THE SECOND CREATION

THE SECOND CREATION

MAKERS OF THE REVOLUTION IN TWENTIETH-CENTURY PHYSICS

Robert P. Crease and Charles C. Mann

COLLIER BOOKS
MACMILLAN PUBLISHING COMPANY
New York

Macmillan Publishing Company
866 Third Avenue, New York, N.Y. 10022
Collier Macmillan Canada, Inc.

Portions of this book originally appeared in somewhat different form
in *The Atlantic Monthly.*

Excerpts from *The Birth of Particle Physics*, edited by Laurie Brown
and Lillian Hoddeson, are reproduced by permission of Cambridge
University Press.

Library of Congress Cataloging-in-Publication Data
Crease, Robert P.
  The second creation.
  Includes index.
    1. Physics—History. 2. Grand unification theories
(Nuclear physics)—History. I. Mann, Charles C.
II. Title.
QC7.C74 1986   539′.09   85-23370
ISBN 0-02-084550-2

Macmillan books are available at special discounts for bulk purchases
for sales promotions, premiums, fund-raising, or educational use. For
details, contact:
  Special Sales Director
  Macmillan Publishing Company
  866 Third Avenue
  New York, N.Y. 10022

First Collier Books Edition 1987

10 9 8 7 6 5 4 3 2 1

Printed in the United States of America

The Second Creation is also published in a hardcover edition by
Macmillan Publishing Company.

Contents

Preface

ACROSS THE GLOBE, THEORETICAL PHYSICISTS ARE ENGAGED IN AN extraordinary enterprise: the attempt to construct, by sheer force of reason, what some refer to, only half-jokingly, as a Theory of Everything. More formally known as a unification theory, a Theory of Everything would be nothing less than a complete description of the foundations of matter, space, and time, a set of linked equations containing the elements of the cosmos. Physicists have always hoped for a unified picture of nature, and now and then individuals have actually sought one. What is remarkable about the present search is that the cream of the present generation of researchers has decided as a group that the time may be ripe for unification, and a task that fifteen years ago was relegated to the lunatic now preoccupies the science as a whole. *The Second Creation* is an account of how this great leap came to pass.

We have taken our title from one of the implications of the search for unification: Working back from the present, the new theories promise to explain the beginnings of the Universe, the awful, minute explosion of the Big Bang, as well as everything that ensued. Never before has science tackled such large questions; the prospect of discovering some answers is thrilling for those who find delight in the accomplishments of the human mind.

Unfortunately for nonscientists, the progress of physics has gone hand in hand with an increasing reliance on abstraction and mathematical representation, a remoteness that has obscured the intense play of style, personality, and thought that permeates this most human of creative practices. Although it is a history of science in this century, *The Second Creation* is a tale of people, not protons—the small community of gifted individuals whose accumulated work has, in the words of Nobel laureate Steven Weinberg, answered all the "easy questions, such as why is the sky blue and what is inside the nucleus of the atom?"

No single person ever wrote a book; for us, the process of writing *The Second Creation* has been a lesson on the object truth of this maxim. Our approach would have been impossible without considerable assistance by professional physicists. We have been fortunate to have had the help of many. Above all, Gerald Feinberg, Murray Gell-Mann, Sheldon Glashow, Abdus Salam, Robert Serber, and Steven Weinberg have tolerated repeated interviews, read many sections of the book, and given us extensive help and criticism. They must have come to dread our reappearance; the book exists thanks to their kindness—a kindness that cost them many hours away from their own work.

"Intellectual debts are odd, for they take hard work to accumulate and are a pleasure to acknowledge." We have many. In the course of research, we interviewed some 125 physicists, many repeatedly, and most at length. Their names are found in the notes; to all we give our thanks. Many scientists whom we interviewed read and commented, sometimes copiously, upon early drafts of chapters. These include Neil Baggett, Ulrich Becker, Hans Bethe, James Bjorken, Robert Brout, Min Chen, David Cline, Martin Deutsch, Samuel Devons, Richard Feynman, Howard Georgi, Gerson Goldhaber, Maurice Goldhaber, Jeffrey Goldstone, Gerard 't Hooft, John Iliopoulos, William Kirk, Willis Lamb, Richard Learner, Leon Lederman, Luciano Maiani, Robert Marshak, M. G. K. Menon, Robert Mills, the late Paul Musset, Yoichiro Nambu, V. S. Narasimham, Yuval Ne'eman, Abraham Pais, Robert Palmer, Francis Pipkin, David Politzer, Charles Prescott, I. I. Rabi, Burton Richter, George Rochester, Carlo Rubbia, Nicholas Samios, Mel Schwartz, John Schwarz, Julian Schwinger, Lawrence Sulak, Richard Taylor, Samuel Ting, Martinus Veltman, Victor Weisskopf, Frank Wilczek, Robley Williams, Frank Yang. Several physicists allowed us to dip into their private papers: Gary Feinberg, Murray Gell-Mann, Sheldon Glashow, Gerson Goldhaber, Richard Imlay, Paul Musset, Robert Palmer, Abdus Salam, Nicholas Samios, Mel Schwartz, Lawrence Sulak, Samuel Ting. In addition, several historians of science resisted the usual hostility of academe to interlopers by giving us suggestions, answering questions, and allowing us to pick their brains: Mara Beller, Stephen Brush, J. L. Heilbron, Andrew Pickering, Silvan Schweber. We have benefitted immeasurably from their writing. If despite the presence of so much eminent help we have introduced errors into this book, we claim full credit for them. We owe a particularly odd debt to Horace Freeland Judson, a man whom we have never met, for his book, *The Eighth Day of Creation*, a work that gave us a glimpse of what is possible in science writing for nonspecialists—and the quotation that begins this paragraph.

We are indebted to Carol Ascher, Mary Lee Grisanti, Dale McAdoo, Peter Menzel, Mark Plummer, Victory Pomeranz, Debbie Triant, Sophia Yancopoulos for a host of personal favors. *The Second Creation* bears the invisible scars from the skillful scissors of a brace of talented editors—Barry Lippman, Debbie McGill, and William Whitworth. Richard Balkin, *agente incredibile*, walked the contractual tightrope with aplomb.

We are also indebted to the American Institute of Physics, the Churchill College Archives, the Oppenheimer Papers at the Manuscript Division of the Library of Congress, the National Archives, the Pauli Archives at CERN, the Archives for the History of Quantum Physics at the American Institute of Physics in New York City (where Spencer Weart helped us), the Science and Physics Libraries of Columbia University, the New York Public Library, the IMB collaboration (special thanks to Dan Sinclair and Larry Sulak), Brookhaven National Laboratory (especially Neil Baggett and Anne Baitinger), the Enrico Fermi National Accelerator Laboratory (where Dick Carrigan looked after us), the Stanford Linear Accelerator Facility (William Kirk and William Ash), the European Organization for Nuclear Research (Roger Anthoine and the incomparable Gwendoline Korda), the Bohr Institute (Eric Rüdinger), the Tata Institute in Bombay (B. V. Sreekantan), the International Centre for Theoretical Physics in Trieste (where Dr. Salam hosted us). One of us (R. P. C.) would like to thank the Department of Philosophy at Columbia University for tolerating his delayed dissertation.

Finally, each of us would like to thank the other for his wholehearted commitment to the sometimes frustrating, always fascinating task of writing this book.

I
Waves and Particles

1
Beginnings and Ends

THE COVERALLS IN THE TRAILER WERE STIFF AND GRAY WITH SALT, CRACK-
ling as we stepped into them. A cool June wind blew in from Lake Erie and
whistled past the corners of the building. Outside were heaps of bluish salt
piled up thirty feet or more into a series of high waves that lapped at the feet
of the four concrete silos that dominate the area, each with an image of the
Morton Salt girl across its face. Hair askew, umbrella bent with rain, the four
Morton girls presided like innocent deities over the entrance to the salt
mine. We put on gloves. We buckled on hard metal shoes and miner's
helmets with lamps and a special wide leather belt with a battery and a gas
mask. It was just after dawn. Fully tricked out, we went outside to the
elevator shaft, ready for the long ride down to where, two thousand feet
beneath our feet, a team of twenty-five sober physicists had filled a tunnel
with sophisticated electronic equipment and spent millions of dollars in
government funds trying to find out when the Universe will end.

A particle physicist named Larry Sulak was taking us down. Tall and
handsome in a lean, country-boy style, Sulak has the easy bearing of a man at
home in the world of objects. The mine entrance proper is a collection of
skeletal girders and cable drums spattered and rusted by the ever-present
salt. We passed through a security post, signed safety releases, then
squeezed into the peeling blue elevator car for the endless slow descent. A
sudden draft whipped through the shaft, carrying a faint chemical odor and a
quick little shower of water. "Watch out for that water," Sulak said. Seeping
through layers of sulfur in its passage through the earth, the liquid becomes
charged with sulfuric acid. It smells quite exactly like rotten eggs and has
chewed holes in Sulak's coveralls. We came to a halt in the semidarkness.

Sulak opened the door and led us into an ugly wonderland of salt.
Brownish stalagmites rose from the floor, wrapped stone tendrils around
ancient equipment, froze ladders to the wall, sheathed machine tools in a
foot-thick casing of stained crystal. Surfaces shimmered with a thin acid film
of slimy water. Sulak pointed his miner's light upward and silently moved it
along the ceiling, revealing a barricade of dripping salt icicles. We inhaled
through our mouths to avoid the stench. "They should drag the theorists
down here once in a while," Sulak said. "Five minutes with a pencil up there
means three years down here for us."

He guided us through an airlock—a huge affair with roaring fans that made our ears pop and moisture fly from our clothing—and into a vast cavern that opened into a set of tall, quiet tunnels lined with crumbling gems of rock salt. To one side was a meeting point for miners, who clambered expertly onto trucks, lunch boxes in their hands, their voices booming with unnatural loudness in the dome. Their equipment bore the logo of Morton-Thiokol, the company that owns the mine and generously donates space to the physicists inside it. A truck bound for the face disappeared down a wide corridor; three miles away, thousands of feet beneath the lake itself, machinery ground into a subterranean cliff of salt. "The commute's not so long if you're a physicist," Sulak said. A hundred yards down a stone ingress was a big steel door behind which a score of scientists wait for the proton to decay.

Along with neutrons and electrons, protons are the primary constituents of ordinary atoms. *Decay* is a physicist's term for the process that transpires when a subatomic particle turns into other, lighter particles—the word comes from *decadere*, "to fall into." Over the last decades, particle physicists such as Sulak have watched hundreds of millions of particle decays, and have come up with rules of thumb for the behavior of the smallest components of matter. One of these principles, or so they thought, was that the proton never, ever decays; it never transforms into some lighter particle. This permanence is a lucky thing: If protons decay, the atoms they comprise will fall apart just as surely as would a house if the stones in its foundation turned to porridge. If atoms fall apart, matter as we know it will some day cease to exist.

We pushed our boots through a little machine that scraped off the salt and then walked into the experiment proper. Inside was a thunder of liquid—the roar from a dozen pumps. The pumps take the water from Lake Erie, a national symbol of filth, and force it at enormous pressure through plastic membranes, sterilizers, scrubbers, and filters until it is clearer than the clearest runoff from an arctic lake, after which it is sent into an underground reservoir the size of a five-story building. Like a sports car aficionado with a new Lamborghini, Sulak's delight in the equipment was expressed by his insistence on explaining to us the nuances of every valve and every reading on every dial. Under the alarmed eye of Ted Darden III, the technician in charge of the pumps, Sulak opened a spigot in the middle of one of the filters. "Here," he said, "try some! Go ahead—taste. It's so pure it's utterly tasteless!" We did. It tasted like nothing.

Sulak had already loped to the other end of the room, where Joe Reese, another technician, had a panel of electronics open on a workbench. "By golly, Joe, using teeth for wire-strippers! Don't tell that to your dentist!" His fingers deftly flicking from switch to dial to gauge, Sulak examined the small computers at the end of the room that monitored the equipment. Satisfied, he said, "Come on, let me show you the tank." His face was taut with pleasure. We took a right turn, put on blue plastic protective booties, then swung left and up a few iron steps to find ourselves on a catwalk above what looked like the largest, cleanest swimming pool ever made—a pool that had taken over two years to excavate, and was the size of a five-story building.

A small knot of technicians, doctoral students, and physicists winched up a long chain of photomultipliers—extremely sensitive detectors of light— that ran along one side of the tank. About the size and shape of the big glass light bulbs used at the turn of the century, the photomultipliers detect the minute flashes of light given off by the passage of subatomic particles and pipe them to computers that analyze their trajectories. Dan Sinclair, the physicist in charge of the day-to-day running of the experiment, crouched on the astroturf that lines the catwalk as the phototubes rose from the water. Power screwdrivers whirred, and a globe was pulled free from its plastic cage. All of the two thousand phototubes that line the top, bottom, and sides of the pool were to be replaced, because the old, obsolete ones had begun to crack. The tube was discolored: A shimmery film of brown and blue ran across its face. Its replacement stood nearby—bigger, cleaner, more sensitive, made in Japan. Sinclair hovered anxiously over the technicians who levered the new photomultiplier into place. "If you get *any* water in there," he said, "everything goes to hell."

Scuba gear littered the far end of the catwalk; divers had just probed the tank walls for leaks. Sulak hopped over the fence to a narrow steel girder that crossed the room a few feet above the water. The pool was covered by a thick black plastic tarpaulin supported by pontoons; Sulak stepped onto a pontoon in the center, threw open a porthole, and pointed the beam of his flash into the tank. The circle of light swept clearly over the rows of photo-multipliers standing at attention on the bottom, five stories below. (Later we learned that the detectors were so sensitive that they require more than a day to recover from such treatment.)

Sulak and his colleagues—physicists from the University of California at Irvine, the University of Michigan in Ann Arbor, and Brookhaven National Laboratory in Upton, Long Island—had been working on the experiment for more than six years, nursing it through the maze of bureaucracy and politics necessary to obtain funding, cutting the hole in the Morton mine, writing computer simulations, developing the detector technology, and, finally, waiting and watching to see what happened after they switched the system on.

The object of so much effort is to prove a theory that draws within one picture all matter and energy, from the hottest supernovae to the whirring fragments of the atom. This goal is known as "unification," and in a sense it defines the discipline of physics as a whole. Physics is a generous term that comprises the study of phenomena as disparate as friction, X rays, the formation of ice crystals, and the evolution of the Universe; it is a canon of faith that these aspects of nature, no matter how dissimilar in appearance and scale, arise from a few unifying principles—perhaps just one—which branch out, interweave, and extend outward like the veins in a leaf, becoming finer and finer until they mesh indissolubly with the cosmos; becoming the stuff of the cosmos itself. Finding these unifying principles is equivalent to writing the equations of the world. Whether they knew it or not, all of the physicists who ever lived spent their days laying down bricks in the road to this goal. History is littered with discarded notions of how to put it all together; Einstein looked for such a theory for years, and failed completely. In view of this dismal record, the progress achieved in physics since World War II is remarkable. Within that relatively short period scientists have taken greater strides toward unification than all of their predecessors put together. When Sulak went to graduate school in the 1960s, physicists were at a loss to account for the constituents of the atom; today he is testing unified theories, and some of his colleagues even have boldly asserted that the end of physics, or at least the beginning of the end, is nigh.

The earliest of the recent spate of unification theories is known as the "grand unified theory." Proposed in the early 1970s, the grand unified theory accounts for almost every known form of matter and force. At first dismissed as implausible, the theory had a compelling clarity that eventually caused many physicists to change their minds and others to propose variations. Despite their magniloquent name, the various grand unified theories do not completely unify, but their construction has inspired physicists to embark on a quest for a theory that does.

Fully unified theories are not yet complete enough to be subjected to experiment, and Sulak and his collaborators have concentrated on the earlier grand unified theories (or GUTs). The most startling implication of the GUTs is that the Universe is mortal. According to these theories, over a long, long period—so long it would take at least thirty-two digits to write the number of years—all protons everywhere will fall apart, and therefore all atoms. If this prediction is true, then humanity is located on a time line in a Universe that is destined to vanish.

"It was a pretty dramatic prediction," Sulak said. "They said, if you're going to unify the laws of physics, the proton must decay in 10^{31} years. [10^{31} is one with thirty-one zeroes after it.] Now there are two ways to find out whether this is true. One is to get a proton and watch it for a billion trillion trillion years to see what it does. The other is to get a billion trillion trillion protons and watch them for one year. Obviously, given funding constraints and the human lifetime, the latter course is more feasible."

Below as we stood on the slowly rocking pontoon were about 10^{33} protons: the nuclei of the water molecules. According to the predictions of the simplest version of the grand unified theories, around two thousand of these protons should decay every year, exploding into two or more smaller particles. These will be traveling so fast that they will actually leave behind wakes of blue light as they course out of the pool. Rippling outward, the light will be picked up by the photomultipliers. By noting the exact time these waves of light reach the side of the tank, the experimenters can re-create the paths of the particles involved and, with care and a soupçon of luck, discern whether they are seeing an extraneous interaction or a true example of proton decay. "I spent several hours lying here with a facemask and my head in the water one night," Sulak said, "hoping that when my eyes adjusted I might be able to actually *see* the flash. I couldn't. It turns out that the light is just a little bit too low in energy for the human eye to pick up. But the new detectors have wave shifters to amplify the light; you might be able to see them light up, so I'm going to try it again."

We watched the students push the winch, a big device that resembled a giant's fishing reel, on a trolley over the catwalk and begin methodically hauling up another string of photomultipliers. A low rumble of voices came over the water; the gentle splashes from the rising phototubes awoke memories of fishing at a lakeside. We asked Sulak if the experiment had observed proton decay. "We're still working over the data," he said carefully. "We have candidates, but as Maurice says—" Maurice Goldhaber, one of his collaborators "—not every candidate is elected. We do know that they decay much more slowly than the simplest grand unified model predicts. So we've killed that one off. The others are harder, but we'll get to them."

Although unification is central to physics, an idea that regulates the discipline, the subject occupies a curiously ambivalent place in physicists' hearts. There is no proof that it is possible, merely the aesthetic conviction that a single, primary cause can be found, that humanity can encompass the basic components of all physical phenomena in a simple, unique, and elegant theory. Moreover, as Sulak told us, the stir caused by the still unproven unified theories has inadvertently masked a more solid, more certain accomplishment: the construction of an experimentally tested theory of matter, a framework that ties together all the bits and pieces of the atom and the forces that play among them. This group of ideas is known as the "standard model," and it is the piecing together of the standard model that makes the physicists today most proud of their work.

Sulak took us to the computer, where viewing screens could be made to show proton decays, candidates and simulations, in swirls of particolored asterisks and stars across the walls of the tank. The images are color-coded, the time the light strikes the photomultipliers indicated by shifts from red to yellow to blue. "We have this kind of event," he said, calling up an image with two rough lighted circles cutting into the walls of the tank. He rotated the image on the screen; unsteady arcs of light swam across our field of vision. "It's a two-body decay. If the proton decays, it's like a stick of dynamite, it explodes from a point. Everything shoots outward in all directions, but everything in this case is just two particles. Now if instead this event were produced by something coming in from the outside, it would have an enormous amount of momentum, and the debris would shoot in only one direction. Sometimes events occur in the corner of the tank and it's hard to tell the difference between a false signal and a real event. That's what we have to decide."

We had first encountered Sulak several years before, and had been instantly captured by his aggressive charm, the rapid staccato of his sentences, and a patience with obtuseness surprising in a man who moves so quickly. Sulak has an enormous, almost greedy enthusiasm for his subject, a love for the process of doing physics that is the hallmark of a fine scientist. His experiments are "discovery" experiments, efforts to discern whether an exceedingly rare phenomenon exists rather than precise measurements of previously known qualities. Both types of work are essential to the science, but the discoverers tend to be brasher, more competitive, and more outspoken than the measurers, more given to communicating the pleasure they take in their work and the pride they feel in the discipline of physics.

There is every motive for Sulak to feel as he does. The history of physics in this century is astonishingly distinguished—who in this age of popularization has not heard of quantum mechanics and relativity?—a steady drive toward knowledge that today pushes at the beginning of time and the end of matter. But there is another reason for the joy good scientists take from their work: They are members of a tradition, a collegial association that continually celebrates its past even as it seeks ever greater novelty. Just as it would be impossible to begin to understand the nature of humanity without some conception of our primitive ancestors, so it is difficult to appreciate contemporary physics without some feel for its past development. Sulak said, "The unification theories are beautiful and fascinating, and may even be true, but they are only the tip of a pyramid built by God knows how many other people. They don't make any sense without the rest of the structure. By themselves they're not *physics*. Physics is the whole thing, the continual process of trying to put the pieces together."

The word *physics* carries a certain weight in Sulak's lexicon; it means the clarity, depth, and beauty with which one can illuminate the workings of nature. He is by no means alone. In the last few years, we have sought out a number of other practicing physicists, and have been struck repeatedly by the passion with which they approach their subject, and the eagerness with which they seek the thrill of the sudden, excruciatingly sharp insight into nature that is the ambition of all scientists. Over and over again, we were surprised—naïvely, perhaps—by the similarities between these participants in the hardest of hard sciences and artists, and how often the growth and flowering of twentieth-century physics resembled the growth and flowering of an artistic movement. Like the best artists, the finest scientists work to bring forth a vision of the world; like artists, their search for that vision is guided by the times they live in and their teachers, their individual tastes, and their personal proclivities. Over time, our conversations with physicists allowed us to see the compelling sweep of the work of the last several decades, the standard model of elementary particle interactions—an artistic landmark that physicists themselves are still discovering how to appreciate.

In the big laboratories and small university offices where high energy physicists do their work, researchers are shaking the surprise out of their heads, bemusedly realizing that they have been through an extraordinary time, and that they have discovered much and learned even more. Because they knew all along that they were in unfamiliar territory, there were no old answers to overturn, and hence this second wave of physicists—Einstein's children, if you will—does not call its accomplishments revolutionary. Rather, these men and women worked for decades in a frightful tangle of confusion, which suddenly resolved itself in a collective moment of insight that bound together the long work of many separate hands. When the standard model fell into place, it did so all at once, after much disorder, and physicists who had expected to spend their careers straightening up a small corner of the immense arena of their incomprehension abruptly found themselves closer to their goal than they had ever dreamed. The atom revealed its secrets only gradually, like a cleverly written play, allowing sparks of knowledge briefly to illuminate corners of the story, but jumbling scenes and inferences so that full realization came only at the end, after many members of the audience had given up hope of understanding. Physicists are pleased, too, that nature indeed is like a good playwright and has closed this particular show with a dramatic promise of unity.

Sulak sat at the small eating table, a cup of coffee steaming unnoticed in front of him. Through the plastic ceiling we could see a rock spar pressing against the border of the air-conditioned bubble that held the experimental station. He was asked what he thought of the current push toward unification. "This is a truly amazing time," he said. "So much has happened in the last fifteen years." Then, rapidly: "It is all unifying—we are seeing the global picture—the forces are coming together. In some ways, the field has not been like this since the development of relativity and quantum mechanics in the teens and early twenties." He gave us a crooked smile, motioned toward the tank of water. "But, really, you don't believe any theory; it's all crap until you've got some real hard facts. That's what we're searching for—two good, hard facts, namely, what the decay rate is and what the decay mode is. That will go a long way to pointing to the right theory."[1]

□   □   □   □   □

Once before, physicists—or some of them, at any rate—thought they were close to a total picture. The certainty with which many physical scientists of the 1880s thought they had the fundamental questions nailed down is today a source of puzzlement to scholars. At Harvard University, for instance, the then-head of the physics department, John Trowbridge, felt compelled to warn bright graduate students away from physics. The essential business of the science is finished, he told them. All that remains is to dot a few *i*'s and cross a few *t*'s, a task best left to the second-rate.[2] In 1894, Albert Michelson of the University of Chicago, one of the most prominent experimenters of the day and the future recipient of a Nobel Prize, told an audience that "it seems probable that most of the grand underlying principles have been firmly established and that further advances are to be sought chiefly in the rigorous application of these principles to all phenomena which come under our notice. . . . [T]he future truths of physics are to be looked for in the sixth place of decimals."[3]

Michelson's timing was comically bad, as it happened. Before the conference proceedings were printed, the first evidence of the previously unknown phenomenon of radioactivity was discovered by one Antoine-Henri Becquerel, the third Becquerel in a row to occupy the chair of physics at the Musée d'Histoire Naturelle in Paris. A balding, irascible man with a fierce little Vandyke beard, Becquerel had spent his twenties and thirties performing undistinguished experiments on phosphorescent crystals. He got his doctorate at the age of thirty-five and almost immediately gave up research, settling into the comfortable respectability of his professorship. Becquerel was, to say the least, an unlikely candidate for celebrity; everything about him suggested that he was destined to be a footnote to future histories of science.

There are few scientific discoveries whose circumstances are known as minutely as those around the almost accidental finding of radioactivity.[4] On January 7, 1896, the great French mathematician Henri Poincaré received a letter containing several astonishing photographs of the bones in someone's hand. The bones belonged to Wilhelm Conrad Röntgen, a scientist Poincaré had never visited. The letter explained that the pictures had been taken with the aid of a new discovery, X rays, that Röntgen had turned up the previous month, and that he was publicizing his findings by mailing off prints all over Europe. Publicized they were: The photographs created a sensation across the globe. Within three weeks, little Eddie McCarthy of Dartmouth, New Hampshire, became a local cause célèbre when his broken arm was set by physicians armed with X-ray images of the fracture.[5] It is easy to imagine

Poincaré's amazement—photographs of the inside of a human being!—and he quickly asked two local doctors if they could duplicate Röntgen's work. On January 20, they showed their own X-ray photographs to the assembled members of the French Académie des Sciences.[6] The reaction was immediate and extreme. In the next fortnight, five members of the Académie presented papers on the new phenomenon.

Becquerel, too, was sitting in the audience when the X-ray photographs were shown. He was fascinated by the strange ghostly images and the mysterious emanations that produced them. Both he and his father had studied the phenomenon of phosphorescence—the museum laboratory was filled with lumps of stone and wood that shone in the dark. The glow of X-ray emission put Becquerel in mind of the light in his study; although he had not done much active research in the last few years, he thought immediately of putting some phosphorescent rock on photographic paper to see if it would darken it in the same way as one of Röntgen's X-ray sources. It would not be all that much work.

What happened next has been recounted many times: how over the next month Becquerel tried a variety of phosphorescent stones, and found nothing; how one day he happened to pick up a chunk of potassium uranyl sulfate, a messy crystalline mix of uranium, potassium, sulfur, and other elements, which he knew from experience glowed under ultraviolet light; how he set the rock out on his balcony to be charged up by the ultraviolet rays in the winter sunlight; how he took a photographic plate, wrapped it up in thick black paper to shield it from the sun, and put it beneath the uranyl sulfate; and how, to his pleasure, in the darkroom he saw that the radiation— he called it "penetrating rays"—from the rock had glided through the paper and produced gray smudges on the plate.

Becquerel was certain that he had shown that X rays were somehow linked to phosphorescence. But he wanted to prove it scientifically—nail it down. Over the next few days he put coins and irregular pieces of metal between the uranyl sulfate crystals and the plates. Sure enough, they blocked the penetrating rays, showing up as coin-shaped spots of white in the darker gray. On February 24, Becquerel told the Académie of his results: Phosphorescence caused X rays.[7]

Becquerel's study was a model of the scientific method. It has come down to us, however, as a textbook of the practical difficulties in applying that method—which Becquerel was the first to find out. By February 26, the weather became dreary, as often happens in Parisian winters. While waiting for the sun, the professor put the plates, paper, and crystals into a file drawer. They lay in the dark for nearly a week; nothing could happen there, Becquerel knew, because the uranyl sulfate was not exposed to light, and hence could not phosphoresce. Nonetheless, on March 1, when the sun came back, he had one of those happy, once-in-a-lifetime thoughts: Why not develop the plates anyway? He had time. In the darkroom, he saw the darkest exposed blotches yet. Becquerel realized to his dismay that the photographic plates were *not* exposed by phosphorescence. There was something in the rock that did it. The uranium, it seemed, was spitting out X rays all by itself.[8]

This, too, was not entirely correct. In fact, the lump of potassium uranyl sulfate was emitting a whole spectrum of radiation, of which only a small portion was X rays. Nonetheless, the discovery caused a sensation, in part because it was so easy to duplicate. Almost every laboratory in the world had construction paper, photographic plates, and chunks of uranium ore. Within weeks, scientists across the Continent were looking in astonishment at the blurred black patches on their photographs. Becquerel became famous; judging by his contemptuous dismissal of rival claims to the finding, he seems to have enjoyed his sudden notoriety.[9] In 1903, Becquerel was given one of the new science prizes established by a posthumous bequest of the late Swedish industrialist Alfred Nobel.

Within weeks, news of Becquerel's findings had spread to Germany, Great Britain, Italy, and the United States, further exciting researchers already stirred by the discovery of X rays. Tests of the two phenomena were often conducted on the same workbench. The consequences of each discovery, however, were far different. X rays were found to be simply pulses of light—light of an intensity and power never before seen, but light nonetheless. Radioactivity, on the other hand, was something entirely new, something that did not fit anywhere. The existence of radioactivity—metal that somehow shot out energy!—was a direct attack on the most ardent beliefs of Becquerel and his colleagues. When the strange behavior of uranium was first noted, Becquerel wrote in his memoirs, "There was no reason to presume that the phenomenon was [anything but] a new example of a known type of energy transformation. Contrary to every expectation, the first experiments demonstrated the existence of an apparently *spontaneous* production of energy. . . ."[10] They had spent many years, those nineteenth-century scientists, establishing the law of conservation of energy: Energy was neither created nor destroyed. But every single piece of uranium seemed of its own accord to produce radiation that fogged photographic plates, electrified gases, and sometimes even burned physicists—and the energy needed to do these things evidently came from no place at all. The metal just sat there, its atoms quietly working away, continuously beaming out penetrating rays in seeming disregard for the conservation of energy.

As it happened, the first clue to the nature of radioactivity, although it was not recognized as such immediately, was found just a year after Becquerel's work, when Joseph John Thomson, an Englishman, deduced the existence of small objects later called *electrons*. The director of the Cavendish Laboratory in Cambridge, England, "J. J." had a gift for designing experiments, although his clumsiness prevented him from actually building the equipment. He was notoriously inattentive in matters of dress—his tie is askew in his official Cavendish portrait. To Thomson's own surprise, he had been chosen to head the Laboratory in 1884, when he was only twenty-eight.[11] By 1897, when Thomson made his discovery, the Cavendish had become the most prominent physics laboratory in the country. It had twenty full-time staff members, and had recently acknowledged the growing women's rights movement by allowing women to enter the Laboratory. Research was steadily turning away from practical matters such as telegraphy to such useless topics as the nature of electricity.[12]

At the time, a chief means of examining electricity was to pump out the air from a long glass tube, insert wires in both ends, and connect the wires to a battery. If one of the ends of the tube was painted with zinc sulfide or some other fluorescent material, a tiny, glowing spot would appear on the paint as soon as the battery was switched on. Obviously, *something*—or maybe a stream of somethings—was passing out of one wire, shooting across the tube, and smacking into the zinc sulfide. (The dot of light was like the one seen on old television sets the instant after they are turned off.) Because once the air was sucked out and there was nothing in the tube, scientists reasoned that whatever was going across and making the paint fluoresce must be a flow of electricity in its elemental form. If one could learn of what that flow was made, it might reveal the nature of electricity. Thomson and his subordinates had spent more than a decade, off and on, worrying about the problem.

After a series of experiments, Thomson gave a talk on Friday, April 29, 1897, in which he announced that he had the answer.[13] (A long paper was published six months later.) The glow, Thomson claimed, was caused by a stream of small particles—*corpuscles*, in the jargon of the day—each bearing a set amount of negative charge. They were, in a way, atoms of electricity: electrons, as they are now called. The corpuscles, Thomson said, were sailing off the wire into the zinc sulfide, and somehow the energy of their collision was causing the chemical to emit light. They were very small—too small to weigh by any known means. And the amount of electric charge on each corpuscle was also tiny—too tiny to measure. But Thomson found one property of these electric corpuscles he *could* evaluate: the ratio of their electric charge to their mass. Only the ratio of these two quantities, however, not the actual value of either. Indeed, for several years, the charge-to-mass ratio of these particles was their only precisely measurable property.[14]

Nevertheless, it was clear that Thomson's corpuscles were smaller than any known object, including the atom. The reaction to this notion was not positive: It is difficult to grasp how startling the notion of a subatomic particle was to nineteenth-century physicists, many of whom did not believe that atoms existed, let alone that they had constituent parts.[15] An influential and primarily German school of thought argued that physicists ought not to truck with creatures they could not see; because atoms could not be observed directly, they therefore should not be made the object of speculation. If this were true for atoms, it was all the more true for pieces of atoms. Years afterward, Thomson recalled his colleagues' lack of enthusiasm for his discovery: "At first there were very few who believed in the existence of these bodies smaller than atoms. I was even told long afterwards by a distinguished physicist who had been present at my lecture at the Royal Institution that he thought I had been 'pulling their legs.' I was not surprised at this, as I had myself come to this explanation of my experiments with great reluctance, and it was only after I was convinced that the experiment left no escape from it that I published my belief in the existence of bodies smaller than atoms."[16]

Atoms are building blocks of nature. Pile up enough of them and you get a lump of stuff that you can see or break or hit someone over the head with. With electrons it is an entirely different affair. No matter how many you gather together, it is impossible to make a lump of something that can be tasted or smelled or held in the hand. Little wonder that Thomson's colleagues did not think he was speaking seriously.

They were forced to accept Thomson's work by the increasing number of scientists who argued that the results of their experiments could only be understood by believing in the existence of subatomic particles. Becquerel, for example, made his last important contribution to science when he demonstrated, in 1900, that the emanations from the uranium in his laboratory included a considerable number of electrons.[17] But this discovery raised more questions; if there were negatively charged bits of matter in the atom, and if ordinary atoms were electrically neutral, there had to be some positively charged matter to balance out the negative. In other words, there had to be more pieces to the atom. Where were they, and how were they arranged? What are the ultimate constituents of matter and what forces play among them? Answering these questions required decades of work, which culminated in the 1970s with the standard model of elementary particle interactions and the prospect of unification.

The first hints were provided almost a decade later by Ernest Rutherford, a Thomson protégé then working at the University of Manchester. A large, confident man with a big red face and a walrus moustache, Rutherford was born and raised in colonial New Zealand. He was good with his hands, clever, ambitious, and hardworking; he did not want to spend his life grubbing about a subtropical farm. In 1894, he won a scholarship to Cambridge, which had just changed its rules to admit graduates of other schools, thereby permitting Rutherford to become the first foreign research student in the Cavendish laboratory. He arrived in England in September of the following year. He was just twenty-four years old.

As soon as Rutherford appeared in Cambridge, he went to Thomson and, according to the letters he dutifully sent his fiancée in New Zealand, "had a good long talk with him. He's very pleasant in conversation, and he's not fossilized at all. As regards appearance, he's a medium-sized man, dark and quite youthful still—shaves very badly and wears his hair rather long." Rutherford reported that Thomson "seemed pleased with what I was going to do."[18]

The Cavendish then was crammed into a Victorian neo-Gothic building, gray and bilgewater yellow, in the center of Cambridge. The experimenters worked cheek by jowl in grubby, crowded laboratories where Rutherford was apparently allowed to hear the mocking comments of Cambridge instructors about scholarship winners from the Antipodes. Nonetheless, he quickly impressed Thomson with his brashness, skill, and extraordinary drive, the same qualities that prevented him from making friends immediately. "One can hardly speak of being friendly with a force of nature," Rutherford's Cavendish colleague Paul Langevin is said to have remarked.[19]

Rutherford was a perfect man for the time, a hardworking, hardheaded scientist who was impatient with mathematical abstraction and complicated equipment. Consciously presenting himself as the epitome of an experimenter, he declaimed that he liked to discover facts the reliable way, through experiments, without a lot of theoretical pettifogging. The experiments themselves should be quick, simple, and performed on equipment scavenged from the basement of the laboratory. Each test should build on the one before. The object was to find good, solid facts—did X happen or not?—to find them first, and to let other people clean up decimal places. He said, "There is always someone, somewhere, without ideas of his own, who will measure that accurately."[20]

Rutherford possessed one of the talents that a good physicist must have, a sense for the right direction to pursue. At any given moment, there are literally thousands of experiments that could be done; a great experimentalist has a sixth sense for which one will lead to something profound, rather than merely informative. Rutherford had early on been intrigued by electromagnetic waves, and actually built a practical radio transmitter a little ahead of Guglielmo Marconi. At the Cavendish, however, Rutherford quickly decided to drop the radio, which was only practical, and chase after radioactivity, which might involve some real *physics*. With characteristic directness, he began to take the phenomenon apart. He learned that radioactive materials boiled with activity; they constantly spewed out huge numbers of particles, and the particles traveled at terrific speed—thousands of miles a second. Moreover, Rutherford found radioactivity had two distinct forms, which "for convenience" he named alpha and beta rays, after the first two letters in the Greek alphabet.[21] The alpha rays could be blocked by a piece of paper, but the beta rays had a hundred times the ability to punch through a shield. (Both had been emitted by the uranium in Becquerel's laboratory. Beta rays were later shown by Becquerel to be composed of electrons.)

In 1898, Rutherford was offered a professorship at McGill University, in Montreal, where he was "expected to do a lot of original work and knock the shine out of the Yankees!"[22] At McGill, Rutherford began what was to be his life's work, the study of alpha rays. For the next few years, Rutherford, Becquerel, and Becquerel's collaborators and friends, the young Marie and Pierre Curie, kept up an intense but friendly rivalry to be the first to comprehend the nature and behavior of alpha particles. The contest was sharpened by the Curies' discovery of radium, an element a million times more radioactive than simple uranium. Radium was fantastically rare; the Curies processed tons of uranium to get microscopic amounts of radium, and by 1916 the total world supply was less than half an ounce, parceled out in minute doses among the score of laboratories investigating its properties.[23] Using a few hot milligrams of the stuff sent to him by the French, Rutherford measured the charge-to-mass ratio of alpha particles, just as his mentor Thomson had done for electrons; after four years of work, Rutherford was certain they were positively charged helium atoms.[24] During this time, most scientists vaguely thought the atom was a buzzing hive of thousands of electrons somehow held together by a positively charged glue. Rutherford decided that his alpha particles were made by knocking out a few electrons from the glue in helium atoms; the absence of the negatively charged electrons gave the helium a net positive electric charge.

At about the same time Rutherford was establishing the identity of alpha particles, Becquerel had performed an experiment that seemed to indicate that alpha particles had the spooky property of increasing their momentum as they pushed through the air.[25] This was bizarre: Alpha particles were spat out of atoms like so many minute bullets, but instead of slowing down as they traveled, they seemed to speed up.[26] Rutherford, on the other hand, found that alpha particles slowed down gradually.[27] The two men challenged each other's findings, and both repeated their own experiments. Rutherford was correct. The argument would today be forgotten except that it sparked Rutherford's curiosity. He derived no particular satisfaction from being the victor in a minor scientific dispute, but he *was* intrigued with the question of why his French colleague had gone astray. When Rutherford turned his attention to the details of Becquerel's experiment, he was impressed with how difficult it was to measure precisely the paths of the alpha rays. The lack of definition was "evidence of an undoubted scattering of the rays in their passage through air."[28] In other words, as the alpha particles sailed on their merry way, at least some of them bounced off the molecules of air in their path. This finding was a critical step towards Rutherford's discovery of the structure of the atom.

He didn't realize it at first. For several years, Rutherford thought that the deflection was only another stumbling block that nature maliciously had put in the way of experimenters who wanted to find facts without any bother. In 1907, he accepted a post at Manchester and moved back to England, where he continued working with alpha rays, using a much stronger source of radioactivity he had obtained after much wrangling with colleagues. A year later, he won the Nobel Prize. To his astonishment, it was for chemistry, not physics.[29] Despite the move and the prize, he continued to work at the same pace. But whenever he was forced to make a precise measurement— despite his dislike, he would do it in a pinch—the scattering kept making the job harder. For example, when he wanted to find out the precise charge of alpha particles, he thought he would fire them one by one into a device capable of assessing them individually. But the various measurements never agreed with each other, and Rutherford and his team, annoyed, realized that "the scattering is the devil" that was plaguing their work.[30] To the scientists' dismay, alpha particles seemed to be ricocheting all over their equipment.

Exasperated, Rutherford told his assistant, Hans Geiger, that to avoid further problems they would have to measure how much the alpha particles were being jostled. Geiger was joined by an undergraduate, a New Zealander named Ernest Marsden. They beamed alpha particles through a thin metal foil into a thin metal screen that gave off tiny flashes whenever it was struck by the particles. Beforehand, the experimenters had to sit in the dark for a quarter of an hour to let their eyes adjust enough to see the flashes. The problem was, of course, the scattering: So many alpha particles were deflected by the air and the walls of the tube that it was difficult to discern what particles were bouncing where. One day in the early spring of 1909, Rutherford told Marsden to see if any particles would actually bounce back from the foil.[31] They quickly found that about one in eight thousand alpha particles would slam into a sheet of gold leaf and rebound.[32]

At first, Rutherford assumed that the alpha particles had, like billiard balls in a complicated shot at snooker, simply ricocheted off several atoms of gold. But over the next year the scattering apparently nagged at him; he simply did not think it likely that a fast little alpha particle could graze a few atoms and end up turning 180 degrees.

On the other hand, Rutherford could imagine an alpha particle rebounding off *one* atom—the guiding metaphor here might be shooting a bullet at an anvil. The problem was that atoms were not supposed to be like anvils. The most prominent physicist in England, J. J. Thomson, argued forcefully that the atom must consist of a ball of positive charge studded with electrons. The whole ensemble was often described, rather vaguely, as a spongy, doughy blob—a "plum pudding," as it was later termed, with electrons standing in for plums. Rutherford liked to have a clear pictorial image in his head of what was going on when he performed an experiment. He knew the alpha particles were drilling through the air at great speed; he could not imagine how such bullets could bounce back from a lump of pudding.[33]

By late November or early December of 1911, Rutherford had the first inkling of the answer. A quick seat-of-the-pants mathematician, Rutherford figured that if almost all of the mass of the atom were concentrated into a little charged node in the center, that would be enough to deflect an alpha particle. He wasn't sure whether the charged center was positive or negative—that is, whether it kicked away the alpha particle or whipped it around like a comet—but the opposite charge had to be in a sort of thin, gaseous sphere surrounding the middle of the atom. Although he was pleased with this image, Rutherford hesitated. He was not really certain of his ideas and, despite a well-deserved reputation for speaking his mind, he was leery of placing himself in the classic Oedipal situation of publicly disputing his mentor.

At this point, Rutherford had a stroke of luck: At about the same time he was chewing the matter over, another Thomson protégé, J. A. Crowther, announced that he had confirmed the plum pudding model by an experiment similar to the one done by Geiger and Marsden, except that Crowther had fired beta rather than alpha particles at metal foils.[34] The experiment gave Rutherford something to react against; it was psychologically much easier to go after Crowther than Thomson, although the end result was the same.[35]

On March 7, 1912, Rutherford first presented his theory at a session of the Manchester Literary and Philosophical Society.[36] He discussed the Thomson model and Crowther's experiment and then bluntly attacked both. The results of Geiger and Marsden, he contended, could not be explained by a plum pudding. Although it was a small effect, the deflection of one in eight thousand alpha particles occurred and must be caused by something; the alpha particles had to have slammed into something very small and very hard in the atoms of the target. Rutherford called that small, hard thing "a central electric charge concentrated at a point." We now call it the nucleus of the atom. Around the nucleus, Rutherford said, is a "uniform spherical distribution of opposite electricity," which today we know to be the orbiting electrons.[37]

The solar system–like picture of the atom has become familiar to us from the covers of numerous high school physics books and from the logo of the old Atomic Energy Commission. But this symbol is a domesticated, smoothed-over version of the real thing. Rutherford found that the atom, and therefore matter as a whole, consists overwhelmingly of empty space. If an atom were blown up to the size of a domed football stadium, the nucleus would be the size of a fly in the center; scattered throughout the enclosure is a sprinkling of even tinier electrons. More strangely still, the nucleus is incredibly heavy: It accounts for almost all the weight of the stadium, while all the grandstands and roof panels are as light as mist. The apparent solidity of everyday objects is due to the play of electrical forces among atoms and molecules, not the substance of the material itself; in truth, substance is one of humanity's most persistent illusions. With the discovery of the emptiness in matter, nuclear physics—indeed, the whole nuclear age—was born.[38] It was Rutherford's greatest accomplishment; we are still reaping the consequences.

At first, nobody paid attention. That spring, Rutherford had written to many of the physicists he knew, telling them about his model with characteristic ebullience. Reactions ranged from polite acknowledgment to indifference; after all, other than one rare type of alpha scattering, there was little evidence that the atom had a nucleus and that matter was mostly void. Rutherford seems to have been a little daunted by the lack of enthusiasm. At any rate, for a while he gave up proselytizing for his idea in favor of writing a big book summarizing the state of knowledge about radioactivity.[39]

But even if physicists had taken the nucleus seriously, Rutherford's model literally could not work. It was like a model airplane whose designer had included elements that would not fit together. In nearly all arrangements, the orbiting electrons would either be sucked into the positively charged nucleus or be ejected by the negative charge of fellow electrons. With considerable mathematical finagling, everything could be balanced, but even then the slightest nudge or disturbance would cause the whole system to go awry. If an atom like Rutherford's had ever existed, it would have torn itself apart in a fraction of an instant.[40] Any reasonable physicist, therefore, would have dismissed the idea after thinking about it for five minutes.

2
The Man Who Talked

ON A GRAY RAINY STREET IN THE CENTER OF THE GRAY RAINY CITY OF Copenhagen is a small cluster of buildings that protrudes into the side of a city park. Neatly tended and vaguely inhospitable in the Continental manner, Faelled Park is a stretch of wet greensward laced by gravel paths that run beneath stands of trees. The park is old, square, pristine, proudly aloof from city life—except where the complex on its edge has gobbled up a meadow and nibbled at the edges of a gathering of oak. As if to disguise their intrusion, the buildings cut into the flank of Faelled Park have inconspicuous slate-colored walls and red tile roofs and curtained windows like their neighbors in Copenhagen proper. They are, however, one of Europe's greatest centers of theoretical physics and a living monument to the torchbearer of the quantum revolution, Niels Hendrik David Bohr.

Bohr's working habits have become legendary among his successors, part of the lore of science along with Einstein's flyaway hair and Rutherford's remark that relativity was not meant to be understood by Anglo-Saxons. Bohr *talked*. He discovered his ideas in the act of enunciating them, shaping thoughts as they came out of his mouth. Friends, colleagues, graduate students, all had Bohr gently entice them into long walks in the countryside around Copenhagen, the heavy clouds scudding overhead as Bohr thrust his hands into his overcoat pockets and settled into an endless, hesitant, recondite, barely audible monologue. While he spoke, he watched his listeners' reactions, eager to establish a bond in a shared effort to articulate. Whispered phrases would be pronounced, only to be adjusted as Bohr struggled to express *exactly* what he meant; words were puzzled over, repeated, then tossed aside, and he was always ready to add a qualification, to modify a remark, to go back to the beginning, to start the explanation over again. Then, flatteringly, he would abruptly thrust the subject on his listener— surely this cannot be all? what else is there?—his big, ponderous, heavy-lidded eyes intent on the response. Before it could come, however, Bohr would have started talking again, wrestling with the answer himself. He inspected the language with which an idea was expressed in the way a jeweler inspects an unfamiliar stone, slowly judging each facet by holding it before an intense light.

His continual struggle with language extended to the most ordinary acts. Bohr was one of the few people on earth to write drafts before sending postcards. His articles were composed with such care and precision that they sometimes verged on incomprehensibility, and were always late. He asked friends to read preliminary versions, and weighed their comments so thoughtfully that he would often begin over again; a frustrated collaborator once snarled to a colleague who had given Bohr a minor suggestion on a draft, prompting a seventh rewrite, that when the new version was produced, if "you don't tell him it is excellent, I'll wring your neck."[1] Bohr studied problems with the slow gravity of an earnest child; he was willing to appear foolish if it meant he might learn. He was utterly unable to tease. He was entirely without malice.[2]

Shy and pensive, he had a long, oval, big-cheeked Danish face and thick hair which he combed straight back from his forehead. Although he had been a brilliant student, Niels was consistently overshadowed by his younger brother, Harald, who was considered the Bohr child with real promise. Niels was an excellent soccer player, but it was Harald who in 1908 played halfback on the Danish Olympic team and brought home a silver medal. Harald was two years younger than Niels, but he entered the university just a year behind him and completed his doctoral dissertation a year earlier. Niels wrote strange, difficult, brilliant physics articles, but at first it took Harald's urging to make scientists look at them. Harald became a distinguished mathematician; even late in life, Niels claimed that his brother "was in all respects more clever than I."[3]

As a schoolboy, Bohr's worst subject had been Danish composition, and for the rest of his life he passed up no opportunity to avoid putting pen to paper. He dictated his entire doctoral dissertation to his mother, causing family rows when his father insisted that the budding Ph.D. should be forced to learn to write for himself; Bohr's mother remained firm in her belief that the task was hopeless. It apparently was—most of Bohr's later work and correspondence were dictated to his wife and a succession of secretaries and collaborators. Even with this assistance, it took him months to put together articles. Reading of his struggles, it is hard not to wonder if he was dyslexic.

Early in 1911, at the age of twenty-five, Bohr defended his dissertation and received a fellowship from the Carlsberg Brewery, which many Danes believe makes the finest beer in the world, to study in England for a year. He was excited about the prospect of working with the famous J. J. Thomson in the Cavendish. Bohr had thought a great deal about Thomson's plum pudding atom, and was sure that it could not possibly be correct; he could hardly wait to discuss his criticisms of it with the master. He came to Cambridge in September of 1911, sixteen years after Rutherford, and like Rutherford was at first elated to be there. A few days after his arrival he wrote to Margrethe

Nørlund, his fiancée, "I found myself rejoicing this morning, when I stood outside a shop and by chance happened to read the address 'Cambridge' over the door."[4]

Like many shy young people, Bohr often had to work up the nerve to speak with strangers, and the accompanying anxiety would cause him to blurt out whatever was on his mind. This failing, coupled with his then-unsteady grasp of English, made his first encounter with Thomson something of a disaster, and his stay at Cambridge discouraging. Unlike Rutherford, he could not learn to get along with the English. Bohr showed up in the lab, bumptiously anxious to talk of plum puddings, and was promptly and rudely dismissed. "I had no great knowledge of English, and therefore I did not know how to express myself," he said later. "I could only say [to Thomson], 'This is incorrect!' He was not interested in the accusation that it was not correct."[5] Thomson seems not to have known what to do about the young, anxious, inarticulate Dane; Bohr soon noted that whenever he managed to catch Thomson's attention "for a moment, he [Thomson] gets to think[ing] about one of his own things, and then he leaves you in the midst of a sentence (they say that he would walk away from the King, and that means more in England than in Denmark), and then you have the impression that he forgets all about you until the next time you dare to disturb him."[6] Fending off the young foreigner, Thomson promised to read his thesis, but never did. Bohr felt stranded in the coldly civil laboratory.

Unhappy, he attended the annual December Cavendish dinner, a boisterous, collegiate affair featuring music hall numbers the scientists wrote parodying themselves and their work. Wine flowed freely, and researchers bellowed physics jokes, like the toast, "To the electron! May it never be of any use to anybody!" Rutherford, who always enjoyed a good party, often came down from Manchester for the occasion; the Cavendish dinner of 1911 was no exception.

Bohr's feelings upon meeting Rutherford for the first time are easy to imagine. Standing up at the dais was a rugged man with the ruddy complexion and thick moustache of a country butcher; laying aside a pipe that spat out smoke and ash at a volcanic rate, he launched into a direct, humorous, even bawdy account of developments at Manchester. His manner was enormously heartening to a young man distressed by English formality. Later Bohr discovered that almost everyone had a favorite Rutherford anecdote; one Cavendish man told him that of all the physicists with whom he had worked, Rutherford was the one who could swear at the experiments most effectively.[7] In March 1912, Bohr moved to Manchester, staying in a little room at Hume Hall. In a short time, his attention drifted to Rutherford's model of the atom. Rutherford told his new assistant not to spend too much time wondering what went on in the nucleus; it was just an idea, and an idea was not worth as much as a fact. Bohr ignored the advice.

In their different ways, both men had the great gift of physical intuition, of being able to picture the doings of the unseen entities they were studying. Bohr was drawn immediately to the nuclear atom—"I just believed it," he said later—partly *because* it would not work; some extra new thing would be necessary to make the model fly. Bohr had the notion that a young man unimpeded by the received wisdom of his elders could provide the answer. Moreover, he seems to have been one of the first to guess that the structure of the atom and the behavior of its parts would be the question that would drive the progress of physics in this century, something not obvious at the time. By the end of spring, he was mulling over the suspicion that the nuclear model might be coherent if it were joined to something from an apparently unrelated idea of physics—the quantum.

A child of the century, the quantum was born on December 14, 1900, when a conservative German academic, Max Planck, reluctantly announced that certain experimental results could best be understood if it were assumed that substances emit light only of certain energies and not others.[8] Born in 1858, Planck came from a family of ministers and lawyers—upright, dutiful, honest people. He attended universities in Munich and Berlin, doing well but not strikingly well. His thesis adviser told him to look into another field, because physics in the 1880s was just about finished. Planck took a teaching job in Kiel and then, to his surprise and pleasure, was offered a prestigious post at the University of Berlin in 1889. He was not certain whether he believed in the reality of atoms.[9]

In the late 1890s, Planck spent six years studying the way substances emit light; an example is the blue glow of the gas in a neon sign. As can be imagined, the connection between light and matter was hard to understand at a time when many reputable scientists did not think atoms existed. Planck tried to avoid the whole question of what was making the light by treating light as if it were emitted by "oscillators" whose oscillation produced light waves in somewhat the way a plucked guitar string makes sound waves. One way of performing such calculations is to divide up the total energy of each oscillator into little pieces of approximately equal size, let them become infinitely small, and then use the techniques of calculus to add them all back up; the sum—or, more properly, the integral—would then be the original energy. Unluckily, in Planck's case it didn't work that way. If he were to make his result fit with the experimental data, the little pieces of energy *could not* become vanishingly small. They had to have some finite size—meaning that their sum, the total energy of the oscillator, could only have particular values.

The more Planck thought about this, the less he liked it. If his oscillators could not vibrate with any energy they pleased, then something made them choose certain values and prevented them from selecting others. This was apparently as senseless as claiming that a guitar string could be tuned to produce C$^\sharp$ and B$^\flat$ but no sounds in between. Planck was absolutely unable to justify his statement; he had to fudge the math to make the equations even *look* right.[10] On the other hand, Planck, like most physicists, was a practical sort, and he realized that his wild idea produced formulas that matched the charts and graphs of the experimenters.

After weeks of uncertainty, he finally performed what he later called "an act of desperation" and asserted that his oscillators could only vibrate at certain specified rates, and that therefore the light they produced came in certain discrete energies, each energy proportional to some primal value.[11] In modern language, Planck's claim is reducible to the expression

$$E = nh\nu$$

where E is the energy of the light source, n is a positive integer (that is, a number like 0, 1, 2, 3, and so on), ν is the Greek letter *nu*, which physicists use to mean frequency, and h is a small, unchanging number now known as Planck's constant.[12]

Simple in appearance, Planck's formula had great resonance. Quite literally, its strangeness threw the world into turmoil—although slowly, for truly great discoveries sometimes acquire their stature only in retrospect. If n must be an integer, then it cannot have a value between 0 and 1. Elementary multiplication thus shows that E, the energy, cannot have a value between νh and 0. Values like $\frac{1}{2}\nu h$, $\frac{1}{4}\nu h$, and $\frac{1}{10}\nu h$ are forbidden. They cannot exist. If light is emitted only in certain selected energies, Planck found himself saying, it must be packaged in little νh-sized units, which he called *quanta*, from the Latin for "how much."

Deeply uncomfortable with his own formula—why couldn't lightwaves have any energy under the sun?—Planck suggested that maybe light actually can have any energy value under the sun, but that somehow it is emitted only in quanta, just as milk is packaged only in pint, quart, and half-gallon containers, but, once bought, can be poured or spilled in any amount.

Although Planck's formula was bizarre, it *worked*. When physicists calculated with it, they got the right answers, and much more beside. Many scientists strove to explain the results in some other way. Planck, in particular, seems to have felt like Epimetheus, the mythical Titan who opened Pandora's box; although he was richly rewarded for his findings, the discoverer of the quantum spent years in a useless battle to remove it from the world. Years after Planck's death, his student and colleague, James Franck, recalled watching his fruitless struggle "to avoid quantum theory, [to see] whether he could not at least make the influence of quantum theory as little as it could possibly be—whether he could not, for instance, say it might be only the emission but not the absorption. I mean, a lot of things he tried out. He was really trained in classical physics, and if ever there was a classicist in character, it was he. He was a revolutionary against his own will. And I remember that he always came with attempts to see whether one could not avoid—with some resignation, but also with looking ahead. He finally came to the conclusion, 'It doesn't help. We have to live with quantum theory. And believe me, it will expand. It will not be only in optics. It will go in all fields. We have to live with it.'"[13]

Physicists did have to live with it. In 1905, Einstein made one of his first and greatest contributions to the field when he took Planck's idea more seriously than its creator. Light not only comes in quanta, Einstein argued, it *is* quanta.[14] This step was even crazier; as late as 1913, when Planck and three other physicists recommended Einstein for membership in the Prussian Academy, they stressed Einstein's great contributions to physics, even though "he may sometimes have missed the target in his speculations, as, for example, in his hypothesis of light-quanta." Even when Einstein's equation incorporating the light-quanta was proven by the American experimenter Robert A. Millikan in 1915, Millikan described the theory behind the equation as "wholly untenable."[15] But Einstein's reasoning was clear and precise, his assertions confirmed, and when the Swedish Academy finally awarded him the Nobel Prize, in 1922, this, not relativity, was the work they cited.

Today's physicists readily accept light-quanta, and call them "photons."[16] But back in 1911, the concept seemed far-fetched indeed. In Rutherford's opinion, the Continental theorists didn't want to explain how their fancy, highfalutin talk of the quantum translated into something real that he could find in an experiment. Rutherford was not satisfied by hypothetical "oscillators." He thought that such theoretical chatter shirked the essential task of physics, and the whole overly mathematical Germanic school of physics was suspect for it. At the same time Rutherford was putting together his model of the atom, he complained to a friend that "continental people do not seem to be in the least interested to form a physical idea of the basis of Planck's theory. They are quite content to explain everything on a certain assumption, and do not worry their heads about the real cause of the thing. I must, I think, say that the English point of view is much more physical and much to be preferred."[17] He would have been flabbergasted if someone had told him then that a Continental physicist would establish the physical basis of the quantum in Rutherford's own lab, and that the basis was, in fact, Rutherford's own model of the atom.

Nobody knows who first told Bohr about the nucleus, or how he learned that atoms with nuclei would be unstable. It is evident, however, that this instability, which made most non-Manchester theorists unwilling to consider the nucleus seriously, was what interested Bohr most. Convinced, perhaps irrationally, that the nucleus must exist, he soon realized that demonstrating its reality would require wholly new ideas. Indeed, he was certain that any explanation of Rutherford's model was impossible *without* these ideas. Sometime in the late spring of 1912, it occurred to him that quite possibly the explanation for the stability of this atom lay in the still growing domain of quantum theory. "It was clear," he said later, "and that was *the* point in the Rutherford atom, that we had something from which we could not proceed at all in any other way than by radical change."[18] *Radical change* was something that interested the young Bohr.

Bohr knew that Einstein had said that light consisted of tiny bundles called quanta; what if quantization was a fundamental property of all energy? Could this somehow be linked to the atom's stability? By June he thought he had it. Excited, he wrote Harald that "perhaps I have found out a little about the structure of atoms. Don't talk about it to anybody, for otherwise I couldn't write to you about it so soon. If I should be right it wouldn't be a suggestion of the nature of a possibility (i.e., an impossibility, like J. J. Thomson's theory) but perhaps a little bit of reality."[19] More confidently, he wrote his fiancée two weeks later, "It doesn't perhaps look so hopeless with those little atoms, even though the outcome of the calculations has its ups and downs."[20]

By this time he was working under some pressure, for he was planning to leave on July 24 for Copenhagen, where he was to be married a week later. Like every busy bridegroom who has tried to plan a traditional ceremony in the faraway hometown of his bride, Bohr had trouble juggling his personal and professional life. Nevertheless, he had time to prepare a little summary of his ideas for Rutherford.[21] In it he stated as a hypothesis the idea that the electrons around an atom neither pushed each other away from the nucleus nor fell into it because they simply could not do so unless something external—a photon, say—intervened. Bohr suspected that the fixed quantities of energy comprising light somehow corresponded to fixed electron orbits in atoms. He could give no reason why, other than that it seemed to ensure that Rutherford-style nuclear atoms did not fall apart.[22]

Rutherford was surprised that Bohr had taken the model so seriously, and drawn such far-reaching conclusions about the nature of the atom from his little picture of how it was organized. "He thought that this meagre evidence about the nuclear atom was not certain enough to draw such consequences," Bohr remarked later. "I said to him that I was sure that it would be the final proof of his atom."[23] Excited, Bohr convinced his wife that they should not, as they had planned, go to Norway on their honeymoon, but instead should spend the time in Cambridge, where he could get some work done.[24]

Bohr expected to write a paper fairly quickly from the memorandum he had given Rutherford. But seven months later, in February, he was still slowly working out his ideas, quietly juggling everything he knew about atoms and quantum theory—all without real satisfaction. His problems were resolved, he often said later, at a stroke, when a friend suggested that he look into a formula giving the frequency of light emitted by atoms. (Recall that the frequency of a lightwave is the number of wavelengths—crests and troughs, if you will—per second.) It had been discovered that heated materials gave off only certain specific frequencies of light, and not any others; gas in a neon sign, say, glows with particular shades of blue that are characteristic of its composition. These colors can be broken up into their individual components in the same way that a prism divides a ray of white light into a rainbow of red, yellow, blue, and green. If an element—for example, hydrogen—is examined in this way, the result is not a continuous spectrum, but a series of colored bands, called spectral lines, that represent the tones and frequencies hydrogen atoms are capable of producing. Spectroscopists stack these lines atop of each other in sequences that look quite like those ugly black marks printed on tin cans that are "read" by electronic cash registers; the pattern of horizontal lines is different for every substance and as individual as a fingerprint. In 1885, a Swiss high school teacher and amateur numerologist, Johann Balmer, had noticed that the frequencies of the light emitted by hydrogen atoms were mathematically related.[25] (Hydrogen is the lightest and simplest element, with just one electron.) The regularity interested scientists, but few were convinced that it was important. It is unlikely that Bohr had never seen Balmer's work before, but he had probably forgotten it. This time, a glance was enough to electrify him. "As soon as I saw Balmer's formula the whole thing was immediately clear to me," Bohr said.[26]

What he experienced at that moment was the single, intensely pleasurable instant of illumination when a great deal of hard thought abruptly coalesces into a vision, a process similar to the abrupt emergence of a painter's style from a morass of false starts and derivative juvenilia. In a sense, Bohr realized that he should take Planck's constant very seriously indeed. Planck's constant h is 6.62×10^{-27} erg-seconds, an incredibly small number with somewhat peculiar dimensions. The dimensions of a number are the units it is written in; for instance, the dimensions of velocity are distance divided by time, miles per hour. Erg-seconds, the dimensions of Planck's constant, are energy multiplied by time, which is identical to those of a quantity scientists call *action*. For this reason, h is often termed a quantum of action. The idea of action was elaborated by eighteenth-century astronomers, who found that they could simplify complicated problems of planetary orbits by introducing a new variable related to energy: action. For example, the action of the earth going about the sun is calculated by dividing the orbit into a series of points, multiplying the earth's momentum at each point by the change in radius from the point before, and adding up the result; engineers have to do this kind of arithmetic today to tell astronauts in the space shuttle when to deploy weather satellites. Later, astronomers learned that the easiest way to work with action variables was by incorporating them into functions called Hamiltonians, after their inventor, the nineteenth-century physicist William Rowan Hamilton.[27] Hamiltonians are a method for finding the minimum value of a given equation. Physicists use them to calculate orbits, trajectories, and the like because objects naturally follow the path of least action. Bohr won the Nobel Prize for the insight that when calculating the orbits of electrons he should stick in h, the quantum of action, every time he saw an action variable in the Hamiltonian.

Because h is a number with a fixed value, this was the same as saying that the electrons in the atom could only have orbits with specified values for the action, and that these values were multiples of Planck's constant. If this were true, Bohr realized, the electrons around the nucleus must exist in fixed arrangements—as if satellites could circle the globe only in certain orbits—and going from one arrangement to another must take or release a certain predetermined amount of energy. These amounts of energy, absorbed or released, come in the form of electromagnetic radiation—that is, light. An electron can absorb light (or, rather, a photon) if and only if the light photon has exactly enough energy to kick that electron from one state to another; no more, no less. Because $E = h\nu$, this is the same as saying that a given orbiting electron can only absorb light of certain frequencies. Similarly, when an electron falls from a high-energy state to one of low energy, it does so by squirting out a photon, again of a definite frequency and energy—the same frequency and energy required to get the electron there to begin with. Quanta of light are the tolls collected or paid by electrons as they jump about their permitted places in the atom. The spectral lines described by Balmer's formula are a set of subatomic hops, skips, and jumps, the jitterbug moves of hydrogen's lonely electron.

Like an artist who dithers over a canvas for years but quickly executes it once inspiration arrives, Bohr wrote up his insight rapidly indeed. He saw the Balmer formula in mid-February; on March 6, he put a draft of a long paper in the mail for Rutherford. In an attached note, Bohr said it was the first of several related articles.[28] Fifteen days later, Bohr sent a second, amplified draft. In the meantime, Rutherford had read the first and found it interesting, if not entirely plausible. He also thought it was much too long. He wrote to Bohr, explaining that "long papers have a way of frightening readers. It is the custom in England to put things very shortly and tersely in contrast with the Germanic method, where it appears to be a virtue to be as long-winded as possible." He kindly offered "to cut out any matter I consider unnecessary in your paper. Please reply."[29]

Bohr replied by taking the first available boat to England, going directly to Rutherford's office, and arguing doggedly against cutting a single phrase. He had spent weeks agonizing over the thing; each stilted sentence had a precise meaning. Rutherford experienced the astonishment, common to editors, caused by an ordinarily quiet writer's apparent willingness to kill over a comma. He relented, and Bohr's three long, historic papers were published in virtually unedited form in the *Philosophical Magazine* issues of July, September, and November of 1913.[30]

The Rutherford-Bohr vision of the atom—a tiny central nucleus surrounded by a pattern of electrons in different states—was confirmed by another Rutherford protégé, Henry Moseley. Bohr had predicted that when the innermost electrons moved from a very high to a very low energy state, they would give off high-energy light—X rays. The precise energy of the X rays would depend on the electric charge of the nucleus, for the stronger the positive charge attracting the electron, the more energy it would use up when it moved outward, or give off when it moved inward, and the higher the frequency of the corresponding X rays. (Because $E = h\nu$, energy and frequency are directly proportional.) Using Balmer's formula and his own ideas, Bohr realized that one ought to be able to work backward and, from the frequencies of the X rays given off, determine the charge on the nucleus. Moseley set out to test this by shooting a beam of electrons at different elements. In an experiment of classic simplicity and elegance, he sealed in a vacuum chamber a sort of toy train, which hauled samples of different elements back and forth in the line of an electron gun; it was as if he had put the samples in the middle of a television tube, blocking off the stream of electrons that generate the image. He planned to try every element, going one by one, step by step from lightest to heaviest. Some of the electrons shot from the gun and crashed into the inner electrons of the target atoms, knocking them free; when others rushed in to fill their place, they gave off light quanta—X rays—in the process. Moseley measured the results and discovered that as the elements got heavier the frequencies increased; as each element was replaced by the next, the frequency rose to match.

The result not only established the Rutherford-Bohr model of the atom, but also illustrated a growing phenomenon in the science of physics, the widening division between theory and experiment. Despite his real brilliance, Rutherford lacked the talent for purposive daydreaming that is the hallmark of the theoretician; if he had possessed it, he would not have been the experimenter he was. And Bohr, despite his enormous powers of concentration, did not possess the skill to overcome the brute intractability of matter necessary for performing experiments. His forte was talk; he liked to stop and think things over, and frequently took advantage of the theorist's freedom to discard or rework a troublesome idea. Experimenters like Rutherford, on the other hand, must invest themselves in a course of action and see it all the way through.

"Science walks forward on two feet, namely theory and experiment," said the American scientist Robert Millikan on the occasion of receiving the Nobel Prize for physics in 1924. He continued, "Sometimes it is one foot which is put forward first, sometimes the other, but continuous progress is only made by the use of both—by theorizing and then testing, or by finding new relations in the process of experimenting and then bringing the theoretical foot up and pushing it on beyond, and so on in unending alternations."[31]

□   □   □   □   □

The Cavendish laboratory, like the whole of English physics—for that matter, like England itself—was shaken by the war. The universities were emptied by the fight: research assistants given artillery commissions, students inducted, professors sent off to do something useful. Just before the hostilities, Geiger left for Berlin; one of Rutherford's best students, James Chadwick, went there to work with him and spent the whole of the war in an internment camp. Marsden served in New Zealand; Moseley refused Rutherford's offer of wartime scientific work and patriotically went to the front. In 1915, he died in the senseless battle of Gallipoli. Rutherford was drawn into working on antisubmarine warfare for the War Research Department, but managed to steal time now and then to perform his own experiments.

Once again, he was thinking about what happened when atoms were bombarded with alpha particles. After Geiger had left for Berlin, Marsden continued experimenting with, again, a target, a screen, and a source of radiation. Alpha particles from the source hit the target and were absorbed or reflected. By this time, people knew quite a lot about alpha particles; they knew, for example, that they had quite a short range—after shooting through a few inches of air, they petered out. What Marsden noticed was that the screen kept flashing even if he moved it farther away than alpha particles were supposed to reach. Whenever he brought a magnet near, the flashes moved in response. Physicists had long known that a magnet bends the paths of charged objects speeding by it—positively charged objects in one direction, negatively charged objects in the other. Just before returning to the Antipodes, Marsden ascertained that the sparks were caused by something with a positive charge, something as light as hydrogen, the lightest element. Rutherford and he quickly had the idea that they were seeing the nuclei of hydrogen atoms. It was logical to suppose that hydrogen nuclei, the lightest known entity of positive charge, might be the positive version of the electron, that is, the particle that had to exist to cancel out the negative charge of the electron. On the other hand, Rutherford realized, the flashes might be due to a previously unknown gas, one lighter than hydrogen. People were still discovering elements; Rutherford himself had found one.[32] He carried on alone, one of the few active physicists in the nation. By the end of 1917, he was fairly sure of the answer: The alpha particles were slamming into atoms of nitrogen in the air, and breaking off chips—hydrogen nuclei. If hydrogen nuclei were *inside* nitrogen nuclei, this was a strong indication that the hydrogen nucleus might be a fundamental building block of matter. It also meant that Rutherford, alone in the Manchester lab except for an assistant, was splitting the atom. Over the next months, he sat in a darkened room, counting minute flashes of light as he varied the experimental apparatus slightly to eliminate the possibility of error. He knew he was onto something important. Nonetheless, he was sufficiently engaged by war research that he did not submit a paper on the splitting of the atom until April 1919.[33]

A year later, he had progressed to the point of calling the ejected hydrogen nucleus a *proton* and arguing that nuclei in general must be made largely of assemblages of protons packed closely together.[34] Protons weigh almost two thousand times more than electrons, which is why the mass of the atom is concentrated in the nucleus. By this time, he had succeeded J. J. Thomson as director of the Cavendish. He ran the laboratory according to his own lights, which fit in well with Cambridge traditions: simple experiments, scavenged equipment, readily understandable ideas, strong facts, sharp inferences. Greedy as ever for quick, cheap, dramatic discoveries, Rutherford tried hundreds of experiments, a scattershot approach that reaped great immediate benefits even as it laid the seeds for the laboratory's long-term decline. The most important Cavendish work stemmed from Rutherford's conviction that even if the proton and electron were the only components of matter, with no more to be found, some close combination of the two might form a neutral particle, and he and his former student Chadwick made sporadic attempts to discover it. In 1931, they had a stroke of luck when the discovery was virtually handed to them on a platter. Irène Curie and Frédéric Joliot, the daughter and son-in-law of Mme. Curie, studied a new type of radiation made by bombarding beryllium with alpha particles.[35] This radiation—high-energy photons, they thought—was uncharged, but so powerful that it could knock protons flying at tens of thousands of miles a second. When Rutherford heard of such photons, he snapped, "I don't believe it!" Chadwick immediately suspected that the protons were reeling from collisions with massive particles, rather than with massless photons. By doing the experiment the Cavendish way, Chadwick was able to show that the radiation indeed consisted of neutral particles about as massive as protons. Neutrons were soon shown to be present in the nucleus along with protons.[36]

By 1932, all three components of ordinary matter—the electron, proton, and neutron—were known. But identifying the parts of the atom told as little about the structure of the nucleus and the forces holding it together as the knowledge that a building is made from brick would say about its architecture. The answers, Rutherford thought, would not be found "in this generation or the next and probably not completely for many years, if at all, or for many hundreds of years, because the constitution of the atom is, of course, the great problem that lies at the base of all physics and chemistry. And if we knew the constitution of atoms we ought to be able to predict everything that is happening in the universe."[37]

The gram of radium in the Cavendish was the heart of the quest, and the institution as a whole. It was kept inside a sort of oven made from lead bricks that rested in a skinny tower on the top floor of the laboratory. Low and heavy, the radium box was treated with the respect of an altar; only the trusted few were allowed near it. There was less than a gram of radium bromide sealed away inside, but that amount produced enough radiation to give the room the sharp, fresh odor of ionization. When an atom of radium splits off an alpha particle, what is left behind is an atom of radon, a heavy, inert, highly radioactive gas that seeps out of the source like fog from a marsh. (The radium was kept in an upper floor so that if the radon somehow got loose it could be fanned out of the window before it collected in the cellar.) Day in, day out, radon gradually accumulated in a sealed glass bottle whose walls were stained purple by radiation; experimenters routinely siphoned off doses thousands of times stronger than the largest permissible measures today. These early experimenters with radioactivity did not have the handicap of knowing how dangerous the phenomenon was; if their precautions had been better, it is unlikely they could have had sources of sufficient strength to get proper results. One of the last people to work with the Cavendish radium, Samuel Devons, told us recently, "I would ask for, say, three hundred millicuries"—a curie is the amount of radioactivity put out by one gram of pure radium—"which today would be considered an absolutely absurd lethal—I mean, you wouldn't go within miles of it! The hullaballoo about it! Christ, I had that in my bare hands. Well, I had rubber gloves on."

We spoke to Devons at the Barnard College History of Science Laboratory, an historical re-creation of early experimentation, which he ran. He has an English manner—tweeds, a white beard outlining his mouth and chin, fiercely hooked nose. His fingers were smudged from working with the equipment in the laboratory. "There was a little flight of stairs going up the tower," he said. "Twenty steps. At the bottom you'd change your jacket for a jacket you kept on a peg and at the top you'd change that jacket and put a coat on and rub your hands with chalk for a little bit and put rubber gloves on. When you came out, there was a place at the top where you were supposed to wash your hands and another at the bottom—progressively getting cleaner. I don't know how much of this was important, but if I forgot to do any of the ritual at all, I could tell when I passed by Maurice Goldhaber's counters. He was doing neutron work with incredibly primitive sources. And as I walked past, just enough stuff would be on me to swamp his counter. If I picked up a Geiger counter after being in the tower and blew on it—wfff! the thing would go *brrrr*." The result, of course, was that much of the laboratory was poisoned by radioactivity. "There were rooms where you could take a Geiger counter and run it down the baseboards and at a certain bench you'd hear that *brrrr*! They were painted over and scraped and painted, but they were never thrown out. It was too expensive. It wasn't all that bad, but oh yes, rooms got contaminated. Certain experiments that required a very clean background you couldn't do in a certain room. And the tower itself was so filthy, it was absolutely shocking."

Rain came down as Devons spoke, making fat wet asterisks on the office windows. His lab was closed for the day, but nobody was going anywhere until the shower was over. On his desk, his copy of the *Collected Papers* of Lord Rutherford lay open to a drawing of the equipment used in a scattering experiment, a simple collection of wire, wood, and brass that could have been, and was, constructed from scrap by a graduate student in a few hours. We asked Devons about the difference between working at the Cavendish and working in one of the enormous national laboratories today. Devons shrugged. The scale of the Cavendish, he said, was so small that people nowadays find it hard to believe. "There are more conferences going on today than there were scientists a hundred years ago. And each conference may have a few hundred, a thousand people in it, say. Physics has just *grown*." He looked up for a moment at portraits of Rutherford and Thomson on the wall. A lamp made from a glass cider jug illuminated their faces. "You might say the underlying basis of it is still doing the same thing. It's like seeing the first Mr. Henry Ford humming away, being his own banker, his own bookkeeper, his own employer, and making cars, right? And the Ford Motor Company of today is still producing cars. But look at the organization today. The people at top, they've got manicured fingernails, and—you see? They've never smelled a furnace."[38]

3
A Children's Crusade

"PEOPLE," HOWARD GEORGI SAID, "STILL HAVE AN EINSTEIN COMPLEX." WE were sitting in his office in Lyman Hall, a lumpy pile of brick in an unlovely corner of the Harvard University campus that over the last thirty years has hosted many of the brightest luminaries in theoretical physics, Georgi among them. His window was open, and the sound of construction drifted across the small, paper-strewn room. Georgi is a tall, limber man with a wide face and an even wider russet beard curled around the edges of his smile. That morning he was wearing tennis shorts and sneakers; his white socks were pulled up nearly to his knees. He was talking about his generation of physicists, the successors to Werner Heisenberg, Bohr, and Planck, and explaining a dissatisfaction, common among his colleagues, with certain aspects of the public legacy of Albert Einstein. As he spoke, he gestured toward a small picture of the savant thumbtacked to a bulletin board. Rumpled, sorrowful, and ethereal, Einstein looked as if he were ready to sink beneath the weight of his own wisdom, whereas Georgi was the picture of American health—he had just come off a tennis court. A racket in its press lay across the papers on his desk. Yellow-green tennis balls dotted the floor, a menace to visitors. "Revolution!" Georgi said, dismissing the subject. "You're always getting asked if physics is having another revolution, if it's like Einstein all over again. But that's not the point at all. The things we've learned are so *interesting*—" here a quick burst of his infectious, high-pitched laughter "—that it's just *foolish* to ask about Einstein. Because that's not what it's about, really—revolution."[1]

In the fall of 1973, Georgi became one of the progenitors of the current drive toward unification when he and another Harvard theorist, Sheldon Glashow, wrote down the first completely unified theory of elementary particle interactions, the first grand unified theory, which started the long chain of events that eventually led to physicists working in salt mines beneath the shores of Lake Erie.[2] Their speculations have attracted considerable public interest. Among other things, they promise to provide an outline history of the Universe, from the first fraction of an instant after the Big Bang to the long cold slumber at the end of all things, as well as a complete inventory of its constituent parts.

Despite the enormous scope and recent impact of his work, Georgi would be the last to call it revolutionary, or indeed to so regard most of the current unification ideas. The reason is that they are all couched in the language known as "quantum field theory," the same theoretical grammar that produced the standard model. Georgi sees the grand unified theories and their successors more as logical culminations of past ideas than as striking novelties. Although they represent significant departures, they are implicit in earlier work.

A couple of years later, we took up the subjects of quantum field theory, the standard model, and unification with Georgi when he came into New York City to give a seminar at Columbia University. He amiably agreed to meet for breakfast at a local landmark, the Hungarian Pastry Shop, near the half-complete majesty of Saint John's Cathedral. The morning was cold, and the cafe's cappuccino machine was busy. Georgi had to lean over the table every now and then to make himself heard over the steam. We asked him about the way theorists approach the art of building unified theories.

Putting such theories together, he said, physicists are strongly guided by the necessity of expressing their thoughts in terms of quantum field theory, the theoretical language that was invented in the late 1920s by melding quantum mechanics and relativity. "What's happened since [that time] is that the lexicon was expanded. We discovered that we could build more theories of this kind. We seem to have found them all now—again, without changing the rules. One can, as I say, almost prove that the lexicon is exhausted, where what I mean by exhausted is that this is all you can do without a dramatic change in the rules."

You need constraints, he said. You need to operate within a framework. "The primary constraint comes from the language which we use. Quantum field theory is, as far as we know, the only way of combining relativity and quantum mechanics, at least without changing the rules in some more complicated way, like assuming there are ten dimensions. You can more or less prove that the only way of building a reasonable quantum mechanical theory that has special relativity in it and that doesn't violate causality and the various other simple principles is to build it in the form of a quantum field theory. So we're constrained by relativity and quantum mechanics to speak this funny language." The waitress called out our order, interrupting the discussion with hot apple turnovers and hotter coffee. "At some level—" Georgi took a healthy bite of his turnover "—it's description. Field theory is a language which you use to describe the interactions of the particles and fields that you see. You fit them into the grammar and syntax of quantum field theory, and then you're sure that you've succeeded in describing the interactions in a way that's consistent with all of the constraints of quantum mechanics and relativity. You can then ask—and sometimes answer—questions such as, why do the interactions have a particular form?"

Using quantum field theory must carry with it some implications, we said. He nodded in agreement. "The language is fundamentally absurd. What I mean by that is this: If you insist that relativity and quantum mechanics work down to arbitrarily short distances, then you're stuck with local quantum field theory down to arbitrarily short distances. What 'local' means is that the interactions which cause everything take place at single points." The cappuccino machine suddenly blasted, and Georgi had to shout. "A point is not a physical thing! A point is an infinitely small mathematical absurdity! So this assumption is crazy. It's clearly an act of hubris to assume that you know what's going on down to arbitrarily short distances." The point business, in fact, caused vast difficulty all the way along; it was only when a definitive end run was made around the problem that quantum field theory became solid enough to base ideas of unification on it.

We mentioned some models and theories we had heard about that had recently been developed in the search for unification—including one with the improbable name of "ten-dimensional string theory"—and asked whether these portended a complete description of nature.

"It depends on what you mean by 'complete,' " Georgi said equably. "I don't think we have the right to talk about a complete description of nature. The point is, one of the things we've learned is that our description of nature is organized by distance scale. Newtonian mechanics, for instance, is fine down to something like a billionth of a centimeter. After that, quantum effects come in. We certainly have the right to say that we have a complete description of nature at distances greater than 10^{-16} centimeters—that's a thousand times smaller than a proton—because we have some information about what's going on at distances greater than 10^{-16} centimeters. We might even guess that we have a complete description of nature down to some shorter scale, and that may not be unreasonable as long as (a) we're really just extrapolating from what we know, and (b) this description has experimental consequences that can be checked some day. That I regard as a sensible sort of physics. Unfortunately, you can also go the other way around and say, 'Look, damn it, the geometry of strings in ten dimensions is so elegant that this must be the unique theory of the world!'—and then try to work your way back to long distances. That seems to me to be rather silly." He laughed, took a bite. "You have to remember that theoretical physicists are parasites. The people that do the real work are experimenters. Theorists, of course, tend to forget that." More laughter; another bite. "A lot of the present speculations I'm really sort of unhappy with. I feel about the present status of grand unification a little bit the way I imagine that Richard Nixon's parents might have felt if they'd been around in the latter days of the Nixon administration. I'm very proud that my creation is so important and that people talk about it all the time, but I'm not really very happy with the things that it's doing at the moment." He finished the turnover. "It's gotten a bit too *theological*."[3]

At Cambridge, Georgi had spoken about the same ideas; they had been much on his mind in recent years. "We've really come an amazingly long way," he said. "I'm sometimes not sure if my colleagues realize that. Everything anyone has ever seen can be described in terms of the standard model. And all of it in some sense stems from relativity and quantum mechanics, plus a few experimental discoveries." He sat cross-legged in his swivel chair, tossing a tennis ball up and down as he considered the question. Above the construction noise was the sound of undergraduate laughter somewhere outside Lyman Hall. "I guess I'm saying that I think we're on the right track," he said finally. "You start with relativity and quantum mechanics, you make quantum field theory, and you proceed from there. If there is going to be unification, I suspect it will stem from that approach. We're not in for revolution. Progress will come from the experimenters, and from things that are already in our hands."[4]

□   □   □   □   □

Physicists today mention them in a single breath, but to their creators, relativity and quantum mechanics could not have been more different.[5] Whereas quantum conditions govern the properties of extremely small bits of matter, relativistic conditions govern the properties of matter traveling at extremely high speeds. Their difficult marriage produced what is called quantum field theory, a sickly child that eventually grew into the robust standard model of elementary particle interactions.

Relativity was developed almost entirely by one man, Albert Einstein, who, like Rutherford, labored alone during the First World War.[6] Einstein had developed some aspects of relativity earlier, in 1905, in a paper that introduced the famous equation $E = mc^2$, although in a slightly different form.[7] In subsequent articles, he struggled to make his ideas more general; he succeeded only at the end of 1915, when the war was in full swing.

Einstein developed relativity in the course of pondering a striking inconsistency between Newtonian mechanics, which describes the gravitational force that makes apples fall from trees and keeps the planets in their orbits, and Maxwell's electrodynamics, which show how electricity, magnetism, and light are different aspects of the single phenomenon of electromagnetism. One of the most impressive aspects of Newtonian mechanics is that its laws do not depend on the location or velocity of the system to which they are applied. Objects fall in the same way whether they are in New York or Tokyo, whether they are in the middle of a stadium or traveling thousands of miles an hour in a jet plane. If someone in the plane were to throw a ball out the window, it would be moving much, much faster relative to the ground than to the aircraft. Nonetheless, the trajectory of the ball could be predicted from Newton's laws, regardless of whether the point of view was the ground or the plane.

The situation is radically different in the four equations that sum up Maxwell's electrodynamics. Electricity, magnetism, and optics can only be tied together if light travels at specific velocities in specific media—186,000 miles per second in outer space, somewhat slower in air. For example, if instead of throwing a ball the airplane passenger shines a flashlight, the light, unlike the ball, would not be moving faster relative to the ground than to the aircraft.

Both theories were enormously successful, but both could not be completely correct. Starting with Maxwell, a number of prominent physicists tried to reconcile them, usually by postulating that the speed of light was not really an absolute quantity, but relative to some sort of invisible medium that carried it. This medium, which was called the *ether*, was everywhere the same, and Maxwell's laws held strictly for light's passage through it. Newton's laws also held for light, but they predicted variations in speed that were too small to be noticed.

This explanation was set back considerably when it was established that there was no ether pervading space. In 1895, the great Dutch theorist Hendrik Lorentz constructed a series of equations on an *ad hoc* basis that attempted to resolve the conflict between Maxwell and Newton by introducing changes in the definition of space and time.[8] This set of rules is today known as the "Lorentz transformations," and quantities that fit into them are said to be "Lorentz invariant," that is, they are not thrown off kilter by the redefinitions. Lorentz introduced his transformations simply to show that a mathematical schema could be built in which electromagnetic phenomena could be described independent of the point of view, as was the case for Newtonian mechanics, although one had to pay the price of doing odd things to space and time. Over a decade later, Einstein decided the Lorentz's redefinitions had much more meaning than their creator knew: Space and time truly were different than had previously been supposed.[9]

Einstein's fundamental reconception of both is the foundation of the theory of relativity, an achievement so great that his successors still marvel that one man could have accomplished it. It was accomplished in two stages, the first, special relativity, in 1905, and the second, general relativity, completed in 1915 after several false starts. With special relativity came a host of strange effects—time slowing down at high speed, the shortening of objects in the direction of motion, and so forth—all of which are mathematically derivable from the necessity of changing Newton to fit Maxwell. Small wonder that when Einstein first put forward his even stranger ideas about general relativity, which included the concept that space is curved, many physicists had not the faintest notion of what he was talking about. In practical terms, Einstein's legacy is the physicist's duty to work only with quantities and theories that take account of relativity, special and general, by being Lorentz invariant.

General relativity had few predictions that could then be tested by experiment. One of its deepest consequences involved the curvature of space. Although it was not widely known at the time, Newton's laws could be interpreted as predicting that strong gravitational fields would bend light in much the same way that strong magnetic fields cause charged particles to swerve off course. General relativity made a similar prediction, but added that the curvature of space around massive bodies would cause light to bend even more. Many scientists withheld their acceptance of the theory until this prediction was examined. The best way Einstein could think of to test this was to see how much the light from stars bent as it passed near the sun. Performing this experiment required observing the sun during a total eclipse; the stars within a few degrees of the sun, ordinarily invisible, would then be both viewable and out of position by tiny fractions of a degree. The fate of relativity depended on how much.

In 1912, Argentine scientists traveled to Brazil to measure starlight during an eclipse; the experiment was rained out. A German expedition to observe an eclipse at the Crimea in 1914 was foiled by the outbreak of war. As it happened, these failures were lucky, for in 1915 Einstein finally understood how the curvature of space bent starlight, and revised his predictions. The war also scotched a 1916 expedition to study an eclipse in Venezuela; an American team was unable to establish anything conclusive from an eclipse two years later. Two British teams finally ran a successful experiment that confirmed Einstein's prediction. They announced their results on November 6, 1919, at a joint meeting of the Royal Society of Sciences and the Royal Astronomical Society. J. J. Thomson himself was the chair of the gathering; although older and increasingly set in his ways, he proclaimed, "This is the most important result obtained in connection with the theory of gravitation since Newton's day," describing the result as "one of the highest achievements of human thought."[10] The following day, the London *Times* headlined[11]

REVOLUTION IN SCIENCE.

NEW THEORY OF THE UNIVERSE.

NEWTONIAN IDEAS OVERTHROWN.

Further in the article was the alarming subhead

SPACE "WARPED"

(In fact, Einstein had not demonstrated that Newton was wrong; he had not even claimed he did. Rather, relativity showed that factors had to be added to Newtonian calculations, and that these factors could be easily ignored except in certain circumstances.) Three days later, the *New York Times* printed its first article on relativity. Crossing the Atlantic, Thomson's praise inflated; he was now quoted as saying the result was "one of the greatest—perhaps the greatest—of achievements in the history of human thought."[12] A second *New York Times* article, printed the next day, had a marvelously grandiose set of headlines:[13]

LIGHTS ALL ASKEW
IN THE HEAVENS

Men of Science More or Less
Agog Over Results of Eclipse
Observations.

EINSTEIN THEORY TRIUMPHS

Stars Not Where They Seemed
or Were Calculated to Be,
but Nobody Need Worry.

A BOOK FOR 12 WISE MEN

No More in All the World Could
Comprehend It, Said Einstein When
His Daring Publishers Accepted It

During the next few years Einstein toured the world, speaking to packed audiences. In the popular imagination, the man typified physics, even science itself. For the rest of his life he would remain "Mr. Physicist," in the mainstream of the popular idea of physics. But he did not remain in the mainstream of the discipline, for he rejected the quantum theory he had done so much to establish. He spent much of the rest of his life on a fruitless quest to unify gravity and electromagnetism, hoping to show that they are two aspects of the same thing, within the framework of what he called a "unified field theory." Einstein's attempt to unify was, to say the least, premature, and he failed utterly. The year 1919 represents the culmination of his career; thereafter, as physicists became increasingly preoccupied with quantum theory, Einstein's views became for his peers a source of puzzlement, sorrow, and finally indifference.

"Already in 1905," Einstein once said, "I realized what a *Schweinerei*—" what a stinking mess "—the quantum theory was."[14] Relativity, like Athena, sprang full-grown from Einstein's head. Quantum mechanics, on the other hand, had a difficult delivery that required dozens of midwives. The notion that many aspects of the atomic domain were quantized proved singularly difficult for physicists to grasp, although the best among them were absorbed by the challenge of interpreting it. They learned and relearned that atoms and their constituents have properties that can be described in terms of simple numbers— +1, 0, and −1, say, or perhaps −½ or +⅔—multiplied by constants like h (Planck's constant), c (the speed of light), or pi (the ratio of the diameter of a circle to its circumference), which implied that the relations among the bits and pieces of the atom were vastly different than the interactions among ordinary, nonquantized objects. Physicists found the puzzles electrifying, but the answers were dissatisfying, abstract, even nonsensical; as hard to understand as they were impossible to visualize.

Early in the 1920s, theorists realized that what a particle is and how it behaves can be completely identified by a small set of numbers. There just isn't anything more to describe; particles are too simple. In this respect, the pieces of the atom are vastly different from objects like tables and cans of chicken noodle soup. The Campbell's soup cans elevated by pop art into symbols of uniformity are in truth not uniform at all; each can has hundreds of physical characteristics that can be measured and described, such as its precise shades of red and gold or the exact dimensions of its label, and the soup cans stacked in a supermarket differ, however slightly, in all of them. Subatomic particles, on the other hand, have less than a dozen attributes. They have no color, taste, nor odor; they are neither hard nor soft, shiny nor dull. They have only a handful of numbers for a few simple properties. All electrons have exactly the same electric charge, −1; protons all have a charge of +1.[15] An electron at rest has a mass of about 10^{-30} kilograms; a proton's mass is about 1,836 times greater. As far as scientists know, there is no reason why electrons couldn't have a charge of, say, −1⅜, and protons a mass, say, nine hundred times larger than electrons. But they don't.

Oddest of all, however, some of the numbers describing how particles behave are not only identical, but also quantized; that is, they can have *only* certain particular values. Niels Bohr showed that orbiting electrons have specified energies. Moreover, electrons whirl about the nucleus with just a few preferred values of angular momentum ("angular momentum" can be loosely thought of as the momentum with which something, in this case an electron, goes around a curved path). Because the electron can be curving any which way, angular momentum is associated with direction. The direction—or, anyway, the part of it that can be measured—is also quantized. It, too, can only have particular values that physicists can write as 0, 1, 2, or 3.[16] The complete set of such numbers for a particle is called that particle's *quantum numbers*. Particles with the same quantum numbers are absolutely identical—they *cannot* be distinguished.

Freshly encountering this situation, it was of obvious interest to physicists in the 1920s to specify a complete list of the quantum numbers associated with the electrons in atoms. A decisive contribution was made by a
twenty-five-year-old Viennese named Wolfgang Pauli. Brilliant, acerbic, and
fat, Pauli was already well on his way to becoming one of the great physicists
of the century; even as a young man, his intellect and passionate style of
argumentation intimidated many of his colleagues. He had a kind heart, an
irascible temperament, and a streak of melancholia that he overcame by
driving at physical questions as if his life depended on it, which perhaps it
did. Like many theoreticians of the day, Pauli was concerned with understanding the spectral lines emitted by atoms. Bohr's original model worked
for the relatively simple patterns emitted by hydrogen, but heavier, more
complex elements were much harder to understand. For example, cesium,
strontium, and barium—several of what chemists call the alkaline-earth
metals—produce spectral lines that upon close examination are seen actually
to be split in two; these lines, called doublets, are made of two almost
identical frequencies. In December of 1924, Pauli suggested that a complete
list of the quantum numbers of an orbiting electron would include its energy,
angular momentum, and orientation in space; in addition, to explain the
alkali doublets, he suggested that there had to be a fourth quantum number,
which he called, rather unhelpfully, *Zweideutigkeit*—two-valuedness.[17]

In the summer of 1925, Samuel Goudsmit, a young Dutch physicist,
was trying to explain Pauli's ideas to a young countryman, George Uhlenbeck, who had been out of Holland for a while and was trying to get back into
physics. Uhlenbeck had been told by his teacher that he should be briefed
by Goudsmit. During afternoon talks, Goudsmit explained Pauli's four quantum numbers to Uhlenbeck. "I was impressed," Uhlenbeck recalled later,
"but since the whole argument was purely formal, it seemed like abra-
cadabra to me. There was no picture that at least qualitatively connected
Pauli's formula with the old Bohr atomic model."[18]

It occurred to Uhlenbeck that Pauli's *Zweideutigkeit* was not really
another quantum number, but simply another property of an electron. He
suggested that perhaps an electron spins on its axis like a toy top; unlike a
top, however, the spin of the electron would be quantized, and it could only
turn at certain speeds. Looking at Pauli's formulas, Uhlenbeck and Goudsmit
realized that if electrons had a second angular momentum associated with a
spin, this would perfectly account for both "two-valuedness" and the double
spectrum lines from the alkaline earths. The amount of spin was one-half $\hbar$,
where $\hbar$ is physics shorthand for $h/2\pi$, that is, Planck's constant divided by
twice pi. Although both men were struggling in the sea of quantization, they
"appreciated right away that if the [spin] angular momentum of the electron

was $\hbar/2$, one had a picture of the alkali doublets as the two ways the electron could rotate with respect to its orbital motion."[19] The two ways of rotation would give the electrons two slightly different energy levels, and electrons with two slightly different energy levels would create two slightly different frequencies of light.

They took their idea to Uhlenbeck's teacher, Paul Ehrenfest, who headed the physics department in Leiden. Ehrenfest made some suggestions, had them write up a short paper about spin, and then told them to take it to Hendrik Lorentz, the grand old man of Dutch physics. In addition to inventing the Lorentz invariance on which special relativity is based, Lorentz was the first man to construct a theory of the electron. In 1925, Lorentz was seventy-two and ostensibly retired, but he still taught a class at Leiden every Monday morning from eleven to noon. After one class, Uhlenbeck and Goudsmit showed Lorentz their paper, which was only a few paragraphs long.

"Lorentz was not discouraging," Uhlenbeck once said. "He was a little bit reticent, [but] said that it was interesting and that he would think about it." *Thinking*, for Lorentz, was apparently an active occupation. "It was so typical of Lorentz that he immediately made very extensive calculations on the classical theory of rotating electrons. I think the next week, but maybe two weeks later, he gave me such a *stack* of papers with long calculations. Large white paper, I still remember. He tried to explain it to me, but it was so learned that I . . ."[20] Uhlenbeck's voice trailed off. Lorentz had explained several problems, one of the most grievous being that if the electron really had a spin angular momentum of $\hbar/2$, this implied that it rotated with a particular velocity. The old man had figured out the speed: ten times faster than the speed of light. Because nothing can go faster than the speed of light, this was, Uhlenbeck and Goudsmit decided, a devastating critique. They were most unhappy.

Uhlenbeck was then a professor at Rockefeller University, a set of isolated buildings of funereal modernity on the Upper East Side of Manhattan. The university is one of the more carefully guarded in the nation, and we were detained by an armed guard for some twenty minutes. It was one of those clear mornings when the gray, boxy magnificence of the cityscape is just slightly fuzzed over by smog. He was eighty-four, a tall man with thinning hair scattered across his head like so much straw. Impatient with his growing deafness, he asked us to sit close. "Talk in my left ear," he said cheerfully. "The other one is primarily for decoration." He kept a pair of glasses in his hand, twisting the frames around his fingers like worry beads. We asked him what he had done with Lorentz's calculations.

"I thought, well, therefore it is all wrong, what we have done. And I went back to Ehrenfest and said, 'You better not publish that paper, because Lorentz has shown that it is not correct.' He said, 'I sent it out right away. It will come out next week.' "

Ehrenfest had mailed it off without waiting to hear from the master? Uhlenbeck roared with laughter.

"He *knew!*" he said. Ehrenfest had immediately seen the difficulties with spin. "But he said to us—he said it in German—'*Sie beiden sind jung genug sich eine Dummheit leisten zu können.*' Both of you are young enough to afford a stupidity!"[21]

Ehrenfest was not being entirely cavalier. It is important for theoreticians not to pay attention at all times to what is wrong with their ideas. They must not be too afraid to follow their intuition; every now and then, the best way to hit a target is to shoot from the hip. Spin, as an example, had been thought of earlier, by one Ralph de Laer Kronig, an American of Hungarian descent. Unfortunately, Kronig asked caustic Wolfgang Pauli for his reaction to the notion that *Zweideutigkeit* was due to the effects of a spinning electron. Pauli tore apart the idea and, to the later regret of both men, Kronig never published it.[22] Bohr, on the other hand, dismissed Lorentz's objections immediately, telling Uhlenbeck that the faster-than-the-speed-of-light problem would "disappear when the real quantum theory is found."[23]

As it turned out, Bohr was correct. When the real quantum theory was found, the spin problem disappeared. But the complete quantum theory, quantum mechanics, could not be established until the various pieces of quantum theory—spin, angular momentum, orbitals, and the rest—were expressed in a common mathematical language. Metaphorically speaking, every physical theory is woven on a frame supplied by mathematics. The frame provides the backing, the warp and woof on which physicists tie and knot their tapestries of ideas. Newton, for instance, could not have put together the laws of mechanics if he had not first invented a language in which to write them: calculus. Similarly, Einstein could not have described general relativity if a sickly German named Georg Friedrich Bernhard Riemann had not developed a strange kind of geometry in which parallel lines could meet: Riemannian geometry. As the realm of quantization grew apace, the need became evident for a mathematical formalism that could handle the eccentric requirements of the new physics.

One of those who felt this need particularly strongly was a young German named Werner Heisenberg. The son of a professor of Greek at the University of Munich, Heisenberg had the sort of good looks usually associated with the word *dashing*. His character was a museum of turn-of-the-century Teutonic virtues: romantic temperament, ironic patriotism, love of intellection and the natural world. He was thoroughly steeped in the Greeks;

he was an excellent pianist; he could recite Goethe from memory. His late autobiographical essays are full of evocations of the special quality of well-being he felt walking across a winter beach, the cold spray across his face, head full of the ticking mechanisms of the world.[24]

Just before Heisenberg's seventeenth birthday, Germany was swept by a leftist rebellion. Munich was a center of the fighting; when the revolution collapsed, the city spent the spring of 1919 in a state of starvation and violent anarchy. The re-formed national government assembled troops to seize control of the city, and Heisenberg, caught up in an adolescent sense of the moment, volunteered to act as a guide. After a few weeks of shooting, Munich was captured, and Heisenberg was assigned to guard the telephone exchange. The city began to calm down. Heisenberg realized that classes would soon start again.[25]

"I had duty during the night," he explained years later, "and it was a nice summer in 1919. During the morning at 4:00, nothing was happening at the [telephone] office, of course, and somehow I couldn't sleep, so I went up to the roof of the house into the sunshine. It was nice and warm. I had Plato's *Timaeus* with me. I studied the *Timaeus* partly to keep up with the Greek, because I had to know Greek for my examination, but partly also because I was really fascinated by atomic theory. You know all of Plato's atomic theory was in the *Timaeus*."[26]

The *Timaeus* baffled Heisenberg. The basic text of Greek cosmology, it asserts that the Universe will never be understood until the smallest components of matter are known, and that these tiniest pieces consist of tiny right triangles, jointed together in various arrangements to form all the regular bodies of solid geometry. Heisenberg thought Plato's ideas were ridiculous. Nonetheless, Heisenberg was left, as he wrote later, with the feeling that "in order to interpret the material world, we need to know something about its smallest parts"; furthermore, he continued, "I was enthralled by the idea that the smallest particles of matter must reduce to some mathematical form."[27] He decided to study atomic theory.

During the 1920s, there were three outstanding centers of theoretical physics: two older German schools, Munich and Göttingen, and an institute at Copenhagen founded by Bohr in 1921, which soon rivaled the others in importance. Each had its own character: Munich was known to be more physically oriented, Göttingen was one of the world centers of mathematics, and Bohr's institute was characterized predominantly by the philosophical attitude of Bohr himself. Heisenberg eventually spent time at all three.

In 1921, he attended Munich, where his father still taught, as an undergraduate, then went to Göttingen a year later. In the last months of 1923, he went back to Munich to obtain his doctorate—and almost flunked. One of his examiners, Wilhelm Wien, was annoyed by Heisenberg's indifference to experimentation, and amused himself during the orals by asking the Ph.D. candidate questions about experimental technique. How clear were the images from telescopes? Wien asked. Heisenberg had not the faintest idea. How does a battery work? Heisenberg didn't know. Wien expressed his incredulity. A lengthy argument ensued, during which Wien announced that Heisenberg had failed, and the other two examiners insisted that he was the best physics student at the school in many years. Compromise was finally achieved in the form of letting Heisenberg pass with a *rite*, the equivalent of the "gentleman's *C*" given to wealthy but indifferent scholars in elite American prep schools.[28] The humiliated Heisenberg then became a *Privatdozent*, a very junior teaching fellow, in Göttingen. At about the same time, he got a fellowship and split his time between Copenhagen and Göttingen.[29]

At Munich, he was taught what was being called *quantum mechanics*. But as a graduate he found that the new laws of Bohr, Planck, and the rest were often nothing but hunches and badly formed approximations—it was a great shock, he said later, learning that theoreticians *guess* at answers—and, in the winter of 1925, twenty-three-year-old Werner Heisenberg set out to create for himself a more rational quantum theory.[30] He was part of a *nouvelle vague* of young theorists, arrogant, self-assured, and convinced that the time had come for children to sort out the confusion bequeathed them by their elders. The quest, he thought, was best pursued by the young—an attitude certain to ruffle feathers in the authoritarian German university system.

Having struggled to make a picture in his head of the orbit of an electron—the fixed states Bohr had imagined, and that he and others had elaborated on—Heisenberg began to wonder if the game was worth the candle. Clearly, the electrons were not simply going around the nucleus. If that were all that was going on, the electrons would (as Bohr had known ten years before) either fall into the nucleus or push each other away. In addition, electrons seemed to be able to jump almost instantly from one orbit to another and back again, absorbing and releasing the necessary energy for the move in the form of a photon. This was like claiming that Venus could suddenly hop close to the Earth, stay a while, and then, without missing a beat, leap back to its original orbit. Heisenberg thought that calling that kind of behavior *orbiting* was stretching the term out of shape.[31]

Moreover, because you couldn't ever *see* the orbit, Heisenberg wondered if one shouldn't junk the idea of electron orbits altogether—temporarily, anyway—and think about what could be seen, namely, the spectral lines.[32] These lines were created by lightwaves of specified frequencies. If the frequency of a lightwave is known, it is an easy task to calculate the wavelength, the distance between two successive crests or troughs. With a bit more mathematical finagling, one could figure out the amplitude of the waves, that is, how high the crests are. Unlike orbits, the frequencies and amplitudes could be directly measured in the lab. Physics being an empirical science, Heisenberg liked basing his ideas on tangible quantities rather than intangible entities like "orbits." If a particular frequency was supposed to be emitted when an electron jumped from one "state"—Heisenberg's replacement for "orbit"—to another, one could make a little table representing all possible states and frequencies:

	S_1	S_2	S_3 $\cdots$
S_1	ν_{1-1}	ν_{2-1}	ν_{3-1} $\cdots$
S_2	ν_{1-2}	ν_{2-2}	ν_{3-2} $\cdots$
S_3	ν_{1-3}	ν_{2-3}	ν_{3-3} $\cdots$
$\vdots$	$\vdots$	$\vdots$	$\vdots$ $\ddots$

S_1 stands for "state #1," S_2 for "state #2," and so on, and ν, again, is physics shorthand for "frequency." When an electron goes from S_1 to S_2, it produces light of frequency ν_{2-1}.

At this point, Heisenberg wrote later, "My work along these lines was advanced rather than retarded by an unfortunate personal setback."[33] He was smitten by such a severe bout of hay fever that he had to ask his adviser, Max Born, for a two-week leave of absence. A dripping, sneezing, coughing wreck, Heisenberg fled for Helgoland, a small island in the North Sea, where he hoped the bracing, pollen-free sea air would help him recover. The landlady of his *Gasthaus* took one look at his swollen face and concluded that he had been in a fight. But she installed him in a quiet, second-floor room with a view of the nearby houses, the beach, and the dark expanses of the North Sea. There Heisenberg made the fundamental discovery that you can accomplish more when nobody bothers you and you don't have a phone.

Working in a solitary fury of excitement, he realized that he could construct equations describing the tables of frequencies, amplitudes, positions, and momenta—"quantum-mechanical series," as he called them— that related the quantities that experimenter *actually observed*. Adding in an extra assumption or two, he came up with an awkward but definitely workable scheme that did not rest on ineffable ideas like nonorbiting orbits. Then he realized that he wasn't sure if his scheme was consistent with the law of conservation of energy.

I concentrated on demonstrating that the conservation law held, and one evening I reached the point where I was ready to determine the individual terms in the energy table, or, as we put it today, in the energy matrix, by what would now be considered an extremely clumsy series of calculations. When the first terms seemed to accord with the energy principle, I became rather excited, and I began to make countless mathematical errors. As a result, it was almost three o'clock in the morning before the final result of my computations lay before me.

The energy principle still held. Sitting in his rented room that June morning, Heisenberg felt that oceanic sense of clarity, of luminous and special insight into nature, that is the greatest joy a scientist can experience. "At first," he recalled,

I was deeply alarmed. I had the feeling that, through the surface of atomic phenomena, I was looking at a strangely beautiful interior, and felt almost giddy at the thought that I now had to probe this wealth of mathematical structures nature had so generously spread out before me. I was far too excited to sleep, and so, as a new day dawned, I made for the southern tip of the island, where I had been longing to climb a rock jutting out into the sea. I now did so without too much trouble, and waited for the sun to rise.[34]

But later, when Heisenberg then tried to calculate something with his quantum-mechanical series, he discovered a distressing asymmetry: No matter how hard he worked, his new quantum theory seemed to violate one of the first mathematical principles learned by every schoolchild, the commutative law.[35] According to the commutative law, the order in which you multiply two numbers does not affect the result: $A \times B$ always equals $B \times A$. But quantum-mechanical series A multiplied by quantum-mechanical series B did not give you the same answer as quantum-mechanical series B times quantum-mechanical series A. Heisenberg did what any theorist would do when a horrible problem marred a wonderful idea: He swept the difficulty under the rug. In the paper he submitted that July, the noncommutativity is only mentioned, rather sheepishly, in a single sentence, which is quickly followed by an example in which, oddly enough, the problem does not show its ugly face.[36]

That summer, Heisenberg traveled to Berlin, Leiden, and Cambridge to talk about his ideas. At the Cavendish, where Heisenberg knew almost nobody, he stayed with Ralph Fowler, Rutherford's son-in-law and the house theoretician. After two months of constant travel, work, and hay fever, Heisenberg was dazed with exhaustion. When Fowler left for a day of meetings in London, Heisenberg went to breakfast and slept through the entire day at the table, terrifying the maid.[37] The next day, speaking to the Cavendish experimenters, Heisenberg merely stressed the need for a new theory of spectral lines; but when he talked privately to Fowler, he pushed his own recent work. Fowler asked him to send over a copy of the proofs of the article as soon as they were available.[38]

Back in Göttingen, Heisenberg's supervisor, Max Born, had in the meantime been poring over the quantum-mechanical series. Fascinated, he felt that there was something important in the paper, but that he wasn't sure what it was. In the middle of July, he suddenly realized that he had seen these quantum-mechanical series before. They were, in fact, what mathematicians call matrices.[39]

Although they are today taught in any linear algebra course, matrices were virtually unknown to physics then. A classic use of matrices is given in the "payoff" matrix depicting the possible outcomes of matching pennies, the children's betting game in which the two participants flip pennies simultaneously, Child #1 winning if the coins land with the same face showing, Child #2 if one penny comes up heads, the other tails. The matrix below shows which child will win for the various combinations of throws.

		CHILD #1	
		Heads	Tails
CHILD #2	Heads	1	2
	Tails	2	1

Ordinarily, mathematicians just write the matrix without the marginal entries. Thus, the penny-matching payoff would be

$$\begin{pmatrix} 1 & 2 \\ 2 & 1 \end{pmatrix}$$

with the parentheses indicating that it is a matrix. Other examples of matrices include

$$\begin{pmatrix} 1 & 2 \\ 3 & 4 \\ 5 & 6 \\ 7 & 8 \end{pmatrix} \quad \begin{pmatrix} a & b \\ b & a \end{pmatrix} \quad \begin{pmatrix} 1 & 0 & 0 \\ 0 & 1 & 0 \\ 0 & 0 & 1 \end{pmatrix} \quad \begin{pmatrix} 1 & 1 & 1 & \dots \\ 1 & 1 & 1 & \dots \\ 1 & 1 & 1 & \dots \\ \vdots & \vdots & \vdots & \ddots \end{pmatrix}$$

The second matrix from the left is identical with the penny-matching payoff matrix written earlier; the numbers have been turned into letters. The matrix on the right is an infinite matrix, with the entries stretching endlessly in all directions.

A French mathematician, Augustin Cauchy, was the first to write matrices in the rectangular form used today. His successors, especially Arthur Cayley, a Cambridge mathematics professor, realized that these tables had mathematical properties. They could be added together, subtracted from each other, and multiplied; they even had inverses. To find the product of two matrices

$$\begin{pmatrix} a & b \\ c & d \end{pmatrix} \times \begin{pmatrix} A & B \\ C & D \end{pmatrix}$$

Cayley multiplied and added the members of the individual rows and columns

$$\begin{pmatrix} aA + bC & aB + bD \\ cA + dC & cB + dD \end{pmatrix}$$

This had a strange implication. If he multiplied the two matrices

$$\begin{pmatrix} 1 & 1 \\ 2 & 2 \end{pmatrix} \qquad \begin{pmatrix} 2 & 2 \\ 1 & 1 \end{pmatrix}$$

the answer depended on the order in which he put the matrices. Despite the similarity of the two matrices,

$$\begin{pmatrix} 1 & 1 \\ 2 & 2 \end{pmatrix} \times \begin{pmatrix} 2 & 2 \\ 1 & 1 \end{pmatrix} = \begin{pmatrix} 2+1 & 2+1 \\ 4+2 & 4+2 \end{pmatrix} = \begin{pmatrix} 3 & 3 \\ 6 & 6 \end{pmatrix}$$

but reversing the order of the two produced

$$\begin{pmatrix} 2 & 2 \\ 1 & 1 \end{pmatrix} \times \begin{pmatrix} 1 & 1 \\ 2 & 2 \end{pmatrix} = \begin{pmatrix} 2+4 & 2+4 \\ 1+2 & 1+2 \end{pmatrix} = \begin{pmatrix} 6 & 6 \\ 3 & 3 \end{pmatrix}$$

Because $\begin{pmatrix} 3 & 3 \\ 6 & 6 \end{pmatrix}$ is not the same as $\begin{pmatrix} 6 & 6 \\ 3 & 3 \end{pmatrix}$, matrix multiplication is not commutative.

Born was one of the few physicists in Europe—perhaps the only one—with a good knowledge of matrix mathematics.[40] He realized that Heisenberg's quantum-theoretical series were nothing more, nothing less, than awkward manipulations of frequency matrices

$$\begin{pmatrix} \nu_{1-1} & \nu_{2-1} & \nu_{3-1} & \cdots \\ \nu_{1-2} & \nu_{2-2} & \nu_{3-2} & \cdots \\ \nu_{1-3} & \nu_{2-3} & \nu_{3-3} & \cdots \\ \vdots & \vdots & \vdots & \ddots \end{pmatrix}$$

Born was delighted. Rewriting Heisenberg's equations as matrices led to a whole new world of applications he could explore. The first thing he figured out was that the matrix q for position and the matrix p for momentum are noncommutative in a very special way: That is, pq is not only different from qp, but the difference between pq and qp is always the same amount, no matter what p or q you chose. Mathematically, he wrote this

$$pq - qp = \hbar/i$$

where $\hbar$, as usual, is Planck's constant divided by twice pi, and i is the special symbol mathematicians use for the square root of minus one.

Born was pleased with this result, but he was tired, and wanted someone else to help him develop the insight. He approached Pauli, a former student; the latter bluntly informed Born that he would end up ruining Heisenberg's lovely ideas with "tedious and complicated formalisms." Born turned to another pupil, Pascual Jordan; by the end of July, the two men had laid out the basic principles of what was latter called matrix mechanics. A skittish, egotistical man, Born was driven by the excitement of the chase into a nervous collapse that lasted through all of August 1926.[41] Jordan had finished most of the work by the time Born returned; their paper was submitted to the *Zeitschrift für Physik* a few weeks later.[42]

Principally, what the two men had discovered was a deep connection between classical physics and quantum mechanics. Just as relativity is an extension of nineteenth-century physics, the new mechanics of Born, Jordan, and Heisenberg stemmed from Newton. "The concepts for quantum mechanics can only be explained by already knowing the Newtonian concepts," Heisenberg remarked in an interview many years ago. "That is, quantum theory is based upon the existence of classical physics. This is the point which Bohr emphasized so strongly—that we cannot talk about quantum physics without already having classical physics."[43] Without Newton, quantum mechanics would have nothing to diverge from; without classical mechanics, the new physicists would not even have words and concepts to redefine. As physicists will, Born and Jordan made the connection in formal terms: After developing the matrix methods they needed for quantum theory, they argued that their matrices could be linked to classical theory by using them in Hamiltonians. (Bohr had plugged Planck's constant into the Hamiltonian; now matrices also went in.) Intrigued, Heisenberg wrote to Jordan, and the three men began a series of discussions that culminated at the end of October, when Heisenberg joined Born and Jordan in Göttingen for a week of intense work. Born had promised to lecture in the United States, and was anxious to publish; Heisenberg thought they had an appalling amount of work to do, and no time to do it. He stole a moment to dash off a letter to Pauli in Zurich: "I am almost entirely occupied just now with quantum mechanics, and I further doubt very seriously whether this problem—writing this three-man paper—really can be solved in a finite time."[44] Nonetheless, after a week the three men had a draft of a long article; the complicated paper was sent out in mid-November.[45]

But by that time, to their chagrin, they had company.

4
Uncertainty's Triumph

ONE OF THE MOST TRYING ASPECTS OF PRACTICING THE ART OF PHYSICS IS
that the shape of the answer is not known from the outset. Although they can
draw upon advanced experimental technology and the wealth of data collected by past scientists, physicists must work in the dark whenever they
proceed close to the frontier of knowledge; they are aided only by a set of
aesthetic prejudices, a few mathematical tools, and the knowledge that whatever they come across is unlikely to contradict directly the conclusions of the
past, although it may modify them. The ideas of theoreticians must be at
least somewhat amenable to being tested by others in the community; experimenters need to make it seem plausible that others could reproduce
their work, and achieve the same results. But these guidelines leave more
than enough room for error, and the scientists who make the most remarkable advances—perhaps especially the most remarkable advances—are almost inevitably haunted by doubt, anxiety, and the fear of being forgotten.

Consider Werner Heisenberg, residing temporarily at Göttingen, writing to his close friend, Wolfgang Pauli, on the very day that the three-man
paper on matrix mechanics is received by the office of the *Zeitschrift für
Physik*. Heisenberg encloses a copy, now apparently lost, of the typescript,
which, like most products of Göttingen at the time, is laced with complex
mathematics. (Heisenberg prefers the air more in Copenhagen, where
Bohr's approach is more conceptual.) The math, which comes mainly from
Jordan and Born, makes Heisenberg a little uneasy. He awaits, a bit ner-
vously, the famously critical Pauli eye. Here is how he feels just after completing the work for which he was later awarded the Nobel Prize:

*I've taken a lot of trouble to make the work physical, and I'm relatively content
with it. But I'm still pretty unhappy with the theory as a whole and I was
delighted that you were completely on my side about [the relative roles of]
mathematics and physics. Here I'm in an environment that thinks and feels exactly
the opposite way, and I don't know whether I'm just too stupid to understand the
mathematics. Göttingen is divided into two camps: one, which speaks, like [the
prominent mathematician David] Hilbert (and [another mathematical physicist,
Hermann] Weyl, in a letter to Jordan), of the great success that will follow the
development of matrix calculations in physics; the other, which, like [physicist
James] Franck, maintains that the matrices will never be understood. I'm always
annoyed when I hear the theory going by the name of matrix physics. For a while,*

I intended to strike the word "matrix" completely out of the paper and replace it
with another [term]—"quantum-theoretical quantity," for example.

Fortunately, Heisenberg did not follow up this last suggestion. He continues, grumbling in an aside: "Moreover, 'matrix' is one of the dumbest mathematical words that exists." He wonders if the whole mathematical base of the theory had not been constructed too hurriedly. In the paper, Born, Jordan, and he have been satisfied with rough, approximate calculations; Heisenberg now tells Pauli that "when one *really* integrates," chunks of the theory reveal themselves to be nothing but "formal garbage."[1]

His apprehensions were more than justified. Matrix mechanics impressed his colleagues, but many, if not most, of them found the equations impossible to solve. In addition, they resisted the notion, propounded by Heisenberg, that the impossibility of visualizing matrices was actually one of the theory's more favorable qualities; this seemed as opaque as the Kierkegaardian dictum that the fact that one cannot comprehend how God could be is evidence for His existence. Other theorists pointed out that if quantum matrices were all based on the light emitted in transitions from one atomic state to another, the theory therefore admitted the existence of steady states—so much for Heisenberg's hope to get rid of the detested orbits!—and therefore should encompass some picture of exactly what these states are. (It did not.) Finally, there was the transition problem. In the everyday world—what physicists tend to refer to as the "macroscopic scale"—there is no need for matrices to describe the position and momentum of, say, the trees on a front lawn or the motorcycles whizzing by them. It was therefore necessary to specify at what point matrix mechanics came into the picture, and how one made the passage from ordinary variables to "quantum-theoretical quantities."[2]

Nonetheless, physicists had few alternatives; with varying degrees of enthusiasm, they gamely tackled matrices. Pauli, an early advocate of the approach, helped the matrix advocates along when he used the new methods to arrive at the Balmer spectrum lines for hydrogen.[3] Inelegantly, perhaps, as Pauli was the first to admit, but he did it; Heisenberg was elated.[4] Two other physicists independently did the same thing shortly afterward.[5] Unfortunately, the matrix methods were so hard that none of these very smart people could then go on to calculate the spectrum for the next simplest atom, helium, which has two electrons.[6]

The impasse was broken by Prince Louis Victor de Broglie, a young nobleman from one of the more illustrious families in France, an amateur in science unknown to the research community, who wrote a doctoral dissertation sufficiently farfetched that the Sorbonne faculty was unable to evaluate its correctness.[7] Whereas Einstein had insisted that light comes in particlelike packets governed by the sovereign relation $E = h\nu$, de Broglie turned the question on its head; insisting with equal vehemence that energy is energy, no matter the form in which it is encountered, he argued that particles have energy, and therefore they, too, are ruled by $E = h\nu$. This meant that, say, an electron is characterized by some frequency and wavelength; indeed, all material objects have both. Refrigerators, baseballs, subatomic particles—all have wave characteristics.*

Such a thesis would have seemed simply ludicrous had not de Broglie quickly made use of it to explain features of the Bohr atom that had puzzled theorists since its creation. Bohr had shown that electrons surround the nucleus in orbitals of particular sizes, but neither he nor anyone else had been able to figure out why nature had chosen those orbitals and not others. De Broglie claimed that the orbits in an atom are of exactly the right size to permit electrons to complete each "lap" around the nucleus in a whole number of wavelengths. This limits the number of possible orbits in a way that is easy for anyone with a handful of quarters to picture. Imagine squatting on the floor, a pile of coins nearby, placing one quarter next to another to make circles of quarters on the carpet. A circle of one size can be formed with, for example, five quarters; one of another size can be made with six; but it is impossible to form circles of some in-between size without overlapping quarters. There is no problem if the coins on the carpet overlap, but if *waves* overlap, they reinforce and interfere and cancel each other out higgledy-piggledy, as everyone who has splashed water in a tub can attest. For an electron wave to avoid becoming entangled with itself, it must whirl 'round the nucleus in a whole number of wavelengths, and only a whole number of wavelengths. Thus de Broglie's simple idea ingeniously explained

---

*The chain of reasoning can be followed by anyone with a few minutes and some knowledge of simple algebra. (We ignore a few factors of pi and the like.) The length between the successive crests of a lightwave is known as its wavelength and called *lambda* (λ). It can be calculated by taking the velocity of the wave (c) and dividing it by the number of waves per second (ν, the frequency).

$$\lambda = c/\nu \qquad (1)$$

In the quantum world, $E = h\nu$. To find the value of ν, divide both sides by h.

$$\nu = E/h \qquad (2)$$

Sticking E/h into the first equation,

$$\lambda = ch/E \qquad (3)$$

Another fact: Einstein demonstrated that $E = mc^2$. Substituting for E,

$$\lambda = ch/mc^2 = h/mc \qquad (4)$$

Mass times velocity—mc—is equivalent to momentum. Particles like electrons and protons also have momenta equal to their mass times velocity. It is well known that a particle cannot travel as fast as the speed of light. Therefore, de Broglie said, to find the wavelength of a particle, one should simply insert its actual velocity v—not to be confused with the frequency, ν—into Equation (4). Young de Broglie argued that this was not an algebraic trick but that particles really do have wavelengths and frequencies given by the relation

$$\lambda = h/mv \qquad (5)$$

what Bohr had only been able to postulate: why electrons have fixed, quantized orbits.

One of de Broglie's thesis examiners knew Einstein and passed on the thesis to the great man, who in turn recommended it to another colleague, Erwin Schrödinger. Few people had paid attention to the thesis; Schrödinger changed all that. He pushed de Broglie's ideas far enough to change the face of physics when, in March of 1926, he published a single equation purporting to explain almost all aspects of the behavior of electrons in terms of de Broglie waves, rather than of matrices.[8] Schrödinger proposed that matter could best be understood as a collection of waves adding up, interfering with each other, creating nodes, and so on; their behavior was described by a new branch of physics he later called *wave mechanics*.[9] A theorist then at Zürich, Schrödinger was a few years older than the matrix physicists, respected in the field as a serious craftsman. He was a gentle, sober, decent man, an incessant chain smoker, competent in six languages and something of a poet. He hated Hitler and is said to have intervened when he saw Jews assaulted by the Gestapo; shortly thereafter Schrödinger fled Berlin, where he had been teaching.[10] Like Einstein, he spent many of his later years in a futile attempt to reach unification.

The Schrödinger wave equation was an enormous if somewhat disturbing achievement. In a series of papers that appeared with the speed of a drumroll, Schrödinger tied his wave equation to almost every aspect of the new physics; in his second paper, for example, which appeared just three weeks after the first, Schrödinger rewrote his equation in terms of the Hamiltonians used by Bohr, thereby converting most theorists to his approach in a single blow.[11] Moreover, the mathematics Schrödinger used was much easier for physicists to understand. In school, they all had *studied* waves. They had taken *classes* about waves. If it was hard to imagine how a solid object like an atom could really be made out of waves—what was making the waves?—many physicists had confidence that Schrödinger, a clever fellow, would figure out the answer. Meanwhile, theorists had formulas they knew how to manipulate. They could *solve* Schrödinger's equation; they could obtain exact answers. They could not do this with Heisenberg's version of quantum mechanics. Moreover, the trick of envisioning quantum phenomena as the result of crests and troughs, nodes and interference patterns meant that physicists could visualize the atom in terms of continuous processes—the ripple and flow of long-standing waves—whereas with matrices they had to deal with Heisenberg's assertion that the nature of the microworld was discontinuous and impossible to picture. Little wonder that many physicists threw away their matrices and started working with Schrödinger's methods![12] Even Max Born, who had helped develop matrix mechanics, wrote to Schrödinger that after reading the first wave paper he had become so excited "that I want to defect—or, better, return—with flying colors to the camp of continuum physics. After my whole course [through quantum mechanics], I feel myself drawn to the place from where I set out, namely, the crisp, clear conceptual formulations of classical physics."[13] Solving an individual atom, he said, was simply a matter of solving its particular wave equation. No need for all this stuff about quantum leaps! As Schrödinger wrote (using the now-obsolete term "vibrational mode" instead of "standing wave"), "It is hardly necessary to point out how much more gratifying it would be to conceive of a quantum transition as an energy change from one vibrational mode to another than to regard it as a jumping of electrons."[14]

Unfortunately, the very success of Schrödinger's approach pushed forward the question of what was supposed to be waving in the wave equation. For those who wanted to know what these waves were supposed to be and how wave theory proposed to explain the evidence, accumulated since J. J. Thomson's discovery of the electron, that the atom was made from particles, Schrödinger had an answer: A particle was in reality nothing but "a group of waves of relatively small dimensions in every direction," that is, a sort of tiny clump of waves, its behavior governed by wave interactions. Ordinarily, the bundle of waves was small enough that one could think of it as a dot, a point, a particle in the old sense. But in the microworld, Schrödinger argued, this approximation broke down. There it became useless to talk about particles. At very small distances, "we *must* proceed strictly according to the wave theory, that is, we must proceed from the *wave equation*, and not from the fundamental equation of mechanics, in order to include all possible processes."[15] This was an implicit attack on the matrix methods—and Heisenberg recognized it as such. Nonetheless, Schrödinger kindly mentioned his hope that wave mechanics and matrix mechanics "will not fight against one another, but on the contrary . . . will supplement one another, and that the one will make progress where the other fails. The strength in Heisenberg's program lies in the fact that it promises to give the *intensities of the spectral lines*, a question that we have not approached as yet."[16]

It is not difficult to imagine Heisenberg's reaction to this last statement, and to wave mechanics in general. He was terribly aware of the failings of the matrix methods he had initiated; although Schrödinger in Zürich may not have known it, the physicists at Göttingen and Copenhagen had already proven unable to calculate the intensities of the spectral lines.

In the first months of 1926, Werner Heisenberg was twenty-four years old. He was already one of the most prominent physicists of the day, a position that, of course, he enjoyed. His reputation was due almost entirely to his work on quantum mechanics, a subject that had occupied the whole of his short professional career. He had spent a year and a half on the train of reasoning that culminated in the three-man paper, and eighteen months is a long time to a young man in a hurry. When Schrödinger's wave equation appeared, Heisenberg reacted in a fury; he must have envisioned all his work being consigned to oblivion. He berated Born for deserting matrices and told Pauli that the wave business was so much "crap."[17] It was, he said later, "too good to be true." He hoped it was wrong.[18]

It was not. On April 12, Pauli sent a lengthy letter to Jordan in which he proved that the two approaches were identical.[19] Schrödinger himself proved the same thing, a little less completely, a month later.[20] (It was the third wave paper in less than two months.) The attitude of the Göttingen-Copenhagen group seems to have surprised and annoyed him; thirty-eight-year-old Schrödinger didn't like their barely suppressed scorn for their elders and their conviction that exploring quantum theory was a quest for young men— a game Schrödinger was already too old to play.[21] In the equivalence paper, Schrödinger mentioned, pro forma, that it was really impossible to decide between the two theories—and then went on to argue fiercely the merits of wave mechanics.

Far from being a good point, he said, the impossibility of picturing matrices was "disgusting, even repugnant." It meant that it was terribly difficult to imagine how one would go from the macroworld (where objects could be seen and measured) to the microworld (where, according to the matrix methods, objects per se might not even exist). On the other hand, Schrödinger said, waves existed in both domains; the wave equation was ideally suited for the basic question of future research—that of the interactions between the atom and the electromagnetic field.[22]

Privately, Schrödinger was even more deprecatory: He suspected— correctly, as it turned out—that Heisenberg, Pauli, and the rest of the matrix people had papered over difficulties in their mathematics.[23] When they calculated the spectral lines for hydrogen, they had fitted their work to the data (unconsciously, to be sure). Later, when the dust had long since settled, it was shown that if the matrix technique then promoted by Heisenberg, Jordan, and their associates was used strictly and carefully, it produced the wrong answer.[24]

To Heisenberg's dismay, yet a *fourth* Schrödinger paper—this one submitted in May—showed how to calculate the Balmer lines for hydrogen *and* their intensities.[25] (The success with the intensities should have particularly galled Heisenberg, for the Göttingen-Copenhagen physicists had been unable to come up with them.) That summer, Heisenberg recalled in his memoirs, there was a large conference in Munich, where he planned to confront Schrödinger directly. After the latter's lecture, Heisenberg leaped to attack his rival. "I . . . raised a number of objections, and, in particular, pointed out that Schrödinger's conception would not even help explain Planck's radiation law [that is, $E = h\nu$]." Heisenberg's reasoning was, apparently, that continuous waves could not produce discrete packets of energy. "For this I was taken to task by Wilhelm Wien," the experimental physicist who had attempted to flunk Heisenberg on his dissertation,

who told me rather sharply that while he understood my regrets that quantum mechanics was finished, and with it all such nonsense as quantum jumps, etc., the difficulties I had mentioned would undoubtedly be solved by Schrödinger in the very near future. Schrödinger himself was not quite so certain in his own reply, but he, too, remained convinced that it was only a question of time before my objections would be removed. My arguments had clearly failed to impress anyone—even [Arnold] Sommerfeld, who felt most kindly toward me, succumbed to the persuasive force of Schrödinger's mathematics.[26]

The measure of Heisenberg's consternation perhaps is most fully demonstrated by his neglect, in this and other autobiographical writings, of the fact that Wien, Schrödinger, and Sommerfeld were absolutely right to wave off his objections. Even half a century later, Heisenberg apparently did not enjoy recalling that after the Munich conference Schrödinger promptly published a paper showing that Planck's radiation law could indeed be derived by his methods.[27] Heisenberg was furiously determined to eject waves from the microworld; light consisted of photons, not waves; electrons were particles, not waves; continuous equations were wrong, discrete and unvisualizable matrices were right.

Dismayed, Heisenberg wrote Bohr that there would soon be nothing left of the quantum theory that had been gradually built up since Bohr's model of the atom thirteen years before. Although tending to agree with Heisenberg, Bohr was troubled by the entire wave-particle question. How could an electron be a particle that bounced and ricocheted like a tiny bullet yet be described by an equation for a wave? Bohr asked Schrödinger to visit Copenhagen in October of 1926. A man of pronounced intellectual honesty, Schrödinger was happy to debate; in addition, he seemed to be on the winning side, and could afford to dally in his rival's headquarters. He was not prepared for the welcome he received. Schrödinger met Bohr at the railroad station and immediately was involved in an argument—a discussion that went on for days, early in the morning until late at night. Bohr had arranged for Schrödinger to stay at his house, so that nothing could interrupt the confrontation. Heisenberg once recalled,

Bohr was an unusually considerate and obliging person, but in this kind of discussion, which concerned epistemological problems which he thought were of vital importance, he was capable of insisting—with a fanatic, terrifying relentlessness—on complete clarity in all argument. Despite hours of struggle, he refused to give up until Schrödinger had admitted his interpretation was not

enough, and could not even explain Planck's law. Perhaps from the strain,
Schrödinger got sick after a few days and had to stay in bed in Bohr's home. Even
here it was hard to push Bohr away from Schrödinger's bedside; again and again,
he would say, "But Schrödinger, you've got to at least admit that . . ." Once
Schrödinger exploded in a kind of desperation, "If you have to have these damn
quantum jumps then I wish I'd never started working on atomic theory!"[28]

With Heisenberg at his side, Bohr was able to browbeat Schrödinger into a
temporary retraction. But it didn't last, and by the end of October, Heisenberg had worked himself into that desperate, creative frenzy that precedes
much of the finest scientific work. Like Rutherford fueling himself with an
attack on the Thomson "plum pudding" atom, Heisenberg had railed against
wave mechanics to the point where he was now ready to entertain crazy
ideas.

The intensity of the debate at the birth of quantum mechanics was
more than a matter of careers, temperaments, and prejudices; the participants were gripped by the conviction, endemic to the science, that they
were arguing about the shape of the Universe itself, and that the picture they
were forming had profound philosophical resonances. In some respects, the
belief that discerning the laws of the quantum world is equivalent to deciphering the most primary code of nature is naïve and reductionist; but in
other ways it is exactly what they were doing, and the physicists of the
time—as well as their successors today—have rightly been caught up in the
breathtaking implications of their quest.

Many physicists have experienced a moment, usually in their childhood, in which they perceive the Universe as a house of law, a domain whose
architecture can be learned by human beings. The American physicist Isidor
Isaac (I. I.) Rabi, for example, remembers such an experience vividly, although it occurred when he was just twelve years old. Born in 1898, Rabi is
one of the few surviving members of the first generation of quantum physicists. He was raised by almost violently orthodox Jewish parents—"God was
in almost every sentence," he once told us—in pre–World War I Brooklyn, a
time and place in which it was still possible for a boy never to learn that the
earth went around the sun, and to believe that the sunrise was a daily
miracle. One day in 1910, Rabi stumbled across a book on Copernicus in the
library. There, in the stacks, he discovered the grand movements of the solar
system and was filled with a perception of the cosmos as a composition of
musical beauty and sculptured precision. He passed through a private version of the Copernican revolution, went home, and bluntly informed his
parents that there was no need for God. "I was too young to appreciate what a
terrible blow it was for them," he said to us.[29]

Rabi is a short, pugnacious man, whose sharp opinions and warm laughter have not been dimmed by age. He and his wife, Helen, still live in New York, near Columbia University, in a fine old apartment building that faces Riverside Park. We sat around a silvery coffee table that had been presented to him upon his retirement from Columbia some seventeen years before; its top had once formed part of a particle detection chamber that operated in Columbia's physics laboratory for many years. Sunlight streamed in through the window over the treetops. Unlike most New York City apartments, this one was almost noiseless save for the wind blowing in through the window and the occasional noise of a bus rumbling by on Riverside Drive. Rabi was uninterested in questions about the chronology of physics developments in the 1920s. "What the hell do you want?" he snapped, waving his hand scornfully. "It's been fifty years!" More important, he thought, was why he or anyone could care so much and for so long about such an abstract subject as physics; he wanted to talk about the role physics had played—still played—in his life.

"Quantum mechanics and relativity affected me deeply—personally. It affected my attitude toward the world. I've always thought of physics as a sort of ivory tower, from which you venture forth into all other human affairs, of all kinds. That's why I became a physicist. I could've earned more money as a lawyer." He laughed a little, hunching forward over the silver table.

We pointed out that the vantage he was speaking about is one more traditionally associated with religion, philosophy, or art.

Rabi nodded in agreement. "I believe in the importance of literature," he said. "The basic things like the Russian novel, for instance. I'm very impressed by philosophy. On the other hand, this is one world, as far as our experience goes. And I never know how much of philosophy as such is a play on words and the moving power of language. Like in German, *Volksgefühl*." *Volksgefühl* means "folk wisdom." Some philosophers only articulate the common wisdom, the *Volksgefühl* of their people or their country, whereas others seek something unchanging and deeper. Rabi had been unsure he could distinguish between them; he turned to physics, where he was confident of his own ability to work on a fundamental level. "With science I felt I could grab on to actual things and try to understand them. And then they turn out to be so extraordinarily *mysterious!* Newton's laws of motion, the laws of the electromagnetic field, relativity—they're so far removed from experience, but yet there it is. It's a measure of all the other things that I look at. It gives you an approach to the human race, apart from these inherited things of nationality and whatnot, which you can't take very seriously. That's what science was for me—a citadel. I know some place where I can find out things which are *so*, and not trivial. Far from trivial." He was speaking slowly, hunting for the right phrases. Behind him, wind shook the treetops of the park. "So relativity can have an enormous effect on how I regard myself in the world. It's hard to communicate to other people who haven't that experience. Since it is, as far as we know, a universal human possibility to investigate nature, and the nature of discoveries is so remarkable, so wonderful—if you want to think of the goal of the human race, there it is. To learn more about the Universe and ourselves. In physics, the newest discoveries, like relativity and the uncertainty relation, uncover new modes of thought. They really open new perspectives." A sudden, sad look passed over his face. "And I thought that, say, fifty years ago, that this would happen, that these revolutions and advances in science would have an effect on mankind—on morals, on sociology, whatever. It hasn't happened. We're still up to the same things, or, well, I think, regressed in values. There's this terrible thing— between the United States and the Soviet Union, we might destroy the world. What differences are there between them, except some ideas—not too well-founded—about the distribution and manufacture of the world's goods? And one side has Reagan, who solemnly believes in the God of his fathers, and the other side, Marx and Engels. And they're such unimportant things in comparison with what actually *exists*—the wonder and mystery of the world."

There may have been no revolution in morals, but the physicists of Rabi's time found wonders and mysteries aplenty. Perhaps the greatest and most widely known of these is the indeterminacy relations—popularly called the "uncertainty principle"—which were discovered by Werner Heisenberg in the months after Bohr's confrontation with Schrödinger. Pushed by the apparent failure of matrix mechanics, his own ambitions, and the drive to find the beautiful and profound secrets in the domain of the quantum, Heisenberg spent the autumn of 1926 in a furious attempt to make sense of the two approaches.

He was not alone. Max Born, too, spent the last half of the year working on the problem. Born had been chagrined when Schrödinger's equations appeared, for he was convinced that he could have found them first if Heisenberg and Jordan had not led him astray with their matrices.[30] But he modified his initial enthusiasm after he began working with wave mechanics; by the fall, he was convinced that the waves in Schrödinger's papers were not real, three-dimensional waves, as their creator thought, but something else. Born suggested they might be probability waves, rising and falling with the likelihood that a particle might be in a given place.[31] His basic idea was right, but it was Wolfgang Pauli, in a letter to Heisenberg from the middle of October, who put it into its final form.[32]

Pauli, too, was preoccupied with the puzzle of resolving the wave and matrix approaches. He, too, had been struck by the noncommutative relationship $pq - qp = \hbar/i$—an equation that hung like an icon over quantum mechanics, full of unfathomable meaning. (Here, as before, p stands for momentum, q for position, $\hbar$ is Planck's constant divided by twice pi, and i is a symbol meaning the square root of negative one.) It was difficult to know what this noncommutativity *meant*. For instance, if physicists followed their natural inclinations and put p and q into formulas, the result depended on the order in which the terms were arranged. This implied that one equation could produce several answers; conversely, if you knew, say, p and the answer, but wanted to find q, you ended up with an array of possible values. In the quantum domain the position and momentum of a particle (or bundle of waves) were inextricably related, but what, Pauli asked, was the meaning of it? What was hidden in $pq - qp = \hbar/i$? Pauli felt the problem like a missing filling; he could not stop tonguing the hole. In October, he wrote to Heisenberg, a lengthy missive in the famous Pauli style, a mixture of the abstruse, the elegant, and the gossipy. "But now comes the obscure point. The p's must be assumed to be *controlled*, the q's *uncontrolled*. That means one can always calculate only the probabilities of particular changes of p's with fixed initial values, [and only when they are] *averaged over all possible values of q*." The more you learned about one, the less you could say about the other. Pauli spent a page and a half demonstrating this rigorously.

"So much for mathematics," he continued. "The physics of this is unclear to me from top to bottom. The first question is why can only the p's (and not simultaneously both the p's *and* the q's) be described with any degree of precision? This is the old question of what happens when you know (with reasonable accuracy) the direction of the velocity and the greatest distance from the nucleus of the orbiting electrons. About this I know nothing I haven't known for a long time. Always the same question . . ."

In a burst of annoyance, Pauli summed up: "One cannot simultaneously hook together both the p-numbers and the q-numbers with ordinary c-numbers [that is, regular classical variables]. You can look at the world with p-eyes or with q-eyes, but open both eyes together and you go wrong."[33] And he still had no idea what this meant.

Nine days later, Heisenberg gratefully responded. He had shown Pauli's letter to Bohr and several other physicists at Copenhagen, where it had stimulated heated discussions. Heisenberg, too, had been intrigued by Pauli's "obscure point." It seemed to him that what his friend was implying was that the $pq - qp = \hbar/i$ relation meant that the individual p's and q's couldn't be precisely defined in the old way. Reaching for an analogy, he wondered if talking about a particle's momentum p at a point q might not be like trying to determine the length of a wave if you only had measured one point. (Less exactly, this is like trying to guess the height of the Empire State Building when you know only that the front door is about ten feet wide.) Perhaps momentum and position were large-scale concepts, qualities that required a little breadth of scale before they took on meaning. Then he confided, "Above all, I hope there will eventually be a solution of the following type (but don't spread this around): That time and space are really only statistical concepts, something like, for instance, temperature, pressure, and so on, in a gas. It's my opinion that spatial and temporal concepts are meaningless when speaking of a *single* particle, and that the more particles there are, the more meaning these concepts acquire. I often try to push this further, but so far with no success."[34]

In the meantime, Heisenberg's friend and collaborator, Pascual Jordan, had been thinking about what it meant when physicists said that they wanted to restrict their theories only to quantities they could observe. Shy, amiable, and apparently bereft of personal ambition, the twenty-four-year-old Jordan was easily sidetracked into helping others on lengthy projects of no particular profit to himself.[35] In school, he had flunked out of elementary laboratory physics; his first paper, which sought to modify Einstein's approach to the light quantum, was quickly demolished by the old master himself.[36] In 1926, he was the youngest of the important physicists in Born's circle; he had studied physics seriously for only four years, and regarded the twenty-five-year-old Heisenberg as his senior colleague. He liked fundamental questions; this inclination meant that he soon drifted from quantum physics into cosmology. (On the other hand, he also wrote a humorous paper about the jaw motions of cows chewing grass.)[37] At the time, a number of physicists were wondering about the nature of measurement.[38] Several had claimed that even if a microscope were developed that was precise enough to observe individual atoms, the endless random tumbling and rattling of the microscope's own atoms would render it impossible to measure exactly what other atoms were doing. Jordan argued that if one could somehow freeze the microscope to absolute zero, the temperature at which there is no molecular movement at all, then one could, at least in theory, use the instrument to make a certain measurement of both the position and the momentum of an atom or its constituent electrons.[39] Therefore, Jordan argued, these variables are not, in principle, unobservable; they are just hard to see.

Heisenberg hadn't thought of this. Jordan's paper excited him because it was a direct challenge on his home ground. In the beginning of February, Heisenberg wrote jokingly to Pauli that he "was still occupied with the logical foundations of the whole phony $pq - qp$ business." Jordan, Heisenberg said, talked about the "probability of finding an electron in a particular space," but nobody had a good definition for the idea of "the place of a particle." Heisenberg thought that this definition was something to think about. A bit unkindly, he signed his letter, "With complete youthful unconcern for continuum tricks, W. Heisenberg."[40]

In Copenhagen Heisenberg stayed in the garret apartment of Niels

Bohr's brother, Harald, a small place with walls that sloped toward the ceiling. Niels liked to come in after supper, pipe in hand, and the two men would argue about physics until the early hours of the morning. Their routine was ordinarily quite fruitful, but the difficulties with the wave and matrix approaches to quantum mechanics proved so intractable that by February of 1927 Bohr and Heisenberg were tired and snappish in each other's company. Bohr wanted to understand the wave-particle confusion; Heisenberg wanted to get rid of the waves in any form. Feeling their friendship under strain, Bohr left to go skiing in the mountains north of Oslo. Late one February night, Heisenberg took a solitary walk in Faelled Park, behind the Institute, before going to sleep. As he walked, Heisenberg was thinking that if one multiplied p and q, you simply could not be sure what the result would be. He was thinking that he did believe in the matrix theory. He was thinking about how one would discover where a particle was and how fast it was moving. He was thinking about Jordan's frigid microscope. And then, in the park, it occurred to him that if his scheme were right—that is, if the laws of quantum mechanics were truly correct—he ought to see how nature fulfilled them.[41]

For example, suppose you did want to look at a subatomic particle cooled to absolute zero. Looking at an object in this way means bouncing light quanta—photons—from it and then catching the reflected photons in the lenses of the instrument. If you are looking at something small, such as an electron, the energy of the photons will shove the electron away from its original position in somewhat the way that a rolling billiard ball will knock aside another, stationary ball. You inevitably change the particle's momentum in the act of measuring its position.

To prevent changing the momentum, Heisenberg realized, would require using a photon with a tiny amount of energy. The energy of a photon is given by Planck's formula $E = h\nu$, where ν is the number of waves per second. If E is small, then ν must also be small. (Planck's constant h cannot be changed.) To have a small number of waves going by every second, the waves must be long, just as fewer limousines packed bumper to bumper pass by in a given time than tiny sports cars. But the longer the wavelength of the photons, the less precisely they pinpoint the location of the electron; this leads to the absurd situation where you are trying to measure something by bouncing off it, not a billiard ball, but an enormous whiffleball.

Enormously excited, Heisenberg spent the next little while writing a fourteen-page letter to Pauli about his idea. He described his imaginary microscope and the "thought experiment" of trying to look for particles with it.[42] "One will always find that all thought experiments have this property: When a quantity p is pinned down to within an accuracy characterized by the average error Δp, then . . . q can only be given at the same time to within an accuracy characterized by the average error $\Delta q \geqslant \hbar/\Delta p$."

Today, this now-famous uncertainty principle is usually phrased

$$\Delta p \Delta q \geqslant \hbar$$

where the triangular symbol Δ is the Greek letter delta, and Δp is read "delta-pee." In other words, the position and momentum of an electron cannot be determined exactly—a small but irreducible margin of uncertainty is unavoidable. If the error in the position (that is, Δq) is very small, then the error in the momentum (that is, Δp) must increase to keep their product, $\Delta q \times \Delta p$, larger than $\hbar$. If the position of an electron is measured with such precision that the error is almost zero, then the corresponding error in the momentum becomes almost infinite—the momentum is completely indeterminate. Not only that, Heisenberg showed Pauli how this was a direct consequence of $pq - qp = \hbar/i$—at last giving the equation some physical meaning. Heisenberg found that now that he could visualize some aspects of the matrices, he didn't mind at all.

Heisenberg later recalled that Pauli's reply from Hamburg was enthusiastic. "He said something like, '*Morgenröte einer Neuzeit*,' "—the dawn of a new era.[43]

With such encouragement, it must have been a shock to Heisenberg when Bohr returned from vacation and showed Heisenberg that his paper contradicted Heisenberg's own interpretation of quantum affairs. Bohr reminded Heisenberg that even in the subatomic domain energy and momentum are conserved. If you bounce a photon from a motionless electron and set it moving in the process, you can still know the precise momentum of the electron by looking at the recoil of the photon. As long as you measure the momentum of the photons bouncing off the electrons, all uncertainty vanishes.

However, Bohr said, Heisenberg's idea was still correct. It works because you *can't* determine the momentum of the recoiling photons precisely. You can't because they spread out and diffuse in a wavelike manner—which is precisely why you need a microscope lens to collect and focus them. Even if an electron was used instead of a photon, Schrödinger's equation would ensure that it would diffuse in exactly the same way. But to acknowledge this meant acknowledging that Schrödinger-style waves played an essential role in the theory—a step Heisenberg was so little prepared to take that he burst into tears with frustration.[44] He could establish the physical basis of quantum mechanics only by firmly entrenching his rival's place in it.

Heisenberg and Bohr had a terrible quarrel. "I did not know exactly what to say to Bohr's argument," Heisenberg said years later. "And so the discussion ended with the general impression that now Bohr has again shown that my interpretation is not correct. Inside I was a bit furious about this discussion, and Bohr also went away rather angry. . . ."[45] The two men met again a few days later; Bohr flatly told Heisenberg not to publish the article.

Ignoring his friend's advice, Heisenberg wrote his paper with the old, incorrect arguments. At the end, Heisenberg tacked on a curious little postscript advising the reader that Bohr had found problems, but not disclosing what they were. On March 23, 1927, the article, "On the Visualizable Content of Quantum Theoretical Kinematics and Mechanics," arrived in the offices of the *Zeitschrift für Physik*. It was published at the end of May.[46]

On July 4, Schrödinger wrote to Max Planck, who had recently retired from his position at the University of Berlin. Schrödinger had taken his place, and in his conscientious way he wanted to keep the old man well informed about the discipline to which he had given his life. It had not escaped Schrödinger that the indeterminacy relations restored visualizability to physics, and that wave functions were an integral part of the theory. He and Planck had talked often, it seems, about waves and matrices; now he wrote, "In a brand new article by Heisenberg, my-much-smiled-at wave packets are said to have at last found their correct interpretation as 'probability packets.' . . . Well, as God wills. I keep quiet."[47]

A kind and honorable man, Schrödinger kept true to his word. And in December of 1933, he and Heisenberg found themselves on the same podium, smiling fraternally as the Swedish Academy awarded them both the Nobel Prize for physics—Heisenberg the delayed awardee for 1932, Schrödinger the co-winner for 1933. Although there is no record of any love lost between the two men, their conjoined work became part of the common lore of physics. Schrödinger's wave methods are used by all the workers in the field, while his premises have been rejected; despite the celebrity of Heisenberg's premises, the matrix methods he pioneered are now rarely used for the purposes he had hoped.[48]

□   □   □   □   □

To this day quantum mechanics has not lost its ability to provoke. Many, if not most, students have trouble swallowing the ideas when they learn them in class; it has been said that physicists never understand quantum mechanics, they just learn to use it. At times, thinking about the subject exhausted our faculties, and we consulted a physicist named Robert Serber. We were impressed by Serber's sure grasp of the heart of a question. Every so often we met him for a few hours in his office at Pupin Hall, in Columbia University. Serber is a patient man, and, not unnaturally, we liked him because he was patient. We would talk for a couple of hours, and then he would look at his watch and say he had to pick up his son from the sitter. A short, wiry man with thinning hair and strong blunt-fingered hands just beginning to bend with age, Serber is one of the physicists who keep the field going, making solid, unflashy contributions that rarely come into the public eye. Typically, he would wear a loose turtleneck shirt, faded jeans, and a Western belt buckle studded with turquoise and wide as any made; his well-used glasses have thick black frames. The kind of man whose presence in a room is not immediately noticed, he speaks softly, in narrow circles, continually curling back to the subject, reformulating, clarifying, breaking off a sentence—maddeningly, in mid-phrase—to erase an earlier equation, to modify a parameter, to restate a definition, to think it over again from the beginning; a kind of deep stammer that involves whole phrases and yards of discourse and is impossible to reproduce on the page. It took a little while to realize that what we heard was the audible record of a man thinking through a physical question, pushing hard as he could at nature. If we said something he did not think was so, he would make a barely perceptible grimace, and then say, "Ah, yeah . . ." Then, about ten seconds later, he would say politely, "Well, I'm not sure." Then he would say, "That's just all wrong," a little apologetically, and get up and draw on the blackboard whatever the right idea was.

On one occasion, we came to talk about a few books that we had read. These books, which are quite popular, assert that there is a connection between quantum mechanics and some forms of Eastern mysticism. Behind the connection, more or less, is the view of quantum mechanics hammered out by Niels Bohr and subsequently known as the Copenhagen interpretation. Because position and momentum cannot be measured simultaneously, the Copenhagen interpretation claims that experimenters choose which one shall have a definite value by the act of measuring it. Physicists are therefore part of the reality they are probing; the element of conscious choice, some theorists believe, means that any picture of an observed phenomenon must include the mind of the observer. The Nobel Prize–winning physicist Eugene Wigner described the logical conclusion: "[I]t was not possible to formulate the laws of quantum mechanics in a fully consistent way without reference to the consciousness [of the observer]. . . . [Remarkably,] the very study of the external world led to the conclusion that the content of the consciousness is the ultimate reality."[49]

The books we had read took this a bit further, insisting reductively that the connection between physicists and the minute particles with which they work demonstrates that we are all floating in a sea of mind.[50] The philosophical resonance of quantum mechanics, it has been said, is, if anything, "psychedelic."[51]

Quantum mechanics is complicated enough without being psychedelic, and we thought we might talk to Serber to shore up our shaky understanding of the subject. Serber learned about the uncertainty principle three years after its formulation, in 1930, as a first-year graduate student at the University of Wisconsin, in Madison, under the able tutelage of John H. Van Vleck, one of the first and foremost expositors of the new physics in the United States. For some years now, Serber has been retired; his office is down the hall from that of I. I. Rabi.

Clean, orderly, seldom used, Serber's office has the blue, melancholy lighting of retired scholarship. A bookcase stands to the side, its glass doors framing the reflections of the Chagall and Klee prints on the opposite wall. Over an empty blackboard hangs the urgent message: *DO NOT WASH.* Tucked into a corner is a frightening souvenir of the atomic era, a piece of wallboard from a school in Hiroshima with the shadow of a windowframe burned into its surface. Although faded by time, the burn is a clear reminder of the practical consequences of abstruse physical theories. We talked with him a while, and then, on another day, Serber had thought of a better way to say what he meant and we returned to the subject. (Some of the remarks that follow are taken from his reformulation on the second day.)[52]

We asked right away if it was true that the uncertainty principle brings the observer into physics. He shook his head; no; shook it again, even more emphatically. "I mean, the whole idea of observers—that's a pedagogical thing. It's very convenient when you're trying to understand something to imagine doing an experiment or think about an observer doing something. But that doesn't have anything to do with physical law, that has to do with the understanding of the law. A physical law is a description of nature, not a description of observers. Using the word *observer* in any place in physics at all—it's irrelevant. It's *never* part of a physical theorem." He took his glasses off and rubbed his eyes, which were slightly reddened from lack of sleep. He had been up the night before with sick children. "Look," he said, "they apply quantum mechanical laws to the Big Bang! There were no observers there!"

What made the uncertainty principle so important, he said, was the way it helped physicists reconcile the two apparently incompatible concepts of particles and waves. Einstein had shown that light waves can behave like particles, de Broglie had shown that particles can behave like waves, and physicists had become terribly confused. "Heisenberg realized that there wasn't necessarily a contradiction between the two views," Serber said. "In any real physical situation, you're never dealing with a single wave—that's a mathematical abstraction—but with a wave packet." He went to the board, found a stub of white chalk, and drew a series of waves whose crests coincided at one point, which he labeled A.

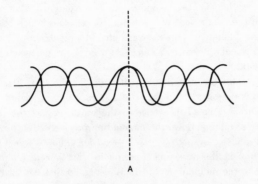

Ordinarily, waves in such a jumble, overlapping each other every which way, would tend to cancel each other out—go out of phase, as scientists say. If circumstances are right, however, they may all pile up for a brief moment, as they do at *A*. "If you add the way these waves interfere," Serber said, "you wind up with something like this." He superimposed the sum on the first sketch.

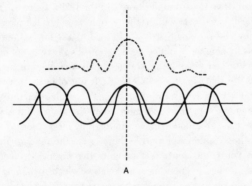

The monadnock in the center of the drawing is very narrow because the various waves rapidly cancel each other out as they spread from Point *A*. The sum of the waves, that is, the wave packet, is therefore only detectable close to Point *A*, and the wave function represented by the picture describes something quite like the tiny dot conjured by the word *particle*. If the waves did not interfere with each other right away—if the central bump smeared out into a mesa or a series of ripples—the Schrödinger function would describe something more and more like a conventional *wave*. The speed with which the component waves of the packet cancel each other out depends on the difference in their wavelengths. A large difference, and one has a particle; a small difference produces a wave. The underlying reality is the same for both.

Now comes the uncertainty principle, Serber said. "The spread in wavelength is proportional to the spread in momentum, right? De Broglie showed that. So a small difference in the wavelength of the components corresponds to pinning down the momentum precisely. But that means the wave is not localized in space very well—you don't know where it is, really. If you want to make the wave packet narrow in space, on the other hand, you have to make the momentum spread very big, so that the waves interfere right away.

"You can't have both at once. Heisenberg said that either you don't have a definite momentum or you don't have a definite position. And the fact that you can't have both at once is what removes the contradiction between wave behavior and particle behavior." The famous formula $\Delta p \Delta q \geqslant \hbar$, Serber said, is simply a mathematical consequence of Schrödinger's equation; all Heisenberg did was to figure out its physical meaning.

What was the role of quantum mechanics and the uncertainty principle in the present drive toward unification? "In the 1920s, there wasn't any push toward unification to begin with, right? People were just learning a lot of new physics. They had no idea that it would lead to any kind of unified theory. Einstein was looking for a unified theory, but, if you really get down to it, it didn't make any sense. All he was talking about was electromagnetic theory and gravity. But that isn't the whole story. It was obvious that you couldn't expect to get a complete solution from an incomplete picture in the first place. You had to understand a lot more about the nature of the world before you could proceed. You had to know about strong interactions, weak interactions, what the symmetry properties were, what the fundamental particles were before you could start. Just to unify the weak and electromagnetic interactions, you had to know an awful lot about electrodynamics that wasn't known until the late 1940s. Then you had to know an awful lot about the weak interactions that wasn't learned until the 1960s. So all this—" the equations for the uncertainty principle he had put on the blackboard "—is the preliminary stages, when you're learning the pieces that you have to put together, that you have to unify. Quantum mechanics is the groundwork.

"It's astonishing, really. Look, quantum mechanics was created to deal with the atom, which is a scale of 10^{-8} centimeters. That was in the twenties, right? Then in the thirties, forties, fifties, they applied it to nuclear physics. Instead of 10^{-8} centimeters, it's now 10^{-13}, and quantum mechanics still worked—that's another leap. Then after the war, they began building the big machines, the particle accelerators, and they got down to 10^{-14}, 10^{-15}, 10^{-16} centimeters, and it *still* worked. That's an *amazing* extrapolation—a factor of a hundred million!"

Serber once showed us the notebooks he used in graduate school in 1930. They were artist's notebooks with neat red spines and inner covers papered with Florentine designs. His handwriting is an impeccable tribute to the rigidity of prewar grammar schools. Van Vleck, his teacher, had an approach to the subject that would seem ludicrously complicated today. He explained Heisenberg's microscope and gave another example, deriving the uncertainty principle from a prism. On one page, nestled in a tangle of integrals, is a reference to a popular essay by the Harvard physicist Percy Bridgman, which appeared in the magazine *Harper's Bazaar* two years after the discovery of the uncertainty principle. Entitled "The New Vision of Science," it is a exposition of quantum mechanics for the nonphysicist as good as any that has been published in the intervening six decades. (The lack of success of scientific popularization may be adjudged by the fact that if Bridgman's article were published today it would be equally illuminating to the general public.) One of the dreams which humanity refuses to let die is that some new breakthrough—usually from the realm of philosophy, art, or political revolution—will one day permanently and unalterably change humanity for the better. For Bridgman, the startling discoveries of quantum mechanics, and especially the uncertainty principle, were proof that this saving power would come from physics. The collapse of the commonly understood law of cause and effect, Bridgman correctly predicted, would at first "let loose a veritable intellectual spree of licentious and debauched thinking"; to fix this, new methods of education would have to be introduced to teach the young the right way of understanding concepts foreign to everyday experience. But once this is accomplished, Bridgman thought "understanding and conquest of the world will proceed at an accelerated pace"; indeed, humanity will improve, acquiring a certain "courageous nobility" in the face of this new wisdom. "And in the end, when man has fully partaken of the fruit of the tree of knowledge, there will be this difference between the first Eden and the last, that man will not become as a god, but remain forever humble."[53]

II

Particles and Fields

5
The Man Who Listened

THE STANDARD MODEL OF ELEMENTARY PARTICLE INTERACTIONS WAS pieced together by three distinct intellectual generations of twentieth-century physicists. Each arrived suddenly, of a piece, in the course of two or three years, a group of young men—women as well, in the case of the third—which emerged with their style of play fully developed, in confident command of the tools of the trade. Like the abstract expressionists, whose intemperate urgency and immediate prominence in postwar New York City stunned their elders, the new physicists startled their contemporaries with the sweep and precision of their attack and the fierceness with which they demanded to be heard. Although, like the expressionist Willem de Kooning, the new workers may actually have labored unrecognized for years, it seems to the community at large that a movement has abruptly formed, fast and bright as a stroke of lightning, and that everything has changed. The construction of quantum mechanics, in the mid-1920s, was the work of the first *nouvelle vague*; names like Heisenberg, Pauli, and Oppenheimer, Rabi, Schrödinger, and Jordan, suddenly appeared on papers that had to be read by every serious practitioner in the field. Meanwhile, the old hands, the Rutherfords and Bohrs, were pushed to new accomplishments. (Some never adjusted; J. J. Thomson was one, and—sadly, grandly—Einstein was another.) The years immediately after the Second World War marked the entrance of another group of scientists—young, hungry, predominantly American; the beginning of the 1970s has witnessed the rise of a third.

In each of these brilliant, collective entrances, there has been one player whose superior insight into the structure of field theory has led his confreres to adjudge him, rightly or wrongly, the most brilliant of the lot. These men are not well-known outside the field—in a related discipline, how many members of the public know of David Hilbert, arguably the finest mathematician this century has produced?—and they may not produce as much physics as others of their generation, but nonetheless the acuity of their vision and the formal tools under their control have awed their contemporaries. Physicists have as much difficulty with complex mathematics as anyone else, and it is little wonder that the legends should accrue around mathematical physicists. In the recent rush to assemble the standard model and move toward unification, the name of Gerard 't Hooft stands out; in

1948, the honor went to Julian Schwinger. During the short, happy heyday of quantum mechanics, the reputation went to Paul Adrien Maurice Dirac.

At a time when young physicists created stirs with their first papers and every dissertation opened a new field, it was Dirac who most shaped the resulting science. If Einstein, with his disdain for the quantum theory, was truly the last classical physicist, then P. A. M. Dirac, as he always signed his work, was the first completely modern one. In an article written soon before Dirac's death in 1984, the physicist Silvan Schweber remarked, "Dirac is not only one of the chief authors of quantum mechanics, but he is also the creator of quantum electrodynamics and one of the principal architects of quantum field theory. All the major developments in quantum field theory in the thirties and forties have as their point of departure some work of Dirac's."[1]

Circumstances seem to have conspired to make Dirac painfully shy and taciturn. He was born on August 8, 1902, in Bristol, England, then, as now, a commercial town on the confluence of the Avon and Frome rivers. His father, a Swiss émigré, was retiring to the point of being antisocial; the family never had guests, and the Diracs never went out. Dirac ate with his father in the dining room, while the rest of the family had dinner in the kitchen; the boy would have preferred to be with his mother, sister, and brother, but there were not enough chairs in the kitchen. At home, Dirac once recalled, "My father made the rule that I should only talk to him in French. He thought it would be good for me to learn French in that way. Since I found that I couldn't express myself in French, it was better for me to stay silent than to talk in English. So I became very silent at that time."[2] Quiet and introverted, Dirac spent much of his time outdoors, taking solitary walks through the English countryside. He liked order and symmetry, the meticulous compression of mathematical relationships. "A great deal of my work is just playing with equations and seeing what they give," Dirac said later. "I don't suppose that applies so much to other physicists; I think it's a peculiarity of myself that I like to play about with equations, just looking for beautiful mathematical relations which maybe don't have any physical meaning at all. Sometimes they do."[3]

Although Dirac's father had no appreciation for the importance of social contact, he did realize the use of a good education, and encouraged his son's mathematical bent. Moreover, this proclivity was fostered by an accident of history: As a teenager, Dirac was pushed into a higher level than was normal for his age, because the more advanced classes had emptied out when older students left to go to war. He liked the Merchant Venturer's School, which shared quarters with the engineering college of Bristol University, partly because the curriculum deemphasized philosophy and the arts, subjects that he found nearly incomprehensible for most of his life.[4] Afraid that there would be no jobs in mathematics, Dirac chose to specialize in engineering when he attended the university. He was a good student, but only the theoretical aspects of the field really interested him; his early on-the-job training ended disastrously when his employer found that Dirac "lacked keenness, and was slovenly."[5] Although his only brother was working in the same factory, the two boys never exchanged a word.

In the fall of 1921, having completed an engineering degree, Dirac found himself unable to obtain employment. The Bristol mathematics professors, who had made known their disappointment that a brilliant mathematician was taking engineering courses, offered Dirac free tuition; because he had nothing else to do, he accepted. The only other student in the honors program of mathematics was a woman firmly intent on studying applied mathematics, especially that which could be used in physics. Because he did not have any firm convictions, Dirac went along with her wishes; in this way the career of one of the great physicists of the century began.

From the haphazard beginning of his career to the end of his days, Dirac was convinced that mathematics is the key to progress in physics. In one of his last addresses, Dirac explained his credo: "[O]ne should allow oneself to be led in the direction which the mathematics suggests . . . [o]ne must follow up [a] mathematical idea and see what its consequences are, even though one gets led to a domain which is completely foreign to what one started with. . . . Mathematics can lead us in a direction we would not take if we only followed up physical ideas by themselves."[6]

At Bristol, Dirac became acquainted with relativity, which electrified him. He won a B.Sc. and, buttressed by two grants, entered Saint John's College, Cambridge, in 1925. By 1927, at the age of twenty-five, his contributions to quantum mechanics had ensured that he was one of the most important physicists in the world.

It should be said that he was not changed unduly by the growth of his reputation; he continued to be so laconic that people who met him often thought him rude. Although an honored member of the Cambridge physics group, he had few students, established no school, and talked rarely to experimenters. Samuel Devons, who spent the late 1930s at the laboratory, once told us, "There was a Cavendish Physical Society meeting, a sort of semiformal gathering once every two weeks. Some lecturer would come in, and Dirac would sit in the front row and listen. Very rarely would he open his mouth. Sometimes he would be prodded by Rutherford, who'd say, 'Now what do you theoretical people think?' Rutherford had this notion that theory was some sort of speculation—the real facts were in the experiments. And Dirac would sit and say nothing."[7]

Dirac spoke so precisely and carefully that he approached the Delphic; when he taught quantum mechanics, he stood behind the podium and read to the class from the book he had written on the subject, believing that he had set down his point of view there as well as he could. In 1928, he gave a series of talks at Leiden, where Paul Ehrenfest, the Dutch theoretician who had quickly submitted the idea of spin before its creators could get scared, was frustrated by the Olympian manner of Dirac's presentation.[8] H. B. G. Casimir was in the audience. Each lecture, he recollected later, "was presented in perfect form. You know Dirac's habit—if you didn't understand things, he would not offer any explanations but would very patiently repeat exactly the same thing. Usually it worked, but it wasn't quite Ehrenfest's way of doing things." Ehrenfest wanted always to see the human being behind the work. "I—" Casimir again "—remember that once Ehrenfest put a question to Dirac to which Dirac had no immediate answer. And so Dirac began to work it out on the blackboard. He covered the entire blackboard with very small [writing]. And Ehrenfest was [standing] right behind him, trying to see what he did, and exclaiming, *'Kinder, Kinder! Schaut jetzt zu! Jetzt kann man sehen, wie er es macht!'* " Kids, kids—look at this! *Now* we can see what he's up to![9]

In August of 1925, R. H. Fowler of the Cavendish received the proofs of an as-yet unpublished paper by Werner Heisenberg that, in the opinion of its author, created "some new quantum mechanical relations."[10] It was Heisenberg's first paper on quantum mechanics, the paper that inspired Born to think of matrices. Fowler gave the proofs to his young research assistant, Dirac. Dirac's interest vanished as soon as he saw that Heisenberg had prefaced his work with several paragraphs of philosophical musings about the importance of considering observable quantities. Noticing that the principal example of the new methods Heisenberg had provided was of no real import, Dirac decided that there was no reason to pay much attention. He came back to it a week or so later, and this time realized that Heisenberg's quantum-theoretical series, the variables with which the equations were formulated, were noncommutative, that is, $A \times B$ did not equal $B \times A$. Dirac saw that this was the key to Heisenberg's entire scheme, and that the reason he had not appreciated the article's significance was that Heisenberg in his anxiety had chosen to demonstrate his idea with a sample calculation in which the noncommutativity didn't show; because noncommutativity was central to the whole conception, he had been forced to provide a trivial example. Dirac remarked later that Heisenberg "was afraid this [noncommutativity] was a fundamental blemish in his theory and that probably the whole beautiful idea would have to be given up. . . . At this stage, you see, I had an advantage over Heisenberg because I did not have his fears. I was not afraid of *Heisenberg's* theory collapsing. It would not have affected me as it would have affected Heisenberg. It would not have meant that I would have had to start again from the beginning."[11] In Dirac's opinion, the originators of new ideas are too protective of their brainchildren to develop them properly; creativity implies a corresponding lack of perspective.

Like Born, Dirac quickly realized that $pq - qp = \hbar/i$. But unlike Born, Dirac did not see at first that p and q were associated with matrices. He regarded them simply as strange versions of position and momentum, and looked for some means to connect Heisenberg's new quantum mechanics with the classical mechanics Dirac felt he understood.

At this time [September of 1925] I used to take long walks on Sundays alone, thinking about these problems and it was during one such walk that the idea occurred to me that the commutator A times B minus B times A was very similar to the Poisson bracket which one had in classical mechanics when one formulates the equations in the Hamiltonian form. That was an idea that I just jumped at as soon as it occurred to me. But then I was held back by the fact that I did not know very well what was a Poisson bracket. It was something which I had read about in advanced books of dynamics, but there was not really very much use for it, and after reading about it, it had slipped out of my mind and I did not very well remember what the situation was.

The French mathematician Siméon-Denis Poisson had introduced this idea in 1809, as a curiosity, to help him make some calculations about the motion of a planet in its orbit. Although Poisson brackets had drawn occasional interest from physicists, until Dirac's insight they were rarely used.

Well, I hurried home and looked through all my books and papers and could not find any reference in them to Poisson brackets. The books that I had were all too elementary. It was a Sunday. I could not go to a library then; I just had to wait impatiently through the night and then the next morning early, when the libraries opened, I went and checked what a Poisson bracket really is and found that it was as I had thought. . . . This provided a very close connection between the ordinary classical mechanics which people were used to and the new mechanics involving the noncommutating quantities which had been introduced by Heisenberg.[12]

A lengthy letter in Dirac's minute, fussy handwriting arrived in Göttingen on November 20, 1925. Written in English, it explained to the astonished Heisenberg how in a few concise steps his version of quantum mechanics could be reformulated in classical terms using a mathematical device he had never heard of. Moreover, working alone, Dirac had come up with a version of quantum mechanics much more general and complete than that produced by Heisenberg, Born, and Jordan in their just-completed joint paper. Stunned at this authoritative communiqué from a complete stranger, Heisenberg replied immediately—in German, a language that Dirac was barely able to follow:

I have read your extraordinarily beautiful paper on quantum mechanics with the greatest interest, and there can be no doubt that all your results are correct as far as one believes at all in the newly proposed theory. . . . [Moreover, your paper is] really better written and more concentrated than our attempts here.[13]

Casually, Heisenberg mentioned that he and others have been working along some of the same lines.

I hope you are not disturbed by the fact that part of your results have already been found here some time ago and are being published independently in two papers, one by Born and Jordan, the other by Born, Jordan and me. [He didn't mention that "some time ago" meant about a month.] Your results, in particular . . . the connection of the quantum conditions with the Poisson bracket, go considerably further.[14]

He then peppered Dirac with questions about the applicability of Poisson brackets to quantum theory. In the next ten days, Heisenberg sent off a postcard and two more letters, all with further questions.

Unfortunately, Dirac's replies are lost. They were among the papers that American military authorities confiscated from Heisenberg at the end of World War II, and, despite all pleas, never returned. But Dirac recalled later that, in one way or another, he was able to answer all of Heisenberg's objections and show that Poisson brackets do, indeed, allow quantum mechanics to be cast neatly into Hamiltonian form by simpler and more classical methods than those Heisenberg used. "That," Dirac said, "was the beginning of quantum mechanics so far as I was concerned."[15]

Three months later, in March of 1926, Schrödinger's wave equation appeared.[16] Like Heisenberg, Dirac was annoyed, but for entirely different reasons. Dirac was hot in pursuit of the mathematical analogies he had unearthed between Heisenberg's quantum mechanics and classical mechanics, and he resented having to turn his attention to a formidable-looking new set of basic ideas whose value was not clear.[17] But he, too, was eventually unable to avoid the wave function.

By the time Dirac had completed his Ph.D. in the spring of 1926, he had completely reorganized quantum mechanics, and was ready to begin pushing its frontier forward. Following graduation, Dirac had the chance to travel to further his studies. His first thought was, unsurprisingly, Göttingen, the home of matrix mechanics. But Fowler, who was fond of Bohr, strongly encouraged him to go to Copenhagen. Torn, Dirac decided to spend several months at each place, going first to Copenhagen. He arrived in the second week of September 1926.

Despite Bohr's garrulity, his overwhelming concern with the philosophical implications of physics, and his notorious difficulties with precise expression, he hit it off immediately with Dirac. "I think mostly Bohr was talking and I was listening," Dirac recalled. "That rather suits me because I'm not very fond of talking."[18] At late-night talkfests in Copenhagen taverns, Dirac met Heisenberg, Pauli, and the rest of the "quantum mechanics" for the first time—a loquacious, snobbish, simpatico bunch that appears to have gotten him to open up a little. Like many of the shy, smart people in that time and place, Dirac had violently colored political views; with passionately lofty detachment, he told his Continental colleagues that there was no reason for the poor to suffer, that he saw little purpose in rewarding the greedy with wealth, and that organized religion was a ludicrous sham.[19] After one such disquisition, Wolfgang Pauli, who had mystical leanings, is supposed to have remarked, "Dirac has a new religion—there is no God, and Dirac is His prophet!"

In Copenhagen, Dirac began to work on some ideas that he hoped would reconcile quantum mechanics and relativity. Quantum mechanics had much to say about the principles of action in the subatomic domain, but it did not take into account the special effects that occur at speeds near that of light. What happens when a flashlight beam hits a wall? The electrons in the wall absorb and emit photons, in the process moving at the enormous velocities where relativity is important. A proper theory of light and matter thus had to describe quantized, relativistic behavior. Dirac set out to create such a theory during the last months of 1926 and the first weeks of 1927, the same period in which Heisenberg wrestled with the uncertainty principle. In the optimism of the time, Dirac had little notion of what a tangle he was to create, and how many years—decades, in fact—it would take to unravel.

Dirac's principal tool was an old idea, that of a *field*. A field is a region of space in which particular quantities are precisely defined at every point. For instance, the pattern of arcs that iron filings form around a magnet illustrates the magnet's field. Each point along each arc is subject to a force of particular strength and direction. The alignment of each filing indicates the direction of the field at that point, and the density of the filings indicates the field's strength. Fields can be defined for such diverse domains as temperature, sound, and matter. The idea of a field was slowly developed by nineteenth-century physicists, and culminated in James Clerk Maxwell's demonstration, in 1861, that light of all varieties could be described as a pattern of electric and magnetic fields.[20] (For this reason, visible light, radar waves, X rays, and infrared beams are all called electromagnetic radiation.) In 1905, Einstein showed that electromagnetic fields are associated with quanta—photons—that act as the agents of the field.[21] If the field is like the domain of a feudal overlord, then photons can be thought of as the soldiers, tax collectors, magistrates, and factota who make the wishes of the ruler felt in that area. By the time Dirac considered the problem of the interaction of matter and electromagnetic radiation, physicists fully realized that to describe the comings and goings of electrons and photons it would be necessary to treat the field associated with the photons. And this is exactly what he set out to do.

Like Planck, Heisenberg, and Schrödinger before, Dirac had recourse to systems of oscillators. But unlike his predecessors, he thought of the atom *and* the field in these terms. Using Heisenberg's quantum mechanics, Dirac was able to come up with a Hamiltonian for the field that was fully compatible with the Hamiltonian for the atom from quantum mechanics.[22] Dirac was thus able to say that the Hamiltonian for the entire process could be found by adding up the separate Hamiltonians for the atom, the field, and the interaction. Moreover, Dirac showed that by juggling the Hamiltonians through an appropriate mathematical procedure, he could prove a law that Einstein had discovered, which gave the probability that a given atom in a given state that sat in a particular field in a particular configuration would absorb or release a photon.[23]

The result was the first real quantum field theory. Because it linked quantum theory with the dynamics of electromagnetic fields, Dirac called it *quantum electrodynamics.* Another way of describing Dirac's accomplishment might be to say that in the beginning of 1927, human beings understood fairly accurately for the first time what happens on the atomic level when someone shines a light in a mirror and the glow is reflected back.

Highly satisfied, Dirac submitted his paper to the *Proceedings of the Royal Society* in the last days of January of 1927, just three weeks before Heisenberg wrote his long letter to Pauli about the uncertainty principle.[24] One of the most influential papers in the history of twentieth-century physics, "The Quantum Theory of the Emission and Absorption of Radiation" begins with the prescient assertion that there is little further work to do in quantum mechanics.

The new quantum theory, based on the assumption that the dynamical variables do not obey the commutative law of multiplication, has by now been developed sufficiently to form a fairly complete theory of dynamics. . . . On the other hand, hardly anything has been done up to the present on quantum electrodynamics.[25]

"Hardly anything" is a classic bit of Oxbridgian understatement. *Nothing* had been done before on quantum electrodynamics.[26]

The questions of the correct treatment of a system in which the forces are propagated with the velocity of light instead of instantaneously, of the production of an electromagnetic field by a moving electron, and the reaction of this field on the electron have not yet been touched.

Dirac said often that he was beset by fears when he came up with his new theory. Perhaps. The tone of the article doesn't betray it.

[I]t will be impossible to answer any one question completely without at the same time answering them all.[27]

In any case, he soon discovered that his fear was justified. Quantum electrodynamics was indeed a great step forward, but it came at a great price. Dirac had set down the beginnings of the modern theory of electromagnetism—the first solid piece of the standard model—but he had also unwittingly let loose an onslaught of conceptual demons that would change our views of space and matter. As a step to quantizing the electromagnetic field, Dirac hypothesized that his oscillators did not disappear when there was no field. Rather, they went into a "zero state," in which they existed but could not be detected. Associated with the zero state oscillators were zero state photons; these, too, could not be detected. It didn't trouble Dirac that his mathematical scheme implied that empty space should contain billions of invisible photons. As long as they didn't show up, their presence made no difference.[28]

But when Dirac's theory was interpreted in light of the uncertainty principle, it turned out that in some sense the zero state photons *do* show up. According to the uncertainty principle, the energy of a field over any given time cannot be determined exactly. (The indeterminacy relations apply equally to particles and fields.) Paradoxically, the less time one spends measuring, the greater this margin of error becomes. It is therefore conceivable that within a time interval on the order of trillionths of a trillionth of a second the zero state oscillators of a field actually might not be at the zero state. In fact, they might have a vast amount of energy that escapes detection. As Einstein's equation $E = mc^2$ demonstrated, mass and energy are two forms of the same thing. Thus the uncertainty principle dictates that if any small area can contain undetectable energy, then, according to Einstein's equation, it can contain undetectable matter. This basic uncertainty is not just a lacuna in our knowledge. Mathematically, there is no difference between this uncertainty and actual random fluctuations in the energy (or matter) measured. Therefore, at least in theory, because any space *might* harbor particles for a short time, it *must* do so.

As strained through the uncertainty principle, quantum field theory exposed a frightful chaos on the lowest order of matter. The spaces around and within atoms, previously thought to be empty, were now supposed to be filled with a boiling soup of ghostly particles.[29] From the perspective of quantum field theory, the vacuum contains random eddies in space-time: tidal whirlpools that occasionally hurl up bits of matter, only to suck them down again. Like the strange virtual images produced by lenses, these particles are present, but out of sight; they have been named *virtual particles*. Far from being an anomaly, virtual particles are a central feature of quantum field theory, as Dirac himself was soon to demonstrate.

At first, however, he does not seem to have realized the full implications of quantum electrodynamics. Instead, he worried about how to make his work jibe with that of his colleagues.[30] There were several major difficulties. First, quantum electrodynamics was based on Dirac's formulation of quantum mechanics, which was itself not entirely in accord with the dictates of relativity. (Dirac's electromagnetic field was relativistic, but his matter field was not.) Deeply sympathetic to the urge for consistency and order that was one of the wellsprings of Einstein's thought, Dirac thought the discrepancy was terribly bothersome. Moreover, in recent months two German physicists, Walter Gordon and Oskar Klein, had produced a relativistic version of Schrödinger's equation for the electron.[31] A little while before his death, Dirac said that the Klein-Gordon equation "could not be interpreted in terms of my general [version] of quantum mechanics and was therefore unacceptable to me. Other physicists with whom I talked at the time were not so obsessed with the need for having quantum theory agreeing [sic] with the general . . . theory, and they were rather inclined to let it go as it was. But I just stuck to this problem."[32] He stuck to his last until the end of 1928, when he returned to England.[33] By then, he had developed yet another equation for the electron's motion—the one used today. To reconcile his own version of quantum mechanics, relativity, Klein-Gordon, and the experimental data, Dirac produced a single equation for an electron traveling through space that had four components, two associated with the spin, and two with the energy of the particle.[34]

The Dirac equation, as it is called, suddenly explained a host of puzzles about the electron. To give one example, it confirmed Bohr's old prediction that the problem of the velocity of electron spin would vanish when the *real* quantum theory was found. According to Dirac's theory, an electron is not localized in a particular place, but has a set of probable locations that are scattered around a point of maximum probability like the cluster of holes around the bull's-eye of a sharpshooter's target. Moreover, these locations circulate around the center as if the target were spinning; according to Dirac's equation, this motion is the spin of the electron. Lorentz had envisioned a little ball rotating when he calculated the electron spin velocity, and found that it turned faster than the speed of light—a flat impossibility. But Dirac said that you should use the average radius of the cluster instead, which is more than a hundred times bigger than the electron envisioned by Lorentz. This made the spin velocity much slower, and removed the contradiction with relativity. Not only that, Dirac's theory predicted that this circulation would create a tiny magnetic field around each electron, and accurately predicted its strength, something previous descriptions of spin had been unable to do.

It was difficult for physicists not to be impressed, even awed, by the Dirac equation.[35] Like children shaking fruit from a tree, they extracted equations describing the collision of two electrons, the interaction of photons and free electrons, and the correct formulas for the spectral lines of the hydrogen atom.[36] The results came so quickly that some members of the field were sure that it was only a matter of time—months, perhaps—before Dirac or someone else would come up with an equation for the last remaining piece of the puzzle, the proton. With photons, electrons, and protons disposed of, scientists would soon knock off the nucleus, and then, except for filling in a few loose ends, all of matter and energy would be explained. Unification would occur, and physics would be over.

"There was one of the regular conferences in Copenhagen," the physicist Rudolf Peierls has recalled. "It may have been the conference of thirty-two or thirty-one; I don't know. The interesting point is that there was a general feeling among some people there, not everybody, that physics was almost finished. This looks ridiculous looking back, but if you look at it from the point of view of the time, practically all of the mysteries had resolved themselves—nearly all. Everything that had bothered one about the atom and molecules and solids and so on, had suddenly fallen into place. . . . Now, I'm not saying this was the common view. I don't think I ever shared it, really. I don't think Niels Bohr, for example, would ever have had any such illusions. But there were sort of over-lunch—or sometimes quite serious—discussions about what we would do when physics was finished. By 'finished' was meant the basic structure; of course, there are all the applications. The majority of people said that that would be the time to turn to biology. Only one person really took that seriously and did turn to biology, and that was Max Delbrück, who certainly was present at these discussions."[37] Delbrück did well in molecular biology, winning the Nobel Prize for medicine in 1969 for beginning the chain of investigations that led to the discovery of the doubled helix of DNA. On the other hand, physics was anything but over.

In the meantime, a small physics community had begun to form in the United States, but it was as yet a paltry thing compared to its equivalent on the other side of the Atlantic.[38] America had the resources to do experimental work, but except for isolated savants like Josiah Gibbs of Yale, the theory came entirely from Europe.[39] More important, the context of physics was set by Continental researchers. Classes at Caltech and Columbia University were assigned textbooks written in Göttingen and Leiden, and graduate students in Cambridge, Massachusetts, angled for the opportunity to travel to Cambridge, England. Aided by grants from the newly formed National Research Council, a pool of European-trained physicists slowly accumulated in this country; from them would later come the leaders of the next generation, one dominated by Americans. I. I. Rabi was among them.

Rabi got his dissertation in 1927, and promptly left for Europe with his wife, Helen. "I found I was actually better prepared than most Europeans of the same level," he told us. "What we lacked—and my generation was to supply it in this country—was a kind of understanding and feeling for the subject, which is hard to get if you're not in contact somehow with a verbal tradition, with people who are making the subject. [You have] to see the living tradition." After a month in Munich and two in Copenhagen, Rabi settled down in Hamburg to work with Wolfgang Pauli. There he met all the creators of quantum mechanics and quantum field theory. He took Dirac to Hamburg's well-known Hagenbeck Zoo, and was startled to discover that the Englishman's love for order extended to an insistence that they see the exhibits *in seriatum*. For Rabi's benefit, Max Born listed all the accomplishments of the past few years. "It was heady stuff," Rabi admitted, a half-smile playing on his lips. "Born said, 'We *have* all that. I think in six months we'll have the proton, and physics as we know it will be over.' There would be a lot to do, but the heroic part was finished. I was just a fresh postdoc [a postdoctoral fellow], and he was foreclosing on the field!"[40]

Not everyone was as sanguine as Born. Heisenberg and Pauli, for example, were troubled by some of Dirac's assumptions, and spent 1928 and 1929 working on their own version of quantum electrodynamics.[41] (To the relief of other physicists, the two versions were subsequently shown to be identical, much as Schrödinger wave mechanics and Heisenberg matrix mechanics were discovered to be different ways of describing the same thing.)[42] Dirac was perhaps the least satisfied of all. It pleased him that two of the four components of his electron equation corresponded to the spin of the particle. But the two other components, those corresponding to the energy of the electron, were more difficult to interpret. When the Dirac equation is solved for the energy of an individual electron, there are two answers, one positive, one negative, in a manner analogous to the way that the square root of a number can be either positive or negative. (The roots of 49, for instance, are 7 and -7, because both 7×7 and -7×-7 equal 49.) The negative answer was trouble: negative energy, itself a puzzling idea, implied, by $E = mc^2$, negative mass—an absurd impossibility. Ordinarily, physicists would assume that the world had started off with all electrons in positive energy states, in which they would remain, and no harm would come from the theoretical existence of negative energy.[43] The problem was that a positive energy electron should be able to emit a photon with enough energy to drop into a negative energy state. In fact, *most* electrons should end up with negative energy.[44]

Alas, "negative energy" is not an easily understood concept. Nobody knew what negative energy electrons could be. One theorist called them "donkey electrons," because they always do precisely the opposite of what you want.[45] Since energy and mass are equivalent, negative energy implies negative mass. What does it mean to say that a subatomic particle can have less than zero mass? As Dirac admitted, the whole subject "bothered me very much."[46] He wrestled with negative energies through all of 1929, trying to see if there was some way to keep the Dirac equation but get rid of the negative energy states.

And then [Dirac recounted later] I got the idea that because the negative energy states cannot be avoided, one must accommodate them in the theory. One can do that by setting up a new picture of the vacuum. Suppose that in the vacuum all the negative energy states are filled up. . . . We then have a sea of negative energy electrons, one electron in each of these states. It is a bottomless sea, but we do not have to worry about that. The picture of a bottomless sea is not so disturbing, really. We just have to think about the situation near the surface, and there we have some electrons lying above the sea that cannot fall into it because there is no room for them.

In other words, we don't notice the negative energy electrons because they are omnipresent—as undetectable to us as water is to a fish. But wait:

There is, then, the possibility that holes may appear in the sea. Such holes would be places where there is an extra energy, because one would need a negative energy to make such a hole disappear.[47]

Because of quantum randomness, in other words, photons should hit some of the electrons in the sea with enough energy to make them jump out of it— that is, become positive energy electrons—leaving a hole in their former location. This hole would appear as a sort of "opposite" electron: positively charged, because it corresponds to an absence of negative charge. Therefore, Dirac realized, even if he accommodated the negative energy electrons, he was forced to predict the existence of particles just like the electron, except with a positive electric charge.

Here Dirac's courage failed him for one of the few times in his career. He refused to follow the naked mathematics. Although logically the hole particles must have the same mass as electrons, because they are created by them, Dirac's paper on the subject said they must somehow be protons— which are almost two thousand times heavier.[48] He was also attracted by the hope of thus explaining both known elementary particles with the same theory. (The neutron would not be discovered for another two years.) Dirac's suggestion was promptly flattened by, among others, the mathematician Hermann Weyl, who reported from Göttingen that whatever the holes might be, they could not be protons.[49] Unafraid of *Dirac's* theory collapsing, Weyl said that if Dirac wanted to salvage his quantum electrodynamics, he was going to have to predict a previously unheard-of type of matter with the same mass but the opposite charge as an electron.[50] Moreover, these new particles should be all over the place. Not one had ever been noticed.

No matter how good a theory might look on paper, Dirac thought in dismay, if it predicts particles that don't exist there is something terribly wrong. Discouraged, he even wondered if all of quantum field theory might have to be scrapped. "I thought it was rather sick," he said afterward. "I didn't see any chance of making further progress."[51]

It is pleasing to report that the answer came quite literally from the clouds. Decades before, Charles Thomson Rees Wilson, a young and impoverished Scotsman with a scholarship at Cambridge, had gone to the highest mountain in the Highlands, Ben Nevis, where there is a meteorological observatory. He had been entranced with the beauty of the clouds surrounding the hilltop, the shimmering rings that suddenly burst into colored existence when the sun burst through the cumuli, or the great shadows cast by the peaks on the foggy masses below.[52] When Wilson returned to the Cavendish with the announced intention of studying the process of cloud formation, J. J. Thomson was willing to let him try to make clouds himself in the Cavendish. The fruit of these studies, made between 1896 and 1912, was the cloud chamber, which for decades was the primary means of studying subatomic particles.

Clouds, Wilson knew, had something to do with water vapor in the air. The amount of water vapor that a given volume of air can hold depends on its temperature; it can contain more when hot than when cold. Wilson also knew that if a volume of air suddenly expands, its temperature drops. When the compressed gas in an aerosol can is released, for instance, it expands out the nozzle and chills immediately—the reason that spray deodorant feels cool on the hottest of summer days. While Thomson and others at the Cavendish puzzled over radioactivity and X rays, Wilson constructed glass tanks with ingenious pistons tightly fitted into the inside walls. Wilson discovered that if he sprayed the air inside his tank with mist until it could absorb no more water, then suddenly pulled out the piston, expanding and cooling the air, the surplus water vapor "fell out" and formed a miniature cloud.[53]

The condensation was started by droplets of water collecting around dust particles in the air. Wilson tried cleaning the particles out of the chamber to see if that prevented clouds from forming. It did not, although the clouds emerged only if the piston were withdrawn a lot more.[54] Now what was the water coagulating around? In February of 1896, less than two months after the discovery of X rays, Wilson shot a beam of them into his glass tank. It was much easier to make clouds. Wilson was sure that the X rays smacked into air molecules, creating the charged objects known as ions, and the ions caused the clouds.

A year later, when J. J. Thomson discovered the electron, it would be clear that the X rays knock loose electrons, which are easily swept onto tiny droplets of water in the air. When several electrons collect on one droplet, the net charge begins to exert an effect on nearby molecules of water vapor. Water molecules (H_2O) are shaped like a broad V, with the oxygen nucleus at the vertex and the two hydrogen nuclei—protons—sticking out the ends. The electrons of the hydrogen atoms are yanked toward the oxygen atom by its much bigger, positively charged nucleus, which means that the tips of the arms of the V tend to end up with a net positive charge. The positive charge is easily drawn toward the negative electrons on the droplet, with the result that in rapid order a lot of water molecules are pulled to the charged droplets, gas condenses to form liquid, and the number and size of the droplets grow. A tiny cloud is born.

With the excitement in the Cavendish caused by radioactivity, it was soon discovered that water vapor would also condense around the electrons left behind by alpha particles speeding through the chamber. For a few instants—long enough for a photograph to be taken—the paths of these particles are visible as slender white lines that arc across the chamber like the contrails of a jet squadron. Alpha particles leave thick tracks, for these slow and doubly charged entities dislodge many electrons; the electrons from beta radiation have a smaller charge, are much faster, and leave light, wobbly tracks reminiscent of scratches from a cat's claw. If a magnet is wrapped around the chamber, particle identification is even easier: The magnetic field forces negatively charged electrons to curve in one direction, whereas positive alpha particles and protons are sent off in the other. The degree of curvature earmarks the particle's momentum. Thus Wilson could put a lump of uranium by his cloud chamber, pull out the piston with a thump, quickly click the camera shutter, and produce a photograph of the tracks left by the spray of radiation. Using a ruler and compass, he could discern that such-and-such a thick line that curved across the photographic plate in such-and-such a direction was left by an alpha particle with such-and-such momentum, whereas a light little line was a beta electron with a different momentum. Cloud chambers soon became standard equipment in laboratories where radioactivity was studied, and Wilson was awarded the Nobel for his invention in 1927.

In 1930, Robert Millikan, the head of the laboratory at the California Institute of Technology, asked Carl D. Anderson, a fresh young Caltech Ph.D, to build a new cloud chamber for use in studying cosmic rays, the recently discovered high-energy radiation that bombards the earth from space.[55] With the cloud chamber, Anderson built an electromagnet so powerful that the laboratory lights dimmed when the cloud chamber was in use.[56] For the next two years, Anderson expanded the chamber at random times, photographed the tracks of any cosmic rays that happened to pass through, and studied the results.

The great majority of the photographs were blank—no cosmic rays came through. But from the very beginning Anderson also started to find that a small number of the photographs had something strange, tracks from light particles that could either be negatively charged particles traveling upward or positively charged particles traveling downward. Anderson wrote later:

> *In the spirit of scientific conservatism we tended at first toward the former interpretation, i.e., that these particles were upward-moving, negative electrons. This led to frequent and at times somewhat heated discussions between Professor Millikan and myself, in which he repeatedly pointed out that everyone knows that cosmic-ray particles travel downward, and not upward, except in extremely rare instances, and that therefore, these particles must be downward-moving protons. This point of view was very difficult to accept, however, since in nearly all cases the [thickness of the track] was too low for particles of proton mass.*[57]

To settle the argument with Millikan, Anderson inserted a metal plate in the center of the chamber. Any particle passing through the plate would lose momentum, slow down, and thus be bent more sharply afterward, thereby indicating its direction. Furthermore, Anderson could calculate the particle's mass from the momentum lost passing through the plate.

On August 2, 1932, Anderson obtained a stunningly clear photograph that shocked both men.[58] Despite Millikan's protestations, a particle had indeed shot up like a Roman candle from the floor of the chamber, slipped through the plate, and fallen off to the left. From the size of the track, the degree of the curvature, and the amount of momentum lost, the particle's mass was obviously near to that of an electron. But the track curved the wrong way. The particle was *positive*. Neither electron, proton, or neutron, the track came from something that had never been discovered before.[59] It was, in fact, a "hole," although Anderson did not realize it for a while.[60]

The identification was made by two chagrined English experimenters—chagrined because they had actually seen the particle in their apparatus before Anderson and had been told by Dirac what it might be, but had waited much too long to be sure.[61] Anderson called the new particle a "positive electron"; *positron* was the name that stuck.

Positrons were the new type of matter—antimatter—Dirac had been forced to predict by his theory. (The equation, he said later, had been smarter than he was.) Physicists soon realized that electrons and positrons would annihilate each other when they met, producing two photons—two flashes of light. Similarly, a photon going through matter could split into a virtual electron and a positron.[62] From an embarrassment the negative energy states were transformed into a triumph for a quantum electrodynamics, the first time in history that the existence of a new state of matter had been predicted on purely theoretical grounds. Dirac won the Nobel Prize in 1933; Anderson went to Sweden three years later.

The canonization of the new developments occurred at the Solvay Conference of 1933, in which the evidence for both the positron and Chadwick's just-discovered neutron was discussed eagerly by Bohr, Curie, Dirac, Heisenberg, Pauli, and the rest of the luminaries of subatomic physics. (The Solvay Conferences were periodic gatherings of prominent physicists that were paid for by a rich Belgian industrial chemist named Ernest Solvay.) With a negative, positive, and neutral particle of real matter and the discovery of antiparticles, the mood was of extraordinary excitement. A form of hydrogen, called deuterium, had been discovered which had a proton *and* a neutron in its nucleus, and scientists were beginning to build machines, called particle accelerators, that promised to unlock even more discoveries. Some physicists thought (again) that the subject was rushing to a close.[63] One of the few sour notes came from Ernest Rutherford, who remarked of the positron that he "would find it more to [his] liking if the theory had appeared *after* the experimental facts were established."[64] Nonetheless, despite the feverish pace of development and the triumph of the positron, there was one idea shared unanimously among the conference attendees: Something was very wrong with quantum electrodynamics.

6
Infinity

WOLFGANG PAULI READ THE FIRST OF DIRAC'S TWO PAPERS ON THE ELEC-
tron with characteristic care as soon as the *Proceedings of the Royal Society*
came in the mail. Pauli understood immediately that an equation for a single
electron floating through space could be nothing more than a starting point,
for most situations in the real world involve many electrons interacting with
each other; he seems also to have realized that the constant presence of
virtual particles in the subatomic domain made the very idea of talking about
one particle by itself unrealistic—unphysical, in the scientist's phrase. Pauli
promptly dispatched a letter to Dirac informing him of the necessity to
formulate quantum electrodynamics without such a dubious assumption. In
the middle of a discourse on the version of quantum electrodynamics that he
and Heisenberg were working on, Pauli broke off to ask, "I would like to ask
your opinion about what is essentially a physical difficulty that Heisenberg
and I have run into and can't get around." It seemed that their calculations
were being thrown off whenever "the problem of 'a particle interacting with
itself' rears its ugly head."

In certain conditions, an electron is affected by the electromagnetic
field it gives off, in somewhat the way that a boat can be rocked by its own
wake, or an airplane shaken by the sonic boom it has created. Working ever
deeper into the thickets of quantum electrodynamics, Pauli and Heisenberg
had become increasingly concerned at the apparent inability of the theory to
account for "a particle interacting with itself." A satisfactory theory should be
able to deal with this phenomenon. As far as Pauli could see, quantum
electrodynamics instead produced nonsense: "If a single electron is present,
and its energy is taken into account, . . . you get an ever-expanding (even
infinitely increasing) equation."

In other words, quantum electrodynamics apparently predicted that
the interaction of an electron with its own field would be infinitely strong—
as if a plane were smashed to bits in its own sonic boom. The same odd
prediction was tucked into Dirac's version of the theory. Pauli believed that if
an equation that is supposed to produce a definite answer instead gives an
infinite result, then something is wrong. Worried, he asked Dirac: "*What do
you think about this?* I don't have a satisfactory way out yet. And I even think
we will have to make fundamental changes in our perspective if we are going
to avoid these difficulties."[1]

In a sense, Pauli's letter was old news. The interaction of a particle with
its own electromagnetic field is known as its "self-energy," and the self-
energy of the electron had puzzled physicists ever since the particle's discovery in 1896. Initially, the difficulties were due to the problems inherent in
applying physics rules derived from ordinary objects with definite shapes
and sizes to the electron, an entity which apparently has neither. As far as is
known, electrons are points in space, without breadth, width, or depth. In
high school physics, students learn that the way to figure out the energy of
the field produced by an electrically charged object—the metal ball of a Van
de Graaf generator, say, which shoots out the sparks in Frankenstein
movies—is to add up the energy of the field at every point in space. Carrying
out this measurement is next to impossible, but nineteenth-century physicists came up with an easy way to get the answer nonetheless.[2] The energy
of an electric field at any individual point is equal to the electric charge
creating the field (which is often indicated by the letter e) divided by the
fourth power of the distance (written r^4) between the point being measured
and the center of the field. The whole thing is multiplied by $\dfrac{1}{8\pi}$. In textbooks,
the equation looks like this:

$$E_{\text{point}} = \frac{1}{8\pi} \frac{e^2}{r^4}$$

Using the techniques of calculus, physicists can add up all the little individual points to get the total energy, which is

$$E_{\text{total}} = \frac{e^2}{a}$$

where a is the radius of the spherical object that is producing the field. With
this equation, it requires only thirty seconds and a hand calculator to find the
total energy of the field—provided you know what a and e are. Simple!
Except that the formula does not work for electrons. Electrons are points,
and thus a is zero. It is simply not possible to divide by zero; the answer
becomes infinitely large. Thus, physicists discovered that their equations,
which work very well for Van de Graaf generators and other big things,
predicted that infinitely close to tiny charged objects are infinitely large
electric fields.

The puzzle of infinite self-energy was augmented by another oddity of
physical theory, electromagnetic mass, which surfaced when scientists still
believed that electromagnetic waves travel through the ether, a mysterious
fluid that was once supposed to permeate all of space. In 1881, J. J. Thomson
noted that the ether would resist the passage of an electrically charged object
in much the way water resists when you swish your hand rapidly around in a

bathtub.[3] Just as your hand feels heavier from the drag of the liquid, the charged body acts as if its mass had increased due to the interaction of its electric field with the ether. This extra mass is today called "electromagnetic mass."

The same year that Thomson did his calculations, Albert A. Michelson began the first of a series of experiments that would establish definitively the nonexistence of the ether. Theorists continued to work on the concept of electromagnetic mass, and around the turn of the century several realized that even without the ether *the same effect still happens*; in a vacuum, a charged particle effectively acquires mass because of its interaction with its own field, that is, its self-energy.[4] Clearly, if the field close by a point electron is endlessly large, as physicists' equations suggested, then its electromagnetic mass must also be infinite. Both conclusions are patently false.

In an ingenious but inelegant argument published in 1904, the great Dutch theorist Hendrik Lorentz tried to dispose of the two infinities by taking the measured mass of the electron, assuming it to be entirely of electromagnetic origin, and feeding it into formulas for electromagnetic mass.[5] Cranking the calculational handle, he came up with what the energy of the field would have to be to generate the electron mass experimenters measured in their equipment. He then plugged this number into the equation $E = e^2/a$, and came up with a radius for the electron of two hundred and eighty quadrillionths of a centimeter, or, in scientific notation, 2.8×10^{-13} centimeters. Lorentz knew enough about relativity to know that objects become pancake-shaped and do other odd things when they travel at high speed, and he fashioned his model accordingly. But accounting for relativistic phenomena made it hard to apply the electromagnetic formulas consistently, and the answers he found depended on his method of calculation. Such mathematical inconsistency is the hallmark of a diseased theory, and for the next several decades the infinite self-energy remained inexplicable.

After the birth of quantum mechanics, physicists hoped that if they worked on other parts of quantum theory, the self-energy infinities might somehow disappear. Such hopes were not as foolish as they may seem. The self-energy complications occurred because of the point nature of the electron, and the idea of a point—a position, in other words—was much changed by quantum mechanics.[6]

Pauli, too, at first nourished this hope. Perhaps, he thought, by juggling around the new equations of quantum electrodynamics, the self-energy could be eliminated.[7] When he discovered that it could not, an alarm sounded: An infinite answer is an infinite answer, and something was terribly wrong. He wrote Dirac that physicists needed to make "fundamental changes in their perspective," intimating that quantum field theory might have to be replaced by something else.

The impact of Pauli's argument had as much to do with the real difficulties with self-energy—difficulties which were to grow ever more intractable the more physicists examined them—as with his remarkably forceful personality. Pauli read everything, knew everyone, and wrote thousands of elegantly phrased, passionate, acid-tongued letters to everybody he knew. His criticism was often brutal, a chill wind that occasionally had the effect of freezing out good ideas but killed off many more bad ones; if an idea was useful, Pauli felt, it would survive despite him. His fierceness was legendary, and no one was spared; as a student, he is reported to have astounded his classmates at a seminar by beginning a question with the words, "What Professor Einstein just said is not so stupid." Pauli told one assistant, "Your duties will be very light. The only duty you will have is to contradict me when I propose anything, with all the facts at your disposal." He terrified Born and badgered Bohr; he sometimes signed his letters, "The Wrath of God."

His intuition was more critical than creative. Although he had many good thoughts of his own, he preferred to place the work of others in perspective, to set the research agenda, to root out confusion and lack of clarity where he found it. For three decades, he was the arbiter of the science, a position that suited a man with a thirst for judgment. In 1958, when Pauli died after a short illness, one of his colleagues called him "the conscience of physics."[8]

While still a twenty-year-old student, Pauli established a reputation by writing a definitive presentation of relativity for the *Encyklopädie der Mathematischen Wissenschaften*, an important encyclopedia of science.[9] Four years later, in 1925, he formulated the "exclusion principle," which says that no two nearby electrons can be in exactly the same state.[10] One of the primary laws of quantum mechanics, the exclusion principle helps dictate the complicated patterns formed by the electrons in the atom. Because the chemical properties of an element derive from the behavior of its orbiting electrons, the exclusion principle is one of the pillars of chemistry. For its discovery Pauli won the Nobel Prize in 1945.

After 1925, Pauli worked almost constantly with Heisenberg, a collaboration that reached one of its peaks while they pieced together their version of quantum electrodynamics. As they wrestled with the theory, the two men kept trying to rid the equations of infinities—"divergences," as mathematicians call them. Heisenberg and Pauli thought it possible that even if quantum mechanics alone had not done the trick, a properly formulated quantum field theory might avoid the divergences of the past. As the two theorists worked on their long and painful proof of the relativistic invariance of quantum electrodynamics they began to suspect the existence of new, more intractable infinities. The prospect was so dark that Pauli threatened to quit physics and live in the countryside, writing utopian novels.[11] Heisenberg left for a lecture tour of the United States. Gloomy and depressed, Pauli explained the problem to the assistant who helped him complete the paper.

The assistant was J. Robert Oppenheimer. Born into a wealthy family, he was a sickly, precocious, over-indulged child with a stutter.[12] He was given the best education money could buy, and he dipped intermittently but deeply into religion, languages, poetry, and music. Eventually he became known in the community for his fascination with Hinduism and the *Bhagavad-Gita*.[13] He was extraordinarily quick—fatally so, for his physical intuition was all too frequently spoiled by arithmetical errors. As an adult, he grew to have a magnetic, rumpled charm; when Oppenheimer became head of the Manhattan Project, he ran the Los Alamos laboratory with such commitment and enthusiasm that many of the physicists there remember constructing the atomic bomb as one of the high points in their lives. He raced through his thoughts, speaking in a rapid tumble of erudite phrases and ironic slang that annoyed his teachers but enchanted his fellow students. When Oppenheimer graduated *summa cum laude* from Harvard in 1925, he went immediately to Europe—the dream of every U.S. physicist—where he studied with Rutherford for more than a year, toured the Continent, and received his doctorate from Göttingen. He went back to Europe in 1928, working under Paul Ehrenfest in Leiden. Ehrenfest soon discovered he could not abide Oppenheimer.[14]

Oppenheimer was an impatient, neurotic young man with brilliant prospects and occasional fits of generosity. Ehrenfest begged Pauli to do something with Oppenheimer. To prevent the situation in Leiden from exploding, Pauli agreed to have the young American come to Zürich to work with him.[15] Pauli thought that Oppenheimer was full of ideas, but all too prone to shoot off on wild, poorly thought out tangents. "Oppenheimer's physics is always interesting," Pauli told his friends. "But Oppenheimer's calculations are always wrong."[16] Still, something certainly could be done with him. He needed a steady problem—like working with infinities.

The exact problem Pauli gave Oppenheimer concerned that favorite subject of physicists, the spectrum of light emitted by a hydrogen atom. When Heisenberg and Pauli produced their version of quantum field theory, they, too, ended up with the Hamiltonian.

$$H_{total} = H_{field} + H_{particle} + H_{interaction}$$

Simply adding the respective Hamiltonians for the field and the particle and ignoring the interaction was enough to account reasonably well for the spectrum of hydrogen. Indeed, this was a triumph for the theory. But there remained the last term, the interaction Hamiltonian, which includes the self-energy. In quantum field theory, the self-energy of a charged particle can be thought of as what occurs when the particle emits a virtual photon and quickly reabsorbs it—a process that happens fast enough to stay within the bounds of the uncertainty principle. Because there are an infinite number of ways in which this emission and reabsorption can occur, the self-energy is not easy to evaluate. Nonetheless, it should have some effect on the behavior of the electron in the atom. Because the spectrum of hydrogen is produced by its electron jumping from one state to another, the energy level changes induced by emitting and absorbing virtual photons should displace the spectral lines ever so slightly. It was this displacement that Oppenheimer set out to calculate.[17]

When Oppenheimer came back to the United States at the end of 1929, he accepted a joint appointment at the California Institute of Technology, in Pasadena, California, and the University of California at Berkeley, outside San Francisco. His first published paper upon returning was the self-energy calculation.[18] Modestly entitled "A Note on the Theory of the Interaction of Field and Matter," the article claimed to show that if quantum electrodynamics was true, the endless number of possible virtual photon interactions should lead to an infinite displacement of the spectral lines of hydrogen—that is, the theory had the ludicrous consequence of predicting that there should be no lines at all because the self-energy effects would knock the electron to kingdom come. Appearing a month after the final formulation of quantum electrodynamics by Heisenberg and Pauli, Oppenheimer's paper seemed both to confirm their worst fears and to expose a terrible flaw in the whole enterprise. A theory may be wrong and still be useful. It cannot be ridiculous.

Few things are more difficult to understand than a brilliant man without a voice. Oppenheimer was strikingly perceptive, but not careful; thoughtful, but not original; decisive on the small scale, but paralyzed by doubt over the long run. Although he accomplished much, this great man never quite made the fundamental contributions to knowledge that his gifts promised. His career was brilliant, but Oppenheimer's talent and energy were such that merely brilliant seemed disappointing, and his very real accomplishments were always shadowed by his even larger potential. When a challenge came from the outside, he met it nobly and well: He built the bomb; he created the American school of theoretical physics; he fought cancer until, in 1967, it killed him. But scientists—good ones, anyway— must have their own visions of nature, and how to tease out her secrets. Oppenheimer did not have his own agenda. It is not easy to say why. The physicist Abraham Pais, a friend for more than fifteen years, told us once that Oppenheimer was in Pauli's company at a time when Pauli was particularly pessimistic and Oppenheimer was particularly impressionable. Clasping

Pauli's received discontent to his bosom, Oppenheimer simply never became his own man.[19]

In any case, Oppenheimer was convinced throughout the 1930s that quantum field theory was not good enough, not deep enough, to penetrate the secrets of matter and light. His concern infected his students, and some of them, later, felt that it slowed the progress of science. Certainly he was despondent in 1930. From his self-energy calculation he could have concluded merely that the infinities were an unexpected blemish on the surface of field theory. Instead, he took them to mean that physics was wholly off track.

Unfortunately, Oppenheimer missed a wonderful opportunity, as did all the other physicists who read his brilliant and influential paper. The dictates of quantum mechanics, and in particular the Pauli exclusion principle, mean that electrons can only reside in particular "neighborhoods" around the nucleus. Hopping from one neighborhood to another is accomplished by emitting or absorbing a photon. Over the years, these neighborhoods, which scientists call "orbitals," have been exhaustively catalogued; tables of orbitals fill fat appendices in atomic physics textbooks. As it happens, some orbitals have different quantum numbers but the same energy. The $^2S_{1/2}$ ("two-ess-one-half") and $^2P_{1/2}$ orbitals of hydrogen, two neighborhoods occupied fairly frequently by the atom's single electron, are examples of this phenomenon.[20] If Oppenheimer had thought about the $^2S_{1/2}$ and $^2P_{1/2}$, he could have subtracted one identical energy from the other, thus getting rid of all the infinite terms in the equation and leaving a finite residue that can be calculated exactly and matched against experiment as a test of the theory's validity. Had Oppenheimer thought of this, he might not have been so distrustful of quantum electrodynamics, and the recent history of physics might have been considerably different.

Three years later, Oppenheimer and a postdoctoral fellow, Wendell Furry, put together a lengthy, overcomplicated proof demonstrating that a simple change in notation would allow one to forget about the infinite sea of negative energy electrons and simply talk about electrons and positrons.[21] But even here, while restoring some degree of visualizability to the subatomic world, the authors are almost gleefully pessimistic about quantum electrodynamics. Getting rid of Dirac's sea, they note, does not get rid of the self-energy difficulties, "which rest ultimately on *an illegitimate application of the methods of quantum mechanics to the electromagnetic field.*"[22] Having disposed of all quantum field theory in a phrase, they claim the divergences "are not to be overcome merely by modifying the [equations for the] electromagnetic field of the electron within these small distances, but require here a more profound change in our notions of space and time. . . ."[23]

Oppenheimer's gloom was more than habitual fatalism; in that paper,

Furry and he had come across perhaps the strangest consequence yet of quantum field theory, the interaction of electrons with the vacuum—that is, with *nothing*.[24] Nine thousand miles away, Dirac had independently explored the subject a month or two earlier, but with even more distressing results. He stunned his colleagues at the seventh Solvay Conference, held in Paris, with a demonstration that quantum field theory seems to predict that the interaction between the electron and the vacuum should have the peculiar result of reducing the total charge of the particle to zero.[25]

Delivered in the perfect French that was a legacy of his father, Dirac's address began by recapitulating the hole theory, which still confused many if not most of the physicists who encountered it. Taking his audience through the whole chain of logic, he demonstrated that his "holes" really did correspond to positrons. He then pointed out that an electron could "jump into a hole, which it would then fill up. We then have an electron and positron annihilating one another," producing two photons with the combined energy of both particle and antiparticle.[26] Similarly, a photon passing through matter could split into an electron and a positron, each with half the energy, momentum, and spin of the original photon. Therefore, the vacuum, which can, according to the uncertainty principle, randomly produce energy in the form of virtual photons so long as it doesn't stay around long enough to be observed, may also create pairs of virtual electrons and virtual photons.

And even the virtual particles, the evanescent electron-positron pairs, can have real effects. Dirac asked his audience to consider an electron floating through empty space—or, rather, not-so-empty space. At any given moment, the particle is surrounded by a swarm of, virtual photons, electrons, and positrons, buzzing like ghostly bees around the particle in its progress. In the brief moment of their existence, the virtual positrons, which are positively charged, are pulled toward the real electron; the virtual electrons are in the meantime repelled. The result is that the electron has a cloak of positrons; it is surrounded by a shimmer of ghostly antimatter. As the distance becomes shorter, the number of virtual particles grows, in accord with the uncertainty principle. The result is that extremely close to the electron, its charge is blanketed by an ever-increasing, indeed endless snarl of, virtual positrons. Dirac called the process of attracting virtual positrons and repelling virtual electrons the "polarization" of the vacuum; it was yet *another* infinity in quantum electrodynamics.[27] (This infinity has a completely different source than the self-energy infinity, but physicists often include the various divergences in quantum electrodynamics under the loose rubric of "the self-energy problem.")

Dirac was far from giving up. The presence of vacuum polarization implied that whenever experimenters measured the charge of an electron, they never measured the "real" charge, which self-energy rendered infinite, but the "real" charge *minus* the offsetting infinite vacuum polarization. Nature, in other words, had subtracted one infinite quantity from another, and obtained a finite result. Although subtracting one infinity from another is fraught with mathematical ambiguity, Dirac proposed that theorists do likewise, and balance the divergences in their equations against each other. It wasn't a solution—infinite quantities were still present, and no one knew how to get rid of them—but theorists were ready to embrace any technique that would even sweep them under the rug.

Although Pauli ridiculed Dirac's "subtraction physics" and the arabesque of infinities in the hole theory, it was the only game in town.[28] By this time, he had a new assistant, Victor Weisskopf, whom Pauli had selected when he could not get another young man named Hans Bethe. After baldly explaining that Bethe had been the first choice, Pauli gave Weisskopf a little problem. About ten days later, he asked for the results. Weisskopf showed them to him. Pauli said, "I should have taken Bethe!" It was the beginning of a friendship.[29] Viennese like Pauli, Weisskopf was an intuitive thinker, a scientist who sought physical reasons rather than correct formal expressions. Pauli must have liked his new assistant's way of working, for he gave Weisskopf a fundamental problem: Calculate the self-energy of the electron, but now take into account the hole theory and the offsetting vacuum polarization discovered by Dirac. Weisskopf finished the figuring early in 1934, and the result was bad news. The self-energy was just as divergent as before.[30]

Shortly after the paper appeared, Weisskopf had the dreadful experience of receiving a letter from Wendell Furry, Oppenheimer's collaborator at Berkeley, that gently informed him of a foolish mathematical error in his work. Once the mistake was removed, the answer changed dramatically; the self-energy was still divergent, but only logarithmically. This was vastly different. The logarithm of a number—say, 100—is the power to which a second number (often 10, because we use the decimal system) must be raised to equal the first. Thus, the logarithm of 100 is 2, because $10^2 = 100$. The logarithm of 1,000 is 3, because $10^3 = 1,000$, and the logarithm of a trillion (10^{12}) is 12. Clearly, a very big number can have quite a small logarithm. Furry pointed out that the self-energy did not depend on x^N, but on $\log_x N$—and that this crawled up to infinity at the pace of the proverbial snail.

Weisskopf today is an emeritus professor at the Massachusetts Institute of Technology, in Cambridge, Massachusetts, a former director-general of CERN, the giant international particle physics laboratory in Switzerland, and one of the founding fathers of nuclear physics. He was, to his sorrow, intimately involved in the effort to build the first atomic bomb, and for years has been a prominent advocate of arms control. A distinguished career did not seem likely to him in 1934, when he received Furry's letter. He was humiliated by his error. "Pauli gives me a fundamental problem to solve and I do it wrong," Weisskopf told us in his small office at CERN. "I went to Pauli and said, 'Pauli, I have to give up physics! It's terrible!' And he said, 'Oh, no, don't worry. There are many people who made mistakes in their papers—I, never.' "

We laughed and asked if Pauli really had said that.

" 'I, never,' " Weisskopf insisted, smiling at the memory. He paused for a moment, thinking; many of his sentences were graced with small, reflective rests. Then he said, "I find it wonderful, you know. It shows the courtesy of physicists at that time, that Furry didn't publish. He merely wrote to me. I published a correction, of course, thanking him.[31] But that wouldn't happen today. Today, the corresponding Furry would publish a paper and say that I was wrong. So this is why on every possible occasion I say how grateful I am to Furry. Because some of my fame comes from this—it's always connected with my name—and actually it was Furry."[32]

Heisenberg, for his part, took little comfort in the shrinking of the divergences.[33] Infinity is still infinity, he thought, and quantum electrodynamics will remain a cheat until the divergences are gone. Heisenberg took his turn at exorcising them. He still hoped that if he set down the full and complete theory of quantum electrodynamics without using any of the approximations dear to theorists, the infinities would cure themselves. The resultant paper, "A Remark on Dirac's Positron Theory," is one of his greatest, but one of his least conclusive.[34] Unlike anyone before, he explicitly wrote down the quantized field equations for the electron and the electromagnetic radiation—both at once—and the interaction. With this article, modern quantum field theory was born, although Heisenberg derived little satisfaction from it.

Long, knotty, awkward, and intense, the article contained in a crude form the rules by which physicists can subtract infinities one from another. There are several such rules, the most important being that the calculations cannot be just a little relativistically invariant; they have to take relativity into account every step of the way.[35] ("Relativistic invariance," again, means that the quantities in an equation are completely faithful to the dictates of general relativity.) Conscientiously subtracting, Heisenberg tried to deal with all of the divergences that grew like weeds in the crevices of the theory: the infinite field produced by the sea of negative electrons, the infinite mass from self-energy, the infinite vacuum polarization, and yet another paradox, the inability of a particle surrounded by an infinite cloud of virtual particles to interact with anything else.[36] But after a lengthy journey through the brambles of the theory, he was still in divergent terrain. When he treated the matter field and the electromagnetic field on an equal basis, he found the same infinities in both—the photon, like the electron, had self-energy problems. The reward for writing down the full theory, treating matter and energy on an equal footing, was discovering the same set of divergences in both. Dismayed by Heisenberg's results, Pauli complained that he was "drowning in the Heisenberg-Dirac hole equations, and would grasp at any straw that was offered."[37]

As striking as the noble mess of Heisenberg's paper is the fact that it took seven years from when Dirac first quantized the electromagnetic field to the first time anyone tried to write down the theory in a complete way. Why the delay?

"It shows how *crazy* the whole theory was considered," Weisskopf said in Geneva. His fingers tapped softly on a copy of a speech he had given recently in opposition to the militarization of outer space. "I mean, the filling of the vacuum and Dirac's theory of the position—nobody really believed it at the time. And the theory was so ugly! All these absurd ideas and this terribly complicated mathematics, it all looked very ugly. In a sense, I suppose the problem [of the infinities] seemed academic because for a long time nobody knew where the physics was in it. And a lot of people just said, 'To hell with it, I'm going to do some nuclear physics.' "[38]

On a summer afternoon not long ago, we walked through the crowded city campus of Columbia University toward the malachite dome of Pupin Hall, the seat of the physics and astronomy departments. Pupin Hall has several floors of dusty laboratories, a small observatory on the roof that is nearly blinded by the lights of Manhattan, and a grim brick facade that stands in crabbed contrast to the Gothic reaches of nearby Riverside Church. Baffled by the plethora of infinities, we turned again to Robert Serber, one of the handful of physicists during the 1930s who came very close to solving the divergences of quantum field theory.

Serber met Oppenheimer for the first time in 1934, at a summer physics seminar at the University of Michigan, in Ann Arbor. Like many of his contemporaries, Serber was immediately enthralled by Oppenheimer's intensity and articulateness. "No one thought quicker than he did," Serber told us. "People used to say that you never finished phrasing a question to Oppenheimer, because he'd answer before you got to the end of your sentence." Hadn't Pauli said that his calculations were always wrong? "That's because he would whip through something in five minutes on the back of an envelope, doing just the essential calculations, and leaving out all the numerical terms. Somebody like Dirac would take a few hours and get all the twos right, and the pis in the right place." Oppenheimer was marvelously, poetically impatient; rushing along in an enthusiastic wave, he didn't like the painful details such as whether there was a three in the numerator or denominator.[39]

For the first few days at Ann Arbor, Serber kept an awed distance.

Then the participants of the summer school were all invited out for drinks to the home of an Ann Arbor physics professor named David M. Dennison. Upon their arrival, the company discovered that Dennison had taken Prohibition seriously, and that "drinks" meant lemonade. In that instant of shock, Serber happened to catch Oppenheimer's glance, and both simultaneously rolled their eyes heavenward. Serber had been on his way to Princeton as a National Research Council Fellow; he decided to turn right around and follow Oppenheimer to Berkeley instead, where he arrived in September of 1934.[40] A decade later, he went with "Oppie" to Los Alamos to build a bomb; a few years after the mushroom clouds over Hiroshima and Nagasaki, he was hired by Columbia, where he has been ever since.

We asked him about Oppenheimer's first self-energy paper. How many of the divergences had he known about at that time?

"*Well*," Serber said. He pulled a cigarette out of his breast pocket. It had been many years. Oppenheimer's paper lay on the desk; Serber looked at it for a minute, smoking, while the traffic rumbled below on 120th Street. "There are three basic divergences in electromagnetic theory. There's the proper energy, and there's the vertex term, and there's the vacuum polarization." He went to the blackboard, hunted around for a piece of chalk. "You have an electron coming along—I'll use a diagram." He drew a line across the board to represent the path of the electron, then a second, wavy line arcing out of the first and then falling back into it.

"This is an electron emitting a virtual photon and then absorbing it. That gives rise to the proper energy of the electron. The proper energy is a big part of what is usually lumped together as 'self-energy.' That's one infinity. Another one is if the electron interacts with a photon—say it gets scattered by a photon." He drew another electron line, but this one bent at an angle when the electron bounced off a photon. "Then you could put in a virtual photon like this." The chalk squeaked as he made a curved, wiggly line joining the two halves of the electron line, pre-and post-collision.

"That's a higher-order term. The electron emits a photon, the scattering takes place, and then the photon is reabsorbed—you see? Now, this has nothing to do with proper energy, although we didn't realize that for an awfully long time. This is the vertex correction. If the electron absorbed the photon before being scattered, it would be—" he tapped the first diagram "—part of the proper energy.

"The third type is vacuum polarization, which is . . ." Serber paused; vacuum polarization is not as easy to diagram. "A photon comes along and creates a virtual pair." Another wiggly photon appeared on the board, this one divided by a circle. "The polarization is that if you have a charge—" he quickly put a fat dot on the board to symbolize an electron "—its electromagnetic field produces not only direct photons, but photons that make pairs." For good measure, another few photons and pairs were added to the picture.

"So if this central point is a negative charge, the net effect is that you have a bigger—it will bring a positive charge closer and push a negative charge further way. So it's just like the vacuum were behaving like a dielectric." A dielectric is the material that keeps the positive and negative plates of a capacitor from discharging into each other. "That is, the negative charge here induces a polarization in the vacuum, the positive charges being attracted and the negative charges being pushed away." Properly speaking, only the first two are "self-energy," but in what follows we shall follow the sloppy habit of physicists and lump the vacuum polarization under the same name.

"In the complete theory, they are all mixed up with one another, so it took some time for everyone to understand that there are just these three primary processes, and many variations of them." To calculate the full self-energy in quantum electrodynamics, then, requires adding up the likelihood of these three interactions and each of their infinitely many variations and combinations, because all have a slight chance of occurring at any given moment. As there are an endless number of ever more complex self-energy diagrams, with ever-branching trees and loops of electrons and photons, evaluating their total contribution to the theory is a horrendous mathematical quagmire.

"That was the problem," Serber said. "Nobody knew how to evaluate these things directly—we still don't." Above his three diagrams, he wrote

$$H_{\text{total}} = H_{\text{field}} + H_{\text{particle}} + H_{\text{interaction}}$$

He put his hand across the final term, blocking it from view. "This is the one that caused the trouble," he said. "At the beginning, they just ignored it;

they really didn't know what else to do. You could actually calculate fairly precisely if you pretended the interaction term wasn't there, so you knew that even if it *looked* infinite the interaction term had to be fairly small. But the whole business was a mess. We knew that. And when we finally started to work with the whole theory, we said, 'Okay, we can assume that the interaction term, the term with the infinities, is really only a small part of the total. Then you can calculate the whole thing [H_{total}, in other words] as a slight change in the sum of this and this' "—the Hamiltonians of the particle and its electromagnetic field.

The calculations were and are extremely cumbersome. They can be done by using what is called a perturbation expansion, which works by viewing each possible self-energy interaction as a small change, a perturbation, of some known state. By adding up the infinite number of such small changes, a final answer can be obtained. However, this answer is finite only if the successive terms become smaller, or, more exactly, if the series of terms *converges* toward a particular number. If the successive terms become bigger, the series *diverges*, and you end up with infinity. In the early 1930s, the series seemed to diverge off the map.

When Serber came to Berkeley, Oppenheimer gave him his self-energy paper, the article he had written with Furry that got rid of the sea of negative electrons, and Heisenberg's just-published "remarks" on Dirac's subtraction theory—the papers everyone in the group was talking about. Much as the presence of a single energetic play director is sometimes enough to create an entire theatrical community in a city where there had been little before, Oppenheimer created a theoretical community in Berkeley, and lent his weight in addition to the Caltech and Stanford departments. In his office was always a coterie of awed graduate students and National Research Fellows, many of whom imitated the light cough, mumble, and penchant for incredibly hot food that were Oppenheimer trademarks. Oppenheimer had no desk; he worked on a table in the center of the room covered with paper and cigarette ash. A blackboard covered one wall, densely packed with the equations whose constants Oppenheimer never could remember. There were another six or seven people at nearby Stanford with Felix Bloch, a Nobel Prize-winner. Everyone congregated in 219 Le Conte Hall, Oppie's office, intent and informal young men with the dark narrow formal ties and thin-collared shirts of the time. A hundred miles outside of town was *Grapes of Wrath* country and the Depression at its fullest; the air in Berkeley bore the whiff of radical politics and Oppenheimer's blackboard chalk.[41]

Once a week, the Journal Club discussed what had come out during the past week in journals; at these meetings, the theorists and the experimentalists who were working on the first, beginning particle accelerators could make close contact. In late 1934 and early 1935, the Berkeley theorists heard reports at the Journal Club that some spectroscopists at Caltech were finding slight deviations from the predictions of quantum electrodynamics for the Balmer series of the hydrogen atom; it looked as if some of the lines were not where they were supposed to be. Oppenheimer suggested that Ed Uehling, another National Research Fellow, should use Heisenberg's subtraction techniques to see if vacuum polarization could account for the new data. Serber joined the discussions, and eventually found himself writing a companion paper. The two articles were published back-to-back by the *Physical Review* in July 1935.[42]

Uehling took Heisenberg's theoretical framework as a given, but he wanted to provide some of the numbers he thought Heisenberg had been too busy to get around to. Because of the polarization of the vacuum, as Uehling noted, "the energy levels for the [hydrogen] electron are slightly displaced. This displacement may be calculated . . . as a small perturbation," and Uehling then did the mathematics. The answer was anything but satisfying. Both he and Serber found that there would be a change in the spectral lines, but it was much too small and in the wrong direction from the experimental results.[43]

A few months later, the Berkeley group was delighted to host P. A. M. Dirac on one leg of a round-the-world trip. Knowing that Serber and another Oppenheimer protégé, Arnold Nordsieck, were working on a variant of the subtraction technique originally suggested by Dirac, Oppenheimer arranged for them to meet the great man. Elated at the opportunity, the two young Americans explained their work for over an hour in a small Berkeley conference room. Dirac uttered not a word. At the end of the talk, a lengthy silence ensued. Finally, Dirac said, "Where is the post office?"

Exasperated, Serber asked if he and Nordsieck could accompany Dirac to the post office. Along the way, perhaps, Dirac could tell them his reaction.

"I can't do two things at once," Dirac said.[44]

Weisskopf, on the other hand, was extremely interested in what Serber and Uehling had done. In 1936 he left Pauli to work with Niels Bohr in Copenhagen, where he tried to further generalize the calculations for vacuum polarization.[45] He argued that if quantum electrodynamics is right, the infinite polarization obviously has to be there; equally obviously, it is not seen directly. In most situations, Weisskopf said, the "electron" seen in experiments actually consists of an electron with an infinite cloak of virtual positrons that shields most of its charge. But if another particle bores in very close to the electron, shouldering through the mass of positrons, it discovers that the real charge of the "bare" electron is much higher than it usually seems. In fact, the "bare" charge is infinite. The infinite charge is almost always hidden beneath an infinite shield of vacuum polarization, for, like the slight tremors that constantly ripple across the junction of the two straining sides of a geological fault, the small observed charge of the electron is just the visible result of two opposite and almost equal infinities. In quantum field theory, the world is full of infinite quantities that nearly negate each other, and nature is a balance of divergences.

Although published in a relatively obscure Danish journal, Weisskopf's paper attracted much attention. The arguments were in many respects similar to those of Heisenberg, Serber, and Uehling, but Weisskopf's clear style and simple presentation made them accessible to colleagues still uncomfortable with the complicated formalism of field theory. The spectre of vacuum polarization that Dirac had conjured up three years before at the Solvay Conference in Paris had been reduced to the trivial role of eliminating another, otherwise unobservable infinity.

One of the three infinities on Serber's blackboard had been killed off, but the other two were as alive as ever. In an article that sharply criticized Heisenberg's subtraction program, Serber set down the rules for calculating the proper energy.[46] He carried the calculations through at the lowest level, but did not publish them because he could see so many other divergences coming down the pike that he concluded the task was hopeless. Something else would have to be tried.[47]

We asked Serber one day why he had given up canceling infinities when from today's vantage he was so close to the answer. There were many reasons why, he said, shrugging, and he had not been the only one. People had been distracted. The formalism was difficult. And he, at least, had not clearly understood the difference between the proper energy and the vertex correction, between his first and second diagram. If the problem isn't identified correctly, it's insoluble. Did anyone make the distinction?

"Well, yes and no," Serber said. We were at a restaurant high above the wintry slopes of Morningside Heights. Glasses tinkled, dishes scraped, and Serber's meal lay almost untouched before him. "Heisenberg's paper at the time didn't describe the problem of the vertex infinities, although he may have been aware that they existed. Then the next thing was that [Felix] Bloch and [Arnold] Nordsieck solved the [low-energy vertex] difficulties, which really went a long way to show how you could do things.[48] And then Bloch and Oppenheimer tackled the vertex corrections on the high end. They didn't do the actual calculation themselves, they gave it to Sid Dancoff to do. Sidney had a fellowship with Bloch. We had some of the right ideas, distinguishing between proper energy and vertex infinities, but Dancoff made a mistake."[49] Serber pointed out the error, but Dancoff insisted it made no difference. It *did* make a difference, and Oppenheimer always felt afterward that he might have nailed all the divergences and maybe won a Nobel Prize if Dancoff had corrected his paper. "We were getting discouraged too easily.

Essentially two-thirds of it was solved by '37, but we didn't really understand that. Then there was still the vertex, and Sid Dancoff almost got that right. A lot of it was *there*. It just wasn't put together properly."

The process of shaking the divergences out of a theory is now known as *renormalization*—getting a theory back to making ordinary, finite predictions. The word may have been coined by the Berkeley-Stanford group; its first appearance in print is apparently in Serber's papers on proper energy.[50] "The interesting thing about renormalization," Serber told us over lunch, "is that the thing never happened all at once. Sometimes science works like that, sort of staggering along. There were a lot of steps and developments, but the *real*" —he laughed, interrupting himself—"the earlier steps didn't make much difference. The people who did it right later on didn't even know about them. Or I don't think they did; it's very hard to tell about things like that. Some things you hear in the summer and haven't understood, and then all of a sudden it'll come back to you three years later and you finally get the point. It may even be that the people who did it later heard our ideas—in the background, in a speech—and it didn't sink in until they used them later. I don't think I consciously had much influence on someone like Schwinger, when he was working it out."[51]

By 1937, European physicists had other things to think about than self-energy. Jewish scientists had to protect their lives and their families; many non-Jews were preoccupied with saving their colleagues. Bohr ran an "underground railroad" for young Jewish theorists, piping them from Copenhagen to England or the United States; Weisskopf, who went to the University of Rochester that year, was one of "Bohr's refugees."[52]

But physicists like to do physics, and a few tried to keep working on renormalization despite the war clouds scudding overhead. In retrospect, they were handicapped by their belief in the radically new. With a kind of hopeful despair, Heisenberg and the other founders of quantum physics hoped the infinities indicated the need for another revolution, another set of apparently crazy ideas to save the theory. They did not recognize that the science had shifted, and needed a different sort of originality, a different sort of perseverance. Heisenberg, for example, tried to resolve the divergences by arguing that the electron was not a point because space itself was quantized. There was a "universal length," and the continuum was grainy, like the dots in a half-toned newspaper photograph.[53] Pauli spent years trying to figure out another way to solve the equations of quantum field theory, but got nowhere.[54] Dirac convinced himself that trying to rid quantum field theory of its divergences was a case of working on too many questions at once, and came up with an infinity-free classical theory. To do it, he had to introduce "negative probabilities," events which had less than no chance of occurring, a concept that even Dirac could not make sense of.[55] Other theorists speculated that the self-energy effects might be canceled out by a second, previously unknown field invented by nature to do just that.[56]

By and large, physicists washed their hands of the problem. Something was wrong, but the theory seemed to produce more or less correct answers if the infinities were simply ignored. The difficulty was oddly abstract; the mathematics was convoluted; and many scientists were seduced away from the problem by the emergence of nuclear physics, and the exhilarating, appalling discovery that the atom could be split.

The last man to deal directly with self-energy before the war was Victor Weisskopf. In a speech before a New York meeting of the American Physical Society and a subsequent paper, Weisskopf demonstrated that the "peculiar interaction between the electron and the vacuum" reduced the self-energy problems to logarithmic divergences not only in simple situations, but in the most complex interactions.[57] The divergence was incredibly weak, but it was still there. Like Lorentz four decades before, he tried to make an end run around the problem by using the infinity to tell him how big the electron radius would have to be. It was 10^{-58} times smaller than the classical electron radius—but it was still not a point. Four months after the article appeared, the Germans invaded Poland, and physicists, like everyone else, had other things to worry about than the size of the electron.

We asked Weisskopf one time what he had been missing in 1939. He was so close to renormalizing the theory. What had he lacked?

"Persistence." He laughed. "Persistence! I was much too much interested in other things. I always say, you know, because I really believe it made me miss the Nobel Prize, that I think I'd gladly pay the Nobel Prize for having this general overview of physics which I have from those years. I was interested in too many things. I prefer it this way. I prefer to know nothing about everything rather than everything about nothing. At that time life was not so easy, and there was a lot of time I had to spend helping refugees and things like that. If I had sat down and said, 'This is my purpose in life [renormalizing the theory],' and forgot everything else—all the nuclear physics and the other things—I probably could have done it. But I haven't got that persistence.

"I don't regret it at all. I always tell my students when they say they cannot work on physics in these terrible days that physics is what kept our equilibrium then. Physics was a wonderful thing! If it were not for physics, my character would have come *apart* at that time. I was a Jewish refugee, my family lived in terror—and physics was the great thing that kept us *human* at that time." He thought a moment, his face turned to the ceiling with a curious mixture of emotion. "But I would say, look, I'm not the type who can work on one thing. I'm just not persistent enough."[58]

7
The Shift

EVERY NOW AND THEN, A FACT, A SINGLE ASPECT OF NATURE, ASSUMES AN importance to physicists far outside of its intrinsic significance, and it makes or breaks theories, careers, reputations. After the contretemps is over, the fact becomes a footnote in future papers, and the irregular means by which it was brought to light is forgotten. The next generation of physicists learns the fact in graduate school as an accepted part of the world; it seems to have appeared in experimental equipment when required, its importance never in doubt, and to have been there always—had past scientists only possessed the wit to seek it.

Such a fact is the exact energy level of the $^2S_{1/2}$ ("two-ess-one-half") electron orbital in atomic hydrogen.* One of the lowest and simplest energy states of the lightest and simplest atom, its location is a corollary of quantum theory. The study of this orbital was the subject of a dozen experiments during the 1930s, all of which bore directly on the worth of quantum field theory in general and the renormalizability of quantum electrodynamics in particular. But, because science is a human enterprise of fallible people, ideas were not put together, connections short-circuited, and the result was a slow tragicomedy on the experimental side that accompanied the contortions of the theoreticians. While theorists spent the 1930s alternating between pretending the infinities weren't there and fruitlessly trying to grapple with them, experimenters passed the same decade painfully trying to decide if the predictions of quantum electrodynamics could be confirmed. Self-energy was the root of both problems. At the end, quantum field theory seemed to be vindicated, and yet again physicists had the hope that this time they were at last on the road to a complete picture of nature. They were, but in a different way than they had imagined.

Many of the students of the hydrogen spectrum and the $^2S_{1/2}$ wanted to measure a magic number known as alpha (α), which had nothing to do with

---

*The terminology for spectral lines is somewhat antiquated and specialized to the point of incomprehensibility. Orbitals are referred to as S, P, D, or F, after the old spectroscopist's language for the type of lines they create—"sharp," "principal," "diffuse," and "fundamental." The left-hand superscript is related to the total spin s by the formula $2s + 1$, and the right-hand subscript is associated with the angular momentum. Thus the $^2S_{1/2}$ is the sharp line from an orbital with spin ½ and a total angular momentum of ½ $\hbar$ (its full name is actually $2^2S_{1/2}$, but we have dropped the initial 2 for convenience). We apologize to our nonspectroscopist readers for the necessity of inflicting them with this language.

alpha particles but everything to do with quantum mechanics and the atom. Like parsley in cooking, alpha appears everywhere in quantum electrodynamics, for the number pops up in the perturbation expansions necessary to calculate the self-energy terms in quantum electrodynamics. These equations express the probability that the electron will emit a virtual photon, which in turn depends on the strength of electromagnetism as a force. The relative strength or weakness of any force is measured by experimenters and expressed by a number called a *coupling constant*. For no special reason, the coupling constant for electromagnetism is called alpha, and thus every calculation of the interaction of photons and electrons has one or more αs in front of it. By 1930, many experimenters had measured alpha. It was almost exactly 1/137.

The constant 1/137 was, as it happens, first described by Arnold Sommerfeld, who supervised Heisenberg's dissertation at Munich. He gave the coupling constant alpha yet another name, "the fine structure constant," because its size plays a role in determining the separation of the fine, bunched-together lines in the hydrogen spectrum. The Balmer lines of the hydrogen spectrum on which Niels Bohr constructed his model of the atom had subsequently turned out to be composed of many different lines of slightly different wavelength—a fine structure. Their exact placement was described by equations couched in terms of alpha. More important, however, alpha put together many of the mysteries of twentieth-century physics. Alpha is the ratio $e^2/\hbar c$—where e is the charge of the electron, $\hbar$ is Planck's constant divided by twice pi, and c is the speed of light. It was hard not to think that the magnitude of alpha must somehow be *necessary*, must be a consequence of deep and hidden connections among e, $\hbar$, and c.

"People were *fascinated* by that number," remembered Markus Fierz, Pauli's assistant and co-worker in the late 1930s. "Now, e is electrodynamics, $\hbar$ is the quantum theory, and c is relativity. So in this one constant all the fundamental theories are related. The hope was that if one could figure out why this number had its particular value—1/137—the whole thing would be solved. It was a magic number!

"It's hard to realize how important this all was just from the publications. They made many attempts to get it that were wrong, and they didn't publish those. Great physicists don't publish their abortions."[1] Heisenberg and Pauli spent years trying to understand why the fine structure constant was 1/137 and not, say, 1/136.[2] Convinced that unexplained factors of 137 have no role in a proper theory, they believed that quantum electrodynamics could be renormalized only if they understood alpha; at the same time that they struggled in the slough of hole theory, they chased up blind alleys in search of the fine structure constant. They were not alone. Sir Arthur Eddington, a well-known astronomer whose exposition and testing of general relativity were partly responsible for its rapid rise to fame, viewed alpha as the key to the way nature hangs together. At first, the fine structure constant was thought to be 1/136, and Eddington said the Universe was an "*E*-matrix" with 136 parameters. Later, when subsequent measurements moved the number to 1/137, he added another parameter, restoring agreement with alpha itself. *Punch* published a cartoon identifying him as "Sir Arthur Adding-One."[3] When Eddington went public with his numerology, he embarrassed his colleagues and exiled himself from the main currents of science.[4]

Nevertheless, Eddington did help stimulate a vogue of interest in the fundamental constants, and a number of measurements transpired.[5] The first experiment after the onset of quantum electrodynamics was announced by Frank Spedding, C. D. Shane, and Norman Grace, all of Caltech, in the middle of 1933.[6] They hoped to use the fine structure of the Balmer lines of hydrogen, which the Dirac equation seemed to predict exactly, as a standard for determining the fine structure constant. To their evident surprise, they found alpha to be 1/138—"distinctly greater," as they noted, than the results from other experiments. "This discrepancy," they noted, "does not necessarily imply erroneous values for *e* and *h* but may arise from an incompleteness of the theory of the fine structure." By "theory of the fine structure," of course, they meant quantum electrodynamics.

A second Caltech team, William Valentine Houston and a visitor from China, Y. M. Hsieh, also examined the fine structure of hydrogen. They, too, found that some of the fine structure lines were about 3 percent off the predictions of the theory.[7] The discrepancy was "large," they said. It was caused by "a deficiency in the theory." Houston had become an assistant professor of physics at Caltech in 1928, and was widely regarded as a man to watch. When he said the experiments might not show what the theory predicted, the theorists listened.[8]

At the same time, R. C. Gibbs, an experimenter at Cornell University, in Ithaca, New York, and one of his graduate students, Robley Williams, were measuring the same thing. Unlike the Californians, Gibbs was not initially interested in determining the fine structure constant. He had merely noticed that an earlier measurement of the fine structure by three experimenters from Boston University was riddled with technical errors.[9] The two men decided to redo the experiment and see if the spectrum was where theory said it should be.[10] At first, the readings seemed to agree with theory, but as the experiment went on the Cornell group, too, saw discrepancies. They gave an initial report on their work at the same time as Spedding, Shane, and Grace, but weren't certain of their results for another six months.[11]

Theory said that when you plugged in all the numbers and solved all

the equations—or, rather, all the equations except the interaction terms—
the primary line in the Balmer series was supposed to be split up into two
major and three minor components formed by different transitions among
the ^2S, ^2P, ^3S, ^3P, and ^3D orbitals.[12] The two most intense components (#1

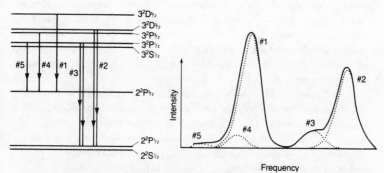

A portion of the hydrogen spectrum (*left*) gives rise to experimental values for intensity
and frequency (*right*). Because we are wholly concerned with the bottom three energy
states—the $2^2S_{1/2}$, $2^2P_{1/2}$, and $2^2P_{3/2}$—we have in the text taken the liberty of dropping
the initial numeral 2, which is physics shorthand for the general energy level.

and #2 in the drawing) were the ones that spectroscopists could see with
relative ease. Gibbs and Williams knew the theoretical value; they got numbers about 6 percent lower. On Epiphany Day, 1934, they sent off a little note
to the *Physical Review*; it was published on February 1.[13] More was coming,
they told their readers.

They were beaten by Houston and Hsieh, whose long and excellent
paper on the fine structure of the Balmer lines appeared in the following
issue of the *Physical Review*.[14] Claiming that their measurements were sufficiently accurate that they had "attained to that degree of precision in which
the theory is no longer satisfactory," they make what is in hindsight an
astonishing suggestion: "One possible explanation of this [discrepancy] is
that the effect of the interaction between the radiation field and the atom
[that is, the self-energy] has been neglected in computing the frequencies."[15]
This explanation, which is now thought entirely correct, had come from two
theorists: another member of the Caltech physics department, J. Robert
Oppenheimer, and a distinguished guest, Niels Bohr, who was visiting the
United States at the time. Both men emphasized to Houston and Hsieh that
all the theoretical predictions they used ignored the self-energy, which nobody yet understood. Few experimenters were aware of such theoretical
niceties, but, once informed, Houston and Hsieh argued that "it seems to us

very probable that this [self-energy] is the cause of the discrepancy we have observed."[16]

Everyone interested in field theory in those days was talking about the newly discovered polarization of the vacuum. Oppenheimer, one of its co-discoverers, was particularly excited; he allowed his new National Research Fellow, Ed Uehling, to see if the separation found by Houston and Hsieh could be explained by vacuum polarization.[17] Unfortunately, as mentioned, Uehling found that the vacuum polarization gave an effect, but it was of the wrong sign and more than ten times smaller than the observed discrepancy.[18] Having already given up on proper energy infinities, Oppenheimer gave up on vacuum polarization as well, and began to doubt whether the experimental results were real.

In Cornell, Gibbs and Williams published *their* long paper two weeks after Houston and Hsieh.[19] They picked out the cause of the discrepancy, a slight shift upward of the $^2S_{1/2}$ level, but not the cause of the shift, self-energy. Gibbs was not theoretically minded; if he had even heard of the divergences of quantum electrodynamics, he would have been unlikely to connect them with something as physical as spectrometer readings. Nonetheless, by mid-1935 there was a rapidly developing consensus that something was wrong with the predictions of quantum electrodynamics. Then, at the end of the year, Spedding, Shane, and Grace recanted.[20]

Having, to some extent, started the fuss, the three men now said the theory was all right. Their experimental data had not changed very much, but their interpretation was radically altered. This time, they found the "right" result for the fine structure constant, 1/137; when corrected by a "more detailed method" of analysis, their readings fit theory exactly. Perhaps wishing to avoid antagonizing their colleagues at Caltech, Spedding, Shane, and Grace said little about Houston and Hsieh, but attacked Gibbs and Williams for insufficient statistical expertise. The three men also suggested that the Cornell experimenters might not have carefully controlled the pressure of the hydrogen that produced the spectrum.

Gibbs and Williams were by now in contact with the California groups.[21] They responded immediately and indignantly, telling the New Year's meeting of the American Physical Society that the separation of the line components was "remarkably uniform" no matter what the experimental conditions.[22] Later, Houston, too, stuck to his guns.[23]

The three groups used the same equipment in similar conditions to test for the same phenomenon, but came to opposite conclusions—Houston, Gibbs, and Williams in favor of the line displacements, Spedding, Shane, and Grace against. Looking back, one sees that the disagreement was partly due to the minute size of the shift and, more important, the nature of the hydrogen atom itself. Like everything else, hydrogen does not emit light in ordinary circumstances. It does so only when stimulated by energy in the form of something like an electric current, which kicks the electrons into higher orbitals. When the electrons fall back to the ground state, they emit light. The discharge is a kissing cousin of the familiar glow of a neon sign.

Unfortunately, there is a conflict between brightness and clarity. To make the light bright and easy to measure, you pass a large current through the gas—but this heats up the little hydrogen atoms so much that their furious jiggling smears out the spectral lines you are trying to measure.[24] To get sharp lines, the hydrogen must be cooled to the temperature of liquid air, about $-380°F$ ($-190°C$), which means using a minimal current and therefore getting a dim, hard-to-photograph line. If you try to circumvent the trouble with a long exposure, you run into other problems—temperature drifts, vibration, and so on begin to swamp the tiny effects you are looking for. In sum, making cold hydrogen emit bright light was no easy task.

Physicists had to make compromises if they were going to work with hydrogen. They would get results, but the numbers would necessarily be affected by the conditions in which they were obtained. The argument among the spectroscopists was over how the data could be corrected for these perturbations—how close to reality the raw measurements were.

The spectrometer used by the three experiments was called a Fabry-Pérot interferometer, after Charles Fabry and Alfred Pérot, the turn-of-the-century French physicists who invented it.[25] The device takes light emitted by chilled hydrogen and bounces that light through a collection of mirrors and lenses in such a way that the crests and troughs of the waves reinforce each other; once amplified, the crests and troughs create a series of glowing concentric rings that fall on photographic paper. The intensity and position of the rings are determined by an electric eye. After processing through a Fabry-Pérot interferometer, the hydrogen alpha line comes out to be a slightly asymmetric double curve that looks like the back of a camel. During the 1930s, spectroscopists like Houston, Gibbs, and Williams spent a lot of time staring at these double curves. The curves are net readings of all five parts of the primary Balmer line; by carefully examining the slight ripples and deformations of the curves, physicists could infer where the two major and three minor components fell. Experimenters would make plausible guesses, raising and lowering this curve or that one, comparing one series of curves to a second and a third, until they came up with something that felt right. How one team fussed with the data could radically alter its conclusions, as Spedding, Shane, and Grace showed when they withdrew their earlier findings.

By this time, Gibbs had become chairman of the Cornell physics department, a post he took seriously enough that it nearly ended his active career as an experimenter. Working alone, Williams redid the experiment, obtaining essentially the same measurements. At this point, however, he was waylaid by an unfortunate accident—the arrival of the distinguished physicists James Franck and Hans Bethe at Cornell. Franck was the first to directly measure the quantized nature of energy, an achievement for which he won the Nobel Prize in 1926; Bethe, a Nobel laureate in 1967, was the author of a definitive and widely quoted theoretical article discussing the fine structure of hydrogen.[26] Interested by the new spectroscopic findings, Bethe and Franck quickly suggested several reasons that Williams's shift might be explainable by theory after all.[27] "They really had discovered a shift in the energy level," Bethe told us. "But I was too stupid to recognize it. Their initial explanation was even the correct explanation, but there was no theory at the time which predicted a shift of the $^2S_{1/2}$ level, so I was looking for other reasons."[28] Williams went back and rechecked his data on the basis of Bethe's ideas. The shift was still there. But by this time he had finished his Ph.D., and was ready to go back to his first love, astronomy. He went to the University of Michigan observatory, and his supervisor, Gibbs, was left with the paper. Gibbs was loath to publish a measurement which he could not explain. He was not a theoretician, and did not know if the discrepancy was really important.[29]

Over two years later, in mid-1938, Williams's paper finally saw the light of day.[30] The tone was more cautious than that of the first paper, but he had found that a third component of the line, in addition, was not where it was supposed to be. The interval between line #1 and line #2 was about 2.7 percent smaller than the value predicted by quantum electrodynamics. All the discrepancies, although Williams did not say this, were associated with transitions from the $^2S_{1/2}$ orbital.

By then, the theorists had become tired of banging their heads against the wall of quantum electrodynamics. The spectroscopists seemed to be disagreeing with each other, and few people outside the experimental world could tell whether Spedding, Shane, and Grace or Houston and Williams were right. "I don't know whether my paper was actively disbelieved or whether it was mostly ignored," Williams says now. He shifted out of astronomy shortly after the war, and became a distinguished molecular biologist. Now a professor emeritus at the University of California at Berkeley, he has little but fond memories from his prewar spectroscopy. "Don't forget the water had been real muddy up till then. Spedding, Shane, and Grace came up with something that was just disgraceful. They didn't *begin* to show the kind of detail they should have seen with this material—it was just lousy spectroscopy. Why? I guess their mirror surfaces weren't very good. And then this [Boston University experiment] was *terrible*. Houston and Hsieh, that was good work. So the situation had been kind of muddy. I suppose that helped our work not to be believed. I don't think anyone insisted we had made a mistake, it was more that it was thought to be kind of marginal.

"And also don't forget, we and the others had a formidable bunch of competition, namely fellows like Heisenberg and Schrödinger and Dirac and a few other types, who predicted what that interval should be—and we damn well better get that interval."

What do you mean, we asked him, that you had to get that interval?

"Those guys had it," Williams said ruefully. "They were very capable, and who was going to doubt the theory?"

In the first paper, though, he *had* doubted the theory. Did he drop the idea because of Bethe's prestige?

Williams was still chuckling. "No, I don't think so," he said. "In a way, our experiments were just marginally ahead of their time. Had you been able to do them on a calculating machine with no fuss or muss and no curve fitting to do, maybe they would have been accepted or believed. It's like any research, I guess, and we were somewhat on the edge of what could be done—I hope so. And we got a result that didn't agree with theory and the result then must be wrong. Well, gee, that's not unprecedented, not at all."

We asked Dr. Williams how he felt a decade later when his measurements were repeated and the Nobel Prizes were being given out.

"Then I was mostly measuring tobacco virus and nucleic acids and stuff. I had dropped all this entirely. So I was pleased that they found something agreed with me."

That's all? Simply pleased?

"Oh, have I ever said that a good idea for the Nobel Prize would have been to share it between the two people who found what the real answer was? Yes, I've thought that at times. But not very hard."[31]

One of the few people interested in Williams's result was Simon Pasternack, a theorist who had recently arrived at Caltech. Pasternack talked to Houston, read the papers, and arrived at the same conclusion Gibbs and Williams had come up with in 1934: the $^2S_{1/2}$ level must not be where Dirac said it was supposed to be.[32] Pasternack came across the discrepancies in 1938, when theorists had largely given up trying to understand the infinities. Instead of invoking self-energy, he ascribed the shift to "some perturbing interaction between the electron and the nucleus." His note caused a little stir, but neither he nor anyone else knew how seriously to take it or the experiments it was based on. Williams, who first met him after the war, remembers him as a curious man who had never "worked with his fingers." They sat for hours in the University of Michigan observatory, talking about the fine structure of hydrogen. The experimenter explained to the theorist why the measurements were good; the theorist explained to the experimenter that the results were very, very important, and that the validity of quantum electrodynamics hung on them. Williams was astonished. It was the first time that he learned of the implications of his own work.[33]

In London, a big name weighed in. Sir Owen Richardson, a pupil of J. J. Thomson, a colleague of Rutherford and Dirac, and the recipient of the 1928 Nobel Prize in physics, was an internationally recognized expert on the spectrum of hydrogen. By 1939, he had become aware of the new controversy over the hydrogen fine structure; he decided to redo his measurements.

Richardson was joined by a colleague, William E. Williams (no relation), who had designed a new type of interferometer that worked in a vacuum, unlike the Fabry-Pérot machine. Williams built the equipment, and Richardson assigned one of his graduate students, John Drinkwater, the task of performing the experiment. Drinkwater was not particularly interested in arcane theoretical problems; he wanted to go into industry, where his work would have immediate, practical results. A man with an Olympian view of collaboration, Richardson was rarely in the lab with Drinkwater, and told him little of the issues involved. Richardson said nothing, and wrote the paper entirely alone.[34]

The result is a most curious document.[35] For sixteen pages, Richardson carefully, even laboriously, explains the motivation and means of the experiment and lists the spectroscopy readings. At first, the paper seemed to promise excitement. Pasternack has suggested a shift in the $^2S_{1/2}$ level; Drinkwater's observations "apparently support this hypothesis." However, Richardson writes, "there are certain other factors that appear to rule out the possibility of this explanation."[36] And, incredibly, he spends the last four pages of the paper on a farrago of absurd suppositions that wish away the discrepancy. Although he spent little time in the laboratory, Richardson calmly assumes that Drinkwater must have used hydrogen that was an improper mix of its atomic (i.e., H atoms) and molecular forms (H_2). The taint of H_2 generated "secondary" spectral lines that gave rise to the apparent discrepancy. "We conclude," Richardson writes, "that no real evidence has yet been obtained to show that the fine structures depart substantially from the values calculated from Dirac's equations."[37]

"That killed it off," Dick Learner, one of Richardson's successors, told us in the University of London. We had gone there in the hopes of finding out what had happened in the experiment; Richardson has been dead for more than thirty years, and his surviving collaborators remembered little of the experiment. At Imperial College, we met Learner, who uses the paper in a class on errors in measurement as an example of what *not* to do. A heavy, genial, dark-haired man, Learner had a clear picture of what had happened to Drinkwater, W. E. Williams, and Richardson. "It's the clearest example I know of somebody refusing to believe his own data," Learner said during our long conversation. "I lecture on this in my course, and the odds of a bad result, according to his own figures, are something like ten thousand to one. He had to work *hard* to explain the effect away."

He was asked why Richardson would have worked so hard.

"Reading between the lines, I would say that he thought Dirac had achieved an outstanding synthesis of atomic physics and relativity—a theory of seductive elegance. Richardson knew Dirac personally, remember." Learner shrugged. Easy to see what was happening there. A few minutes later, he added, "In those days, the professor and the head of the department was barely distinguishable from God. England was still in the Germanic tradition, where the professor said do it and you bloody well did it or else. Richardson was really one of those old despots. Now Drinkwater's in the position of a grad student whose supervisor talked him out of a Nobel Prize."

Why did Richardson's experiment succeed in killing off the others?

"The general feeling after that experiment was that Dirac was right, oh, good, thank God. After that experiment, if you'd applied for funds to re-measure the fine structure, they'd have said, 'Forget it!' It wouldn't have got past the referees on the first chucking-out level. Particularly when it's a Nobel Prize winner whose conclusion is that there is nothing to find. The NSF would blow you a big, gray raspberry, wouldn't it?"

Is that all there was to it?

"People always forget an absolutely fundamental point, and that is that physics has got to be *paid* for. It's always assumed that critical experiments just get done, but they only get done if they get funded. After Richardson, this was obviously an unfundable experiment. Nobody in their right mind would have pushed for it.

"I teach this experiment to remind the people that the theorists are often wrong. If you're an experimenter, you get the illusion that the theorists are all such *smart* bastards. But many theorists have no idea what's going on in an experiment. If you stand a theorist next to an apparatus, it breaks. Some members of the group here, for example, are good in all directions, but among the rest are a few really banana-fingered gentry." Learner abruptly swung to face us. "The point to all this," he said, "is that you must remember that you know more about what you are seeing and how you are seeing it than they do. Experimental physics, alas, has an inferiority complex."[38]

Not every good physicist went to Los Alamos or the other centers of military research during the war. One of those who didn't was Willis Lamb, an Oppenheimer protégé who taught at Columbia. Lamb had a rocky relationship with Oppenheimer; he was fascinated by Oppie, but the two men did not really get along. While at Berkeley, Lamb read his mentor's 1930 paper on self-energy and the displacement of spectral lines; he also heard discussion of Houston's measurements of the fine structure of hydrogen. But his dissertation subject lay in another area of physics, and he never really studied the question. After finishing his thesis in 1938, Lamb met I. I. Rabi, who arranged an instructorship at Columbia for the then-princely salary of $2,400 a year. There he taught beginning courses in engineering physics, worked on theoretical nuclear physics, and came across Pasternack's explanation of Houston's spectroscopic data. He'd also read the paper by Drinkwater, Richardson, and W. E. Williams, and concluded from it that the level shift envisioned by Pasternack was an intriguing possibility, but one unlikely to exist in the real world.[39]

Rabi was then working on experiments involving beams of ions and molecules. Lamb did various calculations for the molecular beam. As the war work began, the molecular beam shut down because most of the experimenters, including Rabi, went off to build weapons or design radar stations. Because Lamb's wife, Ursula, had fled Europe in 1935—"for all the usual reasons," according to Lamb—she was classified as an enemy alien.[40] This prevented Lamb from obtaining the requisite security clearance, and he continued to teach at Columbia.

In November 1943, Lamb received a phone call from Oppenheimer. Security problems were over, Oppenheimer said. Did Lamb have any idea what they were working on in New Mexico? Lamb said he had a fairly good idea. Would Lamb like to work on a very important and very secret project? Lamb said no. He did, however, join the Columbia Radiation Laboratory, which was housed in Pupin Hall, blocked from the rest of the campus by an armed guard. In the Radiation Laboratory were magnetrons, the devices that produce the microwaves used in radar tracking. "We had a rule there," Rabi recalled, "that everyone who worked there had to make a magnetron. So that's how Lamb got introduced to experimental work."

Lamb became interested in magnetrons, but quickly discovered that if he wanted to have his own ideas about them tested, he would have to do the experiments himself. He knew none of the necessary sophisticated metal fabrication and vacuum techniques; he learned. He also taught atomic physics, which made him think about what are called the selection rules for transitions between orbitals in the hydrogen atom. Because the spin, momentum, and other aspects of the electron are quantized, jumps between one orbital and another can only take place if the orbitals involved possess the right quantum numbers. Lamb noted that for selection reasons it is difficult for the $^2S_{1/2}$ level to fall into the $^1S_{1/2}$, which is the lowest energy state of the hydrogen atom. The $^2P_{3/2}$, however, can do so readily. Because the $^2S_{1/2}$ takes a long time to drop down to a lower energy level, it is called a "metastable" state—almost stable.

In the summer of 1945, three thoughts came together in Lamb's mind. First, he realized that if he could get a large number of atoms in the $^2S_{1/2}$ state and popped them into the $^2P_{3/2}$ state, they would fall so quickly into the $^1S_{1/2}$ level that there would be a sudden reduction in the percentage of $^2S_{1/2}$ atoms. Second, he noted that going from the $^2S_{1/2}$ to the higher $^2P_{3/2}$ involved a photon of almost the same microwave frequency as that emitted by the magnetrons in the radiation laboratory. Third, he could delicately fiddle with the energy of the hydrogen atoms by running them through a tunable magnetic field. Combining these insights, he was sure, would allow him to measure the difference between the $^2S_{1/2}$ and $^2P_{3/2}$ energy levels with unparalleled precision. (The size of the gap was predicted by quantum electrodynamics, but it is important to recall that the calculations neglected self-energy.)

Lamb was kept busy by the radiation laboratory and his students at Columbia, but every now and then went back to the idea of making a microwave measurement of the fine structure. By July 1946, he had a vague idea of what to do, and put an order into the machine shop for the parts. His experience was with magnetrons, so his design looked like a magnetron. Even before he finished putting it together he realized it would not work.

That September, Lamb met a graduate student named Robert Retherford, who had decided to return to academics after a spell of working for Westinghouse. Retherford knew a lot about the techniques necessary to build the apparatus; delighted, Lamb and he quickly put together a proposal for a somewhat different method.[41]

The two men planned to bombard a jet of hydrogen with electrons from an electron gun similar in principle to the cathode ray tube used by Thomson to discover the electron half a century before. Gaseous hydrogen usually consists of H_2, two hydrogen atoms linked into a molecule. The idea was that the incoming electrons could have just enough energy to break the molecules into atoms and kick the atoms into the metastable $^2S_{1/2}$ state.

The detector took advantage of the special properties of metastable atoms. The energy they cannot release can be thought of as being like water in a child's water balloon: Under most circumstances, the water can't escape, but if the balloon smacks into something, the water splashes out. In the 1930s, it had been discovered that a metastable atom will fall to the ground state if it comes close enough to a metal atom to knock out one of its electrons. When the ejected electron is close to a positively charged plate, the electron, which has a negative charge, will be attracted to the plate. If there was a large number of collisions producing a large number of electrons, the result would be a small electric current. Thus Lamb knew that if he beamed microwaves at the stream of $^2S_{1/2}$ hydrogen atoms, he would be able to tell when he had popped them to a ^2P state because the electric current from the detector would suddenly drop.

The two experimenters spent the next few months building the experiment on a metal stand in Pupin Hall. When they finally switched it on for the first time, the detector registered no current at all. It was not clear the hydrogen was even hitting the plate. They painted the plate with a sooty yellow mixture of molybdenum oxide, which combines with atomic hydrogen and turns blue. They switched the oven on again. The plate turned blue. They were getting hydrogen, but their method of exciting them into the $^2S_{1/2}$ state was failing.

They decided to rebuild the apparatus. This time, they proposed blowing a thin stream of H_2 into a small oven made from tungsten and heated to a temperature of over $2,000°C$.[42] The heat would rip the hydrogen molecules apart, and the individual atoms would be spat out of a tiny slit in the side of the oven. Most but not all of these atoms would be in the ground $^1S_{1/2}$ state. The electron beam would now only have the task of pushing the atoms into the $^2S_{1/2}$ state. Still, the experiment had to be fiddled with until it worked. They had to keep adjusting the electron gun until they started getting a halfway decent reading.

On Saturday, April 26, 1947, the experiment finally succeeded. Lamb and Retherford set the magnet and microwaves to the energy which conventional quantum theory predicted would make the hydrogen atoms jump into the $^2P_{3/2}$ state. The instrument registering the detector current—a common galvanometer with a beam of light that fell onto a long paper scale—stayed motionless. Surprised, they fiddled with the microwave beam. At a frequency level about 10 percent less than they started, the galvanometer beam dropped. The $^2S_{1/2}$ was going to $^2P_{3/2}$—but at the wrong place. Which meant that the $^2S_{1/2}$ was not where it was supposed to be to begin with. Excited, they realized that the discrepancies fitted Pasternack's analysis. Houston, Gibbs, and Williams had been right.

An experienced theorist, Lamb was fully aware of the implications of the result. Late that night, after Retherford had gone home, Lamb went over alone to the laboratory and tried to confirm what he had seen earlier. He discovered it was impossible for one person to operate the machinery alone and called his wife Ursula to help. They could see the galvanometer spot moving. He woke up the next morning feeling good indeed. Unless he was badly mistaken, the spot of light in Pupin Hall meant that something was going to have to be done about quantum electrodynamics.

A few months after the atom bomb was dropped on Hiroshima, Victor Weisskopf was offered a job by the Massachusetts Institute of Technology. His hiring was in itself a testament to the dramatically changed fortunes of physics in the United States. Before the war, American universities had an informal quota system for the hiring of Jews. The development of the bomb and the subsequent end of the war made heroes out of physicists; the Holocaust made anti-Semitism increasingly less respectable. Backed by the greater prestige of science, Jewish physicists were sought by the academic temples that had previously looked askance at their applications. Once in the Ivy League, the physicists' new stature was enhanced by their address.

Weisskopf immediately went to work on quantum electrodynamics. He decided to go back to the beginning, retracing Oppenheimer's footsteps, and see what the self-energy of the electron in the hydrogen atom would do to its atomic spectrum. About Halloween of 1946, he gave the problem to his graduate student, J. Bruce French, and suggested that he pay particular attention to the $^2S_{1/2}$ and $^2P_{1/2}$ levels, which were supposed to be coincident.[43] They worked at the problem very slowly, hampered by the extreme difficulty of the perturbation expansion, matching up one infinity with another and trying to see if they canceled each other out.

As the work went on, Weisskopf received an invitation to a small conference on the foundations of quantum mechanics hosted by Duncan MacInnes of the Rockefeller Institute for Medical Research, in Manhattan, and Karl K. Darrow of Bell Telephone Laboratories, in Murray Hill, New Jersey.[44] MacInnes, the president of the New York Academy of Sciences, had arranged for the National Academy of Sciences to provide five or six thousand dollars for two conferences, one of which was to be on theoretical physics. Darrow, who had been secretary of the American Physical Society for years, offered to help MacInnes arrange it. After thinking about it for a while, they decided to arrange a sort of physics retreat—twenty to twenty-five physicists, mostly young and promising, for three to four days in an isolated spot. The subject would be "the foundations of quantum mechanics." It would be held on Shelter Island, off the eastern end of Long Island, New York. There would be no formal papers, no fixed agenda, no published proceedings. To ensure the conference would have some prestige, Darrow and MacInnes first invited Oppenheimer. Oppenheimer thought the whole thing sounded peculiar, but was willing to go along.[45] This settled, Darrow and MacInnes invited Weisskopf, who found "the whole idea of a few quiet days in the country together with Heisenberg's 'uncertainty relations' . . . exceedingly attractive."[46] Einstein, too, was invited, but he was unable to attend, and in any case was out of the mainstream of physics.

Weisskopf, Oppenheimer, and a Bohr disciple, Hendrik (Hans) Kramers, were to be discussion leaders. All three men were asked to prepare outlines of what they thought should be talked about. Weisskopf's deserves to be quoted at some length, for it represents the viewpoint of a leading theoretician on the state of play in the spring of 1947.

The theory of elementary particles has reached an impasse. Certain well-known attempts have been made in the last fifteen years to overcome a series of fundamental problems. All these attempts seem to have failed at an early stage. An agenda for a conference on these matters contains, necessarily, a list of these attempts. After returning from war work, most of us went through just these attempts and tried to analyze the reason of failure. Therefore, the list [of problems] which follows will be well known to everyone and will probably invoke a feeling of knocking a sore head against the same old wall. The success of this conference can be measured by the extent it deviates from this agenda.

Weisskopf listed the problems of quantum electrodynamics:

1) *Self Energies. Attempts to remove infinite self energies. . . . Why do logarithmic divergences defy most of these methods?*
2) *How reliable are the "finite" results of quantum electrodynamics derived by means of a subtraction formalism? Polarization of the vacuum and related effects. Is there a high energy limit to quantum electrodynamics?*[47]

He could have written the same list in 1934, thirteen years before.

Kramers and Oppenheimer, too, wrote outlines for the discussion. Oppenheimer focused primarily on cosmic rays, which interested physicists at the time. Kramers, on the other hand, turned to electrodynamics. But, unlike Weisskopf, he was optimistic. The reason was that he thought he was on the track. Kramers had collaborated with Bohr since 1916, and had occupied the chair of theoretical physics in Leiden since 1934. Before the war, he had become interested in the divergences.[48] But he, like Dirac, was convinced that avoiding the infinities in quantum field theory required solving too many problems at once, and he put his efforts into a classical theory, with no quantum mechanics. Along the way, however, Kramers had come to the conclusion that the infinite proper energy of a particle was mostly invisible; just as the observed charge of an electron is the finite residue of the "bare" charge and the vacuum polarization, so the observed mass is really the combination of a "bare" mass and the mass due to self-energy. Therefore, Kramers thought, if one could write a theory wholly in terms of the experimental mass and charge, rather than the bare mass and charge, the infinities would be avoided from the outset, and the theory be renormalized.

When the Nazis invaded Holland, Kramers remained in Leiden, secretly aiding the Jewish physicists whose locations he knew. In broken health, he came to the United States in late 1946, where his ideas began to generate interest in theorists returning from the war. Weisskopf, who was discussing his self-energy work with Bethe, wanted to hear what Kramers was up to. There were rumors floating around that one of Rabi's people had measured the hydrogen fine spectrum and got results that strongly deviated from theory. Weisskopf was pleased to learn that both Rabi and the experimenter, Lamb, were coming to the Shelter Island conference on the foundations of quantum mechanics. At the end of May 1947, Weisskopf took the train to New York with another conference attendee, one of the young comers whom the meeting featured—Julian Schwinger, an astonishingly gifted theorist who had been made a full professor at Harvard at the age of twenty-nine. If the tales from the experimenters were right, Schwinger and Weisskopf thought, then the deviations very likely had their origins in the self-energy.[49]

They met Darrow, MacInnes, and Oppenheimer at the American Institute of Physics, then on East 55th Street in Manhattan. There most of the twenty-five physicists piled into a bus and headed for Long Island. Lamb drove his old friends from California, Serber and Nordsieck.[50] Bethe went to New York City, where he borrowed his brother-in-law's car for the very long trip to eastern Long Island.[51] He missed a spectacular ride: Oppenheimer's prestige ensured that the last part of the trip was made in a police motorcade, sirens a-whirl, whizzing through the village stop signs.

Low, dry, and gnarled, Long Island runs parallel to the Connecticut shore for more than a hundred miles. Its eastern tip branches into two long peninsulas, known respectively as the north and south forks. Between the tines is a small, sandy triangle known as Shelter Island. Unlike the rest of Long Island, Shelter Island is still much as it was fifty or even a hundred years past. In marked contrast to the parade of boutiques that now crowd the fashionable Hamptons on the south fork, Shelter Island still has the plain, unornamented quality of the Quakers who first settled there. There are weathered rural stores, small roads, few cars; the air seems guarded, as if the island were poised to ward off inundation by the tides or, more probable, affluent summer people. Shelter Island is edged by long promontories—"heads," in the local parlance—that spike outward like the curiously curved swords of the old Ottoman Empire.

One of the most remote of these spits is Ram's Head, accessible only by a causeway that winter flood tides leave awash in brine. The telegraph poles lining the road are covered by the untidy evidence of osprey nests. Debouching from the causeway, one quickly sees the Ram's Head Inn, an unpretentious clapboard country hostel. The roses were out when we visited, curling over a split rail fence in front of the establishment; oak and maple trees swept invitingly over the back porch. The entranceway, edged with white trim, leads to a small hall with the eponymous stuffed ram's head on prominent display. The lounge behind has a piano, a fireplace, and a bar; its walls are cluttered with a comfortable jumble of maritime foofaraw.

Beside the door we came across a large scroll of the sort invariably found in rural hostels, celebrating forgotten football championships and the long-ago strikes of a prodigious bowling team. Made curious by the yet-untarnished condition of the metal frame, we approached the placard and discovered it to be the only physical memento of a landmark in twentieth-century intellectual history.

FIRST SHELTER ISLAND CONFERENCE
ON THE FOUNDATION OF QUANTUM MECHANICS

2–4 June 1947

The first Shelter Island Conference on the Foundations of Quantum Mechanics, held at the Ram's Head Inn in 1947, is remembered as the starting point of a series of remarkable developments in physics that have changed our views of the basic structure of matter and given us a new cosmology. . . .

To present-day physicists, the name Shelter Island is synonymous with a sense of remembered excitement; the congress was one of those rare, jeweled occasions where the machinery of the scientific imagination ran faultlessly, and with speed and quiet ideas were put together whose connections had eluded so many minds in the decades before. Years later, Richard Feynman, one of the younger physicists present at Shelter Island, said, "There have been many conferences in the world since, but I've never felt any to be as important as this."[52] Oppenheimer thought it was the most successful scientific meeting he had ever attended. Its total cost was just $850.[53]

The conference opened early in the morning of Monday, June 2, 1947, with Lamb's presentation of his experimental work. By that time, he and Retherford had succeeded in determining that the $^2S_{1/2}$ level was about 3 percent off where it was supposed to be—an even greater separation than found by Williams. Rabi then explained the work of his junior colleagues, John E. Nafe and Edward B. Nelson, who used a somewhat similar apparatus to examine the magnetic behavior of the electron in hydrogen and its isotopes.[54] Here, too, they had found discrepancies.[55]

Kramers, Oppenheimer, Schwinger, and Weisskopf argued that the discrepancies were not due to actual errors in the theory of Dirac, Heisenberg, and Pauli, but to an inadequacy in the way the theory was *used*. That is, the theoretical values had been arrived at by ignoring the last term of the total Hamiltonian. If one calculated the way one ought to, using the full equation, one should arrive at something like the values found by the Columbia experimenters. Unfortunately, as everyone present knew, this was more easily said than done.[56] Getting rid of the infinities—renormalizing the theory, in the jargon—would be a lot of work.

Until late in the night the men hashed out the question, gulping down meals in a fury of technical discussions, wandering through the corridors in groups of two or three. They were the only guests in the inn, which had opened early to accommodate the famous J. Robert Oppenheimer. For many of the participants, the meeting was the first chance since Pearl Harbor to dive into physical waters untainted by weaponry; little wonder they took to it with a giddy pleasure. As Schwinger said later, "It was the first time that people who had all this physics pent up in them for five years could talk to each other without somebody peering over their shoulders and saying, 'Is this cleared?' "[57] A mark of a good physicist is a sort of surprised delight at the working of nature. Every man there was experiencing again the deep pleasures of curiosity. Why didn't the galvanometer beam in the tenth floor of Pupin Hall slip down when it was supposed to?

On the second day, they talked of cosmic rays. Here, too, the discussion was frenetic, animated, sharp.[58] The third day swung back to renormalization, and the group splintered off, its members filled with a sense of mission. Oppenheimer left on a seaplane to take an honorary degree from Harvard, bringing with him Schwinger, Weisskopf, and an Italian cosmic ray expert, Bruno Rossi, on what proved to be an alarming voyage. The pilot was unhappy at Oppenheimer's casual instructions to land at the New London Coast Guard Station, which is not open to civilian aircraft—so unhappy, in fact, that he almost sank the plane. They were met by an armed and infuriated coastguardsman. In Schwinger's recollection, Oppenheimer hopped out of the cockpit and calmed the enraged officer with the magic words, "My name is Oppenheimer." "*The* Oppenheimer?" was the gasped response. They were escorted to the train by an honor guard.[59]

Bethe, meanwhile, drove back to New York. He stayed overnight with his mother and took the train to the General Electric laboratories in Schenectady, New York, where he was then working. The Lamb shift excited him enormously. For years he had believed that the complexity of quantum electrodynamics made it unlikely that he would ever contribute anything.[60] Now, he had an idea of how he might account for the level shift in a simple way.

We asked him once to tell us about the Lamb shift. "It was a wonderful new effect—completely unexpected!" he said. "It tied in with the old puzzle of the infinite self-energy, and especially with a talk by Kramers on the idea of renormalization of the electromagnetic interaction. Kramers hadn't done any calculations, but he had the idea of renormalization. He had the idea that the self-energy of a free electron traveling in space was not observable, but part of its mass. To calculate the shift in the spectral lines, one should only consider the difference between the self-energy of a free electron, and the self-energy of an electron in a field, like the electron in the hydrogen atom." The difference is all we can see, Bethe thought, and that may be finite. For a quick calculation, he simply ignored the relativistic effects. "A relativistic calculation was far beyond my ability to do in any short time," he told us. "I knew—and said so in the article—that the self-energy diverges logarithmically in the relativistic case. That had been known since 1934. So I said, all right, I'll get the nonrelativistic part. I finished the calculation in the course of the train ride."[61] Actually, Bethe didn't quite finish: He had to go to the library the next morning to look up a factor and a logarithm. Forgetting about relativity made the calculation much simpler, and, to his joy, he was able to account for 95 percent of the effect. Five days after the close of the conference, on Wednesday, June 9, 1947, he was ready with a preliminary draft.[62]

The calculation chagrined his colleagues. "I instantly saw what Bethe had done," Lamb remembered, "and mentally kicked myself for not being clever enough to do it first."[63] Crestfallen, the physicists saw that (1) Bethe was obviously right; (2) Bethe had done something that any of them could have done fifteen years before. "Any of the Berkeley people—Serber, Uehling, Oppenheimer, and the rest—could have done what Bethe did," Lamb said on another occasion. We asked him why they had not. "They weren't thinking of the right thing at the right time. It was also because of the sheer simplicity of the calculation. Why do something nonrelativistically when you knew how to do sophisticated things?" In the 1930s, physicists knew a correct theory had to obey the dictates of relativity. It hadn't occurred to them that a simple nonrelativistic calculation might be enough to point the way. "Then there was the experimental fact of the shift," Lamb said. "If you knew there is a substantial effect involved, then there is much more of an incentive to calculate it than if the effect is tiny. It's like the atom bomb—it was a much better secret before anybody made one than after."[64]

8
Killing the Hydra (Part I)

JULIAN SCHWINGER WAS BORN IN NEW YORK CITY ON FEBRUARY 12, 1918. HIS remarkable abilities became quickly apparent. At the age of fourteen, he attended a lecture by Dirac on the hole theory; two years later, he wrote but did not publish his first article on quantum electrodynamics. His high school, Townsend Harris, was a teaching adjunct to City College. The teachers there soon suggested that he should be going to City College itself.

By chance, Lloyd Motz, a Ph.D. candidate at Columbia, became friendly with Schwinger's brother, Harold, who was at Columbia Law School. Motz had already heard through the grapevine that there was a high-school-age prodigy at City College. When he found out the prodigy was a friend's younger brother, he made it his business to look up Julian Schwinger. Motz discovered that although Schwinger had only completed his freshman physics courses, he was completely familiar with relativity and quantum mechanics and nuclear physics and hole theory and the rest of it and was doing advanced computations for his professors. Impressed, Motz asked Schwinger if he would like to collaborate on a paper. They started talking together, and Motz discovered that despite Schwinger's unquestioned talent, he was not doing well in school. Part of the reason was that he was working too hard to pay much attention to his classes. Motz decided he'd better introduce Schwinger to his own thesis adviser, I. I. Rabi.[1]

"The circumstances were rather romantic," Rabi said recently. We were sitting in his apartment, watching Rabi use a portable phone to fend off demands for his time. "I'd read a paper by Einstein, Podolsky, and Rosen. My way of reading a paper would be to get a [doctoral] student to explain it to me. So Lloyd Motz walked by, and I called him in and talked about it. After a while, he said, 'There's somebody waiting for me outside.' And there was this kid, Schwinger, who was a student at City College at the time. Lloyd brought him in. I said, 'Sit here,' and he sat there, and we went on with our discussion. Then Schwinger *settled* the argument by an application of the completeness theorem. I said, 'Who's *this?*' This kid! Then I heard about how he had difficulties at City College, and I suggested he change to Columbia. I got him a scholarship—not without difficulty."[2]

Schwinger said he would transfer to Columbia if he did not get into an honors graduate program at City College. Motz, who was teaching part-time at City, met with the department chairman. A sophomore in an honors graduate program? Out of the question! In the spring of 1935, Schwinger entered Columbia to the accompaniment of a lecture by Rabi on the importance of going to his classes.

Ordinarily less than impressed by mathematical dexterity, Rabi knew that Schwinger was a special case. He quietly brought eminent visitors—Pauli, Bethe, Fermi, and Uhlenbeck among them—to Motz's office, where they listened in stupefied silence to the words of an undergraduate physics major. Bethe wrote Rabi afterward that Schwinger already knew ninety percent of all physics, and that the remaining ten percent would only take a few days. In the summer of 1937, just after graduating from college, Schwinger wrote his Ph.D. thesis; Rabi told him that he should stick around and fulfill his residence requirements. "One always receives this bad advice," Schwinger said later.[3]

Schwinger amazed the physics department but annoyed the mathematics department. Although it was understood that he knew the subject material thoroughly, Schwinger's math professors refused to pass him unless he showed up for his classes. Rabi asked George Uhlenbeck, then spending a semester at Columbia, if he would do something about it. Uhlenbeck taught statistical mechanics, one of the many courses Schwinger never attended. "He never came," Uhlenbeck recalled with amusement. "At the end of the semester I had to give the usual exams. I asked Rabi, 'What should I do with Julian? Because he never came! I'm willing to give him an A, because I know already that he's the best.' But Rabi said, 'No, no, no, you give him an E!' I was appalled. Then he thought and said, 'No, Julian will take an examination from you at the end of this week.' It was three or four days away. Julian got my lecture notes from other students and studied them. And after three days he came, and I gave him the regular exam. He of course knew everything, except that he could do several of the derivations in a simpler way than I had shown in the course."[4]

"He had this notebook that was just *filled* with results and calculations," Motz said recently. "Every time a new paper came out, he'd show me he'd already worked it out in his notebook. He used to give seminars that were absolutely perfect. There was no one I knew in physics like him, including Fermi—and I knew Fermi very well, we wrote a book together. Schwinger was to physics what Mozart was to music."[5]

Bearing his new Ph.D. and a National Research Council fellowship, Schwinger arrived in Berkeley in the fall of 1939, and was soon deep into nuclear physics and collaboration with Oppenheimer. Two years later, in 1941, his grant expired, and he went to Purdue University, in Lafayette, Indiana. The Japanese attacked Pearl Harbor on December 7. Soon thereafter, Hans Bethe swept through the Purdue campus on a physics recruiting mission for the war effort. Schwinger joined the M.I.T. Radiation Laboratory. In 1943, he was asked to follow Bethe and join Oppenheimer's team in Los Alamos. Schwinger had a good idea of what was transpiring there, as did most nuclear physicists in the United States. Hesitantly, he asked if he could first put his toes in the heavy water by joining the group at the University of Chicago, which was designing reactors. After a few months, he began to comprehend the proposed size and yield of the bomb. He backed out and returned to M.I.T. "I give myself high marks for gut reactions," he said to us.

We were talking in a restaurant a few minutes' drive from the campus of the University of California at Los Angeles. Schwinger is a small man with heavy, leonine features and an almost Middle European air of elegance. He drove us to the restaurant in an immaculately maintained Italian sports car; his suit was cut with an attention to style rare in a physicist. A perfectly knotted silk tie hung straight down before his shirt. Almost inevitably, he reacted to his status as a prodigy by becoming hard to reach in his maturity, a practice his colleagues attribute variously to shyness, aloofness, or the simple wish for quiet. While at Columbia, Schwinger began to work at night, a habit that Rabi, with affection and respect, surmised was due to the desire to avoid his mentors—chiefly Rabi himself. While waiting for our order, we asked him about Shelter Island.

"Shelter Island was the first gathering of the clan of research physicists after the war," he said. He spoke readily and quickly, in concise sentences. "It was not a gathering of theorists proclaiming their theories. It was centered around the experimental discoveries that came from Willis Lamb and Rabi and so on." Lamb's work had been circulating about the grapevine for several weeks; before Shelter Island, Weisskopf and Schwinger talked about the level shift. They agreed it was an electrodynamic effect, and that the correction should be done relativistically. "The theories should account for it," Schwinger said. "There was this effect. It wasn't infinite and it wasn't zero. The reaction to the divergences was to say that the theory was wrong, let's throw out these wrong things, so everything that gave an infinity was set equal to zero. It was just thrown away. So here was an experiment saying it's not infinite, it's not zero, it's small, it's finite—we've got to understand it."

He was asked if the Lamb effect forced people to take their own theories seriously.

No, Schwinger said. "The point is, you don't get rid of history as dead wood. Electrodynamics was conceived to be *wrong*. There was a whole generation dedicated to changing it." He listed a few of the ways theorists had tried to rectify quantum electrodynamics. "All this had run wild," he said. When Lamb's work came out, theorists immediately rushed to their cutoffs and classical models and fundamental lengths. "People began immediately, of course, to say, 'Oh, I have a finite theory, now let me calculate this effect.' Very few people—and I count myself first and foremost—said, 'Let's go back and look at the original theory.' I mean, yes, people said it, but it wasn't the first thought of the majority of them. Not at all." He was modest about his own special qualifications. "What I brought to the problem was a physical instinct and an ability to do calculations. That's all that was needed."

Unlike Bethe, Weisskopf, and most of the other people at the Shelter Island conference, Schwinger's imagination was captured not by the Lamb shift but by the discrepancy in the magnetic behavior of the electron. Moving electric charges create magnetic fields; the spin and orbit of the electron ensure that it behaves like a tiny magnet. Speaking exactly, Rabi's doctoral students Nafe and Nelson had established that the electron's magnetic moment—which is related to the strength of its magnetic field—was not precisely what theorists had calculated.

"That was much more shocking," Schwinger said. The Lamb effect, as Bethe showed, could be accounted for almost entirely without the use of relativity. "The magnetic moment of the electron, which came from Dirac's relativistic theory, was something that *no* nonrelativistic theory could describe correctly. It was a fundamentally relativistic phenomenon, and to be told (*a*) that the physical answer was not what Dirac's theory gave; and (*b*) that there was no simpleminded way of thinking about it, that was the real challenge. That's the one I jumped on."

Were the others as interested in this?

"Not as much. That was my particular hangup. You know—holy cow!" He laughed. "But it was the right one, because you had to have a relativistic theory to get that right. Whereas any basically nonrelativistic theory would give the Lamb shift to better than an order of magnitude."

We asked Schwinger how long it had taken him to renormalize the theory. He had, we knew, spent two months after Shelter Island touring the country with his new wife.

"Well, I presume the little gray cells were ticking even as I was driving through California and various other places," he said. "But I began to work on it, I would say, in September. The first thing I began with was the magnetic moment, as I said. That was a real challenge. I had the answer in three months. I then turned to the Lamb shift—I couldn't stop doing calculations!—and I got the wrong answer. I got the wrong answer because in my calculation the magnetic moment that you recognize when the electron is at rest and the magnetic moment that you recognize when it's moving in the atom under the action of its electric field turned out not to be the same. Which meant that the methods I was using violated relativistic invariance. An electron in motion and an electron at rest are the same thing, and they better be that way in the theory."

Working with his student J. Bruce French, Weisskopf had also calculated the Lamb shift. They arrived at an answer and compared notes with

Schwinger and Richard Feynman, another young theorist looking at the question. Unfortunately, both Schwinger and Feynman had made the same mistake and got the same erroneous answer. A flurry of cross-checking ensued. Trusting the math abilities of the younger generation more than his own, Weisskopf postponed publication. The discrepancy between the answers remained, and Weisskopf wrote Oppenheimer just before Christmas, 1947, that he was beginning to wonder if they had at last reached the true failure of quantum field theory.[6] He needn't have worried: Schwinger soon came up with the right answer. Feynman, too, found his mistake and handsomely apologized by juggling the footnotes of his paper until the admission of error appeared in the "appropriately numbered" footnote 13. But in the meantime, Lamb and his student Norman Kroll had scooped everyone with the right calculation.[7]

"The technique I was using before was primitive," Schwinger said in Los Angeles. "I was driven to invent a new method that was explicitly relativistic so I would get the same numbers no matter when I looked at the electron. I had the first ideas of that by the very end of December." His thoughts came together all at once; Schwinger worked night and day. "I gave a talk at Columbia at the end of January 1948, in which these results were announced[8]—the agreement with experiment as it stood then for the magnetic moment, the Lamb shift, and the first suggestions of this relativistic theory. I was on Cloud Nine, or whatever the unrenormalized number was back then."[9]

On Tuesday, March 30, 1948, a second Shelter Island conference convened, this one at Pocono Manor Inn, in Pocono Manor, Pennsylvania. Most of the members of the previous conference were there, but among the newcomers was Niels Bohr. The high point of the occasion was an extraordinary five-hour talk by Julian Schwinger. Covering the portable blackboards in the hotel lounge with equations, he led the assembled physicists through a dazzling and complete reformulation of quantum electrodynamics in which every infinity was subsumed into the observed charge and mass of the electron, and which avoided entirely the concepts of the "bare" mass, the "bare" charge, and the electromagnetic additions to them. Having recently been burned by relativity, he began with an explicit demonstration that every expression in his equations completely fitted the dictates of relativity, in the process taking some of his audience into byways of geometry and differential calculus they would just as soon not have traveled—and then derived the theory from there. Some of the physicists present reacted ambivalently to the glittering array of mathematical skills that Schwinger marshalled into play. They described his talk as the epitome of virtuosity, more technical flash than music, a beautiful but cold piece for a solo voice.[10] Most of the stuff he had been scribbling on the blackboard was merely a record of how he came to the theory, they said, not really *physics*. (Schwinger didn't see it this way.) Nonetheless, every man there was conscious of being present at a historic occasion. A new generation of physicists had finally taken over the reins. Before Niels Bohr's eyes, a young man—Schwinger was then not quite thirty years old—had vindicated field theory and renormalized quantum electrodynamics.

Delighted as he was by Schwinger's speech, Oppenheimer was even more pleased to find on his return to Princeton a missive from Japan informing him that a young theorist named Sin-itiro Tomonaga had mostly renormalized the theory some years before but had been prevented by the war from making the work known in the West. Working in appalling conditions— "perfectly isolated from the progress of physics in the world," as Tomonaga wrote—the Japanese had essentially duplicated Schwinger's reasoning in 1943.[11] When a copy of *Newsweek* arrived with news of the Lamb shift, Tomonaga, too, had been electrified; like the Americans, the experiment galvanized him into a fury of calculation.[12] Oppenheimer sent a copy of the letter to every participant and fired off a cable to Tomonaga by return post.[13]

GRATEFUL FOR LETTER AND PAPERS STOP FOUND MOST INTERESTING AND VALUABLE CLOSELY PARALLELING MUCH WORK DONE HERE STOP STRONGLY SUGGEST YOU WRITE A SUMMARY ACCOUNT OF PRESENT STATE AND VIEWS FOR PROMPT PUBLICATION PHYSICAL REVIEW STOP GLAD TO ARRANGE STOP MOST CONSTRUCTIVE DEVELOPMENT HERE APPLICATION BY SCHWINGER OF YOUR RELATIVISTIC FORMALISM TO SELF-CONSISTENT SUBTRACTION TO OBTAIN SEVERAL DEFINITIVE QUANTITATIVE RESULTS STOP BEST GREETINGS ROBERT OPPENHEIMER

By mid-May, Tomonaga had the summary ready for Oppenheimer, who got it to the *Physical Review* by June 1.[14] Although it was published shortly thereafter, Tomonaga was still six months behind Schwinger.[15] Nonetheless, American scientists, including and especially Schwinger himself, hastened to give Tomonaga credit. Physicists were in a generous mood; twenty years after its initial formulation, quantum electrodynamics was at last on its feet. Victor Weisskopf wrote later that

The war against infinities was ended. There was no longer any reason to fear the higher approximations. The renormalization took care of all infinities and provided an unambiguous way to calculate with any desired accuracy any phenomenon resulting from the coupling of electrons with the electromagnetic field. It was not a complete victory, because infinite counter-terms had to be introduced to remove the infinities. . . . It is like Hercules' fight against Hydra, the many-headed sea monster that grows a new head for every one cut off. But Hercules won the fight, and so did the physicists.[16]

To be sure, the show was not entirely over. Schwinger's calculations, so elegant in their initial impact, proved resistant to practical usage—or, rather, many theorists found them that way. There seemed little purpose to renormalizing the theory if the formalism was too difficult for anyone but a Schwinger to use. Pauli eventually subjected the work to severe criticism; Dirac let it be known that he thought the whole business was a cheap trick designed to paper over fundamental problems.[17] (He held that opinion until the day he died.) And everyone, Schwinger included, knew that checking every possible infinity in every possible interaction would be a wearisome job.

Most particularly, the show was not over because an entirely unexpected thing occurred at the second Shelter Island Conference: a *second* renormalization theory was displayed, this one with such a wholly different approach that Bohr, among others, thought that its creator, Richard Feynman, had not understood elementary quantum mechanics. In fact, the superiority of Feynman's methods was so enormous that the generation of aging young iconoclasts who had created quantum mechanics at first found them incomprehensible.

Feynman, too, was something of a prodigy, although one of a diverse stripe. Born in Far Rockaway, Queens, famous throughout the city as the end of the A train, Feynman was brought up by his father to be a scientist. He was a puzzle nut, an inventor, a practical joker, the local fixer of radios and typewriters; he didn't like the symbols in his high school mathematics texts, so he made up his own.[18] Some of his difficulty at the second Shelter Island Conference was due to the fact that his theories were couched in his own, brand-new system of notation. He had given up his teenage efforts to create new formalisms because he realized nobody else would know what he was talking about; by 1948, he was determined to *show* them what he meant.

He was and is a man without an internal censor, a formidable, hawk-faced presence who says exactly what is on his mind the instant it occurs. In conversation, he is unnervingly present; a prankster, he seems constantly on the point of making faces and has recently composed an autobiographical set of anecdotes strangely disserving to this original and serious thinker—it is devoted to the curious notion that a thrice-married Nobel laureate in particle physics is a regular old hell-raiser, a sort of Lucy-like figure who gets into madcap scrapes with amazing regularity. While a young man in the atomic bomb project, he met Niels Bohr, whose presence awed everyone else at Los Alamos. Bohr made some remarks about a technical problem. Feynman said flatly, "No, it's not going to work," or something to that effect. Bohr said afterwards that it was the only honest reaction he heard during the day, and asked Feynman to be his talking partner while they were together at Los Alamos.[19] Feynman acquired some notoriety on the project for picking locks and blithely informing people that top-secret materials could be found in easily opened safes. He was, at the time, a curiously tragic joker: His first wife, Arlene, was sick with tuberculosis in an Albuquerque hospital. (She died in 1946.)

Feynman has almost always been rewarded for plain speech and going his own way. "I have an impatience and great difficulty in reading other people's papers," he said to us. "For me it's much easier to work a thing out from the beginning than it is to read another paper. Especially if I understand the idea. If I read the paper to give me an idea of the thing, then I prefer to work it out than to follow the equations, in most cases."

He was sprawled in a creaking chair, cradling his head in one arm, his loose plaid shirt sagging, eyes as bright and wild as a bird's. From the years before Los Alamos, he had been interested in the divergences besetting quantum electrodynamics. Feynman is not the type of person to let a problem go. He had some mathematical tricks—in particular a technique called "path integrals"—and after the war, while occupying a professorship at Cornell, he set out to use them. Or tried to, at any rate. Perhaps he had not recovered from the death of his wife, perhaps the destruction of Hiroshima had put a stain on science, perhaps any number of things, but he couldn't get to work.

Depressed, he turned down an offer from the Institute of Advanced Studies, in Princeton, New Jersey, because he felt he had nothing to contribute. Then he turned it around, he said, realizing that *he* was not to blame for *their* misconception about his abilities. He was just going to do whatever he liked in physics. "And only that afternoon," he said once, "while I was eating lunch, some kid threw up a plate in the cafeteria which had a blue medallion on [it]—the Cornell sign. And as he threw up the plate and it came down, it wobbled and the blue thing went around like this." He made a fluttering motion with his hands. "And I wondered—it seemed to me that the blue thing went around *faster* than the wobble, and I wondered what the relation was between the two. You see, I was just playing. No importance at all. I played around with the equations of motion of rotating things and I found that if the wobble is small, the blue thing goes around twice as fast as the wobble goes round. And then I tried to figure out if I could see why that was directly from Newton's laws instead of through some complicated equations. I worked that out for the fun of it." Pleased, he approached Hans Bethe, who asked him what was the use of finding the rate of rotation of cafeteria plate wobbles. None, Feynman said. He was reminded of another, similar problem, related to the spin of an electron, which in turn took him back to quantum electrodynamics. "It was just like taking the cork out of a bottle," he said. "Everything poured out."[20]

After that, of course, Hans Bethe's calculation of the Lamb shift gave him, he said, "a kick in the pants." Instead of a Hamiltonian, Feynman used another type of equation known as a Lagrangian, which was familiar from classical physics.[21] The Lagrangian was what physicists had tended to use before quantum mechanics resuscitated the Hamiltonian. In Lagrangian form, the basic electron-photon interaction in quantum mechanics consists of a term for the electron, a term for the photon, and a term for the interaction, each of which is made up of several wave functions and matrices. For complicated interactions, the equations got very long, and Feynman fell into the habit of making little doodles to help him keep track of the terms. For each electron term in the expansion, for example, he drew a line:

To make photons, he put down a wiggly line:

When they interacted, he had them meet at a point.

"I don't know when I started to make such pictures," Feynman said. He couldn't seem to keep still in the chair. "I do remember at one particular stage, when I was still developing these ideas, making such pictures to help myself write the various terms—and noticing how *funny* they looked." He was sitting on the floor of his room in Telluride House, a residential house at Cornell, late one night, a young man with a pencil and a head full of stick-up hair. (Early photographs of Feynman show him dressed in what today would be called "modified punk"—skinny little tie, white shirt, and thick hair erupting an inch and a half from the skull.) There was no room on the desk to work, almost no room on the floor: The room was awash with calculations and doodles. Pictures everywhere of complicated crossings among electrons and photons, positrons represented by reversing the arrow of the electrons. Thus, in Feynman's little diagram, an electron and positron collide to make a photon, which in turn briefly becomes a pair:

"I was sort of half-dreaming, like a kid would, that it might someday be interesting, that it would be funny if these funny pictures turned out to be useful, because *the damn Physical Review* would be—" he was laughing incredulously "—*full* of these odd-looking things. And that turned out to be true."

Feynman discovered that his little bookkeeping device had mathematical properties. By sorting through the diagrams he could quickly guess whether the associated expression, the sum of the lines and vertices, would be convergent or not. There were simple rules of thumb one could evolve. Highly excited by the ease and proficiency this brought to otherwise dreadfully complex calculations, he took his new methods to the conference in the Poconos.

Feynman spoke after Schwinger, and his exhausted listeners were anything but receptive to hearing more ideas they could not follow. Compounding the difficulty, Feynman has arrived at his conclusions through guesswork, intuition, unfamiliar mathematical tricks, *ad hoc* rules, and hours of cut-and-try calculation. He had invented a branch of mathematics called "ordered operators" to contain his ideas, but the only way he knew it was right was that he always got the right answers. Unable then to deduce his ordered operators from established theory, he tried to present his prescriptions without explanation and proceed by solving examples. This landed him in deeper trouble, because in each case Feynman had already done the problem and knew he didn't have to bother about this principle or that—but his audience was composed of the very physicists who had invented the principles, and who wanted to ensure that they were respected. These men were not satisfied by *ex cathedra* promises from a junior professor that things would turn out all right in the end.

Forced into a dreadful tangle of exposition, Feynman started to explain with diagrams. But as soon as he started drawing his pictures, Niels Bohr interrupted to say that the little line of the trajectory in Feynman's diagram was *exactly* the sort of thing that the uncertainty principle showed was impossible. Feynman replied that the diagram wasn't intended to show a real trajectory, but was, you see, in fact, a bookkeeping device for terms in the Lagrangian—

Bohr would not listen; he loathed Feynman's pragmatic, seat-of-the-pants approach to the sacred ritual of the Copenhagen interpretation. He sternly informed Feynman that there was a lot more to physics than what you get off the top of your head.[22] "They didn't understand it very well," Feynman said. "It was too much stuff to explain. So Bohr said that we already knew in nineteen-whenever that we can't talk about trajectories and the quantum mechanical laws don't permit that any more. I realized—sort of, I mean—I suddenly said, 'That's enough.'" He laughed, sharply and intensely. "They said that this idiot didn't understand quantum mechanics at all. My ideas were consistent. I knew that his objection was wrong. I realized that I'm going to have to write it for them. That's the way I remember it, okay?

"Now, Hans Bethe tells me that I was very depressed, and that he had to hold my hand and put me back together and all this kind of stuff. That psychological trauma I don't remember at all. I just decided—" his voice parodied bureaucratic resignation "—I'd just have to write it up, that's all, because they're not understanding." Then he hunched forward, leaning over his desk. "On the other side, I talked to Schwinger in the halls. And we had *no difficulty* with each other. We had *faith* in each other. If he would say something, I wouldn't argue that his equations might be wrong. If I'd say, 'I'd do it this way,' he didn't bother. He didn't even try to understand it. And I didn't try to understand his equations. But we could understand each other, if not the math. Because we had done the same problem. We talked about different kinds of terms that came in, compared notes, had a nice conversation. So I knew that I wasn't crazy."

It is there in the notes of the participants, this sense of two young men leaving the old guard in their wake. Physics was changing indeed when Niels Bohr, the man who personified the taste for the radical solution, found himself in the role of the ossified fogy. Feynman and Schwinger sat down with their notebooks—they each kept notebooks full of unpublished work— and compared notes. Each man's methods was opaque to the other; to their mutual pleasure, they still kept getting the same answers. "We talked to each other," Feynman said. "We compared notes, we exchanged ideas, I helped him, he helped me, I stole—not stole, I *used*—some of his clever tricks for doing integrals, but I referred to it, I liked it, I thought he was clever, I think he's a smart guy, and vice versa. I don't feel like I felt strongly competitive. In an amusing way, yes. In a mildish sort of way, it's hard to explain, an *amusing* competition like when two friends are racing."[23]

The competition may have become a bit more serious over time, as physicists read Feynman's papers, learned to draw pictures, and forgot many of Schwinger's methods. Freeman Dyson, a physicist now at the Institute for Advanced Studies, gave many theorists much more confidence in the whole enterprise when he proved what had been only hoped, that the methods of Feynman, Schwinger, and Tomonaga were mathematically equivalent.[24] They all led to the same result, all proved the same thing, all demonstrated that quantum electrodynamics made sense. Quantum mechanics and relativity could become the cornerstones of a field theory; at long last, physicists had a place to stand on. For their contributions to quantum electrodynamics, Richard Feynman, Julian Schwinger, and Sin-itiro Tomonaga shared the Nobel Prize for physics in 1965.

☐ ☐ ☐ ☐ ☐

And what about alpha, the fine structure constant, $e^2/\hbar c$, the famous 1/137? The latest tables of the physical constants give its magnitude as 1/137.03604, but we know no more than before *why* it has this value, and not another. In token of the mystery, Julian Schwinger's fine Italian sports car has vanity license plates with the number 137; Wolfgang Pauli died in a hospital room with that same number on the door.[25] Sir Arthur Eddington thought that the cosmos contained exactly $(137 - 1) \times 2^{256}$ protons and $(137 - 1) \times 2^{256}$ electrons; he may even be right, although it would be difficult to find a physicist nowadays who supports this notion.[26]

The fine structure constant remains, tantalizing and seductive to those theorists afflicted with the old physicist's dream of explaining the immensity and beauty of the cosmos on the basis of pure and simple whole numbers. A few years ago, Victor Weisskopf met Gershon Scholem, a prominent scholar of Jewish mysticism. Scholem told Weisskopf that one teaching of the Kabbalah is that every Hebrew word has a matching number with a meaning that can be decoded. Later, when Scholem asked about some of the questions that perplexed physicists, Weisskopf immediately thought of the fine structure constant. "His eyes," Weisskopf said later, "lit up in surprise and astonishment: 'Do you know that one hundred thirty-seven is the number associated with the Kabbalah?' " To this day, Weisskopf likes to remark, it is the best explanation that he knows.[27]

There is a hoary old particle physics joke about Wolfgang Pauli's first day in Heaven. At the pearly gate, Saint Peter says, "Pauli, God wants to meet you right away," and shows him the proper direction. At God's palace, the Deity says, "Pauli, you've been a good man, and I want to reward you in some way. Ask me any question you like." Without hesitation Pauli says, "Explain the fine structure constant." So God goes to the blackboard and starts writing, and Pauli listens in pleasure. But after two minutes Pauli stops smiling; after five minutes, he's shaking his head—and suddenly he's up on his feet, hissing, *"Das ist ganz falsch!"* That's *completely* wrong!

□   □   □   □   □

We could not count ourselves among the old baron's friends; we met him only once, at his apartment in the old town of Geneva, on a frigid gray day in February, 1984. A discreet brass plate outside the door bore the words: E. C. G. Stueckelberg, 20 rue Henri Mussard. Stueckelberg's wife ushered us into a close, dark, oppressively hot office, and left to tell her husband, whom she referred to as the "professor," of our arrival. Lighted only by a window facing the sleet-gray foothills of the Jura Mountains, the small room was jammed with the impedimenta of Victorian life: grand piano, rolltop desk, oversized sofa piled with wooden boxes. Portraits of Stueckelberg's titled ancestors marched across the walls, their frames done in faded gilt and black. Scattered everywhere were hundreds of books: philosophy, physics, biology, theology, and genealogy; Goethe, Swinburne, and Pauli; English, French, German, and Danish; a life's accumulation of learning, heaped in voluptuous and dusty confusion.

Stueckelberg appeared in the doorway, supporting himself on two canes. His sport coat hung on him slackly; plastic bags of tobacco and pipe cleaners were taped to his canes and on the armrest of his chair; his hair was mostly gone. The window light played across his long, angular face, the features thinned by his long fight with gravity. He fumbled in the bags and, after a few moments, extracted a fat, solid meerschaum pipe. "I'm living entirely on medicaments," he said suddenly. "I have terrible arthrosis, arthritis, whatever you call it." He had spoken English rarely since he taught at Princeton a half-century before. From his shirt he withdrew a thick pair of glasses. A match flared; he sucked the pipe into smoky life with evident satisfaction.

Ernst Carl Gerlach Stueckelberg, Baron Souverain of the Holy Roman Empire of the Teutonic Nations of Breidenbach at Breidenstein and Melsbach, professor of particle physics at the universities of Geneva and Lausanne, and the man who just may have first renormalized quantum electrodynamics, was born in Basel on February 1, 1905.[28] His father's family had been citizens of the canton since the fourteenth century; his mother was the baroness of a minute, forgotten fiefdom in central Germany. Stueckelberg acquired his Ph.D. in Munich at the age of twenty-two, under the aegis of Arnold Sommerfeld. Sommerfeld's name was sufficient to win him a post at Princeton University, in Princeton, New Jersey, where he taught until the Depression forced the school to let him go.

Once back in Switzerland, Stueckelberg had the first bit of what was to be a long run of misfortune. He discovered that even though he had been an associate professor at Princeton, he did not have the academic qualifications to teach in Switzerland, and thus was obliged to write another thesis. For many years after, he could find a job only as a *Privatdozent*, a poorly paid teaching assistant, at the University of Zürich. To make matters worse, he foolishly invested and lost the considerable wealth that belonged to his first wife, whom he married in 1931. Facing bankruptcy, he was forced to go into the military service to earn his keep, further delaying his academic career and making it difficult to leave Switzerland, something of a backwater, to meet his colleagues. Under considerable financial and personal pressure, Stueckelberg began to exhibit symptoms of what would today be called manic-depressive behavior. Most of the time he was rational, even brilliant, but occasionally he would feel a fit coming on and pack himself off to the asylum for a few weeks. Over the years, he had a score of different treatments, including electroshock therapy. Nothing helped.

Personal troubles notwithstanding, he gradually managed to acquire a small reputation for the originality and difficulty of his work. Unfortunately, the reputation had to be spread by word of mouth, because many of Stueckelberg's most important thoughts were dismissed out of hand by his colleague in Zürich, Wolfgang Pauli. He predicted the first of the hundreds of subatomic particles discovered shortly before and after the war, but did not publish the idea after Pauli told him it was ridiculous. (Later, the Japanese physicist Hideki Yukawa received a Nobel Prize for this idea.)[29]

When Stueckelberg did publish, his papers were written in a convoluted style that not even Pauli could understand; they were further complicated by his habit, not uncommon among the most mathematically inclined theorists, of inventing a special notation to replace the helter-skelter of mathematical symbols that is the common language of physics. Stueckelberg switched the ordinarily used terms for variables with those for parameters, put the indices on the opposite side of the symbols, and filled his equations with an incomprehensible forest of curved arrows and colored letters. Moreover, his papers were usually in French and published in the venerable but not widely read Swiss journal *Helvetica Physica Acta*. "I practically always published there," he told us in the course of a long conversation. "It was easy—also, my secretary knew only French. This was one reason that my papers were never read. At that time, German and English were common languages for physics, but French was not. I also must admit that when I reread my papers later on, I saw that they were very complicated. I don't know why, but I had a very complicated style. By the way, my friend—he has since died—Professor [Jean] Wiegle always put on the introduction and the summary, and these are understandable."

Sometimes, too, he courted obscurity by his enthusiasm for questionable research programs. Convinced that all reality should be described by real numbers such as one, three-sevenths, or the square root of two, Stueckelberg devoted years to a quixotic attempt to eliminate imaginary numbers, such as the square root of minus one, from the equations of quantum theory. Unfortunately, imaginary numbers, whatever the difficulty one has in picturing them, are a central feature of contemporary physics, as firmly embedded in modern theory as pi in geometry. In the midst of such dubious schemes, he would sometimes toss out another, almost unrelated idea of fundamental import. As an aside to a paper on the atomic nucleus, for example, he postulated that the number of "heavy particles," by which he meant protons and neutrons, in the Universe never changes. If they could decay into other, lighter particles, matter itself would be unstable, ever so slightly radioactive, and the world as we know it would eventually disintegrate. Testing the validity of the old man's postulate is, in a sense, the purpose of the experiment in the salt mine beneath the shores of Lake Erie.

In the mid-1930s, he began to consider the divergences in electrodynamics.[30] The infinities became a focal point of Stueckelberg's career; he lavished years upon their removal. Even decades later, when we met him, his face came alight when he discussed his theoretical strategems, the short-cuts he had devised, the cutoffs and approximations he had brewed. Reciting equations from memory, he traced their symbols in the air with small move-ments of his long, tobacco-stained fingers. Profligate with his ideas, he fol-lowed two separate tracks, a quantum and a classical approach.[31] Each was idiosyncratic; neither was understood; both were ignored. He explained his ideas to Pauli and Weisskopf; neither understood the presentation. They left Switzerland, and Stueckelberg, who was still in the army, was almost totally isolated from physics. Nonetheless, he apparently wrote up a lengthy pa-per—in English, for once—that outlined a complete and correct description of the renormalization procedure for quantum electrodynamics. Sometime in 1942 or 1943, he apparently mailed it to the *Physical Review*. It was rejected. "They said it was not a paper, it was a program, an outline, a proposal," Stueckelberg remembered. "Afterward, I was told that our friend and teacher, Gregor Wentzel—he was the expert [referee]—he got my paper." He rejected it? "Oh, it was done in an extremely obscure style," he said. Stueckelberg was not a bitter man. "Later, he took the manuscript and wanted to have it published to show that I got it before." We asked if he had the manuscript, which would help him establish priority. "I never cared much about that question," he replied. "I don't know what happened to the original copy. I lost it, it completely disappeared."

War swept over Europe. Even in neutral Switzerland, Stueckelberg was mobilized, although he obtained special dispensation from the army to teach his seminar every other week. It was his only contact with science. Nonetheless, he struggled to carry out the program rejected by the *Physical Review*. By the end of the war, in 1945, he seems to have done it.

The triumph, if there was one, was short-lived. His wife divorced him a year later. Long before, he had agreed to her family's demand for a marriage contract; now he courted ruin when he was forced to restore the fortune he had lost. When there was time between his need to scare up money and his sessions in the hospital, he wrote up bits and pieces of his ideas.[32] Eventually they were presented in a complete form in a chapter of the thesis of one of his students, Dominique Rivier.[33] But by then Schwinger had come out with his program, and Stueckelberg, who had the ideas first, published afterward.

He continued to do important work. In 1951, for example, he and his student, André Petermann, invented something called the renormalization group, which is now essential to the construction of grand unified theories.[34] Stueckelberg brought his dog, Carlo III, to seminars at CERN, the new particle accelerator laboratory outside Geneva. When Carlo barked, people would turn expectantly to Stueckelberg. He would survey the blackboard—

"There's always a mistake on them," he said—and point out the error. The rumor grew that somehow Carlo spotted the problems. As Stueckelberg grew older, he appeared less often at seminars. By the mid-1960s, years of experimental medication had slurred his speech, interfering with his thinking. Crippled by arthritis, he was carried to colloquia in the arms of his former students—a painful procedure that Stueckelberg described to us in detached, ironic detail, chuckling every now and then at his own frailty. He married again. He turned to the embrace of the Roman Catholic Church.

After we had talked for a couple of hours, he abruptly extended a hand as light and dry as a dead leaf. He was tired; the interview was over. The array of barons on the wall glowered in the deepening twilight. The old man gathered up his two canes and painfully lifted himself out of his chair. A clock ticked away. Stueckelberg nodded toward it and said, "I look forward every day to my eventual journey to Heaven." A heavy gold cross dangled from his thin neck. He was trembling slightly from the effort of standing. "We live too long," he said.[35]

Seven months later, on September 4, 1984, Ernst Stueckelberg was buried in Geneva at Plain Palais, the cemetery where Calvin had been laid to rest three centuries before.

Strange Interlude

9

From Deepest Space

ON MARCH 30, 1910, A COLD DAY IN PARIS, FATHER THEODOR WULF OPENED the elevator door at the top of the Eiffel Tower and pulled his equipment onto the platform.[1] A physics instructor in a Jesuit high school at the Dutch town of Valkenburg, Wulf was also an amateur scientist. A thousand feet above the Champ de Mars, Wulf spent the day measuring with glass and metal instruments how well the air conducted electricity. The results surprised him no end.

Like many other turn-of-the-century scientific dilettantes, Wulf was fascinated by radioactivity, which had remained novel and mysterious since its discovery fourteen years before by Antoine-Henri Becquerel. In the course of his plodding, methodical inventory of the properties of radioactive materials, Becquerel had found that the lumps of uranium in his laboratory made the air nearby conduct electricity—something that the atmosphere does not do under ordinary circumstances.[2] It happens because the alpha and beta ray emissions from radioactive materials knock out electrons from air molecules, creating *ions* that conduct electricity. The higher the level of radioactivity, the greater the air's conductivity.

The instrument then used to measure electric charge was the electroscope, a device consisting of two little strips of metal foil hanging like two stiff flags from a central rod. Strips and rod are sealed inside a glass jar, and the whole ensemble looks something like a TV aerial in a bottle. Electric charge spreads apart the two lower ends of the strips—a similar but opposite effect is "static cling"—with the distance apart indicating the strength of the charge. When the electroscope is charged up near some uranium, the conductivity of the air around the two strips causes them to discharge, and their ends fall limply back together. Scientists realized that this effect made it possible to press electroscopes into service to measure radioactivity, with the rapidity of discharge indicating the strength of the source.

Early radioactivity buffs had found, however, that their instruments "leaked," slowly discharging even when no uranium was around.[3] Called *residual discharge*, this slow dissipation interfered with measurements.[4] Annoyed, experimenters tried to get rid of it—and couldn't.[5] Even insulating the electroscope with five tons of pig lead could not stop it from losing its charge.[6] By 1905, it was clear that something was going on. The physicists

147

who had used electroscopes to measure radioactivity were now staring in confusion at the electroscopes themselves.

Father Wulf made his first contribution to the puzzle in 1909, when he invented an electroscope so sensitive that it was rapidly adopted by scientists everywhere. The precision of the Wulf electroscope of course highlighted the fact that it inexplicably and inevitably lost its charge, and the hunt for residual discharge began in earnest. Meteorologists and geologists as well as physicists took part, because for all anyone knew the cause might have to do with climate or the structure of the earth. Scientists took Wulf electroscopes all over the world to measure the discharge in different places, climates, and times. Wulf himself took one on a tour of Germany, Austria, and the heights of the Swiss Alps.[7] Commander Robert F. Scott's ill-fated expedition to the pole in 1911–12 included a meteorologist who took measurements over the open sea and on the Antarctic continent with a Wulf electroscope.[8] Residual discharge showed up everywhere, in different amounts, and scientists concluded that it must somehow be caused by low-level background of radioactivity in the crust of the earth.[9] Wulf wanted to make sure, by going to the top of the bizarre, skeletal steel structure that was then the highest building in the world: the Eiffel Tower.

Wulf knew that the thousand feet of air between him and the ground would absorb nearly all the earth's radioactive emissions, and that radioactivity from the Tower itself was almost nonexistent. But throughout his four days there, the electroscope still discharged. Something was causing the loss, but it was not in the Tower, the ground, or the electroscope itself. By August, Wulf concluded that there must be "either another source [of radioactive emissions] in the upper portions of the atmosphere, or their absorption by the air is substantially weaker than has hitherto been assumed."[10]

The remarks of the good Jesuit intrigued Victor Hess, a recent arrival at the newly founded Vienna Institute for Radium Research. Like many early students of radioactivity, Hess was careless about radium, eventually losing his thumb to radiation burns.[11] After checking to see that the air really should have absorbed the radiation from the ground before it arrived at the top of the Eiffel Tower, Hess began to "believe that a hitherto unknown source of ionization may have been in evidence in all these experiments. . . ."[12]

A tenacious and stubborn man, Hess decided that the only good way to test Wulf's results was to go even higher. In 1911, that meant taking a balloon, a dangerous procedure at best. The tiny and unpressurized baskets of fin-de-siècle balloons were easily buffeted by wind; worse, a single spark could blow up the hydrogen in the gas bag. Indeed, the two previous attempts to take electroscopes up in balloons had been plagued by instrument failure.[13]

Still, Hess could think of no other way to proceed. From 1911 to 1913, he went up ten times. Like Father Wulf, he found that the residual radiation decreased with altitude, but not as much as it ought to if it were caused by terrestrial gamma rays.[14] (Gamma rays, a third type of radioactive emission along with alpha and beta rays, are high-energy photons.) An hour or so after dawn on August 7, 1912, Hess began his ninth flight by taking off from Aussig, outside Prague. Six hours later, the orange-and-black balloon landed in a pasture about thirty miles east of Berlin. There is a photograph of Hess peering out through the ropes of the balloon, smiling against the sun, a wave of farmers lapping against the side of the basket. He was just twenty-nine, and had discovered something important.[15]

As in previous flights, the ionization of the electroscope had slowed down until the balloon reached a few hundred meters and then leveled off. At six thousand feet, however, the electroscopes began to discharge more rapidly than ever before. At fifteen thousand feet, the rate of discharge was more than twice ground level. Hess came to the unsettling conclusion *"that rays of very great penetrating power are entering our atmosphere from above."* At first Hess guessed that these penetrating rays came from the sun. But half the balloon flights had taken place at night, and there was no difference between the results of the day and night flights. Clearly, the sun was not the source of the rays.[16] They must come from outer space.

Hess's ideas were greeted with derision. The idea that interstellar rays were continually bombarding the earth with enough strength to shoot through several feet of lead was too wild for many scientists. It was intimated that the low pressure of the heights had confused Hess's instruments; that high-altitude electrical effects overwhelmed the electroscope; that Hess was simply incompetent.[17] While Wulf worked to improve his electroscopes, Werner Kolhörster, a German, made five balloon flights, culminating in a marathon ascent to thirty thousand feet, slightly higher than Everest, on June 28, 1914. He found the ionization level there twelve times that at sea level.[18] Hess was right.

Unfortunately, the same day that Kolhörster took his electroscopes to record heights, the heir to the Hapsburg throne was assassinated on the ground, igniting a chain of events that led with giddy speed to warfare throughout Europe. The balloon flights ceased; the mountain laboratories were abandoned; and all research into the strange radiation from above stopped as Western civilization turned its attention to destroying itself.

Thus began the tale of what would later be called *cosmic rays*, a powerful emanation which bathes our planet from deep space and that scientists are only now beginning to understand. The story of cosmic rays, in a sense, is a tale apart; beginning almost accidentally, even trivially, with the inability of laboratory equipment to stop leaking electricity, cosmic ray physics grew into a hybrid discipline of its own, populated by adventurous sorts who were unwilling and even unable to confine themselves to the laboratory.[19] Traveling across the globe and into the gelid heights of the stratosphere, they discovered that cosmic rays smash into ordinary atoms with enough power to change energy into matter and create dozens of new particles with strange new properties. Ultimately, cosmic rays provided the raw material for the standard model—the matter it classifies, the interactions it describes. Karl Darrow, the organizer of the Shelter Island conference, once wrote, "The subject is unique in modern physics for the minuteness of the phenomena, the delicacy of the observations, the adventurous excursions of the observers, the subtlety of the analysis, and the grandeur of the inferences."[20]

And, Darrow might have added, for the depth of the confusion. Right ideas were often held for wrong reasons, wrong ideas for right reasons; accidents, feuds, and ideologies rather than logical deduction shaped the scientific conclusions. After the war to end all wars, when cosmic ray studies resumed, the greatest share of woes befell a newcomer to the field, Robert A. Millikan, the most famous American scientist of the day.[21]

The son of a Congregationalist minister, Millikan won the Nobel Prize for physics in 1923—only the second bestowed on an American.[22] (He was the first to measure the charge of the electron accurately.) Much of his celebrity was due to his ardent attempts to mediate between the religion of his father's generation and the scientific accomplishment of which Millikan was an exemplar at a time when conflict between them seemed unresolvable. He was a deeply Christian man of science in the 1920s, a time when the nation was preoccupied by Prohibition, the Ku Klux Klan, and the Scopes trial. He believed that religion without science had in the past produced "dogmatism, bigotry, and persecution," but he also argued that science was less important than "*a belief in the reality of moral and spiritual values.*"[23] To the pillars of the social order this was welcome news; the *New York Times*, for one, suggested that Millikan's moral stance was "even more significant" than his physics.[24] A rigid, uncompromising man, Millikan lacked the essential quality of doubt; he treated scientific ideas as if they were tenets of a creed. The lapses caused by his dogmatism were magnified by his genius for self-publicity, which ensured that he was always surrounded by a virtual cloud of reporters.

Millikan had read the work of Hess and Kolhörster in 1914, and decided to look for penetrating radiation himself. Believing that manned flights could not reach sufficient altitudes, he set out to develop extremely light, self-recording Wulf electroscopes for unpiloted balloons. His work was stopped by America's entry into the war, but when the conflict ended he and a graduate student built tiny, seven-ounce instrument packages, each with a Wulf electroscope, temperature and pressure gauges, and a camera to record the readings. In the spring of 1922, Millikan launched them from Kelly Air

Force Base at San Antonio, Texas; one flew to fifty thousand feet, almost twice as high as Kolhörster's highest flight. Millikan found that the rate of discharge was much less than the Europeans had reported. Although he had been prepared to confirm the existence of penetrating rays, he concluded that he had "definite proof that there exists no radiation of cosmic origin having such characteristics as we had assumed."[25] A second experiment, conducted during a snowy week in a hut atop Pike's Peak, 14,100 frigid feet above sea level, produced similar results.[26]

Millikan's data were right, but his conclusions were wrong. As scientists discovered later, cosmic rays are distributed unevenly over the earth. By chance, Millikan performed his experiments in regions of the American West with unusually low levels, and he incorrectly concluded that the Europeans were mistaken. Annoyed, Hess argued that Kolhörster's data were more trustworthy than Millikan's, while Kolhörster carried out further experiments on Alpine glaciers to confirm those of his balloon flights.[27]

Not one to tolerate even the suspicion of error, Millikan looked a third time in 1925. Cosmic rays, if they existed, were assumed to be gamma rays from the stars. Millikan knew that a gamma ray able to travel through, say, a foot of water before being absorbed would also be able to traverse 1,116 feet of air, because water is that many times denser. In August that year, he took his equipment to two deep, snow-fed lakes in the San Bernardino range of southern California: Muir Lake, 11,800 feet above sea level, underneath the brow of Mount Whitney; and Arrowhead Lake, three hundred miles south and 6,700 feet closer to sea level. In terms of gamma ray absorption, Millikan calculated that the big difference in altitude should correspond to six feet of water. If the penetrating radiation really comes from the stars, Millikan reasoned, he should detect the same amount of radiation six feet beneath the surface of Lake Muir as he did at the shore of Lake Arrowhead, for the cosmic rays would have traversed an equivalent amount of matter. The levels were the same. Millikan reversed himself, and asserted that he had never *really* come out against cosmic rays.[28] (In hindsight, he was right for the wrong reason, because cosmic rays are not gamma rays, do not behave like them, and are not equally absorbed by equal masses of air and water. Had Millikan's equipment been more sensitive, he would have discovered the different absorption levels, and once again drawn the wrong conclusion.)

Millikan's results excited the public, convinced the American scientific community, and pleased his European colleagues—at first. But Hess and Kolhörster were angered when Americans soon began to call Millikan the *discoverer* of cosmic rays. Although Millikan himself had coined the name "cosmic rays," the *New York Times* suggested editorially that they be called "Millikan rays" to honor "a man of such fine and modest personality"; American journals spoke of "M rays."[29] There ensued a ludicrous feud between

Millikan and the European cosmic ray workers as to who had produced the first definite evidence of their existence.[30] Although most American scientists came to realize that Millikan had been the latecomer, a history of cosmic rays published in the United States as late as 1936 described Hess and crew as a "nationalistically minded but misguided group of scientists."[31] But that year, almost a quarter century after his epochal balloon flight, the Nobel Prize in physics was awarded to Victor Hess, "for his discovery of cosmic radiation."

Throughout the 1920s, cosmic rays were assumed to be a more powerful form of gamma rays; Millikan in particular was a champion of this view. The first experiment to challenge this assumption was made by Kolhörster and another German physicist, Walther Bothe. Bothe had been initiated into physics by Hans Geiger, who taught him about a new instrument for studying radiation, the Geiger counter.[32] Named after its developer, the Geiger counter was a vast improvement over the old particle detectors—Wulf's electroscopes and Rutherford's scintillation screens. It consists basically of a gas-filled metal tube with a thin wire down the axis. The wire is positively charged; the tube walls have a negative charge. When a charged particle shoots through the tube, it strips away some of the electrons in the gas atoms. The negative electrons can be pulled toward the positive wire with such strength that they actually collide into more gas atoms on the way, knocking off additional electrons. Domino-style, the cascade spreads, until by the time it reaches the Geiger counter wire the flood of electrons causes a noticeable jolt in the current. In some models, the current surge is translated into the "click" now famous from movies and television shows. Geiger counters can detect charged particles with impressive efficiency but are not nearly as good at detecting uncharged photons, meaning that the counters are useful for studying only certain types of interactions.

In the spring of 1929, Bothe and Kolhörster used a brilliantly simple arrangement of Geiger counters to test the nature of cosmic rays.[33] They stacked one counter on top of a second, and counted the number of coincidences, or times the two registered at precisely the same instant. Although there would be an infinitesimal number of "sheer" coincidences caused by two simultaneous cosmic rays, nearly all the signals would be due to single particles, which Bothe and Kolhörster assumed would be electrons struck by gamma rays with enough force to be thrown through both counters. Bothe and Kolhörster knew that materials with lots of protons in their nuclei, such as gold, are particularly good at catching electrons. To eliminate penetrating electrons, the experimenters therefore inserted a two-inch slab of gold between the Geiger counters. If, as was generally believed, cosmic rays are photons, the coincidences should vanish; it seemed unreasonable to imagine that an incoming photon would hit an electron in one counter, tunnel through two inches of gold into the second counter, and then hit a second electron.

Unexpectedly, the slab barely put a dent in the number of coincidences: Fully three-quarters remained. To Bothe and Kolhörster, this strongly suggested that the Geiger counters were not picking up recoil electrons, but the high-energy cosmic rays themselves. This meant the conventional wisdom was wrong: Cosmic rays are not gamma rays. They are charged particles powerful enough to punch through gold without blinking an eye.

Many cosmic ray specialists thought the experiment somehow must be wrong—especially Millikan, who insisted cosmic rays were photons. When reporters who followed Millikan asked whether cosmic rays might be charged particles after all, he snapped, "You might as well sensibly compare an elephant and a radish."[34] What Bothe and Kolhörster observed, Millikan argued, was not *primary* cosmic ray particles, but *secondary* particles, particles that had been hit by cosmic rays.[35] Bothe and Kolhörster could not disprove this. But they did know that if primary cosmic rays consisted of charged particles, their paths would be affected by a magnetic field. The problem was to find a magnet that could reach out into space far enough so that physicists could study its effect on incoming cosmic rays. There was no magnet big enough on earth except one, the earth itself.

In the 1920s, the Norwegian Fredrik Størmer had discerned that charged particles spat out in solar flares are bent by the earth's magnetic field to the poles, where their shivering aerial dance creates the aurora borealis.[36] If cosmic rays were indeed charged particles, they, too, should ride the magnetic lines and come to ground in greatest numbers at the poles. A Dutch physicist, Jacob Clay, took electroscopes on boat trips between Holland and Java in 1927, and found that there was half again more radiation in northern Europe than near the equator.[37] Millikan had not found any latitude effect on his 1926 journey by boat to Bolivia—one reason for his conviction that cosmic rays were photons.[38] Bothe tried to resolve the disagreement himself with a sojourn in Spitzbergen, a large Norwegian island a few hundred miles away from the North Pole. He found no evidence of any latitude effect, but ascribed his failure to his inability to make headway with Størmer's calculations.[39]

The controversy inspired European scientists to measure cosmic ray intensities in the most diverse locations and the highest altitudes. Hess and some colleagues set up electroscopes in mountain observatories in the Alps.[40] Kolhörster took his equipment to the salt mines in Stassfurt, now part of East Germany.[41] The development of pressurized balloon cabins allowed physicists to bob up to unprecedented altitudes in search of cosmic rays. The first balloon flight to the stratosphere occurred on May 27, 1931, when Auguste Piccard, a former pupil of Einstein, and Charles Kipfer, his assistant, took an electroscope up from Augsburg, now part of West Germany. When the hot sun of the heights baked the seven-foot cabin, the crew survived by licking drops of water off the walls. Italian poet Gabriele d'Annunzio, moved by the thought of courting death in the name of knowledge, begged Piccard to be taken on a flight; never one to miss the opportunity for an extravagant gesture, D'Annunzio announced his readiness to be thrown overboard as ballast if necessary. Far better to die the noble death of being tossed from a balloon, the poet said, than to pass away "shamefully between two sheets."[42]

Such flights contributed to little except the romance of science. The practice of randomly measuring cosmic ray intensities in picturesque settings was an impossibly inefficient way of resolving the confusion over the latitude effect. Ultimately, these studies of the upper atmosphere became the absurd pretext for an early version of the space race in which European, American, and Soviet aeronauts battled feverishly to fly ever more advanced stratospheric balloons. Inevitably, disasters occurred. Three Russians soared to a record height of thirteen miles before their bag burst and they plummeted to the ground; according to some reports, their last words were, "We have studied the cosmic rays." They were buried in the Kremlin Wall.[43]

In the meantime, the venue for cosmic ray research had shifted definitively to the United States, where Robert Millikan was locked in yet another widely publicized fight, this time with a former student, Arthur Compton. The third American physicist to win a Nobel Prize, Compton earned his award in 1927 for describing what happens when a photon and an electron collide, a phenomenon now called the Compton effect. Bothe and Kolhörster's experiment had strongly impressed on him the possibility that cosmic rays were charged particles. Like Bothe, Compton believed that the issue would only be settled by discovering whether there was a latitude effect.

In 1930 Compton asked the Carnegie Institute, in Washington, D.C., for money to mobilize a worldwide study of cosmic rays. Once funded, Compton divided the world into nine regions and sent a different expedition to each. In all, over sixty physicists took part. Allen Carpé of the American Telephone and Telegraph Company laboratory in New York City was sent to Alaska, where he and another man climbed the slopes of Mount McKinley, fell into a crevasse on the Muldrow Glacier, and died. (Their equipment was found and the readings used in the paper.) Compton himself filled in all the gaps by embarking on a round-the-world trip with his wife and teenage son in March 1932.

Millikan, too, applied to the Carnegie Institute in late 1931 for money to perform his own series of experiments. Now working with Henry Victor Neher, a young Caltech Ph.D., Millikan developed a new and more sensitive type of electroscope, which they tested for steadiness by giving it a ride down some bumpy roads in a 1928 Chevrolet. During one test drive, Millikan peered into the eyepiece while Neher struggled with the unfamiliar clutch; a lurching shift into first smashed the electroscope into Millikan's nose, barely missing his eye and bequeathing a permanent scar. In September 1932, Neher set sail with his electroscopes on a ship going from Los Angeles to Peru.

Millikan missed the latitude effect on his 1926 trip to the Andes because just south of Los Angeles there is a cosmic ray "shelf" where the intensity abruptly dips and levels off; he happened to switch on his electroscope after the boat had passed it.[44] By sheer bad luck, Neher missed the "shelf" on his trip, too. Neher had two electroscopes with him, but only one was working properly at the start of his journey. That one malfunctioned within forty-eight hours of the launch, and became completely useless shortly thereafter. To his dismay, Neher discovered that the seas were too rough to allow him to repair the other until the boat anchored at Mazatlan, on the Mexican coast. By then, he had already passed the critical point. When Neher arrived in Panama, he sent a telegram to Millikan with the news that there was no latitude effect.

On September 14, 1932, Compton announced that cosmic rays varied considerably in intensity from the equator to the North Pole. Speaking before a crowd of reporters, Compton said, "Obviously if the north magnetic pole has any effect on the rays, they must be electrical in nature instead of a wave, as Dr. Millikan contends [that is, charged particles and not uncharged photons]. The difference shown by my experiments will be a severe blow to Dr. Millikan."[45] Millikan refused comment, saving his fire for the annual meeting of the American Association for the Advancement of Science (AAAS). Three days before his talk, he received a telegram from Neher: SEVEN PERCENT CHANGE RETURNING STOP CONCEALED BEFORE BY BROKEN SYSTEM AND DIFFERENT SHIPS STOP NEHER. Coming up to New York, Neher had again passed the cosmic ray "shelf"; this time, however, his electroscope was working.[46] Faced with overwhelming evidence, Millikan still blasted Compton in his speech. But hours afterward, reason returned. He conceded the point, and sent a long and savage telegram of denial to the New York Times, which had correctly reported his earlier declaration that cosmic rays are photons. In the telegram, Millikan insisted that Compton and he were in complete agreement. When published, the AAAS speech was completely rewritten, thus hurting Millikan's reputation further. Indeed, in his later autobiography, he made but passing reference to cosmic rays, the subject that had absorbed twenty years of his life.[47]

□ □ □ □ □

There were two main topics of conversation at the cosmic ray section of the International Conference on Physics at London in 1934, both with profound implications for the future of physics. The meaning of one passed almost unnoticed for forty years; that of the second was seized upon immediately. Five years before, a Franco-Soviet collaboration discovered that cosmic ray collisions sometimes occurred in connected bursts that sprayed across cloud chambers like shotgun blasts. By the London conference, people had counted showers of up to twenty particles and realized that the energy needed to produce the initial cosmic ray is literally astronomical, the kind of energy found in the sun and stars.[48] In fact, the figure was so high that most physicists considered the relation between subatomic particles and cosmic phenomena to be out of the purview of physics altogether.[49] Only in the 1970s, forty years later, would the connection between the smallest constituents of matter and the greatest reaches of the Universe at last be made.

Much more readily acted upon was the puzzling absorption of cosmic rays by various amounts of matter. Several conferees pointed out that cosmic rays seemed to come in two groups, with markedly different energies.[50] Cosmic rays of the first group, dubbed the "soft" component, were easily absorbed by a few inches of heavy metal, and were evidently a mixture of electrons and positrons. The other component was much "harder," and could pass happily through several meters of lead without slowing down. Nobody in London knew quite what to make of the "hard" particles. They were not heavy enough to be neutrons or protons, yet they punched through lead more readily than light little electrons. A little while later, Caltech experimenters began the tradition of irreverent nomenclature in subnuclear physics by announcing that the "solution" was to say that the soft component consisted of "red" electrons, and the hard of "green" electrons.[51]

Some physicists felt sure that both soft and hard cosmic rays were electrons, but that the latter were moving at such great speeds that their behavior fell outside of the purview of quantum electrodynamics. This was an eminently reasonable guess: Charged particles radiate energy when they accelerate or decelerate, affecting the motion of the particle, in a way reminiscent of self-energy.[52] Negligible at low speeds, the effect rises with the energy of the particle. Because physicists were then wholly unable to explain self-energy, many expected that as experimenters studied particles moving at higher and higher energies, at some point the theory would begin to give wrong predictions. Theorists even thought they knew where—when the electron energy reached $137mc^2$ (m is here the mass of the electron, and mc^2 is the energy of the electron when it is at rest).[53] Seeing the approach of alpha, some researchers were sure it signaled the end of the theoretical road.

After the London Conference, Carl D. Anderson, the discoverer of antimatter, and a graduate student, Seth Neddermeyer, set out to nail down the identity of the hard "green" electrons, using the same cloud chamber

and magnet that Anderson had employed to discover the positron. Once
back at Caltech, they inserted a thin metal plate in the chamber, with Geiger
counters above and below that set off the chamber and a camera whenever a
charged particle passed clean through the plate. The trick of using Geiger
counters to trigger cameras was invented in 1932 by P. M. S. Blackett and
Giuseppe Occhialini; by the time Anderson and Neddermeyer looked for
hard electrons, experimenters routinely took thousands of photographs. The
magnet's field would make the incoming particles curve to one side. By
studying the degree of curvature, the range, and the level of ionization in the
tracks, it was possible to determine precisely a particle's mass and charge. If
particles were pushed one way, they were positive; if they arched in the
other direction, they were negative; and the heavier the particle, the less it
was affected by the magnetic field.

The Caltech team spent two years photographing the chamber and
gradually convinced themselves that the particles seemed to be disobeying
the laws of quantum electrodynamics, not because the theory failed, but
because the experiments were finding something that had never been seen
before—"particles of a new type," as Anderson put it—heavier than the
electron but lighter than the proton.[54] They were dissuaded by J. Robert
Oppenheimer, who flatly informed them that they were just looking at fast
electrons; his superior mathematical acumen and confident lack of confidence intimidated them. In August of 1936, Anderson and Neddermeyer
published several photographs that might have been left by new particles,
but, to their regret, they refrained from claiming discovery.[55]

The article created a small sensation across the Pacific Ocean, in Japan,
where a theorist named Hideki Yukawa had actually predicted the existence
of such a particle. Moreover, Yukawa had predicted that this particle, which
weighed less than a tenth as much as the proton, should be the agent that
kept the nucleus together despite the mutual electrical repulsion of the
protons. Convinced immediately that Anderson and Neddermeyer had
found his particle, Yukawa told his Japanese colleagues that they had to find
some of the things themselves before Anderson and Neddermeyer tumbled
onto what they had.

Yukawa was born in 1907 as Hideki Ogawa. Shortly after his first birthday, his family moved from Tokyo to Kyoto, where he spent the rest of his
childhood. He was a conservative, painstaking, methodical child, insisting in
grammar school that he could not work unless his desk was aligned parallel to
the stripes of the tatami mats on the floor.[56] A short boy with a shy round
innocent face, he spent so much time in solitary silence that his father, a
professor of geology at Kyoto Imperial University, considered recommending
that Hideki, alone of his five sons, not go on to higher education.[57] In 1932,

Hideki's father arranged his marriage to Sumi Yukawa, the daughter of a local physician; following a Japanese custom, Hideki was adopted by his wife's father and assumed the family name.

Soon after entering the university, Yukawa disappointed his father by dropping the practical study of geology and embracing theoretical physics of the most abstract variety. One of his university classmates was Sin-itiro Tomonaga, who later played a major part in renormalizing quantum electrodynamics; the two young men helped each other through quantum mechanics, and stayed on afterward as unpaid assistants in the physics department. In the spring of 1931, Yukawa attended a lecture by Yoshio Nishina, a researcher at the elite Institute of Physical and Chemical Research of Tokyo (usually known as Riken, the abbreviation of its Japanese name). A towering figure in Japanese physics, Nishina worked at both the Cavendish and Göttingen, and spent six years at Copenhagen with Bohr; returning to Japan at the end of 1928, he was a one-man incarnation of the spirit of the new physics.[58] Much of Nishina's lecture concerned quantum electrodynamics, and the half-revealed intricacies of photons and electrons made Yukawa decide to pursue the subject himself. Immediately he ran into the problem of the infinities.

Sitting in an office overlooking the pastures of the agriculture school, Yukawa struggled futilely with the divergences in quantum mechanics; while he worked, mountain sheep bleated mockingly beneath his window. His account of that frustrating time is an echo of his European colleagues' struggle:

As I fought daily with the devil called infinite energy, the call of those sheep sounded to me like the sneers of that devil. Each day I would destroy the ideas that I had created that day. By the time I crossed the Kamo River to my home in the evening, I [would be] in a state of desperation. Next morning, I would leave my house feeling renewed strength, but again would return at night thoroughly discouraged. Finally, I gave up that demon-hunting and began to think that I should search for an easier problem.[59]

Yukawa turned his attention to the atomic nucleus.

Few physicists of the time found the nucleus to be "an easier problem." Heisenberg had made the most recent attempt to tackle it, in a lengthy, three-part article appearing in the *Zeitschrift für Physik* in 1932 and 1933.[60] The chief theoretical difficulty with the nucleus was to describe why the mutual electrical repulsion of the protons did not make it fly apart. Physicists since Rutherford had argued that something else must hold it together—electrons, most likely. Nobody had been able, however, to figure out how the proton-electron attraction could actually cancel out the proton-proton and electron-electron repulsion. In the last few years, theorists had taken to speaking vaguely of a "nuclear force" that held the protons together. Having no influence on the everyday world, the nuclear force must have an extremely short range; beyond a distance of 10^{-13} centimeters, it has no effect at all—as if the gravity that holds people to the surface vanished the instant they climbed a footstool, allowing them to float into the clouds. In his three papers, Heisenberg treated the attraction among protons and neutrons as analogous to the relatively common situation in which two hydrogen nuclei form a molecule by passing an electron back and forth. Although he did not have a clear picture of the process, Heisenberg imagined the nucleus as held together by a game of intranuclear "catch" in which the protons and neutrons rapidly toss a rather strange sort of electron among themselves.

Yukawa knew that the two known fundamental forces, gravity and electromagnetism, were both described by field theories (although, to be strictly accurate, general relativity is a classical field theory and quantum electrodynamics is a quantum field theory). He applauded Heisenberg's decision that the nucleus was held together by another fundamental force and that it, too, needed a field theory; Yukawa even translated part of Heisenberg's work into Japanese, adding his own introduction.[61] He liked the scheme well enough to worry about the problems that cropped up if one took it literally. In the model, neutrons can emit electrons, turning into protons in the process. (The total electric charge, zero, is the same before and after.) The electron is attracted to a nearby proton, which absorbs the electron and turns into a neutron. With a constant flow of virtual electrons flickering between them, the two particles are bound to each other. Neutrons and protons have the same spin, however, so creating electrons involves making a quantum of spin appear out of thin air—violating the laws of conservation of energy and momentum. For this reason, Heisenberg knew the exchanged particle could not possibly be a real electron. He was willing to entertain the possibility of a new "spinless electron" to get around the problem, but Yukawa thought this preposterous. Some other tack was needed.

The obvious strategy was to think more carefully about the type of particle that would be required to carry the nuclear force. Stated this way, Yukawa wrote later, the answer almost jumps at you, "but my brain did not work so quickly. I had to take a wrong path first, before I could arrive at my destination. Those who explore an unknown world are travelers without a map; the map is the result of the exploration. The position of their destination is not known to them, and the direct path that leads to it is not yet made."[62]

Yukawa spent two years on blind alleys. Depressed and harried, he exacerbated his loneliness by his formal, withdrawn mien and the necessity of commuting between two lecture positions in Kyoto and Osaka. In Kyoto he had a new office overlooking the road to the railway station; the bleating of sheep had been replaced by the rumble of heavy trucks. All the while, he struggled to construct a theory of the nuclear force that would not compel him to make up the existence of some particle that nobody had ever seen.

His spirits were raised unexpectedly in the summer of 1934, when he chanced upon an Italian journal containing an article by Enrico Fermi on beta decay, the form of radioactivity responsible for beta rays.[63] Beta decay occurs when a neutron emits an electron and turns into a proton. Although beta decay had been known to exist since the discovery of radioactivity in 1896, all theoretical explanations of the process had been plagued by precisely the same difficulties that troubled Heisenberg's theory of the nuclear forces. During beta decay, particles in the nucleus actually seemed to create electrons with apparent disregard for the conservation of energy and spin. Fermi championed a solution that Pauli had thought of three years earlier but had been reluctant to commit to paper.[64] Pauli had wondered if the spin of the created electron might be canceled out by the simultaneous emission of another particle, electrically neutral and previously unknown to science, with the opposite spin. He, too, had been unsettled by the thought of making up a fundamental constituent of matter solely to resolve a theoretical difficulty, and had never quite wanted to publish it. This particle was called, variously, the "neutron," "Pauli particle," and "neutrino." The last name, coined by Fermi as a joke, is the one that stuck.[65] Hoping that the new particle might also play a role in the nuclear force, two Soviet theorists rushed into a futile effort to produce a neutrino theory that would account for the stability of the nucleus. They ended up by demonstrating that the nuclear force could not be linked to the neutrino, a conclusion that disheartened them—but not Yukawa.[66]

Learning about the neutrino, Yukawa wondered if he were not being unimaginative. He had been trying to arrive at the field by looking at the known particles; perhaps he should start from the field and see what particles were necessary to transmit it. But how did one go about this?[67] Sitting one sleepless October night in his small room, Yukawa realized that with some simple calculations he could figure out the mass of the particle that held together the protons and neutrons in the nucleus.

The calculation proceeded from the known range of the nuclear force. If the particle responsible for that force takes a while to decay, it can travel a long way in that time, and the force will be felt over a considerable distance. Because the range of the nuclear force is very short, its particle must quickly wink out of existence. Yukawa could thus come up with an approximate lifetime. He then recalled Heisenberg's uncertainty principle, which can also be phrased in terms of energy and time.

$$\Delta E \, \Delta T \geq \hbar$$

If the time (ΔT) approaches zero, then the energy (ΔE) must rise to giddy heights to keep the product near $\hbar$. Knowing the lifespan (in other words, ΔT) and the uncertainty relation, Yukawa could calculate the energy of the virtual particle (ΔE). If one considered the famous equation

$$E = mc^2$$

it is easy to find m, the mass of the particle at rest. Given that the range of the nuclear force is about 10^{-13} centimeters, Yukawa thought it should be transmitted by a particle that weighed about two hundred times as much as the electron, or, equivalently, about a tenth as much as a proton.

At the beginning of November, Yukawa began to put together a paper in English, which was published, many drafts later, early in 1935.[68] With its heavy sentences and slightly skewed syntax, the finished article bears the mark of Yukawa's imperfect mastery of the language. He begins with the assertion that it is natural to model the nuclear force on electromagnetism, and thus to describe it as a field, the "U" field. "In the quantum theory this field should be accompanied by a new sort of quantum, just as the electromagnetic field is accompanied by the photon."[69] This newly minted U quantum is the nuclear "glue" that shuttles electric charge among the protons and neutrons in the nucleus. Unlike the neutral photon, the U must come in two varieties, positive and negative. Thus, when a proton emits a positive U, it becomes a neutron; when a neutron emits a negative U, it becomes a proton. The scheme is tidy, but, as Yukawa admitted, "As such a quantum with large mass and positive or negative charge has never been found by the experiment, the above theory seems to be on a wrong line."[70] The answer to this objection, Yukawa said, is that this particle decays too fast to be spotted without great effort. Next, Yukawa brilliantly tried to kill two birds with one stone by arguing that beta decay is nothing other than the decay of a virtual U particle into an electron and neutrino. Almost in passing, Yukawa mentions one place the U quantum might show its face: the highly energetic interactions found in and around cosmic rays.

Yukawa gave a few lectures on the paper, but almost nobody but Nishina took it seriously.[71] Optimistically, Yukawa sent reprints to several European and American physicists, including Oppenheimer, and awaited their comments. Few came. When Niels Bohr came to Japan, Yukawa told him about the U quantum. Bohr's only answer was to ask if he enjoyed making up particles.[72] In the meantime, Yukawa wrote later, "I felt like a traveler who rests himself at a small tea shop at the top of a mountain slope. At that time I was not thinking about whether there were any more mountains ahead."[73]

He began to think again about the new particle when he came across the photographs by Anderson and Neddermeyer in the *Physical Review*.[74] Yukawa identified their photographs as his U quantum even before they were ready to claim they had discovered a new form of matter. Nishina and two other experimenters quickly tried to come up with some of the new particles themselves. During the 1930s, Nishina's Riken lab was a private organization largely supported by the industrial patents its engineers had won for inventing new methods of distilling sake.[75] It could not afford an electromagnet strong enough to bend cosmic rays. Discovering that he was competing with richer foreign groups, Nishina, an ardent nationalist, convinced the Imperial Japanese Navy to let him use a generator that charged submarine batteries.[76] Taking thousands of photographs at a frantic speed, the Riken team managed in the summer of 1937 to turn up a single cosmic ray track that stopped in the chamber, allowing them to estimate the mass of the particle; it was between 180 and 360 times that of an electron, exactly as Yukawa had said. But despite their efforts, they were beaten to the punch.

A few months before, Anderson and Neddermeyer had traveled to MIT, where they learned that two Harvard experimenters, Jabez Street and E. C. Stevenson, had used the same plate-in-the-cloud-chamber technique and were thinking of declaring that they had found the same new particle. His hand forced, Anderson dashed off a paper to the *Physical Review* claiming that "there exist particles of unit charge, but with a mass (which may not have a unique value) larger than that of a free electron and much smaller than that of a proton."[77] He was very annoyed with Oppenheimer.

The paper was published in May. Three months later, the disappointed Japanese sent their single picture to the *Physical Review*, which promptly sent it back. The editors refused to print the photo unless the accompanying text was cut drastically. A squabble ensued, delaying the appearance of the article and allowing Street and Stevenson to come in ahead of the Japanese.[78]

Until the discovery of the new particle, Yukawa's paper—a speculative article by an unknown Japanese theorist published in the obscure *Proceedings of the Physico-Mathematical Society of Japan*—received scant attention in the West. Anderson and Neddermeyer's announcement jogged a few memories. Oppenheimer pulled Yukawa's article off the shelf where he had put it two years before, and with Serber, sent a note about it to the *Physical Review*.[79] Reversing himself, Oppenheimer argued that the new particle did indeed exist, and, moreover, was Yukawa's U quantum, the particle that transmitted the nuclear force, although he and Serber did so with much hemming and hawing. (The hems and haws, as we shall see, were entirely justified.) The next issue of the journal included an article by Stueckelberg, who made the same identification in a more positive tone.[80] (Unlucky as ever, Stueckelberg had independently invented the U quantum, but was apparently talked out of it by Pauli. When he finally published, the paper was ignored.)[81] Other papers equating Yukawa's particle and the cosmic ray particle soon followed; Yukawa's shares soared on the physics bourse.

Yukawa's name for the new particle—the U quantum—never caught on, and a variety of bizarre terms began to circulate, including "penetron," "dynatron," "heavy electron," "X-particle," "barytron," "Yukawa particle," and even "yukon." Taking advantage of their status as discoverers, Anderson and Neddermeyer attempted to end the confusion by sending a letter to *Nature* proposing the name "mesoton" (*meso* is derived from the Greek for "middle"). Anderson's mentor, Robert Millikan, soon objected. Years later, Anderson recalled,

At the time, Millikan was away, and after his return we showed him a copy of our note to Nature. *He immediately reacted unfavorably and said the name should be mesotron. He said to consider the terms* electron *and* neutron. *I said to consider the term* proton. *Neddermeyer and I sent off the* r *in a cable to* Nature. *Fortunately or not, the* r *arrived in time, and the article appeared containing the word* mesotron. *Neither Neddermeyer nor I liked the word, nor did anyone else that I know of.*[82]

Because of the general lack of enthusiasm for "mesotron," the matter was put to a vote the next year at an international meeting of cosmic ray researchers in Chicago. Millikan, a classicist manqué, stridently complained about the popular contraction *meson*, because it is an ordinary Greek word meaning "middle one."[83] No name received a plurality, although *mesotron* and *meson* led the pack.[84] Confusingly, scientists used both, although after the war *meson* eventually prevailed.

For a brief, shining instant—a few months at the end of 1937—cosmic rays made sense. The soft component was comprised of electrons that followed the laws of quantum electrodynamics, whereas the hard component consisted of mesotrons whose actions were described by an emerging branch of physics called "mesotron theory" or "meson theory." The state of the art of meson theory was presented in a lengthy paper by Heisenberg and Hans Euler in 1938.[85] They said that mesotrons are created at high altitudes by collisions between the still-mysterious incident cosmic radiation and the molecules of the atmosphere. A fraction of the mesotrons reach sea level, forming the hard component observed five years before. En route, some of the hard component interacts with nearby atoms, producing cascades of electrons—the soft component.

Most important, physicists believed that mesotron theory would account for the interactions among protons and neutrons, as quantum electrodynamics had taken care of interactions among electrons and photons. Just as the photon was the quantum of electromagnetism, the mesotron was the quantum of the force that held the nucleus together. If one chose to ignore the divergences in quantum electrodynamics—and many did in 1937—it was easy to believe once more that a grand synthesis accounting for all known particles and forces was not far off.

Einstein is famous for the remark that the Creator is subtle, but not malicious. Many physicists think this true, but their belief was sorely tested in the 1940s, when the hoped-for triumph of mesotron physics came apart in their hands. Again, cosmic rays were not what they seemed. Far from being Yukawa's particle, the mesotron was an undreamed-of mystery now called the *muon*. The mesotron was not the key to the nucleus, not the solution to beta decay; it was what Oppenheimer, in a bitter tribute to nature's un-kindness, called a "ten-year joke."[86]

The joke was exposed under dramatic circumstances. In July 1938, Mussolini passed "racial laws" that prohibited Jews from holding such gov-ernmental positions as university professorships. Stripped of his job, the physicist Bruno Rossi left in October, and after some turbulence was offered a position at the University of Chicago by an old friend, Compton.[87] Bohr broke the traditional secrecy surrounding the Nobel Prize to tell Fermi that he was to be the next recipient, allowing Fermi, whose wife was Jewish, to plan his escape. After collecting his award in Stockholm, Fermi went directly to Columbia University.

The Italian physics community was devastated by the departure of Fermi, Rossi, and practically every other good physicist in the country.[88] As war approached, Edoardo Amaldi, the last remaining professor of physics at Fermi's Istituto Guglielmo Marconi, realized that if physics was to continue, he would have to band the remaining researchers into a single group. In that small, depressed group were Marcello Conversi and Oreste Piccioni, two young men who viewed themselves with the arrogance of their years as part of a "new generation" of physicists. Priding themselves on their skills with vacuum tubes and electronic components, they scorned the older genera-tion, which venerated the arts of glassblowing and carpentry.[89]

They knew of the mesotron theory, and also knew that it had to be checked with precise measurements of the properties of the new particle. Believing that previous attempts to ascertain the mesotron's lifetime were flawed by inadequate electronics, Conversi and Piccioni threw themselves with almost religious zeal into the task of producing a circuit that could measure time differences to a ten-millionth of a second. Though one of the glassblowing older generation, Amaldi gave them whatever help he could.

Conversi avoided conscription due to amblyopia of his left eye; Piccioni was drafted but managed to be stationed in Rome; Amaldi gave lectures at six-thirty in the morning so that physicists in the service could attend before boot camp. At night, Conversi and Piccioni bought RCA tubes on the black market and started to work on their circuits, building the fastest electronic circuits on earth from scratch and stolen wire.[90] Their first workshop was on the university campus, where they could collaborate with their colleagues.

But the university was located alongside an important military target, the San Lorenzo freight station. After the Allied invasion of Sicily in July 1943, American warplanes appeared above the Palatine hill for the first time. Dozens of bombs rent the university campus. Conversi and Piccioni spent the next day carrying boxes of resistors and oscilloscopes to a deserted high school close to the Vatican, where the aura of holiness and political neutrality offered some measure of protection to those nearby. Working between air raids in a city on the point of starvation, they rebuilt their equipment in a basement used by the antifascist resistance as a weapons cache. They watched over underground transmitters and guarded rifles; grateful resistance leaders contributed scavenged tubes.

Early in September, with the Allies in Calabria, the Italian government fell and the Nazi occupation began. The two men pressed on—"Our work was the only pleasure we had," Piccioni explained—constantly afraid of being caught by the Germans.[91] (He was, once, but was ransomed by a friend's father for a pile of silk stockings.)[92] They stacked up counters and layers of metal in a score of arrangements to determine how much matter it would take to stop a mesotron. Just before the Allied liberation of Rome in June 1944 they showed that mesotrons lived slightly more than 2.2 microseconds before falling apart, a short lifetime but one that was still many times longer than Yukawa had predicted.[93] In their basement in the ruined city, Conversi and Piccioni realized that something was amiss.

At the cessation of hostilities, Conversi and Piccioni were joined by another young man, Ettore Pancini, whose nascent physics career had been interrupted by the necessity of getting out of the fascist army and into the partisan groups in the north. The three men planned a new experiment to study the puzzling mesotron. If the particle discovered by Anderson and Neddermeyer was indeed the particle in Yukawa's theory, the positive and negative mesotrons should behave differently when they were slammed into matter. The positive mesotrons should be repelled from the positive nuclei by the electromagnetic force, but attracted by the nuclear force; because electromagnetism has a long range, it would win out, and the particles pushed away and left to decay. On the other hand, negative mesotrons should be attracted by both forces, and would in almost all cases be absorbed by nuclei before having a chance to decay.[94] Using a set of magnetic "lenses," the three experimenters deflected cosmic ray particles into a carbon rod. In the carbon, all the positive mesons decayed with the usual rate—but so did many of the negatives. In other words, the negative mesotrons were not absorbed by the nuclei, but captured and put into orbit until they decayed.[95] The three men knew that Yukawa's nuclear force particle should plummet directly into the nucleus; the mesotron was therefore not the transmitter of the nuclear force.[96]

In the summer of 1939, as war loomed, the English government sent a group of physicists from Manchester to man a secret radar installation, one of several that had been built along the eastern and southern coasts.[97] Among the scientists was George Rochester, who had switched to cosmic ray work at the instigation of P. M. S. Blackett, a Rutherford student who did not share his master's predilection for the small and the cheap. Blackett had brought to Manchester a cloud chamber with an electromagnet that weighed eleven tons.

Recently we spoke with Rochester, now an emeritus professor at the University of Durham, in northern England, about that hectic time at the dawn of particle physics. He hadn't lasted long at the radar station, he said. They wanted physicists for other purposes. A slightly built man with a mild round face, Rochester was dressed in warm tweeds against the cold day. He showed us the Durham cathedral, which was visible from his office window. High, stately, immaculately preserved, the church had never been bombed; the entire north had suffered little from the air war. "At some point," Rochester said, "the Manpower Commission decided that scientists would be needed as the war effort went on. They'd made a frightful mistake in the First World War by pushing everyone out onto the front, and lots of the best young scientists were killed. So they sent me back to Manchester to run the honors school in physics.

"Now, each large establishment with a lot of buildings, like a factory or a university, was asked to form its own fire brigade, because the city fire brigade would have too much to do in the event of a real air raid. I led the university squad every sixth day and on many weekends. As there were few air raids on the city of Manchester, we had much spare time. Also on duty, but in Civil Defense, was a Hungarian physicist called Lajos Jánossy, whom Blackett had brought to Manchester. And whilst most of the other people on duty played bridge and billiards, Lajos and I talked about cosmic ray physics the whole time. Oh, there were training periods, and I would get my squad out and we'd run them up some ladders—that was quite a lot of fun—but then Jánossy and I would talk about the creation of mesotrons. Jánossy had already come to the conclusion that mesotrons must be created in large explosions—nuclear events, not electron showers—and had devised a method of selecting such events by a clever disposition of Geiger counters in a great mass of fifteen tons of lead. The lead cut out spurious events, like large electron showers." After 1940, they investigated what they called "penetrating showers," showing that they were produced by protons and neutrons, rather than photons. "All sorts of strange things appeared," Rochester said. "Penetrating charged particles, strange deflections in the gas—but we could not identify the particles because we had no magnetic field. We weren't allowed to use the Blackett magnet because it took too much power." Rochester and Jánossy stared at the cloud chamber as if they were watching a movie with the sound switched off, certain that something fascinating was happening but helpless to tell what it was.[98]

Japanese physicists, too, were conscripted for the war effort. Nishina led the Japanese drive to build an atomic bomb, which was severely hampered by material shortages and the conviction that the thing could not be built.[99] Although the Japanese had no uranium and few of the requisite technological skills, Nishina struggled to put together a bomb until the destruction of Hiroshima announced that the conflict was over. Tomonaga was employed by the Imperial Navy to develop microwave techniques for radar. Solitary Hideki Yukawa, a man with pacifistic leanings, avoided military duty and spent the war doing what he had done in peacetime, commuting between his teaching positions in Kyoto and Tokyo. The trains were often halted by bombing raids, and Yukawa would don a helmet and shoulder pads and quietly walk along the tracks to the next station.[100]

The last issue of the *Physical Review* entered Japan in July 1940; thereafter, the fledgling group of Japanese theorists was completely cut off from their Western confreres.[101] As the war exhausted the country, some scientists ran out of paper to write on, and were forced to calculate on the backs of previously used scraps.[102] Even before Pearl Harbor, most cosmic ray workers had realized that something was wrong with the mesotron: It wasn't absorbed easily enough and it lived too long.[103] The emerging discrepancies worried Yukawa, and he organized a "Meson Club" in Osaka to work on the problems.[104] In June of 1942, one club member, Shoichi Sakata, made the simple and ultimately correct proposal that there were *two* mesotrons, the one that experimenters saw in cloud chambers and the one that Yukawa had predicted, which Sakata assumed had not yet been seen. He wrote up the suggestion in Japanese that year, but it took a long time to reach the rest of the world.[105]

An atomic bomb fell on Hiroshima. Terrified by the devastation, the Army command asked Yoshio Nishina, Yukawa's mentor, to tour Hiroshima to decide if Truman's claim that there was an atom bomb was correct. Nishina arrived the day after the bomb into a scene that no human being has ever described successfully. With shocked calm, he went about the macabre business of estimating the size of the blast by inspecting the burns on corpses. In a smashed hospital he found some unused photographic plates for X rays. Like Becquerel a half-century before, Nishina developed them. They were black from radiation. He used a Geiger counter to test the bones of the dead men in the street. If this deeply patriotic man was humiliated that he had lost the race to build the bomb, he never talked about it. He told the American investigators that they should study the mutations in the area. He never admitted trying to make such a device himself.[106]

During the American occupation, Japanese physicists were closely supervised, access to journals was limited, and there was little chance for dialogue with foreign scientists. Yukawa managed to start publishing an English-language journal, *Progress in Theoretical Physics*, on the brownish spotted paper that was all that he could scavenge. The first issue contained both Tomonaga's demonstration of the renormalizability of quantum electrodynamics and a paper by Sakata and a colleague on the two-meson hypothesis.[107] The journal did not arrive in the United States until December 1947.[108] But by then the Americans had figured both ideas out for themselves.

In addition to the experiments on self-energy, the participants at the Shelter Island conference in June of 1947 spent a day debating the findings of Conversi, Pancini, and Piccioni in Italy. At the height of the discussion, Victor Weisskopf proposed the rather peculiar idea that an encounter with a cosmic ray might somehow make a proton or neutron "pregnant," capable of emitting mesotrons; the period of gestation, he suggested, might explain some of the discrepancies involving the lifetime of the particle.[109] The awkwardness of this solution prompted Robert Marshak, a theorist from Cornell, to make the simplest explanation he could think of, namely, that there are two kinds of mesotrons with different masses. One is produced in the upper atmosphere by cosmic ray collisions, while the other results from the decay of the first.

Marshak was pleased with the reception to his idea at Shelter Island, but was startled upon his return from the conference to find that the most recent issue of *Nature* contained a photograph of precisely the process he had described.[110] The picture had been taken by a team of four British experimenters with a specially developed photographic emulsion that was sensitive to most of the charged particles in cosmic rays. The resultant thin black lines could be measured in exactly the same way as the tracks in a cloud chamber, with the blackening and scattering of each track providing clues to the particle's mass and energy. The British had exposed many emulsions in this way, and on one of them, a mesotron had slowed to a halt in the middle of the film, whereupon it decayed into another, somewhat lighter particle. Marshak immediately set to work on a new paper, enlisting Hans Bethe's help because of the latter's extensive knowledge of cosmic rays.[111]

The first and heavier of the two particles is today called the pion or pi-meson. It corresponds to the Yukawa particle, and is exchanged by the protons and neutrons in the nucleus. It is about 273 times heavier than the electron and has a lifetime of about 10^{-8} seconds, roughly what Yukawa estimated it to be. Experimenters were shortly to discover three pions: positive, negative, and neutral.

The second and slightly lighter particle, created in the decay of the pion, is an entirely different matter. It is the particle Anderson and Neddermeyer discovered. It plays no role in the nuclear interaction, nor in any process essential to ordinary matter. It is about 207 times heavier than the electron; its accidental similarity in mass to the pion created the confusion between the two particles Oppenheimer described as a "ten-year joke." This second particle is nothing more, nothing less, than a plump, short-lived version of an electron—a second electron, that exists for no known reason and has no known function. Unpredicted by theory, unneeded by nature, the muon, as it is now known, turns into an electron a millionth of a second after it is made. It may not last long, physicists thought, but why is it there at all? I. I. Rabi summed up the general feeling of bafflement in the physics community with the snarl, "Who ordered *that?*"[112]

The muon was just the first surprise. After Rochester and C. C. Butler in Manchester finally were allowed to switch on the Blackett magnet, they began to photograph the so-called mesotron showers in a new cloud chamber. With this setup, they found, in October 1946, the traces of what was apparently a new particle. They happened on a second set of tracks seven months later. The first was interpreted as the decay of a neutral particle; the second was charged; both were about a thousand times the mass of the electron. Because both particles made forklike tracks in the cloud chambers when they decayed, they were christened "V particles."[113] Many scientists were initially inclined to dismiss the two photographs as the sort of glitch that occurs in experiments; before endorsing the existence of new states of matter, they waited for Rochester and Butler to produce some more of them.[114]

"Unfortunately," Rochester said later, "although the dear Lord was very kind to us in sending two of these, on the other hand He was extremely unkind in sending no more than two."[115] The two men spent another year unsuccessfully trying to snare more V particles, and growing ever more embarrassed when none showed up. In the summer of 1948, Rochester crossed the Atlantic to attend a conference in honor of Robert Millikan's eightieth birthday. He spoke about the V particles to a polite but unenthusiastic audience. "Then," he said to us, "at Pasadena I met Carl Anderson. And Carl was extremely interested, and after this conference Carl put his best chamber *straight on* to this problem on top of the White Mountain." At such an altitude, V particles, if they existed, should materialize much more often because there were many more cosmic rays to make them. To

Rochester's considerable relief, he and Blackett received a letter from Anderson saying that he had turned up "about 30" V particles, and that he agreed with their analysis.[116]

The discovery of the V particles, like most scientific discoveries, prompted those who missed it to reexamine their evidence. Sure enough, earlier photographs of high-energy cosmic ray showers were found to be sprinkled with V particle tracks. The discovery of yet another exotic type of matter—besides the pion and the troublesome muon, sent researchers in Europe, Japan, and the United States to the heights. By the beginning of the 1950s, cosmic ray laboratories dotted mountaintops across the globe. Conditions were rugged when they were not impossible, and experimenters had to stay for weeks at a time in tiny mountain huts. Nonetheless, some physicists found a certain Hemingwayesque thrill in their splendid isolation. Louis Leprince-Ringuet, a French physicist with the aristocratic mien of a nineteenth-century explorer, spent years eleven thousand feet up the frigid Aiguille du Midi, snowbound in tiny rooms stocked with scientific equipment and emergency brandy. There is a certain Gallic fluidity to his account of the virtues of high-altitude life:

> The work goes on day and night, without stopping, without the slightest disturbance, without any of the quotidian annoyances of ordinary life: in such splendid isolation, one can do a year's work in a month. All the while the storm blows ceaselessly; for two solid weeks, we could not step over the threshold of the door, remaining shut up in the snow, unable to see out the window, for days on end. But after our seclusion was over, we had the rapturous pleasure of putting on our skis and darting, heedless of the crevasses, across the fine, crystal-sparkling powder snow of the heights. On beautiful days, the mornings . . . were magnificent: there was the mist rising up from the plains, and the perfectly calm and level sea of clouds, real seas whose billows were set and animated by a slow rocking motion, creamy bluish seas whose depths were tinted pink by evening. The great peaks were simply parts of our world; Cervin, Mont Rose, and Mont Blanc seemed so close—almost near enough to touch—that the rest of the earth seemed to collapse like Atlantis beneath the sea of clouds.[117]

Others, however, were less entranced, as we discovered when we talked to Antonino Zichichi, an Italian experimenter with a photogenic mane of white hair. In 1953, when he was twenty-four, Zichichi was sent from Amaldi's institute in Rome to the Jungfraujoch in the Swiss Alps.[118] On his arrival in the wasteland of wind and snow, Zichichi was greeted by the lab supervisor, who introduced himself as Hans Wiederkehr. A tyrannical sort, Wiederkehr kept the laboratory functioning at the price of total obedience from the intimidated physicists in his charge. When Zichichi first arrived, Wiederkehr pointed to an exit and explained at considerable length that it should never be used. Zichichi said that he could read and that there was a large *DO NOT EXIT* sign on the door. Grumpily, Wiederkehr indicated that the new arrival was to accompany him to the library, a splendid wood-paneled room with a large bay window that overlooks a glacier. Journals, charts, and maps rested in cozy disorder on the shelves, hauled up by an ancient funicular from the ski town below.[119] Wiederkehr indicated a gallery of neatly framed black-and-white portraits on a side wall. The subjects were physicists who had worked at the laboratory. The conversation, in Zichichi's recollection, went as follows:

WIEDERKEHR (snarling): *See these gentlemen? They're all dead.*
ZICHICHI: *What happened?*
WIEDERKEHR: *They didn't* obey the signs!

"And then," Zichichi said, "he explained to me that one physicist a year died on the average there. In fact, when I was there my group leader, Dr. [Anthony] Newth from Imperial College, went out one day—one nice day—on his skis. And he fell into a crevasse. He survived because at 5:30 Mr. Wiederkehr said, 'Where is Dr. Newth?' So he went out, followed the tracks, and pulled him out of the crevasse. Dr. Newth told me he thought he was dead. I was absolutely terrified! One physicist a year dying, on the average! And all their pictures were in the library. And the incredible thing is—" he leaned closer "—nobody understood a thing they were finding in all that time!"[120]

□ □ □ □ □

Between 1947 and 1953, cosmic ray researchers in a score of cramped mountain laboratories discovered a whole new world of physics. Opening and closing the pistons of their cloud chambers or the switches of their electronic counters, they photographed the tracks of muons, pions, and V particles; beyond that, they found half a dozen *more* particles, all born from the evanescent shimmer of high-energy cosmic rays. The most interesting photographs were made by standing up parallel sheets of photographic film and letting cosmic rays shoot through them, leaving little white trails in their passage. As more people worked with this method, they began to find— inevitably, perhaps—rare, spectacular events. (In the jargon, a particle interaction is referred to as an "event.") Scratched across the film, the lines showed *something* that broke up into three widely separated pions; piecing together another series of photographs, a second set of researchers might see that an unknown entity had disintegrated into a scatter of muons and electrons; yet a third might find that a nucleus in the film emulsion had apparently been hit by a cosmic ray particle, creating a starlike splash of light with subatomic tracers flying away in mysterious profusion. Baffled by the unexpected profusion of particles and decays, physicists might apply different names to one particle, misidentify a second, and deny the existence of a third—only to reverse themselves when they saw data from other laboratories. Within five years of the first observation of V particles, there were thought to be half a dozen subcategories for V particles and a sprinkle of objects loosely called theta, tau, and K particles—all known only from a few hundred photographs and all of which might be anything or nothing. Each lab had only a few examples of every type of event, and their piecemeal publication in widely scattered journals made it impossible to gather a global picture. As a result, nobody knew what they were seeing or why it was there; they did not know how they should refer to it or even if anyone else had seen the same thing. Physics here was completely in the hands of the experimenters; the theorists who had been so confident after Shelter Island were now utterly lost, reduced to vague supposition and unsupported guesswork.

Among the many mysteries, the puzzle of the V particles stood out most sharply. Found at the beginning of the new era, these particles seemed to be abundantly produced, once there was enough energy, which suggested that the force responsible for their existence was powerful. A strong force's strength is demonstrated by the likelihood it will manifest its effects, and the plethora of newly formed particles therefore testified to the might of the force creating them. Such a powerful force should cause the V particles to decay quickly—"easy come, easy go," is the rule of thumb. By looking at cloud chamber photographs, physicists were able to make guesstimates of the time that it should take this strong force to pull the V particles apart. The calculations were somewhat awry: V particles had a lifetime *ten trillion* times longer than expected.[121] When a theoretical prediction is off by thirteen orders of magnitude, it is an indication that something strange is in the mix.

The riddle's solution lay in an old idea called *isotopic spin*. The concept of isotopic spin, although not the name, was invented by Heisenberg as an almost extraneous sidelight to his theory of the nuclear force. As a matter of convenience, Heisenberg imagined the proton and the neutron to be different states of the same particle, in somewhat the way that the atomic isobars carbon-12 and nitrogen-12 have the same mass but different charges. (By right, the name should really be "isobaric spin," not "isotopic spin," because the reigning metaphor has nothing to do with atomic isotopes, such as carbon-12 and carbon-14, which have different mass but the same charge.) To describe this relationship formally, Heisenberg proposed that each particle be thought of as spinning like a top in some imaginary space. If the axis of spin points up, the particle has a positive charge and is a proton; if the axis points down, it has no charge and is a neutron. Heisenberg said that the proton and the neutron could then be assigned the same value—½—of a hypothetical quantity later baptized isotopic spin.[122] One recalls that ordinary spin, too, is associated with a direction, and thus an electron with spin ½ has an axis of spin that is indicated by saying that its spin orientation (S_z in the diagram) is $+\frac{1}{2}$ or $-\frac{1}{2}$. Similarly, isotopic spin is associated with a direction, and thus the upward proton is said to have an isotopic spin orientation (I_z in the diagram) of $+\frac{1}{2}$; the neutron, $-\frac{1}{2}$.[123]

As put forth by Heisenberg, isotopic spin was not only complicated and confusing, but also of little real use; most of his colleagues paid little attention. Four years later, however, a single issue of the *Physical Review* con-

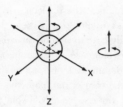

An upward axis of spin

Ordinary Spin	Particle	Orientation of Spin in Magnetic Field (technically known as S_z because it is the z axis on graph.)	
$S = \frac{1}{2}$	Electron	e^- $S_z = +\frac{1}{2}$	e^- $S_z = -\frac{1}{2}$

Isotopic Spin	Particle	Orientation of Isotopic Spin in Imaginary Charge Space (technically known as I_z, as above)		
$I = \frac{1}{2}$	Nucleon	p $I_z = +\frac{1}{2}$	n $I_z = -\frac{1}{2}$	
$I = 1$	Yukawa particle, or pion	π^+ $I_z = +\frac{1}{2}$	π^0 $I_z = 0$	π^- $I_z = -\frac{1}{2}$

ORDINARY SPIN AND ISOTOPIC SPIN

tained three papers that dramatically resuscitated the idea. The first gave an experimental value for the force that clamped together protons in the nucleus; the second analyzed the data to show that the force was equal between two protons and a proton and a neutron; the third used the first two as an excuse to relaunch isotopic spin, arguing that the nuclear force, whatever it was, could not "see" the difference between the proton and neutron, and that therefore they should be regarded as two states of one particle called the *nucleon*.[124] In 1938, one of Pauli's assistants, a Russian emigré named Nicholas Kemmer, extended the idea past the nucleon. He showed that if one were to take seriously Yukawa's notion of a particle that transmitted the nuclear force, one would actually have to assume the existence of three such entities—three pions, in today's language—that could be regarded as states of one particle, with three different axes of isotopic spin: up, down, and sideways.[125] (Note that, as in the case of the pion, the isotopic spin of the particle may have a value of 1, whereas its axis of spin may be some other number, such as 0.)

In an important paper, Oppenheimer and Serber then postulated that the total amount of isotopic spin stays the same in particle interactions, in the same way that the total energy, momentum, and electric charge do not change.[126] All four are said to be "conserved." The difference is that the conservation of isotopic spin is approximate; it is conserved only in interactions involving the nuclear force, whereas energy, momentum, and electric charge are conserved no matter what.

A hitherto unknown law of nature had shown its presence, but briefly, like a flash at the corner of an eye. Enter Abraham Pais, vigorously. Pais was a Dutchman then at the Institute for Advanced Study. In a remarkable talk at a 1953 conference commemorating the hundredth birthday of Hendrik Lorentz, Pais indicated the course that elementary particle physics would follow for the next two decades.[127] Pais was thirty-five; much of his twenties, ordinarily among the most productive years of a physicist's life, had been spent hiding from the Germans in Amsterdam. Eventually he was caught and put into a Gestapo prison. He was lucky enough to survive. Earlier than most Occidental theorists, he had the hunch that the new cosmic ray particles were important.[128] In the years before the Lorentz conference, Pais and several Japanese had come independently to a conclusion that now seems baldly obvious: The creation and decay of V particles are not attributable to the same agent. Unable to think of a real reason why this should occur, Pais complained about the lack of clues: "The search for ordering principles may indeed ultimately have to be likened to a chemist's attempts to build up the periodic system if he were given only a dozen odd elements."[129] Speaking at the Lorentz conference, he began by listing "a number of questions in meson physics that seem to be of outstanding interest."

The first question concerns the isotopic spin. Ever since it became clear that proton and neutron can transform into each other and in many instances are exchangeable in nuclear systems, we have been faced with the question whether a theoretical foundation could be given for the fact that these two particles seem to behave like different states of one entity now called [the] nucleon. The formal shorthand of isotopic spin takes cognizance of the situation but explains nothing of course.[130]

He then argued, in general and by example, that great discoveries were yet to be made, for new variables like isotopic spin would be the key to the study of both the nuclear forces and the cosmic ray particles. More such intrinsically conserved qualities should be sought, and their description merited the introduction of new quantum numbers, the first in twenty-five years. (Using today's language, Pais was the first to urge the search for higher symmetries.) By pushing and pulling these properties, twisting them into mathematical knots, one could rank the elementary particle interactions into a hierarchy and peer inside their inner workings.

Although encumbered by a half-dozen other ideas and a wholly incorrect model, a program for the next generation lies swaddled within the nervous evidence of Pais's creativity. The paper had little direct influence, however, because it was almost immediately superseded by a concrete example of the kind of inquiry Pais advocated, in the form of a new quantum number proposed by a twenty-two-year-old named Murray Gell-Mann, who was then beginning a meteoric ascent into prominence. A young man who gave a rich new layer of meaning to the term "brash," Gell-Mann knew almost nothing about Pais's ideas; he had been led to the same considerations independently.[131] Whereas Pais's style was more discursive and conversational, Gell-Mann was terse, oblique, clipped. One article was a hint, the other an announcement: Another era had begun.

Gell-Mann's paper—"Isotopic Spin and New Unstable Particles"—is remarkable, a pocket symphony in the modern style. It begins with a quick fanfare:

[L]et us suppose that both "ordinary particles" (nucleons and pions) and "new unstable particles" [here a list of V particles] have interactions of three kinds:

(i) *Interactions that rigorously conserve isotopic spin. (We assume these to be strong.)*

(ii) *Electromagnetic interactions. (Let us include mass-difference effects in this category.)*

(iii) *Other charge-dependent interactions, which we take to be very weak.*

This music, familiar to the cognoscenti, had never sounded so clearly. In modern terms, one speaks of a hierarchy of interactions: *electromagnetism*, which binds the electrons around the nucleus, conserves electric charge, and is fully described by a field theory, quantum electrodynamics; the *strong force*, which holds the nucleus together, conserves isotopic spin, was thought to be transmitted by pions, and had no theory to describe it whatsoever; the *weak force*, which causes particles to decay into lighter particles, does not seem to conserve anything at all, was not known to be transmitted by any particle, and was imperfectly covered by Enrico Fermi's theory of beta decay; and *gravity*, which holds planets around suns and suns in galaxies, is fully described by the theory of general relativity, plays no role on the subatomic level, and until recently seemed unreconcilable with quantum mechanics. Today, electromagnetism, the strong and weak interactions, and gravity are thought to be the basic forces of the Universe, from which all others derive.

This ordering of forces was matched at about the same time by an ordering of the particles they affected. Particles that feel the effects of the strong force are now known as *hadrons*. Hadrons in turn are split into two subcategories, the heavy *baryons*, like the proton and neutron, and the lighter *mesons*, like the pion. Those particles, such as the neutrino and electron, which do not feel the effects of the strong force, are called *leptons*.[132] Four interactions for leptons and hadrons: the cosmos in a clause.

Having explicitly laid out the ladder of interactions, Gell-Mann considers papers by Pais and another theorist, David Peaslee.[133] In connection with their work, Gell-Mann writes, "The author would like to put forward an alternative hypothesis that he has considered for some time. . . ." The entering theme is deceptively modest: Assume that the V particles have an isotopic spin half a unit different from that of ordinary particles. If the nucleon and pions are set at ½ and 1, respectively, then the heavy and light V particles are 1 and ½, respectively.

Now Gell-Mann laces together the initial brassy chords and the theme: The stronger interaction of type (i) can produce V particles with isotopic spins of, say, +1 or −1, only if it makes *two* of them at once. Because (−1) + (+1) = 0, no new isotopic spin is created, and the net isotopic spin is unchanged. And because the heavy and light V particles are ½ a unit off the neutron and proton, there is no way the strong force can make Vs into nucleons without changing the net isotopic spin by ½—which cannot occur. Therefore, Gell-Mann notes, "[A]s far as (i) is concerned, the decay of new unstable particles into ordinary ones is forbidden. . . . Only interactions of type (iii), which do not respect isotopic spin at all, can lead to decay." Shifting the isotopic spin values neatly solved the riddle of the V particles. Simple addition and subtraction showed how particles could be created by a powerful force that was unable to make them decay.

Simultaneously, Gell-Mann wrote another paper that cut more deeply into the V particles.[134] Coming across the manuscript now is like reading the private drafts of an author, such as Mark Twain, who was chary of revealing his most ardent beliefs; widely circulated and widely influential, the article was held back by its author until it was obsolete, and eventually never published. An important idea thus slipped in quietly, through the back door, for in that paper Gell-Mann set out what is in effect a formula, good for any interaction or single particle, that reads, in its modern form,

$$Q = I_z + B/2 + S/2$$

where Q is the electric charge, I_z the axis of isotopic spin, B the number of baryons, and S a new quantum number, to which Gell-Mann attached the whimsical and provocative label of "strangeness." Particles with non-zero values of S thus became "strange particles." Strange particles are V particles, that is, those new particles with odd values of isotonic spin. (The same idea was developed independently by a Japanese physicist, Kazuhiko Nishijima.)[135]

Strange indeed. Strangeness was a wholly new phenomenon man-ifested only in the particles created at high energies, although the presence of Q indicated its links with old properties such as charge. The formula was couched exclusively in terms of elementary particle properties, without ref-erence to atoms or nuclei. It accounted concisely for the observed data, but stood mute on their causes. It told physicists that they had reached a plateau, and that there was much to learn. It marked the full emergence of a new discipline: elementary particle physics.

We once asked Abraham Pais to think back to the early 1950s, when humanity first saw what at this moment appears to be the shape of the fundamental constituents of the Universe. What would the physicists of that era have thought if somebody had told them that thirty years later there would be the first attempts at a unification theory? Would that have shocked them? "Oh, absolutely!" he said. There was a nostalgic softness in his tone. "It was a *wonderful* mess at that time. Wonderful! Just great! It was so *confusing*—physics at its best, when everything is confused and you know something important lies around the corner."[136]

In the summer of 1953, cosmic ray workers from all over the world gathered in Bagnères-de-Bigorre, a small Basque town in the foothills of the Pyrenees a few thousand dry feet beneath the Pic du Midi laboratory. Com-pared to today's conferences, which are split up into sub-sub-categories, eagerly covered by the press, and attended by thousands of researchers and job-hunting graduate students, the gathering at Bagnères was a modest affair, a matter of a hundred and fifty cosmic ray enthusiasts, *vin ordinaire*, and cots in the town schoolroom. It was the first time these cosmic ray experimenters had been in one place since the war, the first time many of these scientists felt certain that they were not the only ones seeing peculiar stuff in their cloud chambers. Moreover, these Vs and Ks and thetas—whatever they were, and whatever they should be called—were not flukes or unimportant details, but something of vital interest.

Many of the experimenters at Bagnères remember the conference as one of the greatest ever held. They came to take stock of where they had been; they left, six days later, with the feeling that the physics of elementary particles was a scientific domain every bit as rich with implication as the study of the nucleus. Committees were organized to classify the known particles and to lay down the rules for the nomenclature of future discoveries. Greek letters were assigned to each new particle: small if light, capital if heavy. The conventions were published simultaneously in journals across the globe, and the Bagnères attendees spoke excitedly of the onslaught of new particles to come.[137]

The new particles were discovered, but not by cosmic ray physicists. At Bagnères, an experimenter named Marcel Schein created a sensation with the report that a new and enormous machine in Chicago called a cyclotron, one of the first modern particle accelerators, had successfully thrown a spray of protons at a metal plate with enough power to create a burst of V particles. (Schein's claim later turned out to be mistaken.) The conference members realized that accelerators in laboratories could easily match the energies of cosmic rays in the upper atmosphere. No longer would experimenters have to seek exotic forms of matter on mountaintops. They would be made in the tens of thousands, spat out from machines like so many cans of soup. At the conference closing, Leprince-Ringuet exhorted his fellows in the common room of the Bagnères schoolhouse. "We must hurry," he cried.

We must run without slackening our pace: we are being pursued—pursued by the machines! . . . We are—I think—a little in the position of a group of mountain climbers climbing a mountain. The mountain is very high, maybe almost indefinitely high, and we are scaling it in ever more difficult conditions. But we cannot stop to rest, for, coming from below, beneath us, surges an ocean, a flood, a deluge that keeps rising higher, forcing us ever upward. The situation is obviously uncomfortable, but isn't it marvelously lively and interesting?[138]

It was too late. To explore the new realms of matter found in cosmic rays, cosmic rays themselves would be virtually abandoned by physicists. The euphoria at Bagnères was the high-water mark of cosmic ray physics; although the town dedicated Place Hyperon in honor of a cosmic ray particle, the scientific interlude that began at the beginning of the century with leaky electroscopes had come to a close.

III

The Weak Force

10
Symmetry

AT HALF PAST FOUR IN THE AFTERNOON OF DECEMBER 10, 1979, THREE men—Sheldon Glashow, Abdus Salam, and Steven Weinberg—entered the auditorium of the Stockholm Concert Hall to a flourish of trumpets. They were to be awarded the Nobel Prize for physics, for their construction of a single theory incorporating weak and electromagnetic interactions, an achievement that began, almost unnoticed, in the years following the discovery of the strange particles. The laureates walked down the wide aisle—Glashow and Weinberg in tails and studs; Salam in full Pakistani formal regalia, including shoes with toes that curled several painful inches into the air—and onto a platform, where they were introduced and individually extolled by a middle-aged member of the Swedish Academy of Sciences. At the end of each peroration, the two thousand people in the audience applauded as the prizewinner walked to center stage to meet King Gustav XVI, who presided over the ceremony. Each physicist shook the king's hand and received a leather-bound diploma, a hefty gold medal, and a letter informing him of when and how to collect his share of the prize money.[1]

After the rite, the three physicists, the laureates in chemistry, medicine, literature, and economics, the royal family, and about half the people in the concert house were bundled into a fleet of chartered limousines and buses that conducted them through the bitter cold to Stockholm City Hall. (The Peace Prize is awarded in Oslo; the economics prize, which was established later, is technically the Nobel Memorial Prize.) A turn-of-the-century waterfront building constructed at lavish expense and frequently said by Swedes to be the Continent's finest, the City Hall features an enormous Gold Room lined with the largest single display of twenty-four-karat gold mosaics in the world. There the laureates were subjected to a series of toasts and pronunciamentoes and then led by the king onto a balcony overlooking yet another gigantic space, the Blue Room. The high bare walls of the Blue Room are composed of individually chiseled red bricks whose rough beauty convinced the original architect to change his mind about painting them blue, as had been planned. An ornate marble staircase debouches onto the gray floor of the room, which on Nobel night is jammed with students from the University of Stockholm, who have their own banquet. At the sight of the royal family and the rather stunned laureates, the students jumped to their feet and lustily bellowed a Swedish drinking song that nobody seemed to feel needed translation. Dancing commenced, and spread to the Golden Hall. After two o'clock, some students abducted the physicists for a Swedish ritual known as a *vickning*, a postprandial event featuring, mainly, more eating and drinking, the champagne and caviar now being replaced by schnapps and herring. At dawn, the revelers were released—but had to be at the bank early in the morning to pick up their checks.[2]

Winning a Nobel Prize involves a week of festivities that is the Swedish World Series, World Cup, and Olympics all in one. The laureates are shuttled back and forth among so many cities and so many elaborate dinners that the actual prize ceremony, impressive as it is, is often reduced to a splendid blur in the memory. At some point, however, each prizewinner is asked to give a formal address. The speeches by the awardees of the Peace Prize and the Prize for Literature often attract global attention; those given by the winners of the science prizes are usually ignored, because they commemorate discoveries that are already well-known in the research world. The speeches in 1979 were something of an exception, for the physics prize celebrated the most important advance in particle physics for a generation, and a step toward the full unification of nature that is the dream of every physicist.

Weinberg's speech, like those of Glashow and Salam, was a historical stroll through the path he had taken toward unification. He began in the simple, clear style that is the envy of his colleagues and the wellspring of a parallel career as a popularizer of science.

Our job in physics is to see things simply, to understand a great many complicated phenomena in a unified way, in terms of a few simple principles. At times, our efforts are illuminated by a brilliant experiment . . . But even in the dark times between experimental breakthroughs, there always continues a steady evolution of theoretical ideas, leading almost imperceptibly to changes in previous beliefs. In this talk, I want to discuss the development of two lines of thought in theoretical physics. One of them is the slow growth of our understanding of symmetry, and in particular, broken or hidden symmetry. The other is the old struggle to come to terms with the infinities in quantum field theories. To a remarkable degree, our present detailed theories of elementary particle interactions can be understood deductively, as consequences of symmetry principles and of a principle of renormalizability which is invoked to deal with the infinities.[3]

The importance of the criterion of renormalizability was appreciated in the late 1930s and early 1940s; the understanding of symmetry and its role in the Universe came afterward, and resulted in the current attempt to cover all of space and time in one theory.

Several years after his Nobel, we talked with Weinberg about renormalizability and symmetry. A tall man with wavy red hair exploding into gray curls at the back of his neck, Weinberg is now the head of the theory group of the department of physics at the University of Texas, in Austin. His dark eyes stand out in his pale, slightly flushed features. Half-moon reading glasses protrude from his shirt pocket. A certain measure of gravity, solidity, and dignity is present in his bearing; he speaks well, in slow, deep, concise sentences. On the right side of his office window is tangible evidence of the rewards of accomplishment in theoretical physics: ceremonial photographs of the Weinberg family with the king and queen of Sweden, Queen Beatrix of Holland, and Jimmy Carter, then president of the United States of America.

He hadn't been working there much, he said. Having the writer's habit of pacing while he thought, he preferred to do physics at home, where he could keep a television on his desk. It is a pleasing image: a man in his study, trying to strain sense through a screen of equations, *The Edge of Night* flickering unheeded on the desk as he asks himself, Now what is important in all this here?

We asked him how he had become involved in quantum field theory. "When I was a graduate student, that was the latter part of the fifties," he said. "It was a somewhat confused time, because there had been a great triumph of quantum field theory with the development of quantum electrodynamics—and then a complete failure at extending it beyond the area of quantum electrodynamics. I once took a look at Fermi's Silliman lectures, which were written in 1952, because I had to give the Silliman lectures in 1977, twenty-five years later. These are a lecture series at Yale. I looked back at what Fermi had said, and boy, was that depressing! Attempts made to understand the physics of nuclear forces, mostly based on conservation laws and guesswork and really weird ideas about weak interaction. It just didn't seem to be going anywhere. The one thing that seemed like a fantastic and indisputable accomplishment was the success of quantum electrodynamics. You could calculate till your eyes fell out of your head and you would get the right answer that agreed with experiment to umpteen decimal places. There was nothing else in physics like that. It was an ideal model for the way physics should be.

"By the way, I think in that respect I was not at all untypical. In most graduate schools in the 1950s, the people who were most ambitious at doing theoretical physics—not of a phenomenological kind but a fundamental kind—took quantum electrodynamics as their ideal. There was one thing, I suppose, that mattered more to me than it might have to other people. And that is that I was impressed with the *uniqueness* of quantum electrodynamics. You know, the theory had these infinities, and then it was discovered that the infinities would cancel, provided you took an appropriate definition of what you meant by mass and charge. The thing that fascinated me was the fact that the whole infinity cancellation would only work if the theory was limited to be just the original, simplest possible quantum electrodynamics. If you tried to muck around with quantum electrodynamics—

the way, for example, Pauli had suggested, adding additional terms—then the whole application of renormalization theory would break down. You just couldn't do it.

"Now, to many people, that just showed that renormalization theory was just—it *accidentally* worked in this special case, but didn't really mean that much, and you shouldn't take it that seriously, and perhaps you shouldn't take quantum field theory that seriously. But to me, that always seemed the most exciting thing there was. I never liked these arguments of simplicity. I don't think nature chooses simple equations just because they're simple. I think there are some simple principles underlying everything, and then the equations are just what they are. So merely saying that quantum electrodynamics is just the very simplest possible minimum theory never appealed to me. But here was a principle—the infinities had to cancel—that was a golden key that explained why the theory was the way it was. It was a tremendous expansion of the power of pure thought."

In other words, going from a criterion of simplicity to one of necessity?

"Right, right," Weinberg said. He went to the coffee dispenser and began pouring coffee. Beyond the window rose the hundreds of miles of Texas hill country, gentle green ridges oddly reminiscent of the foggy slopes of northern California. Weinberg said that one still looked for simple equations, but that this was not something to celebrate. "Now, Dirac had taken the attitude that—he just wrote down the equation [for the electron] that seemed to him the simplest, and that it was beautiful, and therefore that was why it was true. I've never liked that attitude. I find that very antithetic to the way I think about things. But renormalizability—obviously the theory, in order to make sense, had to have no infinities. That was a requirement that was reasonable to impose on a physical theory, and that seemed the key that would explain why the physics was the way it is."

One of Weinberg's first papers after his Ph.D. was a tough proof that dotted the last i's and crossed the last t's in the demonstration of the renormalizability of quantum electrodynamics.[4] Published in 1960, the article is still the only piece of Weinberg's oeuvre that he thinks would be regarded with respect by a professional mathematician. In fact, Weinberg asked a mathematically inclined physicist, Arthur Wightman of Princeton, to check through the argument; when Wightman handed the paper back, he commented that there was blood on every lemma. "I'm very proud of it, and it is often quoted," Weinberg said. An expression of amusement crossed his face. "I think it's very rarely read."

At about that time, Weinberg became explicitly interested in the question of symmetries, a subject that had been present in physics for many years but whose importance was not fully appreciated for much of that time.[5] Used loosely, symmetry means harmony or balance. Physicists and mathematicians, however, define the term more precisely. They say that something is symmetrical if one or more of its aspects is indifferent to a change. A rubber ball can be turned freely and its appearance won't be altered; therefore, as a physicist might say, the ball is "symmetric with respect to rotation." More important, space and time themselves have certain symmetries. Space, for example, has a kind of symmetry called by the clumsy name of "translational invariance," which means, simply enough, that space is exactly the same no matter where you are in it. The laws of physics are the same whether you are in New York or Nagasaki, the moon or the bottom of the ocean. Time has "time translational invariance," an even more cumbersome expression for an equally simple idea, namely, that no particular instant in time is inherently different from any other. In other words, physics is the same whether it is yesterday or tomorrow, a thousand years ago or a week from last Sunday. Because of these two symmetries of space and time, business executives in Toronto can know today that next week, when they attend a conference in Brussels, water will still flow downhill and their morning *café au lait* will not crawl out of its cup.

If translational invariance did not hold true, a rocket ship thundering through the heavens might suddenly slow down, losing its momentum, just because it had entered a different part of space. Conservation of momentum is, so to speak, wired directly into space, its presence implied by the symmetry of translational invariance. And if time translational invariance did not hold—that is, if the laws of physics could change from second to second—it would be impossible to assure that energy would be conserved. Similarly, the conservation of angular momentum is linked to the fact that it does not matter which way you turn an object around in space.

Newton implicitly based his laws on another kind of symmetry known as Galilean invariance, named after Galileo Galilei, from whose work it is derived.[6] Imagine you are a passenger in an automobile and that you are drinking a cup of coffee. The simple fact that the coffee stays in the cup when you are cruising along at a steady fifty-five miles per hour just as it does when you are stopped means that the behavior of physical laws does not depend on the velocity of the system to which they are applied. Strictly speaking, this is only true when the velocities are constant and do not change, as is shown by the fact that the coffee tends to slop out when you accelerate out of a stop light or brake to a halt.

Einstein's theory of special relativity stems from the recognition that Maxwell's equations say that the speed of light is always the same no matter how fast you are going. In other words, electrodynamical laws are symmetric with respect to the velocity of light. Unfortunately, it is mathematically impossible for the two symmetries of Maxwell and Newton to coexist. Einstein came up with a way of reconciling them, but only at the price of supposing—in flat contradiction to Galilean invariance—that the behavior of some physical properties such as mass and length *does* depend on the velocity of the system to which they are applied. The special effects predicted by relativity only show up at enormous velocities, which means that you can go right on assuming that the coffee in your morning cup will not be affected if you drive on the freeway. Because the equations were actually invented first by Hendrik Lorentz, the new kind of symmetry is called Lorentz invariance. Einstein, however, had never heard of this work by Lorentz, and was the first to understand the physical meaning of the equations.[7] But despite his use of symmetry it is not at all clear that Einstein realized then or later that the symmetries were the fundamental entities he was dealing with. This is the modern point of view expressed by the physicists who write textbooks on relativity, such as Weinberg, whose treatise on the subject is widely read.[8]

"It's sometimes hard for a physicist today to appreciate the frame of mind of the physicist of the 1930s," Weinberg said to us. "For example, take isotopic spin. Today, if a physicist who had never heard of atomic nuclei was suddenly exposed to a lot of information about them, he or she would immediately begin to ask questions about what are the symmetries of the interactions that hold them together?" He sketched the mathematical reasoning that would lead a physicist to the answer. "That's not the way people thought in the thirties," he said. "As far as I can understand it—and I wasn't doing physics then—their attitude was, what is the *dynamics* of the interactions? What is the theory of the nuclear *force*? And in the course of making a theory of the nuclear force, it might have a symmetry between protons and neutrons. If it did, you could form conclusions, like the conclusion that the nuclei form multiplets, but that would be a side issue. It would only be relevant as an incidental thing that you would use to test a particular theory of nuclear forces. The idea of separating off the question of symmetry as a completely separate subject which you study in its own right—and put off the dynamics for later—is more modern." Caught up in his own thinking, Weinberg had described a full circle around the neat expanse of his desk. He recalled that Einstein had used symmetries to develop special relativity. "The mood of 1905 was, the electron has just been discovered, the next thing to do is to make a model of the electron. And the obvious thing is that the electron is pure electromagnetic self-energy. People like Lorentz, [the great French mathematician Henri] Poincaré, [Max] Abraham, and so on made such models. . . . Einstein's real contribution was precisely to say—he didn't say it in these words, but this must have been his attitude—'We don't know what an electron is! Don't bother me with your models of the electron!' Many people—[J. J.] Thomson—thought that the electron was the only fundamental particle. Einstein said, 'This is not the time to say what the electron is. Let's just look at one aspect of it—what is the symmetry governing the laws of nature?' And he said, 'Well, it's Lorentz invariance. It's a different symmetry. It's not the one you thought it was.'

"I would say that symmetry again and again plays two roles. One is that very often symmetry is the thing to think about because it's the only way you can make progress. Einstein could only make progress by thinking about symmetry because in 1905 it was premature to make a model of the electron. Einstein understood—I don't know how!—but somehow or other he knew that that was not a time to make a model of the electron, that was the time to think about symmetry. In the beginning of the 1960s, it was not the time to make a theory of the strong interactions. The pieces were not in place. It was the time to think about the symmetry of the interactions, and Gell-Mann and [Yuval] Ne'eman did that and made tremendous progress. The other way that symmetry comes in, though, is more characteristic of Einstein in 1915 [general relativity] and [Chen Ning] Yang and [Robert] Mills in 1954 and also the development of the electroweak theory in 1967. And that is that the symmetry isn't just the only thing you can get a handle on, it is the thing that actually *drives* the dynamics. It's the central problem."

He was asked if this second kind of symmetry was what he had in mind when he talked about the simplicity of nature.

"Yeah. At the deepest level." He qualified the statement a little. "Well, I think it is. I don't know why nature has these symmetries, but Lorentz invariance, for example, the most important symmetry of all, is not only a symmetry which governs the form of the equations, it tells us what the equations are about.

"Particles are bundles of energy and momentum. What are energy and momentum but the quantum numbers defined by [time and space] translations? What is angular momentum but the quantum number which is defined by rotation? In a sense, if you have an elementary particle, and you describe how it behaves under various symmetry transformations, including translations, rotations, gauge transformations, then you've said everything there is to say about the particle. The identity of the particle is fixed by its symmetry properties. The particle is nothing else but a representation of its symmetry group." He laughed. "The Universe is an enormous direct product of representations of symmetry groups. It's hard to say it any more strongly than that."[9]

The role of symmetry in modern physics was established by Amalie Emmy Noether, one of the most important mathematicians of this century.[10] Emmy Noether was born in the German university town of Erlangen in 1882, where her father was a mathematics professor. Although university rules ordinarily prohibited women from matriculating, Noether obtained permission to attend lectures; of the 968 students at the University of

Erlangen, there were two women, neither of whom had been legally allowed to matriculate. In 1904, when the law changed, Noether took her degree; four years later, her doctoral thesis was accepted, *summa cum laude*, by the same university.[11] There was little of the rebel about her. She simply loved mathematics of the most abstract kind. From 1908 to 1915 she taught, without pay, at the Mathematical Institute of Erlangen and occasionally substituted for her father at the university; in 1915, she moved to Göttingen, where she became an important but still unpaid part of a team that worked on calculations for Einstein's formulation of general relativity.

Despite her evident brilliance and the support of the mathematicians, the Göttingen philosophical faculty, which encompassed disciplines from philology to physics, refused to accept her application for a salaryless position. The great mathematician David Hilbert, however, had her lecture under his name. (Deeply interested in the mathematical aspects of physics, Hilbert held the conviction that "physics is much too hard for physicists."[12] Indeed, he derived much of the theory of general relativity on his own.)[13] At the same time, he pressed for her promotion, angrily declaring at one meeting of the Philosophical Faculty, "I do not see that the sex of the candidate is an argument against her admission as *Privatdozent* [teaching assistant]. After all, we are a university and not a bathing establishment."[14] Evidently finding Hilbert's remark more provocative than persuasive, the assembled professors rejected Noether's application. She was finally given a teaching position in 1919. Three years later, she was promoted from teaching assistant to "unofficial associate professor" with the grudging admonition from the Prussian minister of culture that "this designation does not carry with it a change in her duties. In particular, her position as *Privatdozent* and her relation to her faculty remains unchanged; neither is she to receive the salary due an official position."[15] A year later, in 1923, she began to be paid for her work. She had been in the field full-time for over a decade.

Noether did her best work in her later years, which is rare for a mathematician. She was neither a manipulator of formulas nor a virtuoso calculator. Rather, she looked for the ideas that lay at the heart of entire disciplines, and strove to discover the conceptual bedrock upon which real mathematics could be based. Because of the abstraction and generality with which Noether worked, she became widely respected among her colleagues— which didn't save her from ejection by the Nazis. The Göttingen school of physics and mathematics was destroyed in 1933 when Noether and the other Jewish members of the Philosophical Faculty, such as Max Born and James Franck, lost their jobs. Many refugees found jobs in the United States, although not necessarily ones suited to their eminence. After fleeing Germany, Noether could only find a temporary position at Bryn Mawr College, outside of Philadelphia. In 1935, she died unexpectedly after surgery.

Noether's friend, the mathematician and physicist Hermann Weyl, himself a
refugee, delivered her memorial address. Near the end, he remarked, "She
was not clay, pressed by the artistic hands of God into a harmonious form,
but rather a chunk of human primary rock into which he had blown his
creative breath of life."[16]

Had Noether been a man, her appearance, demeanor, and classroom
behavior would have been readily recognized as one of the forms that absent-
minded brilliance frequently assumes in the males of the species. Fat, loud,
and disorganized, Noether cared little for how she looked or what she ate.
During lectures, she repeatedly withdrew the handkerchief she kept tucked
in her blouse, waved it about furiously to demonstrate a point, and tucked it
back in. She neither backed down from arguments nor easily relinquished
the speaker's role. Weyl confessed that he and his colleagues sometimes
referred to her unkindly as "der Noether," with the masculine article.[17]
There is a photograph of Noether taken at the Göttingen train station just
before she fled to the United States. Her humorous mien, touched with
melancholy, her thick glasses and baggy rumpled overcoat, and the plain flat-
brimmed hat pulled down over her forehead make her seem indeed like a
female version of Einstein.[18]

The theorem on which much of today's particle physics is based forms
only a small part of her total oeuvre, and is usually ignored by mathematicians more interested in her fundamental ideas on commutative ring theory
and algebraic number theory. Developed for her *Habilitationschrift*, the
second thesis that people in Germanic school systems had to write before
joining the faculty of a university, the work was an offshoot of David Hilbert's
interest in relativity.[19] As in her more mathematical papers, Noether went
straight for the most fundamental statement about the kind of relations with
which Einstein wrestled. Briefly, Noether's theorem demonstrates that
wherever there is a symmetry in nature, there is also a conservation law, and
vice versa.[20] In other words, the symmetries of space and time are not only
linked with conservation of energy, momentum, and angular momentum,
but each *implies* the other. Conservation laws are necessary consequences of
symmetries, and symmetries necessarily entail conservation laws.

The simplicity, power, and depth of Noether's theorem only slowly
became apparent. Today, it is an indispensable part of the groundwork of
modern physics, and scientists have compiled a list of over a dozen important
conservation laws and their associated symmetries. They have also discovered different types of symmetries, the most important of which is a
gauge symmetry.

Gauge symmetry was introduced to physics by Noether's friend Hermann Weyl in 1918, the same year that Noether proved the theorem that now
bears her name. A pupil of Hilbert, Weyl had also worked with Einstein in

Switzerland. Like Einstein, Weyl was afflicted by the urge to unify. In a series of three papers, he tried to show that electromagnetism, like gravity, is linked to a symmetry of space, which would be a step toward showing that the two forces are aspects of the same thing.[21] He based the argument on what now has the name of "gauge invariance," a term sufficiently undescriptive that even many physicists admit they wish it could be replaced.

A gauge, quite simply, is a measuring standard. When measuring most things, the size of the standard can be changed at will, and the results remain the same. The length of a board, for example, is the same whether a carpenter's measuring tape is marked in inches and feet or centimeters and meters; as long as the proper conversion is made, identical pieces of wood can be cut in both systems. Weyl called this kind of symmetry "measuring stick invariance," then changed it to "gauge invariance," apparently thinking of the metal devices used by railroad engineers to measure the distance between train tracks.[22] Weyl also realized that gauge invariance comes in two varieties, *local* and *global*, and that local gauge invariance was an extremely important idea.

Local gauge symmetry is difficult to visualize or describe in simple terms, and may perhaps best be approached by contrasting it to global gauge symmetry. A happy example of global gauge symmetry is the earth itself, which has been divided by cartographers into longitude and latitude lines. Zero degrees is arbitrarily set as the longitude of an observatory in Greenwich, England. If the whole system were turned ninety degrees, zero would fall in the middle of Memphis, Tennessee, and the international dateline would be outside Dacca, Bangladesh. Despite the changes, pilots would be able to navigate perfectly well. They would just add ninety degrees to their charts—changing their gauge, so to speak. Mathematicians might say that navigation on earth is symmetric about a longitude change of ninety degrees. This is a *global* gauge symmetry, for all longitude readings are changed in the same way. They are all ninety degrees west of what they were before.

Another example of global symmetry can be drawn from the world of high finance, or at least that portion of it described in elementary economics textbooks. If the income of every person and the price of every commodity suddenly increased tenfold, nothing would happen. People would still spend their wages on the same things in the same way, except that now paychecks and prices would have an extra zero. The supply and demand of every commodity would be undisturbed and, in the jargon of economists, all markets would still "clear." The forces of supply and demand can be said to be globally symmetric with respect to a change in the economic measuring stick—money.

Now imagine that incomes and prices were to jump around *randomly*, some rising, some falling by different chance amounts. The situation should then change dramatically: Supply and demand would be out of whack. Some people would be clamoring to buy more than is currently offered for sale, while others would be unable to afford their previous standard of living. This state of affairs, according to classical economists, brings into play the "invisible hand" of market forces. Prices adjust, decreasing in one place, increasing in another, until once again all markets clear. Automatically compensating for each random change in the system, the invisible hand maintains the original symmetry, which in this case is a *local* one. (In fact, economists make such claims only for small stochastic perturbations, but the principle holds true.)

Given the arbitrariness of the changes in local symmetry, it is hard to believe that any physical system could have this property. And indeed, gauge symmetries cannot occur unless the random changes in one aspect of a system are precisely compensated for by a change in another aspect, so that a quantity related to both remains conserved. Such compensation cannot take place unless a force intervenes. Indeed, it presupposes the existence of one. Maintaining local gauge invariance is not an easy, passive task; the Universe must act to preserve it. Thus understanding a force means tracing it back to an originating symmetry. In physics, the forces are real entities that act in the world, whereas economic forces are the half-understood sum of many individuals' actions.[23]

Weyl indeed postulated that space does have local gauge symmetry. He argued that the behavior of space and time in fact can vary randomly from place to place and instant to instant, but these chance alterations are canceled out by the activity of the electromagnetic fields that permeate the cosmos.[24] The result of this local gauge symmetry, by an application of Noether's law, is the conservation of electric charge. Soon after receiving a draft of Weyl's paper, Einstein appeared to kill the idea by pointing out that such local gauge invariance led ultimately to the erroneous prediction that if one took a clock around the room it would not show the same time as an identical clock in the original place because of all the gauge changes.[25]

In 1927, another theorist, Fritz London, revived Weyl's work with the suggestion that he had the right idea but applied it to the wrong symmetry.[26] Electric charge is indeed conserved, he said, but the associated symmetry is not the scale of space and time, but rather a symmetry of a much more abstract property, the phase of the Schrödinger wave equation. The phase of a wave is that part of it one is looking at—foamy crest, briny trough, or something in between. When two waves are *out* of phase, crest matches with trough and they cancel out; two waves *in* phase, on the other hand, reinforce each other. One of the enduring lessons of quantum mechanics is that particles have wave characteristics, and are described by wave equations. (Recall that Bohr, Heisenberg, and Schrödinger had long arguments about exactly this point.) Among the wavelike features of any elementary particle is its phase, which is intimately tied to such traits as electric charge. Changing the phase of a wave equation therefore would usually lead to changes in the electric charge. It is known that the electric charge of, say, an electron absolutely *never* changes. As Fritz London showed, the reason the phase can shift and not affect the charge is because of the gauge symmetry, which compensates for such changes by creating virtual photons—and hence an electromagnetic field—whose actions ensure the conservation of electric charge. Canceling out the effects of any phase change, the field protects the charge with single-minded drive. Thus the gauge symmetry of the phase implies, even *creates*, the dynamic actions of the electric field. From this protective action on the subatomic level stems, incredibly enough, all the phenomena of electromagnetism that we see about us, from the small shocks of static in carpeted rooms to the jagged arc of lightning across a storm sky. A single symmetry gives birth to them all.

Fittingly, it was Weyl himself who put the final formulation on this idea; it is too bad that he retained the name "gauge symmetry" for what was now a symmetry of phase.[27]

Just as someone who had never played chess would have to watch for a while to pick out the usually untouched king as the most important piece on the board, it often takes scientists a while to put their fingers on the central aspect of any phenomenon. Although physicists were well aware that electromagnetism could be described as a local gauge field, they did not know if this was particularly significant. Indeed, the concept stayed nearly dormant for twenty-five years.[28]

Modern gauge field theory was largely created at Brookhaven National Laboratory in early 1954 by Chen Ning Yang and Robert Mills. In a single short paper published that year in the *Physical Review*, "Gauge Invariance and Isotopic Spin," Yang and Mills built the frame upon which modern quantum field theory is woven. At first their ideas were treated with skepticism and even derided as being pure mathematics, without any physical significance; now, theories of Yang-Mills gauge symmetries are powerful enough to account for almost every particle and force in existence.

The thrust of the paper occurred to Yang in 1948, when he was a graduate student at the University of Chicago. He had recently left war-torn China, where his father was a mathematician, to study with Enrico Fermi.[29] Under the theory that his given name, Chen Ning, might be too difficult for Americans to say, he awarded himself the nickname "Frank," after Benjamin Franklin, a man whom the youthful Yang admired. He was impressed with the idea that gauge symmetries could be used as a base to construct the entire theory of quantum electrodynamics. "When I was a graduate student," he said recently, "people talked a lot about gauge invariance. People

would say that a calculation was not right, for example, because the result
was not gauge invariant. It was a kind of check on calculations. But it was not
appreciated that gauge invariance is a principle that can *generate* forces.
Today we recognize it as the only principle that can generate forces. It had
been understood that way for electromagnetism in the 1920s, only somehow
people were not paying attention. They also didn't realize the principle could
be moved to new situations."[30]

Yang began to suspect, in other words, not only that such a description
might be essential to an understanding of electromagnetism, but that there
might be other kinds of local gauge fields. He was unable to carry his postulate further than this. Theorists with incomplete ideas are like people with
songs in their heads that they can't identify; they can't stop trying to place
the tune. Yang finished his thesis and went to the Institute for Advanced
Study, in Princeton, New Jersey. He kept picking up and dropping gauge
invariance. In 1953, when he spent some time at Brookhaven, the idea still
bothered him, and he told Mills, his office mate, about it. "There was no
other, more immediate motivation," Mills says. "He and I just asked ourselves, 'Here is something that occurs once. Why not again?' "[31]

To find a conserved quantity like electric charge, Yang and Mills seized
upon isotopic spin. The paper begins with celerity:

*The conservation of isotopic spin is a much discussed concept in recent years.
Historically an isotopic spin parameter was first used by Heisenberg in 1932 to
describe the two charge states (namely neutron and proton) of a nucleon. The idea
that the neutron and proton correspond to two states of the same particle was
suggested at that time by the fact that their masses are nearly equal, and that
[many] nuclei contain equal numbers of them.*

They note that the strong force does not care whether the axis of isotopic spin
is up (proton) or down (neutron).

*Under such an assumption one arrives at the concept of a total isotopic spin
which is conserved in nucleon-nucleon interactions. Experiments in recent years
on the energy levels of light nuclei strongly suggest that this assumption is indeed
correct.*

Saying that the strong force conserves isotopic spin is the same as arguing
that it can not see the difference be .ween neutrons and protons. Yang and
Mills said that therefore, as far as the strong force is concerned, "the differen-
tiation between a neutron and a proton is then a purely arbitrary process."

The point made, the two men then busied themselves with the argument of the paper. Reminding the reader that in electromagnetism the phase
of the wave function can be shifted arbitrarily in space and time because the
action of the electromagnetic field will invariably cancel out the alteration,
they proposed to do the same thing with isotopic spin. After randomly

spinning the third axes of isotopic spin—that is, after imagining that they had haphazardly rearranged the labels "proton" and "neutron" throughout the Universe—they hypothesized the existence of a "B field" to counteract the change. Just as the *raison d'être* for the electromagnetic field is to ensure the gauge symmetry of electromagnetic interactions about the phase, so the B field maintains the gauge symmetry of strong interactions about the orientation of isotopic spin.

Yang and Mills then spent three pages working out what this B field would be like if it exists, and what the characteristics would be of the virtual B particles that transmit it. The paper concludes:

The quanta of the B field clearly have spin unity [i.e., one] and isotopic spin unity. We know their electric charge too because all the interactions that we proposed must satisfy the law of conservation of electric charge, which is exact. The two states of the nucleon, namely proton and neutron, differ by charge unity. Since they can transform into each other through the emission or absorption of a B quantum, the latter must have three charge states with charges ±1 and 0.

Today, the B particle is called a *vector boson*, which is jargon for a particle with a spin of one. (The name is redundant: A boson is a particle with integral spin, whereas any particle with a spin of one is called a vector particle.) Yang and Mills hoped that the equations would work out in such a way that this purely speculative B particle would convey a force identical to the strong force. They would thus have an explanation for strong interactions that would be in the same language as quantum electrodynamics, the model theory.

Unfortunately, when Yang and Mills did the mathematics, the figuring seemed to suggest that their conjectured vector boson had electric charge but no mass. This was most discouraging: Massless particles, such as photons, are easy to make in experiments, and charged particles, such as electrons, are easy to detect in them. The two men could not understand why, if charged, massless vector bosons existed, they would not already have been discovered. "We do not have a satisfactory answer," they admitted.[32]

On February 23, 1954, Yang spoke to a small audience at Princeton about the work he and Mills were doing. The audience included luminaries such as Pauli and Oppenheimer, who was now in the midst of his battle over his security clearance. Although Yang did not know it, Pauli had explored the possibility of generalizing gauge invariance a year before, but had decided not to publish because of the mass problem.[33] Yang recently recalled the occasion:[34]

Soon after my seminar began, . . . Pauli asked, "What is the mass of this [vector boson]?" I said we did not know. Then I resumed my presentation, but soon Pauli asked the same question again. I said something to the effect that this was a complicated problem, we had worked on it, and come to no definite conclusions. I still remember his repartee: "That is not sufficient excuse." I was so taken aback that I decided, after a few moments' hesitation, to sit down. There was general embarrassment. Finally Oppenheimer said, "We should let Frank proceed." I then resumed, and Pauli did not ask any more questions during the seminar. I don't remember what happened at the end of the seminar. But the next day I found the following message:

> *Dear Yang:*
> *I regret that you made it almost impossible for me to talk to you after the seminar. All good wishes.*
>
> *Sincerely yours,*
> *W. Pauli*

Despite the unresolved question about the mass of the vector boson, Yang and Mills decided to publish their work. "We did not know how to make the theory fit experiment," Yang said. "It was our judgment, however, that the beauty of the idea alone merited attention."[35]

Although it would not be fully realized for fifteen years, Yang and Mills had transformed the role of symmetry in quantum physics. They had given symmetry muscle, shown how it could create forces and set particles in motion. In essence, their message was this: "Look at particle properties. See if you can find any unchanging qualities and try to imagine a gauge field that might account for them. Work out the properties of that fictional field and its associated virtual particles. Are they like anything in the real world?" If they are, then it would be a major step toward unification, toward the day when scientists would be able to trace out the laws of the Universe from a single principle, one that reaches out to cover the whole of nature in a single explanatory web. Yang and Mills had raised the hope that the clue to the diversity of the world lay in symmetry.

In the meantime, however, physicists had to occupy themselves with the question of how to use this new insight. Yang and Mills, and many of their colleagues, chose to look at the strong force and the heavy particles it affected, about which a wealth of data was pouring in from the first particle accelerators. Although they couldn't know it, that was a blind alley. Most of the speculation about Yang-Mills gauge fields—the term was soon applied to any local gauge field—stayed shut in physicists' notebooks, and gauge symmetries lay dormant until years later, when Steven Weinberg, Abdus Salam, and Sheldon Glashow applied them to the weak interaction.

11
Weakness

WHAT ARE NOW KNOWN AS THE WEAK INTERACTIONS HAVE HISTORICALLY been associated with a striking degree of experimental confusion ever since their accidental discovery by Becquerel in 1896. The *rayonnements invisibles* that elided through the heavy black paper in which he wrapped his photographic plates largely consisted, we say today, of the electrons emitted by the beta decay of neutrons in his uranium crystal. The process of beta decay was named and distinguished from its confreres by Rutherford; Becquerel himself made the hypothesis that the "beta ray" was a speeding electron, although this universally accepted belief was in fact not completely proven until much later, in 1948, by an experiment performed by Gertrude and Maurice Goldhaber, in which they showed the identity of beta rays and atomic electrons.[1] No one had any idea why every now and then an atom should spit out an electron. After the alpha-scattering experiments of Rutherford, Geiger, and Marsden, the location of the phenomenon was thought to be the nucleus, but beta decay was in no way less mysterious.

The most puzzling aspect of beta decay was that the electrons came out of the atom with a dizzying variety of speeds. Radioactive atoms generated electrons in a continual rain, but there seemed to be no preferred speed or direction to the flow.[2] From the law of conservation of energy, one would expect that the energy of the emitted electrons would be equal to the energy lost by the nuclei that ejected them. Measurements showed that this was not the case.[3] In 1927, two experimenters at the Cavendish totted up the full heat energy of a small amount of radioactive material and compared it to the energy of the electrons it created. There was little connection between the two figures.[4] Other, more precise experiments discovered the same result.[5] Niels Bohr said that these experiments represented irrefutable proof that the law of conservation of energy must not apply to radioactive decay.[6]

Rather than give up conservation of energy, Wolfgang Pauli proposed— reluctantly, perhaps, and certainly with less than usual vigor—that beta radiation actually consisted of *two* particles, the already known electron and a previously unimagined entity later dubbed the "neutrino." Pauli had an ambivalent attitude toward publishing radical ideas; he announced his guess at the end of 1930 through a letter to some colleagues—"Dear radioactive ladies and gentlemen," as he called them—begging off from a conference.

(He wanted to go to a dance contest instead.)[7] The neutrino was small, light, electrically uncharged, very hard to detect; it carried off the missing energy.

At the same time, there was the puzzle of where, exactly, the electrons came from. It was generally assumed that they must somehow be in the nucleus until the question was reopened by the development of Heisenberg's uncertainty relation. Recalling that one formulation of this principle links the uncertainty in the position of a particle to that of its momentum, it is a simple matter to show that pinning down an electron's location to the nucleus of an atom, which is extremely small, implies an enormous uncertainty as to the particle's momentum, and the corresponding likelihood that the latter quantity is extremely large. A large momentum means a high velocity. If electrons indeed whiz around inside the nucleus at great speed, why do they not shake loose anyway, long before beta decay has a chance to kick them out? On the other hand, where else could the beta decay electrons come from if not the nucleus?

(Heisenberg doubted that the electrons were in the nucleus at all, a skepticism that he shared with Victor Weisskopf on a hot day in the spring of 1931 as the two men sat outside the entrance to a swimming pool. In Weisskopf's recollection, Heisenberg said, "These people go in and out all very nicely dressed. Do you conclude from this that they swim dressed?")[8]

On the theoretical side, the first measure of order was established in 1933 by Enrico Fermi, who took advantage of the recent discovery of the neutron and Pauli's neutrino suggestion to create the modern picture of beta decay: neutrons turning into protons, giving out electrons and neutrinos as they do so.[9] Later, when it was more clearly realized that the creation of a particle is always balanced by the creation of an antiparticle, the picture would be amended again to say that the newly manufactured electron is accompanied by an *anti*neutrino.

Even so, the picture discomfited some physicists. It was odd to imagine that the electron and neutrino were not originally in the nucleus at all, just as a bubble is not in a bubble pipe until one blows through the mouthpiece.[10] Hard, too, to picture an elementary particle, one of the bricks that comprise the Universe, falling apart. Maurice Goldhaber, who was at the Cavendish at the time of the first experiments with the neutron, recently described his alarm. "We found that the neutron was definitely heavier than the hydrogen atom, which is a proton and an electron," he said. "So I could easily see that a neutron could decay into a proton and electron." In fact, as was found later, the average time before beta decay hits a free neutron is about eighteen minutes. "We decided, here is a fundamental particle which should be radioactive. I remember being shocked by that idea—that a fundamental particle could *decay*. This is now taken for granted. But such a basic concept, that a particle can decay— One knew of course that nuclei decay, that particles in the nuclei change, but elementary particles seemed to me . . ."

He put his finger to his upper lip, considering. A cold wind keened at the window. His office is at Brookhaven National Laboratory, where Goldhaber is involved in such things as examining the unification theories that sprang from such earlier work as the measurement of the neutron lifetime. Once dismayed by the end of a particle, Goldhaber is now a partner in the salt mine collaboration near Cleveland that is testing for the end of all things. "Maybe I shouldn't have been shocked. But it seemed to me a shocking idea then, that a fundamental particle could *decay*."[11]

The theory of the weak interaction was first exposed to the world between Christmas and New Year's Eve of 1933 from a small, bare, cold hotel room in the minute Italian hamlet of Selva, nestled in the Alps about twenty miles east of Bolzano, where four physicists from the University of Rome— Enrico Fermi, Franco Rasetti, Edoardo Amaldi, and Emilio Segrè—had decided to spend their Christmas vacation. One evening, after a full day of skiing, Fermi invited the others into his hotel room to talk about the paper he had just sent off for publication. There were not enough chairs, so the four men crowded together on the edge of the bed. Fermi bent over, scribbling calculations on a sheet of paper he had propped against his thighs, while Rasetti, Amaldi, and Segrè craned their necks to read his drawing. Sore from repeated tumbles on the icy snow, Segrè kept changing his position restlessly. But the three were impressed, although a bit confused, by the bold theory that was emerging on Fermi's knees. Beta decay was not only associated with the unheard-of transmutation of neutrons and a never-seen neutrino, Fermi told them, it was the handiwork of an entirely new force of nature.[12]

When theorists are confronted with a phenomenon they don't understand, they often cast about for something similar they *do* understand, borrow its theory, and squeeze as much of it as possible into the new phenomenon. This had been precisely Fermi's strategy. The most promising similar theory was quantum electrodynamics, despite the infinities then driving theorists to distraction. Fermi therefore set out to model beta decay on electromagnetism.

An electron emits a photon when it changes from one state to another. This change of state is brought about by a familiar force of nature: electromagnetism. In the process of electron scattering, for example, two electrons approach each other, exchange a virtual photon, and fly away like so many billiard balls. This interaction is said nowadays to be composed of two "currents," which might loosely be described as the two halves of the diagram below, that is, the two V-shaped electron lines. Fermi pictured the process of beta decay likewise as a combination of two currents, a neutron-proton current (n to p in the diagram) and an electron-antineutrino current (e to v̄, with the bar indicating that it is an antiparticle). Avoiding all speculation as to what, exactly, happened at the point where the two currents intersect, he argued merely that the changes of state are brought about by an extremely feeble force, trillions of times weaker than either electromagnetism or the nuclear force responsible for strong interactions. Its effects are almost always swamped by the other two forces, like the political course of a tiny nation caught between two superpowers. Nevertheless, the actions of this small force can sometimes make a decisive difference when the effects of the other two offset each other, as in unstable nuclei. The result is beta decay. (There is also a form of inverse beta decay, in which a proton changes to a neutron, absorbing a neutrino and releasing a positron.)

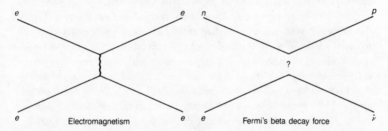

Electromagnetism   Fermi's beta decay force

The strength of the electromagnetic interaction between two objects is described by its coupling constant, alpha, the infamous 1/137, which in a sense indicates the absolute value of the force. Fermi wrote down his theory of beta decay to include such a coupling constant, and was able to deduce its value from the tables collected by experimenters. This Fermi constant is extremely small, as befits such a feeble interaction. Although its exact value depends on the system of notation employed, physicists sometimes say the strength of the weak force relative to electromagnetism is on the order of 10^{-13}, a number whose tininess is best appreciated when written out in full: 1/10,000,000,000,000.

Using the analogy with electromagnetism and the value of the beta decay coupling constant, Fermi was able to calculate the range of energies the emitted electrons should have, when beta decay could take place, and many other features of the process. His rapid series of papers on the subject is extraordinary: The theory was his first and only work on beta decay and was so complete that its description of the phenomenon has remained essentially unchanged to this day. At a stroke, Fermi first accurately described beta decay, first identified the separate nature of its cause, and first predicted most of its essential features. "Seldom was a physical theory born in such definitive form," Rasetti has written.[13]

The editor of *Nature*, to whom Fermi sent his first paper, was less appreciative. Fermi's note was bounced immediately, with the remark that its speculations were too remote from physical reality to be of interest to the practicing scientists who made up the audience of the journal. The idea was therefore published first in a small Italian review, *Ricerca Scientifica*; follow-up articles appeared in Italian and German.[14] (Maurice Goldhaber, in fact, explained Fermi's Italian paper to physicists at the Cavendish.) By the time they appeared, Fermi, the last twentieth-century physicist to make great contributions to both theory and experiment, had turned his attention to the experiments on which much of his fame rests—bombarding the nuclei of all the elements with neutrons, creating radioactive forms of some of them in the process. These experiments were eventually to make him one of the fathers of nuclear physics and the atom bomb. Certain British experimenters found this to be a more promising route to good science; at the end of April the following year, after Fermi had published some preliminary data, he received a letter from Rutherford whose parting salutation is almost a conditioned reflex: "I congratulate you on your successful escape from the sphere of theoretical physics! You seem to have struck a good line to start with. You may be interested to hear that Professor Dirac also is doing some experiments. This seems to be a good augury for the future of theoretical physics!"[15] (Needless to say, experiments did not form a major part of Dirac's oeuvre.)

One small theoretical adjustment was made to Fermi's thesis. For mathematical reasons, Fermi argued that beta decay would take place rarely unless the summed spins of the constituent protons and neutrons in the original and the final nucleus remained the same.[16] (Here the discussion is of ordinary spin, not isotopic spin.) Two theorists, George Gamow and Edward Teller, showed that beta decay was also mathematically possible if the spins differed by one, and developed the appropriate equations.[17] Ever since, beta decays where the spin of the nucleus remains unchanged have been called Fermi transitions; those where the spin of the nucleus changes by one are called Gamow-Teller transitions. Although Gamow and Teller developed their equations only as a theoretical possibility, actual instances of Gamow-Teller transitions were soon discovered.

Despite its theoretical success, Fermi's theory soon ran afoul of experiment. His articles predicted that the emitted electrons would come out with varying energies, but that if experimenters plotted the number of beta electrons against their velocity, the resultant curve would have a particular shape, somewhat like the famous bell curve used to grade students in large classes.[18] Two British and three Soviet experimenters promptly measured this curve with the electrons emitted from radium and radioactive phosphorus. They found many too many slow electrons.[19] To fit the data, George Uhlenbeck, the co-discoverer of spin, and Emil Jan Konopinsky, both of whom had recently moved to the University of Michigan, formulated a brand-new, modified theory, which became known as the K-U theory after their initials.[20] Discerning the velocity of the electrons was no easy task, and the work was made even harder by the fact that the weak force is so feeble that beta decay does not happen very often. In order to scare up enough electrons to measure, the British and Soviets had used thick chunks of radioactive material. Unfortunately, if the source is thick, electrons emitted in the middle of the stuff must travel through many layers of atoms, ricocheting as they go, before emerging into the open air. The resultant deflection and loss of energy distorted the readings and the graphs based on them. When thinner slivers of radioactive material were used, the experimenters found fewer slow electrons than before.[21]

Although the new set of experiments invalidated the K-U theory, the curve for low-energy electrons still did not agree with Fermi, leaving the understanding of beta decay in what was beginning to be a customary state of confusion. The war intervened, and most of the players in nuclear physics and electrodynamics became preoccupied with building weapons or hiding from Fascism. An exception was Willis Lamb at Columbia, who thought about measuring the fine structure of hydrogen; another was Chien-Shiung Wu, who spent most of the war years teaching physics at Smith College and Princeton University. Born in China in 1912, Wu came to the United States twenty-four years later to win her doctorate. Although the skill and care with which she worked were immediately apparent, her sex guaranteed her anonymity. At the end of the war, she was finally taken in at Columbia University's Division of War Research. The hostilities concluded, she began to consider the question of the low-energy electrons from beta decay. Slow electrons are particularly susceptible to scattering effects, the way, for instance, slight ripples on a putting green may make a barely moving golf ball swerve round the hole but have no effect on a faster one. Even the thinnest of radioactive sources would throw out slow-moving electrons in odd ways if its surface was uneven; the electrons could also be deflected by the backing material and the detector itself. In 1949 Wu experimented with an intensely radioactive film of copper—the technology to make this kind of thing was a legacy of Los Alamos—and was able to show with convincing thoroughness that Fermi's theory held true through the gamut from fast to slow electrons.[22]

During the sixteen years experimenters tried to sort out the electron end of beta decay, they attempted in addition to discover whether the neutrino existed.[23] At the time Fermi proposed his theory, the plausibility of the scheme depended on the near-impossibility of detecting the neutrino, which explained why it had not yet been observed. It is unlikely that Fermi realized at the time just how elusive this particle was going to be. Hans Bethe and Rudolf Peierls calculated that neutrinos of average energy could pass through fifty billion miles of water without hitting a single atom.[24] Unless a particle interacts with matter, it cannot be found. As the decades went on, the undetectability of the neutrino became an embarrassment. Massless, chargeless, hardly present but supremely necessary, the neutrino was an enormous lacuna in the physicist's image of matter. By the beginning of the 1950s, scientists were beginning to regard it as a sort of epistemological boondoggle, a poltergeist particle that belonged in its own special category of being.[25]

In 1952, a thirty-three-year-old Los Alamos scientist named Frederick Reines took a leave of absence to think about questions that would be worth devoting a lifetime to answer. After some months of staring at his empty office, he said later, "all I could dredge up out of the subconscious" was the recollection that the frightful by-products of an atomic explosion include a big blast of neutrinos—if they exist. Given enough neutrinos passing through a detector, some of them might strike a proton, causing inverse beta decay, that is, producing a neutron and positron.[26]

Reines talked about blowing up an atom bomb to find neutrinos with a friend, Clyde Cowan, and together the two men realized that if the detector was sensitive enough the experiment could be done in the considerably less hazardous surroundings of a nuclear reactor. (When informed that Reines had decided to try the reactor, Fermi drily commented that the new experiment was much better if only because it did not require setting off an A-bomb every time somebody wanted to check the results.)[27] The idea was that the neutrinos from the reactor would pour into a vat of dilute cadmium dichloride. Reverse beta decay would take place. Positrons would quickly strike and annihilate electrons, releasing photons; neutrons would be absorbed by cadmium nuclei, which acknowledge the action by emitting more photons. The two parallel bursts of photons would travel outward, passing into a layer of highly flammable triethylbenzene, a smelly substance somewhat like household cleaning fluid that has the happy property of flashing when struck by high-energy light. Amplified, the light is picked up by photomultipliers, devices like the sophisticated uptown cousins of the phototubes that keep open elevator doors. The task of the experimenters was thus to find and measure those twinned sparks of light.

The first run, at Hanford Engineering Works, in Hanford, Washington, was a nightmare of stacking and restacking lead shielding, cleaning dirty pipes, and trying to prevent the liquid detector from making the paint peel. Nonetheless, Reines and Cowan thought they saw a hint of the neutrino's presence.[28] Nearly three years later, using a much larger system that weighed ten tons even before it was surrounded by lead blocks, Reines and Cowan found the signals that enabled them to wire Wolfgang Pauli that the neutrino had been discovered three and a half decades after its invention.[29] (It took another eight years before the find was independently confirmed.)[30] Reines set to work on another problem, that of trying to see if a neutrino could scatter off an electron. Twenty years later, he proved it could.[31]

Millennia of intermittent thought went into deciphering the character of the gravitational force, into realizing that whatever pulled objects to the ground was the same as whatever bound the moon to the earth. Lightning, static electricity, and magnetism were known for centuries before humanity became aware that they were actually different aspects of one thing—electromagnetism. The weak force, too, was recognized only after scientists learned that what they thought were several different phenomena were in fact part and parcel of just one. But this new force, its actions mostly hidden from our eyes but no less fundamental than the other two, came to light in a time of frenzied activity that telescoped the scientific process. The weak interactions were discovered, named, and incorporated into a unified theory within the span of a single generation, but only after the most appalling experimental confusion as to exactly what they were.

Linking together the various aspects of the weak force began when clever investigators suspected that the Fermi beta decay force also played a role in the emission and absorption of muons.[32] (Muons, one recalls, are the "ten-year joke" particles in cosmic rays that were mistaken for pions.) In 1948 and 1949, four groups of theorists independently claimed that the coupling constants of the force responsible for muon decay, muon capture, and beta decay were identical, and thus all three phenomena were controlled by the same force.[33] The implication of each paper was that beta decay was merely one example of a whole class of actions in the subatomic world, a thesis which became known as the "Universal Fermi Interaction." Strange particle decays, too, were soon assigned to the same class. If the Universal Fermi Interaction hypothesis was true, several disparate interactions would be unified as the doings of one force; if it was not, particle physics would be more hopelessly complicated than anyone had ever dreamed.

The Universal Fermi Interaction would have to survive two tests: First, this new force must be shown to have its own unique *properties*; second, the *form* of its alleged manifestations must be proven identical. Remarkably, the Universal Fermi Interaction—what physicists now call the weak interactions—survived both tests in a single year.

As it happens, the property that characterizes the weak interactions is related to a symmetry called *parity*. Like symmetry, parity has a loose, general meaning and a much more restricted scientific definition. Ordinarily, parity means equality; two individuals are on a par or have parity if they are equal in some respect like position or stature. To scientists, however, parity means the kind of equality that results under a particular kind of transformation—when all spatial coordinates are reversed, and right becomes left, down up, and backward forward. This complicated-sounding process is quite simply what happens when you look in a mirror, except that up and down are not reversed in a mirror. The sameness that results is a special form of symmetry called reflection symmetry or parity.

Physical laws were assumed to be indifferent to reflection transformation. What works for left, up, and forward works equally well for right, down, and backward. Apparent asymmetries in nature, such as the right-spiraling shells of crustaceans or the left-sided position of the human heart, are the result of historical evolution rather than physics; there is no reason why shells could not spiral left, or why a "mirror-image human" could not exist whose heart is on the right. The heart may have its reasons for its left-sided position, scientists thought, but physics does not.

When physicists speak of the parity of elementary particles, they have in mind what happens when minus signs are placed in front of the spatial variables (the Xs, Ys, and Zs of position), for the Schrödinger wave function of a stationary particle, an operation equivalent to flipping the equation in the parity mirror. If the wave function is not changed, the particle is said to have even parity and given a quantum number of $+1$; if the wave function reverses sign when its orientation in space is reversed, the particle is said to have odd parity and given a quantum number of -1. (If the particle is moving, a few extra complications enter.) Over the years, experimenters observed that if one multiplied together the individual parities of all particles in a system, the total parity, like electric charge, was conserved.

The parity of elementary particles came into sudden prominence as a result of the decays of the strange particles, which were candidate members of the Universal Fermi Interaction. While grouping together the various particles found in cosmic rays, the Bagnères conference attendees took the time to note a small peculiarity concerning two strange particles, the theta and the tau. These had almost identical masses and lifetimes and could only be told apart by the manner in which they decayed: The theta particle changed into two pions, whereas the tau became three. A stationary pion has a parity of -1. Ignoring a few additional factors, the two-pion decay has a total parity of $-1 \times -1 = +1$. The three-pion decay has a parity of $-1 \times -1 \times -1 = -1$. In this way the tau and theta were distinguished. As Australian physicist Richard Dalitz showed after the conference, such calculations left physicists in a terrible bind.[34] On the one hand, describing two different particles with almost identical physical properties required a theory so complicated that nothing remotely like it had ever been seen in nature. On the other, asserting that the tau and theta were the same thing meant that this particle could blithely decay as it pleased into products of even or odd total parity—in short, that parity was not conserved in its decay.

In April 1956, Frank Yang attended the 1956 Rochester Conference, in Rochester, New York. Organized by Robert Marshak to carry on the Shelter Island tradition, the Rochester Conferences were the principal meeting places of physicists in the United States, yearly landmarks when the luminaries of the field would gather together to discuss where they were. Unlike Shelter Island, the Rochester Conferences had equal mixes of theorists and experimenters. Some even roomed together; Richard Feynman's roommate at the 1956 conference, for instance, was a Duke University experimenter named Martin Block. Feynman is notorious for arguing heatedly with whomever he is in closed quarters, and he and Block began quarreling about the tau-theta puzzle. At one point Block suggested that parity was not conserved. Block has recalled his roommate's reaction: "Feynman, in his usual gracious way, was ready to say to me, 'How *stupid* you are!' And by the time his jaw got over to the word 'stupid' he had thought a little bit—I'm quoting Dick on this—and he said, 'Well, maybe you have something there.' We spent one week, every night, arguing until the wee hours of the morning."[35]

The last day of the conference began with a session on the "Theoretical Interpretation of New Particles." Oppenheimer held the chair; Yang gave an introductory review.[36] Several talks followed, each wrestling with the problems involved in making the tau and theta one particle, or separate particles. Delphically, Oppenheimer opined, "The tau meson will have either domestic or foreign complications. It will not be simple on both fronts."[37] A few minutes later, Gell-Mann inventoried his own possible approaches to the problem, but he, too, was perplexed. He was unable to bring himself to advocate his own ideas of dealing with the problem—only to mention that he had some. The proceedings then report: "Pursuing the open mind approach, *Feynman* brought up a question of Block's: Could it be that the theta and tau are different parity states of the same particle which has no definite parity, *i.e.*, that parity is not conserved. That is, does nature have a way of defining right or left-handedness uniquely?"[38] After Yang said that he had looked at parity violation and several other possibilities without reaching a conclusion, Oppenheimer remarked that the time had come for everyone to close their minds. Afterward, Block asked Feynman what Yang had answered. "I don't know," Feynman said. "I couldn't understand it."[39]

Two weeks after the conference, Yang moved to Brookhaven for the summer.[40] Nearby at Columbia was Tsung-Dao (T. D.) Lee, another Chinese emigré, frequent collaborator, and long-time friend. The two met twice a week to work together, with the tau-theta puzzle occupying most of their attention. One morning in late April or early May, Yang made the long drive in from Brookhaven and picked up Lee at his office. The two parked near 125th and Broadway, where there were, until recently, several fine Chinese restaurants. Upon discovering they were not yet open, Lee and Yang retired to the nearby White Rose Cafe. Over coffee, they discussed the possibility that parity could be *conserved* in the strong interaction that produced the tau and theta, yet *violated* in the weak interaction by which they decayed. Although they couldn't imagine why parity might be violated, they agreed that this possibility should be further explored.

Lee went to the office of the local expert on beta decay, Chien-Shiung Wu, on the thirteenth floor of Pupin Hall, and asked whether she knew of any experiments that had definitely shown anything one way or the other about parity in the weak interactions. Wu said that Lee would have to look through the literature. She lent him the "literature," the recently printed *Beta- and Gamma-Ray Spectroscopy*, a massive tome of almost a thousand pages.[41] Crammed with graphs and tables in tiny type, the book summarized more than forty years of work by hundreds of physicists. All Lee and Yang had to do was make sure that not one of the cited experiments had directly tested for the conservation of parity.

Several weeks of intense work later, they knew the answer. Incredibly, nobody had ever proved that parity conservation was valid for the weak interaction. On June 22, 1956, the results of their labors, "Is Parity Conserved in Weak Interactions?" arrived at the offices of the *Physical Review*. The paper was published under the title of "Question of Parity Conservation in Weak Interactions," because Samuel Goudsmit, the editor of the journal, believed that question marks in titles were a blot on the escutcheon of physics. The work is an exceptional document. It begins simply, with a confident statement of the tau-theta puzzle: The tau particle and the theta particle are known to have nearly identical properties except for their parities, and thus are considered to be separate particles.

One way out of the difficulty is to assume that parity is not strictly conserved, so that theta and tau are two different decay modes of the same particle, which necessarily has a single mass value and a single lifetime. We wish to analyze this possibility in the present paper against the background of the existing experimental evidence of parity conservation.

What is this evidence?

At first sight it might appear that the numerous experiments related to beta decay would provide a verification that the weak beta interaction does conserve parity. We have examined this question in detail and found this to be not so.

Lee and Yang did not argue that parity was actually violated in weak interactions. Pointing to a surprising gap in our knowledge of the world, they merely tried to convince experimenters to settle the question.

To decide unequivocally whether parity is conserved in weak interactions, one must perform an experiment to determine whether weak interactions differentiate the right from the left.[42]

Few experimenters were willing to spend the effort of working in the notoriously difficult area of beta decay to check an absurd idea. Some theorists, too, tended to dismiss the paper as clever but irrelevant—scientific virtuosity displayed to little apparent purpose. Freeman Dyson, for example, read the article twice, thought "this is interesting," and filed the journal back on the shelf.[43] Gerald Feinberg, who got his Ph.D. from Columbia that year, heard about Lee's struggles with parity during Tuesday night theory seminars. He was interested by the idea of parity violation, but not convinced. "People didn't know what to make of it," he recalled. "That's true of any genuinely new idea—you don't know what to think of it at first. Lee had gone through a lot of different ideas. This paper did not come with an arrow pointing to it, saying, 'This is it!' "[44]

Wu had been the first to hear about Lee and Yang's work, and was the first to consider testing it. For months she and her husband, Chia-Liu Yuan, had planned to visit the Far East on the twentieth anniversary of their exodus from China. With passage booked on the *Queen Elizabeth*, Wu abruptly canceled and left her spouse to make the sentimental journey alone. She wanted to start an experiment before anyone else realized the paper's importance. By the beginning of June, three weeks before Lee and Yang were ready to submit the paper, Wu was already lining up her collaborators.

She was fully aware of the difficulty of the experiment, because she herself had informed Lee and Yang of the possibility of its execution. Briefly, the task was to align the spins of the separate particles in an atomic nucleus, thus making them, so to speak, face the same way. The experimenters would then watch for beta decay. If the Fermi force respected parity, the electrons could have any orientation in space whatsoever and would come out in a random, symmetrical spray. If beta decay violated parity, the emission would be asymmetrical. The neutrons needed to be lined up so that their own random motions would not hide any asymmetry.

Aligning the neutrons is known as "polarizing" the nucleus. At the time, polarizing nuclei was, to understate the case, something of a challenge. The only known technique involved cooling a substance to almost absolute zero. With the nuclei nearly still, a carefully applied magnetic field could then rearrange each atom's cloak of electrons in such a way that they in turn pushed and pulled the frozen protons and neutrons into a sort of lockstep. Even at incredibly cold temperatures, the alignment could not be maintained very long because the atoms would slowly bang into one another. With luck, the polarization might last a few minutes. Wu intended to use Co^{60}, an intensely radioactive form of cobalt that has sixty protons and neutrons and the right configuration of electrons.

Not only the cobalt but the electron detectors would have to be cooled, for the electrons emitted in beta decay could not possibly penetrate the heavy insulation around the equipment. These detectors were thin anthracene crystals that gave off tiny flashes of light whenever struck by electrons; the flashes would be carried through the insulation by glass windows and a four-foot Lucite pipe to a bank of photomultipliers. Knowing that the small low-temperature laboratory at Columbia could not do the job, Wu called Ernest Ambler, a pioneer in nuclear alignment, at the National Bureau of Standards in Washington, D.C., in early June to propose a collaboration. He agreed; they began to construct the equipment. By September, Wu commuted regularly to Washington to set up the apparatus. She and Ambler also recruited Ambler's division head, R. P. Hudson, and two nuclear physicists, R. W. Hayward and D. D. Hoppes.

The cobalt had to be layered in a thin film—Wu knew all about the importance of thin films—onto a small crystal of cerium magnesium nitrate (CMN), which was then shielded in a housing made from larger crystals of the same material. Halfway into the experiment, the team discovered that nobody knew how to make the big crystals. Consulting a huge, dusty, forgotten nineteenth-century German text on the crystalline properties of CMN, they set up beakers under an array of heat lamps and dumped in as much ground CMN as the water would hold. They slowly lowered the temperature and watched the CMN crystallize on the bottom of the glass like so much rock candy. Once the crystals were glued together, Wu and her colleagues froze the mass of CMN and Co^{60} to a fraction of a degree above absolute zero. The whole ensemble fell apart. In this manner they learned that Du Pont cement loses its sticking power at cryogenic temperatures.

Even after they tied together the crystals with nylon thread, they could only keep the nuclei aligned for fifteen minutes at a time. Nonetheless, at the end of six months' work the team was almost certain that more electrons were being emitted in one direction than any other. A period of intensive checks of the equipment followed; Hoppes slept on the floor next to the equipment in a sleeping bag for fear of something going awry. About two o'clock in the morning of January 9, 1957, all checks were finally complete. Hudson uncorked a bottle of Chateau Lafite-Rothschild 1949, ceremoniously poured it into paper cups, and everyone drank to the overthrow of parity.[45]

Before New Year's Eve, Wu had told Lee and Yang of the asymmetry, but said the team was not ready to go public. But the word leaked out anyway. Someone informed Leon Lederman, another Columbia experimenter whom Lee and Yang had unsuccessfully urged to investigate parity. Lederman had a colleague, Richard Garwin, who was creating beams of muons in order to repeat the Conversi-Pancini-Piccioni effect on his own. Over a meal at the Shanghai Cafe on January 4, Lee, Lederman, Garwin, and half the Columbia physics faculty discussed the muon beam. If parity nonconservation was a general feature of weak interactions, the electrons created by the decaying muons would also emerge more in one direction than another. The experiment was a gamble, because the process of stopping the muons might wreck the alignment of their spins. Despite all the difficulties, Lederman and Garwin went ahead. Lederman telephoned Lee before breakfast on Tuesday, January 8, with the words, "Parity is dead." The effect was huge, unmistakable, even more pronounced than Wu's. Incredible that nobody had seen it before. As a matter of course, Lederman's team waited for Wu's group to finish before sending off their paper.[46]

Valentine Telegdi, an experimenter at the University of Chicago, also had an assistant making muon beams. Having run across a preprint of the Lee and Yang paper in August, Telegdi, unlike most of his colleagues, was immediately interested. Not knowing of Wu's work, he and the student began an experiment that was similar in many respects to Lederman's. He might have finished it first, but illness in the family delayed the run. His parity paper arrived two days after the other two at the offices of the *Physical Review*. It was bounced. A rewritten article was published two weeks after the others.[47] By then the news had already thrown the physics world into turmoil.

Typical of the reactions was that of Wolfgang Pauli, who with amazingly bad timing put the following remarks in a letter to Victor Weisskopf just three days before he was told the news:

I do not believe that the Lord is a weak left-hander, and I am ready to bet a very large sum that the experiments will give symmetric results.[48]

Ten days later, he wrote Weisskopf again.

Now, after the first shock is over, I begin to collect myself again (as one says in Munich). Yes, it was very dramatic. On Monday 21st, at 8:15 P.M. I was to give a lecture on "the past and recent history" of the neutrino. At 5 P.M. the mail brought me three experimental papers. . . . Now, where shall I start? It is good that I did not make a bet. It would have resulted in a heavy loss of money (which I cannot afford); I did make a fool of myself, however (which I think I can afford to do)—incidentally, only in letters or orally and not in anything that was printed. But the others now have the right to laugh at me. What shocks me is not the fact that "God is just left-handed" but the fact that in spite of this He exhibits Himself as left/right symmetric when He expresses Himself strongly. . . . How can the strength of an interaction produce or create symmetry groups, invariances or conservation laws?. . . . Many questions, no answers![49]

Pauli was not alone. Yang informed Oppenheimer with a telegram to his vacation retreat in the Virgin Islands. Oppenheimer cabled back, "Walked through door." I. I. Rabi explained to the press, "In a certain sense, a rather complete theoretical structure has been shattered at its base and we are not sure how the pieces will be put together."[50] That fall, the Nobel Committee awarded Lee and Yang the Nobel Prize barely a year after the work which it canonized.

Revolutions are seldom accomplished without some legacy of bitterness. Telegdi was sufficiently angered at the rejection of his paper by the *Physical Review* that he resigned from the American Physical Society.[51] Later, several experimenters were chagrined to realize that they had observed parity violation as far back as the 1920s but had not understood what they saw.[52] Block believes that his question at the Rochester Conference started the entire process, and that the physics community does not recognize his contribution.[53] The most lasting divisions came between Lee and Yang, whose friendship was a casualty of their increasing celebrity. In 1962 they formally ended their decade-long collaboration. Physicists took care to invite one only when the other was absent, and watched in distress when these two remarkable thinkers used the handsomely printed volumes of their collected works to attack each other.[54] Science is ordinarily a collective enterprise, with the flux of ideas as impossible to monitor as it is to measure. Thoughts and concepts flow freely, and the most solitary of thinkers is influenced by the talk in the corridors and the rumors from experiment. When the stakes are high, however, worries about priority and reputation creep in; the relationship between Lee and Yang could not withstand their own success.

Parity nonconservation was soon found to characterize all manifestations of the proposed Universal Fermi Interaction. But to clinch the case for its existence, physicists would have to do more than show that the interactions—beta decay, muon decay and capture, and strange particle decay—had similar properties and strengths. ("Similar strength" means that the various decays had apparently identical coupling constants.) A sleigh can be pulled with an equal force by, say, a small pickup truck, several Clydesdale horses, or a large pack of Siberian Huskies, but that does not imply that the nature of the entity pulling the sleigh is identical. To show that, experimenters would have to demonstrate that the *form* of the interactions is in all cases the same.

Physicists have classified elementary particle interactions into five kinds, which bear the labels scalar, vector, tensor, pseudoscalar, and axial vector. These imposing names—customarily abbreviated as S, V, T, P and A—designate the only ways that one particle's wave function can transform into another during an interaction (as the wave functions for the neutron and

proton do in Fermi's beta decay theory) and satisfy the demands of both
relativity and quantum mechanics. Physicists can learn much about an interaction by discovering its form. Quantum electrodynamics, for instance, is a
vector (V) interaction, which implies that the transition from one wave function to another is always accompanied by the creation of a virtual particle, the
photon, that has a spin of one and negative parity. For this reason, any
particle with a spin of one and negative parity is called a "vector" particle.
Similarly, an axial vector (A) interaction is associated with a spin-one,
positive-parity particle. And a scalar (S) interaction is one mediated by a
particle with a spin of zero and positive parity.

Copying electromagnetism, Fermi had hypothesized that beta decay is
a vector interaction. But the beta decays of Gamow and Teller could not be
described in this manner. From measuring the velocities of the emitted
electrons, Fermi-style beta decays were known to be either V or S, whereas
Gamow-Teller decays were either A or T. This left only four possibilities for a
Universal Fermi Interaction: VA, VT, SA or ST. And there the matter stood
when Wu started aligning frozen cobalt nuclei.

The Seventh Rochester Conference opened on April 15, 1957, just
three months after the discovery of parity violation. A feeling of elation was
still in the air. One of the most eagerly awaited sessions was the discussion on
weak interactions, in which the chair had been given over to Yang and the
introductory remarks to Lee. Lee quickly threw cold water on the hopes of
the crowd by pointing out that although "there exists a large class of interactions characterized by coupling constants of order 10^{-13}," and that "all weak
interactions seem to have striking features [such as parity violation] in common," there was no reason to believe that they were all caused by one entity,
and indeed some reason to think that they were not.[55]

He drew a triangle representing the particles affected by the proposed
Universal Fermi Interaction, with each vertex representing a current—
again, the top or bottom half of the appropriate Feynman diagram—and
each leg the interaction between the two currents. The left-side leg (proton,
neutron—electron, neutrino) was beta decay, which one remembers has two
different forms, Fermi and Gamow-Teller transitions. Each of the two had to
be examined separately to determine which letter applied to it.

Two years earlier, experiments had been done on neutrons, Ne^{19} (an
isotope of neon with 19 particles in the nucleus) and He^6 to determine their

form of beta decay, with the He^6 especially noteworthy because it was performed under the aegis of Chien-Shiung Wu. He^6 pointed unambiguously toward T for Gamow-Teller transitions, whereas the other two indicated S for Fermi transitions.[56]

All that was fine, Lee said. ST fitted perfectly well into the Fermi theory. But the parity violation experiments of Telegdi, Lederman, and Garwin indicated that muon decay—the bottom leg of the triangle—was V. A single force could not be scalar and tensor in one case and vector in another; beta decay was not the same as muon decay. The similarity in coupling constants appeared to be one of nature's nasty coincidences, like the similarity between the masses of the muon and pion that led the two particles to be confused for a decade.

The case against the Universal Fermi Interaction was hammered home a few minutes later by Wu, who described some as-yet unpublished experiments with another type of radioactive cobalt. The new data contradicted the two experiments that indicated S for Fermi decays. Wu said she believed that they were vector. Combining this with the He^6 results for the Gamow-Teller transitions, she argued that beta decay was a combination of V and T, which was as compatible with Fermi as S and T. This removed the contradiction with the experiments by Telegdi, Lederman, and Garwin and should have been pleasing to advocates of the Universal Fermi Interaction.[57]

Instead, her speech threw the conference into confusion. V and T behave completely differently under parity violation. In practical terms, parity violation means that the neutrino emitted during beta decay always has a spin that went in one direction. A neutrino with "left-handed" spin rotates clockwise to the direction of motion; "right-handed" neutrinos go counterclockwise. (The names come from the act of sticking a thumb out and curling the fingers into the palm; the left hand curls clockwise, the right counterclockwise.) When one plugged V and T into the now parity-violating equations, the vector terms predicted a left-handed neutrino whereas the tensor terms came up with a right-handed particle. In other words, if Wu was correct, the neutrinos emitted in the two types of beta decay were completely different. Not only might there be no Universal Fermi Interaction, but beta decay itself might be caused by a mess of different forces.

One of the more startled conference attendees was its organizer, Robert Marshak.[58] His graduate student, E. C. G. Sudarshan, had reviewed all available experimental data about the "class of interactions characterized by coupling constants of order 10^{-13}." If there really were a Universal Fermi Interaction, Sudarshan and Marshak concluded, it had to be a mixture of V and A, specifically V minus A $(V - A)$. This was a vector interaction with some axial vector thrown in to account for parity violation. Sudarshan and Marshak had intended to present their ideas, but Wu's remarks threw them completely off kilter; moreover, there were rumors floating around that two other experiments were coming out that would back her up. To Sudarshan's annoyance, Marshak, the senior collaborator, said that they should keep quiet until the new data appeared.

One of the rumored experiments was done by Felix Boehm at Caltech. By chance, Marshak ran into Boehm's colleague Gell-Mann in late June at the RAND Corporation in Santa Monica, where both were consultants, and asked him to arrange a meeting. Over lunch, Marshak and Sudarshan outlined the $V - A$ idea to Gell-Mann and Boehm and asked the latter about his experiments; Boehm replied that his data contradicted V and T, but were compatible with $V - A$. Reassured, Sudarshan and Marshak completed their paper in a matter of days and decided to unveil $V - A$ in September at a conference in Padua, Italy.[59] In Padua, Marshak boldly declared that the weak interactions must all be caused by the same thing, and that therefore $V - A$ was the only possible way to go. "I said," Marshak recalls, "that it had to be $V - A$ for a universal interaction, and that to be so will require that four experiments be murdered—I used the word *murdered*—including the He^6 experiment. One physicist then said to Lederman during a coffee break, 'Marshak's *crazy*. How can He^6 be wrong?' "[60]

Gell-Mann, too, had been toying with $V - A$. At the Boehm luncheon, he had been interested to hear Sudarshan and Marshak also pushing the idea.[61] Gell-Mann then talked up $V - A$ with a colleague, who in turn described it to Richard Feynman. (Such is the operation of the physics grapevine.) Feynman had been thinking about weak interactions for some time, but had been unable to understand how nucleon beta decay could be S and T and fit in with anything else. He spent that summer in Brazil. When he returned to Caltech, he asked some experimenters to fill him in on beta decay.

Finally they get all this stuff into me, and they say, "The situation is so mixed up that even some of the things they've established for years are being questioned—such as the beta decay of the neutron is S and T. Murray says it might even be V and A, it's so messed up." I jump up from the stool and say, "Then I understand EVVVVVERYTHING!" They thought I was joking. But the thing I had trouble with at the Rochester meeting—the neutron and proton disintegration: Everything fit but that, and if it was V and A instead of S and T, that would fit too. Therefore I had the whole theory!

Delighted, he went home to work.

I went on and checked some other things, which fit, and new things fit, new things fit and I was very excited. It was the first time, and the only time, in my career that I knew a law of nature that nobody else knew. The other things I had done before were to take somebody else's theory and improve the method of calculating. . . . I thought about Dirac, who had his equation for a while—a new equation which told how an electron behaved—and I had this new equation for beta decay, which wasn't as vital as the Dirac equation, but it was good. It's the only time I ever discovered a new law.[62]

Feynman wrote a paper on $V - A$ with Gell-Mann, and they sent it to the *Physical Review*. It had some features not considered by Marshak and Sudarshan, but, like theirs, claimed that the He^6 experiment must be wrong. Due to the vicissitudes of publishing, the Sudarshan-Marshak article appeared months after the Feynman–Gell-Mann article.

Within a few months of the Padua conference, experimental evidence began to collect that $V - A$ was indeed correct.[63] The question was settled when Maurice Goldhaber and two collaborators performed an elegant and simple demonstration showing that the neutrino was left-handed and the antineutrino right-handed.[64] After this experiment, $V - A$ was established as the form not only of beta decay but all other manifestations of the weak interaction as well. The Fermi force that had grown into the Universal Fermi Interaction was now an accepted fundamental force of nature: the weak force.

All participants in the $V - A$ theory had many reasons to be pleased, but Gell-Mann in particular felt as if the plums had at last fallen into his lap. There had been years of experimental wrestling with the phenomena, but at last theorists had enough solid ground to stand on; the new picture of the weak interactions represented, at least in part, a hope that physics could make giant strides indeed. The vector and axial vector nature of weak interactions was deeply compatible with the vector form of electromagnetism. Two vector interactions—could they somehow be the same thing? Gell-Mann spent a fair amount of time in 1958 and 1959 thinking about a principle that would account for the existence of the weak force, something like the way Yang and Mills had shown how fields can be generated through symmetries. He never came up with a workable model, and so didn't publish any of the scribbling on his sketch pad, but he was sure that a true understanding of the weak interactions would entwine and embrace them with electromagnetism.[65]

12
Steps to Unification

ONE CASUALTY OF THE CONFUSION OVER THE FORM OF THE WEAK INTERAC-
tion was Julian Schwinger, the prodigy who facilitated the renormalization of
quantum electrodynamics. During the 1950s, he built up a school of quantum field theory at Harvard that drew its inspiration from his extraordinary
thrice-weekly lectures on the subject. He drove, in the recollection of his
students, the largest and most elegant Cadillac in Cambridge, a purring
baby-blue sedan that pulled into the lot by Lyman Hall moments before class
began.[1] "Cast in a language more powerful and general than any of his
listeners had ever encountered," one of Schwinger's advisees recalled,
"these ceremonial gatherings had some sacrificial overtones—interruptions
were discouraged and since the sermons usually lasted past the lunch hour,
fasting was required."[2] After a short break, students would gather in a clump
outside Schwinger's office in the hope of a private meeting with the master,
whose fundamental insights were written in a wholly idiosyncratic system of
symbols that avoided all use of the Feynman diagrams that by then pervaded
theoretical physics.

Schwinger navigated a solitary course that paralleled the path of the
rest of the field, independently duplicating when he did not anticipate the
summed efforts of every other person in particle physics. His great physical
intuition and enormous breadth of interest made him into that rarest of
creatures, a one-man community of thought—a forbidding role that eventually isolated him from his fellows. Like Yang and Mills, Schwinger had
been impressed by the way that the theory of quantum electrodynamics
could be associated with a symmetry of phase. Characteristically, he derived
a mechanism somewhat like that of Yang and Mills in his own way, and with
his own notation. (He tried a global, rather than a local gauge symmetry.) But
instead of using the idea to look only at strong interactions, as most of his
colleagues were doing, Schwinger turned also to the weak. And in October
and November 1956, while Sudarshan and Marshak at Rochester were putting together their first notions about the $V - A$ form of the weak interactions, Schwinger dazzled his students at Harvard with a series of lectures
attempting to derive the very existence of the weak force from the symmetries of its interactions—an extension of Noether's theorem and a much more
ambitious project. As if this were not enough, Schwinger introduced, for the

first time anywhere, the idea that the weak and electromagnetic interactions might be two facets of the same phenomenon.

Schwinger had been thinking about unification since before the Second World War, when he worked with Oppenheimer at Caltech. "In 1941," Schwinger once told us, "when I was Oppenheimer's assistant, we talked about all parts of physics. And Fermi's theory of beta decay was one of the hot theoretical topics because obviously it had divergences that were even worse than those we were already conscious of [in electrodynamics]. It was really messy. And I had a thought. Fermi's theory was then expressed in terms of a basic coupling constant. The basic coupling constant in electrodynamics is 1/137. It's a pure number, which is why the whole thing works. Now, Fermi's coupling constant is a dimensional quantity, and since it arose in the days when—well, it's concerned with electrons, so the electron mass was used as a unit of length." (Oddly, by fiddling with the ergs, centimeters, and so on used in the equations, it is possible to use a mass to express an associated length.) "It turns out to be some absurdly low number, 10^{-13}, something or other. Now, I was then a nuclear physicist. And I knew that while the electron was involved in beta decay, it was something that the proton and neutron also engaged in. So I said, 'First of all, why isn't the proton or the nucleon mass the important thing?' When I introduced that, the coupling constant became 10^{-5}, a more reasonable number." Using the much larger mass of the nucleon as a standard pushed the coupling constant up by a factor of a hundred million. "Then, being very aware of electrodynamics, I said, 'Hey, could it be that if the relative mass [of the virtual particle in beta decay] was a couple of hundred proton masses, the coupling constant would become the fine structure constant—1/137?'

"It was numerology," Schwinger said dismissively. "Numerology. But—that's the whole idea. I mentioned this to Oppenheimer, and he took it very coldly, because, after all, it was an outrageous speculation."

We asked if this had all happened in the 1940s.

He nodded. "In 1941. But it's the idea of the intermediate particle, with the mass coming from making the coupling constant to be the fine structure constant. And so that had always been on my mind. In 1956, when I was hit with parity—it was a blow to me, parity violation—I began to rethink the whole subject. And at that point, I introduced the idea explicitly, that there was a vector particle which carried weak interactions, and all the rest of it."[3]

Parity violation was indeed a shock to Schwinger. When the rumors about Wu's experiment began seeping out of the Bureau of Standards, Schwinger urged his Cantabrigian colleagues not to jump to the idea of parity violation before the results were known. In one of the discussions that took place everywhere in the physics world in the first days of 1956, Schwinger told his fellows that parity was too lovely a symmetry to be easily abandoned. The argument was interrupted by a call from Schwinger's mentor, I. I. Rabi. Schwinger returned with the sad words, "Gentlemen, we must bow to nature."[4]

Like many elementary particle physicists, Schwinger was pushed by parity violation into rethinking the nature of the weak interactions. The thoughts grew into a series of lectures, and out of the lectures came a paper, provocatively entitled, "A Theory of the Fundamental Interactions," which was published in the November 1957 issue of the *Annals of Physics*.[5] An expansive piece of writing, it is filled with Schwinger's trademarks: an evident love of mathematical ingenuity, an utterly eccentric notation, and a bold, imperious style that makes almost no reference to the work of other physicists. (This last has offended many of Schwinger's colleagues, whose reputations depend on such citations.)

The most important part of the paper is a section at the end, on the class of particles called leptons, which are unique in that they respond only to the weak force and not to the strong. In 1957, there were three known types of lepton: the familiar electron; the neutrino; and the muon—the "who ordered *that*?" particle—which is exactly like the electron except that it is about two hundred times heavier. All are as impervious to the strong force as a tree stump is to an electromagnet. In his paper, Schwinger simply assumed the existence of two types of neutrino, one associated with the electron and one with the muon. True to form, he did not identify his conjecture as such, but wove it into the logic of his argument.

("Schwinger always pretended that everything he wrote was a direct revelation from God," one of his colleagues told us. "You'd be reading along and suddenly there's two neutrinos—a direct revelation from the Lord!"[6] On the other hand, other physicists had independently made the same guess beforehand.)[7]

Schwinger needed this first speculation to make a second bit of speculation come out right. He postulated that the weak force is carried among the leptons by three particles—the photon and two hypothetical vector bosons, which are now called W^+ and W^-, and are analogous to the B particles in the paper by Yang and Mills.

There had to be *two* vector bosons to account for the way that weak interactions convey electric charge. When the weak force transmutes an uncharged neutron into a positive proton, a virtual particle is emitted, and in its decay an electron and an antineutrino are created. Schwinger argued that when the weak force transfers a positive charge, that charge is carried by a positive version of the vector boson, now called the W^+, whereas when the weak force transfers a negative charge, it is carried by a negative version of the vector boson, the W^-. The ordinary photon would be the intermediary when no charge is involved.

Schwinger told his classes that this picture of the weak force had far-reaching consequences. Because the photon and the Ws make up a triplet, Schwinger said, electromagnetism and the weak interaction should be treated as part and parcel of the same phenomenon; that is, he had unified them. But he was more cautious in print, as is proper. He never used the word "unification" in the paper, referring instead to the weak force as "a partner of the electromagnetic field."[8]

There was no compelling reason for his supposition; Schwinger merely wanted to demonstrate that his pet notion could be used to construct a coherent theory. In addition, he found the result aesthetically pleasing. His colleagues did not. During a visit to New York City, Rabi told him bluntly, "They hate it."[9] Upset by the reaction, Schwinger was further dismayed to learn that the experimental data he was relying on were wrong. By artful juggling, he had managed to make the description of the combined weak and electromagnetic interactions fit the data on the then-believed V and T form. (Schwinger, too, had been at Rochester when Wu argued in favor of that interpretation of the experimental findings.)[10] Between the time he submitted the paper and the time it was published, $V - A$ was in. By the time Goldhaber and others proved that $V - A$ was right, Schwinger had thrown up his hands and turned to other areas of physics.

Before he did, however, he asked a student, Sheldon Glashow, to look at the possibility of a connection between the weak and electromagnetic forces—to see if a connection could still somehow be established. "It was a vague request," Glashow recalled later. "He said, 'Think about intermediate vector bosons as agents of weak interactions.' My problem was to think about intermediate vector bosons. And that's what I did for two years—think about it."[11]

Sheldon Glashow's path toward unification began at the extreme northern tip of Manhattan, where a little side street, Payson Avenue, runs uphill for a long block beside Inwood Park. Facing the grass and trees is a row of small two-story houses, one of which—number sixty-five—was Glashow's boyhood home. Glashow's father, Lewis Gluchovsky, had come from the town of Bobruysk, in White Russia, where the family had owned a Turkish-bath house ("*the* public baths of the city," according to Glashow). When he arrived here, the immigration authorities changed his name; shortly thereafter, Gluchovsky/Glashow became a plumber in Manhattan. He married another immigrant, Bella Rubin. They had two boys, aged fourteen and eighteen, when the third, Sheldon Lee, was born. One of Glashow's brothers became a doctor, the other a dentist. Sheldon decided at the age of eight or nine that he wanted to be a scientist.

"It was the beginning of the Second World War," Glashow said to us. "People were interested in those days in identifying airplanes and understanding how bombs are dropped, and I got curious about the ballistic problem. My brother explained to me that if a plane drops a bomb and doesn't take some evasive action, the bomb is going to blow up right underneath where the plane is *going* to be. That of course struck me as a very odd fact, because the assumption is that when the plane drops the bomb, the plane is moving forward and the bomb is not.

"I had a little notebook, this black-and-white speckled notebook that kids still have. My brother would write little lessons—two or three or five such lessons—about things he had learned. Not long before, Harold Urey had discovered heavy water, and Urey was his professor in school, so he was a little turned on about it, and he communicated that to me."

Lewis Glashow was not entirely pleased by his son's interest in science. "My father told me that science would be all right, but why didn't I do it in my spare time and go into medicine, like my brothers. I differed with him slightly about that. He relented when my brothers came back safely from the war. When I entered high school, he helped me build a chemistry lab in the basement, where I loved to perform long and dangerous experiments."[12]

Glashow's passion for science was not directed to physics until 1947, when he entered the Bronx High School of Science. After classes, the boys in the Science Fiction Club—there were no girls—would cluster around an old laboratory table littered with surplus hospital Bunsen burners and the tracework of test-tube racks. The preferred topics of conversation included the contents of a magazine called *Astounding Science Fiction* and the ideas of L. Ron Hubbard and Immanuel Velikovsky. Two of Glashow's best friends were Gerald Feinberg and Steven Weinberg. (Feinberg became a distinguished physicist and has written several respected books; Glashow and Weinberg shared the Nobel Prize in physics in 1979 for their contributions to a theory incorporating the weak and electromagnetic interactions.) As boys, all three contributed to the Science Fiction Club fan magazine, *Etaoin Shrdlu*; the name is a term from typography.

The founder of Bronx Science, Morris Meister, was a pioneer of fast-track education and a tireless advocate of the notion that if bright, science-oriented students are brought together, certain ill-defined but nonetheless valuable learning processes will occur. As far as Bronx Science is concerned, he seems to have been right. Since its founding, in 1938, it has produced three winners of the Nobel Prize for physics—more than any other high school in the world ever has, and as many in those forty-eight years as most countries, including France and Italy.

Considering that statistic, it is surprising to learn that the physics taught to Glashow, Weinberg, and Feinberg was of dubious merit.[13] In Feinberg's recollection, the instructor who taught his introductory course, "for all we could tell, did not believe in the existence of atoms."[14] The textbook was written by a man named Charles Dull, who seems to have lived up to his name. Students interested in pursuing advanced physics were offered a choice between "automotive physics," in which the kids soberly took apart and put together an old Curtiss-Wright airplane engine each semester (a good course "for those who believe the automobile is here to stay," the 1950 yearbook states); and "radio technology," in which students put together shortwave kits of the sort advertised on the back of *Boy's Life*.[15] Glashow took neither.

Despite the modest course offerings, a kind of *samizdat* science took place outside the classroom. Glashow persuaded his mother to buy him college outline texts whenever she dragged him downtown to shop; at lunch Feinberg often led excited conversations about quantum mechanics. Most important, however, was the give-and-take among the Science Fiction Club members; the boys competed with each other to reach the most complete understanding of the latest physics discoveries they had encountered, but they also helped stragglers to catch up. Feinberg said, "After the meetings, Shelly and I would spend hours on the phone talking about science. An hour, hour-and-a-half a day—it drove our parents crazy. We were sure we were going to find wonderful things."

After graduation, Feinberg chose to stay in New York City and go to Columbia. Glashow and Weinberg went upstate, to Cornell. Although Cornell has become an internationally important center for physics, in the early 1950s it had the nickname "Moo U" and was known chiefly for agricultural science. Cows grazed on the lawns, and campus lore had it that professors were not allowed to eject dogs that might stray into their classrooms. Although some prominent physicists—Richard Feynman, for example—were at Cornell when Glashow and Weinberg entered, in the class of '54, their courses were restricted to graduate students. Those who taught undergraduates regarded most forms of collaboration among students as dishonest, and after Bronx Science Glashow found that attitude hard to take. "There was a style mismatch. Some of us were thrown out of school for cheating. We would collaborate on their dumb problem sets and they didn't like that, for example. One of us would show up with six copies of homework—sort of identical homework—and the teacher would simply ask if one of us wanted the 100 percent credit or if we wanted to divide it up. It got to the point where I flunked a course taught by a solid state professor. I just found the problem sets absurd and refused to do them."[16]

In his senior year, Glashow was exposed for the first time to quantum field theory. He was the only undergraduate in the class, and the subject baffled him. At exam time, the professor, Silvan Schweber, was nonplussed to learn that he could give all the graduate students pass-fail grades, but had to give an undergraduate a numerical one. Glashow recalled: "So he said, 'How 'bout an eighty-five?' Which was not too low, not too high. I wanted to try for something better, so I said, 'No, give me an exam.' Schweber asked me a bunch of questions and I got every single one of them wrong. At the end, he said, 'How 'bout an eighty-five?' I said, 'Great!' "[17]

("I still remember giving him the oral," said Schweber, who is now at Brandeis University, in Cambridge, Massachusetts. "It was clear that he was a very unusual undergraduate. It was also clear that he thought he could do the work with his left hand. My memory plays tricks on me—I can't be sure now why I did it—but by giving him that [very hard] test I may have been trying to tell him you can't do physics that way.")[18]

Despite his trials, Glashow learned some physics, most of it in the same way he had at Bronx Science—by shooting the breeze with his friends. He argued with Weinberg, dashed off letters to Feinberg, and dropped down to New York City on weekends to hear lectures; back in Cornell, he latched on to classmates who could teach him something. Perhaps the most important material he absorbed came from a graduate student, Harold V. McIntosh, who had decided to pursue physics after getting a degree from the Colorado School of Mines. Brainy and eccentric, McIntosh had a little coterie of enthralled undergraduates whom he introduced to his current obsession, the mathematical theory of groups. Glashow said, "He was always coming up to me with problems involving vibrating bedsprings and jiggling ozone atoms and things like that which he claimed group theory was capable of solving. Actually, what I learned from him was as relevant as any course I took there."[19] (The last Glashow heard, McIntosh was teaching physics in Mexico.)

Glashow chose to do his graduate work at Harvard, where he might be able to write his thesis with Schwinger, one of the heroes of the effort to renormalize quantum electrodynamics. At first, Glashow found Harvard graduate school little better than undergraduate life at Cornell. "Nothing is quite as dull as being here," he wrote to Feinberg in November of 1954.[20] He saw Schwinger rarely, and considered the classes insufferably tedious.

By Glashow's second year, however, he was taking courses directly from the master, an experience that he found both electrifying and peculiar. A charismatic teacher, Schwinger spoke at lightning speed, covering the blackboard with equations, many of which were expressed in his own "Schwingerese." Regardless of the description in the catalogue, Schwinger's advanced graduate courses were always about whatever he was working on at the moment. He sometimes began class by saying that he had just realized that what he had told them the day before was utterly wrong. Glashow loved him. A shy man, Schwinger had a reputation for aloofness outside the classroom; his students claimed that all they ever saw of him was the flash of his florid silk tie as he ducked into the men's room to avoid them. ("He suffers from an uncertainty principle," Glashow said.)[21] Nevertheless, at the beginning of his third year, Glashow asked Schwinger to be his thesis adviser. So did eleven other students. To Schwinger's consternation, he found all of them waiting in his office one day. Thinking to test their abilities and avoid dealing with them as a pack, he devised an enormously complicated problem and instructed the multitude to work on it and return, one by one, with their solutions.

"So *of course* we collaborated," Glashow recalled. "The twelve of us came back a few days later, all at the same time, having solved the problem. He was happy with the solution—it was very elegant—but not so happy with the complete failure of his scheme to sort out the masses. So he then came up with a problem for each of us to work on individually, and in that way, against his will, got us all started on our theses."[22]

The question Schwinger awarded Glashow was one close to his current concerns. He asked Glashow to explore the behavior of the virtual particles Schwinger had hypothesized to be the carriers of the unified weak and electromagnetic forces—but this time to look at them in the light of the new experimental data on weak interactions. Glashow did not know he was being put on the track that would eventually lead him to a grand unified theory; he knew only that Schwinger was asking him to look at the sort of wild idea that was dear to his heart.

Today, Glashow's thesis—the product of those two years of thought— sits on his bookcase, stuck in a black spring binder on a neglected top shelf. He had to hunt for it, wading through the stacks of preprints and journals that clutter his office, when he wanted to show it to us. On a bulletin board beside the door were many strata of notes and memos, including a letter from physics students in Beijing claiming to have discovered antigravity ("Is *most* important result!"), newspaper advertisements showing how Transcendental Meditation can harness the power of the Grand Unified Field, and a number of clippings about the Soviet physicist Andrei Sakharov, in whose defense Glashow has been active. "Here it is!" Glashow said, standing on a wobbly stool. He hefted the volume appreciatively. "These things are always long, to show you have lots of bright ideas, and filled with tons of calculations—student showboating. Mine is a complete parade of crazy digressions. I haven't looked at this in *years*. But there's one part I'm still proud of. Here, wait, it's in the appendix."

This was no surprise. Graduate students of all stripes traditionally hide their most radical statements in footnotes and appendices, where they can be disavowed if necessary. Glashow was now sitting at his desk, wreathed by cigar smoke and shaking his head at the maunderings of his youthful self. When he found the page, he stuck an admonitory finger in the air.

"*Yes.* 'It is of little value to have a potentially renormalizable theory of beta processes [i.e., weak interactions] without the possibility of a renormalizable electrodynamics.' That's because the Ws are charged, so you had to get into quantum electrodynamics to talk about them. Here we go. 'We should care to suggest [his voice rises here; his pleasure in the younger Sheldon Glashow is apparent] that a fully acceptable theory of these interactions may *only* be achieved if they are treated together.' "

He snapped the binder shut, kicking up a little cloud of dust. "That's the basic idea of the electroweak theory right there, that the weak interaction'd never make any sense by itself because it's always mixed up with electromagnetism. So you have to look at them together. It sounds simple, but that kind of realization is a big deal. Before Newton could work out his theory of gravitation, he first had to realize that the force that moved the planets around the sun and the force that dropped apples on people's heads were the same thing. So, in a certain sense that sentence won me the Nobel Prize. It took thirteen years more to show that I was right, but damn it, I was."[23]

While working on his thesis. Glashow won a National Science Foundation fellowship, a prestigious award that entitles its recipient to study abroad. Glashow spent two years, from 1958 to 1960, at the Niels Bohr Institute of Theoretical Physics, which invites a small number of physicists every year to participate in the discussions held there. Glashow attended lectures, gave talks, met the great Bohr, and in general lived the life of which he had dreamed years before at Bronx Science.

At the Institute, Glashow expanded the appendix of his thesis into a full-fledged paper, "The Renormalizability of Vector Meson Interactions," which he submitted to the journal *Nuclear Physics* in the fall of 1958.[24] (Maddeningly, physicists for years used "vector meson" and "vector boson" interchangeably in this context.) Unlike Schwinger, Glashow thought that a Yang-Mills theory—that is, a local gauge theory—could be the basis for a unified theory. Moreover, Glashow wanted to prove that such a unified Yang-Mills theory of the weak and electromagnetic interactions was indeed renormalizable, as he had alleged. By November, when the paper was done, he thought he had it.

Pleased with himself, he went to England in the spring of 1959 to present his work. "I lectured in London or somewhere. They all sat there smiling very politely, told me that they liked what I was doing, and went home to tear it apart."[25]

"They" in this case refers mostly to Abdus Salam, who later shared the Nobel Prize with Glashow and Weinberg. Salam and his collaborator, John Ward, had also been inspired by Schwinger's paper to look into the question of linking the weak and electromagnetic interactions. Although the two men were skilled at the techniques of renormalization, they had been unable to rid the theory of infinities. When Glashow claimed he had, they were appalled. Salam said to us, "Ward and I were in London, and he was just passing through. To our surprise, he was working on the same thing—having been put on it by Schwinger. He had the W^+, W^-, and the photon in one group—that part was common. But then he also did a quite *amazing thing!* He said the theory was renormalizable! My God! this young boy was claiming that this theory was renormalizable! It cut me to the quick! Both of us considered ourselves *the* experts on renormalization. So there we were, the two great experts on renormalizability, wrestling for months with the problem—and here was this slip of a boy who claimed he had renormalized the whole thing!" Salam laughed. "Naturally, I wanted to show he was wrong— which he was. He was completely wrong. As a consequence, I never read anything else by Glashow, which of course turned out to be a mistake."[26]

Glashow had made mathematical and physical errors; his theory was not, after all, renormalizable. In fact, as Salam delightedly pointed out, if one did the math correctly one actually proved that the model of two Ws and a photon was not and could not be renormalizable, exactly the opposite of what Glashow had asserted.[27] For Glashow, the episode was embarrassing and annoying—a physicist's nightmare. ("They don't make graduate students that dumb any more," he said, groaning.)[28]

He was now faced with a choice more dependent on his personality than on the rules of scientific inquiry. Young theoreticians have to come up with bold ideas if they are to acquire a reputation, yet their papers cannot be half-baked. Glashow's idea was bold, but it was wrong. If he kept going, he risked being labeled an eccentric; if he went on to something else, it would be an admission of defeat. Glashow chose not to abandon unification; instead, he went further out on a limb, advocating an even wilder, more speculative extension of his original position. He threw himself into a sequel, dashing off notes to Feinberg and Weinberg and receiving encouraging criticism, turning over his ideas with the bemused care of a raccoon washing a shiny ring.

In March 1960, after Glashow gave a talk in Paris, Gell-Mann, whom Glashow had never met, invited him to lunch. Gell-Mann was at that time a visiting professor at the Collège de France. As he was then in the middle of an extraordinary decade in which he totally dominated particle physics, Glashow was more than happy to be his guest. The two went to a seafood restaurant. Glashow mentioned an aversion to fish; his dining companion explained that serious people liked seafood, that Glashow was a serious person, and that therefore Gell-Mann would order fish for both of them. Over the entrée, the two men talked about vector bosons. Loathing the odd notation, Gell-Mann had not read Schwinger's paper, but he had himself been thinking for several years about ways to reconcile the weak and electromagnetic forces. He urged Glashow to press on, and invited Glashow to join him at the California Institute of Technology in the fall to work on field theory. "What you're doing is good," Gell-Mann told Glashow. "But people will be very stupid about it."[29]

"Partial-Symmetries of Weak Interactions," the paper that established Glashow's claim to a Nobel Prize, was published in the November 1961 issue of *Nuclear Physics*—one month before its author's thirty-first birthday.[30] The paper has the dry, confident tone that characterizes the best scientific prose. "At first sight," Glashow begins,

there may be little or no similarity between electromagnetic effects and the phenomena associated with weak interactions. Yet certain remarkable parallels emerge with the supposition that the weak interactions are mediated by unstable bosons.

Next, Glashow squares up to a problem that had arisen first in the work of Yang and Mills and was still unsolvable: how much these *W* bosons weigh.

The mass of the charged intermediaries must be greater than [zero], but the photon mass is zero—surely this is the principal stumbling block in any pursuit of the analogy between hypothetical vector [bosons] and photons. It is a stumbling block we must overlook.

Glashow simply flags the question in passing with a "don't know" sign, and turns to the rest of the theory. This time, he argues that the theoretical problems can be resolved if one assumes not three virtual particles, but *four*.

In order to achieve a . . . theory of weak and electromagnetic interactions, we must go beyond the hypothesis of only a triplet of vector bosons, and introduce an additional, neutral vector boson, Z_S.

As he confesses,

The reader may wonder what has been gained by the introduction of another neutral vector [boson].[31]

What had been gained was this: The Z_S allowed Glashow to separate the weak and electromagnetic interactions, and to deal mathematically with each in turn. (Today a very similar entity is called Z^0—"zee-zero"—and without getting into unnecessary complications we shall henceforth identify the Z_S as the Z^0.)[32] In the first Schwinger-Glashow scheme, the photon had worked overtime—both for electromagnetism and the weak force—with the result that the two functions interfered with each other in the theory. The new model was far neater.

Unfortunately, the neatness came at the expense of unification. Glashow could only resolve the difficulties with renormalization by "detaching" the weak and electromagnetic forces; they now worked hand in hand, but were no longer fully unified. In this model, the weak and electromagnetic forces within the atom are like two children around an elaborate Lionel train set; each has a separate control panel, and frantically switches the gates, toots the whistle, and twists the throttle without consulting the other. The motion of the train is the result of the actions of both, and depends moment by moment on what each is doing. So it is for subatomic particles. Surrounded by a haze of photons, Ws, and Zs, their movements are a synthesis of the effects from all of them.

The contribution of each force to the total interaction is expressed in a ratio, so much weak to so much electromagnetism, that Glashow describes in the paper. It is customarily known as a "mixing angle" because of a principle from high school trigonometry that most students immediately forget: Any ratio or percentage can be expressed in terms of a trigonometric function of some angle. Today, to Glashow's chagrin, the mixing angle is sometimes called the "Weinberg angle."[33]

It is the Weinberg angle because seven years after Glashow's paper was published Weinberg reinvented it. No one had paid much attention to Glashow's paper: It had appeared in an unknown journal and, if that had not been enough to guarantee its obscurity, there was the fact that it was written in Schwinger's odd notation.

A deeper reason was the sheer ungainliness of the model. Glashow had jammed together two gauge theories. The first was ordinary quantum electrodynamics. The second was a theory of the weak force based on the use of isotopic spin by Yang and Mills. But instead of using ordinary isotopic spin, as they had, Glashow postulated the existence of an isotopic spin-like quantity that applied to electrons and neutrinos; its conservation generated the weak force. But because parity is conserved by electromagnetism and violated by weak interactions, half of the model applied equally to left- and right-spinning particles, whereas the other part applied only to left-spinning particles—which meant the model looked like an unholy mess.

"When $V - A$ was established," Gell-Mann said, "everybody began working with electromagnetism, the weak force, and Yang-Mills, trying to unify. But it never worked. The crucial missing step was Shelly's idea. He pinned down the screwy, ugly, disgusting, stupid, horribly asymmetrical model nature had picked out to work with." Indeed, Glashow's paper was the first rough sketch of a triumph of modern physics, a theory tying together the weak and electromagnetic forces. This electroweak theory, as it is called today, does not fully unify the two forces. Nevertheless, it ties them together so firmly that most scientists refer to it as unified. Furthermore, it became a crucial step on the road to the more speculative fully unified theories that dominate the field today.

Intrigued by Glashow's work, Gell-Mann took it before publication to the 1960 Rochester Conference.[34] "I gave the first clear description of his ideas," Gell-Mann said, meaning that he translated the theory into a form his colleagues could understand. "The reaction was very unenthusiastic."[35]

One reason for the lack of enthusiasm was that in addition to the theoretical peculiarity of Glashow's model, it had run into problems with the experimental data. By adding a third vector boson, the Z^0, to make the model look like a gauge theory, Glashow had created a new problem having to do with what are called weak neutral currents. In general, particle interactions are described in terms of currents; when the neutron of beta decay turns into a proton, this is known as a "weak current," because the current is the result of a weak interaction. In Glashow's hypothesis, the weak current occurs when the neutron emits a vector boson. The neutron-proton current can be further typified as a "charged weak current," because the uncharged neutron becomes a positively charged proton—a change of charge occurs. In terms of Glashow's electroweak model, the neutron emits a negatively charged vector boson, the W^-. If there were a weak interaction in which the neutron did not change identity, this would be a "weak neutral current." Such an interaction would occur, Glashow argued, if the neutron emitted a Z^0. The problem was that experimentalists had looked at millions of weak interactions and were certain that weak neutral currents did not exist. Not once, in hundreds of careful experiments, had they ever observed a weak interaction that might be the passing of an uncharged vector boson.

Unfortunately for Glashow, in this case nature was playing a trick on experimenters. Because weak interactions are usually swamped by those of the strong and electromagnetic forces, experimenters had done almost all of their work in this area with the particles in which the effects of the weak force were easy to see—strange particles. Strange particles only decay via the weak force, and physicists had used them since their discovery as a laboratory to study that interaction. For reasons that Glashow would only unravel a decade later, strange particles are just about the only type of particle in which weak neutral currents of the kind experimenters were looking for do not exist, and thus their years of research did not bear on Glashow's theory.

At the time, of course, he did not know this. Like his colleagues, he believed that the data on strange particles held true for all forms of matter. To save his idea, Glashow invented a reason why the Z^0 had never shown its face. He asserted that it must be a lot heavier than its brethren, the W^+ and W^-. To create a Z^0, experimenters would therefore have to bring to bear much more energy than any existing accelerator could generate. No wonder the particle had never been seen. (Such theoretical gambits infuriate experimenters, by the way.)

Alas, this strategy threw the baby out with the bathwater: In order to explain how weak neutral currents could exist but not be observed, Glashow had hypothesized a Z^0 so heavy that it would hardly ever put in an appearance. The paper concludes, "Unfortunately our considerations seem without decisive experimental consequence."[36]

The problem of the mass of the vector bosons and the question of weak neutral currents dogged the electroweak theory from the outset, and these difficulties were not resolved for years.

In 1960, after Glashow finished his paper on weak interactions, he moved to Pasadena, where he became part of a team scrambling to keep up with Murray Gell-Mann. One of the true *enfants terribles* of twentieth-century science, Gell-Mann entered Yale at fifteen and received his doctorate at the age of twenty-one. A polymath, he is sometimes said by his colleagues to have no particular talent for physics, but to be so smart that he is a great physicist anyway. Although only three years older than Glashow, he had been on the scene almost a decade longer. By 1960 he had the run of the California Institute of Technology and was spewing out ideas at a dizzying rate. He is a stocky man, whose short gray hair grows in tight curls; he moves with bulletlike determination, his shoulders thrust forward, his gaze fixed ahead. Notorious for his drive and asperity, he is also quick and generous with praise. Like Glashow, he loves conversation. The two men enjoyed each other's company and soon began to work together on an ambitious article that they hoped would pave the road to unification.

The result—"Gauge Theories of Vector Particles"—appeared in the *Annals of Physics* in September 1961.[37] It is a curious document; Gell-Mann has described the first half as a "classic, a landmark," and the second part as a "mess."[38] The first half sets up a kind of grammar for quantum field theory—a catalogue of possibilities. Drawing upon group theory (Glashow's introduction to it at Cornell, under the tutelage of Harold McIntosh, had not been a waste of time), the two physicists noted that the sets of hypothetical virtual particles described in Yang-Mills theories corresponded to the generators of a particular type of group.

A group is a set of mathematical objects related by a symmetry. An example of a group is the six sides of a die, which has the same shape no matter what side you are looking at. The symmetry is shown when the cube is rotated ninety degrees in any direction. Other groups are the sets of all ways one can turn a round rubber ball or a book. Interestingly, the latter two groups are quite different. It makes no difference whether a ball is turned to the side then flipped upward or the other way round. On the other hand, if this book were to be turned to a side and then flipped up the cover would be in a different place than if these movements were done in the other order. Mathematicians say that the second group, the set of book rotations, is not commutative, because the outcome of several successive rotations depends on the order in which they are performed. Because, as Heisenberg discovered in the 1920s, quantum variables are often noncommutative, this kind of group might be expected to be of interest to physicists, and indeed it is. Moreover, as Gell-Mann and Glashow discovered, all possible variations of such groups had already been laid out nearly thirty years before by the great French mathematician Élie-Joseph Cartan. Cartan had given the groups names. For example, Cartan's group U(1) corresponded to the gauge field for quantum electrodynamics, and his group SU(2) × U(1) to Glashow's electroweak model. (It is usually pronounced "ess-you-two-cross-you-one." SU stands for "special unitary" group and U for "unitary." SU(2) is the name for the group defined by all possible ways of turning something like a book; it is also the group for all possible rotations of the axis of isotopic spin.)

In the first half of their paper, Glashow and Gell-Mann showed that every one of the groups Cartan defined corresponded to a Yang-Mills gauge field, and that if one constructed a Cartan-style group of virtual particles there would always be a Yang-Mills field associated with it. The difficulties arose in the second half of the paper, in which Glashow and Gell-Mann tried to do something practical with the classifications borrowed from Cartan. They proposed a theory of the strong force to go along with Glashow's electroweak theory, and expressed them both in terms of Cartan's groups. With this trick, they sought to treat the strong and electroweak forces as Yang-Mills gauge fields. In this way, the strong and electroweak forces would be explained in the same language as electromagnetism, and thus would be linked to the only proven theory in elementary particle physics: quantum electrodynamics. But the attempt was like a game of basketball and a game of hockey played in the same arena at the same time. Gell-Mann explained, "We wanted to construct a Yang-Mills theory for the weak interaction and one for the strong interaction. But to do this, we had to have them both in the same space. And they *fought* each other. It was *terrible*. They would get mixed up with each other in some ghastly way, and the weak interactions would turn strong—it was *bad*, it was *sick*. Finally, I gave up on it. I said the hell with it. This one, I said, we're not going to solve now."

Glashow abandoned the effort, too. He was discouraged not only by their failure to put the electroweak and strong forces into a single theoretical framework, but also by the still unresolved problem of the mass of the W and

Z particles and by the lack of experimental evidence for weak neutral currents. For almost a decade, he turned his attention to the strong force, consigning all his earlier work to one of the dusty piles of paper in his office.

He returned to unification only once during this time, in the summer of 1962, when he spent a few months in Turkey. Roberts College (now the University of the Bosphorus), where he taught a seminar with Abdus Salam, is in the town of Bebek, which overlooks the Bosphorus. It is not far from Istanbul, where Glashow journeyed on the young American's ritual pursuit of hashish. Salam, who disapproved, accompanied Glashow to Iki-Bucuk Street, where a criminal named Ceri sold them a considerable quantity of greenish-brown Bursa hashish. After smoking much of it, Glashow is reported to have babbled incoherently about unification into the early hours of the morning. According to participants, Glashow resisted all efforts to quiet him, and pontificated with the volume and conviction of a Biblical prophet.[39]

As a testament to his concern, "The Secrets of S. L. Glashow," a notebook he kept during that summer, has a page labeled "For Fun." On the top is the question, "How does finite mass come about?" There was no answer. But the bottom half of the page and part of the next contains a sketch of an $SU(2) \times U(1)$ model that is, in retrospect, almost exactly correct. At the end, the mass of the Z^0 is predicted to be twice what it actually is ("I didn't do my arithmetic carefully"). The notes conclude with an almost audible sigh: "So much for this model—rather offbeat—just intended as a curiosity."

13
Broken Symmetry

GLASHOW WAS BY NO MEANS THE ONLY PHYSICIST TOYING WITH GAUGE theories, nor was he the only theorist with a model of how one might work. From 1957 to 1964, he executed an unknowing intellectual *pas de deux* with the man who later joined him in Stockholm, Abdus Salam. Often physicists happen more or less independently upon the same conclusions, but the temporary harmony between the separate styles of Glashow and Salam is as striking as the sudden, simultaneous appearance of African motifs in the work of Picasso and Matisse. Salam today is head of the International Centre for Theoretical Physics in Trieste, a stark modern cluster of cement-and-glass slabs pocketed in a hillside with a view of the Adriatic Sea. The Centre— Salam follows the British spelling—is funded by two United Nations agencies, the International Atomic Energy Agency and the United Nations Educational, Scientific, and Cultural Organization. Its aim, purely and proudly, is to promote basic research in theoretical physics, to foster thinking unfettered by the demands of practicality, in the belief that in the long run such scientific work is just as valuable to any nation, no matter how poor, as education to fit narrow and immediate needs, no matter how pressing. Salam suggested the idea of creating the Centre in 1960, and fought relentlessly for its creation at United Nations fora. He has given much of his life to the cause of Third World education, and, as shall be seen, some of his science. The Centre won new life after he received a Nobel Prize for the $SU(2) \times U(1)$ model of electromagnetic and weak interactions.

He has a position there and another at Imperial College, in London; he is on the board of directors for a score of Third World and other committees; he advises and consults and in between finds time to do physics. When we first met him, he had just flown to Trieste from the United States and was soon to go to Geneva for a meeting of the CERN Science Policy Committee. A consul from Yugoslavia was expected to arrive with a plaque announcing Salam's induction into that country's Academy of Sciences. Despite the frantic schedule, Salam himself seemed as calm as the surface of a summer lake. Beneath his chair was a white shag rug, a small space heater, and a favorite pair of worn slippers. His face was thoughtful behind twin barriers of beard and glasses. There was a Dickensian sense of coziness in the room. The feeling of comfort was heightened, perhaps, by the racket outside of the gelid *bora*, the notorious Triestino winter wind which gusts so fiercely that ropes are strung along the sidewalks to prevent pedestrians from being blown into traffic. Salam had on a thick brown suit of the sort that looks incomplete without a watch fob. A blackboard filled one wall with the calligraphy of grand unification. On another wall at eye level: a framed quotation from the Koran.

Salam was born and raised in what was then the Punjab province of British India. A gifted child, he created a sensation in his hometown of Jhang by having the highest marks ever recorded in the entrance examination for the Punjab University. Through a series of lucky coincidences, he was enabled after graduation in 1946 to leave the riot-torn subcontinent for England a year before partition would have irrevocably wrecked such an arrangement.[1] He finished an undergraduate degree in mathematics at Cambridge but found himself at the end with another year's scholarship and a hankering to try some physics. He went to his supervisor, the astronomer Fred Hoyle, explaining that he wanted to be a theoretical physicist and had one more year of scholarship money. Hoyle informed him loftily that proper physicists had to take actual experimental courses; otherwise, he told Salam, they could not look their colleagues in the eye. "I said, 'Look, first of all, this physics course which you are asking me to take lasts two years. I have funds for one year only. Secondly, I have not done physics for about six years—proper experimental physics. How the *hell* do you expect me to cope with the lab work?' And he said, 'Never mind, it's a challenge to you.' " Salam went to see his college adviser, a geologist, who rubbed his hands in glee at the prospect of learning whether someone who had won a first-class undergraduate mathematics degree in two years could make a first in the undergraduate physics course in one. "You see," he said to Salam, "[N. F.] Mott and G. P. Thomson [both later Nobel winners] tried, but they failed."

"So he just wrote my name down for physics," Salam said. "I went to the lab in physics and *by God* it was *hard!*" His voice is soft, rather quiet, the vowels held in the throat as if he enjoyed them; he has the English habit of whispering for emphasis. We had come upon him in another place—Geneva, as it happened—six months later and asked him what it was like as an undergraduate in the Cavendish. Salam had just given a talk on Kaluza-Klein theories, the type of unification ideas that then interested him most, and folded, ten-dimensional spaces made a chiaroscuro of equations on the blackboard behind his head. By coincidence, he was wearing the same suit; it had a watch fob, or something like one. "In the Cavendish, there was the old equipment, ancient equipment, and nothing but. Rutherford's own equipment—and you were supposed to make it work. You had to blow glass tubes yourself and carry them three flights of steps. It was a torture. They wanted it to be torture, and they succeeded. I remember the first experiment which I was given. It took me four full days to complete. Basically rather simple, the experiment—you had to measure the difference of two sodium D spectral lines, the wavelength difference, by an interferometer method. I took three days to set up the equipment, set it up properly, and then I took three readings. Three readings on the principle that I wanted to get a straight line—two points to determine a straight line, and the third to prove it. I took this piece of work to Sir Denys Wilkinson, who is now vice chancellor of Sussex University and one of the brightest experimental physicists in the U.K. He was one of the supervisors who awarded you marks on your write-up. He looked at this with a quizzical look on his face. He said, 'What's your background?' I said I came from mathematics. He said, 'Oh, I can see that. You realize that you have to take *one thousand* readings before you have a straight line. This is just not worth grading. Go back!' " Salam laughed. He had not gone back. He never showed Wilkinson his lab work again. In another experiment, he said, "I had these Poiseuille's tubes, through which you look for the laminar flow of water. But I couldn't get the laminar flow. My tubes, which I had blown myself, were probably all clogged up with God knows what. So I invented a new theory to fit my facts." He almost failed the experimental exam, but did so well on theory that he got a first. "The day the results came out, I met Wilkinson in the Cavendish. Wilkinson called me and asked me what class I had gotten. I said, 'I got a first'—rather modestly. He was facing me. He turned full round in a circle. You know, like this—" he stood up, miming his teacher's little dance of astonishment, the vents of his wide suit a-flap "—and then he said, 'Shows you how wrong you can be about people!' "[2]

At Cambridge it was the tradition that first-class graduate students became experimenters, whereas lesser lights went into theory. Because he had done so well, Salam's reward was a promotion to the experimental halls he detested. He was put to work under Samuel Devons, firing hydrogen isotopes at each other and measuring the resultant collisions. He loathed it. "It was hard for me," he told us in his office in Trieste. "I lacked the sublime quality of patience with experimental equipment, which—certainly in the Cavendish—never functioned."

Salam asked Devons if he could leave to join a theoretical department. Devons agreed, provided he could find a supervisor. Salam approached Nicholas Kemmer. After Yukawa had invented the theory that a virtual particle—then called the mesotron, now called the pion—was responsible for the strong and weak interactions, Kemmer had described all possible forms of interactions that could be mediated by mesotrons. After the discovery of the pion and the renormalization of quantum electrodynamics, the natural question was if Yukawa's meson theory could also be renormalized. "The question was, was the pion a renormalizable theory?" Salam said. "At our first interview, Kemmer said to me, 'Well, all the problems have either been solved by

Feynman, Schwinger, Tomonaga, and Dyson, and [P. T.] Matthews is going to solve all the problems concerning mesons. So go to Matthews and ask if he has any problems left.' I went to Matthews, and I said, 'What have you been up to?' He said he had spent two and a half years trying to renormalize meson theories. He had found that only spin-zero mesons would work." This was encouraging; the pion has a spin of zero. "He had done the calculations to one-loop order and shown that the theory of spin zero was renormalizable up to the second order. Matthews said to me that I should read Dyson and look at the general problem of renormalization of meson theories. Following Dyson, in a couple of days I produced a general scheme of renormalizing all meson theories of spin zero."

That quickly?

"Very quickly," Salam said, chuckling. "I went to Matthews and I said, 'Look, is this the problem?' He laughed and he said, 'This is *not* the problem. You've done dimensional analysis, which only shows that various factors fit and everything would be fine—*if* one can show that the infinities *really* can be removed, each to its proper place. That's the problem. That problem devolves to a rather obscure point in Dyson's papers about overlapping infinities. It's so obscure: we really have to work on that.' " In complicated interactions, the reaction can go more than one way along the tangle of loops and vertexes in the graphs. The question was whether each subinfinity had to be removed individually, as if the others were not present. Dyson claimed to have solved this for quantum electrodynamics, but had not given his proof. In the meson theory, Matthews had encountered even more virulent snarls of infinities, many of which overlapped in a vicious manner. Would Dyson's claim hold for these also? "Matthews said, 'I'm taking a vacation. You can have this problem until I come back to work in October. If you don't solve it till then, I'll take it back.' That was the sort of gentlemen's agreement we had.

"This was probably April or May [1950]. I rang up Dyson at Birmingham, where he then was, and said, 'I would like to talk to you. I am busy renormalizing meson theories and there is this problem of overlapping divergences which you claim to have solved.' He said, 'I'm leaving tomorrow for the U.S.A. If you want to talk, come tonight to Birmingham.' "[3] Salam took the train to Birmingham and early the next morning asked Dyson to show him the solution to the overlapping infinities.

"I have no solution," Dyson said. "I only made a conjecture."

"It was as if the earth had opened and I had sunk into it," Salam said on another occasion. "Here was my hero, Dyson, blandly confessing to me that he had no proof and that he had just made a conjecture and gone on!" Salam laughed heartily. "It turned out to be right, of course." Using some hints from Dyson, Salam soon showed that the overlapping infinities indeed could be accounted for; he then wrote a long, yet in some ways oddly sketchy demonstration that the theory of the strong interaction with spin-zero particles could be renormalized to all orders. (Indeed, the complete method was not published until 1983.)[4] "We believed at that time in Yukawa's theory, which said that pions were the final secret of nuclear forces, and we went on believing it till 1951 or 1952. The expectation was that meson theory, once renormalized, would completely solve the strong force problem. And that's why I became famous, because I had proved that the theory was renormalizable. In 1950, when my proof was completed, we believed this was the end. We expected it to be *the* theory. The end! And we lived in that fool's paradise, euphoric for a year or so, until the excited state of the nucleon, the so-called delta, was discovered. Then the bottom fell out of the business. At the same time, the strange particles started to be discovered. So the situation became even more chaotic."

Although Salam was not permitted by Cavendish regulations to submit the renormalization as a thesis for another few years, he was immediately offered a position at the Institute of Advanced Study. Upon going, he discovered his reputation had preceded him, and was horrified to learn that he was regarded as an expert theorist. "All I had was one year of research," he said to us. "The result was disastrous for me. I knew *nothing* of physics except what I had done myself." He was afraid to reveal his ignorance. As a result, he said, "I could learn nothing. In learning, you have to ask questions which are sometimes exceedingly foolish if you don't know the subject. That's why people stop producing when they become very famous, because they cannot bear to write papers that may prove to be trivial. Once that happens, that's the end. You are no longer able to make stupid mistakes. Personally, since then I have never subscribed to this view that I should stop writing articles that might be trivial."

Before finishing his doctorate, Salam returned to Pakistan, where he began teaching at the University of Lahore. To his dismay, he learned that he was the only practicing theoretical particle physicist in the entire nation. No one cared whether he did any research. Worse, he was expected to look after the college soccer team as his major duty besides teaching undergraduates. Just before Christmas of 1952, he received a cable from Wolfgang Pauli, who was on a tour of India. Overjoyed, Salam flew from Karachi to Bombay. When he knocked on Pauli's door at the Tata Institute after traveling all night, Pauli let him in and without a word of greeting snapped, "Schwinger's wrong! I can prove it!"[5] Upon returning from Bombay, Salam was "charge-sheeted" for leaving Pakistan without permission.

Kemmer kept in touch with Salam. When Kemmer decided to leave Cambridge for Edinburgh, he managed to get Salam appointed in his stead. Salam went with his wife and daughter to England early in 1954. One of his first graduate students was a moody young man named Ronald Shaw, who, while working for Salam, independently invented Yang-Mills theories for his Ph. D. thesis.[6] In Trieste, we asked Salam what had been his reaction to the difficulty of assigning mass to the vector bosons. "Well," he said, "I think the . . . " He stopped and pushed around the espresso cup on his desk for a few seconds. "Mass is a very important thing, something whose origin we must explain," he said at last. "But it's also the most difficult problem in physics. Just look at the range of masses. You start with the electron and go to the Planck mass, which is twenty-two orders of magnitude larger. My reaction has always been, don't worry about the mass problem—take it the last. So although Pauli shouted Yang down for ignoring mass, I personally have never let myself be put off by it, especially when the primary interaction and its basic symmetry is yet to be determined." Sipping his coffee appreciatively, he added, "To me at that time, the establishing of the symmetries was the primary motivation."

Like many other particle physicists, Salam was drawn into the field of weak interactions by the tau-theta puzzle. Unlike most others, he was enthusiastic about Lee and Yang's proposal that parity violation might be at work. Appalled at this violation of symmetry, Salam could not understand how nature could prefer left to right. Then he was the first to realize that parity violation could be tied to the existence of a neutrino that spins in just one direction—what is called, in the jargon, a "two-component" neutrino.[7]

When Salam proposed this idea in 1956, before experiments confirmed the left-handedness of the neutrino, Pauli was scornful. "Give my regards to Salam and tell him to think of something better," was his message. And when Salam went on to claim the existence of the Universal Fermi Interaction, he was treated to a second blast from Pauli:

For quite a while I have [made] for myself the rule: "If a theoretician says 'universal,' it just means 'pure nonsense.' This holds particularly in connection with [the] Fermi interaction, but otherwise, too." And now you, my dear Brutus, come with this word! . . . So I have not seen your paper, but I have some small hope . . . that you have already withdrawn it.[8]

We asked Salam why Pauli was so dead set against universality. "Pauli was the oracle for my generation," he said. "So it was fun to see how wrong he was on both counts. But regarding universality, I think he was so angered by Einstein's claims of the unification of electromagnetism and gravity that he disliked anything that resembled it. We, the young people, liked universality, but our seniors did not. They—I'm speaking of the generation of Pauli and Dirac—had also wanted in their time to do everything with one stroke. You see, the greater physicists are, the greater their desire to finish off the subject in their own lifetimes. For them, the proton was going to be the antiparticle of the electron, and this was to be the end. Everything wrapped up in one tidy package. When those ideas didn't work, when new physics came, they became discouraged with the whole idea of grandiose unifying schemes.

"What really shocked them was the muon. There was no reason for it to be there. I remember telling my seniors it was just a heavy electron and getting indignant denials. 'How dare you! It weighs two hundred times more than the electron!' The mass syndrome again! In that sense, the muon was a giant step backward for particle physics. There is the famous remark of Rabi about the muon: 'Who ordered *that*?' Instead of being seen as a replicated electron, it made them think that everything was lost and there was no chance of ever coming back to a unifying picture. It took decades for us to recover our ambition."[9]

Curiously, Pauli's wish was granted. After writing his paper on the Universal Fermi Interaction in the beginning of 1957, Salam went to the seventh Rochester Conference. Like Marshak, he, too, was alarmed by Wu's talk of V and T. Salam's version of the Universal Fermi Interaction was premised on the supposition that the weak interactions dealt only with left-handed particles, as is the case for $V - A$. To his later chagrin, he stopped at the offices of the *Physical Review* on his way back from Rochester, and pulled his article from the editor's hands.[10]

Schwinger's hypothesis later in the year that the weak and electromagnetic interactions might be mediated by a triplet consisting of two vector bosons and the photon delighted Salam, who has a fondness for the imaginative leap. He had already imagined the possibility of a virtual particle for the weak interaction. Why not add in electromagnetism for an elegant and unified theory? With John Ward, a theorist who had renormalized the meson theory at about the same time, Salam proposed a new unified model that directly linked Yang and Mills, universality, and unification.[11] The paper describes the same model as in Glashow's first paper, and is contemporaneous in publication.

"Ward and I were trying to make Schwinger's idea into a gauge theory. We estimated that the mass of the W would be at least thirty times the proton mass—huge, it seemed to us. We were very ashamed of calling this heavy object an elementary particle. And we tried to diminish it as much as possible. We tried our best to keep the mass down." Salam laughed again, swinging the chain on his glasses. "We found reasons to insert the square root of two in the appropriate places—a standard theoretical ploy! We wrote a second paper in '64." He rummaged in a drawer. "This is the manuscript of that one," he said, displaying a tired mass of yellow legal paper. He flipped through it. The pages were covered with scratched-out phrases, canceled equations, inserts, additions, and erasures. Hidden in the mire of second thoughts was the complete SU(2) $\times$ U(1) model; like Glashow, they had found the Z^0. Also like Glashow, they had not resolved the mass question or the difficulties with neutral currents. Salam dropped the scribbled pages on the table with a gesture of disgust. "I did not know until '72 that Glashow had done this already."[12]

After 1964, Salam, too, set aside unification, partly because he was distressed by a general change in the theoretical climate. Frustrated by their decade-long inability to formulate a theory of the strong interactions and disheartened by the phenomenological character of the $V - A$ theory, a prominent group of scientists began to argue that quantum field theory had failed, and that new ideas would have to take its place. But there was nothing ready to supplant it, and most physicists kept right on using field theory to do calculations even as they decried its shortcomings. "Many of our colleagues tried to throw quantum field theory out, keep only relativistic quantum mechanics, and confront its predictions with experiment," Salam said. "I mean, all those people who believed in field theory were badgered. I didn't doubt the thing. But there were others, like Geoffrey Chew, who believed that quantum field theory was dead." Salam was reminded of the only time he had ever met the Soviet physicist Lev Landau, at the 1959 Rochester Conference, which was held in Kiev as a gesture toward Sputnik-era détente. "He was flamboyant, wearing a shirt of extraordinarily vivid colors. He saw me from a distance in the crowd and said, 'Oh, you're Salam! The man who scooped me in inventing the two-component theory of the neutrino!' This was flattering, so I said yes. He said, 'Come here, come here!' I went to him. He said, 'Aren't you *ashamed* of yourself?' This was a shock. I said what for. He said, 'Aren't you a believer in field theory?' I said yes. He said, 'I have just shown that the Hamiltonian in field theory is zero. Aren't you ashamed of yourself?' This was the so-called Landau ghost which he had discovered in quantum electrodynamics, portending that the theory must possess a ghost. Basically, a ghost is a particle that violates the laws of cause and effect. Awful things, those ghosts. Some years later, however, Landau was shown to be wrong. There are no Landau ghosts in local gauge theories." Shaking his head, Salam said, "But in 1959 there were so many people who thought that field theory was rubbish, and only *fools*—" he pointed to himself "—talked about something like *gauge* field theory."[13]

Part of the reason Salam was unworried by the mass problem in his $SU(2) \times U(1)$ model was that he had at least a rough notion of how to solve it. In a lengthy footnote to the $V - A$ paper he yanked from the *Physical Review* is a sketch of an idea now called spontaneous symmetry breaking, which shows how asymmetrical systems can end up being described by symmetrical equations.[14] One of the first and most well-known examples of broken symmetry is the theory of ferromagnetism, which was worked out by

Werner Heisenberg in 1928. Magnets have north and south poles, which means that they have a preferred orientation in space; they are not symmetric, as physicists use the term. If a bar-shaped iron magnet is heated in a forge, it loses its magnetic properties and becomes symmetric—there is no way of distinguishing one end from the other. When the metal cools down, however, it spontaneously recovers its magnetization and again acquires a north and south pole. The symmetry, physicists say, is "broken" or "hidden"; it is manifest when the bar is red-hot, concealed when the magnet is cold.

In an intuitive way, Salam thought the real, asymmetrical world might be described by a theory that was itself perfectly symmetrical. He hoped that he might be able to construct a gauge invariant theory, that is, a theory with the same kind of symmetry as quantum electrodynamics, which described the asymmetric real world in which vector bosons had to have masses if they existed at all. Salam's thoughts were worked out for electrons, not for W or Z bosons, but he had the hunch he was on the track. Although he can be very careful in his formalism, he cares most about broad ideas and general concepts, about the gist of things; after he has skimmed the cream, someone else can look after the mathematical niceties. "I am an intuitive physicist," he says. "Give me the basic thrust and let me get on to the next problem."[15]

In the summer of 1961, Salam attended a conference in Madison, Wisconsin, with Steven Weinberg. The two men had met eighteen months before, when Weinberg, then a junior member of the faculty at Berkeley, told Salam about his proof of the last loopholes in the work on overlapping infinities in quantum electrodynamics. They had hit it off immediately. In Wisconsin, they were both excited and appalled by a series of conversations about broken symmetry with a young Cantabrigian theorist, Jeffrey Goldstone.

Goldstone, too, had studied field theory, but had become interested in superconductivity, the phenomenon in which certain substances lose all resistance to electricity when they are cooled to a few degrees above absolute zero.[16] Discovered at the beginning of the century, superconductivity resisted explanation until 1957, when John Bardeen, Leon N. Cooper, and John R. Schrieffer of the University of Illinois developed what is called the BCS theory of superconductivity.[17] (Cooper, like Glashow and Weinberg, is a graduate of the Bronx High School of Science; Bardeen, Cooper, and Schrieffer won the Nobel Prize in 1972.) When first formulated, the BCS theory was attacked in many quarters because it did not seem to be gauge invariant; by 1960, however, at least some physicists realized that superconductivity was in reality an example of the spontaneous breaking of the gauge symmetry of electromagnetism. When the temperature gets cold enough, the symmetry breaks, and strange things happen.[18] The few particle physicists who understood superconductivity wondered if perhaps the trick of breaking the symmetry could be carried over into some other area of inquiry. Jeffrey Goldstone was one of these physicists.

The most obvious asymmetry in particle physics is the difference in mass of the various elementary particles. Goldstone himself was intrigued by the electron and muon, which are identical except for their masses. From his work in superconductivity, Goldstone, like Salam, hoped that some kind of symmetry breaking could account for the difference. Unlike Salam, Goldstone actually tried to construct such a model during the summer of 1960. He sketched out a simple unrealistic model of particles somewhat like pions interacting with particles like protons (the hot magnet, so to speak), broke the symmetry, and tried to figure out what would happen (the cold magnet). To drive the symmetry breaking, Goldstone hypothesized the existence of a special field permeating all of space that alters the conditions under which the symmetries hold. In Goldstone's conception, the vacuum of outer space is, in short, not really empty; it is everywhere saturated by the symmetry-breaking field. Assuming the existence of this field broke the symmetry, all right, but *en passant*, it created a bizarre problem. The equations with symmetry and the equations without symmetry were equivalent only if he threw in another, massless particle into the stew. In other words,

$$\text{symmetry} = \text{broken symmetry} + \text{extra particle}$$

This was not good. The new particle would presumably be strongly interacting. "Massless particles of that type, like the photon, lead to long-range interactions, like electromagnetism," Goldstone said. "The strong force has a range of 10^{-14} centimeters or so. So you were in trouble."[19]

Goldstone spent that summer in the theory wing of CERN, where his office was near that of Sheldon Glashow. The two men talked over symmetry breaking, but could not see how it fit into anything, including and especially Glashow's own work on $SU(2) \times U(1)$. Goldstone told Glashow that he was completely confused about what this extra particle meant. In Goldstone's recollection, Glashow raised his eyes heavenward and said, "Oh, come on! Publish it anyway!" Somewhat reluctantly, Goldstone did.[20] "It's a ragbag of remarks and techniques," Goldstone told us, "including, thrown in at the end—on Glashow's encouragement because it was unsupported—the statement that you *always* got zero mass particles [if you broke a gauge symmetry]. I would have sworn it was true, but I certainly didn't have any kind of proof at that point."[21]

Despite the lack of proof, this "ragbag of remarks" snapped Salam and Weinberg to attention when Goldstone explained it the next year in Wisconsin. Salam had just shown that the one in Glashow's first theory, in which the masses of the Ws were put in "by hand," were unrenormalizable. Spontaneously broken theories just possibly might run around this problem. They might start with massless particles, and end up with massive ones. Weinberg, too, was intrigued. He had recently come across another paper, by University of Chicago theorist Yoichiro Nambu, that talked about the same analogy with superconductivity, but had not really understood it. "I think that week with Jeffrey Goldstone in Madison was really the first time I began to think about these things seriously," Weinberg told us. "I fell in love with symmetry breaking. It was clear that the old style of looking for symmetries was pretty well played out. It just seemed that it was going nowhere." The way Goldstone thought about symmetry struck a chord in Weinberg's heart. "That I thought was just beautiful, and it gave me the feeling that it wasn't just a bright idea of Nambu about an analogy with superconductivity, but that there really was a whole body of theory that one could master and apply in various ways."[22]

On another occasion, Weinberg returned to the subject of spontaneous symmetry breaking. "The point is," he said, "there clearly weren't a lot of symmetries left to be discovered, like isotopic spin, which were clearly manifest in nature. The wonderful thing about broken symmetry is that there still may be a lot of deeper symmetries left that are still hidden. It wasn't the *breaking* that excited me, but the potentiality that there are still symmetries left for us to discover. It's as if you were mining for some kind of metal—gold, say—and you'd done all the surface mining and you can't find any more, and then you find out that gold is also found underground in certain kinds of rock formations. You get excited not because it's deep underground, but because there's still some gold you haven't found yet."[23]

Thinking this, it is no wonder that Weinberg and Salam were dismayed by Goldstone's conclusion that symmetry breaking must always produce particles of a type that do not exist. Salam couldn't use it to save $SU(2) \times U(1)$; Weinberg couldn't find more symmetries. "Both Weinberg and I disbelieved Goldstone," Salam said. "We christened his particle 'the snake in the grass'—always ready to strike. After the talk, we argued a little bit with him—a good physics discussion. Weinberg and I came away feeling it *had* to be wrong. There *couldn't* be this particle." That fall, Weinberg went to Imperial College, where the two continued to look for the flaw in what is now called the Goldstone theorem. Not only were they unable to find any, they demonstrated the theorem was true in no less than three separate ways. "We started working and ended up proving Goldstone was right," Salam said. "More elegantly, or so we claimed. Then, after the paper was written, we began to wonder if maybe some of the ideas in it had somehow not really come from our talk with Goldstone. He was at MIT at the time, and we cabled to ask him if he wanted to be in on the collaboration."[24]

Goldstone telegrammatically agreed. "Basically, I didn't write a single word of that paper. They wrote it. But some of the results in fact were things we had discussed at length in Wisconsin that summer."[25]

The Goldstone-Salam-Weinberg paper left no doubts about the unwelcome massless particle; it was there, and everyone concerned was exceedingly glum about it.[26] The three men had invented the mathematics for studying the possibility that the vacuum might be saturated with special fields—the prospect that nothingness might have, as physicists say, structure—but were unable to do anything but produce unwanted particles with their idea. Eighteen years later, when Weinberg won his Nobel Prize, he recalled the dismay he felt then: "I remember being so discouraged by these zero masses that when we wrote our joint paper on the subject, I added an epigraph to the paper to underscore the futility [of trying to make the vacuum beget mass]: it was Lear's retort to Cordelia, 'Nothing will come of nothing: speak again.' Of course, the *Physical Review* protected the purity of the physics literature, and removed the quote."[27]

We once asked Glashow why nobody involved had thought of what seems obvious in retrospect, making a connection between the massless boson in $SU(2) \times U(1)$ and the massless bosons in spontaneous symmetry breaking. You have to recall the time, he said. "Goldstone showed that if a symmetry broke down by itself, there would necessarily exist certain massless particles," he said. "Now, there are none of these massless particles. They don't exist. This was just a little exercise in abstract quantum field theory. What Yang and Mills showed was that if you had this kind of crazy gauge theory, then the particles that would carry the force would be massless. So here we have two types of theories, quite independent—one [being] gauge theories, the other spontaneous symmetry breaking—both of which produce massless particles. Two different types of massless particles, neither of which exists in nature. Two sets of guys doing crazy stuff. How could you know any of this would be relevant to any other part?"[28]

In the early 1960s, evading the Goldstone theorem and, as it became known, the Goldstone boson, became a sort of cottage industry in the coterie of physicists still interested in field theory. In the United States, Great Britain, and Belgium, half a dozen theorists tried their hand at it, proceeding less from the conviction that the theorem was empirically wrong than the belief that spontaneous symmetry breaking was too pretty a concept to be fouled with such ugly consequences. The first to find a loophole was Julian Schwinger, who pointed out that broken local symmetries and broken global symmetries might be different animals in a paper that few people read and even fewer understood.[29] One of those who tried to comprehend was Philip

Anderson, a condensed matter physicist at Bell Laboratories, then in Murray Hill, New Jersey. (Condensed matter physics is the somewhat grandiose name for the study of the behavior of atoms in solids.) Anderson had taken courses from Schwinger at Harvard and tended to follow his papers. He picked up on Schwinger's main thrust and argued from it that Goldstone, Salam, and Weinberg were obviously wrong.[30] Superconductivity is an example of broken symmetry in which no extra massless particles appear, and hence is an exception to the Goldstone theorem. Anderson suggested that if electromagnetism can escape Goldstone, then other locally symmetric gauge fields might do the same thing.

It is likely, then, considering the superconducting analog, that the way is now open [for a theory] without any difficulties involving either zero-mass Yang-Mills gauge bosons or zero-mass Goldstone bosons. These two types of bosons seem capable of "canceling each other out" and leaving finite mass bosons only.

What happened, he said, is that in a gauge-invariant theory the Yang-Mills bosons "become tangled up" with the Goldstone bosons and end up with a mass. The massive particles could exist because they might be heavy enough not to be seen.

We conclude, then, that the Goldstone zero-mass difficulty is not a serious one, because we could probably cancel it off against an equal Yang-Mills zero-mass problem.[31]

In retrospect, Anderson's suggestion that these two theoretical embarrassments would take care of each other was incredibly prescient. His idea is precisely what is now believed, which makes it all the more remarkable that almost nobody looked at his paper although it was printed in the *Physical Review*, then the most important journal in the world, and was written in anything but an obscure style. Anderson's article went unread because he was a condensed matter physicist and therefore in some sense not a member of the club.

"Everyone in condensed matter physics knew about Goldstone bosons," Anderson said to us. "If you break symmetries you get what we call 'collective excitations,' which are the equivalent. All Goldstone did was put it into a theorem. That sort of thing—" he chuckled scornfully "—impresses particle physicists."[32]

Among the few to read Anderson were Walter Gilbert of Harvard, and Abraham Klein and Benjamin Lee of the University of Pennsylvania. As often happens in theoretical physics, reasoning from the same base of data led to opposite conclusions. Klein and Lee supported Anderson; Gilbert said that superconductivity and relativistic quantum field theory were totally separate domains.[33] The dispute helped trigger a spate of more or less independent activity by two Belgians, Roger Brout and François Englert; and a

Scot, Peter Higgs. (In addition, Goldstone himself had done some work on the topic a bit earlier.) Each discovered that, in a sense, all the conflicting claims about symmetry breaking were correct. All relativistic theories with broken symmetry had massless particles except—the *sole* exception!—Yang-Mills theories, that is, theories based on local gauge symmetry. Exactly as Anderson had suggested, the massless particles that were a problem for theories such as Glashow's SU(2) × U(1) do become "tangled up" with the massless particles created by spontaneous symmetry breaking. In a sort of mathematical fratricide, the two sets of particles eat each other up and the vector bosons that transmit forces magically acquire mass. Left over at the end is another, as yet unseen particle called the Higgs boson. The entire process is a theoretical Rube Goldberg device that only works for a very small class of theories. Some physicists, such as Glashow, think the hurlyburly of spontaneous symmetry breaking is so awkward and ugly that it is a gravy stain on the tie of physics. Others, such as Weinberg, regard symmetry breaking as a wonderful limiting device, for it means that nature and humanity are forced to work with local gauge invariance. It is a necessary condition; the symmetries again constrain us.

In a perfect world, a world in which science moved logically and scientists were not swayed by fashion or blinded by intellectual prejudice, the discovery that spontaneous symmetry breaking can endow vector bosons with mass would have produced a flurry of theoretical activity.[34] As it was, Brout and Englert were too identified with condensed matter physics for anyone to pay attention; Goldstone was told by Gilbert that this ludicrous scheme could not possibly be correct and therefore, to his regret, did not write up his ideas; and Higgs had trouble even getting people to publish his papers. His first piece showed how to evade the Goldstone theorem; the second, which gave a very simple example, was bounced by the European journal *Physics Letters*, on the grounds that it was irrelevant to particle physics.[35]

Brout phoned Salam and tried to explain the exception to the Goldstone theorem, and discussed it in his solid-state textbook, *Phase Transitions*. He knew that a vector boson with mass spoils the renormalizability of weak interaction theories, and with Englert began to sniff around the question of whether symmetry breaking would be the answer. They were sure it was, but could not prove it, although they published the suggestion in 1966. As Brout recalled:

Perhaps the most thrilling moment of my life as a physicist was when, in conversation, we conjectured that the photon and $W^{\pm}$ were gauge bosons of a broken isotopic spin group, thereby unifying weak and electromagnetic interactions into a common renormalizable theory. We were conversant enough with the weak interactions to realize that one isotopic spin group alone would not do. So *we worked—and very hard indeed—to get a good group. And we failed to find one [i.e., SU(2) × U(1)], and so we dropped the matter.*

After working on symmetry breaking, Weinberg joined a number of other physicists in an attempt to develop chiral symmetry, which treated the isotopic spin of left- and right-handed particles independently. (*Chiral* comes from the Greek word for "hand.") The symmetry is broken by the weak force, which only interacts with left-handed particles; the massless particle associated with the broken symmetry is the pion. Or, rather, because chiral symmetry is only approximate, the pion is approximately a Goldstone boson, and has a small mass. "The fact that there was a way of avoiding Goldstone bosons simply no longer seemed very exciting," Weinberg said. "We didn't need to avoid them. There they were—it was the pion. I didn't pay a great deal of attention to the work of Higgs at all for that reason. I think most theoretical physicists didn't pay much attention to it for that reason. The Higgs mechanism seemed like a peculiarity that didn't have any relevance to the real world. In the real world, there were Goldstone bosons, and pions the example. The general idea of using symmetry breaking, though, captured the attention of a large number of physicists in the mid-1960s."[36]

Weinberg then tried to develop a theory of the strong interaction based on the combined effects of unlimited numbers of low-energy pions. "It wasn't successful. I wrote some papers about [large numbers of low-energy] pions which are *totally* ignored, and justifiably so. They have *no* importance. I remember I gave a talk about it at a conference in Amherst, and nobody paid the slightest attention. Very often my colleagues show good taste like that."[37]

Doggedly, Weinberg tried another tack, constructing a model in which the intermediate particles were a triplet of mesons. "I tried to make a Yang-Mills theory for pi, rho, and A1 mesons, with the rho and A1 being like photons but getting their mass through broken symmetry and the pion being the Goldstone boson. It was not a beautiful theory, and I was not very happy with it. And then, at a certain point [in the fall of 1967]—I have some memory of driving to my office at MIT—I suddenly understood that I'd been applying the right idea to the wrong problem. I realized that the whole mathematical apparatus I was constructing for the strong interactions was *just* what I needed to understand something I hadn't really been thinking about, namely, weak interactions and intermediate vector bosons. Like everyone else at the time, I knew that the weak interactions were carried by intermediate vector bosons. I suddenly realized, 'My God, this is the answer to the weak interaction!' Now I was starting out with an exact gauge theory, for the Goldstone boson had been eliminated by this mathematical formalism in which I had re-created the Higgs mechanism. I knew that Yang-Mills theories with the mass added by hand were not renormalizable—everyone knew that. Because it was gauge invariant like electromagnetism and the massive particles got their mass by spontaneous symmetry breaking, I thought the theory just might be renormalizable."[38]

Weinberg's paper came quickly.[39] Just three pages long, the article reads as if written in a single surge of confident thought. With quick, sure strokes, Glashow's SU(2) × U(1) model is reinvented and twined together with symmetry breaking. The arguments of Yang and Mills are sketched; at long last, the virtual particles have mass. With massive vector bosons, Weinberg now had an answer to the first of the two difficulties that had defeated Glashow. Next came the problem of neutral currents, the second of the two puzzles. Weinberg had no idea of what to do about it. Nobody had ever seen anything that might be construed as being caused by a Z^0. Like Glashow, Weinberg dodged the snag. He restricted the model to leptons—electrons, muons, and neutrinos—by fiat. Having banged his head against the strong force for years, he didn't want to complicate his model with the hadrons that could feel its influence. Moreover, he knew that neutrinos were so hard to work with that no experimenter could possibly have seen a neutral current involving them, and that therefore the purely leptonic neutral currents in his model might still exist. If SU(2) × U(1) could be somehow renormalized, he would have taken an enormous step forward. The last words of the paper convey the special sense of scientific excitement as clearly as anything in the literature:

Is this model renormalizable? We do not usually expect [Yang-Mills] gauge theories to be renormalizable if the vector boson mass is not zero, but our Z and W bosons get their mass from the spontaneous breaking of symmetry, not from a mass term put in the beginning. Indeed, the [equation] we start from is probably renormalizable. . . . And if this model is renormalizable, then what happens when we extend it to include the . . . hadrons?[40]

What happens is the biggest stride forward in elementary particle physics since 1947. Full of excitement, Weinberg explained his electroweak theory to the Solvay Conference of 1967—and was greeted by a resounding "So what?" Faced for the first time with the model that now forms the backbone of high-energy physics, the assembled theorists could not have been less interested. It did not fit in with the aesthetic of the time; it was restricted to leptons; it did not account for neutral currents: For all these reasons, Weinberg, like an accepted painter whose first works in a new style are ignored by the art world, was left to tinker alone with his sketch pad as long as he continued in this line.

In the early months of 1967, Salam was tutored in the techniques of evading the Goldstone boson by a fellow theorist at Imperial College, Thomas Kibble, who had examined the Higgs mechanism in considerable detail.[41] Salam saw immediately the application to the SU(2) × U(1) model he had developed three years before with Ward. Every fall, Salam gives a series of lectures at Imperial College. The subject is always a rough presentation of whatever he is working on at the moment. In the fall of 1967, he presented his electroweak theory. Like Weinberg, he did not know what to do about neutral currents or renormalization, but he didn't mind. He had the general idea, and thought nobody else cared enough to worry much about the details. Toward the end of the year, a colleague told him about Weinberg's preprint. Salam checked Weinberg's paper. The two models were identical. "It was something I had been working toward little by little for many years," he says. "With the new stimulus of spontaneous symmetry breaking, all these things I had worked on for years—local gauge theory, chiral symmetry, $V - A$, unification—even possible renormalizability—just came together. To me, it was mother's milk."[42]

Nonetheless, he decided against publishing. Instead, he planned to extend the model to baryons and mesons, thus superseding Weinberg, who had only considered leptons. Unfortunately, he couldn't do it. He tried a dozen schemes, but could not avoid predicting neutral currents. His exasperation was increased by having no time to think. The International Centre for Theoretical Physics was moving out of temporary quarters to a permanent location in Trieste, and needed transfusions of money on an almost weekly basis. Salam spent most of his day on fund-raising. In addition, he was organizing a conference to commemorate the establishment of the Centre, and had managed to convince twelve Nobel laureates, including Bethe, Dirac, Heisenberg, and Schwinger, to attend. There was no room for physics.

He presented the SU(2) × U(1) model at a small congress held outside Göteburg, Sweden.[43] Impressively titled the Nobel Symposium despite its relative insignificance, the gathering was postponed twice. It took place only two weeks before the large Trieste conference, and Salam barely had time to attend. He left in the middle of the Symposium for London to pick up a grant for the Centre, then returned to Göteburg to give his talk. As a result, the speech was half-prepared to the point of near-incomprehensibility. SU(2) × U(1) appears in the middle, flickers into visibility, and then is shrouded by a screen of other notions. Nobody paid attention; few understood it; some were critical.[44] Overwhelmed by preparation for the forthcoming Trieste conference, Salam never wrote out a paper; the only record of his thought is the published transcript of the talk. Salam did not go back to unification for four years.

The resistance to tying together the weak and electromagnetic forces was in large measure aesthetic. Not only was quantum field theory as a whole regarded as old hat, as passé and overtaken by time as Lawrence Welk, but

SU(2) × U(1) was in some views a startlingly nasty-looking model. Thomas Kibble, for one, recalls reading Weinberg's article and thinking that "it was very intriguing that one could do something like this, but it was such an extraordinarily *ad hoc* and ugly theory that it was clearly nonsense."[45] His colleagues evidently agreed: In the next two years, the only physicist to mention Weinberg's work in print was Salam.[46]

Consider what Glashow, Salam, and Weinberg were offering to the community of physics. The weak and electromagnetic interactions are said to be tied together. The photon, two W particles, and the Z^0 form a family, even though the photon has no mass or charge and the W particles weigh eighty times as much as a proton and have charge. In addition, the huge masses of the W and Z come from an incomprehensible and suspect mechanism involving unseen Higgs bosons. The weak and electromagnetic interactions are said to be tied together despite the fact that the weak interactions violate parity and the electromagnetic interactions do not. Perhaps the strangest twist of all is that the ordinary isotopic spin upon which Yang and Mills based their theory is for leptons replaced by "weak isospin." Before, the proton and neutron were treated on an equal footing; in weak isospin, it is the electron and neutrino. The weak portion of the electroweak force is supposed to be generated by the symmetry between these two wildly dissimilar particles. Incredible! Weinberg and Salam asked the community to believe that the weak interactions were based on a Yang-Mills symmetry that had never been seen; that the force was carried by three vector bosons that had never been observed; that these bosons formed a family with the photon; and that the whole business could in some way be made renormalizable. "Rarely has so great an accomplishment been so widely ignored," Sidney Coleman of Harvard wrote later. It is easy to see why; in 1971, Weinberg called his own SU(2) × U(1) model "repulsive."[47]

On the other hand, the flagrant awkwardness of the scheme could be seen as a testament to its significance. The world around us is not symmetrical, and the untidyness of SU(2) × U(1) is a reflection of the asymmetry of nature itself. Behind the clumsy jamming together of unlike objects in the model are the simple, powerful rules of creation—a source of satisfaction, elegance, and beauty that the best theorists look for in their work.

Weinberg took us to lunch one day at the faculty club of the University of Texas. The afternoon was crisp, dry, almost cold; wind made students huddle into their sweaters. There were trees, tall buildings, sloping asphalt paths: the usual impedimenta of modern higher education. The faculty club rang with conversation; we had to lean forward a little to hear Weinberg. He talked about what he had thought when he had proposed the spontaneously broken gauge theory of SU(2) × U(1). "I was tremendously excited," he said. "I do remember very clearly that my point of view at the time was, 'This is the way that theories of the weak interactions must be.' But not, 'I know that SU(2) × U(1) is right.' I regarded that theory as illustrative, and I still—I can't quite get over the fact that it turned out to be right. I was sure that was the *way* the theory had to be. It had to be a spontaneously broken gauge theory." It did not have to be that particular model, he said.

"You can't test general ideas experimentally. You can only test concrete realizations of general ideas. If you have a general idea—broken symmetries are important, or we should reformulate physics in terms of dispersion relations, or something like that—you can't test that. All you can do is test concrete, specific theories. So when it comes time for an experimentalist to prove a theory is right, it's always a concrete, specific model that he is proving right. And that's where all the attention is focused on. That's what is regarded as the greatest success. To me, the underlying ideas are much more important, and often they are the real breakthrough. In that particular case, I regarded the real breakthrough as this understanding that these spontaneously broken gauge symmetries would work. And that they would be a natural way of understanding the weak interactions. And the specific SU(2) × U(1)—" he spread his hands in amazement "—it's just fantastically lucky that that was the right theory."[48]

IV
The Strong Force

14
The Eightfold Way

BEFORE THE TWENTIETH CENTURY, PHYSICISTS GENERALLY WORKED BY themselves, presented their ideas before small societies or clubs, and had them promptly printed in journals that mixed scientific papers, minutes from meetings, and birthday greetings to distinguished colleagues. Many physicists were gentlemen amateurs; there wasn't much money in science. In England, the first university physics laboratories appeared only in the middle of the nineteenth century, when the invention of the telegraph made it desirable to have people around who knew something about electricity.[1] These early laboratories were dusty, high-ceilinged rooms whose brick walls were lined with bottles of chemicals. A spiderweb of primitive wiring connected thumping vacuum pumps, smelly wet batteries, and glowing cathode ray tubes. Scientists wore wing collars and frock coats on the job; they spent much of their time trying to get the vacuum pumps to work. If the equipment did function, a physicist could perform an experiment in a few days. James Clerk Maxwell, the eminent Scottish physicist who showed that electricity and magnetism were two aspects of a single phenomenon—electromagnetism—did his thinking by himself, in the time (according to legend) he was able to snatch away from the social obligations of his demanding wife. Einstein, too, fashioned the theory of relativity single-handedly, one of the reasons he has often been called the last of the classical physicists.

All this has changed. The construction of the standard model and the unified theories that rest on it has gone hand-in-hand with the transformation of physics itself. A tapestry woven by many hands, as Glashow once put it, the standard model is the work of hundreds of people, an act of sustained, collective creation which recalls the effort that produced the great Gothic cathedrals.[2] It required the construction of enormous machines—the "atom-smashers" of the Sunday supplements—and huge centers to house them. Whether they work in blue jeans or business suits or white laboratory coats, physicists now work *together*, chewing over scientific problems with their colleagues. Experiments are done by scores or even hundreds of Ph.D.'s in concert, designing, building, and operating million-dollar machines run by the fastest computers on the planet.

Brookhaven National Accelerator Laboratory, in Upton, Long Island, has a campus of over 5,000 acres and a staff of 3,200—all for research. When we went there one recent winter afternoon, the campus stretched for miles

253

in splendid, snowy isolation. A ring that could encircle Monaco had been dug out of the woods, and earth-moving equipment lay in the ice about it. Fine cold rain pelted the snow. Their luncheon concluded, a throng of researchers streamed in their winter clothing through the fields around the cafeteria building. Some were going to offices where every available inch was covered with blackboards and bookcases; these scientists had spent the morning scribbling equations about fermions and bosons, leptons and quarks, and all the other arcanae of the subatomic world. Other workers were about to enter quite a different world—an experimental hall the size of an aircraft hangar but crammed so densely with computers, cables, and concrete shielding that only one small pathway was available for people to go from one end to the other. Earlier that day, they had supervised the operations of the yellow bridge crane overhead, which spanned the entire width of the room and carried giant spools of cable and electromagnets the size of a human being.

Brookhaven is one of three national particle physics laboratories in the United States: the other two are the Enrico Fermi National Accelerator Laboratory (Fermilab), in Batavia, Illinois, and the Stanford Linear Accelerator Facility (SLAC), in Palo Alto, California. Other high-energy physics laboratories include CERN, in Switzerland; DESY, in Germany; KEK, in Japan; and UNK, in the Soviet Union. Each is dominated by one or more particle accelerators, machines that break atoms apart and drive the fragments to high speeds. The stream of particle "bullets" thus created is shunted out of the accelerator through various channels into special components known as detectors. There the protons or electrons slam into other particles or into the nuclei of atoms. Pieces fly all over the place, and the detector tracks their trajectories for clues to what is taking place. (The whole process has been likened to firing a gun at a watch to see what is inside.) In the early 1950s, accelerators supplanted cosmic rays as the chief means of studying particles; they could reach higher energies, and could produce many more particles in the same amount of time. Today, these machines are the biggest pieces of scientific equipment in the world.

We had come to Brookhaven to talk about the era of accelerators with Nicholas Samios, the director of the laboratory and a participant in several of the chief discoveries of the past three decades. The discoveries concerned the nature of the strong force and the particles that feel its influence. They took place at about the same time as the unraveling of the weak force that has just been chronicled, but completely overshadowed the evolution of the electroweak theory. Little wonder: The strong interactions are the most powerful in the universe, holding sway over the subatomic world like an inflexible dictator. The desire to understand the strong interactions mandated the accelerators among which Samios's generation of physicists has spent its active years; the successful outcome is at once a capstone of contemporary physics and a landmark on the road to unification. It was early February, budget time on the Potomac, and Samios met us between conferences with federal officials who wished to know why the taxpayers should pay hundreds of millions of dollars to make new forms of matter that have no apparent use. During allocation season, he frequently takes the shuttle to the Capitol, pleading the cause of science before half a dozen oversight committees. His job had recently been made especially difficult: In addition to lobbying for his own laboratory, he was one of many particle physicists promoting the construction of a gigantic new accelerator called the Superconducting Supercollider, a machine some sixty miles in circumference to be constructed somewhere, anywhere, in the empty American prairie. If built, the multibillion-dollar ring would collide protons together at energies that have not been reached since the first fractions of an instant after the Big Bang.

Samios invited us to sit down at a round table in his office that faced a blackboard covered with scrawled Lagrangians and budget figures. He has longish black hair combed straight back over his head that nonetheless tends to fall over his face when he speaks vehemently, which he often does. Samios loves to talk physics, and enthusiastically rattles off particle masses, meson lifetimes, and quantum numbers the way a baseball fan might talk of batting averages, ERAs, and RBIs. Simply put, he relishes the game; like a baseball fan, a certain nostalgic softness filters into his voice when he talks of the 1960s, when there was money for all, nobody had heard of the deficit, and the scale of science was not so grand. The United States had just introduced a new, aggressive style to the sport and American physicists were snatching discoveries right and left. We asked him about the American style of physics.

"America is a place you *do* things," he said. "It comes from the West, okay? It's important to get on with the job instead of setting up committees to study the best possible way. So the style in American physics in the late forties and early fifties was—after the war many places wanted to get into science again. And depending on the drive of different places, each person wanted their thing, to get on with it. And the thing at that time was [a type of particle accelerator called] a cyclotron. Cyclotrons were at Chicago, at Columbia, Carnegie-Mellon—called Carnegie Tech at the time—at Caltech, and at Harvard. People were just interested in *doing* things—in great contrast to today. Before you do anything, you need five blue-ribbon panels to review things. And so as a result you *study* things to death. In those days, you didn't justify it to the utmost, you just *did* it. For instance, there's a document which is the justification for the AGS." The Alternating Gradient Synchrotron, built in 1960, is still the largest accelerator at Brookhaven. "It's a five-page letter, written by the then Brookhaven director, Leland Haworth, to the Atomic Energy Commission, and *that* was the AGS proposal. It's a five-page letter![3] There was no detailed theoretical justification. No theorist is mentioned. He says, 'We believe that as you raise the pion energy interesting things will happen, and this machine can do that.' And the AGS turned out to be one of the most productive accelerators ever built."

Cyclotrons began in the early 1920s, when a few experimenters wanted to find a way to split atoms without relying on radioactive materials. The first to succeed even partly in artificially kicking electrons to high energies was Rolf Wideröe, a Norwegian-born engineer at a university in Germany. Wideröe wanted to put a lot of electrons into a disc-shaped container permeated by an electric field. The positive end of the field would naturally attract the negatively charged electrons. By moving the field in a circle, one could induce the electrons to rotate faster and faster, until they were traveling nearly the speed of light. Unfortunately, when Wideröe actually built the device, he could not prevent the electrons from flying off to the side.[4]

Ernest Orlando Lawrence, an American of Norwegian descent teaching at Berkeley, accidentally picked up Wideröe's paper. Although unable to read German, he studied the drawings and built his own accelerator, blissfully ignorant of Wideröe's problems. Lawrence thought up his own variant of the design, a sort of circular metal sandwich. The top and bottom layers consisted of two coils of wire: electromagnets. The middle layer resembled a steel and brass pillbox and was made of two electrically charged metal semicircles—or *D*s, as they were called—with a small space in between. If electrons were squirted into the middle separation, they would rush toward the positively charged semicircle. Passing through a port in the bar of the *D*, they would feel the influence of the electromagnets, which snapped the particles around the semicircle and back toward the second *D*. This time, the other *D* was positively charged. Picking up speed, the electrons raced toward it, passed through, were turned around, and so on and so forth, until they received so many small "kicks" that they traveled at huge velocity. Drawn on, disappointed, and attracted again, the particles spiraled out until they reached the edge of the machine, where they flew through an open window and smacked into a target, ripping its constituent nuclei asunder. The thing seemed too complicated to build, but Lawrence and a student went ahead and did it anyway. The electromagnet was four inches across; the whole device fitted easily on a tabletop.[5]

Lawrence was a bull of a man, a driven experimenter whose temperament was volcanic and habits were erratic. He was well aware that John Cockroft and E. T. S. Walton were working on a rival accelerator under the doubtful and somewhat disapproving eye of Ernest Rutherford. Despite their chief's attitude toward spending the necessary money, Cockroft and Walton were the first to succeed in smashing atoms with a machine, prodding the reluctant Rutherford to admit that the whole business "seems to be worth the expense and the effort."[6] It was Lawrence who built the largest machine. He literally danced with glee when the speeding protons broke the one-million-volt mark.[7]

The measure of energy used by particle physicist is not the ordinary volt, but the electron volt. An electron volt is a unit, like a calorie.[8] A cup of chocolate mousse contains eight hundred diet-shattering calories, which means that every mousse molecule has an energy of a few electron volts. The molecules in a candle flame, by contrast, burn with a few hundred electron volts. When the enormous particle accelerator in Fermilab reached a trillion electron volts in 1984, the particles inside, small but furious entities, had the energy of hundreds of billions of candle flames. Lawrence's 1932 machine went to one million electron volts, or 1 MeV, but he wasn't satisfied: He wanted 5 MeV.

The Berkeley laboratory where Lawrence worked had no money, no professional engineers, no tools more complicated than a lathe. Accelerator parts were scavenged from radios, bought secondhand, or made on-site. Nothing worked exactly the way it was supposed to, and nobody really understood what was going on. The electrons or protons to accelerate came from a tank of hydrogen that was subjected to an intense electrostatic field that pulled the protons one way and the electrons the other. Sometimes the flammable hydrogen leaked into the accelerator, where it blew up. If the machine was on when physicists descended into the bowels of the enormous magnet, it created electric currents in their brains—a form of self-induced shock therapy. The physicists saw fireworks and strange colors. After discovering they were giving themselves jolts of the sort used to "treat" the clinically insane, they stopped going down if the power was on. When the particles hit the target, they knocked out a dangerous spray of neutrons. To protect themselves, Lawrence and the Berkeley physicists filled dozens of jerry cans with water and stacked them around the target area. They leaked, dripping hot water on the physicists' heads. There were no gauges. To find out how much electricity was loose inside the accelerator, physicists taped a nail to a stick and gingerly approached the machinery, waiting for a spark to leap across to the nail. If the voltage was high, the electricity sometimes jumped from the nail to the hand of the physicist holding the stick; if the voltage was very high, the stick holder went to the hospital.[9]

The war changed everything. Before Hiroshima and Nagasaki, Lawrence had begged for private money to build his machines; he used his Nobel Prize, which he won in 1939, to convince the Ford Foundation to give him a million dollars to build an accelerator that could go up to 100 MeV.[10] He ended up working on the Manhattan Project instead. After the war, atomic energy and physics were big news. People touted fission as the energy source that would power the nation into the future, producing electricity so cheap that it would not be cost-effective to meter it. Physicists convinced the new Atomic Energy Commission, the Manhattan Project's successor bureaucracy, to pay for some accelerators as well as reactors.[11] In addition, the U.S. military, which had just beaten Japan by exploiting the practical consequences of the previously arcane study of nuclear structure, was willing to fund more research that would lead to newer and better weapons. A passionate anticommunist, Lawrence argued before congressional committees that new weapons like particle-beam death rays and radiation bombs capable of killing the populations of cities while leaving the buildings intact could best be devised by building particle accelerators.[12] Although the efficacy of such testimony is difficult to assess, it is certain that the Office of Naval Research, the Atomic Energy Commission, and other federal agencies spent hundreds of millions on the construction of ever-larger particle accelerators in the 1950s and 1960s, initially unaware that they were going to have to keep them running at a cost of *more* hundreds of millions. Some physicists think that when Congress discovers that accelerators will not produce death rays and superbombs, Congress will stop funding particle physics. It is the task of Samios and his colleagues today to make the case for pure science, basic research untroubled by application to war.

In 1952, the first multibillion-volt accelerator, the Cosmotron, was dedicated at Brookhaven. A year later, Nick Samios got his B.S. from Columbia University, one of the universities associated with Brookhaven. Blissfully unaware of the hoped-for military benefits from experimental physics, he received his Ph.D. from the same school.

"Columbia was really a center of physics at the time," Samios said, "because of the people who were there. T. D. Lee was there, Feinberg, Weinberg, Steinberger, Lederman, Rubbia—and then people came by for a year, or to give a lecture series, people like Gell-Mann and Pais. Lots of people around, lots of excitement." The influence of Lawrence and Fermi was strong, he said. Although the accelerators kept costing more, the cost per MeV went down. Samios took off his sports coat, pushed up the sleeves of his white turtleneck. "Strange particles had been found in cosmic rays, but I came in at a time when the machines were just starting to operate. In fact, my early work was done at the Cosmotron. We worked on lots of games— finding particles, looking at their decays, their spins, the beta decays. There were lots of interesting questions because everything was a puzzlement. Today, in the texts, everything is nice and neat, but at that time the whole thing was wide open. The first great breakthrough was the Gell-Mann– Nishijima scheme for strangeness. But then, after that, the particles kept increasing, both mesons and baryons."

Particles, particles everywhere. Neutrinos, pions, K particles, lambdas, sigmas, and xis, all with their own spins and masses and lifetimes and typical decays and isotopic spin and strangeness, different states, different resonances—a wilderness of data that nobody could interpret. Most of these particles are seen infrequently in cosmic rays, because the flux of cosmic rays is not enough to produce many instances of the rare processes studied by particle physicists. In addition, cosmic rays are not under experimenters' control, making them difficult to use systematically. Instead, the accelerators would fire a burst of protons at a small metal strip; energy would change to mass, weak and strong interactions would take place, and in a fraction of an instant a spray of exotic particles would be produced. As physicists increased the force of the collisions, they discovered that at certain energy levels the numbers of pions, K particles, and so forth that appeared in the detector suddenly increased. Just as more patrons pour out of a crowded concert hall when another exit is opened, the jump in particle production showed that a new channel for interactions had opened up. The channel was an additional way that particles could decay; physicists realized that this extra path indicated the existence of a novel type of short-lived particle which quickly changed into the floods of pions and K particles that were observed in detectors. The first particle to be found in this way is now called the delta. It was discovered in 1952, when Fermi and his colleagues slammed together protons with just enough power to create a few deltas.

The energy needed to make the particle can be employed to calculate its mass, through the famous relation between energy and mass discovered by Einstein; as a result, physicists tend to speak of particles having a mass of a certain number of electron volts. The proton, for example, weighs in at 938 MeV, the pion at 135 MeV, and the light electron at 0.5 MeV. Similarly, the quantum numbers of new particles could be inferred by studying the quantum numbers of the decay products. In this fashion, experimenters discovered and named dozens of new entities, which they variously and not always consistently referred to as "states," "resonances," and "new unstable particles." The work was not easy, and errors were frequent. A table of meson resonances was published in the *Review of Modern Physics* in the spring of 1963. Twenty-six were listed, of which nineteen are now no longer believed to exist.[13]

Nobody knew what to do with the particle population explosion, or how to relate the new discoveries to each other. Berkeley published a handbook with information about the sixteen known particles in 1958; two years later, so many more had been found that a second issue, with a handy wallet card, was printed. Samios has the latest edition in his office: the size of a small telephone book, it catalogues some two hundred particles, not counting all of their various avatars, resonances, and states.[14]

"You could go back in the literature [of the '50s and early '60s] and find dozens of classificatory schemes," Samios said. He flipped idly through the pages of the *Review of Particle Properties*; the summed work of hundreds of experimental physicists sped through his fingers in the form of reproduced computer tables. There was an amused lilt to Samios's voice. "They'll explain *lots* of things. They'll give you a whole spectrum of particles. They're all ambitious," he said, shutting the book. "And they're all wrong."

□ □ □ □ □

Taxonomy is the study of the principles of classification, especially the classification of objects based on common characteristics. A proper taxonomy of the subject material is essential for the progress of a science. Chemistry was a jumble of cookbook recipes and unrelated facts until Dmitri Mendeleev put together the periodic table in the nineteenth century, thus ordering the elements and their study at a stroke. Natural history began to change into biology with Carl Linnaeus's lifelong attempt to classify and name every species of plant and animal in the world. Darwin's *Origin of Species* can be regarded an attempt to explain the relationships established in Linnaeus's *Systema naturae*.

The diversity of chemical compounds and biological species was always apparent; the variety of subatomic particles was an unwelcome surprise. Particle classification was at first regarded as an uninteresting chore. Looking back, however, taxonomy was the key to understanding both the fundamental constituents of matter and the strong force. A successful scheme for categorizing the diverse forces of matter is also a prerequisite for unification, which seeks to weave together both forces and particles into a seamless whole. Theorists today reach for fully unified theories by standing on the solid base of the taxonomy of matter established in the 1960s.

The earliest classification of elementary particles assigned them to three "weight classes," leptons, mesons, and baryons. Admittedly crude, this scheme did have things to recommend it. For instance, all the leptons were immune to the strong force, so it made sense to lump them together. Moreover, there was believed to be a conservation law called "conservation of baryon number," which meant that the same number of baryons entered and exited any interaction, although the species of baryons might change. (Stueckelberg, once again, had first proposed this idea, and, once again, had been ignored.)[15] Everything that was neither a lepton nor a baryon fell into the class of mesons. This classification scheme was tidy but arbitrary. Like separating dogs into breeds by weight, it revealed nothing about their nature and the genetic relationships among them.

As particles proliferated, the lepton-meson-baryon triad showed its limitations. Physicists turned to quantum numbers, the handful of numbers that describe a particle's dynamics. By the 1950s, they had grown to six, and included a particle's spin, parity, electric charge, baryon number, strangeness, and isotopic spin. Some quantum numbers, like electric charge and baryon number, are conserved in all known types of elementary particle interactions, whereas others, like isotopic spin and strangeness, are conserved in only some. As Emmy Noether had shown, each conservation law is connected with a symmetry, and particles can therefore be arranged in groups according to the various symmetries that resulted from the way they conserved quantum numbers. For instance, isotopic spin is conserved in all strong interactions, which are therefore said to possess isotopic spin symmetry. Scientists put particles with the same isotopic spin into small groups called *multiplets*. An example is the nucleon, which is the multiplet formed by the proton and neutron. Each particle in a multiplet has roughly the same mass and is identical in all other features except electric charge, which is described, one recalls, as the orientation of the third axis of isotopic spin. This "isospin" scheme had helped Gell-Mann and Nishijima develop the notion of strangeness.

A second attempt to order the particle zoo involved trying to conceive of some of the new particles as composites of the old ones. The first to try out this idea were Enrico Fermi and Frank Yang, who hypothesized that the recently discovered pions might be made up of nucleon-antinucleon pairs.[16] Figuring that "the probability that all [the new] particles should be really elementary becomes less and less as their number increases," they offered their calculation "more as an illustration of a possible program . . . than in the hope that what we suggest may actually correspond to reality." Although at first blush it may seem absurd to suppose that a combination of two heavy particles (the proton and antiproton, in this case) could produce a third (the pion) that weighs one-sixth as much as either of its two purported constituents, Fermi and Yang pointed out that the excess mass could turn into the energy that bound together the proton and antiproton. They were unable to show how this would actually occur, and their paper is today noteworthy principally for introducing the notion that some subatomic particles might be made from others.

In the early 1950s, several physicists expanded on the Fermi-Yang approach.[17] The most noteworthy of these attempts was performed by an unusual group of Marxist-inspired Japanese physicists led by Shoichi Sakata of Nagoya University. While a high school student, Sakata came across *Dialectics of Nature*, a posthumously published work by Karl Marx's collaborator Friedrich Engels that attempts to extend the methods of dialectical materialism to the realm of natural phenomena.[18] Just as dialectical materialists perceived society as an endless sequence of interacting social levels, so Engels viewed nature as an endless sequence of physical strata. The task of the dialectical scientist is to discover the interconnections of these levels, recognizing that any particular stratum of matter, such as atoms or particles, is but one in a teeming panoply of material forms. Sakata kept a quote to this effect from *Dialectics of Nature* on his desk as a "precious stone" to guide his work; until his death in 1970, he regarded Marx, Lenin, and Mao as the founders of a "true science" of society upon which physics should be modeled.[19] An extremely capable theorist, Sakata used dialectical materialism to great effect for many years—which most physicists today would take as an example of how far one can go with a nutty idea. Sakata took the Fermi-Yang nucleon-antinucleon idea and, because strange particles had just been discovered, added a third constituent, the lambda (the post-Bagnères name for the V particles discovered by Rochester and Butler).[20] By combining protons, neutrons, and lambdas with their respective antiparticles, Sakata was able to account in an approximate way for all of the particles then known.

The population explosion of hadrons—particles affected by the strong force—brought both good news and bad news to the Sakata group. In 1959, four younger colleagues—Mineo Ikeda, Shuzo Ogawa, and Yoshio Ohnuki, and, independently, Yoshio Yamaguchi—demonstrated that the mesons, whose number had grown to seven, could be constructed by a slight reworking of Sakata's model.[21] They also predicted the existence of an eighth, which was duly discovered.[22] Simultaneously they tried to build up the baryons, supposing that each was made from two of the basic triplet, together with the antimatter version of the third. They ended up with a mess, a loose gaggle of fifteen possible baryons, with particles like the sigma, xi, and proton, which have similar masses and the same spin and parity, treated as totally different entities. And there the composite approach of Sakata seemed to stall.

In retrospect, the separate searches for symmetries and fundamental constituents were two sides of a coin, but they were not recognized as such at the time. Sakata warned against the "inverted viewpoint of believing the ultimate aim to be a discovery of the symmetry properties" as a type of "theology," whereas his critics ridiculed the Japanese group's sometimes dogmatic insistence that proton, neutron, and lambda begat everything else.[23] In any case, the question was resolved with unexpected speed, largely through the efforts of Murray Gell-Mann, whose role in strong interaction physics is something like that of Linnaeus and Darwin combined.

Murray Gell-Mann has published less in refereed journals than his influence might indicate. He suffers now and then from writer's block brought about, at least in part, by the belief, uncommon among theorists, that publishing a wrong idea leaves an indelible stain upon one's career. A theorist's insight, he has said, "should be gauged by the number right minus the number wrong, or even the number right minus twice the number wrong."[24] Even by this lofty standard, Gell-Mann's score is remarkably high, for he has a gift for recognizing symmetries and patterns in places where they are not apparent to his fellows. Time and time again, he has plucked from the chaotic tables of data about particle properties a connection that illuminates the entire discipline and sends his colleagues scurrying to follow him. When he received what he calls "one of those Swedish prizes," in 1969, the Academy cited just this aspect of his work: The award was given in recognition of Gell-Mann's "contributions and discoveries concerning the classification of elementary particles and their interactions."

Gell-Mann is a busy man these days. In addition to his position at Caltech, where he holds the Robert A. Millikan Chair of Theoretical Physics, he is on the board of the MacArthur Foundation, the enormously wealthy endowment that funds many charities, including what have become famous as the "genius" awards. Had the awards existed in the 1950s, it would be hard to imagine a better candidate than Gell-Mann himself, a brash, forceful, sometimes intolerant, and remarkably cultivated man who rapidly rose to prominence in a difficult and competitive field. His father was a teacher of languages, and Gell-Mann has retained the habit of pronouncing foreign names in an impeccable accent. He is an alarmingly knowledgeable amateur birdwatcher and collector of pre-Columbian pottery. The range of his interests is sufficiently large that some of his colleagues have remarked that he is the world's greatest physics dabbler—an image that Gell-Mann has fostered by proclaiming that he only entered physics after his father informed him he would starve as an ornithologist.[25]

When we first spoke with him several years ago, he was on a lightning cross-country tour and could only speak to nonphysicists on the plane between stops. We found him on the aircraft already buckled in and tapping his fingers with impatience; the only trace of the exhausting series of lectures and seminars that he had just delivered in half a dozen cities was a slight darkening under his eyes. His California tan had not faded from the gray spring skies of New York City. "It's a long flight," he said. "I hope you're interesting." We asked him a question. He groaned, looked out the window. "People talk a great deal of nonsense about science," he said. "And you have *obviously* reaped the consequences of that." The cabin doors clicked shut, and we flew from New York to Los Angeles.

In the air, we asked him about his introduction to physics. He went to Yale at fifteen, he said, but hadn't learned much there. The real physics came at MIT, where his adviser was Victor Weisskopf. "There's a joint Harvard-MIT seminar," he said. "I attended that, but I didn't have any idea what a theoretical seminar was. I didn't have any idea what research was, really. I thought it was some sort of class, and that the young physicists who spoke there were trying to make an impression on the older ones. At college, what goes on mainly is pleasing the teacher, not really understanding anything for its own sake." The stewardess was bustling about with a cart of tea and coffee.

"I didn't know what it meant to have a critical understanding of the theory and to know what was going on at the frontier. And so at an early Harvard-MIT theoretical seminar that I attended, I listened to a student who was just receiving his Ph.D. at Harvard talk about a calculation he had done on the ground state of a certain nucleus, boron-10." Bo^{10} is a boron nucleus with a total of ten neutrons and protons; like all nuclei, it has a certain intrinsic spin which is related to the summed spins of its constituent neutrons and protons. "Everybody knew that the spin of boron-10 was one unit. Of course he had to find that the ground state had spin one. He did, and he told how. While he was talking, I was speculating whether some of the more experienced physicists sitting in the front row would think it was good or bad. At the end of the talk, the people in the front row said nothing. But a grubby little man with a three days' growth of beard sitting next to me, who had crawled out of some lab in the basement at MIT, got up and said in not very cultivated English, 'Say, dey mehsured da spin of boron-10, and it ain't one, it's t'ree.' And suddenly I realized that the whole point of theoretical science was to understand what was *really happening*. It was not to please the teacher or older colleagues, but this grubby man from the basement who was reporting somebody's measurement."[26]

Gell-Mann's Ph.D. was delayed for six months because he was slow about writing his thesis. After a short stay at the Institute for Advanced Study, he spent a few years at the Institute of Nuclear Studies in Chicago, working under the aegis of Enrico Fermi. The weather in the South Side of Chicago is appalling, frigid in winter and searing in summer; Gell-Mann was distressed to discover that the university regarded machinery, not human brains, as the proper target of air-conditioning. He had a few thoughts about the V particles which he had trouble writing up in the heat. His concentration was helped wonderfully by the arrival of a draft notice; his occupational deferment form had not been properly completed. Facing the prospect of KP in Seoul, he began to write furiously. A year earlier, at the age of twenty-two, he had solved the riddle of the V particles by inventing strangeness but only now was writing it up. In 1955 he moved to Caltech, where he has remained ever since; there, two years later, he and Richard Feynman, a fellow Caltech physicist, helped to straighten out the $V - A$ form of the weak interaction. Fascinated by Yang-Mills fields as soon as the idea was presented, Gell-Mann was one of those who, like Glashow and Salam, spent the end of the 1950s trying to extend the idea to the weak force. He called the vector boson the X. He still does, although everyone else in the field refers to it as the W.

While working on the forces, he also tried to figure out a taxonomy for the ever-growing multiplicity of subatomic particles. He first tried working with the symmetries, proposing, then quickly withdrawing, a global symmetry for the strong interaction much like the suggestion floated by Schwinger that started Glashow on his work.[27] As did the Japanese, Gell-Mann then switched to the composite approach. He, too, guessed that the proton, neutron, and lambda were somehow the basic entities of the Universe, and that all other observed particles were created simply by mixing and matching those three. "It was a big mess," he recalled. The xi, sigma, and nucleon were not in the same family. "I didn't like it, and I didn't publish it."[28]

In 1959, he spent a year at the Collège de France, where he tried to use Yang-Mills theory to account for the strong interaction and the strongly interacting particles. As it turned out, he was hampered by his ignorance of mathematics, especially group theory. Although he knew that electromagnetism demonstrated one kind of gauge invariance and Yang and Mills had come up with another for isotopic spin, he did not know that the two symmetries were described respectively by the groups U(1) and SU(2). A group, one remembers, is a collection of items related by specific mathematical rules; to take a simple example, all the integers,

$$\ldots -3, -2, -1, 0, 1, 2, 3 \ldots$$

comprise a group. Every group has one or more "generators," which are entities that give rise to the group when certain mathematical operations are performed. The integers, for instance, are generated by the number 1, because if you add or subtract 1 any number of times from any integer you can generate the entire group. The generator of the U(1) group of quantum electrodynamics is electric charge, with the photon the agent that ensures the gauge symmetry. Similarly, the three generators of the Yang-Mills SU(2) group imply the existence of three vector bosons. In modern terminology, Gell-Mann was trying to find a larger, more complex structure formed from a greater number of generators that could generate the full range of hadrons; at the time, knowing no group theory, never having heard of terms like SU(2) and U(1), he put it in terms of looking for what he called "another Yang-Mills trick," only bigger.[29]

A worrier, a fretter, Gell-Mann slogged through calculations every morning at the Collège de France until he felt as if he were banging his head on a wall. Discouraged, he met his French friends for a good lunch and a couple of bottles of wine, and returned to his office in a much better frame of mind but with his computational facilities impaired by alcohol. Glashow visited Paris in March of 1960 and showed his model of SU(2) × U(1), one Yang-Mills and one electromagnetism. Seeing a concrete example of how one could push Yang-Mills further, Gell-Mann was enthusiastic. Unfortunately, SU(2) × U(1) was bigger, with four generators (three vector bosons and a photon), but not *different*. As he later put it,

I worked through the cases of three operators, four operators, five operators, six operators, and seven operators, trying to find algebras that did not correspond to

what we would now call products of SU(2) factors and U(1) factors. I got all the way up to seven dimensions and found none. . . . At that point, I said, "That's enough!" I did not have the strength after drinking all that wine to try eight dimensions. Unfortunately I did not pay sufficient attention to the identity of one of my regular companions at lunch. It was Professor Serre, one of the world's greatest experts on Lie algebras [the kind of mathematics for the groups with which Gell-Mann was working].[30]

After his return to Caltech that September, he combined forces with Glashow, worked out the examples again, and again got nowhere. Glashow went to the East Coast for Christmas vacation, leaving Gell-Mann on his own. By chance, he talked to Richard Block, an assistant professor of mathematics at Caltech, who informed Gell-Mann that he had spent the last six months reinventing the wheel. Gell-Mann looked at Cartan's table of Lie groups and felt like a fool. Eight generators, the one that he had not bothered to work out, corresponded to SU(3), the simplest group that was not a composite of SU(2) and U(1).

"Of course," he has said, "if I'd had my mathematical wits about me, I would have realized it much earlier. It's trivial." He had actually sat through courses and lecture series on group theory. "But, you know, the way they teach math is so abstract and peculiar, it's very hard for a student to know what's going on. Mathematicians tend to present things in such a—what is the word?—such a *nonconstructive* way. They like to prove that there is something, but not actually *show* you what it is. When they give examples, they are so trivial that you don't learn anything from them."[31]

What Gell-Mann perceived about SU(3) has the simplicity that is the retrospective hallmark of good physical ideas; fitting the thought to the data was a matter of detail work, of algebra, of tidy manipulation and setting up tables. The group SU(3) has eight generators loosely corresponding to a collection of Yang-Mills vector bosons. Two of the generators represent isotopic spin and strangeness; the other six are rules for changing the value of the first two.[32] Metaphorically, the arrangement of the generators can be thought of as being like a chessboard, with the rows and columns representing different values of strangeness and isotopic spin. The six other generators are like the rules that dictate where a piece can move on the chessboard. A knight in the center of an empty board, for example, can move to any one of eight possible squares. In this case, the generators are rules like "Move the knight up two rows and over one column to the left." And the "representation" of the group is the set of squares on which the knight can land plus the one that it sits on. For SU(3), the six generators alter isotopic spin and strangeness during elementary particle interactions. Two of them change isotopic spin by one unit more or less without changing the strangeness; the other four alter the isotopic spin by one-half and the strangeness by one. In

Gell-Mann's original, Yang-Mills–style formulation, the changes in isotopic spin and strangeness were the work of no less than eight vector bosons.

The simplest and smallest representation of SU(3) has just three members (see page 268), and the six generators are the six possible transitions from one member to another. Other, larger representations are created from aggregates of the smallest, which is why the latter is called the "fundamental" representation of the group. Sakata and his followers argued that the three objects in the fundamental representation are the proton, neutron, and lambda, and in fact produced an SU(3) model based on this idea. Gell-Mann did not believe in the primacy of the three particles, but after considerable study was unable to decide what should be in this representation.

Gell-Mann was dismayed, but he had no choice but to continue if he wanted to use SU(3). He simply skipped over this representation, and turned to the next simplest. There he saw a fit with the quantum numbers for the baryons with the lowest mass: the familiar neutron and proton, the lambda, and five other particles—the positive, negative, and neutral sigma, and the recently discovered xi minus and neutral xi. All have the same spin (½) and the same parity ($+1$).

The eight particles can be plotted neatly; six of the particles are at the points of a hexagon, and the other two particles are at the center. Seven mesons with the same spin and parity fit neatly into another hexagon; to fill the empty place Gell-Mann predicted the existence of a new particle, the eta. Gell-Mann ran into more serious difficulties, however, when it came to the next largest representation, which has ten members and is called a decimet; only four baryons with the right spin and parity were known. Perhaps the other six indeed existed, but Gell-Mann thought that he was getting into dangerous terrain with the decimet; he decided for the moment not to talk much about this representation and stick to the octets.

Working furiously in the quiet of a campus deserted by Christmas vacation, Gell-Mann finished his particle periodic table in the first days of 1961. With some gaps and a few problems, he had successfully assigned most of the known particles to families. He honored the family of eight baryons, the representation that fit SU(3) most exactly, by calling his idea the "eightfold way," in joking homage to the teaching of the Buddha.[33]

Now this, O monks, is noble truth that leads to the cessation of pain: this is the noble Eightfold Way: *namely, right views, right intention, right speech, right action, right living, right effort, right mindfulness, right concentration.*

To Gell-Mann's considerable annoyance, his jest has fed the notion that quantum physics has something to do with the mysteries of Eastern mysticism. The only mystery to Gell-Mann was *why* the thing worked.

After typing up the eightfold way as a Caltech preprint, he began to

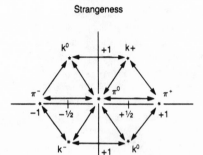

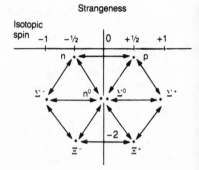

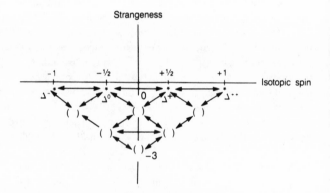

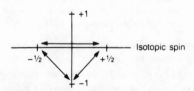

The eightfold way. Here the arrows represent the operators of the group SU(3). The smallest representation (*bottom*) is the fundamental representation, from which the other three are built.

worry about his inability to explain the fundamental representation. The inexplicable disparity in the masses of the particles in the octets was also troubling. In addition, he was at the same time collaborating with Glashow on their attempt to derive both weak and strong interactions from Yang-Mills theories. When that collapsed, he withdrew the Yang-Mills theory that had inspired the eightfold way, and finally presented it as an abstract symmetry whose origin he did not know. Just as Mendeleev had discovered the regularities of the periodic table without the faintest notion of their source, Gell-Mann had invented a taxonomy for the hadrons but had no answer to the obvious question: What caused it?

A similar model was presented independently by Yuval Ne'eman, a colonel in the Israeli army and an amateur physicist who was, during the Begin government, a right-wing member of the Cabinet in Tel Aviv; he is still in the Knesset, where TV cameras sometimes zoom in and catch him doodling equations during dull speeches.[34] A living refutation of the belief that scientists are absent-minded dreamers, Ne'eman may have had the most unorthodox career of any physicist in this century. He is a *sabra*, a Jew whose family lived in Israel before the partition of Palestine in 1948. Ne'eman joined the Haganah, the Jewish underground, at the age of fifteen; about half his four years as a college student at the Israel Institute of Technology, in Haifa, were spent outside the classroom, fighting for the creation of a Zionist state. A child prodigy, he has grown into a short, pugnacious man with thinning hair and intense, slightly squashed features hidden behind the thick black frames of his glasses. We met him in a conference at Fermilab, where he moved around the expansively gesturing physicists with his briefcase clamped securely at his side, watching the proceedings with an intelligence officer's shrewd eye. By 1958, Ne'eman said, he had set up the national defense policy, risen to the rank of colonel, commanded an infantry battalion during some of the heaviest fighting in the 1948 war, and served as the acting head of the Israeli secret service. He was also tired of not working with his first love: physics. Ne'eman asked his friend and superior, General Moshe Dayan, for two years' leave to study physics in Haifa. Dayan agreed, provided Ne'eman did the work between appointments at his new job—defense attaché in London. Ne'eman went to London, armed with yellowing twelve-year-old recommendations from his teachers in Israel and a crisp new note from Dayan.

"When I decided to do physics, I was going to do relativity," Ne'eman said. "But because of the London traffic, I ended up doing particle physics. It turned out that general relativity was centered on King's College. The embassy was in Kensington. It was impossible to be in both places because London traffic is so jammed that I could never get from one to the other. I went looking for something else closer. I found Imperial College." Ne'eman found a professor in the Imperial College catalogue under "theoretical physics," and astonished him by walking into the office in an Israeli colonel's uniform and asking where people studied unified field theories. "He said, 'I don't know about *unified* field theory, but Salam over there is doing *field* theory.' " It was indeed fortunate that Ne'man was directed to Salam, one of the few physicists interested in unification at the time. Salam laughed at the letter from Dayan and told Ne'eman to bring a recommendation from a physicist. Ne'eman never did, but Salam accepted him anyway—partly, he has said, to repay a debt incurred by Islamic science, which in its medieval heyday owed much to Jewish scholars.

We met Ne'eman late one moony night on the steps of the central administration building at Fermilab, a tall curvilinear structure that rises fifteen stories above the empty Illinois prairie. The conference was over for the evening, the blackboards wiped clean, and physicists trickled from the building in animated knots of two and three, talking of symmetries and string theories. A bus was late; waiting scientists dotted the steps. Ne'eman stood erect in the slight breeze, one foot on a higher step, talking in long, proud, harshly rolling sentences. Oddly romantic moonlight played on his square face and heavy glasses. "I worked on evenings and weekends," he said. "And I taught myself quantum mechanics properly from books. After some time, Salam gave me some calculations to make, and I made the calculations, so I learned field theory. Then there was a gap of a year, because in July of 1958—the end of that school year—there was a revolution in Iraq. The Iraqis assassinated their king and their prime minister, and there was a feeling that the whole Middle East was going to collapse and fall in the hands of [President Gamal Abdel] Nasser [of Egypt] and the Russians. The British wanted to overfly Israel with two battalions of paratroopers to protect King Hussein [of Jordan]. And I found myself having to negotiate with them what they would pay for that overflying of Israel. We ended up with the possibility of purchasing two submarines and fifty Centurion tanks. So between the summer of fifty-eight and the end of fifty-nine I barely had time to listen to a seminar. I was completely busy organizing the training of two submarine crews that had never seen a submarine in their lives and had to be taught from the beginning." He described with some amusement the process of driving in his uniform from the base to a seminar, changing his clothes in the Imperial College parking lot. "It was very bad for my physics. I kept sending letters to the chief of staff—Dayan had meanwhile been replaced—telling him that this was not the original deal I had made with Dayan."

A van pulled up. Ne'eman threw the door open and found a seat without interrupting his story. "He regarded all of this as some kind of intellectual escapade. He wanted me to create the national defense college in Israel. So I did him a favor. I prepared a draft for the national defense college, which was afterward created, but I told him that I was going to stay in physics because I liked it. We agreed, finally, that he would send me a replacement for a year, and the replacement arrived in the spring of 1960. On May 1, 1960, I was liberated from my army responsibilities and was given the possibility to stay in London until the next summer at the expense of the state of Israel." The former child prodigy was thirty-five and just beginning his career in the world of science.

Ne'eman heard about Yang and Mills from Thomas Kibble, a theoretician who later helped develop spontaneous symmetry breaking; at the same time, he learned the rudiments of group theory from Salam. Like Gell-Mann, Ne'eman thought of combining group theory and Yang-Mills to produce a taxonomy of the hadrons. Unlike Gell-Mann, he had the advantage of knowing some group theory, but he had the disadvantage of having to learn the rest of physics. "I started coming up to Salam every time with a new model, trying to produce a model for the particles that were known at that time. He would look at it and say, 'Oh, you've again been wasting your time on these things. Why don't you do this calculation I gave you? What you've just done is something that was suggested by [J. C.] Polkinghorne and me ten years ago.' Then I would come up the next time and he would say, 'Oh, that's global symmetry. That was invented four years ago, five years ago, by Gell-Mann and Schwinger.' So he was becoming impatient and saying, 'You know, you told me you just had one year, and you're wasting it on these things. Why don't you start working on a proper calculation that you know where you start and where you finish?'

"I, on the contrary, was gaining confidence, because if I was reproducing serious things that others had thought about then I was not on such a bad track. I told him, 'After all, I'm doing physics only because I love the material, so I want to do what I love to do.' He said, 'Okay, but you're—' he used these words '—you're embarking on a *highly speculative search*. But if you do it, do it properly. Don't be satisfied with the group theory I taught you here. That's nothing. Go in depth, study the thing properly, and then come up with something.' " Salam mentioned that he had heard of a Soviet mathematician named Dynkin who had supposedly done something useful with groups. After searching with some confusion through both the *American Mathematical Society Transactions* and the *American Mathematical Society Translations*, Ne'eman found a thesis by one E. B. Dynkin in the *Translations* that consisted of a modified extension of Cartan's catalogue of groups. Ne'eman was looking for a group big enough to contain strangeness and isotopic spin but small enough to avoid extra complications. From the list, he worked out a set of five groups that satisfied his criteria: SO(5), SO(4), Sp(4),

SU(3), and G_2, the last being one of five "exceptional" groups for which Cartan could not find mates. "Emotionally, I felt that maybe G_2 should do it," Ne'eman said. "It was one of the exceptionals, and I didn't know what that could be. I worked, and I drew the diagram, and it came out to be a Star of David. I said, 'That must be the finger of God!' " He laughed. "But then it turned out that I didn't like the transitions that G_2 gave. I had myself an entire series of criteria, and they didn't obey my criteria. Whereas SU(3) did obey them."

Delighted, Ne'eman waited for Salam to return from the Rochester Conference in October 1960. Salam was impressed, but had heard Sakata disciple Yoshio Ohnuki expound on the Japanese SU(3) model. Disliking the proton-neutron-lambda triplet for the same reasons as Gell-Mann, Ne'eman wanted to publish his own SU(3) anyway. In February, he asked his former secretary at the embassy to type up the article.

A month later, he found Salam in the office at Imperial College with a fat mimeograph from Caltech in his hands and an astonished look on his face. He admitted sheepishly that seeing Gell-Mann's name attached to SU(3) made the whole thing more plausible. At about the same time, Ne'eman's paper was bounced by the editor of *Nuclear Physics* because the Israeli embassy had single-spaced the manuscript, making the welter of equations nearly impossible to read. After retyping, the article was published in July of 1961. In the meantime, Ne'eman began to worry about the same thing that bothered Gell-Mann. If SU(3) was indeed a symmetry of nature, what was the fundamental representation? Could he really pretend it was not there?

The van arrived at the hotel, a bright oasis of conviviality in the midwestern night. Jukeboxes hammered in the bar; the restaurant bustled with the last dinner seating. Physicists thronged the lobby, stumbled over piles of luggage. Ne'eman made directly for a vinyl-covered chair and sat down. He had only a few minutes more to talk before his midnight telephone appointment with an Israeli journalist doing a story about Ne'eman's periodic abdication of his responsibilities as a member of the Knesset to play at theoretical physics. Although he has a special dispensation from the Israeli supreme court to leave the country for up to a month every year on physics-related matters, Ne'eman attracts considerable media suspicion that he is somehow pulling a fast one. "They always find me," he said, checking his watch. He sat hunched over his briefcase in the middle of the room, a small and fiercely solitary man in a rumpled suit, utterly untouched by the roar and sentiment and laughter echoing off the plastic walls about him.

Theoretically, the eightfold way was beautiful: One simple group gave rise to octets and decuplets that contained all the known particles and could be presented in simple diagrams. Experimentally, SU(3) was a disaster.[35]

Gell-Mann had started with two octets and been forced to add a third when a new batch of mesons was discovered. Even so, there were more holes in Gell-Mann's multiplets—almost a third of the total—than there had been particles ten years before. Years later, when we visited Brookhaven, Nicholas Samios recalled the time from the privileged view of hindsight. "There were many people who had many schemes," he said. "In fact, there were some people who said it was a crock of shit, SU(3) couldn't be correct. People tend to say, 'Oh, my God, Murray came up and showed it, and everybody said "fantastic!"' But that's simply not true."

We asked why SU(3) was disbelieved. Samios produced a blank sheet of paper and quickly put a set of dots into an inverted pyramid; four dots in the top row, followed by rows of three, two, and one—ten dots in all. The decimet he said, is very neat. The particles in the top level have a strangeness of 0, in the second a strangeness of -1, in the third a strangeness of -2, and the single particle at the bottom, a strangeness of -3. "The mass difference between the different levels is the same, about 150 MeV," he said. "It works out magnificently! But it was incomplete, okay? Murray had *this*"—pointing to the top layer of the diagram, the quartet of delta particles—"but nothing else. There was not enough information to say whether he was right or wrong. He needed more data, okay?"

That additional data came soon enough.[36] In June of 1962, Gell-Mann, Ne'eman, Samios, and about a thousand other physicists attended a Rochester conference held, in a gesture of international amity, at CERN. Some CERN experimenters had measured the by-products of proton-antiproton annihilations. The data contradicted the Sakata model, whereas they fitted the eightfold way. Elated, Ne'eman asked to speak, only to discover that the chairman of the symmetry session was Yoshio Yamaguchi, an ardent partisan of the Sakata SU(3) model. Yamaguchi refused to let Ne'eman talk about the rival model, having at first misunderstood the calculation. "There were all kinds of funny papers at that session," Ne'eman recalled. "There was another Japanese who explained 137—things like that!—but the octet was not allowed."

In addition, at the conference another team of European experimenters announced the discovery of two new particles: the xi-star, by Samios and collaborators, and the sigma-star. (The star refers to the particles' similarity to other sigma and xi particles.) Although little was yet known about them, their approximate masses fit the decimet. Ne'eman guessed that the sigma-star and the xi-star would fill in the middle two levels of the pyramid, leaving one hole at the bottom. Below the xi-star should be a single particle that is about 150 MeV heavier and has a strangeness of -3.[37]

The plenary session on strange particle physics took place in one of the rather barren auditoriums in the main CERN complex. The rapporteur,

G. A. Snow of the University of Maryland, cautiously went through the still-confusing tangle of data about the competing SU(3) models, then turned the session open for comments. Full of his prediction, Ne'eman raised his hand impatiently. The chairman looked directly at him and said, "Professor Gell-Mann."

Gell-Mann, too, had understood the significance of the experimental findings. He strode to the platform and took the microphone. If the spin and parity of the xi-star and sigma-star are "really right," he said, "then our speculation [about the decimet] might have some value and we should look for the last particle, called, say, omega minus." Because the masses of the new particles fitted the eightfold way, Gell-Mann also predicted the mass of this new particle: 1685 MeV. On the way up the stairs to his seat, he noticed Ne'eman's name tag; the two proponents of the eightfold way met for the first time.

Minutes afterward, the convention broke for lunch in the CERN cafeteria, a sunny L-shaped room surrounded by picture windows that open out onto a view of patios, rusty lawn furniture, and, very occasionally, the Alps that surround cloudy Geneva. Samios and another Brookhaven experimenter, Jack Leitner, approached Gell-Mann and asked him to predict how the omega minus would decay. Gell-Mann jotted down a prediction on a paper napkin in his small neat handwriting. The three talked about how one would find the omega, if it existed, and Samios put the napkin in his pocket to show to Maurice Goldhaber, then the head of Brookhaven.[38]

For about a year Samios had been planning to do an experiment on Brookhaven's spanking new accelerator, the AGS. The AGS speeded up protons to nearly the speed of light, and smashed them into a dense target material. By using magnets, the various fragments of this collision—muons, pions, K particles, and the like—could be sorted out, and split into secondary beams which were in turn directed into particle detectors. Early in his career, Samios had decided that the future lay in beams of strange particles, because nobody had ever seen what happened when a lot of strange matter interacted with ordinary matter. "Since the name of the game was strangeness," Samios said, "it's best to try to make a beam made out of Ks [a type of strange particle], because if you start with strangeness you could leap one more strangeness, or two more strangeness to begin with."[39] The beam of K particles would be directed into a vat of liquid hydrogen, which was inside a particle detector called a bubble chamber.

Bubble chambers were then the most sensitive detectors available. Invented by Berkeley physicist Donald Glaser in 1953, bubble chambers are large and somewhat dangerous devices whose care and feeding preoccupied experimental physicists for years. We once asked Robert Palmer, Samios's longtime collaborator and an accelerator expert, to describe how a bubble chamber worked. A British subject who has been at Brookhaven for a quarter century, he is now working on designs for the sixty-mile Superconducting Supercollider. "I like to say a bubble chamber is like a pressure cooker," Palmer said. "In a pressure cooker, you can heat up the water inside to well above the boiling point, and it won't boil, because the pressure inside is so high. Now, you take the lid off, and what happens? Funnily enough, nothing—not right away. Then it starts boiling. Where does it start boiling? On the surface. It doesn't start boiling in the middle of the liquid—it doesn't know how.

"But in a bubble chamber—in which liquid hydrogen is used, to make it more sensitive—it can. If a charged particle races through the middle of the liquid, the friction heats it up. The liquid hydrogen starts to boil in the wake of the particle, and a string of tiny bubbles forms.

"Then you wait about a millisecond to let the bubbles grow. All of a sudden you *slam* the lid back on the pressure cooker—in the bubble chamber, it's a big piston—flash your lights, and *click*! the cameras peering through the portholes snap a picture of all the little strings of bubbles. Then you advance the film, reset the chamber, and do it again about a second later."[40]

Pouring through the thousands of photographs taken every day in a bubble chamber is a tedious process that is usually assigned to lowly graduate students. Knowing that the AGS experiments would produce tens and even hundreds of thousands of photographs, Brookhaven physicists hired a team of people from the outside. Mostly Long Island homemakers who needed extra income, they worked around the clock, three shifts a day, seven days a week, with special projectors that blew up the seventy-millimeter film onto white tabletops. The scanners, as the women were known, copied the tracks of the events onto magazine-sized sketch books, and then measured the various angles and lengths.

When the mass of tracks is blown up onto a tabletop screen, the result is astonishingly lovely—a thicket of white lines skirls across a dark gray field, intersecting, brachiating, looping like the lines in a Ptolemaic drawing of the heavens. Each type of particle leaves a characteristic track called a signature, and the scanners were taught how to recognize them. Electrons, extremely light particles, produce faint, wiggly tracks, for they are easily knocked about amid the heavy protons of the hydrogen nucleus, like a ping-pong ball attempting to fly through a roomful of baseballs. Protons leave thick, straight lines that run like pencil strokes across the slide. As in a cloud chamber, a magnetic field is placed in the bubble chamber to help identify the particles; it causes the positively charged particles to bend one way and the negatively charged particles to bend the other. The amount of deflection indicates their mass, for the paths of lighter particles bend more than heavier ones. Neutral particles leave no tracks.

"When I was a student," Samios said, "we started off with a chamber that big." He held his hands a few inches apart in front of his face. "Then we went this big." His hands spread about a foot. "Okay? But this was *big*! A thousand liters. So here we were, with a new beam and a new chamber, and we were going to push both to the limit. And along comes Murray with his prediction of the omega minus. That provided us with a focus. We wanted to look for it."

Samios and Palmer built a pipeline from the target that was hundreds of feet long and lined by huge gray magnets that pulled and pushed at the stream of particles until everything but the kaons was filtered out. At the end of the tunnel, the K particles struck the hydrogen in the bubble chamber, producing showers of new particles that had to be precisely analyzed. The experiment started in November of 1963, and promptly ran into trouble. The particles going into the chamber acted nothing like K particles. After a month of figuring, the team discovered that the kaons were being swamped by pions. They spent weeks adjusting the magnets used to select out the Ks—and got nowhere. Early one winter morning when Samios and Palmer were on shift together, they decided that the Ks had to be hitting something somewhere inside the beam line, creating a secondary burst of pions. From the geometry of the situation, the two men were able to calculate approximately where the obstruction must be. They turned everything off and walked down the line of magnets. A machine part called a collimator was sticking out close to the target region, a small piece of metal that had cost them two months' time.

Samios and Palmer were not the only ones seeking the omega minus. Gell-Mann had made his prediction at CERN, which had its own accelerator. A British team built a K beam just like the one at Brookhaven, except six months earlier. Unluckily, their bubble chamber didn't work. A French group did have an operating bubble chamber, but the English refused to use their detector unless the experiment remained entirely British, an idea that the French group rejected. The British kept on struggling with their own detector in proud isolation.[41]

The Brookhaven bubble chamber, too, had trouble. A set of black strips about an inch wide hung from the wall of the chamber like a vertical venetian blind to cut down the reflective images from the camera and lights. Soon after the beam was introduced into the chamber, the repeated jolting of the piston knocked ten of the slats loose. They fell forward and came to rest against the glass window.

"There we were, in the middle of the night, [bubble chamber designer

Ralph] Shutt, myself, W. Fowler, Palmer," Samios said. "And we looked at the window. The question is, Did you damage the glass? Because if you damage the glass, and put pressure on it, then the glass breaks, and you have a thousand liters of liquid hydrogen coming out—you have a real catastrophe. So there we were, with ten of the hangers down, and Shutt looking in. He asked all our opinions, but he had to make the decision. He said, 'Expand.' " The three-foot piston thudded into the liquid hydrogen. "Nothing happened. And he made the decision to continue.

"The other option was to dump the chamber, open it up, fix the slats, and lose a month. The logical thing would have been to stop, to do it right. But we wanted to get on with it. My feeling is, you get five of these things, then you've got to stop. But if you always stop at the first fix, there are usually two or three problems further on. If you fix the first one right, when you get started again, you get the next one. But you've wasted a month before you even *knew* about the other ones! So you go as far as you can without jeopardizing things. We ran, and we kept taking data."

During the months at the end of 1963 and the beginning of 1964 that the Samios-Palmer group could not get their apparatus to work, they were the target of the wrath of other Brookhaven experimenters, who wanted the time and the equipment that were being granted to Samios's team. "We had first priority for a certain window, okay?" Samios said. "But whatever time we took, other people didn't get, so we didn't have carte blanche. We had to perform. If we made mistakes, people would say, 'Why don't you put these guys off for a while and let us do our things, and when they're in better shape, have them come back.' We had big debates. And in fact, I lost some."

In January 1964, Maurice Goldhaber left for a conference in Coral Gables, Florida, where he told the physicists present, including Gell-Mann and Ne'eman, that nothing interesting had been found in fifty thousand bubble chamber exposures; theorists began giving up on the eightfold way, and the Sakata group took new hope.[42]

In the meantime, Samios, Palmer and their team continued to analyze what eventually became more than a million feet of film. "We arranged the scanning so that there would be a physicist on shift at all times," Samios said. "In case any candidate events or something queer showed up, a physicist would judge whether it was good or not. We took data, December, early part of January . . ." He waved his hands to indicate the passage of time. "The physicists would scan, and then, as you're scanning, if someone had questions, you'd stop and you'd go over and look at what they had and tell them yes or no. And then you'd continue, day after day."

Samios was on duty, spooling through bubble chamber film, measuring angles and particle tracks, when a striking negative slid across the white surface of his scanning table. The Ks usually swept in lines from left to right across the frame, slowly spreading out to either side like sheaves of wheat held in an unseen hand. In this frame, one stalk was snapped off. A K particle had come in, hit something, and created a new particle that veered violently downward. A foot away, a thin V appeared in mid-air, the two arms crossing lazily after a few inches.

An old bubble chamber rule of thumb is that the line of flight of the invisible particle that created the V can be estimated by laying a ruler between the two intersections. If it misses the point where the K struck the hydrogen atom in the chamber, that is a sign that some interesting, albeit invisible physics took place in the interval. "So I said, 'This is a real candidate. Looks interesting.' But I didn't do very much with it. I said, 'Measure it.' Early the next morning—I think I was on the night shift—they were measuring it. And while it was on the measuring table, there was a group around looking at it, and somebody noticed a gamma ray."

More exactly, they noticed that an electron-positron pair suddenly burst into existence a few inches away from the candidate interaction. The speeding K had·struck a hydrogen atom, creating a minute cornucopia of particles. One of the products was a photon, which then turned into an electron and positron that spiraled away from each other like watch springs. This was a stroke of luck: By measuring the curvature of the tracks of the electron and positron, the experimenters could determine their mass, and hence the energy of the originating photon. Subtracting that from the energy of the K they would know how much energy was available to create particles, such as an omega minus.

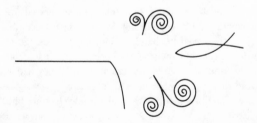

"They noticed one gamma ray first. And then someone said, 'Hey, could there be a second?' This was on the measuring table. They said, 'Hey, there *is* a second!' I said, 'That's *crazy!*'" The likelihood that *two* photons had come out and turned into easily measurable pairs of electrons and positrons was quite literally a million to one. "I said, 'Measure it quickly! Get it off and put it on a regular scanning table!' We had templates—" calibrated French curves for particles "—so I could measure these two gamma rays by hand." In about two hours, Samios was almost ready to declare that they had found an omega minus.

Some thirteen particles were involved in the event. The only reason that the physicists could be certain they had an omega minus was the sheer dumb luck that all the final products—the two particles in the V and the two electron-positron pairs—left easily measurable tracks. This good fortune allowed them to work out the mass exactly. Samios and his team finished a letter to the *Physical Review* and sent it out.[43] By the time it was printed they had come across a second. The Europeans published a year later.

Samios called up Gell-Mann, who gracefully pretended to be surprised by the news, although he had already heard it through the scientific grapevine, which travels at nearly the speed of light. The Brookhaven group later sent him a picture that still hangs on the wall of his office. The following winter, the laboratory mailed out a Christmas card bearing an artist's rendering of the first omega minus event.

The omega minus discovery clinched the case for Gell-Mann's SU(3) scheme. The eightfold way not only arranged particles into patterns, but it successfully predicted where hitherto unknown particles should be found. Unfortunately, nobody had the faintest idea *why* it worked. The person who figured this out was, once again, Murray Gell-Mann.

15
The King and His Quarks

THROUGHOUT THE LAST CENTURY *QUARK* WAS A RARE, POETIC TERM FOR A particular type of animal call, the cry of a heron or gull. Nobody is ever likely to use the word in this sense again, for during the past two decades it made an abrupt transition in meaning from bird caw to subatomic particle fragment, surely one of the most bizarre etymological twists in the history of language. This movement began in 1939 with the publication of James Joyce's last novel, *Finnegans Wake*. Its hero is a Dublin pub owner named, variously, H. C. Earwicker, Here Comes Everybody, and even "Heinz cans everywhere." He is asleep throughout the book; his dreams, which are recounted in its pages, are the expressions of a collective unconscious, reenacting myths, historical incidents, and even the rise and fall of civilizations. As befits a dream, however, the events and scenes of the book do not unfold with the sequential logic of a television miniseries, but through a dense collage of puns, repetitions, misspellings (which begin with the deliberate omission of an apostrophe from the work's title), allusions, and other linguistic highjinks.

The fourth episode in the second section of *Finnegans Wake* begins with the hoots and babbles of four old men, who represent the four authors of the Gospels, the four ancient historians of Ireland, and anything else that comes in fours, as they chortle over the old Celtic romance of Tristan, a young nobleman, and Iseult, the wife of Tristan's uncle Mark, King of Cornwall. The sequence opens with the huzzah:

> *Three quarks for Muster Mark!*
> *Sure he hasn't got much of a bark.*
> *And sure any he has it's all beside the mark.*

Like everything else in *Finnegans Wake*, *quark* has many meanings, all of which have been fiercely argued by scholars. The term is said to be a kind of Bronx cheer for the poor cuckolded king: Three raspberries for the pigeon! It is also a play on a demand that one might have heard in Earwicker's tavern for three hearty Irish quarts of ale: Buy that man a drink! In German, *quark* is a variety of particularly runny cheese: The four men describe Mark as beleaguered by screaming gulls, and the sentence therefore also refers to a deluge of bird droppings—bombs away!

Quark might have passed into the etymological limbo reserved for the obsolete words resurrected by modernist writers had not Murray Gell-Mann accidentally revived the word in connection with his thoughts about hadron taxonomy. On leave from Caltech, Gell-Mann passed the winter of 1962–63 at MIT, from where he traveled, one blustery March day, to present a lecture at Columbia. A guest speaker at Columbia is customarily treated to lunch at the faculty club by a member of the department, and this time the honor fell upon Robert Serber. Walking up Broadway in a freezing wind, Serber, Gell-Mann, and several other physicists began talking about the eightfold way. Serber asked Gell-Mann about the different representations of SU(3), not all of which can be matched with elementary particles. In particular, the scheme had no room for the smallest SU(3) family, which has just three members. Mathematically, such an absence is peculiar, because the larger representations are built up by fitting together an array of the smallest. For this reason, the three-member representation of SU(3) is known as its "fundamental" representation. Serber did not understand how nature could use the octet and decuplet without taking some account of their basic constituent, the fundamental representation.

The physicists entered the faculty club dining room, a large, elegant space in a building overlooking Morningside Park and South Harlem. In Serber's recollection, he asked Gell-Mann why he had not considered suggesting that the SU(3) families were indeed formed from the fundamental representation, which was equivalent to saying that the particles in the octets were made out of littler particles. The eightfold way might therefore be viewed as the set of all possible means of mixing and matching several of these subunits, analogous to the way the more than one hundred chemical elements are formed out of protons, neutrons, and electrons. In this way, taxonomy would be explained by structure. "I pointed out that you could take three pieces and make protons and neutrons," Serber said. "Pieces and antipieces could make mesons. So I said, 'Why don't you consider that?' "[1]

"So I *showed* him why I hadn't considered it," Gell-Mann told us later. "It was a crazy idea. I grabbed the back of a napkin and did the necessary calculations to show that to do this would mean that the particles would have to have fractional electric charges— $-\frac{1}{3}$, $+\frac{2}{3}$, like so—in order to add up to a proton or neutron with a charge of plus or zero. [No such fractional charges had ever been seen.] And he said, 'Oh, I see why you don't do that.' But then, thinking about it the next day or so I said, 'So what if they have fractional charges? Maybe they're permanently trapped inside. Maybe they don't ever come out and they don't ever cause any problems.' I said, 'What the hell, why not?' Speaking the next day at Columbia I may have mentioned it." The talk was in a seminar room at Pupin Hall. Afterward, over coffee, Gell-Mann used an odd word for these subunits: *quork*, to rhyme with *pork*.

"It seemed somehow appropriate," Gell-Mann said. "A strange sound for something peculiar. When I was going to publish the idea eight months later or whenever it was, late in sixty-three, I was paging through *Finnegans Wake* as I often do, trying to understand bits and pieces—you know how you read *Finnegans Wake*—and I came across 'Three quarks for Muster Mark.' I said, 'That's it! Three quarks make a neutron or a proton!' Joyce's word rhymes with *bark*, but it was close enough to my funny sound. Besides, I told myself, in one of its meanings it is also supposed to rhyme with *quart*. So that was the name I chose. The whole thing is just a gag. It's a reaction against pretentious scientific language."[2]

Gell-Mann took almost a year to write a short paper about quarks. Entitled "A Schematic Model of Baryons and Mesons," the article appeared in February 1964, the same month that the Brookhaven team discovered the omega minus and clinched the case for the eightfold way.[3] A landmark in contemporary physics, the eight-paragraph note is a model of scientific prose: brief, logical, achingly clear, so tightly and modestly drawn that its full scope may elude the reader. In the first line, the author sets forth his intention: "If we assume that the strong interactions of baryons and mesons are correctly described in terms of the broken eightfold way, we are tempted to look for some fundamental explanation of the situation."

After pointing out the algebraic necessity of the fundamental representation in the second paragraph, Gell-Mann begins to state the case for subunits in the third. Guessing the reaction to fractional charges, he at first avoids them; he therefore proposes only that all hadrons are composed of three subunits, which he calls "up," "down," and "strange," with charges of $+1$, 0, and 0, respectively. In addition to this triplet, there is an oppositely charged antimatter "anti-triplet" of "antiup," "antidown," and "antistrange." Mesons are made from one member of the first triplet and one of the second; the heavier baryons consist of a meson plus another neutral particle, which Gell-Mann calls b. First suggested in the earlier Sakata model, the neutral b particle has the job of maintaining the conservation of baryon number. Passed along through interactions, it prevents the ups, downs, and stranges in the baryons from all recombining into mesons.

The bit with the b is awkward, as Gell-Mann was the first to admit.

A simpler and more elegant scheme can be constructed if we allow non-integral values for the charges. We can dispense entirely with the basic baryon b. . . . *We then refer to the members* $u^{+2/3}$ *[up with a charge of* $+2/3$*],* $d^{-1/3}$ *and* $s^{-1/3}$ *of the triplet as "quarks" q and the members of the anti-triplet as antiquarks* q̄.

Exactly as Serber suggested, baryons are now made of three quarks; a proton, for instance, consists of two ups and a down clamped together by the strong force. The lighter mesons are composed of one member of the first trio (a quark) and one of the second (an antiquark).

How did this explain why the eightfold way successfully ordered particles? Gell-Mann argued that the quarks and the antiquarks that make up baryons and mesons divvy up each particle's electric charge. Thus the proton—two ups and a down—has a total charge of $(+2/3) + (+2/3) + (-1/3)$ = $+1$. Similarly, the neutron, which Gell-Mann supposed to be made of an up and two downs, has a total charge of $(+2/3) + (-1/3) + (-1/3)$. The particles in the $SU(3)$ baryon families consist of every combination of three units of $+2/3$ or $-1/3$; the result is always an integer.

Quarks not only explain $SU(3)$'s usefulness, they make sense out of strangeness and isotopic spin. A particle's strangeness is given by the number of strange quarks inside it; a strangeness of 0, for instance, means that no strange quarks are inside. Easy![4] And isotopic spin is determined by the number of up and down quarks in the particle: A proton, which has more ups than downs, has an upward axis of isotopic spin; the neutron, with more downs than ups, spins on a downward axis. Values of strangeness and isotopic spin are constant in strong interactions because, as it happens, the strong force cannot change one kind of quark into another. It can shuffle them around like cards in three-card monte, or hold them together in an iron grip, but it cannot alter the proportions of up, down, and strange.

Gell-Mann suggested that although the weak force cannot hold quarks together, it might get them to change types, or "flavors." And that trait would account for the long lives of strange particles, which can decay into ordinary particles only when the feeble weak force finally asserts itself and turns a strange quark into an up or down.

Up, down, strange: Just three quarks explained phenomena that theorists had been puzzling over for years. Nevertheless, Gell-Mann said, "Quarks went over like a lead balloon." The reason was that the fractional electric charges $+2/3$ and $-1/3$ contradicted one of the most thoroughly established rules in physics. Over the half-century during which anyone had been looking, no particle with, say, two-thirds the charge of a proton had ever been spotted. Such particles simply could not exist.

Gell-Mann asserted that quarks had never been seen because they cannot be seen. For reasons he could not explain, quarks must always be chained together, like the hero and heroine of Alfred Hitchcock's *The Thirty-Nine Steps*. ("You can't even pull one out with a quarkscrew," Glashow has said.)[5] This, too, did not go over well. Experimenters did not like the notion of something that *in principle* could not be found with the aid of their machines. Their frustration was not assuaged by Gell-Mann's oracular pronouncements that the failure to see fractionally charged quarks meant that "they exist but are not 'real.'"

(Gell-Mann explained once that he used those terms "because I dreaded philosophical discussions about whether particles could be considered real if they were permanently confined. While a colleague of mine falsely claims to have a doctor's prescription forbidding him to engage in philosophical debates, I really do have one, given to me by a physician who was a student in one of my extension courses at UCLA.")[6]

To others, at least, it seemed that Gell-Mann hedged when it came to the question of the ontological status of quarks—whether they were actually *there*. Although he closed the paper by proposing experiments "to reassure us of the nonexistence of real quarks," Gell-Mann did not expect that experimenters might find them. "Even I thought the idea of observable fractionally charged particles was crank," he said. "My intuitive idea was that maybe they were permanently stuck inside and couldn't get out and were therefore what I called 'mathematical' quarks—that was very hard to explain to people. They thought it was some sort of cop-out. So I asked, 'Why don't you look for them experimentally if you want them to be real?' Then people said, 'That's crazy, too. Everybody knows that there won't be any experimental particles that have fractional charges." About the time he finished the paper, he called his former teacher, Victor Weisskopf, who had been appointed the director of CERN. "I was in Caltech, he was in Geneva. In the course of the conversation, I said, 'By the way, Viki, I have a very good idea for how to account for all the mesons and baryons; they are made of particles of charge $-1/3$ and $+2/3$, three kinds of them.' And he said, 'Oh, nonsense, Murray, don't waste time on a transatlantic call talking about stuff like that.' And I said, 'Well, look, maybe they exist as observable particles. Do you think it would be worth trying to look for them at CERN? How about a search for observable particles of this kind?' And he said, 'Murray, let's talk about something important.' "

The quark paper appeared in the CERN journal *Physics Letters*, whose editor at the time was the Franco-Polish theorist Jacques Prentki. "We were just starting out," he said, "and nobody from the States was submitting papers. I said, 'Murray, why don't you submit something?' So I got the quark paper, which he did not think would be accepted by the *Physical Review*. Now, everybody knew that if you took the hadrons and got them broken up, you ended with fractional charges. That is arithmetic. But it took Gell-Mann to make a very beautiful paper from it. I accepted it immediately. The blame for such a crazy paper would fall on Gell-Mann, not the editor of *Physics Letters*."[7]

Many physicists *were* thinking about triplets. Any theorist with a rudimentary knowledge of group theory could wonder about the fundamental representation of SU(3). Yuval Ne'eman had worried about the problem from the beginning, finally deciding that the regularities of the eightfold way, like those of the periodic table, would be understood fifty years later.[8] Unable to bear the thought of waiting half a century, Ne'eman took a stab at explaining the structure within a few months. There were three known leptons;

Ne'eman asked Salam if a baryon could be made by combining neutrinos, electrons, and muons. Salam laughed outright, then said that no idea is too stupid if it works.[9] With another Israeli theorist, Haim Goldberg, Ne'eman worked out the mathematics of the quark model, but shied away from hypothesizing the existence of quarks themselves. Their hedging, awkwardly written paper was sent to *Nuovo Cimento* in February 1962; it passed almost unnoticed, and is still hard to read.[10]

Somewhat later, Julian Schwinger and T. D. Lee independently came up with fundamental triplets of their own. Respecting fifty years of experimental practice, they, too, were leery of fractional charges. Schwinger scoffed at Gell-Mann's up, down, and strange; in his article, mailed the month after Gell-Mann's article appeared, he said they could presumably be detected solely by "their palpitant piping, chirrup, croak, and quark."[11]

The saddest story belongs to another Caltech theorist, George Zweig, who duplicated the quark model exactly at almost the same time. He called the quarks "aces," and strenuously insisted that they could be detected. Unfortunately, he was at CERN during the year of the quark, and resisted a laboratory regulation that he had to publish his papers in a particular form in the laboratory journal, *Physics Letters*. Zweig had worked out an elaborate scheme of diagrams for the aces, but the editors would hear none of it. Despite heated argument, Zweig never published his long, careful article on aces.[12] (One man's meat: Gell-Mann sent his article to *Physics Letters* voluntarily because he was sure he would be hassled by the American *Physical Review Letters*.)[13] Word got around, though, and the reaction to aces was, in Zweig's words, "not benign." It was all right for someone of Gell-Mann's stature to advocate the lunatic notion that most of matter was made up of ineffable entities that were invisible to experiment; having no reputation to protect him, Zweig was denied an appointment at a major university because the head of the department thought he was a "charlatan."[14]

Nevertheless, there was a certain elegance to the idea of three quarks. For one thing, the three quarks matched up neatly with the three leptons, and physicists were beginning to discover that symmetry begets unification. But no sooner was this pattern pointed out than it was disrupted again through the discovery of a fourth lepton, in what amounted to the last hurrah of Columbia University as the dominant center of particle physics in the United States, and thus the world. Whereas the eightfold way arose as a frontal assault on a discipline-wide problem, the fourth lepton was first suspected by theorists worrying over small paradoxes, the sort of tiny irritants whose slow consideration can with luck produce pearls of insight. There is a dictum known as the totalitarian theorem—Gell-Mann took it from a line in *1984*—which states that every particle interaction not forbidden is compulsory; it is up to experiment merely to determine the rates. By 1960, physicists believed that the electron and muon were exactly, precisely, absolutely identical, except that the latter was more than two hundred times heavier than the former. By the totalitarian theorem, muons should be able to turn into electrons through the simple expedient of emitting gamma rays that take away the extra mass in the form of the photons' energy—unless some rule forbade it. Experimenters had duly searched for evidence of such decays, and found they occurred, if at all, less than one ten-thousandth as often as ordinary decays. Theorists had a little puzzle.

The late 1950s was also when many physicists began to seriously consider the possibility of an intermediate vector boson that plays a role in weak interactions.[15] If the vector boson is admitted to the theory, then the muon-electron decay should occur as follows: the muon turns into a W^- and a neutrino, the W^- releases a photon, and the neutrino and W^- recombine into an electron. Having set up the chain of interactions, Gary Feinberg calculated the probability that it would take place: It was five times more than the experimental limit. Therefore either the boson did not exist or the decay was expressly forbidden for some other reason. Feinberg guessed the other reason might be that the neutrino from the muon simply could not become an electron because it has what one might call an essential "muness." Thus, two neutrinos exist, one each for the electron and muon. Although such ideas had been bruited about before, they were not respectable, and Feinberg, then a young postdoc at Brookhaven, was sufficiently mindful of the climate of opinion that he tucked his solution into an obliquely written footnote.[16]

Despite Feinberg's hesitation, the footnote stirred discussion, which subsequently foundered on the absence of satisfactory means of studying neutrinos. Because neutrinos experience only the weak force, they rarely interact with matter. Floods of neutrinos from the sun shoot as easily through the earth as if the planet were a sheet of tissue paper. Reines and Cowan had barely been able to discover the neutrino by examining the trillions given off every second by a nuclear reactor; although most physicists believed that the particles exist, a smaller number found their demonstration convincing.[17] In the next five years not a single direct measurement of neutrinos was performed anywhere in the world. For this reason, Feinberg did not even consider the possibility that the question of two neutrinos could be settled by experiment.

The answer to the dilemma came in a roundabout way, a product of the afternoon coffee breaks on the eighth floor of Pupin Lab at Columbia University. During these hour-long respites, students and faculty alike gathered before the blackboard, tossing out pet ideas and watching T. D. Lee run through their consequences with chalk in hand. One day in November 1959,

the *koffee klatsch* was attended by a Columbia experimenter named Mel
Schwartz, who by chance found himself on campus without a class to teach.
The subject of the coffee break was the experimental study of weak interactions. Schwartz is a square-shouldered man with the brash, tough manner
that is the hallmark of what Europeans mean by the American style. Infuriated by bureaucracy and committees, he left high energy physics in the
1970s for the greener, more independent pastures of Silicon Valley. He still
loves physics, he said when we met him. He just hates big science. There are
certain people who love to talk about their passionate interests. Schwartz is
one of them; when we asked about the discussion that day in Pupin Hall, the
phrases tumbled out. "Everybody was sort of gathering around, throwing
around crazy ideas on how to measure the weak interactions at high energies," he said. "Somebody was discussing how to scatter electrons at very,
very high energies and looking at the deviations—you know, the cross-
section as a function of angle. There are a whole lot of other ways you can
look for weak interactions, but they're all ugly, because they're all terribly
masked by other things that are taking place." The much more powerful
effects of the strong and electromagnetic forces tend to swamp weak interactions. "At the end of the hour there was no conclusion. I think T. D. and
Gary Feinberg and a few other people were there. I went home, and then
that evening I was pondering the thing, and it became suddenly obvious that
the right way to do that would be to make use of neutrinos. The reason
being, of course, that neutrinos have *only* got the weak interactions. So if you
wanted to study the weak interactions, the question became, could you make
enough neutrinos? Once you think to yourself, in a very simple-minded way,
that you can make a neutrino beam, it's all a matter of calculating how many
you can possibly get, and whether there'd be enough to do something."

By the end of February 1960, Schwartz worked out what would be
necessary to produce and detect a beam of neutrinos. (In Europe, Glashow
had just worked out his SU(2) × U(1) model.) At the same time, Lee and
Yang, whose partnership was not yet defunct, produced a list of theoretical
questions about neutrinos that should be resolved. First in the series was
whether there are two neutrinos. The pair of two-neutrino papers were
published back to back in the *Physical Review Letters*.[18] As difficult in practice as it is simple in theory, Schwartz's method for studying neutrinos
involved shooting protons into a chunk of metal, creating pions, which then
decay into floods of muons and the neutrinos that accompany their creation.
All slam into a thick steel wall. Only the neutrinos make it through the wall
and into the detector on the other side. To discover whether two types of
neutrinos exist, the experimenters must identify the particles produced in
the detector when the neutrinos hit something. If they include muons but no
electrons, then there is some conserved quality of "mu-ness," and both

muon and electron neutrinos exist. If muons and electrons are created in equal abundance, then only one type of neutrino exists, and theorists had some rethinking to do. Schwartz's Letter, published in March 1960, claimed neutrino experiments "should be possible within the next decade." In fact, the first design work began within months.[19]

"There were two machines being built at that time," Schwartz said, "both of which had the capability of doing this experiment: the AGS and the CERN machine. Now, the CERN machine was about six months ahead of the AGS. In the spring of 1961, a group at CERN began looking into the experiment, and we began looking into the experiment at Brookhaven." The rivalry between the two teams was overt and intense from the first: Both were composed of ambitious young men aching to make names for themselves in what looked like wide-open territory—the completely unexplored terrain of neutrino interactions. Moreover, the Brookhaven physicists were led by Schwartz and his Columbia colleague, Leon Lederman, and the CERN group included yet another Columbia experimenter, Jack Steinberger. "There was a certain period of resentment because of the fact that we were in competition with somebody from Columbia who had decided to go off and do the experiment elsewhere. In any case, one day we got the exciting news that the experiment at CERN was a disaster, because Jack had made a mistake in his calculations. Somebody by the name of von Dardel had gone over Jack's calculations and discovered an error and found that you couldn't do the experiment operating out of a five-foot straight section." The straight section was the segment of pipe between the ring of the accelerator and the target. CERN physicist Guy von Dardel discovered that the straight section was too short and that the magnets on the ring were therefore close enough to defocus the pions to one side, which would diminish the neutrino intensity to the point where the experiment became impossible. Physicists are rarely kind to one another, and experimenters in competition rejoice over their opponents' misfortune. "That became of course very exciting to us. The so-called von Dardel effect meant that we were going to be there first. Number one, we had a ten-foot straight section; number two, there was no opposition at Brookhaven toward doing the experiment, whereas at CERN there was very strong opposition."

We asked Schwartz why CERN hadn't installed a longer section, which was, after all, only a matter of five feet of pipe. "Well, the place at that time— I'm not sure how it is now—was not overly full of cooperation. Different teams from different countries were always at each other's throats. If one guy got screwed, the other guy would jump up and down with joy." At a different point in the conversation, he came back to the subject. "Everybody else who was not doing the experiment was in competition with the people trying to do the experiment. And so the minute that von Dardel discovered this problem, the first reaction should have been, 'Well, let's go and switch the experiment to a ten-foot straight section'—right? In fact, the reaction was, 'Can the experiment.' Which made us very pleased. Of course it was about as stupid a thing as CERN could have done, in retrospect."

Discouraged at the nationalistic battling within the international lab, Steinberger returned to Brookhaven, where the hatchet was buried and Steinberger invited to join the experiment. The team spent the rest of 1960 and all of 1961 building the apparatus. The plan was to send pulses of protons from the AGS down the ten-foot straight section into a beryllium target, creating a cloud of pions, which were in turn directed into a seventy-foot pipe. About one-tenth of the pions decayed to produce, among other things, billions of neutrinos. Everything then smashed into a pile of steel forty-two feet thick that only the neutrinos could penetrate. The steel was scrap, rusty chunks cut from the sides of old Navy cruisers; some of the pieces were even stamped *U.S.S. Missouri*. Piling it around the detector took Lederman, Schwartz, and Steinberger months of hard labor. On the other side of the steel was a ten-ton detector consisting of many thin aluminum plates set close together; the plates were electrically charged, and the passage of charged particles created sparks. By photographing the sparks, experimenters could obtain approximate pictures of the particles' trajectories. Once a day, they calculated, one out of the billions of neutrinos passing through the steel would strike a proton or neutron in the detector, producing an interaction.

The experiment began in early 1962 and, as is the case with all genuinely new techniques, something went wrong immediately. In this case, it was the AGS itself, which leaked protons and gave rise to contaminating radiation. Because the team was expecting only one neutrino event a day, it would not take much to hide what they were looking for. After some time, Lederman recalled later, "we traced it down to part of the beam escaping the target and hitting the wall of the vacuum chamber so that neutrons would come down into the concrete and underneath and scatter from the concrete up into the detector." The team figured they could stop the leak if they piled lead right along the wall of the brand new accelerator. "At the time there was a line beyond which Mr. G. K. Green, who is in charge [there], said we could put shielding only over his dead body; this would have made a small but unsightly lump in the shielding. We compromised."[20] Seven days a week, twenty-four hours a day, physicists sat outside the detector, watching the dials, consuming sandwiches and coffee, playing chess, and sometimes napping. The experiment ran eight months until July 1962, during which time over a hundred trillion neutrinos passed through the detector. Just fifty-one of them hit something on their way. Not a single electron was found in the fragments of these collisions, whereas muons were found in all fifty-one.

The American team was elated. With pluck and luck, they had soundly beaten CERN, and they popped open the bubbly to celebrate. We asked Schwartz about the pleasure of being first, of having clear priority to the discovery. "People are very selfish of their priorities in physics," he said. "You know, now I'm in a business [computer systems] where the measure is very simple. It's how many bucks can you bring in, right? If your company makes enough profit, then you're a big man. If it makes a little profit, you're a small man. If it makes no profit, you're miniscule. So it's very simple to make a measure of a man. In physics, the only measure you make is general recognition. In that situation, you have an awful lot of people fighting for the only money that exists, which is the money in recognition. It's a big poker game, with a certain amount of zero-sum, so to speak. In other words, if I win, you lose. If I get the priority for that particular thing, you haven't got it." Experimenters often ask the question: Who was the *second* person to say "$E = mc^2$?"

The tally now stood: Leptons 4, Quarks 3. The symmetry between them could be restored by tinkering with the number of quarks, but this seemed pointless to most physicists, who regarded them as Gell-Mann's fractionally charged folly. Not Sheldon Glashow. Hypothetical particles perpetually locked inside all the particles in the nucleus was just the sort of loony supposition he thrived on, and he was determined to play with it. Glashow quit Caltech in a fit of annoyance in 1961 when the college bookstore refused to cash one of his checks. Exasperated at the lowly status awarded to postdoctoral fellows, he wangled a junior faculty position at Berkeley and moved upstate, where he spent some years energetically propagandizing for the eightfold way.[21]

In the spring of 1964, Glashow returned to the Bohr Institute. Bohr had died, but little else had changed. The Institute was still full of bright young theorists talking up their ideas in tiny offices. One of the other newcomers was James Bjorken, a tall, sandy-haired theorist who had arrived from Stanford a few months earlier. The two men had met in northern California a little while before. Soon Glashow hit Bjorken with his latest pet notion, namely, that there might be a *fourth* quark. He called it *charm*.

Charm excited Glashow, because it arranged the Universe back into an elegant pattern. With the discovery of the second neutrino, the existence of four leptons was universally accepted. If a fourth quark could be summoned into existence, Glashow reasoned, then the subject matter of physics could be divided into two families. The first family makes up ordinary matter; the second occurs in high energy processes.

quarks	leptons
up	electron
down	neutrino

ORDINARY MATTER

quarks	leptons
charm	muon
strange	neutrino

WEIRD MATTER

Unfortunately, he was unable to find any justification for charm other than his certainty that God could not have been stupid enough to decree four leptons and only three quarks. Nonetheless, he managed to persuade the reluctant Bjorken to send the idea off anyway.

It is hard to imagine two theorists with more divergent styles. Bjorken has long arms and long legs and the slight stoop of a very tall man. There is a certain outdoorsy ranginess to the way he occupies a chair; one suspects immediately that he has spent a fair amount of time on hiking trails. His voice is a quiet, almost avuncular tenor. Glashow is tall, too, but with thick graying hair that falls in a tangle over his forehead, needing less to be combed than to be subdued; his glasses are often askew; he speaks with a pure New York inflection untouched by years in Harvard Yard. Glashow has spent his career producing ideas in a hot-brained hurry, tossing one aside the instant another occurs to him, whereas Bjorken is slower, more methodical, closely tied to the phenomena. Bjorken has spent his career working closely with experimenters, first at SLAC and now at Fermilab; he likes trying to guess what the machines will find before they are turned on. The article with Glashow is perhaps the most speculative in his oeuvre.

"At the time, it was not a particularly earth-shattering paper," he said. "The idea of a fourth quark had been floating around among many people, who also wrote papers about it at the same time as we did. You have to remember the flavor of the time. It was an easygoing time for models. Models came and went. Salam was particularly prolific; he would explore every idea, build a model for it, and most all of them were wrong. A mountain of papers proposing various models were published by many authors, including Shelly, that went straight into the trashcan. There was no reason to think that this paper would fare any better. The idea turned out to be right. But then, I've seen Shelly equally enthusiastic about things that didn't pan out. We made a lucky hit, but we weren't the only ones. The one thing unique about that paper was the name we gave the fourth quark: charm. That is what stuck."[22]

When they submitted the charm article, Bjorken indicated the spirit with which he took the whole enterprise by signing a fictionalized version of his name: "B. J. Bjørken," "B. J." being the monicker he is known by, followed by the Danish spelling of his Swedish father's name. In the indexes of scientific publications, there is a mysterious "B. J. Bjørken" who has written only one paper. It appeared in *Physics Letters* in August 1964—just seven months after Gell-Mann's original quark article.[23]

As it turned out, charm worked more magic than either Glashow or Bjorken knew. When incorporated into $SU(2) \times U(1)$, charm explained why experimenters had never observed weak neutral currents.

In many cases, the greater the number of ways in which a particle can decay, the more readily its decay will occur. This much is a familiar part of our world. For example, if two doors on a bus are open rather than one, twice as many people can exit at any given time. But in the quantum world such "exits" can interfere with one another, and opening an additional "door" may have the paradoxical effect of reducing rather than increasing the frequency with which particles decay.

So it is with strange particles—the kind that physicists had always used to study the weak interactions. According to the $SU(2) \times U(1)$ theory, weak neutral currents exist, which means that strange particles should often emit virtual Z^0s and turn into ordinary particles without any change in electric charge. But experimenters had never seen any such event, let alone the number likely if $SU(2) \times U(1)$ were correct. Charm reconciled Glashow's lovely theoretical model to this ugly experimental fact. By a quirk of mathematics (and nature), the probability that a strange particle will emit a Z^0 and become an ordinary particle and the probability that it will emit a Z^0 and become a charmed particle are almost equal. It is as if the strange particle stands irresolute between two equally tempting options and cannot bring itself to choose either. (Glashow wrote later, "As it happened, a sign in the equation that defines [the charmed reaction] is negative, and the two interactions cancel each other.")[24] Thus, although weak neutral currents should be able to occur in strange particles, they never do. Experimenters would have to look at other particles for evidence of such currents.

Unfortunately, when Glashow and Bjorken wrote their paper they didn't see the power of charm. "There it is in black and white, and we missed it!" Glashow said recently, smiting his brow to theatrical effect. "If you read my 1961 paper with Gell-Mann, you'll say, 'These people say the theory doesn't work and something is missing.' And if you read my 1964 paper with Bjorken, you'll exclaim, 'By God, that's it! A fourth quark!' But did we notice it? No, not for six years."[25]

In 1964 Glashow was not even thinking about weak neutral currents. He was preoccupied by how poorly quarks fitted into field theory. If no theorist could come up with a field that would hold quarks together in a way that would explain why they were never detected in experiments, it was difficult to understand how adding a fourth quark would help matters. Troubled, Glashow gave up on charm, as he had earlier given up on $SU(2) \times U(1)$, because he didn't know how to show that it was right. For the next six years he published inconsequential papers, threw others away, and grew cranky. In 1966, he became a full professor at Harvard, and arrived on campus to find other members of the physics department in the same funk.

□ □ □ □ □

"There's a long tradition of theoretical physics, which by no means affected everyone but certainly affected me, that said the strong interactions are too complicated for the human mind." The speaker was Steven Weinberg. We were in his living room, and he had been inveigled into speaking, in his synoptic way, about the strong force and the many attempts made by physicists to understand it. "When I was a graduate student at Princeton, I worked with Sam Treiman, who was my thesis adviser. I don't know remember whether Sam said it to me or we said it to each other, but that was the feeling. You just couldn't get anywhere trying to understand them. What you could do was sort of walk around the outside of them—" look for small effects and simple situations "—which gave you such a feeble interaction that you really could study it mathematically and learn about its symmetry properties." Other people, of course, had waded into the fray, directly confronted the strong force. "I never got involved in that," Weinberg said. "I thought, that's not—maybe their minds can do that, but mine couldn't."[26]

Although both Gell-Mann and Ne'eman based SU(3) on Yang-Mills, the inability of local gauge field theory to account for the eightfold way eventually made them both shy away. The subsequent explication of the representations in terms of quarks removed Yang-Mills theories one step further; by 1965, gauge field theory, which today dominates particle physics, was sufficiently out of favor that even Weinberg, one of its more ardent proponents, could remark that "no one would ever have dreamed of extended gauge invariance if he did not already know Maxwell's theory."[27]

Between 1964 and 1969, there were no major experimental discoveries concerning the strong interactions. Without pace-setting experiments, theoretical ideas tend to branch out, wide, ramose, and shallow, splitting, like the roots of a palm tree, into endlessly fine distinctions, ready to be torn out by the first strong wind. In a repeated, faddish pattern, new wrinkles and clever techniques spark frenzies of speculation that subside as physicists run into blind alleys. Considering the theoretical work of the 1960s, one wants to retreat before the barrage of calculation in the journals, the thousands of articles that nibble at the corners of large, poorly understood questions, and come to small conclusions.

An attempt to evoke the prevailing mood runs into the historiographic difficulty that many if not most high energy physicists spent their time on pursuits that seem remote from what are now considered central problems. Some theorists, for example, put SU(3) and the two orientations of spin together to form SU(6); after two years of frenetic calculation, it was demon-strated that the theory was not Lorentz invariant.[28] When that failed, the-orists turned to symmetries like U(6), SU(12) and E_7, another one of the "exceptional" groups. All were eagerly embraced and readily discarded. Gell-Mann made a habit of building new techniques on field theory but removing the particular wrong version of field theory from the final product, likening the process to the French chef's trick of cooking pheasant between two slices of veal, then throwing away the veal. He had done this with the eightfold way; he did it again with "current algebra," which treated the currents of the interactions as the fundamental entities, rather than the particles comprising them.[29] Other researchers tried to link the spins and masses of elementary particles with what were called "Regge poles," after the Italian theorist Tullio Regge. Some Regge enthusiasts believed that their poles, together with calculations of scattering probabilities, would displace quantum field entirely, leading eventually to the reign of "nuclear democ-racy," in which no particles would be fundamental and every entity in the subatomic world was a transitional form of every other.[30]

In the huggermugger of currents and Regge poles and groups, good ideas vanished if not advertised. Even as the majority of theorists cast about for some clue to the strong interaction and the quarks it held together, if it held them together, the later key to success lay entombed in the back issues of the *Physical Review*. If ideas are not picked up quickly, they are generally forgotten unless their authors continue to promulgate them energetically. In this case, the author was Yoichiro Nambu, a soft-spoken, quiet man who has earned the dubious distinction of being described as the John the Baptist of elementary particle physics—"there in the wilderness before everyone else," Bjorken once put it.

Nambu's manner is self-effacing, almost retiring; his features seem to recede from view, his words to fall into the back of his throat. He doesn't parade his accomplishments.[31] Born in 1921, he is one of the last members of the great brief flowering of Japanese theoretical physics that began with Nishina and Yukawa in the glory days of quantum electrodynamics. At the beginning of the 1940s, Nambu attended the University of Tokyo, where the curriculum did not include particle physics. He taught himself about cosmic rays, relativity, and quantum field theory with a group of friends. When he could, he attended the seminars at Riken given by Tomonaga and Nishina, including one that discussed a letter from Sakata proposing for the first time the two-meson hypothesis. The university was exceedingly formal, comatose with tradition and bureaucracy, the last legacy of an aristocratic style of education that was about everything but learning. Students did not question professors. Professors did not fraternize with students. Hints of change came from the direction of the physics department, but only small indications.

Within a year after Pearl Harbor, Nambu was drafted into the Imperial Army and put to work in Osaka researching the microwaves used in Japan's recently developed land-based radar. One of Nambu's tasks was to steal a document from Tomonaga, working for the Imperial Navy; the two branches of the Japanese military had as little to do with each other as possible, and the Army got word that Tomonaga had written a paper in which he used some of Heisenberg's ideas to understand microwaves. Through a professor who worked for both sides, Nambu managed to get a copy of Tomonaga's work.

After Hiroshima and Nagasaki, Nambu was demobilized. After a brief stint at the University of Tokyo as a research assistant, he was appointed to a post at Osaka City University, a new school that was trying to rise up from the wreck of postwar Japan. Living in the confused desperation of a beaten nation and with their leaders under suspicion, the Osaka group nonetheless managed to keep abreast and even ahead of their victorious American colleagues. With two other Osaka theorists, Nambu wrote one of the first papers on the production of strange particles.

Nambu's office today is on the second floor of the Enrico Fermi Institute of Nuclear Studies at the University of Chicago. One of the long row of neoclassical buildings at the school, the institute is directly across from the site of the old football stadium, where, in a basement squash court, Enrico Fermi supervised the creation of the first man-made nuclear chain reaction. These days the football stadium is gone; in its place is a set of tennis courts and a tall, rounded monument to atomic energy by Henry Moore. Nambu's window overlooks the courts and the sculpture, which were dimpled with snow on the cold February day we visited him; he told us that he and the rest of the particle and nuclear physicists had watched the princes and potentates at the dedication ceremony from their office windows.

We asked him how it was that Japanese science had been so strong despite so many obstacles.

"That's a curious thing," he said, and then waited a while with his hands flat on a long table in his office. "Yukawa had this great vision. He was a self-made man, in the sense that he never went to study abroad. He was a very highly independent person. And Sakata and Taketani, who were his collaborators, they had their own ideas. . . . I don't know how they arrived at those views. I just don't know. I mean, because they were based on Hegelian, on Marxist philosophy."

By all accounts Sakata was a powerful man whose smooth, sociable demeanor masked an overbearing streak. He rebelled against the authoritarian tradition of Japanese academics only to use it to his own benefit when he became established. Holding that real but unseen particles ran counter to the tenets of existence discovered by Engels and Lenin, Sakata refused to consider quarks as a serious possibility. Nambu subscribed to this dialectical view, and to some extent still does. It proved fruitful for him, but unfortunately for most others in his native land the ideological pressure to conform was too high. "No great independently minded physicists have appeared since," he said.

After the war Oppenheimer invited Yukawa to Princeton one year, and Tomonaga the next; Nambu followed in 1952. Somewhat to his surprise, he never returned to the country of his birth except for visits. He learned English quickly and well, and went to the University of Chicago a few years later. His style is one of reasoning from first principles, much as a coloratura might constantly return to the primary scales. A good portion of his time has been occupied with the task of attempting to reconcile currently fashionable theoretical notions to fundamental physical principles. In the late 1950s, for example, Nambu listened to a lecture by a graduate student, John Schrieffer, on the "BCS" theory of superconductivity that he had recently helped to develop. In the middle of the lecture, Nambu realized that the BCS theory violated gauge invariance. Schrieffer came soon afterward to the University of Chicago as an assistant professor, and Nambu talked with him often. He soon showed that superconductivity was an example of broken symmetry, that the U(1) of quantum electrodynamics did very strange things at low temperatures. Then, with a student, Nambu developed a method of symmetry breaking for strong interactions.[32] The audacity of ascribing to the vacuum—empty space!—a structure was remarkable; just as bold was their assertion that when SU(3) was broken, a massless particle popped out. They claimed it was the pion, even though the pion is not massless, and said that it was not "the primary agent of strong interactions, but only a secondary effect." Since Yukawa's invention of the meson a generation before, the pion had been assumed to be the agent of strong interactions. Now it was relegated to the role of a by-product. The argument later became the basis for the surge of work on chiral symmetry in the 1960s.

We asked Nambu whether he had been attracted to the quark model. "Oh, yes. Except for the fact that these quarks had to have these funny charges—a strange fractional charge. *That* I didn't like. My belief in general was that what you assume as constituents cannot be fictitious, but must be real." His reaction, he admitted, was perhaps influenced by the Sakata school.

We said that the Marxist method did not sound too different from the traditional way of doing physics. "Maybe it's a matter of abstraction," Nambu said. Then, quickly: "But when you start building the particles out of quarks, it didn't seem realistic or correct to me just to regard these quarks as symbols." He made a rare gesture by tracing a little symbol in the air. "I had some difficulty with physicists who could easily swallow that—you know, quarks as just another symbol."

Did he have other objections? Nambu nodded. "This again had to do with one of the basic principles of physics, the Pauli exclusion principle. In a very simplified language, it says that two particles with spin one-half in the same state cannot occupy the same place at the same time. The same state—actually, to be rigorous, the electron has to be specified by its position and spin direction. So if you put two electrons in the same place with the spins pointing in the same direction, that is just not allowed. And this quark model just ignored it completely!"

Gell-Mann had said the quarks had to have spins of plus or minus one-half if three of them were to make a baryon with a spin of one-half or three-halves. (The plus or minus indicates the direction of spin.) To make, for example, a delta, which has a spin of 3/2, requires three up quarks, and 1/2 + 1/2 + 1/2 = 3/2. Three quarks with the same spin are thus packed together inside a delta, which flagrantly violates the Pauli principle. The problem was referred to as a problem with "the statistics." Oscar Greenberg from the University of Maryland worked out an alternate type of statistics—what he called *parastatistics*—in which the spin one-half particles could avoid the Pauli principle. He applied it to Gell-Mann's quark model, but not in a way most physicists understood.[33] Nambu was not satisfied with that route, either; he thought he could fix the statistics and have integer charged quarks to boot.

"You cannot deny that the quark model can explain a lot of things," Nambu said. "I tried to save two things: one, the statistics and two, to make them observable." He was slowly pushed in the direction of making the quarks different from each other. "There is always a hesitancy to increase the number of fundamental constituents. But I was slowly driven to this idea: If there is one kind of quark, why not two?[34] In other words, two types of ups, two types of downs, two types of strange. I tried that first. Then just for the heck of it I tried three—three kinds of quarks. And I wrote a preprint in which the possibility of three was mentioned in passing."

In 1965, Nambu was invited to give a lecture at Carnegie-Mellon, called the Carnegie Institute of Technology in those days, and there he looked at a letter sent to him by a young graduate at the University of Syracuse, Moo-Young Han. Han had used group theory to elaborate on Nambu's three-triplet idea, and, like a jazz player pulling unexpected riches from a popular tune, had discovered treasures within. Through the mail, Han and Nambu put together a paper proposing three types of quarks; it was published in the *Physical Review* without its authors having met face-to-face.[35]

Nambu hypothesized at the start that every quark carries one of three charges, in much the same way that a pion can bear a positive, negative, or neutral electric charge. The quark charge and electric charge, however, have nothing whatsoever to do with each other. Rather, quarks have a new, never-before-seen charge that years later was christened *color*, although it is not ordinary color. There are three "color charges," today called red, green, and blue.[36] (Confusingly, Nambu called his new charge *charm*, even though Glashow and Bjorken had already used the term for the fourth quark.) Baryons are composed of three differently colored quarks; a proton, for instance, might be a red up, a blue up, and a green down. Mesons, on the other hand, would be colored quarks and "anticolored" antiquarks; a pion might be a red up and an "antired" down.

Exactly as positive and negative electric charges attract one another, so do red-blue-green and color-anticolor combinations. Furthermore, like the photon associated with electric charge, the color charge has eight particles—"gluons," in current parlance—that hold colored quarks together. The common proton, in this picture, appears as an extraordinarily complicated object: three differently colored quarks crowded into each other like people on a rush-hour bus, surrounded and permeated by a swarm of gluons.

An odd view, perhaps, but one compatible with integrally charged quarks. Three colors also avoid the statistics problem, for the quarks inside the proton are no longer in the same state; they are differently colored, hence in different states, and can happily coexist.

There was more. Implicit in the scheme—and here was Han's contribution—is the startling suggestion that there are two types of SU(3): one, the eightfold way, is a symmetry of particles that feel the strong force; the other is a color symmetry that pertains to the force itself. Moreover, the intense strong interaction that holds together protons and neutrons in the nucleus is merely a secondary by-product of the second SU(3), the enormously powerful "color force" that grips the quarks. The exchange of pions that clamps together the constituents of the nucleus, and thus every atom, is a feeble reflection of the forces among quarks; the pions should be regarded, Nambu wrote later, "as perturbing forces rather than the decisive factors in the physics of hadrons."

Up, down, and strange quarks were far out enough for most physicists; the Han-Nambu triplets of each seemed even more outrageous, especially when the two men were forced to admit at the end of their paper that at the energies available to experimenters it was "difficult to distinguish" their model from the regular quark model. Unhappily, Nambu realized that arguing that the proton is really a collection of colored quarks and gluons was equivalent to saying that the strong interactions could best be explained *entirely in terms of particles that had never been seen*, and that the particles that could be observed were nothing but secondary manifestations of these invisible entities. Although he was predisposed to believe that matter was composed of a series of levels, such dealing with ineffable entities bordered on the metaphysical, if not the crank. Nambu did not know if his guess was profound or simply a clever new way to be wrong. In any case, nobody seemed to be interested. Convinced that the *Physical Review* would never accept further work on color, Nambu turned over the idea a little more, briefly considering making a Yang-Mills theory out of it, and then gave up. He let the paper with Han slide into the morass of failed speculation on the strong interaction, where it remained for another five years.[37]

□   □   □   □   □

"It really was an extraordinarily boring time in particle physics," Howard Georgi once said of the late 1960s. He was in his office at Harvard, leaning back in his chair, in a good-humored, discursive mood. "Hardly anybody knew what they were doing." Small bursts of laughter punctuated his sentences, as if thinking over the confusion of the past inevitably threatened to spill into simple guffaws of amusement. "Witness the fact that the $SU(2) \times U(1)$ model—Weinberg's model of leptons—was written down and completely ignored by everyone. Him, too, really. No one even blinked at it in 1967. I remember looking at it as a graduate student, and I said, 'Well, it doesn't look renormalizable to *me*.' And dismissed it.

"The reports of friends of mine who were here—" he swept out a hand, indicating Harvard "—at the time were not encouraging, because this place . . ." He paused. "Well, Shelly [Glashow] tends to get depressed during boring periods of physics, and when he gets depressed he sort of gets grumpy and wanders around not doing much of anything for long periods of time. And that tended to happen during this period." He paused again, reconsidered. "Actually, I should say it wasn't obvious that anything was going on, although in fact there was a ground swell of what turned out eventually to be physics coming from the experimenters on the West Coast." We asked Georgi what had been so unexpected about that experimental work. He laughed. "Really, who would have ever thought then that somebody would show that *quarks* existed?"[38]

The ground swell came from the Stanford Linear Accelerator Center, a brand new national laboratory south of San Francisco in the Santa Clara Valley. Begun in 1962, SLAC is centered on almost five hundred acres of countryside—ridged, verdant, marked by forest—that belongs to nearby Stanford University. The heart of the facility is a tunnel a dozen feet wide and two miles long that cuts from the campus laser-straight toward the Pacific: the linear accelerator. Controversy over the size and use of the machine accompanied the laboratory's parturition. For the years before its construction, the accelerator was known as Project M—*M* for monster—and regularly derided by East Coast experimenters. The two-mile length of the machine, they whispered, had been chosen only because that was the longest straight path the planners could make on Stanford property (which was true); electrons, they fairly shouted, were thoroughly understood by quantum electrodynamics and could never produce much new physics (which turned out not to be true). What could be less interesting than slamming electrons down a big sewer pipe into a vat of hydrogen?

The monster now lies rigid across the Santa Cruz foothills like the backbone of a huge extinct beast. Tunneling relentlessly through sharp loess hillocks, under orchards, and even under Interstate 280, the accelerator starts a quarter mile from the San Andreas Fault. (The laboratory has an earthquake committee that reviews all experiments and supervises construction; even the cafeteria vending machines are bolted to the wall.) Although the beam itself is forty feet below the grass, its path is marked by a sort of corrugated iron shed that runs the entire length of the line and contains almost two hundred and fifty man-sized devices known as klystrons. A klystron is an amplifier for microwaves; low-power microwaves come in one end and high-power microwaves go out the other. The high-power microwaves push the electrons down the pipe in the tunnel. When we visited the klystron gallery, as the long shed is called, the monster was on, and small flashing red lights played in mock alarm over the walls and ceiling. It was an unseasonably warm day in February, and the faces of the joggers who regularly use the gallery as a track were glossy with sweat. The klystrons buzzed amiably, noisy but not uncomfortably loud, sending short bursts of microwaves down precisely engineered copper tubes that disappeared into the earth; each tube connected to the beam pipe, and the quick pulse was timed to meet with a bunch of electrons as they shot toward the target, whipping them to greater vigor.

The electrons come from a piece of hot metal oxide subjected to an electric field of intensity sufficient to rip the light little electrons free of their nuclei; shooting out one end of the "gun," they enter the accelerator and are almost immediately pushed to within a whisker of the speed of light. All the further "kicks" they receive from the klystrons do not significantly increase their velocity. They go at almost the same speed, but accumulate energy, storing it like misers, ready to slam into a target with minute, awful violence. Joggers need twenty minutes to run the course of the gallery; the electrons whip through it in about one hundred-thousandth of a second. At that speed, they could travel 'round the world seven times in a single second.

SLAC is open to anyone who cares to enter, a rule that Wolfgang Panofsky, the founder and first director of the laboratory, fought for with considerable vigor. Short, plump, bespectacled, and articulate, Panofsky, who is nicknamed 'Pief' after a German cartoon character, has been involved in political struggles over arms control and scientific budgets since the advent of fission. He is said to be in many ways responsible for the ban on atmospheric testing of nuclear weapons, a treaty that has helped reduce the level of radioactivity in the air. To create SLAC, Panofsky had to negotiate a course through both the bureaucracy in Washington and the unpredictable politics of the San Francisco Bay area. The laboratory was designed so that the base of the accelerator faces a small grassy hill from whose summit we could see the earthmoving equipment for the next SLAC machine, the linear collider. Designed to take one beam of electrons and one beam of positrons from the original linear accelerator and send them down opposite sides of a circular track until they hit each other head on, the collider is an attempt to explore the ramifications of the electroweak theory quickly and cheaply. Most of the new work will be underground; the largest structures on the surface are the old experimental halls, factory-sized concrete and metal boxes where the first evidence of quarks was found.

When the long accelerator switched on in 1967, one of the first experiments was a SLAC-MIT-Caltech collaboration that measured what happened when electrons smacked into protons at high speeds. (Actually, the electron and proton don't directly hit each other, but interact through virtual photons.) The expectation was that the paths of most electrons would be only slightly bent by coming near a proton, but that a few would come close and be bounced off at a wide angle, the way a few of Ernest Rutherford's alpha particles had been knocked back by the atomic nucleus. If there were something hard inside the proton, something with its own charge and mass—something like, say, a quark—the electron would recoil, much as Rutherford's alpha particles had recoiled from the concentrated electric charge of the hard little nucleus in the center of the atom.

"The Rutherford thing was self-evident," James Bjorken recently recalled. He was a student at Stanford when the new machine was proposed. "I certainly remember Leonard Schiff [the head of the Stanford physics department] saying something about it the first time that Project M was talked about at the colloquium at Stanford." Bjorken had been sitting in the seminar room, still marveling at the concept of a machine with dimensions described in terms of miles. "Something two miles long was really *off scale*. It was just *enormous*. The first real geographic machine.

"But anyway, it was announced there'd been some planning done by the senior people and then there was a general colloquium in the physics department to explain this to the community there. And Schiff was the guy who talked about the theory. I remember him talking in those terms, that Project M would give enough energy to look at instantaneous charge distributions inside of a proton." At least to Schiff, that much was clear: One could hunt for constituents in this way. The problem was going to be interpreting the results. "Nobody knew at all the right descriptive language for that, because this was clearly a very relativistic situation, whereas all the previous applications were nonrelativistic in nature. So the barrier was not the generalization of the Rutherford idea to looking inside of a proton. That much was self-evident to anyone who thought at all about it." Later, after his return from Copenhagen and charm, he joined the theory group at SLAC and quickly became preoccupied with the experiments about to be run on the machine. "The real problem was how to handle the problem that this was a very new situation, because whatever the explanation was, it had to include relativistic motion. Extreme relativistic motion, in fact. In a very essential way. Whatever the constituents were inside of a proton, they were moving around at the speed of light, sort of, and the old-fashioned way of doing things just didn't work. That was the barrier that was so hard to overcome."

From the point of view of an electron in the SLAC tunnel, the protons in the hydrogen target rush toward it at close to the speed of light. At such speeds, as Einstein showed, peculiar things happen; from the vantage of the electrons, the protons are flattened out into pancake-shaped objects that move with nearly infinite momentum and seem to be frozen in time. The bizarre relativistic environment made for circumstances that many theorists thought to be both hard to predict and of little intrinsic interest: hard work, no profit. The experimenters were sure that they would come across odd things, but nobody quite knew how they should interpret the results when they came in. Bjorken had many friends on the MIT-SLAC-Caltech experiment; as the beam line neared completion, he started making calculations. He didn't look up for more than two years. "I got really hooked on the problem," he said ruefully. "Worked relatively obsessively on it. I can remember saying at some point or another, 'There's all sorts of other physics around, why the heck am I stuck on *this* thing? I can't get loose from it.' "[39]

He had been trained to approach physics linearly, to build up a deductive chain of reasoning, a series of "ifs" leading to a small number of "thens." Calculating electron-proton scattering didn't work like that at all. On the research frontier, all the half-certain theoretical notions that were jumbled in the foreground had to be applied at once, a little dubiously, but none could be relied upon. The yardsticks were quantum mechanics and relativity. Gell-Mann's current algebra gave a solid foundation, but one of limited extent. Bjorken soon got down to the slippery precepts of common sense. Don't get stuck on a model; strip away assumptions and see what you can believe if you start off believing nothing.

The first batch of low-energy trial runs started in January of 1967. They went well. So well, in fact, that the Caltech members of the group dropped out of the experiment because the results had been completely in accord with expectations. The electrons ricocheted off the protons; a few were sent off at wide angles. We told you so, the Caltech experimenters were informed by their colleagues on the East Coast. The remaining experimenters— Henry Kendall and Jerome Friedman of MIT, and Richard Taylor of SLAC— pressed on in the second half of the year, looking for cases where the electrons had scored more direct hits on the protons, creating a splash of new particles, and considerably complicating the situation. By the beginning of 1968, the group was ready to start looking at the data.

Here a distinction must be introduced. The first set of runs, in which electrons and protons simply bounced off each other, measured what is known as *elastic* scattering, interactions in which the particles knock each other about like so many billiard balls. When, in the later runs, the electrons had lost enough energy in the collisions to create new particles, this was *inelastic* scattering. "The whole East Coast establishment didn't believe that electron scattering of any kind was interesting at all," Richard Taylor recalled. "Ever! They *still* don't." Inelastic scattering was regarded merely as an uninteresting way of producing resonances and measuring their properties, work that had in most cases already been done; elastic scattering, in which particles retained their identities, was thought of as the way to study the structure of a proton. The electron came in, bounced off whatever was inside, and left, without confusing the issue by making more particles. Several elastic scattering experiments had already been performed, and Taylor liked to quote a Harvard physicist's remark: "The peach didn't seem to have a pit."

When the later, inelastic data was processed by the computers, however, the experimenters found themselves staring at an odd batch of numbers. A relatively large number of electrons seemed to fly off to the side even when the proton exploded in the collision. Theorists had predicted rather vaguely that the likelihood of such wide-angle interactions would be small; instead, the probability of an electron careening off to the side was a hundred times what they had imagined. It was unclear what this meant.

Bjorken thought he knew—"sort of," as he put it. From half a dozen sources—principally Gell-Mann's current algebra, but also from work on neutrino scattering using various mathematical rules—he had come to the conclusion that the possibility of scattering at the large angles would depend on the ratio of two simple quantities: the energy loss of the electron and the momentum transferred from the electron to the proton. Above a particular energy threshold, the electrons would stop behaving as if they were hitting big protons and start acting as if they were ricocheting off something as small as another electron. That small something might be a quark, although Bjorken was anything but certain about it.

A model of caution, the experimental group was afraid that the photons radiated by speeding electrons sufficiently complicated the interaction to make any calculation dubious.[40] They did their analysis two different ways and released their data in slow bits and pieces. As the numbers dribbled out, Bjorken made his own rough graphs. The curves more or less did what he thought they would. Although he was young and leery of pressing his views, he showed his graphs to the puzzled experimenters. They were interested in the fit, but baffled by his abstract style of exposition.

Nonetheless, they felt they had enough to go on. In September 1968, Panofsky told a large conference in Vienna about what was becoming known as "scaling."[41] The presentation was lengthy, evenly phrased, and sufficiently encyclopedic that no member of the audience seems to have noticed that Panofsky ended the talk by mentioning "the possibility that these data might give evidence on the behavior of point-like, charged structures within the nucleon."

At that point, Richard Feynman paid a short visit to SLAC. After working on the weak interaction, he drifted out of elementary particle physics temporarily, because he thought there were not yet enough experimental clues to make good guesses, and into the quantum theory of gravity. Little research had been done into quantum gravity since the flowering of interest in the subject created by Einstein's theory of general relativity. Feynman struggled for a few years and made some small headway, but decided to return to high energy physics.[42] Tall and sparely built, with thick long gray hair combed straight back over the skull, he has a face marked by years of vivid internal experience and a bold, precise, intense, and almost imperious manner of speaking. His office in Caltech is by Murray Gell-Mann's; one overworked secretary serves them both. Feynman is an enthusiastic amateur percussionist; he announced his presence on the floor by rapping drum patterns on the corridor walls with his knuckles as he approached. He walked with the jaunty, hips-out stride of a young blade checking out a bar the day after payday.

Sunlight filled Feynman's large, cluttered office. A pair of glasses rested at an angle in his shirt pocket. His dislike of interviews has increased since the Nobel Prize made him willy-nilly into a media figure; his voice was edgy, suspicious, ridged with theatrical intonation. He is scornful of history and philosophy. We asked how he had become interested in electron-proton interactions. Staring straight into his interlocutors' eyes, he waited for three long seconds before answering. "I had not been in physics much," he said. He spoke with an urban snap, the fast emphatic cadence of a city dweller. "I mean, I'd been doing gravity, waiting for phenomena to grow so that it was more interesting. Then I thought I'd better get back into hadronic physics. I asked some friends where a problem was and they said that high energy collisions are peculiar between protons. About the way the cross-section behaved and some other things." In the jargon, cross-sections are measures of the likelihood that particular types of interactions will take place. "I worked it all out and didn't see anything peculiar. And then I got interested when I realized that the machines were going to be for high energies, so I started to work on the theory of what happens at relativistic energies, trying to guess what the behavior would be."

He didn't want to imagine that Gell-Mann's quarks were inside the protons, and that this would affect the interaction between two protons. Instead, he guessed agnostically that there was an indeterminate number of unspecified objects he called *partons* inside each proton, and tried to proceed from there. "Partons were nothing but the field quanta of some future relativistic field theory, whatever it may be. The method I had of thinking was to not decide yet which it is but to see what would happen in general. I gave it a word, so that meant I didn't have to say quarks or whatever. I could say, 'Whatever they're made out of, when we find out later, they'll behave like this.' We can find out what they are and how they're distributed by experiments. I was noticing I could deduce something irrespective of what they were made out of." He went to SLAC to find out what the experiments were doing. We asked him if he had heard of scaling beforehand.

"No. They said that Bjorken had proven by some way that the data should be plotted a certain way and that it fitted pretty nicely, that it behaved like he said it would behave. They asked me why did it behave like that. He [Bjorken] at the moment was out of town. I said I didn't know a cause, why don't you ask him? They said something about that he had tried to explain it, but it wasn't very clear or something. They showed me the data. They showed me that it fitted this thing, this function business that he had. And then I went home after they showed it to me—" *home* being his motel room "—and I began to think about it." Abruptly he comprehended that scaling was caused when an electron was going fast enough to interact with an individual parton, not the proton as a whole. The complicated, messy interaction with the proton is then suddenly replaced by a much cleaner collision between two pointlike objects. "I suddenly realized that I was very *stupid*, that I'd been working on a doubly hard problem, that the [collision of] two protons is harder [to understand] than one proton being hit by an electron. I thought I was rather dumb. I had not thought of the easy experiment to analyze, I had always tried the harder one. I had some half-assed ideas about it, but when they showed me this experiment, I started to analyze it from my other point of view, which was the partons. And believing that that was what Bjorken had done, I returned to them and said, '*This* is why Bjorken says there should be scaling.' "

It wasn't, exactly. Bjorken returned the next day and discovered that Feynman, whom he had never met, was in a state of great and alarming enthusiasm over scaling. "We got together," Bjorken recalled, "and he said, *'Of course* you know this,' *'Of course* you know that.' And I said, 'Uh, sort of.' Some of the things I knew very well and other things I didn't and other things I knew better than he did, but in any case he just went away."

The laboratory was thrown into a state of excitement by Feynman's visit. His analysis was intuitive and transparent, and, as Henry Kendall said, "Feynman being Feynman, if Feynman says that you fellows are observing pointlike constituents in the nucleus, then you pay attention."[43] Whereas Bjorken expressed his ideas in the language of current algebra, Feynman had a simple picture: The proton actually consists of a swarming mass of particles. We don't know what they are now, but they move around freely inside the proton. When an electron is going fast enough, it sees the partons as almost motionless because of the slowing of time associated with relativity. Whereas under ordinary circumstances, the electron interacts with many partons— the proton as a whole—it now goes by fast enough to hit just one. Thus by carefully watching how the electrons rebound, you could deduce something about the nature and distribution of these partons. Chiefly, what was learned was that the partons moved around inside the proton as freely as balloons at a birthday party.

Pleased, Bjorken, too, went to work on partons. With a colleague, E. A. Paschos, Bjorken started working on trying to understand other aspects of the phenomenon: what happened when photons or neutrons hit partons, and then how the struck parton turned into a lot of hadrons. In the middle of the work, they ran into a minor question of etiquette: Feynman never got around to publishing anything about partons. It seemed *comme il faut* for the SLAC team to wait for Feynman to announce the idea himself before printing work based on it. Bjorken was too shy to get on the phone and ask Feynman. Eventually, they published the parton model first, with a prominent and somewhat embarrassed acknowledgment to Feynman.[44]

A more serious disagreement arose when Paschos asked Feynman if, as seemed plausible, the partons were quarks. No, Feynman said. "I did not feel too sure about this, basically because I was worried by the fact that my theory was for partons, which if hit hard enough would come out with only little further interaction. Quarks, the theory went, couldn't escape the nucleus—if they went too far they would find large forces bringing them back." Feynman told Paschos not to jump to conclusions.[45]

Down the hall, Gell-Mann thought this was ridiculous, and said so. Frequently. "The whole idea of saying that they weren't quarks and antiquarks but some new thing called 'put-ons' seemed to me an insult to the whole idea that we had developed. We'd put in so much effort inventing quarks and antiquarks and so on, and here Feynman and other people were talking about these put-ons as if they were something different." A coolness developed between the two men. "It made me furious, all this talk about put-ons. The implication was that there was something fundamental about the scaling behavior of put-ons, and this was not true. It's only a relatively good approximation to the correct theory of quarks and gluons." Gell-Mann paused, reflected; his voice grew amiable. It was getting late in the afternoon. Soon he would startle us and a waiter in a Chinese restaurant by ordering dinner in what was evidently serviceable Mandarin. "But there was a very important physical point that was made by Bjorken," he said. "And then explained, I guess, in a somewhat more—ah, what shall I say?—popular manner by Feynman. And that was that deep in the interior of the nucleon the quarks were almost free."[46]

In the long row of candles that had to be lighted before the strong force emerged out of darkness, the scaling experiments set the match to the one closest to the center of the mystery. Scaling emerged slowly, through a series of meticulous experiments that, one by one, checked and cross-checked various aspects of the phenomenon until the early 1970s.[47] That they have not been granted a Nobel Prize is a testament only to the Swedes' fallibility. In their quiet, tenacious, matter-of-fact way, the experimenters overturned previous dogma and established with icy rigidity that there were freely moving *somethings* inside the neutron and proton, little wheels inside bigger ones.

Somewhat paradoxically, scaling did not create a conversion en masse to the quark model. A curious dilemma: The simplest, most naïve quark model, the picture of quarks as little objects floating inside the particles we see, was amazingly good at accounting for many of the peculiarities of the hadrons and was backed by an impressive number of experiments at SLAC. The sole feature the quark model failed to account for was the quarks themselves, which stubbornly persisted in their improbability. How could they bobble helter-skelter inside a proton yet be stuck so firmly together they could never be seen as individual entities? Weinberg, for one, was extremely skeptical. "The only way you could understand why we weren't seeing physical quarks directly was to imagine that quarks were really very heavy, and somehow inside the nucleon the binding energy almost exactly canceled their rest mass, so that the neutron and proton weighed much, much less than the quarks inside them. That seemed absurd dynamically; I couldn't understand how that could be. And if that were true, then the success of the quark model wouldn't be explained. You know, the successes of the quark model had to do with, for example, calculating the magnetic [behavior] of the neutron and proton on the basis that they're just bound states of three fairly independent quarks. And if the quarks really were very heavy and very tightly bound, then their existence wouldn't account for the success of the quark model." He laughed good-humoredly. "It was a paradoxical situation. In fact, I remember Murray Gell-Mann at a conference saying that the discovery of [observable] quarks would be very exciting and it would be very revealing and very important, but the one thing it wouldn't cast any light on is the success of the quark model."[48]

V
The Great Synthesis

16
Killing the Hydra (Part II)

IN THE FALL OF 1969, HARVARD ACQUIRED TWO NEW POSTDOCTORAL FEL-
lows, Luciano Maiani and John Iliopoulos, from Rome and Paris, respectively. The campus was quiet, resting between spells of activism; police and
petitioners were gone from the streets, replaced by the usual Cambridge
panhandlers. Lyman Laboratory, which houses the physics department, is a
few minutes' walk north of Harvard Square, tucked into a corner by Oxford
Street. A brick building undistinguished by any particular style, Lyman
emanates the decaying Victorian gentility that suits academia. Harvard's
physics department was then among the second rank, but near the head of
the second rank; Maiani had chosen to go there, rather than a top university,
because he feared he would never have access to the physicists in a place like
Caltech. Iliopoulos came because he knew Glashow. Both of the new men
were field theorists; both were interested in weak interactions; and both
knew that their approach to physics made them members of a vulnerable
minority. Like young men everywhere, they had hopes of getting into something big, something important. And—it cannot be put more exactly—their
dreams came true.

That Iliopoulos and Maiani would fire the first volley in a barrage which
would rapidly transform particle physics seems all the more remarkable
when one considers the state of play as they entered Lyman Laboratory that
fall. Physicists spoke of four interactions: gravitation, electromagnetism, and
the weak and strong forces. Each was described by a separate body of theory
expressed in a distinct mathematical language; passing from one to another
was like shutting a book of Balinese folk tales and picking up a juridical code
in medieval French. Gravity had been brilliantly accounted for by Einstein
but remained resolutely mismatched with quantum mechanics. Quantum
electrodynamics, the first and most successful quantum field theory, explained electromagnetic phenomena to umpteen decimal places, but its insights did not seem readily applicable to any of the other interactions. Although a $V - A$ theory described the weak force, the physical mechanism of
the interactions was not understood—did they occur via virtual W particles?—and the equations remained stubbornly unrenormalizable. Finally,
the study of the strong forces was in approximately the same state as that of
contemporary art: a tidal wash of fads and short-lived movements, band-
wagons driven by briefly dominant personalities, an uncertain jittery market

311

of anxious theories with little agreement on the simplest assumptions. Such confusion did not seem unusual to Iliopoulos or Maiani; it was the normal state of the science they had grown up with. They hoped to see a few cobwebs swept away in the course of their careers. The amazing fact is that within three years of their arrival in Harvard the first modern unification theory was proposed.

Iliopoulos is Greek by origin, a man whose thick, wavy hair is black enough to make his skin seem pale.[1] He did his undergraduate work in the early 1960s. There being no place in Greece to obtain a good graduate physics education, he moved to France, where tuition was free. He did not speak a word of the language. Annoyed with the relentlessly practical engineering classes he had been forced to take in Greece, he opted for theory, rather than experiment. His faculty advisers were Claude Bouchiat and Phillipe Meyer, two of the more important figures in the slow postwar resuscitation of French physics. The French school was known for being mathematically oriented—excessively so, perhaps. Bouchiat and Meyer were more interested than most in applications; Bouchiat's wife, Marie-Anne, is a careful experimenter with the European sense of craft.[2] Bouchiat taught Iliopoulos to sail close to the solid shore of the experimental data. He also interested his student in the weak interactions, which the successes of the eightfold way and current algebra had reduced to a minority interest. Because foreigners then could not hold tenured positions in French universities, Iliopoulos went to CERN for a couple of years before getting his *doctorat d'état* in 1968.

At CERN, he worked with Jacques Prentki, the house theorist, who was also interested in weak interactions. A nervous, fidgety man with a heap of white hair and an extraordinary Franco-Polish accent, Prentki shared Iliopoulos's lack of interest in Regge poles and enthusiasm for field theory. They decided that something should be done about weak interactions. In Prentki's small, bare, smoky office, they set to work. "Until that time, weak interaction theory was very phenomenological," Iliopoulos remembered in the course of a long conversation. He was speaking in an office at Rockefeller University, where he was paying a two-months' visit. It was early afternoon; Iliopoulos, a late riser, was sipping his morning cup of coffee. He wore his sweater buttoned Gallic-style along the shoulder. "You had this Fermi theory, which you could use to compute to the first order, and then you ask questions about what happens. And we had learned all these things about symmetries and current algebra and all that. A small number of people felt it was the right time to really ask the right questions about what is the *theory*. First of all, we had run out of processes to compute. All the weak interaction processes were more or less computed already—this was the late sixties— and people really felt that we now had enough experience to go beyond.

Since we had no idea what would be the right theory, this involved a kind of step-by-step approach."

("We believed physicists should have their heads hanged in shame," Prentki said. "Forty, fifty years since Fermi, and what have they learned fundamental about weak interactions? *Nothing!* V − A is wonderful, beautiful, but that is really just making Fermi violate parity." What about the SU(2) × U(1) of Glashow, Salam, and Weinberg? "We knew about that. Shelly did his work in Europe, some here in CERN. I remember seeing Weinberg's paper the first time. Whoof! I not only read the paper, I lectured on it in Paris. But these papers, they do not come with ribbons saying, 'Take me, I am correct.' I saw there was something very fine about it, but the situation was also not very satisfactory. He had a nice way of calculating SU(2) × U(1), but it was not renormalized. There's something at the end about hoping it can be renormalized, but that is not physics, that is *hoping.* Iliopoulos and I said, 'We are going to do physics. We are not going to hope.' ")[3]

The blessing and the curse of the Fermi theory and the V − A theory erected on its back was that they artfully avoided the question of how, exactly, the weak force did its work. As soon as V − A had been announced, Gell-Mann and others had realized that the form of the interaction was compatible with the existence of a positive and negative vector boson, but this had not led to substantial progress, still less a resolution of the endless infinities. Iliopoulos, Prentki, and Bouchiat decided a first step would be to sort them out.

"We were trying to see if we could arrange those divergences," Iliopoulos said. He spoke quickly, sentences pouring out in a soft urgent rapid flow. "We classified them into leading divergences, next-to-leading divergences, next-next-to-leading, and so on and so forth." Others had taken the road before them.[4] The idea was to set up an energy cutoff, and see what happened if they said the theory was only good to a certain energy level. "From a mathematical point of view, it didn't make much sense, because all these things were infinite anyway, but they were less infinite when we put a cutoff. So you have the things that would diverge most strongly with the cutoff, then the things that would diverge with a power less." Strange things started to happen. "It was a phenomenological theory. You would say, suppose that the cutoff tells you up to which energy you could trust the theory. You can't trust it at energies that are much more than the cutoff. Then you try to guess the order of magnitude of the cutoff. If you do nothing to the [V − A] theory, the cutoff comes out to be very, very small—so small that you can't trust the theory *anywhere."* What happened when they just went ahead nonetheless was that the equations worked out so that the weak interactions became much more powerful. "We looked at that and we said, 'Well, suppose that they *are* strong interactions.' " In other words, what if the weak and strong forces were somehow the same thing? "What kind of bad effects would that have? You would produce parity-violating strong interactions, strangeness-violating strong interactions, which don't exist." They discovered that if they put in a clever symmetry-breaking mechanism, they could sweep the parity violation under the rug. This also got rid of the effects of the leading divergences.[5] "We were left with the next-to-leading divergences. They are equally bad, but you sort of say, 'Well, it's one power less, who cares?' " But even that didn't work: The next-to-leading divergences, although much smaller, made rare processes occur frequently. "So you had to get rid of those, too, hoping that somehow this would get inside, let you see the right thing." The whole procedure was intellectually untidy, mathematically unrigorous, but physics at the edge is frequently that way.

Glashow was visiting CERN, and Iliopoulos showed the half-complete work to him. Glashow, too, was pushing around the infinities in the weak interaction, working by brute force, trying to squeeze them into the edges where they wouldn't show. The two men decided to collaborate when Iliopoulos came to Harvard, and agreed that they could use the criterion of renormalizability to try to figure out a real theory. At the beginning of November, they were joined by Luciano Maiani, a young Italian field theorist. The three men's interests meshed immediately: All were working with the same level of ignorance on similar phenomena. Maiani brought some suggestions from Europe; Glashow and Iliopoulos rejected his ideas with the bluntness customary among physicists. Within two months, they had come up with a pivotal piece of the standard model.

The atmosphere around the University of Rome has the stillness of shell shock; the days when tear gas and terrorism accompanied students to class are not long gone, and the gateway to the school on Piazzale Aldo Moro is flanked by sullen *carabinieri*. Five years before, tanks were a feature of campus life. The university is large, desolate, creakingly underfinanced; the heavy neoclassic buildings of the science wings are sprayed with political graffiti and surrounded by indifferently tended islands of grass. Luciano Maiani's office is on the second floor, in the middle of a twist of dusty corridors. A graduate student guided us through, silent as Charon, turning imperturbably this way and that in the dim light.

Maiani was waiting for us, a cigarette burning in a sixteen-millimeter film canister that served as an ashtray. He has a large head, expressive dark blue eyes, and black hair that is swept back from a high forehead sticking out to the side like the cartoon image of an orchestra conductor. His voice is deep, penetrating, an unmistakable peninsular bass. "Want one of these?" he asked, pushing the cigarette box across the desk. They were MS, the state-owned brand. "MS, *morte sicura*," he said. "They are disgusting."

Maiani's family is from San Marino, the minute city-state in northern

Italy. He was born, however, in Rome, and studied there as an experimenter. In the midst of writing a thesis on solid-state detectors, he decided to move to theory, and in particular the theory of weak interactions. There was something of an Italian tradition of weak interactions, a chain of work that started with Enrico Fermi, was interrupted by Mussolini, and reimported by Raoul Gatto from the United States. In 1963, Nicola Cabibbo, an Italian then at CERN, figured out a general formulation for hadronic weak decays. The strange particles, as usual, did not behave exactly like the ordinary particles, and Cabibbo was obliged to introduce a parameter that, speaking crudely, related the probability the weak force would change the strangeness of a particle to the likelihood that it would not. This parameter became known as the "Cabibbo angle," because of the trigonometric law that ratios can be expressed in terms of some angle. (The measured value of the angle is about thirteen degrees.) The angle was absolutely necessary to calculate weak interactions, but its existence was one of the mysteries surrounding them.[6]

Five years later, Cabibbo and Maiani decided that there should be some way to compute this angle, rather than simply letting experimenters find it. "We thought we had a solution," Maiani said. "It was a very strange idea, which was connected with some ideas that Gell-Mann had. That is, that in fact the weak interactions are strong." He had his feet on the desk, his head resting against the cool metal wall of a file cabinet. In unknowing parallel to Iliopoulos, Bouchiat, and Prentki, hundreds of miles north in Geneva, Cabibbo and Maiani had hit upon the notion that the divergences in some way tied the weak and strong interactions together. "Now, you worry whether the parity violation which is in the weak interactions would propagate to the strong interaction improperly. But this can be exorcised. We came out with a theory of the Cabibbo angle which gave a good result.[7] But this theory had a problem, which we became aware of just in the summer of sixty-nine, while I was packing and moving to Harvard." The problem was that a consequence of their ideas was the prediction that a long-lived K meson should decay often and easily into two muons. It does not. Maiani arrived in Cambridge and was delighted to find Glashow and Iliopoulos had similar interests—so similar, in fact, that they immediately informed him that his ideas could not possibly be correct.

"We started discussing furiously," Maiani said. "Because I was defending my work, and they were attacking it, there was always a lot of discussion in which two people were attacking a third one." He laughed. "I was a fool."[8]

"We finally convinced him that what he had done with Cabibbo was not relevant," Iliopoulos recalled. "I don't know how long it took, but he finally admitted it." Nonetheless, Iliopoulos and Glashow liked the essential idea of computing the Cabibbo angle by getting rid of divergences. From his thesis, Glashow also knew that using a Yang-Mills formulation also got rid of infinities; he had even claimed, erroneously, that local gauge symmetry eliminated all of them. "Every day," Iliopoulos said, "invariably, one of us would come up with an idea. Then the other two would prove to him he was wrong. We tried all sorts of different recipes, and nothing worked. There were long calculations and trying to—I mean, there was this angle that we tried to untangle—it was impossible. This took us some time, and we got frustrated. And then we really convinced ourselves—we didn't have any rigorous proof—that there was no solution in the then-standard model of three quarks and four leptons. We couldn't renormalize it. So then we started to change things in other ways. The first thing we tried to do was put in more leptons."

After playing with leptons for a while, they gave up and plugged in a fourth quark, which Glashow, remembering his 1964 paper, again called charm. Almost at once they realized they were onto something. All three men liked the obvious tidiness of four quarks and four leptons. Moreover, adding in an extra quark to the equations made many more divergences vanish; the theory became much neater. Finally, Glashow at last realized that charm canceled the possibility of neutral currents for strange particles, neatly removing one of the major problems of his $SU(2) \times U(1)$ model. He told us, "What I don't understand is why I forgot that charm was relevant for so long. But how often can you forget something for six years and have no one take you up on it? The next paper, with Iliopoulos and Maiani, uses that observation to solve a major problem of physics. But nobody else was aware of the *problem*, let alone finding the solution."[9]

The *problem* came from Maiani's work with Cabibbo. Their Lagrangians predicted K decays that didn't happen, but, the three men realized with astonishment, *so did the regular* V − A *theory*.[10] Every now and then, strange particles, such as Ks, should emit both a W^+ and a W^- and through a complicated interaction decay into two leptons. Such an interaction would mimic the final result of a weak neutral current. (The weak neutral current would arise when the K emits a Z^0 that in turn decays into a lepton pair.) Glashow, Iliopoulos, and Maiani calculated that experimenters should already have seen such decays, and took their absence as an indication that charm was on the job. They were hugely pleased; physicists routinely predicted new particles, but it was not often that one got a chance to postulate the existence of an entirely new state of matter comprised of whole families of charmed particles.

Excited, the three theorists took their work to the office of MIT theorist Francis Low, and talked it over with anyone who came in. They mentioned that charm might fit into an $SU(2) \times U(1)$ model. One of the physicists who wandered by was Steven Weinberg. Glashow laid out the case for charm, mentioning that it might fit into some kind of Yang-Mills scheme. A spectacular failure in communication ensued; Weinberg did not see the relevance of charm to his work three years before. Indeed, in Maiani's recollection, he reacted with instinctive dislike to the extreme lack of economy implicit in introducing whole families of unknown particles to shrink the reaction rate of an obscure class of particle decays. His momentary lapse—together with those of Glashow and Iliopoulos, who knew of Weinberg's paper—meant that the three men continued to put the masses in by hand, instead of employing symmetry breaking. Which meant in turn that the first draft of their paper, which contained the statement that an $SU(2) \times U(1)$ Yang-Mills theory with four quarks might be renormalizable, was bounced by the referee at the *Physical Review*. They excised the offending sentence, inserting instead the remark that the "more daring speculation" of a Yang-Mills theory at least "does not make the theory more divergent." Although the new phrasing was almost as unsupported as the old, it was allowed, and "Weak Interactions with Lepton-Hadron Symmetry" appeared on October 1, 1970.[11]

All three were convinced of the existence of charm, but Glashow's certainty was the most ebullient. Glashow, Maiani, and Maiani's wife went to a Cambridge hangout with the curious name of Legal Sea Food. Over the meal, Glashow informed Maiani's wife that he was extremely pleased with what has come to be known as the GIM mechanism for avoiding strangeness-changing neutral currents. She asked if it was important. Glashow replied, "It will be in the textbooks." They gave a seminar on charm to the experimenters at the Cambridge Electron Accelerator, a small machine run by Harvard and MIT. Glashow flatly informed the audience, "From now on, everything is chemistry. We understand everything now. All you have to do is go out and find neutral currents and charm, and then everything is wrapped up."

The experimenters didn't see it that way. Charm seemed like the worst sort of theoretical fantasizing: In order to make sense out of an as-yet unformulated theory of the weak force, the GIM paper reached into the domain of the strong force and created a fourth quark—when few people felt sure of the existence of the first three. Nonetheless, charm was to Glashow's taste; it matched his physical intuition. When Maiani decided to return to Italy after his wife became sick, he was able to obtain passage for the family on a luxury ship from Boston to Rome. They decided to have a farewell party. Glashow arrived with a trash basket full of ice and champagne, a brace of cigars in his pocket. Toasts were drunk. Talk got loud; it was another one of the expansive collegial gatherings that make the science go round. At a certain point in the proceedings, in Maiani's memory, Glashow, Iliopoulos, and he cornered MIT experimenter Sam Ting. They told him about charm and liquidly informed him why it was important. If electrons and positrons collided with sufficient energy, they said, the burst of energy from matter-antimatter annihilation should produce charmed particles. Ting smiled skeptically and said nothing. Maiani told him charm was the next principal item on the experimental agenda; the words "Nobel Prize" seem to have been mentioned. And Ting smiled and said nothing. Experimental physics is a serious business, and he had no interest whatsoever in spending years chasing after some lunatic speculation that would be disowned by its theoretical parents a few weeks later.[12]

After Maiani's departure, Glashow and Iliopoulos determined to show the referee that the divergences canceled even in a Yang-Mills theory. They spent the first six months of 1971 matching infinities against counter-infinities. The work was arduous, painstaking, deeply mathematical, exactly the kind of highfalutin abstract work Glashow and Iliopoulos loathed. They proceeded without a clear plan, trying to match up each individual tangle in the thicket with another: the head-banging approach. Eventually they showed that the situation was not automatically worsened by going to a Yang-Mills theory.[13] "It was the most intelligent paper I ever wrote," Iliopoulos told us. "The most intelligent that either Shelly or I have ever written. It's a completely obscure paper. Nobody ever read that paper and nobody ever will read it because it's so difficult."

In June, Glashow presented the work thus far to a small conference on renormalization at the Centre de Physique Théorique, in Marseille.[14] During the meeting he received a rude shock. "There we are, laboriously canceling out infinities, when this guy Veltman comes up to us. And he says, 'My young student 't Hooft has solved the problem of renormalizability of gauge theories. All the work you've done for the past year is a waste of time!' And he was right. His student had spontaneously re-created the whole SU(2) × U(1) theory, and renormalized it to boot."[15]

□   □   □   □   □

The Dutch physicist Martinus Veltman was one of the few who had clung unswervingly to quantum field theory throughout the 1960s. Now at the University of Michigan in Ann Arbor, stubborn, independent, isolated "Tini" Veltman labored on renormalizing Yang-Mills theories at the University of Utrecht for five years despite the almost complete indifference of his fellows. Veltman is a proud man with a full rack of dark hair and a direct manner who identifies whatever he doesn't like as "baloney" or "crap." We met him in Robert Serber's office after he gave a seminar about alternatives to the Higgs boson. Unlike some of his colleagues, Veltman thinks that spontaneous symmetry breaking, the Higgs mechanism, and the rest of it is, not to put it too plainly, a lot of hooey—at best an approximation to some more elegant truth. When we entered the office, Veltman was slumped in Serber's old red leather chair, half glasses riding down on his nose, a *toscano*-style cigar nearly finished in his mouth. His beard was trimmed in the

European manner around the adam's apple, patches of white marking the jawline. His hair rose back from a large sloping forehead. He spoke in an English transfigured by a Dutch accent, shimmeringly pure vowels shining like brass through his beard. His affect was melancholy; he began by regretting that he had agreed to speak with us at all.

"I don't like to talk about this. It opens up old wounds. I just want to close it off and continue working. You say it and say it and nobody listens, so after a while. . . ." He looked distractedly about the office. *"No matter what I say,* it sounds like sour grapes," he said, earnestly, believably. "You try to solve a problem and sometimes you don't even know that you've solved it. In your innocence, you do things and you aren't even aware of what are doing, and nobody—" Veltman waved the cigar disgustedly. "Ach, here I am, talking about it again.

"I started this business in the spring of 1968. I spent a month at Rockefeller University, doing nothing but thinking about the problem of weak interactions. Where did you go for the theory?"[16] He had arrived at that position in a highly idiosyncratic manner. After writing his thesis on formal properties of field theories, Veltman became interested in current algebra, which Gell-Mann had abstracted from field theory and then thrown away the field theory. (One recalls Gell-Mann's use of a metaphor from haute cuisine: cooking a slice of pheasant between two slices of veal, then discarding the veal.) The residue was a set of relations among currents that supposedly could be manipulated to produce many of the rules and predictions of field theory without the attendant ambiguities and mathematical difficulties. In fact, however, field theory is so beset with technical complexities that careful calculations often bring unwelcome surprises; the current algebra taken from field theory had problems of its own. Years before, Julian Schwinger showed with characteristic force that two methods of doing the same somewhat obscure calculation in field theory produced different answers unless additional restrictions were factored in.[17] The same problem showed up in current algebra in the form of extra terms called "Schwinger terms." Mathematics being a consistent subject, one should not get two different answers to the same question. No matter how you add, two plus two should always equal four, not three or five. In 1966, Veltman showed that adding in two extra fields to the current equations canceled out the Schwinger terms and restored mathematical sanity, albeit at the cost of sacrificing some of the useful features of current algebra.[18] That was not the end of it, however. When Veltman sent a preprint to the theorist John Bell, a friend at CERN, Bell realized almost immediately that Veltman had proceeded by "avoiding as much as possible the creaking machinery of field theory and simply *imposing* gauge invariance" on the equations.[19] Although Veltman hadn't known it, the fields that he had introduced as abstract technical devices were in fact Yang-

Mills fields applicable to weak and electromagnetic interactions.[20] Gell-Mann had based current algebra on Yang-Mills and then discarded the Yang-Mills; coming round the other way, Veltman had deduced the veal from the flavor of the pheasant without knowing how it had been cooked.

Easy, in retrospect, to state with a single phrase. In fact, Veltman didn't know what to make of Bell's paper for almost a year. Confused, Veltman left Utrecht to spend a month at Rockefeller, where he decided to take advantage of the peace and quiet to figure out the article. One day in his Rockefeller office, while staring out the window at the 59th Street Bridge over the East River, Veltman suddenly grasped the heart of Bell's argument: that local gauge invariance was *the* key to a successful theory of the weak interactions. Current algebra, Bell was saying, is just a mask for a more fundamental Yang-Mills theory.[21]

Why hadn't other people realized the significance of Yang-Mills to the weak interactions? Because, Veltman said, nobody could show the theory was renormalizable. Why couldn't they show that? Several reasons. One was that in addition to the ordinary self-energy-type infinities, the intermediate vector bosons had electric charge, and thus could interact with each other. This meant that calculating involved horribly complex Feynman diagrams with three and four bosons looping and intertwining. Self-coupling was not the only cause for despair. The intermediate vector bosons must have mass, and they do not in a pure Yang-Mills theory. When one simply awarded them a mass, as Glashow had done, the picture became even darker.

Veltman decided that if nature had chosen to express herself in terms of local gauge invariance, then Yang-Mills theories could not truly contain infinities. At this point he was faced with the choice of examining pure, massless Yang-Mills theories or theories with extra mass terms. Massless theories were obviously not realistic, so Veltman elected the latter. He didn't know at the time that Salam and Komar had shown seven years before that massive Yang-Mills theories were nonrenormalizable; Veltman's ignorance was fortunate, for their proof was incomplete.

In a local gauge theory many of the infinities match up neatly with ones of the opposite sign. The trick is to show that in the vast sea of possible diagrams not one divergence is left uncanceled, and that no hidden infinities spoil the game. Veltman realized that pairing infinities the conventional way—lining up leading divergences, next-to-leading divergences, and so on—was going to be fruitless. Radically new methods were needed. He found them, but at the price of making himself incomprehensible to almost every other physicist in the business.

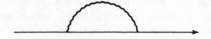

Progress in physics often depending on discerning the right hair to split, Veltman first drew a vital distinction. In the diagram pictured above—a "one-loop diagram," as it is called—an electron emits and reabsorbs a virtual photon. In the brief interval of the photon's life, it can have any momentum whatsoever. Its existence is ruled by the uncertainty principle; standard rules of cause and effect do not apply; calculations seem to produce an infinity. Such a virtual photon is said in the jargon to be "off the mass-shell." Before and after the virtual interaction, the electron is on the mass-shell. "You try to play this renormalization game," Veltman said, "and you realize these diagrams have infinities that disappear on the mass-shell, but not off of it. These off-shell divergences are utterly hopeless, unless you change the rules."

We asked how one changed the rules.

"It's complicated," Veltman said. He stood up, moved heavily to the blackboard. "I think in diagrams, not equations. You have a loop, like so—"

"—you subtract a negative infinity—"

"—the result appears as mass. You don't follow. In gauge theory, you have a freedom in the choice of variables—yes, you know this. Good.

"In classical physics, a gauge transformation takes you from one formulation of the theory to another. That is gauge invariance, in a way. All the formulations are the same, they give you the same result. The question was, How does this take place in quantum field theory? This people could not do. People thought that the diagrams were slightly different for each gauge transformation, that you just fiddled it, but really they are totally different diagrams." Revolution, not reform. "You can't just adjust the Feynman diagrams for gauge theory. The rules are different. Each transformation means that you have to have a totally different set of rules. You had to derive a technique to go from unitary rules, where every line corresponds to physical particles, to others, where diagrams have fewer divergences off the mass-shell." *Unitary rules* are the rules of the real world, where particles are physical objects and all probabilities add up to one hundred percent. Veltman went from unitary rules to a new gauge, a gauge of his choice, where the divergences matched neatly off the mass-shell but very odd things happened in the Feynman diagrams. Chief among the strange things was the appearance of objects of negative probability, "ghost" particles with the impossible quality of having less than zero chance of being around.

In the early 1960s, the apparent prediction of such entities in quantum electrodynamics contributed to the climate of disbelief in field theory. Here, however, Veltman decided that they didn't matter if they disappeared by the end of the calculations. If the ghosts always stayed off the mass-shell and never made it into the final state of an interaction, it didn't matter whether they appeared in the calculation. In the early summer of 1968, he wrote a paper in which he used these ghosts to get renormalized rules for the simplest infinities in massive Yang-Mills theories—one-loop diagrams.[22]

"The immediate reaction was essentially zilch. I was told I was nuts. My colleagues told me I was disappearing into a black hole. Sidney Coleman [of Harvard] told me, 'Tini, you're sweeping an odd corner of the weak interactions.' Most other people were doing fancy things—even Mr. Weinberg was doing no work on gauge theory. They were busying themselves with Regge poles, phenomenology, all kinds of funny things." What made him stick to field theory? "Field theory was my own belief. I have always gone my own way. It is my natural habitat, it is as if you ask a fish why he likes water."

We once asked David Politzer, a Caltech physicist who was a first-year graduate student at Harvard at the time of GIM paper, if the thought of constructing gauge field theories had really been as dead as those in the discipline recalled. In addition to being an active physicist, Politzer actively watches the field, and has many enthusiastically expressed notions about what has happened to elementary particle physics. "I took a course from Glashow in 1970 on weak interactions," he said. "Aside from it being quite inspiring, and aside from the fact that I didn't understand very much, I do remember that in a whole year course on weak interactions, he briefly addressed the question of renormalizability—he mentioned an absolutely disgusting model put together by Gell-Mann, Low, someone, someone, someone, and someone—and did not even *mention* the work of Weinberg. This was 1970, and this was Shelly Glashow! He didn't even mention his own work, for which he supposedly got the Nobel Prize. What had happened was that people had been sort of burned by their lack of understanding of gauge theories." He named a few seared physicists. "They wouldn't touch it with a ten-foot pole! And it is true that Tini Veltman was the only one who kept slugging away at it seriously. Then things turned around."[23]

In August 1968, Veltman went to the Laboratoire de Physique Théorique et Hautes Energies, at Orsay, outside of Paris. The city was still reeling from the student revolution, and Veltman found the intellectual climate interesting if remote from his own views. He was told in Orsay that several Soviet physicists had managed to renormalize massless Yang-Mills to any order—two loops, twenty loops, one hundred loops, whatever they pleased. Although Veltman had chosen to ignore such theories as physically irrelevant, the Russians' resounding success raised his hopes for the massive case.[24]

Veltman revamped his one-loop renormalization proof using the Russians' methods, and commenced working on the next level—loops within loops, two-loop diagrams. These were much more complicated, and Veltman found himself engaged in the sort of sprawling computation that induces math anxiety. He became an expert in little-known mathematical techniques, especially a trick called "path integrals." He prepared computer programs to work on two-loop diagrams and let the machine take over—but the printouts showed that at two loops the massive theory was damningly unrenormalizable. His hopes were evidently dashed; years of work seemed to have reached a dead end; he had indeed swept clean an odd corner of the weak interactions.[25]

Events from this juncture were shaped more by the bent of Tini Veltman's character than compelling scientific logic. This supremely stubborn man kept going, his hopes resting now on a paradox. When he tried the massive theory with an intermediate vector boson of zero mass, the answer was not the same as in the massless theory. One equation had a particle of no mass, the other a massless particle. Zero is zero, and surely the two cases should be identical. They weren't. One was full of infinity, and the other wasn't.

By this time, he had acquired a graduate student named Gerard 't Hooft, a trim, slightly built man whose finely cut youthful features are today filled in by a prominent, barely reined-in moustache. In his inexpensive gray suit, 't Hooft might easily pass for a salesman of electrical appliances, an impression dispelled the instant he opens his mouth: 't Hooft is a very clever fellow. (The 't is an abbreviation for *het*, which corresponds loosely to the French *le*.) When 't Hooft completed his undergraduate studies at Utrecht, Veltman was the only theoretical particle physicist around, and 't Hooft asked him to be his adviser. Veltman agreed, and gave him lecture notes to write up.

Gerardus 't Hooft was born in 1946 and raised in The Hague. Physics ran in the family: His great-uncle, Frits Zernike, won the Nobel Prize in 1953 for inventing the phase contrast microscope, and his uncle, Nicholas van Kampen, was a well-known professor of theoretical physics at the University of Utrecht. While in high school, 't Hooft often asked his uncle questions about physics. "He usually forced me to rephrase them many times until I was more precise," 't Hooft later recalled. "In this way, I learned more about physics than if he had given me straight answers." (On the other hand, being related to van Kampen had its disadvantages; 't Hooft first had to convince Veltman that he wasn't just his uncle's nephew.)

In college, he learned as much particle physics as he could; a classmate did an undergraduate thesis on spontaneous symmetry breaking, and through him 't Hooft picked up a few vague notions on the subject. In the summer of 1970, 't Hooft went to a physics summer school in Corsica where with impressive celerity he learned a great deal about renormalization and spontaneous symmetry breaking from the theorists Benjamin Lee and Kurt Symanzik. "I was very shy in those days, but I did approach both Lee and Symanzik with one question: How do you do this in a Yang-Mills theory? They both said they knew nothing about Yang-Mills theories."[26]

In the meantime, 't Hooft contemplated thesis topics. Veltman made several suggestions, none of which 't Hooft found particularly exciting. He was intrigued, however, by what Veltman himself was working on: Yang-Mills theories. In a first paper, 't Hooft tackled the massless case. He believed the Soviets had only argued that the theory was renormalizable, but not actually done it. This he set out to do, and finished early in 1971.[27] The calculation was the first explicit statement of the renormalizability of massless Yang-Mills theories, although the result was already accepted by the cognoscenti.

"When I had my first draft ready," 't Hooft told us, "Veltman was in Paris for the year. I thought I had done something important, and it was difficult to get his attention at that time. He came back several times and, you know, we had big fights even about this first paper, because he did not agree with the way I formulated things. In fact, my first version did not contain anything about the unitarity proof." That is, 't Hooft had not shown that he ended up with only physical particles and not ghosts.

Once he had done that, Veltman was still suspicious. There was yet another hurdle. Working together, John Bell and Roman Jackiw had discovered, in 1967, what was to be called an "anomaly" in pion decay. An anomaly shows up when, again, two different means of calculations produce two different answers. The anomaly in pion decay was discovered independently by Stephen L. Adler, and the resultant trouble is still called, alphabetically, the Adler-Bell-Jackiw anomaly.[28] Basically, any theory in which something like the Adler-Bell-Jackiw anomaly could occur is not gauge invariant—that is, it is wrong. Veltman had also come across the anomaly himself, but had not clearly understood its significance for a while. By 1969 he had realized its importance to the point of assigning it to 't Hooft as a subject for an undergraduate thesis. Previously, 't Hooft had known the anomalies were there, but thought they were of secondary importance. Now Veltman informed him they were paramount. There could be no anomalies hiding in the computations. Once 't Hooft had shown that, Veltman was ready to be impressed. The massless case would be licked. Of course, the theory would still have to be massive to apply to the real world.

Early in 1971, Veltman had a conversation with 't Hooft that he has never forgotten. In Veltman's recollection, the interchange went as follows:

M.V.: *I do not care what and how, but what we must have is at least one renormalizable theory with massive charged vector bosons, and whether that looks like Nature is of no concern, those are details that will be fixed later by some model freak. . . .*
G.'t H.: *I can do that.*
M.V.: *What do you say?*
G.'t H.: *I can do that.*[29]

"And this he could not believe," 't Hooft said years later, "because he had been working on the massive case himself for so long, and he was convinced that it couldn't work. He may disagree, but I think for me this was the very important moment. He was certain the massive case wouldn't work. And I said, 'Yes, but you need this extra particle. There will be an extra scalar particle around, and then it will work. I am convinced I can do it.' " The extra particle is the Higgs boson; *scalar* is a technical expression meaning that it has a spin of zero. On the surface, the suggestion was foolish. In most situations, tossing in a scalar particle ends up by producing ghosts. "He said, 'No, no, no, it doesn't work because you will get the wrong statistics.' Then I said, 'You *can* do it. I am convinced one can do it.' And so he said, 'Write down the Feynman rules, and I'll check.' By that time he had his computer program ready to check such things. He said, 'Actually, it's rather easy for me to check, if you have a scalar particle.' I wrote down the Lagrangian—the Higgs Lagrangian—but to make it acceptable to him I just wrote down that the mass was there, the interaction, and so on. It looks rather strange. It's not really recognizable as coming from a spontaneous symmetry breakdown. So it just looked like a lengthy Lagrangian with a bunch of extra terms in there which looked rather arbitrary and crazy."

To Veltman, 't Hooft had just shown the result, not the derivation? He had hidden the Higgs mechanism?

"The point was that he didn't want to hear about the Higgs mechanism. So he said, 'Just forget about the Higgs mechanism. Just give me the Lagrangian that you think works.' I ignored where it came from. I just wrote down the whole theory in the broken representation with the extra scalar particle and its self-interactions." Among the many odd-looking terms that flowed from 't Hooft's pen was a factor of four that came from the Higgs boson. "He said, 'Well, I'll take this Lagrangian. I don't believe your factors of four, they look crazy, but I'll take this Lagrangian and put in it my program.' "

Veltman traveled to Geneva to put the paper into the CERN computer. He went down in a skeptical frame of mind, worried that his student did not seem to understand the consequences of making an error in a paper that was sure to be widely discussed. In 't Hooft's recollection, Veltman soon called up in a state of excitement and said, " 'It nearly works! You just have some factors of two wrong.' But that was because he had not copied my factors of four, because he didn't believe them. So then he realized that even the factor of four was right, and that all canceled in a beautiful way. By that time he was as excited about it as I had been."

A few weeks later, David Politzer was at a summer school in Erice, Sicily, run by the Italian experimenter Antonino Zichichi. The school is on the rugged western coast, and the delights of the Sicilian countryside, the sea breeze, and the excellent cuisine have ensured its popularity. In quest of the perfect snorkel, Politzer one day went rowing with Sheldon Glashow. In the middle of the ocean, Glashow suddenly told Politzer that some Dutch student claimed he had renormalized Yang-Mills theories. Politzer asked if Glashow understood the work. He still recalls the response. "No," Glashow said. "Either this guy's a total idiot or he's the biggest genius to hit physics in years."[30]

Initially, 't Hooft had thought of Yang-Mills in terms of a theory of the strong interactions, and had actually played with a model in which the rho mesons, the lightest mesons with a spin of one, played the role of intermediate vector bosons. The rho's interacted—"mixed" is the term of art—with the photon. The relevant group was, by rank coincidence, $SU(2) \times U(1)$.

When Veltman took 't Hooft's proof to the CERN computer center, he also asked a resident theorist, Bruno Zumino, if anybody had done anything else with an $SU(2) \times U(1)$ model. Zumino is known for being one of the rare physicists who actually reads all the journals; he recalled Weinberg's 1967 paper, and Veltman phoned Utrecht with the citation. A small communications failure transpired, the wrong reference was copied down, and 't Hooft couldn't find the right paper, although he did find another Weinberg article that he thought was pretty good. Having been provided with the basic idea, he sat down and started to duplicate the model. He was fairly well along when Veltman returned with a photocopy. Turning through the pages, the two men realized they were in possession of something lovely: a fully renormalizable model of the weak and electromagnetic forces.[31]

As luck would have it, a big conference was scheduled in Amsterdam that August, and Veltman happened to have been given the task of arranging the theoretical talks. It pleased him no end to be able to schedule a special session devoted to renormalization. "Talk number one," Veltman said, "was given by Salam, who was trying to remove the infinities by using gravitation. Talk number two was T. D. Lee, who was introducing physical particles with bad properties. So first I let Salam talk about his baloney, then T. D. Lee about his attempts, and then came the real thing. I got up and I remember announcing to the audience that here was a renormalizable theory 'every bit as good as quantum electrodynamics.' That phrase is not in the proceedings, but I remember the moment very well."[32]

Alas, Veltman's pleasure was quickly to sour. His relationship with 't Hooft did not survive their joint rise to prominence; the teacher and student ultimately could not be together as equals. A sensitive, private man with strong views about the nature of physical truth, Veltman became irritated that the specific $SU(2) \times U(1)$ model became the focus of attention rather than the proof that an entire genre of theories was useful; unwilling to advertise that the student's success was based on his master's techniques, Veltman drifted into bitterness. He withdrew from the limelight, moving across the ocean to the University of Michigan, where with unaltered tenacity he followed his intuition that the Higgs mechanism is an ugly tear in the otherwise unmarred tapestry of field theory.

To Steven Weinberg, 't Hooft's proof just seemed like hand-waving. Then he heard that his friend, the Korean-American physicist Benjamin Lee, was working on it. Lee had taken Veltman's course on his path integral techniques in Paris three years before, and was one of the few people in a position to understand what 't Hooft was doing. Besides lending 't Hooft's work his considerable prestige, Lee spent most of August translating it into a form other theorists could comprehend.[33] "I was really impressed with that," Weinberg recalled. "I thought if Ben Lee takes this seriously, I've got to take it seriously." Working on the problem himself, he slowly grasped that an extraordinary period in the history of elementary particle physics had begun, and that in the middle of the tumult and celebration was nothing other than his old $SU(2) \times U(1)$ model.[34]

□ □ □ □ □

Gell-Mann, in the meantime, was thinking of anomalies. If properly understood, the Adler-Bell-Jackiw study of pion decay might at last make some progress on a small but interesting problem that had defeated theorists for some time. According to the $V - A$ theory and its extensions, the neutral pion ought never to decay into two photons. It does. Adler, Bell, and Jackiw argued that this discrepancy appeared because of their anomaly. Gell-Mann hoped that properly disentangling the anomaly would lead to understanding pion decay.

In late 1971, he teamed up with a young, enthusiastic colleague from Germany, Harald Fritzsch. The collaboration began auspiciously, several hours after a predawn earthquake knocked all the books off the shelves in the Caltech library and set askew the paintings in Gell-Mann's office. They were soon joined by William Bardeen of Stanford (who is not the Bardeen of the BCS theory). Bardeen had recently worked at Princeton with Stephen Adler (who *is* the Adler of Adler-Bell-Jackiw) on a further calculation of pion decay, and brought with him a piece of news from New Jersey.[35] In a talk at Columbia, Adler had shown that pion decay could be computed well enough to test various models of hadron constituents. Three fractionally charged quarks came off very badly, predicting low by a factor of three. Such disagreement, Adler said, means that *"the quark hypothesis is strongly excluded."* (Emphasis in original.) On the other hand, Han-Nambu quarks with three colors did reasonably well.[36]

Reluctant to have quarks strongly excluded just as scaling seemed to give evidence of their existence, Gell-Mann, Fritzsch, and Bardeen realized they could fix things up by adding color to the original fractionally charged quarks; that is, by making each of the three "flavors" of quark come in three additional types or "colors." Why not? "We gradually saw that that variable was going to do *everything* for us!" Gell-Mann said. "It fixed the statistics, and it could do that without involving us in crazy new particles. Then we realized that it could also fix the dynamics, because we could build an SU(3) gauge theory, a Yang-Mills theory on it."[37] The new idea was similar to Nambu's old model of color, except with Yang-Mills and without integral charges; it had, Gell-Mann thought, the feel of something very promising.

Just as quarks resolved many of the problems with the strong force but raised the question of their own existence, so color answered questions about quarks, but in turn raised the question of *its* existence. Although every hadron in the Universe was allegedly composed of colored building blocks, nobody had ever seen anything like color. The three men took recourse in Gell-Mann's old solution: Color is not observed because it *cannot* be observed. Only particles in which red, blue, and green or color and anticolor neutralize each other can exist; the world is made of uncolored, neutral, "white" composites. The tidiness of the scheme enchanted Gell-Mann. "I assumed that everybody was working on that," he said. "It turned out they weren't."[38]

In September of 1972, at a conference celebrating Fermilab's opening, Gell-Mann presented an almost complete picture of the strong interaction.[39] Hadons, he said, are composed of three types, or "flavors," of quarks, each of which can come in three colors. Up, down, and strange can be red, blue, or green. The quarks are held together by a gauge field whose quanta, gluons, are the vector bosons that mediate the strong force. Whereas the quantum of electromagnetism, the photon, comes in only one type and is uncharged, gluons come in eight types and are colored, like the quarks they interact with. Gell-Mann called the new theory "quantum chromodynamics"—*chromos* is the Greek word for "color"—in an obvious but loose analogy with quantum electrodynamics.[40]

Although Gell-Mann presented quantum chromodynamics in a bold and forthright manner to the Fermilab conferees, he had acquired a case of cold feet by the time the proceedings were printed, and the new Yang-Mills theory of color remains only as a kind of shimmer over the printed article, a promised theory that is never quite fully described.[41] Quantum chromodynamics was very new, and Gell-Mann was not certain he understood all the wrinkles. Oddly, however, he was not perturbed by the main cause of his colleagues' skepticism, that the quarks, colored or uncolored, could rattle around the proton like marbles in a bag and yet somehow be stuck together in perpetuity. Something that is loose can be dislodged; something that cannot be dislodged must not be loose.

One can say the words *asymptotic freedom* any number of times without causing them to discharge their informational content. The term stands as a kind of perfect emblem of the gap between scientists and non-scientists; rich with association and historical resonance to physicists of a certain age, the words are blank and impenetrable to the lay public. For a little while in 1972 and 1973, a select coterie of mathematically inclined theorists tossed about the notion as a panacea for all the ailments of strong interaction physics, telling seminars and conferences that asymptotic freedom was the wave of the future. All one had to do was prove it possible; when three physicists did, they gained renown. The standard model came together like a card game in which all the early play is but a setup for a rain of trumps at the end; asymptotic freedom was the last trick in the shuffle, when the hidden face cards came out and the round abruptly snapped into its final shape. It was discovered in 1973, after a complicated, quick little bout of competition and near-collaboration.

When we visited David Politzer, one of the three godfathers of asymptotic freedom, it was entirely natural to ask him what it was. Politzer is now at Caltech, a relaxed, loquacious, slightly cherubic man in a small office with a tiny window and a pair of brown slippers under the desk. He answered with a promptitude indicative of the number of times he had been asked this question. "Roughly speaking," Politzer said, "—I'm lying a little—it means that there is a unique class of forces that gets systematically weaker as the separation between particles gets littler. That allows you to have quarks when they're close together to be weakly interacting, and as they get farther apart, their influence on each other gets stronger instead of weaker. The 'asymptotic' means getting things close; 'asymptotic' means in the limit that there's no separation at all. But in the limit that quark separation goes to zero they are free particles. When quarks are sitting right on top of each other, they don't see each other, each doesn't feel the presence of the other one. It's only when you pull them apart that they feel the influence of their neighbors. And there's only this very small, well-defined class of gauge

theories which are asymptotically free among all possible theories that anyone has ever been able to imagine that could have that property." He fiddled
with his red vest. "Yes," he said firmly. "That's one way of saying it." Another
way is to say that asymptotically free theories, and only asymptotically free
theories, have negative coupling constants.

He was asked how he became involved with this problem.

"Ah, that's an interesting example of how science works," he said. "In
1972, I was looking for something constructive to do, having been in graduate
school already three years, passed my exams, done all sorts of things, but
really had nothing to work on." That summer, he went to a summer program
in Sicily where he learned about the Callan-Symanzik analysis, a field-the-
oretic attempt to understand scaling simultaneously invented by Curtis
Callan, Jr., of the Institute of Advanced Study and Kurt Symanzik from the
DESY theory group in Hamburg.[42] Its details need not concern us here;
Politzer merely thought it clever and inventive enough to apply to symmetry
breaking, which was not—is not—truly understood. "I was trying to do the
Callan-Symanzik analysis specifically on gauge theories to understand their
long-distance behavior or low energy behavior, hoping in the long run to
address the question of spontaneous symmetry breaking. I went down to
visit my adviser at Princeton, Sidney Coleman, to ask him what he thought
about it. [Coleman, a member of the Harvard department, spent the year at
Princeton.] He thought it was a good idea. And I asked him, had anybody
done it? He said not to his knowledge, but let's ask [Princeton theorist]
David Gross. So we go next door to ask David Gross. Gross says no, nobody's
done it. And then I discussed briefly with Gross why it wouldn't be so hard
to do. Even though it used to seem terribly complicated, if you have your
wits about you it's pretty straightforward."

David Gross's thoughts were far from symmetry breaking, although he,
too, was interested in Callan-Symanzik. The Callan-Symanzik analysis
sprang from a curiously powerful yet curiously neglected aspect of field
theory called the renormalization group, a mathematical technique originally
developed to relate the structure of the theory of quantum electrodynamics
at high energies to its predictions at low energies. The renormalization group
was first concocted by Stueckelberg in conjunction with a student, André
Petermann, but nearly all theorists know it through an article written by
Murray Gell-Mann with Francis Low in 1954.[43] "[O]ne of the most important
[papers] ever published in quantum field theory," Steve Weinberg recently
called it,

*This paper has a strange quality. It gives conclusions which are enormously
powerful; it's really quite surprising when you read it that anyone could reach
such conclusions: The input seems incommensurate with the output. The paper
seems to violate what one might call the First Law of Progress in Theoretical*

Physics, the Conservation of Information. (Another way of expressing this law is: You will get nowhere by churning equations. . . .)[44]

As the theorists of the 1930s knew full well, when two electric charges approach each other very closely, quantum effects like vacuum polarization must be taken into account. Another way of saying this is that different aspects of quantum electrodynamics come into play according to the distance scale; because high energies are needed to push particles very close together, the energy scale one works with requires one to treat different parts of the theory. The procedure of renormalization can make this process more complicated and even seem to lead to inconsistencies unless one realizes that the coupling constant—the famous alpha, 1/137—is *not* constant, but depends on the distance and energy. Thus, the coupling constant is a "running" constant, and at terribly short distances and terribly high energies the value of alpha changes. The renormalization group describes the way this odd behavior works.

With the general lack of faith in field theory that characterized the theoretical climate of the 1960s, an idea that treated the extremes of distance—distances like 10^{-291} centimeters, arguably the smallest number ever to appear in a serious physics equation—was not going to be discussed fully. An independent-minded physicist named Kenneth Wilson took up the technique, however, and chewed it over in his idiosyncratic way for many years. Wilson worked slowly; he took his doctorate in 1959, and in the next decade published exactly six papers. In 1971, he revealed the first fruits of his thoughts about the renormalization group.[45] Much if not most of his approach concerned solid-state physics, but he also gave a sample calculation of how one could apply the renormalization group to strong interactions. With assistance from Sidney Coleman of Harvard, Callan and Symanzik gave a general prescription for such an application.

Gross had been trying to come up with field theory that would explain scaling, but could only find theories that predicted violations of scaling, which experimenters were not finding. Gross hoped that the renormalization group would help him find a form of field theory that did not predict scaling violations—in vain. By the beginning of 1972, repeated failure had driven Gross to the conclusion that field theory was not adequate to explain strong interactions. "I sort of decided that I was really going to *kill* quantum field theory," he told us. "I was going to prove (1) that scaling really required asymptotic freedom, and (2) I was going to show that there weren't any asymptotically free theories. I started to work on both of those problems in the fall of seventy-two. The first problem I was working on with Curtis Callan. We tried to show that if you assumed ordinary quantum field theory and you assume you have scaling, within the framework of the renormalization group it must be asymptotically free." The first was successful. "We managed to show that for all nongauge theories, if you have scaling, you had to have asymptotic freedom. So that was nail number one. Nail number two was going to be that there aren't any asymptotically free theories."

The second nail was harder to strike. There were many kinds of field theories of varying degrees of difficulty. Local gauge theories were the hardest to work with, so Gross first tackled all the others and showed that they weren't asymptotically free. "The one hole left in this thing was gauge theories, which didn't fit into the same line of proof. So that hole I was going to close with Frank Wilczek, who had started to work with me as a graduate student."[46]

For some time, Wilczek, too, had been fascinated by the renormalization group, partly because so little had been done on it; the standard textbooks on field theory had brief sections on the renormalization group, but then "they just sort of *stopped*." In the months before his collaboration with Gross began, Wilczek tried to calculate renormalization corrections to weak interaction processes. This led him to wade into the technicalities of the renormalization group; from there, it was a short hop to join Gross on his quest to kill off field theory. Asymptotic freedom became Wilczek's thesis topic.

Working through the fall of 1972, Gross and Wilczek thought they were almost the only physicists concerned with asymptotic freedom. They kept committing and catching mistakes in the long, tedious figuring, but they began to wonder if Yang-Mills theories might not be asymptotically free after all. Then, a shock: Soon after Christmas vacation, Wilczek came across a preprint by Symanzik in which he spoke about asymptotic freedom. Wilczek said to us, "Now, the theory he actually used to illustrate it was a diseased theory. It's asymptotically free, all right, but also unstable. It doesn't really exist. But at the end, the very last sentence of this preprint said something like, 'Gee, it would really be interesting to know if Yang-Mills gauge theories were asymptotically free.' I was terrified. I saw my thesis going down the drain."[47]

He should not have worried about Symanzik. He should have worried about 't Hooft. In June of 1972, 't Hooft attended a congress on gauge theories in Marseille, a follow-up meeting to the one where Veltman had first presented 't Hooft's renormalization the year before. Upon descending from his plane in the Marseille airport, 't Hooft recognized Symanzik, who was scheduled to give a talk on field theories of the strong interaction. Symanzik said he had been trying to figure out if it was possible to have a force with a negative coupling constant, that is, a field that got weaker when you went closer to its source. Something like that was probably happening with the quarks. Symanzik had been discouraged, however, by various "no-go" theorems purporting to show that this was impossible. Well, 't Hooft said, he had looked into the question, and there is one class of theory in which negative coupling constants can exist: Yang-Mills gauge theories. The two men commenced to argue, and went at it from the airport to the university. Symanzik told 't Hooft that he had probably made a sign mistake some-where, an easy thing to do given the complexity of equations in quantum field theory; 't Hooft said he didn't think so. The debate continued until Symanzik went to the dais to deliver his talk.

Symanzik told his audience that he knew just what kind of theory one needed for the strong interactions, but he didn't have the slightest idea of how to put it together. It had to be asymptotically free. But as far as phys-icists knew, such peculiar fields could not be; if, for example, gravity had a negative coupling constant, people would float about the surface of the earth and only acquire weight in space. At the end of the talk, 't Hooft got up and announced that he had done the requisite calculations, and that Yang-Mills theories could have a negative coupling constant. Then he sat down. Nobody there, including 't Hooft himself, quite grasped the significance of the re-mark. Unluckily for 't Hooft, he was still embroiled in dotting the i's and crossing the t's of his renormalization proof. To demonstrate asymptotic free-dom, he would have had to write a paper explaining his technique, and another using his techniques to explain scaling, and then another. . . . He did so, but by then it was too late.[48]

One of the many mistakes Gross and Wilczek made while sweating through the calculations was a sign error in the overall result. For a few days, they thought that gauge fields, like all other fields, invariably became stronger at short distances. During that time, they happened to describe their results to Sidney Coleman. They soon corrected their goof and made some more, but they eventually proved to Gross's amazement that Yang-Mills theories could be asymptotically free. "It was," Gross said, "like you're sure there's no God and you prove every way that there's no God and as the last proof, you go up on the mountain—and there He appears in front of you." (The nail, so to speak, had turned into a door.) But in the interim, Coleman had talked to Politzer, who had run smack into a dead end with the Callan-Symanzik analysis as a means of understanding symmetry breaking.

"Okay, so I'm working on it," Politzer said, miming the action of a wizened savant scribbling field theoretical calculations. "It comes out in the end, after all the dust settles, that it's clearly totally useless for the purpose that I had in mind. What I learned as far as my own interest is that the Callan-Symanzik analysis doesn't tell you *anything* about the long distance behavior of gauge theories! Within the next sort of day, it dawned on me that if it's no good for long distances, it *is* just what you need for short distances to tell you about scaling. All of a sudden I realized that I had something that was potentially interesting. I called up Sidney Coleman; I told him I was very excited. I told him why, and he said, 'Um hum. That's very interesting— except for one problem, which is that David Gross and a student of his had worked on the same calculation, and they said it comes out the other way.' I said, 'It's pages and pages'—I could do it now in two pages or one, but at that time it was pages and pages—and I said, 'I checked it, and I think I got it right.' He said, 'Those guys don't make mistakes.' " Politzer left soon there-after with his wife for Maine, where they were staying in the house of a friend. "It rained a lot, and I thought it was important to check my thing, so I spent a lot of time checking the stuff. I got the same numbers, came back, and said, 'Sidney, I got the same numbers.' He said, 'Yes, I know. David and Frank found their mistake, and they've submitted a paper to *Physical Review Letters* because it's important.' " Politzer dashed off an article of his own, and the two proofs were published back to back in the issue of November 15, 1973.[49]

In retrospect, asymptotic freedom was clearly in the air. A Soviet phys-icist, I. B. Khripovich, had done the calculations around 1972 but had made a sign mistake, and thought gauge theories were not asymptotically free. Tony Zee, a visiting professor at Rockefeller, was also looking for negative coupling constants, but happened not to hit upon Yang-Mills theories.[50] Once the discovery was made, influential physicists like Gell-Mann and Weinberg were enthusiastic—Weinberg proposed that if the color force got stronger at greater distances, it trapped the quarks—and their excitement drove the realization through the scientific community.[51] More important, the two teams of Gross and Wilczek and Georgi and Politzer worked for months to obtain predictions of the effects of asymptotic freedom on scaling. They showed up as slight deviations from simple scaling that were duly found. (These deviations were due to the presence of colored gluons, which Bjorken had not considered when he did his figuring. Amazingly, scaling gave an impetus to quantum chromodynamics, as did the violation of scaling.)

Asymptotic freedom came like the opening of a curtain onto a pre-viously hidden stage, revealing the strong interaction in its full dimensions. Many physicists had pictured elements of the scene—Gell-Mann came clos-est to encompassing it in its entirety—but none till then had fully grasped the flawless elegance of nature's conception. Whereas the electroweak theory was clumsy, quantum chromodynamics was pretty—a perfect, unbroken symmetry. Eventually, the excitement around it built like a tsunami, from an initial slight ripple into a tidal wave. The words *unbroken Yang-Mills SU(3) color with asymptotic freedom* rang through a hundred transcontinental wires; preprints of new work came to departments, were instantly photo-copied a dozen times, and then a dozen times more. In such moments of collective ferment, a scientific sodality moves like a single, self-absorbed organism enraptured by the most pleasing and unexpected interior imagery. Emotional valences tip readily at these times, and theorists who previously held back from constituent models found themselves interpreting the evidence with a more kindly eye. At a stroke, the picture seemed both beautiful and real: Inside every proton are three quarks and numberless gluons, all their colors blending into white.

One of the oldest riddles known is the question, "Is there a smallest piece of matter?" Plato held the elemental things to be unbreakable geometrical shapes, whereas Aristotle stated that all substance was infinitely divisible. Numerous other suggestions surfaced in the intervening millennia. Made uncomfortable by the lack of progress, Immanuel Kant asseverated in the *Critique of Pure Reason* that the phrasing of the question makes it unanswerable. The human mind, Kant wrote, can pose certain questions about nature which have contradictory but perfectly logical answers. One is whether everything in the world is made of simple parts; both the affirmative and negative answer, or so Kant thought, can be proven, which means that the way we think about such subjects is inadequate.

One can speculate endlessly about whether there are particles that can be subdivided infinitely. Quantum chromodynamics does not pretend to answer the question. In the manner of science, however, it does provide a definite answer to what happens when you actually go out and try to do so with the basic components of our world, hadrons. Suppose you begin shooting electrons at a proton, trying to knock loose one of its constituent quarks. As the quark is kicked farther away from its partners, something strange occurs; the virtual gluons whirling between the quarks begin exchanging gluons among themselves. The greater the separation, the more intricate and powerful the web of interactions. Eventually, the energy needed to separate the quark still farther from the snarl of gluons becomes sufficiently great that a new quark-antiquark pair is created *ex nihilo* from the vacuum. The antiquark bonds to the quark separating from the proton to create a meson; the new quark meanwhile pops right back into the proton, leaving it with the same number of quarks as before.

"The whole process is rather hard to visualize," Glashow told us, laughing. "It's like a prison where you don't restrain the prisoners at all except to keep them in the jail. Inside, they can do what they bloody well please, but they simply can't get out. Of course, what happens if you try to pull a prisoner out of the jail is sort of curious in this quantum world, because as you tug on the prisoner, you get him out all right, but you produce a new prisoner-antiprisoner pair. The second prisoner will get stuck in the jail and you will end up with a prisoner and an antiprisoner." He paused for a moment, struck by the worry, common to physicists, that his audience may not be following him. "It's . . . ah, a bit counterintuitive."[52]

Quarks are not parts of protons and neutrons in the same way these particles are parts of atoms; they are not just another rung down on a ladder. According to SU(3), quarks and the gluons that bind them are perpetually unseen constituents of the final rung, ghosts in the machine, concealing their existence from the world in the act of comprising it.

□   □   □   □   □

The power and elegance of quantum chromodynamics—SU(3)—and what some physicists have called "quantum flavordynamics"—SU(2) × U(1)—quickly led physicists to splice them together as SU(3) × SU(2) × U(1), a standard model of elementary particle interactions. The standard model had a curious birth, for it was rushed together headlong after much confusion, and then celebrated as complete well in advance of experimental proof. Theorists intoxicated with the thought of containing all elementary particle interactions in a single coherent package regarded the show as almost over, whereas their colleagues on the machines, as is only proper, recognized only that a new fad had overtaken the pencil pushers. At Brookhaven, at CERN, at Fermilab and SLAC, physicists settled down to put SU(3) × SU(2) × U(1) to the rack of experiment.

17
Neutral Currents/Alternating Currents

IT IS IMPOSSIBLE TO LEARN ABOUT SCIENCE—OR ANYTHING ELSE, FOR THAT matter—without asking a lot of stupid questions. Every scientist has a humiliating memory of revealing ignorance by posing a particularly naïve question to a teacher or a senior colleague. Etched into the brain, the sarcastic response is savored, years later, for the lesson it imparted; being caught short may not be the most comfortable way to learn, but one seldom forgets the result.

While poring over the epochal alpha particle experiment by which Ernest Rutherford divined the existence of the nucleus, we had the notion that we would better understand the discovery if we repeated the experiment with a practicing scientist. We had an enticing image of ourselves watching as the physicist set up the radioactive source, the thin gold foil, and the scintillation screens. We envisioned scribbling notes blindly as the lights were turned out and the little telescope adjusted to count the flashes at each angle. Making a few quick calculations on a scrap of paper, our physicist would announce triumphantly that the evidence indicated that atoms have solid, massy, positively charged centers.

An obvious candidate for this signal honor was Samuel Devons, a former Cavendish physicist who had taught a course on the art of experiment at Barnard College in Manhattan. We broached the idea one day to him and had the embarrassing experience of hearing a kind man attempting not to laugh in our faces; we had obviously made his day. He answered simply enough, but a guffaw kept creeping into the edges of his voice. "In principle, the experiment is simple," he said. "In practice, it would be nearly impossible. First of all, there's the problem of working with radioactive materials. You'd have to find a strong radioactive source—do you want to wait a month to see a flash?—and you'd have to make a new source daily and you would need a certification by a health officer to ship highly radioactive materials around New York City. Do you think you'd get one here at Columbia or anywhere else? No way!

"Okay, so that's one problem, getting permits. Maybe you could," he said dubiously. We were beginning to feel exceedingly foolish, a sensation that Abdus Salam once told us opens the mind to the spikes of insight. "The main problem, though, is that experiment is a *craft*, like making an old violin. A violin isn't a very complicated-looking gadget. Suppose you went to a violin maker and said, 'Could you kindly help me make a Stradivarius? I'm interested in violin-making, and I'd like to see how it was done.' He'd smile at you just like I did. Because craft is a knowledge you have in your finger-tips, little tricks you learn from doing things, and they don't work and you do them again. You have little setbacks, and you think, how can I overcome them? And then you find a way. Every time your equipment changes you forget all the old techniques and have to learn new ones. And you have to know them, because when you're pushing your equipment to the limit it's bloody easy to get spurious results. You're scratching at the ground all the time, and you don't know what you've missed. Every experimenter has made terrible errors at one time or another, and knows of instances where friends have fallen on their faces because they got spurious results and published too early. And yet, you've *got* to push what you know to the limit. If you don't, someone else is going to do it first. And that's dreadful, being beaten. Everyone's got a closetful of discoveries they missed because they were too cautious or some other fellow was cleverer. There was a whole Austrian school working on the same things as Rutherford at about the same time, and nobody's heard of them today. Why not? Rutherford was just a little more daring and crafty."[1]

Looking back at experiments from the vantage of the present, discov-eries appear inevitable, drawing experimenters on like beacons, and errors or sidetracks en route seem like evidence of inattention or stupidity. Such subtle, false teleology is hard to purge from the history of science; it is hard not to say, for instance, that *of course* neutral currents were there, and *of course* SU(3) × SU(2) × U(1) was right, and it was only a matter of time before experimenters found out. Graduate students studying today's text-books find few hints of the long and tumultuous period of testing which the theory underwent. Although the results now appear in the *Review of Particle Properties* as a few lines summarizing an apparently orderly series of confir-mations, the trials of the standard model lasted throughout the 1970s and provoked considerable human conflict.

Experiments focused initially on the electroweak theory of Glashow, Salam, and Weinberg, because this part of the standard model was put together first and because it clearly pointed to the existence of new physical phenomena. When SU(2) × U(1) was first proposed, all known examples of the weak force involved some gain or loss of electric charge by one or more of the particles in the interaction, and hence a change in their identities. The presence of Z^0s in the theory, however, implied that there could also be *neutral* weak interactions, exchanges of Z^0s with no alteration of particle identities or charges. Such currents had never been observed, and their existence was routinely discounted; finding them would impressively con-firm the theory.[2] In two clear, urgent, and important calculations during the latter months of 1971, Weinberg carefully redid SU(2) × U(1) and noted that it predicted the ratio of neutral currents to charged currents—that is, weak interactions with Z^0s compared to those with W^+s or W^-s—to be between one-eighth and one-quarter. Such effects, Weinberg said, "are just on the verge of observability." Extant data gave a lower limit of about one-eighth with a hefty margin of error, which "neither confirmed nor refuted" the theory. [3]

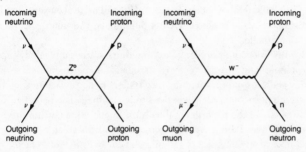

NEUTRAL CURRENT

(No particles change identities; no muon involved in interaction)

CHARGED CURRENT

(All particles change identities; muon inovolved in interaction)

Here was a clear target. According to the electroweak theory, between one-eighth and one-quarter of the time, neutrinos and antineutrinos interacting with protons would emit a Z rather than a W, and no particles would change identities. If SU(2) × U(1) had been renormalized a few years earlier, this prediction could not have been checked, because neutrinos were too hard to work with. By happenstance, exactly the device for such experiments had just become available—a bubble chamber, then the world's largest, which had been built in France over the course of seven years. First proposed in February 1964 by André Lagarrigue, Paul Musset, and A. Rousset of the École Polytechnique in Paris, the new chamber was specifically designed to examine neutrinos. [4] Almost twenty times bigger than any previous European chamber, it consisted of a tank of freon swaddled in a thousand tons of steel and copper. The steel served as a shield from errant particles; the copper consisted of a few kilometers of thick wire spooled around the iron and plugged into a six megawatt generator, which converted the entire assembly into a powerful electromagnet that bent the particle tracks inside. Louis Leprince-Ringuet, the old cosmic ray physicist, dubbed the project "Gargamelle," after the monstrous, big-bellied mother of Gargantua in Rabelais's *Gargantua and Pantagruel*. Gargamelle took six years to construct; the team that planned to use it swelled to more than fifty physicists from eight institutions across Europe.

As Gargamelle came into readiness, the renormalization of 't Hooft and Veltman burst onto the stage. Theorists at CERN learned of the work early from Veltman. In November 1971, Jacques Prentki received a preprint of the first of Weinberg's two papers from an Italian friend at MIT. Written across the front in large letters were the words "MOLTO IMPORTANTE!!!" (Very important!!!)[5] Six weeks later came the second paper. Prentki, Bruno Zumino (the theorist who had told Veltman of Weinberg's 1967 paper), and Mary Kay Gaillard talked about $SU(2) \times U(1)$ and neutral currents to a group of experimentalists in a room of CERN Building 17, the annex constructed to house Gargamelle.

CERN is a complex of long boxy buildings scattered higgledy-piggledy across a mile-long stretch of Swiss farmland. Experimenters are always knocking walls down or implanting new trailers to house facilities, and great cranes stand perpetual watch over the site. Gargamelle had been tucked into the side of one of the experimental halls on Rue Pauli, a few hundred yards southeast of the Proton Synchrotron, then CERN's largest accelerator. Around it were several floors of catwalks and cheap little metal offices, in one of which Gaillard, Prentki, and Zumino talked about electroweak models. They agreed with Weinberg that the most direct proof of the theory would be some evidence of the Z^0 and that the requisite evidence would come most likely from examining neutrino-electron interactions in a bubble chamber.

Such interactions would occur when one of the muon neutrinos in the beam bounced off an electron in a freon molecule; the only way this could happen is if the two had exchanged Z^0s. The neutrino would continue on its merry invisible way, and the electron would, in the subsequent photograph, seem to suddenly erupt out of nowhere, spiraling outward in the magnetic field. (In a charged current, the neutrino changes its identity, becoming a positive or negative muon.) According to the theorists, the scanning team in the basement of Building 17 should be asked to look for the thin, circling tracks of such isolated electrons. The "signature" of the neutral current would be both the ejected electron and, more important, the *absence* of a muon.

Gargamelle's *raison d'être* was to study the general domain of neutrinos, rather than neutral currents in particular, which were dismissed in two short paragraphs of the 194-page proposal.[6] Five members of the Gargamelle team had set low limits on neutral currents as recently as a year before.[7] Despite the theorists' pleading, the experimenters were not impressed with the hand-waving about Ws and Zs. They refused to give high priority to neutral currents, although they did agree to ask the scanners to report any single electrons they saw. Such events would be easy to identify, and looking for them would not be a burden.

Easy to identify, perhaps, but rare. Rare enough, in fact, that Garga-melle project director Musset doubted whether one would be seen in hundreds of thousands of photographs. After spending half a decade immersed in the delicate task of building Gargamelle, the completion of the chamber had left Musset somewhat at loose ends. He decided to occupy his time by looking into neutral currents more closely; his willingness to do so, he told us later, was partly due to his isolation from physics, and his ignorance of the lengthy series of experiments that had failed to observe neutral currents. Musset thought that the team should look for hadronic events (interactions between neutrinos and protons or neutrons) which should occur about two thousand times more often. Unfortunately, hadronic neutral currents—the diagram above pictures one—are much messier than neutrino-electron neutral currents. The reason is that after the neutrino hits the proton, the proton flies off and interacts strongly with any nearby particle, producing a spray of secondary pions, lambdas, and so forth. In hadronic interactions, too, the signature of a neutral current is the absence of a muon. Experimenters would have to be certain that they could identify *every* particle in the melee, and that none of them were muons.

On the other hand, Musset thought that the past experimenters who had not seen neutral currents had probably killed the very signal they sought. They had been worried about the neutrons created by interactions between neutrinos and the hundreds of tons of metal shielding around their detector. These neutrons might slip into the chamber and smack into a proton, creating something that mimicked a neutral current. The initial neutron would be invisible, because, like a neutrino, it is uncharged, and bubble chambers cannot "see" neutral particles. Because no final muon would necessarily be produced, such an event could look very much like a neutral current. To pick out the phony neutral currents, past experimenters had reasoned that these secondary neutron interactions would on the whole have less energy than neutrino interactions, because some energy would be expended making the neutron, the way the second bounce of a tennis ball is always less high than the first. The experimenters had therefore adopted the general practice of throwing out all interactions below a certain energy level.

Musset believed, however, that whereas the escaping muon in charged currents can be photographed and its energy accounted for, the escaping neutrino in neutral currents cannot, because it is not seen. Nobody knew how much energy the neutrino carried off. If it took away even a small amount, the event would seem to have too little energy, and experimenters would consign it to the trashcan as the work of a neutron. Making rough calculations, Musset figured that the energy cutoff had effectively concealed all but 5 to 10 percent of the neutral currents. "It was essential to have the same criteria," he said, "to have a really good comparison, a fair comparison between neutral currents and charged currents."[8] He argued that calculating the energy of charged currents should be done by ignoring the muon, thus putting the two types of interaction on an equal footing. He was unable to convince his colleagues, who thought the idea was ridiculous.[9]

Musset decided to take a look for himself. Together with a few friends—Ugo Camerini, William Fry, Antonino Pullia, and B. Osculati, all of CERN—he went to the scanning rooms in the basement after the scanners had gone home. They turned on the lights, unspooled the film, and let the big black-and-white images play on the formica tables. The five men spent almost a year in the room, working until the early hours of the morning, making drawings and measurements of the tracks that curled soundlessly in the liquid, shooting out like sheaves of wheat held by invisible hands. Musset was helped through this dull task by his myopia, which allowed him to squint closer to the photograph than someone with normal sight. Some time at the beginning of 1972, he came across an event that took his breath away: an interaction well in the back of Gargamelle, but centered on the line of the beam. There was no muon. The event was energetic. It could not be caused by a neutron because on average neutrons could only penetrate two feet of freon without interacting and this event took place over three yards into the chamber. It could not be caused by a cosmic ray because the interaction was right on line with the beam. One picture was not enough to prove anything, but Musset showed it to collaboration head Lagarrigue as a way of exciting his interest.

Years before, in Paris, Musset was on a team that had been beaten to the discovery of the eta meson by physicists at Brookhaven. He had come to the conclusion that the Europeans were too hidebound, too obsessed with precision, and feared that by the time the CERN group understood everything to its satisfaction, it would again be scooped by a few Americans with less regard for niceties. Musset knew American experimenters at Fermilab were examining high energy neutrino interactions. If neutral currents existed, he wanted to find them first.

Designed by Robert Wilson, one of the original accelerator builders on Ernest Lawrence's team, Fermilab is a kingdom of cowboy electronics.[10] A lifelong maverick, Wilson had strong notions about aesthetics and a firm belief in frugality that produced an odd set of priorities in the laboratory's design. When designing Fermilab, Wilson brought with him a graphics specialist named Angela Gonzalez, who was given carte blanche on color— so long as the paint was cheap. The laboratory bought an entire community of tract houses to acquire the site; these became experimental buildings, and Gonzalez ordered them painted solid dark Rustoleum blue, solid dark Rustoleum green, or solid dark Rustoleum pink. In the middle of the campus, the laboratory headquarters rises fifteen stories in a sort of giant H, fat at the base, curving toward the top, an atrium in the middle climbing to the roof—a lone skyscraper in the prairie. The building was named Robert Rathbun Wilson Hall after Wilson retired in 1978. Cooling water for the accelerator flows nearby through canals adorned by small fountains; the water steams in cold weather, fogging the lower windows of Wilson Hall, and is the permanent home to a gaggle of Canadian geese that never migrates from the warm scientific pools.

Despite constraints of budget and bureaucracy, Fermilab was built Wilson's way, making it, in some sense, one of the largest monuments to a personality since Peter the Great laid the foundations for Saint Petersburg. Ten square miles of grassland cut and shaped and molded by a single individual, Fermilab is an utterly eccentric celebration of cerebration. Wilson was from Wyoming, a square-backed, bowlegged man with a hard sunbaked face and square silver-framed glasses. He liked cowboy boots, horses, and buffalo. He liked buffalo so much that eighty of them now live on the Fermilab range, mean-eyed shambling beasts that the physicists avoid when they can. He stood an enormous obelisklike sculpture of his own fashioning in front of the central building and surrounded it with a reflecting pool. He built three experimental halls—the meson, neutrino, and proton buildings—awarded each a distinctive roof shaped like one of the Platonic solids, and placed them to be seen to maximum advantage from the top floor of the laboratory headquarters.

Wilson liked things big and he liked things cheap, and he didn't mind if a few people got scars doing what he wanted. He gave Fermilab the most powerful accelerator in the world for next to nothing, which in physics terms meant about $250 million. He told physicists they could have cranes to move around their equipment or a roof above their heads, but not both. Some of the experimental buildings didn't have floors, because Wilson saw no reason why people couldn't wear boots and walk on dirt. Wilson left no room to tear out experiments; it was cheaper every now and then to knock down a wall.

The main ring at Fermilab is four miles around and marked by an earth berm that resembles, in its simplicity and perfection, the legacy of a Stone Age religion—a technological Stonehenge. Twenty feet beneath the surface is a tunnel with a pipe to carry the protons and about a thousand boxcar-shaped magnets to bend them around the circle. Figuring to save a quarter of a million dollars, Wilson ordered the magnets welded into the beam line without the expensive seals that would allow them to be removed if they failed. Moreover, the tunnel itself was just a tunnel, nothing like the climate-controlled rings at CERN. By 1971, when the accelerator was ready to be tested, Wilson's decisions about the tunnel and magnets had proven catastrophic. When the hot humid summer nights settled over the Illinois farmland, moisture seeped through small hairline cracks in the insulation of the magnets and burned them out one by one. Nobody knew how many magnets had cracks; the technicians just had to sit and wait for the next one to blow; meanwhile, the first experimenters sat on their hands, their detectors dark, their careers on hold, hoping for some beam. For almost a year, Wilson watched powerlessly as his decrees caught up with him. Although the first 200 GeV protons made it through the ring in March 1972, not all of Wilson's colleagues forgave him for the ordeal.

A passionate advocate of unfettered scientific research, Wilson frequently was pressed by Congress to explain why the taxpayer should spend millions of dollars to fund an enormous, expensive gizmo whose sole use was to let physicists chase subatomic particles. One exchange, between Wilson and a skeptical Senator John Pastore of Rhode Island, has become legendary in the research community. Pastore asked, "Is there anything connected with the hopes of this accelerator that in any way involves the security of this country?"

"No, sir," Wilson said. "I don't believe so."

"Nothing at all?"

"Nothing at all."

"It has *no* value in that respect?"

"It has only to do with the respect with which we regard one another, the dignity of men, our love of culture. It has to do with, are we good painters, good sculptors, great poets? . . . It has nothing to do directly with defending our country except to make it worth defending."[11]

One of the teams sweating through Fermilab's rough early days was situated in the neutrino building and composed of a baker's dozen physicists from Harvard, the University of Pennsylvania, and the University of Wisconsin. The team was known as the HPW collaboration from the initials of its component universities; the senior collaborators were David Cline (Wisconsin), Alfred K. Mann (Pennsylvania), and Carlo Rubbia (Harvard), and the initial goal was to find the W. All three men were specialists in weak interactions; Cline and Mann had previously conducted searches for neutral currents with K mesons, and had not found them. Rubbia, who had the hunch that weak interactions were ripe for a theoretical breakthrough, specialized in building large and complicated detectors. The initial round of experiments at Fermilab would be the first to probe a new high energy range, and competition for time slots was, under Wilson's laissez-faire approach, not always gentlemanly. Although the HPW team submitted the first formal proposal to the lab in April of 1970, some members of the group were not certain they were ever going to run.[12]

Cline, Mann, and Rubbia set up their detector at the end of a kilometer-long pile of dirt. After flying into a target, protons from the main ring produced a gush of pions, which decayed mostly into muons and muon neutrinos before slamming into the dirt. The dirt absorbed everything but the neutrinos, which poured into the detector, where the experimenters ardently hoped they would do something interesting. The HPW detector was designed as a long series of vertical slabs somewhat like a fifty-foot-tall Dagwood sandwich turned on its side. Incoming neutrinos first passed through four alternating layers of liquid scintillator (a fluid that flashes when charged particles pass through) and spark chambers (long flat boxes that make sparks when charged particles pass through). Directly behind this segment was the second half of the detector, a magnet consisting of four iron blocks, each four feet thick, interspersed with four more spark chambers. If the neutrinos interacted in the first half of the detector, the various scintillators and spark chambers could record the trajectories of the particles. Of these, all but muons would be stopped by the iron in the second half of the detector. The idea was to use the sparks created by muons in the second half of the apparatus to trigger the detector in the first to record the event. By setting the detector to fire only when muons passed through the iron, the collaboration could avoid scanning the hundreds of thousands of useless events that bubble-chamber physicists had to go through. Unfortunately, because the detector went off only when a muon passed through, the device was not capable even in principle of looking for neutral currents, which are characterized by the *absence* of a muon.

In addition to working on the HPW detector, Rubbia, a fast-talking, ambitious man with an apparently endless capacity for juggling widely separated projects, was running another experiment at Fermilab and one at CERN. After 't Hooft's work, Rubbia said later, Steven Weinberg "took a tremendous amount of time telling [me] how nice his theory was." "Brainwashed," as he put it, into looking for neutral currents, Rubbia insisted that the collaboration modify the detector to trigger even without a muon. He was supported by Larry Sulak, a young Harvard colleague with as much ambition as himself. Sulak assembled the new trigger.[13] Right away, in the small dribbles of the beam they could get before the magnets blew, they started seeing interactions that lighted up the first part of the detector, but not the second—events with no muon.

The HPW collaboration did its work in the Fermilab neutrino building, a boxy affair with corrugated metal walls. One end is a bubble-chamber assembly area with a twenty-sided geodesic roof whose panels were made from soft-drink cans. Somebody suggested to Wilson that you could cut off the ends of Coke cans and sandwich the resultant cylinders between sheets of colored plastic to produce a cute, stained-glass effect on the cheap. Wilson tried it, producing a weird particolored dome that leaked every time it rained. Wilson didn't mind. The floor was dirt, and because the building was sunk into the earth, the place was full of puddles anyway. Come spring, they dried up.

"It was pretty terrible," Sulak said later about his move to Fermilab. "First of all, you had absolutely minimal support. I mean, you were out in the boonies. Wilson didn't believe in any frills at all, where frills meant buildings or sump pumps or cranes. No cranes! He had an objection to cranes. He had an objection to trailers as opposed to his so-called portacamps; we bought at Harvard a super trailer, filled it up with all the stuff to bring out, and then we were forbidden to put it at the experiment because of aesthetic reasons. And had to stick it a quarter-mile away at a farmhouse. Then the watertable was high and the foundation of the building leaked, and the water would come in and it'd get filled with frogs and the frogs would rot and stink, they would die because they couldn't get out. It was just terrible." Sulak spent hours scooping out the corpses of unlucky amphibians from the detector building and wondering what the muonless events were. "But the major problem was waiting forever and ever to take beam and not being able to do it."[14]

Yet a third hunt for neutral currents began at about the same time. It is a measure of the size of contemporary high energy physics that the third search occurred *inside* the first, when Robert Palmer of Brookhaven took a sabbatical at CERN in the fall of 1971. He wanted to work on the Gargamelle experiment, but the software to analyze the scanning data was not yet ready. Realizing that he would have to leave Geneva by the time the computer programs were complete, Palmer and two other American visitors, William Fry and Ugo Camerini, began to wonder what they could do with the scanning data alone. When the scanners went over the first Gargamelle runs, they projected the film on long tables and went over the tracks with a device somewhat like the mouse used in some home computers. By centering the mechanism on each vertex of every track, the scanners created a rough computer "recording" of the event on a keypunch card. Camerini and Palmer realized that they could pore through the cards to search for neutral currents. All they had to do was to figure out what arrangement of holes on the cards indicated highly energetic events with no obvious muon candidate in the rear of the chamber. They found them. Palmer wrote: "I am not sure if Ugo and Fry believed the results at the time, but I was very excited and rushed about trying to interest the other members of the group. I remember having little success and suspected that the already published negative results from the earlier work of the group had a lot to do with it." Incredibly,

Musset was making rough calculations at about the same time with about the same results, but the two team-mates did not run into each other because one man directly measured the photographs themselves while the other examined the computer cards made from them. In any case, both were rebuffed by their colleagues for the crudity of their analyses. As part of a large group, neither man could publish on his own. Palmer returned to Brookhaven in the beginning of 1972 and began telling his friends at Brookhaven that the laboratory should quickly finish its own neutrino experiments, for neutral currents existed and were waiting to be discovered.[15]

It all seems a long time ago now [Palmer wrote]. The originals of [my calculations] are a little yellow, and the angers and frustrations have faded too. Yet I have not altogether forgotten that it was I who first saw these neutral currents and it was I who one night (these moments always come late at night) made that plot and believed, alone in all the world, that God if he exists had decided to have a neutral current.[16]

Palmer's colleagues paid little attention to his claim, partly because of doubts about the accuracy of the measurements on which it was based and partly because a Brookhaven experimenter, Wonyong Lee, had reanalyzed some old neutrino experiments performed at the lab in the 1960s, and concluded, in May 1972, that "there is no evidence for the existence of neutral currents" in the events studied.[17] Moreover, theorists worried by the evident nonexistence of neutral currents had invented ways to save the basic features of $SU(2) \times U(1)$ by making the model a bit uglier or adding extra features.[18] One of the earliest and most ingenious end runs around the apparent absence of neutral currents was made by Glashow and Howard Georgi, then a postdoctoral fellow.[19] By positing the existence of two heavy leptons, they were able to contain the whole electroweak theory in an $SU(2)$ group, making do with just the Ws and the photon and excising the Z. Part of the theorist's sense of elegance comes from doing the most with the smallest number of constituents; the Georgi-Glashow model satisfied this predilection admirably, and many of the growing band of gauge theorists thought, well, maybe one didn't need neutral currents after all. . . .

Perhaps inspired by Musset's superb photograph of a neutral current candidate, Lagarrigue wrote to CERN Director-General Willibald Jentschke in April 1972, informing him that the search for neutral currents was now going to be one of the top priorities of the laboratory neutrino program.[20] Although neutral currents had moved into the forefront, there was no agreement about how best to search for them. Some members of the collaboration focused on the cleaner but much more rare electron-neutrino interactions, whereas Camerini, Fry, Musset, Pullia, and Osculati worked on the hadronic search—formalizing, in effect, an extant informal division of the team. By July 1972, the hadron subgroup had set up ground rules for classifying events.[21]

The HPW group got a bit of beam over the Thanksgiving and Christmas breaks of 1972. Physics lore has it that accelerators work best late at night and during holidays, when the hot shots in the control room who like to tinker with the beam are not around. The great detector in the neutrino building triggered for about one hundred and fifty events with no obvious muon. Recorded on big drums of computer tape, the data was taken first to Wisconsin and then to the fourth floor of Lyman Laboratory at Harvard, where Rubbia, Sulak, and a crew of undergraduates began to analyze it by a process similar to that underway at CERN. The computer singled out events having more than a minimum amount of energy; photographic prints of the pertinent tracks were then developed and blown up. The HPW collaborators built templates to measure the lines of sparks photographed in the detector. By the spring of 1973, they, too, had begun to suspect that the muonless events were for real. They were well aware of what the CERN group was finding. A Mercury of scientific rumor, Rubbia shuttled between his experiments in Cambridge and Geneva, bringing the message to each group that the other was hot on the trail.[22]

Around New Year's Day 1973, a research student named Franz Hasert, a Gargamelle collaborator from West Germany in the subgroup looking for electron-neutrino events, noticed that one of the scanners had discovered a curious interaction that apparently consisted of a muon and a photon, the latter subsequently becoming an electron and positron. Surprised by this odd interaction—three leptons coming from a neutrino!—he went back to the original photograph and realized immediately that the event had been misclassified. The picture actually showed an electron suddenly spiraling out of nowhere across the black background of the chamber, precisely what would occur in the event of a neutral current electron-neutrino interaction. Helmut Faissner, the senior Aachen physicist, was elated by this "picture-book example" of a neutral current. After scanning over 725,000 pictures, luck had allowed them to find the object of their search.

All that remained, Faissner said optimistically, was to check out the background. That took another half a year.[23]

Meanwhile, on January 31, 1973, Musset in the other CERN subgroup presented initial evidence for neutral currents at the annual New York meeting of the American Physical Society, then one of the largest and most important gatherings in the physics world.[24] Upon his return to Geneva, he urged Lagarrigue to help convince the collaboration; no begging was needed, because the discovery of the single electron at Aachen had dramatically changed the climate of opinion. Meanwhile, the hadronic team went through the events one by one, furiously arguing the criteria. They had two hundred and thirteen candidate events, each of which had to be listed and drawn and measured and studied and restudied.[25] To resolve the arguments and doubts, Musset took to carrying big stacks of eighteen-by-twenty-four-inch blowups around from lab to lab, logging thousands of miles as he crisscrossed the Continent.[26] Some physicists couldn't see the kinks; others worried about contamination from strange particles or cosmic rays; nobody wanted to make a false claim that would be subject to much attention and excitement. On the other hand, the Gargamelle group did not want to be beaten to the punch.[27]

Narrowing down the candidates was difficult, tedious, imprecise work, akin to the process whereby a mechanic gains the experience necessary to diagnose the weakness in a motor from the sound it makes. The coils made by electrons and positrons were easy enough to pluck out of the mass of lines in a photograph, but distinguishing between muons and pions was difficult. Both have almost the same mass and charge—one recalls cosmic ray physicists confusing them for a decade in the "ten-year joke"—and the tracks in bubble chamber photographs provide precious few clues to distinguish them. Pions usually end their lives with a small shower, and muons do not, but both sometimes simply stopped in the chamber, absorbed by a freon molecule. Yet distinguishing them was crucial: Mistaking a muon for a pion meant removing the final muon—producing a "neutral current" by an act of misidentification.

We once asked Musset to show us how he went about distinguishing muons and pions. His office was still in the old Gargamelle building, although the bubble chamber itself had been hauled off to a museum of science in Paris. Musset had a few minutes free, so he went across the hall, rummaged in a few old boxes, and emerged with half a dozen yellowing photographs of Gargamelle events. The images were grainy and sometimes washed out by the flare escaping from the edge of the light units. Together with the photograph was a drawing of the interaction and an evaluation sheet telling us that we were looking at Antineutrino Film Roll 592, Camera View 6, Photograph 125. Two tracks emerged from nothing, one short and stubby, the other long and branched. According to the remarks on the evaluation sheet, the event had been moved from the charged to the neutral category several times in 1972 and 1973, changing its identity in the back-and-forth of discussion. Musset's initials were on the analysis sheet.

From the size of the bubbles in the short track, Musset said, he could easily identify it as a heavy proton. "Then there is only one track [besides the proton coming out of the interaction], which is somewhat energetic because it's straight. There is a small kink, here—" his finger poked a point midway along the line "—that you can see, maybe." Kinks indicate where strongly interacting particles glanced off a nucleus. Musset said, "That's important on this particular event, because if you don't take this kink into account, then we have the wrong measurement of the momentum and energy. The energy looks to be less because the kink simulates bending, if you like. We had to measure only up to the kink."

Was that the reason the event had been reclassified?

Yes, Musset said. "This is a good example of an event which can be lost if you are not careful." Spotting the kink had first changed the identification of the particle from a muon to a pion, thus changing the interaction from "charged" to "neutral." Then measuring only up to the kink had boosted the figure for the total energy of the interaction, enabling the interaction to survive the energy cut. Seeing the kink, he said, involved a willingness to lie flat on the scanning table with an eye along the track; interpreting the photographs involved craft, educated judgment, and the kind of knowledge that resides in the fingertips. His myopia helped, too, for he could bring the photographs closer to his face.

At one level, we said, the whole edifice of theory boiled down to whether you saw a bend in a curve when you looked down a line. "Yes," Musset said. "And that was one of the difficulties, because I *never* convinced the scanning girls to put their eye on the table."

He put Photograph 125 atop an old khaki filing cabinet and invited us to look down the tracks. Neither of us could see the kink, though we strained our eyes considerably in the effort.

"It's there," Musset said, amused. He took off his glasses and flattened his cheek against the print. His hair spilled onto the table. "You just have to have the eye," he said, squinting nearsightedly.[28]

Robert Palmer heard Musset speak in New York. Frustrated by his inability to convince anyone on the Gargamelle team with his rough-and-ready calculations, he found himself in the bizarre position of having made a discovery—or thinking he had—from someone else's data. Figuring that the data had been made public by Musset, he tried to publish through the back door—scooping the CERN team on their own results—by writing a *theoretical* article, his first and only, with calculations taken from the Gargamelle data. Unfortunately, he sent it to the CERN journal, *Physics Letters*, where it arrived on May 24. There it sat for months, unpublished, frozen by a bureaucratic deity well disposed to the Gargamelle group.[29]

On April 12, 1973, the Gargamelle neutral current group, which had grown to include members from all the labs in the collaboration, met as an ensemble at CERN to go over every proposed instance of a neutral current. By the end of two days, they had reduced the total to 157 firm candidates.[30] Set down single-spaced on six sheets of paper, the events were listed by roll and frame number with summary comments:

570/253	OUT	1 GeV
570/399	*OK*	E badly known
605/033	OUT	possible mu
436/725	OUT	entering track
571/331	*(OK)*	Check energy (cosmic [ray])
491/263	OUT	error on frame number
443/108	*OK*	
558/316	*OK*	
558/646	*OK*	energy?
795/605	OUT	possible mu kink

The question remained whether they had really accounted for background; this was discussed in a second meeting the next month. To the dismay of some collaboration members, Musset jumped the gun at the end of May. While the team was still counting candidates, he told a French congress that the neutral currents "observed in the [Gargamelle] experiment are at the predicted level. . . . [O]ne may conclude that the current experiments cannot eliminate the Weinberg model."[31] Although the action was bold enough to alarm a few members of the team, Musset's style of presentation was sufficiently cautious that the announcement attracted little attention. Amazingly little, perhaps, in view of the consequences. Backed by Rousset, Musset had the go-ahead by mid-June to write a full paper broadcasting the discovery of neutral currents.

Throughout the spring, Sulak and Rubbia were preoccupied by a single worry, which was known as the problem of wide-angle muons. Although it filled almost the entire building, the HPW detector was long and thin, and it seemed quite possible that muons produced in charged current neutrino interactions could shoot out the sides before reaching the back half of the apparatus designed to detect them, making charged currents look like neutral ones. (Meanwhile, Mann and Cline were principally concerned with understanding the charged currents and the other research subjects the experiment had been built to study.) Years before, at a conference in Europe, Musset had argued with Rubbia about whether the detector would miss wide-angle muons, and Rubbia had said there would be no problem.[32] Quantifying that assertion involved setting up a "Monte Carlo," an elaborate computer simulation of the experiment, and seeing how many wide-angle muons should be produced. The trouble with such modeling, as the team well knew, is the immutable data processing law of GIGO—Garbage In, Garbage Out. If the assumptions they fed in were wrong, the predictions would be worthless. Similar questions beset the CERN group's simulations of neutron interactions. Both groups could therefore be making the same false claim. Nonetheless, Sulak felt good enough about the signal that by mid-June he, too, began to write a paper announcing the discovery of neutral currents.

Meanwhile, the two CERN subgroups prepared to publish. On July 3, the single electron event was submitted to *Physics Letters* from Aachen. The paper finished on a positive note: "We conclude that the probability that the single event [described in the paper] is due to non-neutral current background is less than 3%."[33] Still, the collaboration was uneasily aware that one picture rarely proves anything.

In Geneva, Musset and other teammates finished the rough draft of the hadron paper on July 4. Worrying about background continued until the last possible minute, with Musset bombarded by suggestions, arguments, counterarguments.[34] During a stopover in Europe, Rubbia wrote to Lagarrigue on July 17 that the HPW group had "approximately one hundred unambiguous [neutral current] events" and that they were in the final stages of writing a paper. Rubbia offered to ensure that the European team's work was mentioned in the American paper if the Europeans reciprocated. In a burst of chauvinism, Lagarrigue refused point-blank. He sent Rubbia a letter on July 18 informing him that by the time he opened it the Gargamelle collaboration would already have made a formal announcement.[35] The honors fell to Musset, who spoke in a large lecture hall on the CERN campus on July 19, 1973. The hadronic paper was submitted a week later.[36] For the first time, CERN had soundly beaten the Americans.

"I am not a theorist," Musset said to us. "So I am not able to understand all the beauty of the theories. But I can understand *facts*. It was to me so marvelous—to discover something which had been hidden for centuries."[37]

On July 25, the same day that the hadronic paper appeared in the office of *Physics Letters*, Rubbia received a notice from the district director of the Boston office of the U.S. Immigration and Naturalization Service, explaining that Rubbia's failure to fill out a routine visa extension form meant that he must depart the United States on or before July 29.[38] Notwithstanding his professorship at Harvard, Rubbia could not placate the federal bureaucracy;

he left on July 27, unable to return for three months. The ejection had fatal consequences, for it removed the chief link between Sulak, the junior collaboration member who had done most of the neutral current analysis, and Cline and Mann, the two senior members of the widely scattered group.

At first there was little problem. After amending the draft of the paper to suit comments from team members and others, Sulak took it personally to the offices of the *Physical Review Letters* on August 3. The preprint, distributed to every high energy physics laboratory in the world, announced a ratio of neutral currents to charged currents that was high but not inconsistent with $SU(2) \times U(1)$.[39]

As the news of neutral currents redounded through the physics world, Cline and Mann began to get cold feet. To begin with, Cline had done a half-dozen experiments with strange particles that showed that neutral currents did not exist. He could not understand how he could spend years not finding them in strange particles and then suddenly turn them up with neutrinos. Identifying real muonless events seemed so difficult that they were not at all sure of the analysis. Moreover, both distrusted Monte Carlo simulations.[40] Finally, another collaborator had just examined a second batch of data from Fermilab, taken that spring at a higher energy, and found that the total number of neutral current candidates was small enough to cut the ratio of neutral to charged currents in half. It was hard not to wonder if further work would make them completely disappear.[41]

"It's similar to the situation we are facing now," Cline said to us later. He was then working with Rubbia on an enormous experiment at CERN, smashing together protons and antiprotons to discover the W and Z. "We have some Z^0 decay results that are interesting, and are faced with a tough decision. Do we wait and run the experiment in the same way, or do we change the detector to improve our perception of the results? It's not easy. If you change something in the detector, you introduce new parameters. We did two things in that experiment. First, we modified the detector to pick up the wide-angle muons. Second, we got a more intense neutrino beam. These changed the parameters more than we were able to judge, and that led to what I have to call a disaster."[42]

The modification of the detector seemed harmless enough; Cline and Mann simply stuck in a foot-thick piece of iron before the last spark chamber in the first part of the detector. They hoped that this would block many strongly interacting particles, turning the spark chamber perforce into a muon detector. Although they would have liked to put in more iron, they were cramped by the small buildings that were part of Wilson's legacy. Still, because the spark chamber was only a foot behind the upstream end of the detector, rather than four feet, it would pick up more wide-angle muons.[43]

Although Rubbia was out of the country and thus unable to participate fully, he went along with the changes. However, he was confident enough to discuss the first results at a conference in Bonn at the end of August, and another at Aix-en-Provence two weeks later.[44] The two CERN papers appeared in the same issue of *Physics Letters* on September 3; Palmer's paper came out on September 17.

Even as news of the HPW finding swept Fermilab in early August, a group from Caltech headed by Barry Barish and Frank Sciulli began talking with Wilson about checking the result. Wilson had kept the Caltech group in competition with the HPW team since the start of the laboratory, giving first shot at the beam to the latter only after prolonged skirmishing. As the discovery became the subject of growing incredulity—*Physics Today*, the monthly trade magazine of the physics community, reported that "many experimenters are skeptical that either group has demonstrated the existence of neutral currents"—Barish et al. became, naturally enough, interested in confirming or killing the effect.[45] They were asked to work up a proposal in early September; on October 24, they delivered it. They were happy to state why their detector would not be plagued by wide-angle muons, they wrote, "But the burning question now is: *Is the effect real?*"[46]

Cline and Mann were asking themselves the same question. The first test run on the new detector took place on September 28, and it was quickly apparent that they were not picking up muonless tracks.[47] They were almost relieved when the referees at the *Physical Review Letters* bounced the HPW neutral currents paper, complaining that the wide-angle muon problem had not been handled satisfactorily.[48] Conferring via transcontinental phone lines, Cline, Mann, and Rubbia agreed to hold the paper by the simple expedient of not answering the referees' questions. (Sulak was not informed of what had transpired.) By the beginning of November, Mann had begun work on a "no neutral currents" paper that stated the group's "disagreement with recent observations made at CERN and with the predictions of the Weinberg model."[49]

Still in Europe, Rubbia did not actively participate in the revamped HPW experiment. A certain malicious glee nonetheless informed his encounters with the Gargamelle team, which was alarmed by the report that their overseas colleagues were not going to confirm their discovery. A powerful figure in his own right at CERN, Rubbia transmitted his views to Jentschke. The CERN director was appalled at the prospect that workers at his laboratory might have committed a grievous faux pas, which could have unhappy budgetary consequences on an international body zealously protective of its public image. He summoned Lagarrigue, Musset, Rousset, and the rest of the Gargamelle team to a meeting and grilled them about the validity of their experiment.

"Carlo came in from CERN," Sulak remembered. "I was at the mailbox [in Lyman Laboratory] with Karl Strauch [of Harvard] and Carlo came in, so proud of how much trouble the Gargamelle people were having because of their terrible results in the inquisition that he had arranged, having the Director General check them out and find out where they went wrong in their organization—" here Sulak does a sudden imitation of Rubbia's distinctive voice "—'seence the eh-vents rrreally wer-ren't there.'

"I said, 'What do you mean, they aren't there? The data hasn't changed! What the hell's going on here?' "

The teams were sufficiently large and the experiments sufficiently long that one man, Bernard Aubert, participated in both at various stages of the tale. He spent the latter half of 1973 at Fermilab, from where, on November 5, he telephoned a French colleague, Jean-Pierre Vialle, at Orsay to inform him of the Fermilab group's apostasy. Musset happened to be in Vialle's office when the call came through. "Vialle told me, 'Aubert says they have started to see no neutral currents.' So I said, 'Please, can I have the telephone?' And then I discussed a little bit with Aubert and immediately I found some problem with their experiment that to me prevented their answer from being conclusive. I always had this prejudice—or this opinion— that their experiment was not very good for that subject." On the next day, somewhat to Lagarrigue's dismay, Musset held a previously scheduled press conference at the French Physical Society and strongly laid claim to the discovery of neutral currents. The disbelief in the room, Musset recalled later, could have been cut with a knife. Another CERN experimenter, Jack Steinberger, had informally computed the background and explained away every single instance of the neutral currents. At the press conference, the head of the French Physical Society bluntly asked: "Are you sure this stuff's not junk?"[50]

On November 13, Cline, Mann, Rubbia, and D. D. Reeder of Pennsylvania drafted a letter to Lagarrigue to the effect that the level of neutral currents had basically dropped to zero. Although the letter, on Wilson's advice, was never officially posted, Rubbia showed it around on his next jaunt to CERN: salt in the wound.[51]

Years later, we met Sulak at a hokey Mexican restaurant in Mentor, Ohio, near the proton decay experiment. The hour was late, the waitresses dressed in cowboy outfits and anxious to go. A band thrashed away at some old songs by the Rolling Stones. Sulak had ordered something messy in a tortilla and a Dos Equis; a waitress hovered over his shoulder, asking every few minutes if she could clear the plate. To her dismay, we began to ask Sulak about the neutral currents experiment. He put his food down and answered in a burst. A big physics collaboration has all the tension associated with the cast and crew of a play; sometimes things can inexplicably go sour, dissent wrack the project, and mire the team in anger for a while.

"They wanted to try something," he said, "and they did something crude rather than doing the right job. There was no space to put anything in, so they put the fattest thing they could put in, and that was like a foot thick." That was the reason? we asked. Lack of space? "Yes. The building was too small, so they put in whatever they could. And everyone knows that a twelve-inch piece of iron doesn't shield hadrons for beans. You look in the book." He dug into his pocket, produced the 1984 Particle Properties Data Booklet, a pamphlet four inches high and two-and-a-half inches wide with over a hundred pages of tiny print. "You don't have to be smart. The book's been around for ages, and people know how to calculate the shielding power. You look up a standard table that everybody's had for years that has shielding power—wait'll I find the doggone thing. Here's iron, Fe." He squinted in the dim light. "You look in here, 'nuclear interaction length,' eighty-two centimeters—almost a yard." He threw the book on the table in disgust. "So if eighty-two centimeters is what you have to do to filter something out, and someone puts in something *this* thick"—spreading his hands a foot apart— "you can imagine that any hadron, any pion or whatever, that comes in there is going to make a shower and something has a good chance of coming out the back."

He took a bite, cooled down a bit. "It's very difficult, even today, to understand the propagation of hadronic showers through matter. The thing is very complicated, with lots of pions and neutrons and stuff that you don't really know how to model. To understand, for every pion going into a foot-thick block, how many come out the back, that is nontrivial. So you have to do all kinds of tests, which is what in the end [Richard] Imlay [of Wisconsin] and company finally did.

"But smart people don't put themselves into a situation where they have to understand something which is un-understandable. The reason the original magnet was four one-meter-thick pieces of iron was just to avoid this problem from Day One, and never have to calculate what happens in the middle of the iron because you never looked in the middle."

On November 29, Imlay completed a preliminary study of what is called "punch-through"—hadrons that pierced the iron shield. He combed the literature for examinations of the likelihood that pions would set off counters after going through given lengths of iron, found two measurements, and made from them a rough-and-ready calculation of the punch-through. The level seemed a lot higher than Mann and Cline had guessed. Imlay, Aubert, and T. Y. Ling then scanned thousands of frames of film to look for charged currents—events with muons in the rear half of the detector—that also had secondary tracks in the back. These secondary tracks would be due to hadrons that punched through the thin iron slab. Presumably, this would also have happened in the case of neutral currents, with the hadrons now fooling the detector into thinking they were muons, and making neutral currents appear as charged currents. Imlay, Aubert, and Ling soon saw that their first estimate of punch-through was, if anything, low.[52]

Cline was aware of Imlay's work but didn't think the punch-through would add up to much. On December 6, a week after Imlay's first memorandum, Cline gave a Friday afternoon wine-and-cheese talk at Fermilab about the progress of the experiment. As he wrote on the transparencies illustrating his talk, the collaboration's "very preliminary results!!!" showed that the level of neutral currents was "very likely too small to be consistent with [the] Weinberg model—also CERN data, if due to [the] Weinberg model. . . ." Neutral currents were "not confirmed by the present experiment" and the few surviving candidates simply needed "further study."[53]

In retrospect, the talk was a ghastly public-relations blunder. Playing by the rules of physics, an informal afternoon chat is not a release of experimental data. Nonetheless, it was treated as such by the hundreds of theorists and experimenters who wanted to know what was going on. The HPW team's apparent flip-flop spread with astonishing speed through the physics world even as people within the collaboration maintained that they *were* finding neutral currents. If they stuck to their original findings, the heat of publicity would treat that as a second reversal. The relentless curiosity of their colleagues thus would have translated the routine business of sifting through the data into a set of humiliating about-faces.

Cline's seminar began a week of heated discussion within the group. By December 13, Cline had changed his mind, admitting "the distinct possibility that a muonless signal of order ten percent is showing up in the data. At present I don't see how to make these effects go away." They decided to publish the original paper. It appeared in *Physical Review Letters* on April 8, 1974, six months after the original submission. Another CERN paper came out in *Nuclear Physics* on June 24. When the Caltech group announced that it, too, had observed neutral currents, their reality was at last accepted.[54]

Nobody ever won a Nobel for the discovery. Lagarrigue and Musset were logical choices as the heads of the CERN search, but Lagarrigue died suddenly in 1975 and Musset was killed in the summer of 1985 in a climbing accident near Mont Blanc. In addition, both the HPW and CERN groups had lost a measure of credibility. The lengthy brouhaha, which provided cliff-hanger entertainment to the physics community, was mocked as "the discovery of alternating neutral currents." But now that neutral currents existed in neutrino interactions, the most pressing question became: Why hadn't they been found in strange particle decays? Even the most presumptuous theorist was unlikely to argue that dozens of past experiments were incorrect. "Three or four, maybe," Glashow said. "But this was something like *twenty*. People were walking around wringing their hands and asking, 'What's going on?' So I put them out of their misery and *told* them. It was charm."[55]

In April 1974, Glashow spoke before a conference of meson specialists that was held at Northeastern University, in Boston. In a speech entitled "Charm: An Invention Awaits Discovery," he explained to the assembled experimenters that by virtue of the mechanism described in the Glashow-Iliopoulos-Maiani paper, the fourth quark prevented neutral currents from occurring in strange particles. He set down his reasoning on the blackboard, cheering up some in the crowd when he botched the mathematics the first time through. Undeterred, Glashow challenged the meson physicists in the room: They were rightfully the ones to discover charm, because the lightest charmed particles would be charmed mesons, he said; but because they weren't looking, charm would be found by "outlanders"—other sorts of experimenters, such as the neutrino teams at CERN and Fermilab. By the time of the next meson conference, Glashow said, "There are just three possibilities: One, charm is not found, and I eat my hat. Two, charm is found by [meson specialists], and we celebrate. Three, charm is found by outlanders, and *you* eat *your* hats."[56]

Two months later, Iliopoulos threw down the gauntlet before the Rochester conference in London. An excellent orator, Iliopoulos delivered a tub-thumping defense of SU(3) × SU(2) × U(1). There are, he asserted, two fundamental classes of matter: quarks and leptons. There are four members of each class, shuffled neatly into two families. Electromagnetism, the weak force, and the strong force are covered by two gauge theories consistent with each other, the whole being described in a grand synthesis called SU(3) × SU(2) × U(1).

From this, he said, "a very simple and beautiful picture emerges." With the acceptance that Yang-Mills gauge theories described all interactions, an easy step led to a startling insight. *SU(3) × SU(2) × U(1) is nothing other than the broken remnant of a single unified gauge group that existed in the distant past.* Finding the unified theory now depended on plucking out the right group.

The synthesis demands that charm exist, Iliopoulos said, and he bet the assembled physicists bottles of fine wine that the fourth quark would be found before they met again. At stake, he said, was more than just another quark or a couple of models. At stake was the possibility of unification itself.[57]

18
Charm and Parity

ALTHOUGH GLASHOW AND ILIOPOULOS DIDN'T KNOW IT, THEIR BETS WERE safe. Evidence for charm had already been collected, and more would follow. A team of experimenters at Brookhaven was turning up traces of the fourth quark even as Glashow spoke, although they ultimately played only a small role in the drama that charm became. The first definitive traces were found months later by two different teams, one from MIT and one from Berkeley and Stanford. Yet when all was said and done, it took another year and a fourth sighting before charm was accepted. The finding of charm is now described by a few quick lines in undergraduate textbooks, a circumstance that has prompted Nicholas Samios to remark that such volumes should be printed with some combination of sweat and blood.[1]

Since the revelation of the second neutrino, Samios and Robert Palmer, like Musset and his colleagues at CERN, had planned an experiment to study neutrino interactions by shooting them into a big bubble chamber. At Brookhaven they constructed a seven-foot bubble chamber—much smaller than Gargamelle, but still impressively large—and filled it with liquid hydrogen. The nucleus of a hydrogen atom consists of only a single proton, and therefore when it was struck by a neutrino only two particles would be involved. By contrast, Gargamelle used freon, which is a complicated molecule with 159 neutrons and protons. Neutrinos hitting hydrogen nuclei, Samios said, "is not a messy collision. If you saw something happen, you could measure and calculate it to a gnat's ass."[2] Brookhaven approved the experiment in August 1973.

Shortly afterward, Glashow and Samios met at Brookhaven. Glashow, who was then a consultant for the laboratory, proposed to Samios that the new experiment would be a good way to look for charm. A speeding neutrino might interact weakly with a proton inside the chamber—that is, a W^+ would pass from one to the other. Upon absorbing the W^+, one of the down quarks inside the proton could change to a charmed quark, and for a fraction of a second—10^{-13} seconds, according to theory—a charmed baryon would come into existence. The charmed baryon would not survive long enough to be photographed. But when it decayed (and the charmed quark again changed its identity, becoming a strange quark), it would turn into a strange baryon—a lambda particle. The lambda, too, would eventually disintegrate into other particles, but it would hang around long enough to be detected.

As physicists had known for twenty years, strange particles are generally produced in pairs. (The reason is that they come from the production of a strange quark and a strange antiquark, which fly away from each other to produce two strange particles.) Glashow pointed out that if experimenters saw an exception to this rule—the creation of a single, solitary strange particle—it would be evidence for charm.

Physicists are always discerning the existence of things they can't see from things that they can, but Glashow's chain of reasoning here seems particularly oblique. He was assuming that the presence of a single strange particle meant that it had been produced by another particle (the charmed baryon) that did not live long enough to be seen, in whose interior was *another* type of particle (the charmed quark) that *inherently* could not be seen. Invisible pieces of invisible particles; Samios and Palmer were intrigued but skeptical.

The first neutrino interaction in the seven-foot bubble chamber happened in January 1974.[3] Except for occasional maintenance breaks, the AGS shot a billion neutrinos every two-and-a-half seconds into the bubble chamber until March 1975. Twenty-four hours a day, seven days a week, the three cameras atop the detector snapped pictures, capturing on seventy-millimeter film whatever they saw inside. By the end of the run, Samios's team had taken more than a half million slides. According to the theory, neutrino-proton interactions should be recorded on about two thousand of them, less than one percent.

The Brookhaven scanning team was highly regarded, largely because of the work of its supervisor, Milda Vitols. A Latvian emigré who had spent several years in displaced person camps before settling on Long Island, Vitols was working as a YMCA cashier when she spotted an ad in a local newspaper for a scanning job. She soon worked her way up to a supervisor, thanks to her unusual knack for explaining what the physicists were talking about to the other scanners.[4] She also had to encourage the scanners and keep their spirits up in the face of considerable pressure. Worried that the tedium of the job might make the scanners miss events, the physicists had them placed in separate booths so that they couldn't talk to each other while working. Moreover, the scientists would periodically have the scanners' work secretly redone, and assign them "efficiency numbers" that depended on how many events they had not spotted. "That's an important number," said Michael Murtagh, a physicist who worked closely with the scanners. "It goes into the published results. To have an idea of how many events you had, one of the factors is how many events the scanners missed on the film, and to calculate that you needed to know the efficiency of the scanners."[5] Scanners, however, found it humiliating.

Being electrically neutral, a lambda would not leave tracks, but its decay would leave two particles whose tracks would be visible. Vitols told the scanners to look for two rays appearing out of nowhere and shooting off in a V. (This would be the decay products of the lambda.) If the scanners saw a V, she said, they should look to see if the vertex pointed back to the vertex of another set of tracks. (This would be the locus of the initial collision between the neutrino and the proton.) If they found a second group of lines near the first, Vitols said, the scientists would be very excited, because the configuration would mean that their quarry, the lambda, had glided silently between these two points for the brief flicker of its existence.

On May 3, 1974, Helen LaSauce, an ex-switchboard operator and mother of three, advanced her scanning machine to frame 6,967 of roll 27, and sketched a lone strange particle—the artifact of charm. Five tracks led away from one point of impact, while close by a pair of tracks parted from a second vertex. Camera angles 1 and 2 were unclear, so she decided to sketch angle 3.[6] It was near the beginning of the experiment; she had no idea she was seeing anything special. She just started drawing, her mind on the easy listening music pouring from the transistor radio on the desk. "They never told me much about what I was doing," LaSauce said. "The only way I knew I had found something important was when *they* [the physicists] asked me to remeasure it for the umpteenth time."[7]

At the end of the week, LaSauce had filled up her sketchbook. She passed it to Vitols, who glanced through the book, looking for events she thought would interest the physicists. When she came to frame 6,967, she pointed it out to Samios.

"I remember being in the control room," said Palmer, "and Nick came rushing in, waving this sketch and saying, 'It fits charm!' Now, I think this was the first strange particle we'd seen. The probability that we'd find a *single* strange particle the first time around was just nothing. And I remember thinking, okay, Nick's being wild again. Clearly one of those other tracks is another strange particle. I was skeptical because it was a new chamber, and we weren't really sure how to identify with certainty all the tracks. But it *was* a beautiful photograph. Not one but two electrons stopped in the picture, and that's something that gives us the information to calculate the dynamics of the interaction very precisely. Still, it took us seven months of round-the-clock calculation to conclude that, to an infinitesimal degree of uncertainty, the first set of five tracks consisted of four muons and a pion, and the second pair of a proton and another pion. We had found a lambda—and charm.

"But we'd only found *one* of them. We couldn't turn up another. You don't want to say you've found a new state of matter on the basis of one event, so we kept going back and redoing the calculations."[8]

Once before, Samios and Palmer discovered a particle on the basis of a single photograph: the omega minus. Made in a new detector with a new beam, frame 6,967 was nowhere near as clear-cut a case for discovery. Proving that one and only one of the seven tracks was a strange particle took months of endless figuring, while Samios and Palmer kept begging for more beam time to find a second.[9]

While Samios and Palmer and their team pored over frame 6,967, two other groups of experimenters came up with more definitive evidence of charm. Whereas Samios had found a charmed baryon, these teams found mesons made of charmed quarks. One team, whose head was Samuel Chao Chung Ting of MIT, drew protons from the same Brookhaven accelerator that Samios and Palmer had used. The other, led by Burton Richter, was working on the other side of the country at a new SLAC accelerator. The MIT and Berkeley-Stanford teams were run utterly differently; their joint discovery is a study in contrasting styles—open against closed, precision standing opposed to luck, an army versus a commune—brought into irremediable conflict by nature's chariness with her secrets.

The Patton of experimental physics, Sam Ting is famous for driving himself and his collaborators to exhaustion. A tall, stooped man with large features, Ting has the trick of commanding a room by speaking so quietly that everyone in it has to shut up to hear him at all. His manner is mild, almost affectless; his pants are baggy, his feet slightly splayed; but the look is misleading, for Ting is one of the world's great control freaks—a man of fierce ambition and abiding determination who has ruthlessly pursued his vision of physics for a quarter century. His colleagues love to recount tales of Ting sleeping on the floor by his equipment, or being physically carried out of experimental halls by lab officials who have slated his experiment to be replaced by another. The stories are exaggerated, but their circulation is a measure of the man. Renowned both for sudden, towering rages and for prodigies of self-discipline, Ting has demonstrated a gift for inspiration, command, and bureaucratic maneuvering not usually associated with the pursuit of knowledge. Today, he is putting together the largest experiment ever done, a matter of some 400 Ph.D.'s from more than a dozen nations, which is scheduled to begin on January 1, 1989, the inauguration of CERN's new Large Electron Positron machine. (The vast LEP will itself be *sixteen miles* around.) Ting planned to employ satellites to coordinate the reams of expected data to be analyzed by groups from the Soviet Union, China, the United States, and the thirteen member nations of CERN. "There are two kinds of experimenters," he once told us. "The first kind do what theorists tell them to do. The second kind follow their own ideas. I am of the second kind. I am happy to eat Chinese dinners with theorists, but to spend your life doing what they tell you is a waste of time."[10]

Ting was born prematurely in 1936 at Ann Arbor, Michigan, the child of two Chinese university professors on a brief visit to the United States.[11] Despite his accidental citizenship in the United States, Ting did not return to the United States for another twenty years. Knowing little English, he arrived at the Detroit airport with a hundred dollars and the determination to work his way through college. He spent the next six years surviving on scholarships, learning English, earning degrees in mathematics and physics from the University of Michigan. After accepting a post at Columbia in 1965, Ting became interested in a series of experiments on quantum electrodynamics.

Although quantum electrodynamics had never failed an experimental test, many theorists thought the theory would have to be replaced after particle accelerators reached several billion electron volts.[12] The first important high-energy test of quantum electrodynamics had been done at the recently built Cambridge Electron Accelerator, or CEA. Sponsored by MIT and Harvard, the 6 GeV CEA was one of the last of the university laboratories, built at a time when national laboratories such as Brookhaven were taking over all high energy research.[13] In 1964, a group of experimenters led by Francis Pipkin of Harvard used the machine to make high-energy photons and smash them against a target to create electron-positron pairs, which were then directed into a new type of detector called a double-arm spectrometer. This device took the streams of electrons and positrons issuing from the target and passed them through a series of magnets which bent them back to a point, like two streams of traffic that momentarily separate around an island and remerge. A set of counters on each arm act like "traffic flow meters" to measure the passage of each electron, and to ensure each electron is paired up with a positron on the other side. Pipkin's aim was to test the prediction of how many pairs would be produced at particular angles. It was, he said later, "the first opportunity to test the theory in a domain where you might expect the thing to go screwy."[14] At high energies, the pair production did not seem at all as theory said it should be, and Pipkin announced as much at a series of conferences in the spring of 1965. Indeed, quantum electrodynamics seemed to have gone screwy.

"When we first announced the thing," Pipkin said, "people weren't particularly perturbed. But when people began to think about it, they were." Quantum electrodynamics, which had originated in the work of Dirac, Heisenberg, and Pauli, and had culminated in the work of Feynman, Schwinger, and Tomonaga, was *the* successful physical theory. It was one thing to speculate on its inadequacy, completely another to imagine physics without it. Faced with a growing clamor of dismay, Pipkin's team began working on their equipment in the summer of 1965 so that they could redo the experiment and publish a complete paper. On the evening of Independence Day, when most of the rest of Cambridge was down along the Charles River watching the fireworks, Pipkin dropped by the CEA experimental hall, which was on the Harvard campus, across the street and down a bit from Lyman Laboratory. Working late that night was another team that was filling a big new bubble chamber for the first time. Because new chambers usually leak, they tend to be filled in frustrating fits and starts, with delays to seal problem areas. Both experiments were going smoothly.

Pipkin returned to Harvard for his shift at four o'clock in the morning. As he approached the CEA, he was stopped by a police line and a circle of fire trucks around the building. There were flashing lights, crackling walkie-talkies, ambulances with gaping doors—all the trappings of a modern catastrophe. As he learned later, the bubble chamber had caused an explosion. The windows through which the photon beam was to enter had cracked, and the liquid hydrogen had gushed out to the floor. (Because the filling had gone unexpectedly smoothly, the technicians had neglected to put on the metal safety caps for the windows.) Liquid hydrogen is as explosive as dynamite; hot vacuum pumps throbbed nearby. The blast blew the concrete ceiling off the building; coming down, it shattered on the steel girders and rained concrete on the technicians below. One person was killed, five more were badly injured, and all the equipment in the hall was smashed and twisted by the fire and explosion. Pipkin spent weeks with the subsequent investigation. With two team members hospitalized, the experiment was over. Nevertheless, they published their results.[15]

For someone like Ting, who had just mastered quantum electrodynamics, these results were distressing indeed. "It was a very important experiment, and so I thought that I should do it over again," Ting said. "Quantum electrodynamics was the only theory at that time—even now—that's really very accurate, from cosmological distances to satellite communications to very small distances. And so I was very surprised for someone to say that it is wrong." Ting's first thought was to redo the experiment at the CEA, and on November 9, 1965, he drove up to Cambridge to visit Pipkin. That happened to be the day of the great Northeast blackout; the two men met for the first time in the dark. Ting wanted to know what Pipkin had done, what problems he had encountered, and what suggestions he had if a similar experiment were to be performed. Pipkin responded to all these questions in detail. Then Ting asked how he could go about doing the experiment on the CEA. On this subject Pipkin was less encouraging. "It was made very clear to me that a young person with no background had no chance to do an experiment there."[16] That decision rankled for years. A few years later, when Ting was appointed a professor at MIT, one of his first conditions for accepting the post was that he *not* have to work at the CEA.

Annoyed, Ting contacted lab officials at the Deutsches Elektronen-Synchrotron (DESY), a slightly more powerful "lookalike" of the CEA in Hamburg. The director of the laboratory, Willibald Jentschke, knew Ting from his CERN days, and was encouraging. The following March, Ting took a leave of absence from Columbia to do the experiment in Germany. Just before he left, Ting ran into Leon Lederman, then director of Columbia's Nevis laboratory. Lederman expressed skepticism about Ting's ability to carry off an experiment of a sort he had never done before, and bet Ting twenty dollars that it would take him at least three years.

When Ting arrived in Hamburg, he immediately put into play the knowledge he had learned from Pipkin. Like Pipkin, Ting used collisions between photons and nuclei to produce electron-positron pairs, which he then studied with a double-arm spectrometer. The experiment had all the hallmarks of subsequent Ting experiments: meticulously executed, tightly scheduled, and performed by a totally dedicated international collaboration whose members worked up to seventy-two hours straight and slept by the machine. He has used that electron-positron experiment as a template again and again, and worked with some members of that group for the next twenty years. "Electron-positron collisions are very simple systems," he said to us.[17] He spoke carefully, slowly, laconically; clipped sentences with a slight accent. "They are easy to set up and understand. When they collide, they annihilate each other, and for a brief moment all you have is light. Then the photon of light decays into other things." Primarily, electron-positron pairs. "There aren't lots of other particles floating around, so you have a good idea of what you're doing."

Ting and his group carried off the experiment with such speed and precision that it was completed in several months. Quantum electrodynamics was strikingly confirmed. The refutation of Pipkin's CEA experiment was dramatic enough that Ting decided, in September of 1966, to fly to a Rochester conference at Berkeley to proclaim it. But Jentschke, DESY's director, tried to get Ting to soft-pedal the contradiction, arguing that a flat-out attack on the CEA work might wreck the credibility of the laboratory as a whole. "DESY was built with much help from the CEA. And here we had a result which really would do a lot of damage to the CEA. The question was, should I just present my data alone, or should I present my data as a comparison with the CEA data? I mentioned to the director that as long as people have published their data—published it in *Physical Review Letters*—I can quote it, I can put it on the same graph as mine. He said no, that's too impolite. I said, well, it's *not* too impolite, because after all, I may be wrong, too. If somebody presents his data, and publishes it, other people can quote it—that's the *whole purpose* of publication! We had a *very* strong disagreement about that. We had the disagreement in front of the DESY cafeteria. There were a lot of people going in and out, and he and I were talking, and people were surrounding us, looking on. My comment was, 'If you publish a result, you take the consequences for it!' "

Ting decided to go ahead and compare his results and Pipkin's on the same transparency, which he drew on the plane on the way over. At the conference he listened patiently while a Cornell group, working at a much lower energy on the Cornell accelerator, reported results which appeared to support Pipkin's.[18] A few minutes afterward, Ting quietly but firmly said that the Harvard and Cornell experiments were wrong, and that quantum electrodynamics was still valid. The announcement was a shock: An unknown thirty-year-old from a German knock-off of the CEA had baldly claimed that not one, but two, outstanding American teams had stumbled badly. "Science," Ting has remarked with evident satisfaction, "is one of the few areas of human life in which the majority does not rule."[19]

Ting followed up this experiment with more work on high energy photons, and became particularly intrigued by the way they could occasionally and spontaneously turn into three vector mesons, the rho, phi, and omega, and then back again. (The omega meson is not the omega minus, a baryon, but a meson known as "little omega"; *vector* means "spin-one.") They were enormously difficult to observe; out of a hundred thousand rho decays, only *one* produced an electron-positron pair for his detector. Theorists identified them as quark-antiquark combinations; up-antiup (rho), down-antidown (omega), and strange-antistrange (phi). Because they have exactly the same quantum numbers as photons, Ting called them "heavy photons," and has continued to study them throughout his career. "Always the same thing, always a little more precisely."

At the beginning of 1970, Ting began to have the hunch that there were more heavy photons in addition to the rho, phi, and omega. "The world is not so simple that a photon just changes to three vector mesons, and not more." The most suitable place, Ting thought, would be the new accelerator at Fermilab. He drew up a proposal—Fermilab proposal 144—and one of his collaborators, Min Chen of MIT, went to Batavia to design the detector. Construction lagged, however, and Fermilab officials seemed to be dragging their heels about approving the Ting group's proposal. "I think my style was really very different from the style of Bob Wilson, who was then the director," Ting said. "I did a lot of work, designing a beam, and we never got this approved.

"Then I decided I should propose this experiment at CERN. So I showed up at CERN. And the management of CERN at that time said, 'This is crazy; we know there's no heavy photons. Besides, we have more important things to do.' So I was also very discouraged at the CERN PS [Proton Synchrotron]. And then I decided, maybe I should do it on the ISR, the Intersecting Storage Ring [another CERN accelerator]. So I submitted a proposal there. Well, the management of the ISR at CERN said, 'This is fine, but you are a completely American team—we need some European balance.' And so fundamentally I also didn't get it approved. I did not get rejected, but I did not get the thing approved." As the year 1970 wore on, Ting flew back and forth between Hamburg and Chicago trying to get his various proposals approved, with no luck. He became angry and despondent.

"From Hamburg to Chicago you spend a lot of time and effort to design a beam, and design lots of other things, and you get a flat 'no' from these people. And there was *no reason* for it. They had the beam, they had the money, they have the backup. And basically it was a conflict of style. I became discouraged because people were not using a physics reason to disapprove it. You see, if people say 'no' to you, and give you a physics reason, you can understand. You may not agree, but you can understand. If people say, 'You don't have enough Europeans on your team'—this is not a physics reason." On top of his depressed mental state, Ting began to develop an ulcer. On the advice of his physician, and lab director Jentschke, Ting took several months off from physics to relax.

When he returned, it was with a vengeance. His various proposals still unapproved, he decided to make one last try. At the beginning of December 1971, he called Chen from DESY at Hamburg to announce that he had decided to quit Fermilab and do the experiment at Brookhaven. Chen violently opposed the idea; he had just spent six months completing a design for a Fermilab detector, and was ready to build. A few days later, Chen received a letter from Ting instructing him to quit Fermilab immediately; in place of a closing salutation, Ting wrote, "I will not accept any other alternatives or suggestions."[20] Chen was back at MIT within the week. Between Christmas and New Year's, Ting and Chen got together to write a proposal for Brookhaven, submitting it on January 11, 1972.[21]

They planned to build an experiment on one of the five beam lines in Brookhaven's slow extraction experimental hall. The Alternating Gradient Synchrotron (AGS), the lab's biggest accelerator, can deliver its protons in two ways; It can spit them in one short bunch a few microseconds long, or pour them out in a long bundle of about a second's duration. An experiment is usually suited for either the one or the other, there are separate halls for each, and the AGS beam is usually in only one mode or the other. Samios and Palmer's bubble chamber experiment was in the fast extraction hall, because they wanted to make sure that any events that occurred did so during the bubble chamber's brief period of maximum sensitivity. Most detectors, on the other hand, work best with slow extraction, where the events are spread out so they can be counted more easily. Once delivered to a hall, the beam is subdivided into different beam lines so that many experiments can work at once.

Ting proposed to slam slowly extracted protons from the AGS into a target to produce a hail of particles, mostly protons, neutrons, and pions. These would be ignored; the team was interested only in new, heavier cousins of the rho, phi, and omega. If any were produced, some fraction of them would decay into electron-positron (e^+ e^-) pairs. Any sudden upsurge in the number of such pairs at higher energies would give away the presence of new heavy photons. As ever, a highly sophisticated double-arm spectrometer would count the pairs. The e^+ e^- pairs would be produced in the horizontal plane and bent vertically, separating the measurements of angle and momentum, making them more precise and easier to check.

In May 1972, Brookhaven approved Experiment 598, as it was now called, awarding it a thousand hours of beam time.[22] Ting, Ulrich Becker, and Chen spent the next year constructing the detectors, magnets, and beam pipes, and in the fall of 1973 began installing them in the slow extraction hall of the AGS. Because Ting was examining the direct collision of the protons and the target, instead of watching the decay of secondary particles, the beam line was brought into the slow-extraction experimental hall, where it had to be covered with shielding. To Ting's dismay, he needed much more shielding than planned—ten thousand tons of concrete. The CEA had just been shut down due to budget cuts, and, profiting from Cambridge's misfortune, Ting's group was able to borrow all of its old shielding. In addition to the concrete, they used a hundred tons of lead and five tons of Borax soap, which contains boron, a neutron stopper. The team worked on the A beam line, next to Experiment 614, which was led by Mel Schwartz, Ting's former colleague at Columbia, who had since moved to Stanford. The two experiments would alternately extract protons from the A line.

Ting's detector was enormous in every way: in size, intricacy, sensitivity, and cost. Situated near the center of the slow-extraction hall, the detector—an eighty-foot V swaddled in great concrete blocks—rose in a giant-sized jumble to the ceiling. Fist-thick cables coiled from the apparatus to a small shelter against one of the dull green walls; here the experimenters lived for days on end.

In April 1974, Ting, Becker, and Chen asked for beam and started shaking down their equipment by looking at the phi meson. When the number of electron-positron pairs produced by the collision is graphed against the total energy, the phi is indicated by a hill whose summit, or maximum number of pairs, is at 1.02 GeV (alternatively, 1,020 MeV). To

Ting's mind, the detector had to measure the energy of the electron-positron pairs with great accuracy, because the energy was equivalent to the mass of the particle that produced them. Consequently, his detector was fantastically precise—needlessly so, he was told, for the vagaries of quantum physics smeared the mass value over a broad range. For this reason, particles are spoken of as having a wide or narrow "width," depending on the spread in their mass value; typical widths are on the order of several hundred million electron volts. Ting's detractors complained that making a detector able to detect gradations twenty times smaller than the natural width of the narrowest known particles was ludicrous. "What people worried about was the cost," Ting told us. "The cost was over five million dollars. That was more than ten years ago, when five million dollars still was a lot of money. Most of the theoretical physicists said, 'But there's no particle with a narrow width, and we know it. There's no particle there anywhere. You are just wasting your time.' So I did have a lot of hard times to get the money."

Brookhaven shut down for its summer break in May.[23] Near the end of July, the machine was switched back on and ready to deliver protons to the slow extraction hall. Experiment 598 ran for two weeks, taking data at 3.5 to 5.5 GeV. They found nothing. On August 7 the beam was switched over to Mel Schwartz's experiment, which used the beam for another two weeks.

On August 22, Ting got more beam, this time tuning his detector to between 2.0 and 4.0 GeV. The first few days were difficult; the AGS frequently malfunctioned, and on two successive evenings electrical storms put the whole accelerator out of commission for several hours. Nevertheless, the team ran until the last minute when the AGS was switched off for a maintenance break at four o'clock in the afternoon of Wednesday, September 4.

On every other experiment, a single analysis team dissects the results. Ting's fanatic care is such that he always has had *two* data analysis groups independently crunching the same numbers. Each team wrote their own computer programs to find the energies of each electron-positron pair and to plot the results on a graph. Bumps in the curves would indicate a new particle—a heavy photon. Captained by Becker and Chen, the two analysis squads started working on the data right after the AGS was turned off for a maintenance break on September 4.[24]

Sometime in the next day or two, Terry Rhoades of Brookhaven drove from the lab to Cambridge with the 2-4 GeV data tapes to analyze them with Becker. Each had written half of the computer program; Rhoades and Becker stitched the two parts together and ran them through on great piles of cards through the MIT computer. It was late Saturday night.

When the computer center sent the numbers back, Becker said, they stared at them in "intrinsic disbelief." The printout was a rough graph, with asterisks indicating the number of events at each energy, and the energy in gradations of .025 GeV (25 MeV).[25]

2.875	0
2.900	0
2.925	0
2.950	1***
2.975	0
3.000	3*********
3.025	1***
3.050	6******************
3.075	17***
3.100	15**
3.125	5***************
3.150	1***
3.175	0
3.200	0
3.225	0

Nearly all the events were piled up at 3.1 GeV; instead of a hill, they were looking at a needle. No known subatomic object had such a narrow width. Something was obviously wrong with the computer program; if a routine is incorrectly written, the computer often reacts by shelving all the data into one category. Becker and Rhoades began to shout at each other, each sure that the other had screwed up one of the thousands of statements in the program. They argued until midnight, combing through the cards, neither man giving in. "Since we are both not particularly fond of making compromises," Becker said, "we went to the computer center and threw everybody else out." Brazenly yanking everyone else's cards, they rudely inserted their rechecked cards in front of the stack. This brought down the wrath of another experimenter, Wit Busza, who had been waiting for hours to use the machine; Busza shouted that he was going to call the lab director then and there to protest this behavior. Unperturbed, Becker and Rhoades waited for the cards to go through. Paper clattered from the printer: a needle at 3.1 GeV, the sharpest peak ever seen.[26]

At the same time, Chen began his analysis at Brookhaven. Chen's data were in three separate parts, and when he sat down to total the result by hand, he was astounded to see that most of them were right atop 3.1 GeV. Skeptical, he rechecked his analysis, and got the same result. "I ran into Lee Grodzins," Chen told us, "who was teaching the same course as I was at the time. I yelled at him, 'I just found a narrow resonance!' [Resonance is another term for a short-lived particle.] And he *laughed* at me. He said, 'Maybe the whole world is full of narrow resonances!' "[27]

Becker and Chen each called Ting that Sunday to tell him the news. He told them to come down to Brookhaven immediately and called a joint meeting of the analysis groups. They met Monday morning, September 9, in one of Brookhaven's small conference rooms. Afraid of having committed a disastrous error, both groups sat in silent expectation for what is today recalled as a very long time. Finally, someone in Chen's group said, "Our analysis looks peculiar—" and suddenly everyone was shouting at each other.

Afterward, Chen was overcome by excitement. Delighted, he raced into the central cafeteria, his arms full of cards and charts. Experiment 598, he proclaimed, had found a particle like nothing that had ever been seen—a spike. Ting was enraged. He had seen too many experiments get in trouble by prematurely releasing data. It was made known that he would be extremely angry with anyone who talked about 3.1 GeV outside the group.

Ting was unsure of what to do next. The experiment had ended its scheduled run, having logged only 470 hours of the 1,000 approved hours of beam time. To get more beam time, which would give him more data and allow him to test his results, he would have to wait. The AGS maintenance break was scheduled to continue until October 2, and when the machine was switched on again fast extraction experiments would run. After that, he would have to jockey for beam with the other slow extraction experiments, including 614, Schwartz's experiment. It would be at least six weeks before Ting could run again. He was in an exquisite dilemma: Experimenters must push their equipment to the limit, and announce their findings before others get there first. But they thereby court the risk of overestimating their equipment—a lesson Ting knew well, having established his reputation because of Pipkin's inability to recheck his results.

Against the advice of some group members, Ting decided to wait. In the meantime, he wrote a polite letter to lab officials asking for more time, and followed it up with a less polite phone call.[28]

"If I may make a bad example," Becker said, "suppose you go out with a net to catch a butterfly, and you catch an elephant in it instead; how would you move? The answer is, damned carefully, so the net doesn't break!"

Ting's decision was risky principally because SLAC had a new accelerator, known as SPEAR, that could discover the spike in a day—if the operators knew where it was. Whereas the e^+ e^- pairs in Ting's experiment were by-products of collisions between protons and nuclei in a target, the Stanford Positron-Electron Accelerating Ring used electrons and positrons as the beam itself; Ting was painfully aware of the consequences entailed by this difference. Being the shrapnel from explosive impacts, the electrons and positrons in Experiment 598 shot out with a wide range of energies, some flying away violently, others dribbling out to the side. The detector could thus survey a broad energy spectrum, but it could not focus precisely on any given value because the electron-positron pairs came out more or less randomly. Only Ting's obsessive insistence on fine-tuning the detector to pick up every detail had allowed the group to find the spike at all. Indeed, an earlier experiment with a less precise detector had found only a puzzling "shoulder" in the 3–4 GeV energy range.[29]

SPEAR, on the other hand, circulated electrons and positrons around a ring for hours at a time, pushing them to ever-greater energies. The electrons traveled one direction about the circle, the positrons the other; they met in the middle and destroyed each other, producing minute, savage flashes of energy that turned into other particles. (Today we would say that the energy, which is in the form of photons, "promotes" to the status of real particles the virtual quarks and leptons in the vacuum.) Such machines were built because accelerator physicists realized that accelerators might be able to reach higher energies by letting sprays of particles slam into each other instead of into stationary targets. The difference is the difference between driving a car into a motionless tree and having a head-on collision with another vehicle; the head-on is much more violent.

Controlling the particles in their course around the ring was impossible unless they all traveled with the same momentum. By contrast, imagine trying to guide a mix of speeding race cars and lumbering vans around a narrow race track. Because there was little range in the momentum of the electrons and positrons, the energy—or, alternatively, the mass—of the produced particles was restricted to that narrow range. (The reason is the conservation of energy.) Thus, if each beam was given an energy of 1.5 GeV, for instance, the collision energy of the two beams was 1.5 + 1.5 = 3.0 GeV, and only events of that specific energy could be seen. To switch from 3.0 to 3.2 GeV, physicists had to turn off the whole accelerator and reset the magnets and injection beam so that the beams each had an energy of 1.6 GeV. The result is that the new SLAC machine was a superb probe for any given energy, but useless for sweeping broad ranges. Because the designers of the machine had thought in terms of broad bumps, rather than spikes, the limitations of the machine became apparent mostly in retrospect.

Hence the MIT team's fears: If SPEAR was set to a collision energy of exactly 3.1 GeV, the machine would find the spike instantly. The machine operators need only hear the magic number 3.1, and X would mark the spot. If they sat at 3.2 GeV, they would miss it. During the debate over whether to announce the peak, Chen called an MIT colleague who was visiting Stanford and ever so casually asked what energy SPEAR was running. When he heard that it was skipping between 4.0 and 6.0 GeV, Chen breathed freely; SLAC was in the wrong terrain, and had missed it.

Completed in April 1972, SPEAR was the culmination of a decade-long struggle by Burton Richter of Stanford. Initial speculation about colliding beams centered on beams of protons; Richter, on the other hand, thought of electrons and positrons. In the late 1960s, the Italians built ADA, a tabletop model of an electron-positron ring that was so tiny that it was filled by shooting in a burst of electrons, then flipping the accelerator upside down and firing in the positron beam through the same port. The Italians then built a bigger version, as did Russian and French physicists. Throughout, Richter pushed fruitlessly to have an $e^+ e^-$ machine in the United States. He was lucky: SPEAR eventually caught the fancy of John P. Abadessa, then the controller of the Atomic Energy Commission, who figured out a bit of bureaucratic financial legerdemain to smuggle the accelerator past Congress as a particularly expensive experiment. And, in fact, the ring was pasted onto the extant SLAC accelerator for a price tag just two million dollars higher than Ting's detector alone.[30]

After having spent much of his career struggling to build SPEAR, Richter was the obvious choice to take the first crack at using the ring. A plainspoken, easy-going man, he put together a group from Stanford and Berkeley that worked closely and informally on a preliminary sweep of the energy domain. "We wanted to look for new phenomena," Richter said. "And what we decided was to start at 2.4 GeV and move in steps of 200 MeV. So we went 2.4, 2.6, 2.8, 3.0, 3.2, and at that time we could only get up to about 4.8 or 5 GeV. Because we were not—nor was anyone else—thinking of a new kind of quark in a very narrow resonance." (Resonance, again, is another name for an unstable, strongly interacting particle.)

The SPEAR group was primarily examining a ratio known in the jargon as R, which is, roughly speaking, the number of hadrons divided by the number of muons produced in electron-positron annihilations at a given energy. Ordinarily, high-energy $e^+ e^-$ collisions make mu-antimu pairs, but if the energy is equivalent to the mass of a hadron, showers of mesons and baryons can pop out. A sudden jump in R thus indicates the existence of a new particle. From the first, there were odd things about the results. As early as January 1974, a SPEAR team member named John Kadyk first noticed that the value of R for the run at 3.2 GeV was about a third higher than expected.[31] After the completion of the first series of runs, in June, another team member named Martin Breidenbach followed up on Kadyk's observation by taking data around that point, at 1.55, 1.6, and 1.65 GeV per beam—that is, 3.1, 3.2, and 3.3 GeV total energy, respectively—to check whether this was a machine error; he also took additional data in the area of another unusually high spot, around 4.0 GeV total energy. At a first glance, everything seemed normal. "A preliminary analysis of that data was done," Richter said, "and nothing very dramatic showed up. The group had been working twenty-four hours a day, seven days a week, for more than nine months, and we all collectively collapsed for the summer." SPEAR shut down in July for a three-month break. When the accelerator was switched on again, some members of Richter's team began detailed analysis of Kadyk's and Breidenbach's data, while others set about retuning the machine to run at a higher energy, 4.0 to 6.0 GeV, hoping to find something there. To Ting's relief, they spent October hopping around that energy region, far away from 3.1 GeV.

Back on the other coast, Ting's frustration increased when the AGS was restarted on October 2, for problems developed immediately with a newly developed experimental magnet in the fast extraction beam line, and the AGS had to shut down for a week to fix it.[32] Even when, on October 11, the machine was switched on and fast extraction experiments began, the new magnet continued to cause delays.

The wait made Ting rethink his decision to postpone the announcement of what he was now calling the J particle. He had the chance to make a dramatic announcement on October 17 and 18, when MIT hosted a festival to honor the retirement of Victor Weisskopf. A full-blown bash celebrating one of the university luminaries, the gala gave Weisskopf, an enthusiastic amateur musician, his chance to conduct the Boston Symphony. If Ting used the occasion to announce the existence of the narrowest resonance ever discovered in an energy region that had been thought previously to be well understood, the splash would be enormous. He didn't do it. He knew he was sitting on something very odd, and odd results are usually incomplete.

When Becker gave a previously scheduled seminar on Experiment 598 at MIT on October 22, he had to disguise the narrow peak, which he did by cleverly presenting the number of collisions over a sufficiently wide energy range to smear out the peak. Martin Deutsch, head of the MIT's nuclear physics laboratory, was not fooled. A kind of mentor for Ting, Deutsch had gone out of his way to arrange funds for Experiment 598, and had followed its progress carefully. He glanced at Becker's graph, and mentally spread out the data, differentiating what Becker had integrated. Taking Becker aside, Deutsch told him to publish immediately. Becker said he would, once Ting was ready, and asked Deutsch to help expedite publication once it came time. Fancy mathematical skills were not needed to figure out Ting's team had something, Deutsch said later. "All you had to do was to take one look at Min Chen's face."[33]

That same day, October 22, the AGS was finally switched back to the slow extraction experiments, and Ting got his chance to check out the narrow peak. "There were many, many tests, okay?" Ting told us. "The first is to change the thickness of the target, let's say by a factor of two, and see whether the counting comes off by a factor of two. If it's a scattering from the side of the magnet or something, it won't. Or change the magnet current by ten percent. A true peak better show up in the same place, and not move. Or change both magnets to positive polarity, and see what happens. Plug up the magnet to a smaller aperture, and see whether the rate changes. All kinds of tests. We were constantly going in and out of the experimental area, which is doubly sealed because of the radiation; and we drove the Brookhaven people completely crazy."[34]

Ever meticulous, Ting had laid in a reservoir of good will with the AGS control staff by constantly showing his appreciation of their round-the-clock care for the machine. Now they repaid his concern by whipping the AGS through its paces at top speed and with great care.[35] The narrow peak didn't disappear. Chen kept calling his friend at SLAC.

Newly returned from his first trip to China, Mel Schwartz of Stanford, the head of the group that shared Ting's beam, stopped at Brookhaven on October 22 to check the progress of his experiment.[36] Schwartz's assistant, Jayashree Toraskar, filled him in on the progress of Experiment 614, and then relayed to him some exciting news about Experiment 598 that she had heard from a member of Ting's team: There was a bump at 3.1 GeV.[37] Schwartz walked over to congratulate Ting.

The encounter was a classic clash of opposites. Husky, brash, an all-American go-getter who thought of frankness and openness as among the cardinal human virtues, Schwartz cheerfully asked Ting, a private, driven, careful man, about a secret he was not ready to announce and had energetically tried to keep to himself. Moreover, the result was a spike on the kind of accelerator that wasn't supposed to be able to detect them, and Schwartz was on his way to SLAC, where they could find it instantly if the word got out. In Schwartz's recollection, the following conversation ensued:

SCHWARTZ: *Sam, I hear you got a bump at 3.1.*
TING: *No, absolutely not. Not only do I not have a bump, it's absolutely flat.*
SCHWARTZ (insulted by Ting's caginess): *I'll make you a bet. Ten dollars you got a bump.*
TING: *Absolutely. I'll bet.* (They shake hands. Schwartz is furious. Ting goes to his trailer and hangs up a notice saying, "I owe M. Schwartz $10.")[38]

Long after the event, Schwartz explained to us the reason for his annoyance. "If somebody came up to me and said, 'Mel, I hear you got a bump,' I would say, 'Look, I got one, but I really wish you would not discuss it, because I don't honestly know for sure that I got anything yet. OK? Since you somehow or other found out in one way or another, let me tell you what it is. *This* is what I think I see, *this* is why I don't quite believe it, *these* are the tests I'm going to do. That's the way a rational human being reacts to somebody coming in there and saying, 'Mel, I hear you got a bump.' Now, you realize that Sam Ting, *if* he had leveled with me at that point, *if* he had shown me the thing, *if* he had talked to me in those terms, there wouldn't be *any issue* as to who discovered what. He would have got full credit. By doing what he did, he put himself in the position where he ended up having to *share* it. For what did I do? I flew back to SLAC that very same day, and I said to my guys, 'Hey, I just made ten bucks.' 'What do you mean, you just made ten bucks?' 'I just made a bet with Sam Ting that I'm guaranteed to win.' 'What's the bet?' 'Sam Ting tells me he hasn't got a bump. But I know he's got a bump at 3.1 because his people have told me he's got a bump.' "

Three days later, on October 25, Deutsch telephoned Chen to ask whether the results Becker had presented in the seminar had been confirmed by the recent data. Chen hedged. Growing angry, Deutsch began to bellow at Chen that he had better publicize whatever it was he was sitting on before SPEAR got to it. Loyal to his boss, Chen didn't say anything, but when he finished talking, he pressed Ting to publish.[39]

One member of the SPEAR analysis team was Gerson Goldhaber, a good-natured, able physicist whose office is in the Lawrence Berkeley Laboratory, high on the hills overlooking the San Francisco Bay. We drove up there to meet him one morning in February, when Goldhaber had just returned from a skiing trip; his features above his full beard were red from the winter glare. Crowded by books and journals, a self-portrait was hung near the ceiling; foot-thick stacks of papers blocked every chair. Taped to the side of his desk was a fortune cookie prophecy: "You will finally solve a difficult problem that will mean much to you." A bumper sticker on the wall proclaimed: "Physicists do it with charm." Two empty champagne bottles, souvenirs of charm, rested atop file cabinets. Ultimately, Goldhaber played a big role in directing SLAC to 3.1 GeV, and he did it for reasons that, in retrospect, are entirely incorrect.

"In October," he said, "Roy Schwitters [a member of the SPEAR group now at Harvard] decided that we needed to write a paper on this experiment, and he started to work on it. To write a paper you have to look at the data carefully. And then he noticed that there was a very funny point; at 3.1 GeV there was some inconsistency in the data." Although the second pass through the 3–4 GeV region had not picked up a bump, the pattern of decays was a little odd. Schwitters told Goldhaber about the funny pattern on October 22. "We wanted to publish this paper, and with this funny fluctuation you can't say that you've measured something." Goldhaber set out to find the reason for the odd behavior, and thought he saw something exciting. "It looked like there were K^0s [neutral K particles] in these events, more than one would have expected. My student, Scott Whitaker, who was also working on this data, looked at it, and he somehow thought he saw an excess of *charged* Ks in these events. And then we really got very excited because we knew about the charm hypothesis and we knew that charm should decay into strange particles! As it turned out, my observation of excess K^0s was partly statistical [due to random fluctuation] and partly some were misidentified. There was really no great excess in this data. As Schwitters put it, this was the red herring of the century. It got us thinking that we have to go back and look at this data."

Richter was out of town at that moment, giving a lecture series at Harvard. Goldhaber knew it was going to be difficult to convince him, for SPEAR had just been reset for the 4–6 GeV energy range. "And when you decide on a program," Goldhaber said, "you don't just then go and jump around and do other things. You have to retune the beam and all the magnets had to be reset. So this going back to 3.1 GeV was not completely trivial." Talking over the situation with other group members, he decided that the excess of K particles merited turning the machine back. Richter was due to return in the first two or three days of November; Goldhaber would push him then.[40]

By the end of October, Experiment 598 had fully confirmed the existence of the J, a particle with a mass of almost exactly 3.1 GeV. But something else now prevented Ting from announcing it—"He got greedy," in Chen's words.[41] If there was one unexpected bump, there might be more. Ting knew about the previous experiment that had found a gentle shoulder instead of the sharply peaking J. The shoulder seemed to be centered around 3.4 GeV; perhaps it indicated an even larger, sharper spike there.[42]

Ting was not interested in whether the new particle played in existing theoretical speculations. But some of his collaborators made clever guesses; on Halloween, for example, Min Chen casually wrote at the top of log book page 195: "Study trigger rate for charm particle."

Later that day, Ting inveigled the Brookhaven control room head "Woody" Glenn into postponing some scheduled tinkering with the AGS to give him more running time; he thanked them by bringing a large, carved-out pumpkin full of beer. On the following days he prodded them for increases in the beam intensity, and then made several changes in the targets, trying to see whether this would produce more bumps. Meanwhile, Becker, Chen, and nearly everyone else on the team was demanding that they announce the one they already had. "There is an old Chinese proverb," Chen said drily. "Something about a bird in the hand being worth more than two in the bush. Well, Ting wanted those birds, and more."

Goldhaber called Richter's office on Monday, November 4. For the rest of that week Richter, Goldhaber, Schwitters and other experimenters argued the merits of pushing forward or going back. Although Richter was sympathetic to Schwitters and Goldhaber, there were several obstacles in the way of setting back the energy. First of all, another experiment was being run at the time, which would have to be bumped. Second, Richter wasn't even sure that the machine *could* be switched back. So many improvements had been made in the machine—now called SPEAR II—to allow it to run at a higher energy that he was not sure it could be detuned. The final straw that convinced Richter to try it was Goldhaber's discovery that the K particle production was abnormally high.

Goldhaber said, "On Friday the 8th he called me up to tell me that he has figured out how we can run in that region, and we're going to try it that weekend."

The next morning, Saturday, November 9, Gerson Goldhaber called the SPEAR control room to see how the new run was going. A hastily built maze of metal shelving crammed with humming power sources, noisy computer printers, and beeping control gauges, the control room was in the center of the ring, which stood in a pool of concrete at one end of the two-mile SLAC accelerator. Richter was there—Richter was practically always there when the beam was on—and he told Goldhaber that they were going at 1.57 GeV per beam, 3.14 GeV total energy, and that R was unusually high. Goldhaber drove to SLAC that afternoon, where he discovered they had set the beam at 1.5 GeV, and nothing special was happening. He sat in front of a little computer screen and keyed the computer to show him reconstructions of events, tallying the leptons and hadrons. They were not too difficult to distinguish at a glance: Only two charged "prongs" come out of events that produce pairs of muons or electrons, whereas events that make hadrons create a barrage of particles. At 11:15 P.M., the energy was eased up to 1.56 GeV, and suddenly the number of hadron interactions jumped to twice what it had been. Richter and Goldhaber followed the interactions, with the beam being tuned ever more finely, until three that morning; convinced they had really found some sort of peak, Goldhaber retired to his motel room.

Goldhaber came back later Sunday morning to find they had tuned the beam still more finely, and that R had risen still higher. "By noon the cross section [the likelihood of producing hadrons] had gone up by a factor of seven. So I said, 'We have to go and write this up. This is just so, so unbelievable; we have to write a paper on this. This is such a fantastic effect.' And Burt agreed with me, and so I went off and started writing." For the moment, he called the particle SP(3105); SP for SPEAR, and 3105 for its precise location.

"I picked up a piece of computer output and started writing on it. In the meantime, by about one or two o'clock in the afternoon, we had got an extra factor of ten, so the increase was a factor of *seventy*. I had worked with hadronic resonances, and there you usually get a factor of 1 and ½ percent, 30 percent, 50 percent—no factors of seventy. I had never seen anything like that in my physics career!" In the meantime, someone had brought out a bottle of champagne, and celebration commenced with celerity. "There was a very funny scene," Goldhaber said, "where we went on taking data and standing around, but at the same time we started drinking champagne and having cookies."

Clad in jeans and sneakers, Richter sat at a table covered with computer output paper, a single champagne glass perched precariously atop one stack, counting hadrons and leptons. News spread across the campus, and gradually the control room became jammed. The phone calls to other laboratories began. A simmer of euphoria, the clean pleasure of discovery, coursed through the talk and the cigarette smoke.

Later, Goldhaber said to us, "we realized that nobody calls a particle by two letters. So I looked in my little book—" he hunted around his desk in vain for the Particle Properties Data Booklet "—oops. Today I haven't got it. You know what they look like? Okay. So I was looking through that to see if there was any letter that hasn't been used. While talking with George Trilling on the phone I came up with this—psi. I said, 'Here's a letter that hasn't been used, and it resembles SPEAR, sort of close. It isn't in use, and it's one that you could pronounce.' "

They decided to announce the discovery Monday morning.[43]

That same day, Sunday, November 10, Ting left Brookhaven to attend a meeting of the SLAC Program Advisory Committee, of which he was a member. The meeting was scheduled for the next morning, and Ting's trip had been planned for months. Just before his departure, he accidentally ran into Mel Schwartz, who asked him about the ten dollars he was owed. Irritated by Schwartz's constant pressure, Ting said he had nothing.

When Ting's flight landed in San Francisco, he was paged by Chen, calling from Brookhaven. Chen said he had completed a paper on the discovery and had inked in all the graphs, and made one last plea to submit it for publication. Ting said he understood how Chen felt, but told him to go to bed and catch up on some lost sleep.

At the hotel, Ting found another frantic message, this one from Deutsch, who had learned that SPEAR had found something at 3.1 GeV. Horrified, Ting called Brookhaven and said to prepare to announce right away; he would announce it at SLAC the next morning.

Back at the Brookhaven control room, Austin McGeary was the AGS engineer on the graveyard shift. His shift started out quietly; his log entries record the typical small difficulties of maintaining a large and complex machine:

While at Linac, Dely found a very small H_2O leak on the tygon hose on the Buncher in LEBT—It should be checked periodically. . . .

Lost a little over 2 hours starting shortly after midnight when a +15V P.S. in the linac pipe O.T. bucket crapped out. By bad luck we replaced the supply with another powermate 1521 which although checked OK on the bench would not regulate with even the load of 1 P.C. card. . . . This experience and many others of this type point up the real need for a spare locker. . . .

He was a bit surprised when two members of Ting's team asked him for the keys to the Xerox machines at three in the morning; they had to do some copying immediately, they said, and the machines had been switched off. McGeary hunted around in vain for the keys. A few moments later, the women were back to ask for the phone numbers of some scientists who lived out of town. For some hours they refused to tell him why they wanted to wake up everybody they knew.

0600 Exp. 598 has found a new particle!!! Mss Wu and Schultz have been rushing around madly trying to make Xerox copies of some data and calling outside. They are obviously very excited. Computer printout for data taken since start of this running period shows a very prominent resonance at a mass of 3.1 GeV.[44]

A few hours later, on the other coast, Ting arrived early at Panofsky's office for the committee meeting, and Panofsky told him about SLAC's discovery. Richter arrived a few minutes later, and when he heard Ting's news, said, "Sam! It's the same thing! It has to be right!"[45]

Later that day, Ting relayed the news to the director of the particle physics laboratory in Frascati, Italy, where an electron-positron collider had spent seven years running up and down between 1 and 3 GeV. Within two days, chagrined physicists there had pushed up its energy to 3.105 GeV, becoming the third team to confirm the existence of the particle. The three papers by the three groups on the new particle arrived at the offices of *Physical Review Letters* within a week and were published back to back on December 2.[46] In the meantime, on November 21, SLAC had discovered what Ting had suspected was there but which he had been unable to find—a second spike, which Richter called the *psi prime*.[47]

The particle that Ting and Richter discovered is now called the "J/psi" by everyone except Ting and Richter, who refer to it as the J and the psi, respectively. (The two men won the Nobel Prize in 1976.) The J/psi electrified the physics world for many reasons, including its simultaneous discovery on two different types of machines. Unsurprisingly, the simultaneity has often led to speculation in the sometimes claustrophobic high energy physics community that somehow word about the peak spread from Brookhaven to

SLAC and played a role in the decision by SLAC officials to turn back the machine. The most often cited possible channels for this are Schwartz's journey in late October and a visit to Brookhaven that Schwitters made in mid-September; Schwartz does not deny knowing about Ting's result, while Schwitters does. One member of Experiment 598 told us: "They were at 4.2 the week before, and they reported a rise in R. It was a new result! When you discover something new, you explore it. Why go backwards? You can *always* go backwards—next week, next month, next year. Why go backwards *then?*"

Richter's team members deny that they knew anything about Ting's result, and point to the logical series of steps by which they zeroed in on 3.1 GeV. One physicist, not on Richter's team, even angrily pointed out that Ting, in fact, had decided to announce his discovery only after hearing of SLAC's. "If you aren't ready to announce something, then you really aren't sure yet you have made a discovery," he said. Others at SLAC think that Ting would have been scooped entirely if he hadn't taken that trip to SLAC for the committee meeting.[48]

Although such arguments over priority are the private mania of physicists, they reverberate through any historical inquiry. The discovery of the J/psi was the work of two detectives, each on the trail of something, although not quite sure what it was. Each occasionally caught glimpses of the other; Richter learned of Becker's seminar at Harvard from Deutsch, members of Ting's group followed the course of SLAC; and both were immersed in the sea of rumor, speculation, and gossip that is the natural medium of physics. Still, both were astonished to find that they had converged on the same quarry at the same moment.

The argument over priority masks the difference in the two teams' contributions. Ting found the J on the AGS, a machine barely powerful enough to produce the particle; finding it was a virtuoso turn, a feat of unrepeatable expertise akin to watching a jazz musician squeeze out a beautiful solo from a battered and broken saxophone. But Ting could do little more than establish its existence. Indeed, his experiment ended where Richter's began. A superior tool for the task, SPEAR could—and did—hold the psi up to a clear light, examining it with the attentive care of a jeweler. For the next few months, the California group mapped and measured thousands of psis. Ting may have had the particle earlier, but our knowledge of its properties comes from Richter.

We asked Martin Deutsch whether he thought the discoveries were independent. "Oh, the discoveries were independent—absolutely. I'm convinced of that. But the physics—that's something else. Physics is *never* independent. How independent can you be in such a strongly interacting environment?"[49]

The announcement of the J/psi in November of 1974 is now known in the physics community as the November Revolution. Like many revolutions, its meaning was not clear at first; the unprecedented narrowness of the peak fostered a host of theoretical speculations. Some thought the new particle was the Higgs boson; others suggested that it was the long-awaited intermediate vector boson; still others hoped it might be a "colored" particle, evidence that quarks really did have integer charges and were not confined. Chen called Glashow on the morning of the eleventh. Glashow went to MIT and by lunchtime was sure that the J/psi was a meson formed of a charmed quark and an anticharmed quark, which he dubbed "charmonium."[50]

"Around six months after the psi discovery," Richter said, "it was very, very hard not to think that it was charm. Then there was a problem that left us all doubting." The problem was that if the J/psi truly were charmonium, then experimenters should turn up other mesons with charmed quarks. "We set out specifically to look for charmed mesons. It was still the same experiment—the same machine, the same detector, the same people. And we couldn't find the damned things. It was getting pretty disappointing. I was beginning to wonder about charm. Everyone was—except Shelly Glashow. Shelly is one of the most self-confident people I have ever met. He was quite sure at the time that it was the fault of the stupid experimenters. But we couldn't turn up anything. That's how matters stood when I left for CERN on sabbatical."[51]

The case for charm was only slightly bolstered when Samios and Palmer publicized their single bubble chamber photograph at a meeting in March 1975. They stuck their necks out for charm, but the Gargamelle group, which announced a tentative sighting of a charmed meson at the same congress, left the actual description of the event to a graduate student—a sure sign of unease.[52]

When Richter left on sabbatical that summer, various members of the SPEAR group continued the hunt for charmed mesons, and continued to find nothing. In April 1976, Goldhaber attended a conference in Madison, where he gave a talk about the psi. Another collaboration member reported on SPEAR's failure to turn up charmed mesons after an exhaustive search. Charm wasn't there—a message that displeased Glashow, who was in the audience. Following the conference, the two men flew to Chicago together in the same plane, an occasion Glashow seized upon to harangue Goldhaber for the entire trip. "No bets this time," Glashow said, "just imperatives. I insisted he take another look."[53]

"I told Shelly I would," Goldhaber recalled. "I would spend another month looking. Based on the reaction of the physicists I had spoken with, it would be a month well spent even if I didn't find anything. When I arrived back in Stanford, a new batch of data had just come in. I liberalized the criteria a bit, because they had been rather strict. Within three days, I found the mesons. I called Shelly, and he was elated. He knew it meant the Nobel."[54]

Goldhaber's data convinced even the most die-hard skeptics. In Paris, Iliopoulos received his bottles of wine. And at the next meson conference, the program director, Roy Weinstein (now at the University of Houston), decided to repay Glashow. Near the end of the conference, Weinstein reread Glashow's hat speech of two years before. Whereas Goldhaber, Richter, Samios, and Ting were not meson specialists and were thus outlanders, and whereas according to the terms of the bet everyone present had to eat their hats, as head of the honorable assembly he was duty-bound to hold them to it. He proceeded to pass out candied Mexican hats.

"They were extremely peppery," Weinstein said. "Not a pleasant candy. But everyone ate them with relish, and agreed that Shelly's prediction had been magnificent."[55]

□ □ □ □ □

When theorists first put forth their ideas, they frequently point out the experiments that would suffice to prove or disprove them. For a theory to be right, it must, after all, be capable of being wrong. Often, however, these crucial experiments are not the ones that finally sway the scientific community. Erroneous experimental results, theoretical blind alleys, and mistaken interpretations of each by the other can lead physics down a different path than what would seem to be the straight and logical route. A test that at one moment appears to be a mere exercise can suddenly loom as the arbiter of a theory's fate. So it was with the standard model at the beginning of 1977. Oddly enough, this decisive test was a reprise of the one that began to put $SU(3) \times SU(2) \times U(1)$ on its feet; it involved, once again, a prediction about neutral currents.[56]

The discovery of neutral currents showed that Z^0s could be exchanged between neutrinos and hadrons or neutrinos and electrons; to round out the picture, experimenters wanted to know whether Z^0s were exchanged between electrons and hadrons as well. At low energies the strength of the Z^0 effect would be minuscule compared to that of electromagnetism, and directly measuring such a small contribution was out of the question. But the weak force has a sign which talismanically identifies its presence: parity nonconservation. Experimenters considered experimental setups in which the interaction between electrons and hadrons would show an effect if parity was not conserved in the interaction, and no effect if it was conserved.

Most high-energy physicists wrote off the task as hopeless. But the task was taken up by atomic physicists, scientists who studied the interactions of

electrons and nuclei; sometimes regarded as bookkeepers by particle physicists, they would take a certain pleasure in carrying off a tabletop experiment that could do something a huge particle accelerator could not. The
general theory of "atomic parity violation" experiments was first described
by the French husband-and-wife team of Claude and Marie-Anne Bouchiat
in 1974.[57] Electrons surround the nucleus in many types of elaborate orbits
described by Schrödinger wave equations. In the case of extremely heavy
atoms, the charge on the nucleus is so strong that some of these electrons are
pulled close enough to come within the range of the weak force. According to
$SU(2) \times U(1)$, the electrons and the protons and neutrons in the nucleus
should exchange Z^0s during these close encounters. The Bouchiats calculated that the presence of the Z^0s could be detected by their effect on the
light emitted by the electrons.

The fundamental idea is this: All photons have a spin, which has a
direction that can be viewed as a combination of left- and right-handed.
Similarly, if someone tosses a ball, the ball spins as it flies through the air.
Using the direction of motion as a reference point, the actual spin can be
described as a combination of two different spins, left- or right-handed along
the direction of motion. A beam of light that is circularly polarized has all of
its photons spinning in one direction. Linearly polarized light is the special
case in which the spin of every photon is an equal mix of left or right circular
polarization. In some atomic parity experiments, linearly polarized light is
absorbed and then emitted by atomic electrons. Because the electron feels
the weak force and is surrounded by virtual Z^0s, it interacts differently with
the right-handed and left-handed components of the light wave. The outcome is that the linearly polarized light is emitted with its polarization in a
slightly different direction than it would have if the electron-photon interactions respected parity; the spin is tilted, so to speak. The angle of tilt is
proportional to the difference in absorption rates of right and left. This
emitted light in turn interferes with the incoming light, and what the experimenter actually measures is the shift produced by the interference between
these two.

One of those impressed by the Bouchiat's paper was Edward Fortson of
the University of Washington in Seattle, who had read Weinberg's 1967
paper when it first came out. "He [Weinberg] had a certain way of looking at
the weak interactions in analogy with electromagnetism," Fortson said. "I
didn't get much physics out of it, just the concept."[58] He had thought
periodically about testing Weinberg's concept for parity-violating effects for
several years, but had thought it hopeless. When someone showed him the
Bouchiats' paper in 1974, Fortson realized that he could do it.

According to the electroweak theory, the angle of the tilt in the polarization would be minute in the extreme—about 10^{-5} degrees, or roughly the

angle between the two sides of a nail viewed from five miles away. Nevertheless, Fortson felt that he could arrange his equipment to measure it. He would need a finely tunable, intense laser to provide a precise source of light. In addition, Fortson had to use an atom with a large, heavy nucleus to pull the electron close enough to the nucleus to have the weak force come into play. Only one heavy element gave off light that matched Fortson's laser: bismuth, a white, brittle metal, similar to lead in many respects and its next-door neighbor on the periodic table; in the early days of chemistry, the two elements were often confused. But using an element with eighty-three electrons was a mixed blessing, for it made the exact calculations of each orbit sufficiently complicated that it was not easy to know which approximations were justified and which would lead one astray. If one took the figuring seriously, however, $SU(2) \times U(1)$ predicted that the bismuth spectral line should have a shift of about -25×10^{-8} radians. (Radians are a scientific measure of angle: one degree is just over one-hundredth of a radian; and -25×10^{-8} radians is about 1.43×10^{-5} degrees, that is, a few ten thousandths of a degree, with the minus sign here indicating direction.) In the winter of 1975–76, Fortson's team began to perform their experiment and found a markedly lower tilt of around -8×10^{-8} radians. Fortson was interested—experimenters always like to set the theorists on their ears—but unsure whether he was seeing nothing but an error in his own calculations.[59]

Meanwhile, another team of atomic physicists led by Patrick Sandars of Oxford University had begun a similar experiment in England. Sandars received his doctorate at Oxford at the time of the discovery of parity nonconservation and had been fascinated by the subject ever since. The discovery of neutral currents had pricked his interest, and when he ran across the Bouchiats' paper he decided to try to detect a parity-violating neutral current. Like Fortson, he decided on bismuth, although he chose a different wavelength of light to measure, one for which the Weinberg-Salam model predicted a spectral shift that was also about -25×10^{-8} radians. Sandars began to get his first numbers in early 1976. They, too, were much less than predicted, about $+10 \times 10^{-8}$ radians.[60]

In August 1976, Sandars boarded a plane for the United States, where he planned to discuss atomic parity at a conference in Berkeley. He flew to Seattle, for he wanted to talk to Fortson about the latter's experiment. Fortson met him at Sea-Tac airport, and they exchanged numbers. Fortson's results were slightly to one side of zero, Sandars slightly to the other. At the time, what seemed most significant was that both were much less than predicted by the electroweak theory. Pleased, the two men continued together to California, where they announced their results in a spirit of exuberance.

In December the two groups published a joint letter in *Nature* in which they concluded that "the optical rotation in bismuth, if it exists, is smaller than the values . . . predicted by the Weinberg-Salam model."[61] The same issue of *Nature* carried a comment about the implications for unification by British physicist Frank Close, who concluded that "the clear blue sky of summer now has a cloud in it. We wait to see if it heralds a storm."[62] The storm began to break early the next year, when both Fortson and Sandars released updates on their experiments that confirmed their earlier numbers. The Oxford and Washington results appeared to dash the hopes not only of the standard model but of the entire approach to unification that had been growing ever since the renormalization of gauge field theories.

In such circumstances, advocates of unification had only two ways out—pull a fast trick or turn a deaf ear. Neither alternative was palatable. Some tried to incorporate clever ideas to make the $SU(2) \times U(1)$ conserve parity in atoms after all, but in so doing learned that the model was almost fiddle-free. "In an odd way," Weinberg said, "that made it *more* convincing. You really couldn't do anything much to it without dreadful consequences."[63] Nonetheless, some theorists tried, producing, among other things, a batch of "left-right" theories that restored parity symmetry to the weak interactions at the price of invoking the existence of something else.[64] Others hoped that the atomic parity experiments were wrong or that the atomic physicists were confused by the complexity of the bismuth atom. Certainly particle physicists were baffled; most were simply unable to evaluate the veracity of the experiments.

In March 1978 two Russians reported that they had observed optical rotation of the predicted amount in bismuth.[65] But the score was still two to one against Glashow-Weinberg-Salam, and physicists in the West had no way of evaluating the reliability of the Soviet experiment.[66] With the experimental and theoretical situation in chaos, rumors began to percolate that a SLAC version of the same neutral current experiment was in progress. Based on a logic and a set of assumptions that could be followed by particle physicists, the experiment, to their minds, was a proper experiment, done on a proper accelerator.

The idea for the SLAC experiment came from Charles Prescott, a short, precise man who got his Ph.D. at Caltech in 1966. He, too, had first become interested in neutral currents when he read Weinberg's 1967 paper. ("I didn't understand it, but I was aware of why it was interesting.") Prescott was intrigued by the way that it linked electromagnetism, which conserves parity, with the weak interaction, which does not.[67] In 1969 Prescott, then at the University of California at Santa Cruz, approached Richard Taylor, head of one of the SLAC experimental groups, with an idea for testing the general implication of the Weinberg paper that parity-violating neutral currents existed between electrons and hadrons.

The mutual respect between Taylor and Prescott is masked at first glance by the contrast in their personal styles. Taylor is burly, generously proportioned, expansive, a broad and happy man who puts his feet on the desk, scattering papers, and bids his sentences adieu with a great wave of the hands; Prescott speaks softly, in a neat monotone, hands folded in his lap, a tie closely knotted about his neck, a trim, reserved, and thoughtful player in the complex game of physics. In the parity experiment, Taylor had the role of the go-getter, the worrier about precision, the jack-of-all-trades; Prescott brought inspiration and calculation. (But as in any good collaboration the roles often switched.) Taylor had been incredulous—"We said, 'Shee-it! how you going to do *that*, Charlie?' "—the first time Prescott approached him with the idea. His participation gained, Taylor kept riding Prescott and the other members of the group to make the experiment ever more sensitive. Any effect would be little, and they had better not miss it.

Scattering particles from each other is, in a sense, the basic experiment of twentieth-century physics. In Prescott's first proposal, he hoped to demonstrate that the recoiling protons were polarized as a result of parity-violating collisions involving Z^0s. Gradually he became convinced that the difficulties in measuring this small polarization were insuperable.

At about this time a Yale group led by Vernon Hughes proposed experiments at SLAC with polarized electrons and polarized protons. (Polarized electrons, like polarized photons, have all their spins pointing in one direction.) Hearing of this, Prescott realized that the scattering of polarized electrons from unpolarized targets would test parity violation. The experiment was simple in essence, knottily difficult in specific execution. In $SU(2) \times U(1)$, electrons can scatter from protons in two ways, by an exchange of photons or by an exchange of Z^0s. Parity is conserved in the former, violated by the latter. Prescott's experiment would be performed with the spins of the electrons polarized either in the direction of motion or against it. These two configurations are related by parity reflection. If parity were perfectly conserved, the number of electrons scattered from the protons in a target would be exactly the same in the one direction as in the other; if parity were not conserved, the number would be slightly different—very slightly different.[68]

Prescott wanted to employ the Yale team's equipment to make polarized electrons, which he would measure after scattering with the same apparatus that had been used by the scaling experiments. A SLAC-Yale collaboration was formed; Prescott left Santa Cruz to join Taylor's group at SLAC.

The experiment was approved in February 1972, but events of the following year forced Prescott and Taylor to rethink things from the ground up. "It was about this time," Taylor said, "that 't Hooft showed that the Weinberg stuff was renormalizable. That increased the respectability of Weinberg by an enormous factor, and got every bugger and his brother proposing parity violation experiments. We were ahead, but we weren't alone any more." The discovery of neutral currents that next year caused theorists to scrutinize the implications of the electroweak theory in substantial detail. According to the new bout of calculations, the effect Prescott and Taylor were looking for should have a relative magnitude of about one part in five thousand, which was so small that to see it the detector would have to count a hundred times more electrons on every pass than had been planned. The experiment therefore had to be revamped; the Yale source of polarized electrons was replaced. The house expert in solid state physics, E. L. Garwin, teamed up with laser specialist D. C. Sinclair to design and build a new high intensity source of polarized electrons. A second detector was constructed from pieces of the old one. Both had to be more precise than any built before, and both had to be built quickly if the experiment wanted to play a role in the outcome of $SU(2) \times U(1)$.

For some two years, the group sweated to design tests to ensure that any observed variation in scattering was due solely to the polarization of the electrons. A second big effort went into measuring properties of the beam—position, angle, intensity, and so on—to be certain they did not change when the polarization was switched. The remodeled experiment was approved in 1975.

"There was tremendous pressure at every point in the experiment," Taylor said. "To build that source, learn how to run it, and make it work was an awesome thing to have done on such a short time scale. And then there was all of the electronics that went with the measurement itself. I mean, building electronics in such a way that you could measure this thing—you know, it's very small! It's like standing on top of Hoover [Tower, overlooking Stanford University] and watching ten thousand people streaming into Stanford every day and trying to decide whether there was one more or one less today than there was yesterday. And the ten thousand has to be that *squared* to determine that statistically. So you're dealing with a *hundred million* counts, all right? And we had to deal of course with much *more* than that. We needed to detect *ten billion* scattered electrons in order to see what was happening!"

In March 1978, a decade after Prescott's first proposal, the experiment began. Within a few days, it was clear that they had an effect on the order predicted by $SU(2) \times U(1)$. When they changed the direction of the beam polarization, the counting rate changed as well. "You'd run for awhile and be above the line [the value for unpolarized electrons], then you'd change polarization and it dropped below." Taylor held one hand above an imaginary line, and then plunged it below. "I remember that in just three days you went from wondering whether the experiment was going to work to being pretty sure that you knew that there was parity violation. That, I mean, *that's* why you do this business. That feeling of knowing something before anybody else. It's, ah, it's why you're here."

During the final runs a powerful new check was designed, of which Taylor was particularly proud. Because the accelerator beam was bent by magnets to reach the experimental target, the direction of polarization at the target depended on the beam energy. The bending of the electrons caused the polarization to sweep around in direction as the energy changed. "When the energy was changed by twenty percent, the polarization would go from pointing *that* way"— Taylor pointed one way with his hands —"to pointing *that* way. This would reverse the effect that we were supposed to see in the scattering. But it would *only* reverse it if it were due to polarization."

After the experimental runs were over, Prescott and Taylor insisted that the collaboration spend three months rechecking the data before they told anybody what the numbers were. Taylor said, "We were really worried about credibility. We didn't want to come out with a preliminary number and then change it in three months. I mean, we wanted to give one number as the result of the experiment and have that number sit there. We wanted to do the analysis before we talked. So we had to give Charlie three months."

He was asked whether in retrospect the group had worried too much about nailing down the last decimal places. Had they been too precise? The question made Taylor's jaw pop open in stupefaction. A long silence ensued as he groped for an articulable response. Finally, he exploded, "You *never* know the answer to that question! I mean, because you *don't* go ahead and do the—I mean, we *didn't* do a sloppy experiment to find out whether it was going to make trouble [for the atomic experiments]! You do the *very best* you can!" We wondered if any part of the group's concern for precision had originated in the vicissitudes of the numbers from the atomic parity experiments. He said, "Those experiments looked okay. On the other hand, the theory that went with those experiments was not in very good shape. So the fact that they were getting lower answers—it wasn't clear whether that was because they didn't know their bloody wave functions or because there was no effect. Now, some theorists were saying that there is no parity violation. They must have had little to do in those days, because many of them went off and all those left-right theories just *grew*—I mean, there were *acres* of them, like spring flowers." (Later, when the theories were no longer needed, some people referred to the "SLAC massacre" of the left-right theories.)

The three months were over on June 12, 1978. In his quiet, careful way, Prescott stood before a packed auditorium hall at SLAC and took the crowd through the polarized electron experiment step by step, methodically ticking off the checks, the backups, the alternative procedures, the theoretical figuring, the experimental run-throughs. All doubts vanished; after years of un-certainty, the standard model had been proven by an experiment dis-tinguished for its novel approach and delicate execution.

Years after Prescott's exposition, Taylor still savored the memory. "No-body, I mean *nobody*, could see any way in which the answer could be anything other than the answer that Charlie had put on the board. Every time you looked for a flaw, it was clear that we had covered it from one side or another, and the answer was just boxed in. And you couldn't—it was just *there*." Taylor sighed. "It was quite a moving thing. I mean, Charlie had spent ten years of his life and of many of the rest of us—the life of the group, which was my responsibility, had been poured into the experiment for at least three years at full strength. Now it was working."

At the end of Prescott's talk, there was applause, as there is at the end of many lectures. But witnesses recall this show of hands as different: sub-dued, a soft pattern that went on and on, white and powerful and slow, until it filled the room like a fall of water. They were applauding the experiment, the craft and dedication of its executors, but the sound was more than that: It was a recognition of accomplishment; the physicists were quietly saluting themselves—the long, elegaic salute given to the end of an age. When the clapping died down, Prescott asked for questions. There were none.[69]

VI

Unification

19
The Beginning of Time

"AFTER THAT EXPERIMENT BY PRESCOTT AND TAYLOR," FRANK WILCZEK SAID with admiration, "it was no longer a question of changing $SU(3) \times SU(2) \times U(1)$, but of explaining it."[1] In hand were theories of the strong, weak, and electromagnetic interactions—the three elementary particle forces—each written in the same language, and all shown to be in accord with experiment. No infinities or anomalies troubled the waters; experiments poked lights into corners, lifted up carpets, and found nothing amiss, no unexpected cracks—everything in apple pie order. Save for the embarrassing absence of the W and Z, which were too heavy to be produced in extant accelerators, $SU(3) \times SU(2) \times U(1)$ seemed by 1980 to be marvelously robust. (Three years later, Carlo Rubbia's team found the W and Z exactly where predicted, after a long and expensive search with a new accelerator at CERN.)[2] A new orthodoxy, in Bjorken's phrase, the standard model was suitable for rapid inscription into textbooks.[3]

Few things seem as dull to theorists as solved problems, although the definition of "solved" varies from one physicist to another. Even as experimentalists spent the early 1970s struggling with charm and neutral currents, some theoreticians began to relegate $SU(3) \times SU(2) \times U(1)$ to the status of an old, and therefore uninteresting, idea. They started to look for a reason that the trio of elementary particle interactions possessed such consonant structures. An obvious hope was that the similarity indicated some deeper, unifying principle, as yet undiscovered, from which these theories sprang. In the usual practice of contemporary physics, by the time that experimental verification of the standard model had progressed to the point where such speculation became the rational and logical step to take, less methodical physicists had already taken it, however haltingly, for other reasons.

Fittingly, the first attack on unification was made by the generation of physicists that rose to prominence in the 1950s, Glashow, Salam, and Weinberg among them. They did not realize the full implications of initiating such a search, for it continues to the present day and is still transforming the practice of physics itself. Indeed, in the view of some worried scientists, the current full-tilt push toward unification has more chance of derailing physics than accomplishing its goal.

In the spring of 1984, we met Abdus Salam at midday in a New York City hotel; he had promised to talk of unification—at length and carefully, in a quiet room where he could think. Salam was in town for the annual convention of the American Association for the Advancement of Science, a huge affair that seemed to fill the midtown area with laboratory dwellers on holiday. Scores of seminars on every possible subject progressed simultaneously, followed by bewildered members of the press and an amazing number of ordinary citizens with strong convictions about science. Salam stood out in the bustling Hilton lobby, a zone of calm in a billowing brown suit. He was surrounded by representatives of the Pakistani community in New York. It developed that an elaborate luncheon had been arranged for him. Unfortunately, the arrangers had neglected to inform the guest of honor. As a compromise, we were invited along, riding in a limousine at breakneck speed to a Pakistani restaurant near Gramercy Park.

A room had been set aside and specially decorated for the occasion. The banquet table was reserved for Salam and the men, while the veiled women sat to the side. Cooks had been working since the day before to prepare the many kinds of rice, beautifully colored by aromatic spices, that lay molded in bowls across the table. Course after course appeared from the kitchen; the finale was a great heap of soft candies. Salam was given no time to eat; besieged by books to sign, hands to shake, babies to kiss, and youths eager for career advice, he spent the meal administering to each entreaty with unflappable reserve. He was asked to pose for photographs with every man in the room. Batteries of cameras at the ready, three reporters from three different Pakistani papers came to interview Salam, sequestering him for half an hour of popping flashes and whirring tape recorders. "This happens all the time," Salam said afterward, in his hotel room. "It's extraordinary, the number of Pakistanis who have arrived in different places. It's practically impossible to get any physics done." He propped himself sleepily on the bed. "One always eats too much at these affairs," he said.

The 1971 renormalization of Yang-Mills theories came as a godsend to Salam, whose reputation was then at a low ebb. He had been identified, somewhat unfairly, with a program to mix the eightfold way and spin into a symmetry called, variously, SU(6), U(6,6), and SU(12). Great hopes were invested in this program; when it was thoroughly debunked, Salam said later, he was demolished.[4] A field theorist who had lost the respect of even the few other field theorists, he was elated at the sudden and unexpected surge of interest in the old SU(2) × U(1) model he had developed with John Ward. We asked if he had been inspired by 't Hooft's work on gauge fields to look again into using them for unification, as he had once tried in the early 1960s.

"No." Flatly. "I think 't Hooft's work was very nice, but one took it for granted. Even if it had not come, we wouldn't have stopped. If I think that an idea has some sense, then I always act on the premise that the technical difficulties have *already been solved*."[5]

A consequence of such an attitude is that Salam moves ahead of the wave, skipping from idea to idea, embedding half a dozen notions in omnibus papers whose erratic quality often infuriates the more formally inclined. In Isaiah Berlin's term, he is a fox, rather than a hedgehog, whose thought is "centrifugal rather than centripetal"; turbulent, multileveled, and sometimes self-contradictory; seeking to answer many questions in many ways at the same time.[6] Emblematic of his restless and fertile progress is his development, with a colleague, Jogesh Pati—in the summer of 1972, before neutral currents, before charm, before the parity experiments—of the first steps beyond $SU(3) \times SU(2) \times U(1)$.

That summer, as the Gargamelle group held its first formal debates over neutral currents, Salam was joined at his Trieste institute by Pati. Longtime acquaintances, Pati and Salam had first met in the late 1950s, when Pati was finishing his doctorate at the University of Maryland. Pati was born in Orissa, on the Bay of Bengal, and came, in 1957, to the United States, where he has remained ever since, except for the odd year teaching in Trieste or India. Upon his return from a year in New Delhi, he quite naturally stopped in Trieste for a few summer months by the Adriatic Sea.

At the time, a good proportion of the theoretical energy in particle physics went into examining $SU(2) \times U(1)$ and the apparent absence of weak neutral currents; more would soon be used to erect quantum chromodynamics. Like Salam, Pati felt (as he said afterward) "that the heart of the matter [lay] somewhere else. Even if $SU(2) \times U(1)$ was eventually borne out—at least at low energies—fully, by experiment, it seemed to us, in seventy-two, that there was a lot of arbitrariness in the choice of gauge interactions. . . ." In contrast to most theorists, they were dissatisfied with the theory. Although it accounted compellingly for weak interactions it did not shed light on other, perhaps more fundamental, issues. They did not want to ask questions about the details of weak, strong, and electromagnetic forces. They wanted to know why there were three interactions, and not one or two. Why, they asked, is electric charge quantized? Why do quarks and leptons both have spins of one-half if they are so different? Why, in fact, does nature use quarks and leptons, and not just one or the other? The first answer they came up with was that quarks and leptons were somehow the same thing.[7]

"One of us said, 'Why don't we assume that?' " Salam recalled. "And the other said, 'I was just thinking the same thing, but do you really believe it's true?' And we examined it further, tabulating all possibilities; and then for a number of days we were just dazed by the audacity of it. This was 1972, remember." He shook his head in amusement. "So early, summer seventy-two! The audacity of it! That quarks and leptons are not two distinct types of matter!"

Paradoxically, Salam's unfashionable predilection for quarks of integral charge was of decisive help. "I think there is something unaesthetic in putting particles together which have one-third charge and charge one in the same multiplet. Wouldn't you agree? If you tried to teach a child that these were the same particles, he would say, 'What the hell? You have charge 1, 0, -1, and then you suddenly have $-\frac{1}{3}$, $\frac{2}{3}$, $-\frac{1}{3}$, $\frac{2}{3}$—what's going on?' It sounds very ugly. The formalism does allow you to do this, and we said so in our paper. But if you have all integer charges, then you are in a better—well, you can even teach it to a child."

Early enthusiasts for color, Salam and Pati had also dreamed up something quite like quantum chromodynamics, but without asymptotic freedom and with integral quarks. They also believed in charm. By putting all these ideas into a box and shaking them, they came up with a theory that bears all the defects of great originality. At once a marvel of prescience and a jury-rigged tower of speculation, the Pati-Salam model is brilliant in general architecture, but is stuck together with shoe polish and string around the cracks. They realized quickly that the strong interactions change the color of quarks, whereas weak interactions change their type—their "flavor," in the jargon. To match the four known leptons—electron, muon, and two types of neutrino—they brought in charm, the fourth quark. And to put quarks and leptons together, they baldly asserted that leptons were nothing but a fourth quark color, lilac, split off from the rest by some cosmic symmetry-breaking mechanism invoking a set of no less than fourteen Higgs bosons. The full array fit into a four-by-four matrix for color and flavor, or $SU(4) \times SU(4)$. They called the theory "electronuclear unification."

	red (r)	blue (b)	green (g)	lilac (l)
up (u)	u_r	u_b	u_g	u_l (electron neutrino)
down (d)	d_r	d_b	d_g	d_l (electron)
strange (s)	s_r	s_b	s_g	s_l (muon)
charm (c)	c_r	c_b	c_g	c_l (muon neutrino)

Although both men wanted to explore their pet notion further, the work was interrupted by Pati's return to Maryland and a trip by Salam to China. In Beijing, Salam was asked by Benjamin Lee to speak at the September 1972 conference for the opening of Fermilab. His schedule, as ever, tightly packed, Salam could not make it to the United States in time for the session on weak interactions, but he did send ahead a rough outline of the unification theory. James Bjorken briefly described it, without comment; the idea was obscured in the blizzard of theoretical excitement about gauge theories.[8]

Pati and Salam worked more on the idea during the fall, but it was difficult to make headway while on different continents. A disagreement arose over how to handle a trouble they came across in the course of working through their model. As a general rule, particles in the same family can change into each other. Putting quarks and leptons together thus tends to imply that a quark could become a lepton, and vice versa. If a quark could turn into a lepton, then one of the quarks inside, say, a proton, might suddenly turn into an electron or neutrino. The proton would then cease to exist. In other words, unifying quarks and leptons was tantamount to predicting that the basic building block of atoms was unstable. This meant that all ordinary matter, everywhere, would eventually disintegrate. Indeed, the two men had to worry about why any atoms were left.

"It was obvious the thing would happen," Salam said. He did not recall any particular moment when the realization had sunk in. "But we shouted at each other over whether we should put it into this paper or not. In fact, Pati was more impatient, and I was more cautious—in this respect. I said we should work it out properly before writing it up." Eventually the two men compromised by mentioning it in a whisper. They came up with four unified models, singled out the simplest, $SU(4) \times SU(4)$, and discussed proton decay in the appendix for the other three. They admitted that their proton lifetime was "ridiculously low," but thought that proton decay could be avoided in $SU(4) \times SU(4)$—incorrectly, as they later realized the next summer at Trieste they worked out the model. Bad news: Proton decay was inevitable, and the lifetime must be in excess of 10^{28} years.[9]

A first paper was sent off to *Physical Review Letters*, which gave it a polite brushoff. Saying that the paper was anything but urgent, they suggested that a longer article be published at leisure in the *Physical Review* proper. Annoyed, Salam telephoned Samuel Goudsmit, an old friend and the chief editor of the whole *Physical Review* complex, who overruled the referee and printed the paper quickly.[10] (A longer article did appear later in the *Physical Review*.)

Pleased, Salam flew to Paris in August 1973, where he took a train to a meeting in the south. In his Nobel address, he evoked the scene:

I still remember Paul Matthews and I getting off the train at Aix-en-Provence for the 1973 European conference and foolishly deciding to walk with our rather heavy luggage to the student hostel where we were billeted. A car drove from behind us, stopped, and the driver leaned out. This was Musset whom I did not know well personally then. He peered out of the window and said: "Are you Salam?" I said "Yes." He said: "Get into the car. I have news for you. We have found neutral currents."[11]

Neutral currents dominated the Aix-en-Provence conference. Having seen the first HPW preprint, Weinberg said with cautious enthusiasm, "[T]here is now at last the shadow of a suspicion that something like an SU(2) × U(1) model . . . may not be so far from the truth."[12] T. D. Lee asked Salam for his thoughts on the momentous discovery. To the exasperation of the experimenters who had spent more than a year worrying over the interpretation of bubble chamber photographs, Salam blithely ignored neutral currents—they were a settled issue for him—and extolled the virtues of looking for proton decay. "Nobody took it very seriously," he said. "I think it was so outrageous that . . ." He made a gesture of shrugging indifferently.

That December Salam again boosted proton decay at a conference at the University of California at Irvine. In the audience were Feynman, Gell-Mann, and Glashow. There, too, the reaction was unenthusiastic, except for Frederick Reines, one of the experimenters who first tested for the presence of the neutrino. Concurrently with his continuing research into neutrino behavior, Reines had measured the proton lifetime by looking in his detectors to see if any of their protons had decayed. He arranged for Salam, who was flying to meet Pati in Maryland afterward, to be a dinner guest of William Wallenmeyer, the director for high-energy physics of the U.S. Department of Energy. Salam was to take Pati along, and the two were instructed to sell Wallenmeyer on the idea that the DOE should fund a proton decay experiment. "I still remember that we paid for the dinner out of our own pockets!" Salam said. "So the whole thing just drifted in the hole. Nobody took it very seriously."[13]

At the Irvine conference, Salam also learned that he and Pati were not alone in trying to unify the three particle forces. Sheldon Glashow and Howard Georgi had come up with a unified theory that not only put quarks and leptons in the same family but described all the elementary particle forces with the same coupling constant to boot.

A third-year postdoctoral fellow at Harvard, Georgi began collaborating with Glashow almost as soon as they met. Glashow thought their talents were perfectly complementary. Sitting in his office he would spin out one idea after another, and snarl in mock fury as Georgi shot them down. "He comes at you in the morning with ten ideas he's had since the day before," Georgi told us. "You have to tell him what's wrong with them, which means you have to have this kind of automatic computer in your brain to spit back responses. But if you can't figure out what's wrong with an idea right away, then you go to work on it."[14]

Although to outsiders the collaboration can seem anything but harmonious, Glashow and Georgi have enjoyed it immensely. In 1973 and 1974 alone they wrote four important papers together, one of which set out the first grand unified theory.

When SU(2) × U(1) came together, Gell-Mann has recalled, he was fond of needling his colleagues by informing them that it was not a fully unified theory. "It's a *mixing*," he said. "That's why you have a mixing angle, because the electromagnetic and weak forces are not really unified. A truly unified theory would show how both the weak and the electromagnetic forces are different aspects of the same interaction."[15] A real unification theory, in other words, would not stick together two groups like $SU(2) \times U(1)$, but embed both in a single, larger structure.

The notion that the weak and electromagnetic forces could be *directly* unified occurred to Glashow on the day in October 1973 when he first heard about $SU(3)$ of color, the gauge theory of quantum chromodynamics. He saw immediately the rough outline of the theory that he was looking for. He needed to find a group large enough to contain both $SU(3)$ and $SU(2) \times U(1)$ as constituents. Such a group would permit him to put quarks and leptons into one family and would allow for a unified description of the strong, weak, and electromagnetic forces. The obvious difficulty was that these forces didn't *look* as if they could be unified, inasmuch as the strong force is a hundred times more powerful than electromagnetism, and both are enormously stronger than the weak interaction. Somehow the differing strengths would have to be reconciled.

The relative strength of a force between two objects depends in part on such variables as their masses and the distance between them. Above all, it depends on the coupling constant, which describes the force's absolute strength. To produce a truly unified theory of the strong, weak, and electromagnetic forces, one would have not only to put the quarks and leptons together, but find some way to describe all three elementary particle forces by means of a single coupling constant. Glashow laid all this out for Georgi, who is an expert model builder, and the two sat down to work out a theory.

Georgi knew of the two papers by Pati and Salam, and so was prepared for the idea of quarks and leptons in one family, and proton decay. "The introductory remarks in one of them is really extremely prescient," Georgi said, "and more or less right on. They didn't quite understand or connect the $U(1)$, but they understood almost everything, and then the rest of the paper is not really much fun because it's an explicit model that doesn't do the things that they clearly wanted it to." Georgi laughed. "You see, the reason that they didn't get, and perhaps don't deserve, more credit for all of this was that they were at the time extraordinarily confused about the nature of the strong interactions. At the time, Abdus's primary concern was the Quark Liberation Front. He hated quark confinement and would go around with these buttons that said 'Free the Quarks.' However, if you go back and look at the group theoretical structure, you'd see that it can be translated into something more modern. You don't have to do this awful thing to the quarks. And then it really does look like the precursor of grand unification."[16]

They set out to find a fully unified model; some idea of the naturalness of the step may be indicated by the fact that it took them less than twenty-four hours. (By contrast, Einstein spent half a lifetime on his unified field theories.) It was a day that Georgi remembers well. True to form, he and Glashow spent the afternoon arguing furiously about how to proceed. Unable to hammer out a solution, each left Lyman Hall in some distress. Later that evening Georgi sat down at his desk at home to work on the problem further.

"I first tried constructing something called the SO(10) model, because I happened to have experience building that kind of model," Georgi said. "It's a group in ten dimensions. The model worked—everything fit neatly into it. I was very excited, and I sat down and had a glass of Scotch and thought about it for a while.

"Then I realized this SO(10) group had an SU(5) subgroup. So I tried to build a model based on SU(5) that had the same sort of properties. That model turned out to be very easy, too. So I got even more excited, and had another Scotch, and thought about it some more.

"And then I realized this made the proton, the basic building block of the atom, unstable." Georgi, like Salem and Pati before him, realized that he could not avoid proton decay, although for a different physical reason.[17] If the proton was unstable and would eventually fall apart, so would all atoms—and thus all matter. If the model on his scratch pad was true, then the Universe would ultimately disintegrate. He said, "At that point I became very depressed and went to bed."

The next morning he arrived in Glashow's office with "some good news and some bad news." The good news was SU(5), of course, and the bad news was proton decay. His colleague didn't take the bad news hard. "It wasn't shattering," Glashow said. "I mean, we know the sun will burn out in a few billion years. This is *known*. It's a *fact*. Spaceship earth and all that—poof! That matter falls apart a long, long time afterward is scarcely an upsetting idea. It's bad enough as it is."[18]

Glashow *was* troubled by the question of why the protons in the Universe had not already decayed, given SU(5)'s implication that they could have. The two men raced upstairs to the Harvard physics library to look up what Reines had found to be the minimum lifetime of the proton. The figure was 10^{27} years, trillions of times the present age of the cosmos.[19] Georgi and Glashow sat down to figure out how to make the theory predict that protons would hold out that long.

"My immediate reaction was that we had to make the virtual particles that bring about proton decay very heavy," Glashow said. "That meant they wouldn't appear very often. This we could do thanks to Steve, who had taught us in his 1967 paper how to give the intermediate vector particles a big mass. Only in our case they had to have a *much* bigger mass—a thousand trillion times bigger than anything that had ever been seen." In fact, the proton-decay particle was later calculated to have a mass of almost a billionth of a gram—nearly heavy enough to be weighed on a scale.

Glashow loved the idea of monstrous subatomic particles. "It was another way for him to throw rocks at the establishment," Georgi said. "Nobody we knew had ever even *talked* about elementary particles that heavy. Not even within twelve orders of magnitude of that."

Georgi and Glashow sent SU(5) to *Physical Review Letters* in January of 1974. They had decided not to be shy about what they were doing, and had given the paper the most imposing name they could think of: "Unity of All Elementary-Particle Forces."[20] Only three and a half pages long, the paper linked nearly all of the significant discoveries in quantum field theory that had been made in the past quarter-century. The mathematical techniques it employed came from group theory and Yang and Mills; articles by Gell-Mann, Glashow, Higgs, 't Hooft, Salam, Schwinger, Weinberg, and others were cited. Quarks, symmetry breaking, and gauge fields played a role. All were staples of the theoretical larder; Georgi and Glashow were simply the first to combine them into a meal. Historically speaking, the paper is a culmination of fifty years of physics—even if SU(5) is wrong, as it probably is, for the argument gathers into itself everything that came before.

The paper opens with a fanfare as brassy as that in any recent scientific work:

We present a series of hypotheses and speculations leading inescapably to the conclusion that SU(5) is the gauge group of the world—that all elementary particle forces (strong, weak, and electromagnetic) are different manifestations of the same fundamental interaction involving a single coupling strength, the fine-structure constant. Our hypotheses may be wrong and our speculations idle, but the uniqueness and simplicity of our scheme are reasons enough that it be taken seriously.

They then proceed to lay out the ground rules, acknowledging their sources. (The emphasis is Georgi and Glashow's.)

Our starting point is the assumption that weak and electromagnetic forces are mediated by the vector bosons of a gauge-invariant theory with spontaneous symmetry breaking. *A model describing the interactions of leptons using the gauge group SU(2) × U(1) was first proposed by Glashow, and was improved by Weinberg and Salam, who incorporated spontaneous symmetry breaking.*

In the next paragraph they introduce what was then little more than a favorite hypothesis of Glashow's: charm.

To include hadrons in the theory, we must use the Glashow-Iliopoulos-Maiani (GIM) mechanism and introduce a fourth quark [c] carrying charm. . . . The next step is to include strong interactions. We assume that strong interactions are mediated by an octet of neutral vector gauge gluons *associated with local color SU(3) symmetry.* . . .

They now had the entire standard model, which was still years away from being demonstrated in experiment.

Thus, we see how attractive it is for strong, weak, and electromagnetic interactions to spring from a gauge theory based on the group $\mathbf{F} = SU(3) \times SU(2) \times U(1)$. *Alas, this theory is defective in one important respect: It does not truly unify weak and electromagnetic interactions. The $SU(2) \times U(1)$ gauge couplings describe two interactions with two independent coupling constants; a true unification would involve only one.*

Next Georgi and Glashow hunt through the list of groups that Cartan had compiled decades before, coming up with three that could provide the basis for a truly unified electroweak theory. These three have already been used in models made by Weinberg and, later, Georgi and Glashow. None is reconcilable with SU(3). This allows the writers to conclude, with some pleasure, "We see we cannot unify weak and electromagnetic interactions independently of strong interactions." They then declare themselves forced to consider the "outrageous possibility" that a single group could account for all three. Methodically they set down all of the groups of sufficient size. As it happens, there are nine. One by one, these are eliminated, except the last, which was SU(5)—the model that Georgi had created over shots of Scotch a few months before.

The last half page of the paper was devoted to characterizing SU(5) and its implications, which are multifarious. To start, within this grand unified theory, as within the Pati-Salam model, quarks and leptons are kith and kin. Particles within a family can decay into one another, and thus Georgi and Glashow had to postulate the existence of a brand-new, never-before-seen force that would allow for just that possibility. They named this force the *superweak*, and suggested that it is mediated by a set of extremely heavy vector bosons, which Glashow nicknamed "ponderons" or "vector basketballs." These particles weigh too much to be emitted often, but when a quark does emit such a particle, that quark turns into a lepton. If the quark is part of a proton, the proton will fall apart. The pieces of the proton will, in turn, decay into electrons and positrons, which will eventually meet each other to form photons—light. This light will ultimately traverse an utterly empty plenum. Because protons are a nonrenewable resource, the Universe will end as it began, with light. Eventually, the light will dissipate, and the cosmos will settle into a long darkness. *Obeat lux.*

In their paper Georgi and Glashow, afraid of the reaction that this implication would provoke, refrained from mentioning proton decay until the end:

Finally, we come to a discussion of superweak interactions and SU(3)-colored superheavy vector bosons. In addition to mediating such bizarre interactions as $K^0 \to \mu^+ e^-$ they make the proton unstable.

"Unity of All Elementary-Particle Forces" appeared in February 1974 and initially met with little favor. Most theorists didn't notice it in the excitement provoked by $SU(3) \times SU(2) \times U(1)$. Even the select public that did read it found proton decay hard to swallow. Bjorken, for instance, told us he had been increasingly excited and convinced by the paper until he came to the very end. "It is a very tightly constructed and convincing paper," he said. "Then, all of a sudden, they hit you with proton decay out of the blue. 'Oh, by the way, the proton decays.' I lost my enthusiasm and decided I didn't like the paper. Later, after I'd thought about it, I also became skeptical of the difference in scale between the tremendously heavy particle they had to assume and all the other particles we know about. I'm still not convinced, by the way."[21]

"And *that* was the reaction of the *good* physicists," Georgi said. "To most people, it was just another group. You know, 'First it was $SU(2)$ and then $SU(3)$—' " he shifted into a sudden, surprising imitation of the Dumb Skeptic " '—now it's $SU(5)$ and soon it'll be $SU(18)$.' But we, too, didn't know how seriously to take it. Remember, this was before neutral currents, before charm, before any of the components had been found which would make people want to take the theory seriously."[22]

In addition, the $SU(5)$ paper did not provide any clear reason for supposing that the coupling constants could be made equal; this was provided a few months later, in a clever paper that Georgi wrote with Weinberg and another Harvard postdoc, Helen Quinn. "Hierarchy of Interactions in Unified Gauge Theories" was published in *Physical Review Letters* in August 1974. In it the authors used the renormalization group to examine the behavior of the coupling constants of the strong, weak, and electromagnetic interactions. Pointing out that the coupling constants are not true constants at all, but vary with the energy of the interactions, they climbed hand-over-hand up the scale of energy, rising fifteen orders of magnitude. The strength of electromagnetism and the weak interactions increases, while the strong force grows feebler. At 10^{14} GeV and higher, the coupling constants converge. Where the coupling constants merge, the three forces reveal their unity.

"That paper also was sparked by the excitement about quantum chromodynamics," Weinberg said. "We had Politzer at Harvard, and of course we also knew what Gross and Wilczek were doing, and at a certain point it just occurred to Georgi, Quinn, and me that if the strong interaction is getting weaker when you go to high energy [which is equivalent to small distances], then at *very* high energy it might be comparable to the electroweak." The paper was very much in Weinberg's mode of operation; although it spoke of SU(5), it was applicable to an entire class of models. "I find it hard to believe in specific models. Those results, to a certain accuracy—ten percent accuracy—are valid whatever the grand unified theory is, within a broad range."[23]

Now, 10^{14} GeV is a lot of energy for one particle to have. It corresponds to a temperature of about 10^{28} degrees Fahrenheit—one so extreme that it had not been reached in the Universe since the first fractions of an instant after the Big Bang. Thus, whereas Georgi and Glashow had demonstrated that uniting the strong, weak, and electromagnetic forces is theoretically possible, Georgi, Quinn, and Weinberg showed that these forces had actually been united only at the beginning of time, after which they had separated like curds and whey. Whereas Georgi and Glashow could only confess that the proton decays "with an unknown and adjustable rate," Georgi, Quinn, and Weinberg gave a strong prediction of the average proton lifespan: 10^{32}—ten quadrillion quadrillion—years. Beyond the specific number, however, the Georgi-Quinn-Weinberg paper demonstrated to physicists the compelling need to unite particle physics, the study of the smallest objects in existence, with cosmology, the study of the Universe as a whole.

Cosmology began over a century ago, it can be said, with the Czech mathematician Christian Doppler's prediction of the eponymous Doppler effect in 1842.[24] Three years later, the existence of the effect was demonstrated by one Christoph Baillot, a Dutch scientist who persuaded a band of trumpeters to play on an open railway car as it passed a group of musicians with perfect pitch. The speed of the car changed the frequency of the trumpet blast, an effect familiar to anyone who has heard the change in the pitch of an ambulance siren as it zooms by. It turns out that light waves, too, can manifest this effect, a fact of little consequence until Edwin Hubble discovered in the 1920s that there were millions of galaxies in the cosmos, and that the Doppler shift in the light they gave off indicated that they were flying from us at great speed. It looked to Hubble as if the whole Universe consisted of the speeding fragments of a monstrous explosion.

Attracted by the novelty of the idea, some astronomers leaped on it; others were leery of postulating a definite beginning to the Creation, for fear of invoking the question, "Well, what came before *that?*" The fortunes of the theory of the Big Bang, as the initial explosion was soon dubbed, waxed and waned until the early 1960s, when two Bell Laboratories researchers, Arno Penzias and Robert W. Wilson, were assigned the problem of ridding a big radio dish of a strange type of interference. One of the first satellite trackers, the big dish in New Jersey was designed to pick up signals from Telstar. Penzias and Wilson at first guessed that the interference was due to a terrestrial source, possibly the family of pigeons nesting on the antenna. After scraping off what Penzias obliquely referred to as a "white dielectric substance," the interference remained, no matter where they pointed the dish. Eventually they realized they were picking up the residual radiation from the Big Bang, which floods space like the morning heat from a campfire the night before. (Wilson and Penzias shared the 1978 Nobel Prize for their discovery.)[25]

The combination of the stellar Doppler shift and the cosmic background convinced cosmologists of the reality of the Big Bang, but few of them had then a strong idea of where to proceed. In the 1970s, the advent of unified theories persuaded many that there was a great deal of physics that happened in the first microseconds of creation. (They tended to assert their hypotheses loudly, despite the paucity of empirical data; the saying runs that cosmologists are often wrong, but never in doubt.) Although the standard model was constructed to describe the behavior of particles at ordinary energy levels, it contained within it, much to most theorists' surprise, plausible grand unified theories with a vastly longer reach; united with cosmology, elementary particle physics can present a picture of the history and evolution of the Universe even as it seeks to explain the ties among ordinary matter and forces.

On the far side of the Big Bang is a mystery so profound that physicists lack the words even to think about it. Those willing to go out on a limb guess that whatever might have been before the Big Bang was, like a vacuum, unstable. Just as there is a tiny chance that virtual particles will pop into existence in the midst of subatomic space, so there may have been a tiny chance that the nothingness would suddenly be convulsed by the presence of a something.

This something was an inconceivably small, inconceivably violent explosion, hotter than the hottest supernova and smaller than the smallest quark, which contained the stuff of everything we see around us. The Universe consisted of only one type of particle—maybe only one particle—that interacted with itself in that tiny, terrifying space. Detonating outward, it may have doubled in size every 10^{-35} seconds or so, taking but an instant to reach literally cosmic proportions.

Almost no time passed between the birth of the Universe and the birth of gravity. By 10^{-43} seconds after the beginning the plenum was already cooler, though hardly hospitable: every bit of matter was crushed with brutal force into every other bit, within a space smaller than an atomic nucleus. But the cosmos was cool enough, nonetheless, to allow the symmetry to break, and to let gravity crystallize out of the unity the way snowflakes suddenly drop out of clouds. Gravity is thought to have its own virtual particle (the graviton), and so the heavens now had two types of particles (carriers of forces and carriers of mass), although the distinction wasn't yet as clear as it is in the Universe today.

At 10^{-35} seconds the strong force, too, fell out of the grand unified force. Less time had passed since the Big Bang than it now takes for a photon to zip past a proton, and yet the heavens were beginning to split. Somewhere here, too, the single type of mass-carrying particle became two—leptons and quarks—as another symmetry broke, never to be complete again. The Universe was the size of a bowling ball, and 10^{60} times denser than the densest atomic nucleus, but it was getting colder and thinner rapidly.

One ten-billionth of a second after the Big Bang, the firmament reached the Weinberg-Salam-Glashow transition point, and the tardy weak and electromagnetic forces broke away. All four interactions were now present, as well as the three known families of quarks and leptons. The basic components of the world we know had been formed.

"Let me draw the whole picture for you," Glashow said to us once as he scratched a line in white chalk from one edge of his office blackboard to the other. "This represents the Universe from the beginning to the end of time." After thinking a moment, he drew a second line in purple chalk; it was just a bit shorter. "We live in the fortunate era—the era in which there is matter. Matter first appeared 10^{-38} seconds or so after the Big Bang, and will all disappear maybe 10^{40} seconds from now." He hunted around in the tray below the board, finally happening upon a piece of brown chalk, which he used to plot a line considerably shorter than the first two. "Within the fortunate era of matter there is a somewhat shorter period in which atomic nuclei exist, because the Universe is cool enough to permit them to form. And within that"—a red line now, barely a foot long—"an even tinier domain in which there are atoms." He made a dot near one end of the red line. "Then you have a brief ten billion years or so when things are palatable on earth.

"After that, eventually it all winds down. We won't be bothered when, say, ten percent of the protons go. If we're still around, we may not even notice. But when ninety-nine percent go, you won't have enough left in one place to make a person, and it will be unpleasant. After this point there's a long period of decline, and a very boring period it is, too. This is not a view we could have gotten without grand unified theories. We may not like it, but we have no choice."[26]

Is the picture true? At the end of the 1970s, the standard model was well enough established that experimenters and theorists were ready to look at something that pushed unification further. Unfortunately, one of the predictions of SU(5) is that there are no further elementary particles to be found—not until the grand unification energy of 10^{15} GeV is reached. Such energies cannot be reached by any accelerator conceivable today. "It would take an accelerator ten light-years long," Glashow said, "which, given current budgetary constraints, is unlikely." Grand unification in general and SU(5) in particular did have a second major testable prediction: proton decay, a question Glashow phrased as, "Are diamonds forever?"[27]

◻   ◻   ◻   ◻   ◻

Sunrise never seemed to happen at Kolar. The long slow foredawn, shadowed and almost purple, hung in the air with a last promise of coolness for an hour or more—then was abruptly gone. Harshly the Indian sun went up, and the dust rose to meet it. The miners flinched from the heat as they waited for the koepis to rattle them beneath the surface. Straggling workers joined the line, brushing aside goats, chickens, and somnolent cattle. Great steel drums the size of a house turned with a clank, the metal cable smeared with oil. Bunched with the miners were physicists and technicians, distinguishable by the clipboards in their hands. The small elevator arrived; the crowd slowly moved toward it. Seven thousand five hundred feet below, the oldest running proton decay experiment began another day.

Cosmic ray physics has long been an Indian specialty, because India's high mountains and deep mines enabled its physicists to do important research with inexpensive instruments. Since the 1950s, cosmic ray studies have been carried out at the Kolar Gold Fields, an immense and almost exhausted mine in south central India. There, in 1964, M. G. K. Menon, the assistant director of the Tata Institute of Fundamental Research in Bombay, participated on the Anglo-Indian-Japanese team that detected the first cosmic ray neutrino; twenty years later, he was the senior physicist in the proton decay experiment.

In 1966 Menon succeeded the physicist Homi Bhabba as director of the Institute when the latter died in a plane crash. (India's premier research center, the Tata Institute was founded in 1945 by the trust of a steel magnate, Sir Dorab Tata.)[28] A precise, soft-spoken, patriotic man who has devoted much his life to fostering Indian science and technology, Menon also became his nation's delegate to the United Nations Advisory Committee on Science and Technology. He immediately hit it off with the Pakistani representative, Abdus Salam, and when the two met at United Nations conferences in Geneva or New York they would retire to a corner of the assembly rooms during the more long-winded speeches and chat about physics.

At a meeting in the summer of 1973, Salam told Menon about proton decay. By hooking detectors to a box of heavy material, Menon could scrutinize the box to see if its protons decayed. Because proton decay was expected to be quite rare, the detector should be kept far away from background cosmic radiation that might confuse the issue. A deep mine like Kolar would be ideal. Menon replied that testing the lifetime Salam and Pati had predicted—about 10^{29} years—would require at least a hundred-ton detector, large even by accelerator physicists' standards and huge for a cosmic ray physicist. Nonetheless, Menon decided to give it a shot. In 1978, Menon, now a member of the Indian Planning Commission, V. S. Narasimham, and Badanaval Sreekanton, Menon's successor at the Tata, received permission to go ahead. They had decided to collaborate with scientists from the Japanese universities of Osaka City and Tokyo.

The entrance to the Gifford shaft of the Kolar gold mine is in a large, hangarlike structure criss-crossed by steel girders and the cables from the koepi to the hoist. The morning we visited, monkeys played and chattered in the struts. An amiable security guard escorted us down the shaft; as the lightless rattling box hurtled to the seventieth level, more than six thousand feet below our feet, we shared a moment of claustrophobic alarm. At the bottom was a small bright shrine to the Hindu gods. Compressed air from above blew down the tunnels, keeping the temperature to a bearable eighty degrees. "The temperature is ninety-one at the hottest," Narasimham said. "It's bearable, for Indians used to the climate. But it gives problems to our electronics."

We walked a hundred yards to a second shaft, descended another few hundred feet, then followed the tracks of the bandie cars down a tunnel. A cardboard sign hung from the ceiling: FIRST EVER COSMIC RAY NEUTRINO INTERACTION RECORDED HERE IN APRIL 1965. Narasimham had been at the mine then. A slightly built man with saturnine features, long thinning hair, and somewhat protruding eyes, he had spent years of his life waiting for cosmic rays inside Kolar. Two bandies full of chopped stone were wheeled past us by wiry sweaty men with cloths around their heads. The rock above pressed down like a heavy cloak, smothering. Just before the track rounded a corner was a niche blocked off by a chain-link fence.

PROTON STABILITY EXPERIMENT OF
TATA INSTITUTE OF FUNDAMENTAL RESEARCH, BOMBAY
AND
OSAKA CITY AND TOKYO UNIVERSITIES, JAPAN
Site: 80th Level Heathcote Shaft, KFG.
2300 meters depth.

Fifteen feet behind the gate was a wall of long rectangular pipes, each eighteen feet long and five inches square, stacked Lincoln Log–fashion thirty-four layers high. The finished product was an iron grid of rusted pipe ends, each with a dusty valve and a painted number; a passage to the right led to the data-taking room, where the computers were kept much cooler than the people who attended them. Purchased from Japanese construction firms, the pipes were altered into inexpensive proportional tubes—a kind of

Geiger counter. Engineering students had threaded a single wire down the length of each pipe, sealed it full of argon, and linked the array together to a high voltage line. The central wires and the walls surrounding them acquire opposite charges. If a charged particle passes through, a spark is created. Particle interactions are traced and measured by counting the lines of triggered tubes. "We have six candidates, four of which are solid," Narasimham said. "We have set a lifetime of 10^{31} years. We have stuck our necks out. Nobody else has done that."

"Nobody else" meant the other proton decay experiments, with which the Kolar group was in keen competition. In addition to the IMB experiment in the salt mine near Cleveland, there were tests in the Mont Blanc tunnel, on the French-Italian border; in the Silver King mine, near Park City, Utah; and the Kamioka lead and zinc mine, near Takayama, in western Japan. All took boxes of iron or tanks of water—reservoirs of protons, in either case—to secluded areas and watched to see if any decayed. Mont Blanc, Kamioka, and Kolar had publicized candidate events; IMB did not see them, and in particular did not see the decay modes predicted by SU(5). The result was a curious sort of gossipy standoff among the participants, each of whom was keenly aware of the others' strengths and weaknesses. Narasimham itemized them for a while, defending the design of his detector, pointing out faults in the experiments that disagreed—then said, "You want to see the new machine?"

A second, larger detector was near completion. Perspiring freely, we walked down the dimly lit passage; the rock above had grown heavier with the passage of time. We were led into a cavern, roofed with I-beams, eighty feet long and thirty-five-feet wide, excavated at the risk of a hundred lives. Punctured by long supportive bolts and coated with a yard-thick layer of protective concrete, the walls seemed to tremble with the immensity of the planet above them. Dominating the space was a rough iron cube, twenty feet on a side, that rose like a monolith in the crepuscular light. A corona of garish sparks from welding equipment danced across its crown. Ringed by scaffolding, the three-hundred-ton detector was tended by a dozen quickly moving men. Stripped to the waist, workers clambered over its face like ants. Blasts of compressed air thundered into the space, and Narasimham had to shout above them. In the roar, the dust, the splash of water, and the shriek of machinery, he yelled calmly in our faces, his shirt flapping in the hot ventilating breeze.

"Nobody knows when the proton decays, but everybody seems to think it must," he said. "People used to think it was immortal and hardly anyone looked. Now everyone is looking, and they are sure it decays. We are sticking to our candidates. We may interpret them right or we may not. To be sure, we use as much imagination to say yes as they do to say no."[29]

20
The End of Physics

THE LUCASIAN PROFESSORSHIP OF MATHEMATICS OCCUPIES A CURIOUS POSI-
tion at Cambridge University, for it has traditionally been occupied by a
great theoretical physicist. Newton and Dirac were both Lucasian Professors,
and its current occupant is Stephen Hawking, easily the most well-known
cosmologist in the world. Unfortunately, Hawking is famous less for his
considerable accomplishments than for his long struggle with a degenerative
neural disease that has kept him in a wheelchair for many years, impeded his
ability to speak and write, and may eventually kill him. On April 29, 1980,
Hawking assumed the chair with an inaugural lecture entitled, "Is the End
in Sight for Theoretical Physics?" The answer, he said, is probably yes. A
student read the speech for him:

*In this lecture [Hawking's text began] I want to discuss the possibility that the
goal of theoretical physics might be achieved in the not-too-distant future, say, by
the end of the century. By this I mean that we might have a complete, consistent,
and unified theory of the physical interactions which would describe all possible
observations. Of course one has to be very cautious about making such predictions: We have thought that we were on the brink of the final synthesis at least
twice before. At the beginning of the century it was believed that everything could
be understood in terms of [classical] mechanics. All that was needed was to
measure a certain number of coefficients of elasticity, viscosity, conductivity, etc.
This hope was shattered by the discovery of atomic structure and quantum
mechanics. Again, in the late 1920s Max Born told a group of scientists visiting
Göttingen that "physics, as we know it, will be over in six months." This was
shortly after the discovery by Paul Dirac . . . of the Dirac equation, which
governs the behavior of the electron. It was expected that a similar equation
would govern the proton, the only other supposedly elementary particle known at
that time. However, the discovery of the neutron and of nuclear forces disappointed these hopes. We now know in fact that neither the proton nor the neutron
is elementary but that they are made up of smaller particles. Nevertheless, we
have made a lot of progress in recent years and, as I shall describe, there are
some grounds for cautious optimism that we may see a complete theory within the
lifetime of some of those present here.*[1]

Hawking has been but one among many physicists who have hoped
that unification is at hand, although few have been willing to state their hopes
so boldly. Today, unification is the research topic for an entire new generation
of theorists. The subject has split into branches and subdivisions, and there
are unification conferences and unification journals and even unification text-

books. It is the rare theorist who has not tried his hand at putting things together.

Physics has always progressed by drawing together seemingly disparate phenomena into one framework. Newton's recognition that the force that made apples fall was identical to the force that kept the earth in its orbit was a unification, as was the approximately contemporaneous realization that lightning, static electricity, and Saint Elmo's fire were all manifestations of one phenomenon: electricity. In the nineteenth century, Maxwell synthesized electricity, magnetism, and optics into his theory of electromagnetism.

These successes in turn elicited more grandiose but less happily conceived unification schemes. As Freeman Dyson has pointed out, "The ground of physics is littered with the corpses of unified theories."[2] The early nineteenth-century chemist Claude Berthollet spent years trying to demonstrate that gravity was a sort of chemical attraction. Michael Faraday, Maxwell's great predecessor, anticipated Einstein by eighty years when he tried to establish a relation between electromagnetism and gravity in 1850. The beginning of gauge theory was a mistaken stab at unity by Hermann Weyl. Einstein's long failure with unified field theories is well known; in retrospect, a principal stumbling block was his refusal to treat the forces in atomic nuclei as fundamental, a folly that reached its height when he published a paper unsuccessfully attempting to prove "that the elementary formations [i.e., particles] that make up the atom are held together by gravitational forces."[3] Schrödinger, too, spent the last years of his life obsessively pursuing the chimera of a unified field theory. In the 1930s, Yukawa tried to unify the strong and weak forces, but the eventual success of his work owed nothing to unification. Heisenberg launched, two decades later, a unified field theory that started as a collaboration with Pauli. When Pauli withdrew, Heisenberg pressed on. To Pauli's fury, Heisenberg claimed during a radio broadcast in February 1958 that a unified Heisenberg-Pauli theory was imminent, and only a few small technicalities remained to be worked out. Rumors swept the press. Pauli responded by mailing his friends a letter consisting of a blank rectangle, drawn in pencil, with the caption: "This is to show the world I can paint like Titian. Only technical details are missing."[4]

Mindful of the recent record of failure, unification aficionadoes today practice their craft with a measure of irony, for they are as embarrassed by the loose speculation required as they are entranced by the sweetness of the problem. Almost every practicing physicist has been approached by eager cranks with unified theories, sweaty would-be Einsteins with equations of the world that sew up all loose ends; now look at the theoretical legions invading the territory of the nuts. (The presence of the crackpots is in its own way a measure of the attractiveness of the idea.) A complete unified theory would mean the end of physics. Which is not to say that physics equations would vanish, physics experiments would stop, and all physicists would be out of a job. Many millions of loose ends would need tying up; connections would still need to be drawn; effects, to be understood; applications, to be devised and exploited. Science would continue, but all of the fundamental questions that physics can pose would have been answered, and our knowledge of force and matter would henceforth change only in particulars and not in outline.

In the *Critique of Pure Reason*, Immanuel Kant argued that some aspects of nature will remain forever unknown to us, because our minds must impose a structure on our sensations for us to have any experience of the world at all. As a consequence, we are led inevitably to make certain suppositions about nature that are, in actuality, by-products of the organizing activity of our own brains. Such presuppositions—Kant showed their number, and said they apply to more than science—are unprovable in theory but indispensable in practice. He called them "regulative ideas," and listed the unity of nature as a cardinal example. Because of it, the structure of science itself draws physicists toward unification. Given that the standard model contained the answers to all ordinary physical questions, it is then little wonder that an orgy of unifying came after its completion, as theorists followed the impulse built into the science, and no surprise that the sudden explosion of unified theories in the 1970s brought explosive and diverse reactions.

"To my mind," Gerard 't Hooft said, "the most successful programs of unification always came about when there was some urgent question to be answered. In most cases, the urgent question was that something seemed to be wrong in the present understanding, and the correct answer then turned out to be that the only way to get it right was to say that this effect and that effect cancel, and the only way to make them cancel was to put them into one big theory together and show that they were the same force. Then you get something like unification. That's the way I view, say, the unification of the gravitational forces with special relativity which gave general relativity. That was an enormously far-reaching theory, but it arose because there was an obvious question to be answered: How do you reconcile the notion that the effect of two bodies on each other only depends on their masses, nothing else, and on the other hand keep relativistic invariance for the whole set of equations? If you try to answer it, you run into all sorts of difficulties unless you assume, precisely as Einstein did, curved space-time and all that. There was no other solution. So it *had* to be right. That's the way, to me, you should derive a theory. What Einstein did later was try to put electromagnetism together with gravity. However, this time, he didn't have a good motive. All he wanted was unification. That is, he didn't have as a motive that something would be wrong with physics if he didn't do it. And that's why I think he did not succeed."

We asked if this applied to the current unification efforts.

"I'd make the same objection," 't Hooft said. He had been working on putting forces and particles together for a decade, and felt that in many respects he did not have that much to show for it. "There is no obvious physical need. There is an aesthetic need, but not one of purely mathematical logic. And my conviction is that as long as that need is not there, it's unlikely that it will work in this simple way. It may work, so people should continue trying it, but it may well be just like the fate that struck Einstein— that although it looks from an aesthetic point of view to be an obvious thing to try, nature is more subtle than this. The trick is not to try to put things together which do not really belong together, but rather to search for places in the world where there are discrepancies, where different ideas are clashing that ought to be described in one and only one way."

He was asked what kind of discrepancy he had in mind.

"Well, a very important discrepancy I'm interested in, like many other people, is quantum gravity. Because we still don't have a good way of reconciling gravitation with quantum mechanics. There still isn't. I'm sure that whenever somebody finds a way to do that, he'll solve millions of problems in ordinary physics." We remarked the long-standing difficulties in the theory of quantum gravity. "Well, there simply *isn't* a theory. A theory is completely lacking. People claim that they have ideas of theories abut it—Stephen Hawking is doing a lot on it. He has made, you know, some brilliant contributions. But the most fundamental theory, Lagrangian or Hamiltonian, or a proper description of Hilbert space, is missing. And so we just switch on gravity, which is a pain in the neck."

The difference between the domains of gravitation and the elementary particle forces is the difference between the proton and the plenum. The queen of the interactions, gravitation is easily ignored when looking into the microscope of a particle accelerator; at any other time, its gentle, firm, omnipresent pull dominates the firmament. Gravitation takes place on great, epic terrain—the bald ridges and smooth valleys of space-time itself— whereas quantum theory deals in billionths of millimeters, microscopic ferment, entities beneath visibility and beyond visualization. Big/small, classical/quantum, geometry/algebra, the sole obvious point of contact of these most opposite of physical constructs is that their arena of play is the same Universe.

As the standard model and the various grand unified theories drew together the elementary particle forces, the absence of gravitation became more and more conspicuous. In a sense, the very elegance of general relativity has impeded the urgent task of putting gravity into quantum terms.

Too successful to offer much scope for theoretical tinkering, Einstein's monument stood, a seamless wall of mathematics, in proud isolation from the rest of physics. If quantum mechanics and relativity were uneasy spouses in quantum field theory, how much more troubled would be a quantum theory of gravity itself.

In any quantum treatment of gravitation, the force is transmitted by a massless spin-two particle called the graviton. The graviton gives rise to infinities intractable enough to make veteran renormalizers conclude that straight quantum gravity simply is not finite. "Something is missing there," 't Hooft said. "We are all trying very hard to make it all work, but it turns out to be conceptually extremely difficult. One of the big difficulties is that we realize that some of our well-known concepts have to be abandoned, because nature isn't going to be as simple as it appears now. However, we cannot abandon everything at once, because then there is nothing left to work on."[5]

The problem faced by 't Hooft and his coevals is unprecedented in the history of physics. They believe that matter and energy were one just for an instant, at the dawn of time, in the ravening fire of the Big Bang. Thus the phenomena they seek to describe existed only at unimaginable energies that can never be reproduced in the laboratory. Experiment consequently seems almost helpless. Theorists have spent the decade since SU(5) trying to bootstrap themselves to unification without benefit of data, working out the right theory by pure mathematical ingenuity and physical intuition. They court the risk of divorcing theory entirely from experiment, and turning physics, the prototype of an empirical science, into what Georgi has called "recreational mathematical theology."

Notwithstanding such fears, a dizzying variety of unification theories were developed in the 1970s. Some bolder than others but all unproven, they went by various overlapping names: quantum gravity, supersymmetry, Kaluza-Klein, supergravity . . . the list of permutations is as long as the number of theorists working on the subject. Of all available essays at unification, we found ourselves drawn most to what is called "superstring theory," because its predictions are more than usually bizarre, because its history is more than usually chaotic, because it is apparently renormalizable, because it has a large body of adherents, and because, when asked about it, Murray Gell-Mann told us flatly that he believed that some version of string theory some day would be the theory of the whole world. "It's a fantastic thing," he said. "It's a candidate. It's *the* candidate."[6]

Superstring theory arose from the unexpected recent marriage of two wild ideas: (1) the Universe has extra, hidden dimensions; (2) subatomic particles ultimately are not little points but little strings. The idea that the Universe had hidden dimensions was first proposed by the German physicist

Theodor Kaluza in 1919.[7] Inspired by Einstein's four dimensional theory of relativity as well as the Unified Field Theory, Kaluza tried to incorporate electromagnetism into a *five* dimensional form of general relativity. Kaluza's work was brought into accord with quantum mechanics by a thirty-two-year-old Swedish physicist named Oskar Klein; in Klein's version, the fifth dimension was hidden, curled up into a minute circle and playing no real role in our world.[8]

For a while, theorists found the work of Kaluza and Klein exciting, but it didn't seem to go anywhere, and the idea languished for nearly half a century. Then, in the 1970s, it was revived in a strikingly different context, the multidimensional "string theories." Based on work by Gabriele Veneziano, an Israeli, and elaborated by Yoichiro Nambu and a dozen other theorists, the string model at first dealt just with the strong interactions.[9] Its adherents regarded hadrons as little one-dimensional strings rather than points. Mesons were strings with a quark at one end and an antiquark at the other; when the meson forcefully struck another particle, the string snapped, producing two new strings. Isolating a single quark was thus as impossible as creating a piece of string with just one end. The visualizability of the theory broke down for baryons, which had to be imagined as strings with *three* ends. Although the mathematical properties of the string model were fascinating and elegant, its equations seemed to contain a horrific panoply of ghosts, infinities, anomalies, unobserved spin-two particles, and impossible particles that travel faster than light. Many of these could be removed by artful equation-juggling, but only at the price of assuming that space-time has more than the usual number of dimensions—twenty-six, in fact. (*Twenty-six dimensions*!? The physicists who discovered this didn't even *try* to explain what on earth it could mean.)[10] In 1974, two theorists at Caltech, John Schwarz and the late Joël Scherk, who had worked on an alternative string model with only ten dimensions, realized that the unwanted spin-two particle might be the quantum of gravitation.[11] At a stroke, what had been a troubled theory of the strong force was converted into an excellent candidate for a unification theory.[12]

Because gravitation, unlike the strong force, is a manifestation of the structure of space-time, extra dimensions are not necessarily disastrous. Schwarz and Scherk could use a Kaluza-Klein-like device to ensure that the extra dimensions are perpetually hidden from view, squashed into tiny, unvisualizable balls at each point in space. Moreover, string theories naturally could be extended to "supersymmetry," a method of classifying together particles of force and particles of mass. (For this reason, the theory bears the name of *super*string theory.)[13] The ideas of Schwarz and Scherk were sufficiently off-beat that they attracted little interest until 1984 and 1985, when Schwarz and a colleague proved that superstrings were not only completely free of ghosts and anomalies—that is, they are mathematically consistent—but that they are consistent for just two versions of the theory.[14] These immediately became candidates for a Theory of Everything.[15] Physicists found the thought of deriving the Universe from the requirements of consistency alone to be irresistible, enchanting, marvelous; unlike Candide, who lived in the best of all possible worlds, we might live in the *only* possible world.[16] Theorists have descended upon superstrings in droves, despite its penchant for predictions that even physicists consider bizarre, such as the existence of "shadow matter" in the Universe, matter invisible to us, that can only be detected by gravitational effects and nothing else. Although there is as yet not a scrap of experimental evidence for superstring theory, it is completely renormalizable and does not appear to be in conflict with anything we know so far—no mean feat for a physical theory nowadays.

At the very least, superstring theory is a textbook example of a theoretical bandwagon, of how a clever mathematical conceit can suddenly become *démodé*, dominate discussion and conference proceedings for months and even years, ultimately withering for lack of contact with experiment. At the very most, it is, as Gell-Mann put it, "the theory of everything—gravity, weak, strong and electromagnetic interactions plus a lot of other things all together—a completely unified theory of nature." If Gell-Mann is right, the books of future historians of science may well treat the construction of the standard model as a lengthy parenthetical interlude between the first inklings of superstring theory after the First World War and its successful application to nature sixty years later.

Howard Georgi once remarked that there is little need for string theories and other unified theories because they only apply to phenomena like the Big Bang that can never be approximated by experiment. He advocated what are called "effective field theories," the suggestion that at different energy realms different field theories are applicable. Just as it would be foolish for engineers who build bridges and design cars to use quantum mechanics, it is nonsensical for particle physicists in an $SU(3) \times SU(2) \times U(1)$ world to try to go much past $SU(5)$. The phenomena beyond grand unification are so ephemeral, so distant in time, or so heavy that they play no role even in the subatomic domain probed by the largest particle accelerators. Despite his status as a godfather to the unification movement, he professed to find most unification theory unappetizing. "The most interesting question at the moment is what exactly breaks $SU(2) \times U(1)$," he said. "We still don't know what's giving mass to the W and the Z. We just know that symmetry is broken. It's an absolutely open question whether it's a Higgs or a dynamical mechanism or something that we haven't thought of. I regard that as the only question that I can see at the moment that is both

obviously fundamental and obviously physics. Unification is clearly fundamental, but it may not be physics if you can't see any of the effects."

In 1984, Steven Weinberg came to Harvard to give a lecture series on
string theories. Georgi greeted him by writing a limerick on the blackboard
before Weinberg's first talk.

> *Steve Weinberg, returning from Texas*
> *Brings dimensions galore to perplex us.*
> *But the extra ones all*
> *Are rolled up in a ball*
> *So tiny it never affects us.*

"And," Georgi said, "it bothers me a little that it never affects us."[17] In his
view, unification theories in general and string theories in particular may
inherently be concerned with the hows and whys of phenomena seen only
during the unreachable holocaust of the first instants of creation. If reaching
the energy scale of grand unification requires an accelerator whose length is
measured in light-years, reaching the energy of full unification could only be
done in a machine the size of the galaxy. Because such machines are absent
from any version of the future, Georgi has argued that despite their formal
elegance, mathematical rigor, and beautiful complexity, unification theories
may ultimately be no better than attempts to calibrate the end of the world
by examining permutations of the number 666.

Contemptuous of idle philosophizing, practicing physicists tend to be
uninterested in the metaphysical overtones of their craft. They define the
end of physics operationally, as the day when no government will pay to test
further a future unified theory, and resist speculating about why physicists
keep trying to put such theories together. When we asked Glashow one day
why he had immediately jumped to the idea of a larger, unified gauge group,
he responded by reading a passage written in 1927 by one of his Harvard
predecessors, Percy Bridgman.

Whatever may be one's opinion as to the simplicity of either the laws or the
material structures of Nature, there can be no question that the possessors of such
a conviction have a real advantage in the race for physical discovery. Doubtless,
there are many simple connections still to be discovered, and he who has a strong
conviction of the existence of these simple convictions is much more likely to find
them than he who is not at all sure that they are there.[18]

We asked why unification was the necessary outcome of physics.

"It's *not* necessary," Glashow said. "All I can say is that if you have the
faith, you have an advantage. Physicists in the past who have looked for
simplifying, unifying assumptions have done well. Better than physicists
who haven't. But there's no *reason* that things get simpler. They could
become more and more chaotic and more and more complicated. They may,
at some point. But so far things are getting simpler. I can't say simple, but

simpler."[19] Nonetheless, the difference between unification as a long-range goal and unification *now*, was important to Glashow. He found supersymmetry, supergravity, and the whole panoply of 1970s unification not to his taste. And as for string theory, he told us, in the spring of 1985, "It is sociologically interesting as an example of a theoretical bandwagon, and not much else."

He is not alone. Julian Schwinger, for one, told us that unification was a "fad," a "grand illusion" that is not "a theory in the usual sense but an aesthetic and emotional glow about how things would work if only we could compute them." He dismissed the current push toward unification as simple theoretical hubris. "It's nothing more than another symptom of the urge that afflicts every generation of physicist—the itch to have all the fundamental questions answered in their own lifetimes."[20]

"You know," Steven Weinberg said, "there wasn't that much of an intellectual discontinuity from $SU(3) \times SU(2) \times U(1)$ to grand unification." We had just ordered lunch in the Harvard Faculty Club. Around us were open jackets and open wine bottles, loosened ties and clattering silver: the furniture of academic meals. Weinberg talked about going up fifteen orders of magnitude in his paper with Georgi and Quinn, the giddy audacity of cranking through the numbers across such an enormous range. The fire alarm rang. For a moment or two, everyone in the room looked about with the polite incomprehension customarily awarded to signals of disaster. Waiters shooed out the crowd. Weinberg brushed off his trousers and sat on the steps of a nearby building. We asked what he meant by the lack of intellectual discontinuity in grand unification. "I'm not saying this in any critical spirit. What you were really talking about was a new symmetry structure imposed on the good old dynamics of quantum field theory. But with strings, you really have a new dynamics. It's still within the framework of quantum mechanics. But that's almost the only thing that has remained. String theories *look* like field theories over an enormous range of energies, up to the fundamental scale, which is somewhere in the neighborhood of 10^{16}, 10^{17}, 10^{18} GeV. But if you really get up to the fundamental scale, then they stop looking like field theories altogether. They really are a new kind of dynamics."

Mutterings of false alarm; people started to return to the faculty club, although the fire alarm was still ringing because no one could figure out how to turn it off. As we filed in, we asked Weinberg whether strings change our understanding of the birth of the Universe. Despite the hubbub and the jostling, he spoke readily and concisely. "It doesn't really answer any questions, because if you let the clock run backward and imagine what the Universe looks like as you go to earlier and earlier times, you still see a singular state." That is, properties like the energy density shoot up to infinity as you go back in time, and at the beginning is a white-hot point unexplained by current physics.

On another occasion, Weinberg remarked, "I'd say the period from the mid-sixties to the mid-seventies was enormously exciting, progressive, the best time we've had in physics since the late forties. Unfortunately, since then experiment and theory have gotten out of touch with each other. It's not really the fault of anybody, it's just the logic of the way the subject has developed. It's been the most frustrating decade, really, just awful!—in the sense that the thing that the brightest theorists are doing does not directly bear on any experiment that's about to be done or can be done in the foreseeable future. Supergravity, Kaluza-Klein, grand unification, all of that stuff, with a few little exceptions, can't be tested experimentally. And where it can, it hasn't been terribly impressive. Look what's happened with supergravity. The people who've been working on it for the past ten years are enormously bright. Some of them seem brighter than anyone I knew in my early years. They have elaborated these theories of supersymmetry, supergravity, and superstrings in a way that I think is unprecedented in the history of science for a theory that has *no experimental support whatsoever*.

"I've done it myself, I'm not badmouthing them. I think it was the right thing to do, because, as I say, you do what you can, and this was the best that could be done." He was discouraged by the implicit ironies: So many good theorists with so many good ideas who think they're so close to unification—only to find that proving the theories is utterly impossible with any foreseeable technology. "We just can't go on doing physics like this without support from experiment," he said. "The experimentalists do great things—discover the W and Z—and God bless them, it's wonderful. But the theory has moved to the point where these experiments are not helping. I hope that with the next generation of accelerators, we'll get out of this morass."[21]

We walked through the crumbled salt that littered the floor of the mine in Cleveland. Sulak strode ahead of us, his shoulders slumped a little with tiredness. There was a newborn in the house, and hours of nocturnal activity were taking their toll. We had waited for the end of the maintenance break, and the experiment was running again. The computer screens had been on for a while; we saw nothing, but didn't look for long. The miners were finishing their shift, dirty and joking, ready for a beer, ready for dinner. The elevator goes up only a few times each day, and bodies are packed in as closely as possible.

"It's an exciting time and it's a rotten time," Sulak said. "I really think some of these guys might have the right idea. The superstring theory is amazing, tremendously neat. But, my God, what if we can't see its effects? I always think of physics as the queen of sciences because it is rooted in experiment. I can't bear to think that unification might not be subject to verification by experiment." He'd spent seven years on proton decay, he said. Diminishing returns would come in, sooner or later.

A foreman pushed us into the salty cage, and we rose in slow hot silence, surrounded by the breath of tired men and women. Acid water dribbled down; for some reason the flow was a bit lighter than it had been that morning. Above us was the last flicker of daylight in the blustery sky, grayed by the afternoon's accumulation of smog. Feet made scraping sounds on the rusted floor; chains rattled and looped.

Below us, as the shift ended, half a dozen physicists stared at computer screens, waiting, waiting, waiting for the proton to decay.

Notes

ABBREVIATIONS FOR WORKS FREQUENTLY CITED

AdP: *Annalen der Physik*

AHES: *Archives for the History of the Exact Sciences*

AHQP: Archives for the History of Quantum Physics, American Institute of Physics, New York City

AIP: American Institute of Physics

BPP: Brown, L. M., and Hoddeson, L., eds., *The Birth of Particle Physics*. New York: Cambridge University Press, 1983.

CI: Berthelot, A., et al., eds., *Colloque International sur l'Histoire de la Physique des Particules*. In *Journal de Physique* (coll. C-8), 43, no. 12 (supp.), December 1982.

CQ: Pickering, A., *Constructing Quarks: A Sociological History of Particle Physics*. Chicago: University of Chicago Press, 1984.

CR: *Comptes rendus des Séances de l'Académie des Sciences*

DSB: Gillispie, C. C., et al., eds., *Dictionary of Scientific Biography*. New York: Scribners, 1970.

HDQ: Mehra, J., and Rechenberg, H., *The Historical Development of Quantum Theory*. New York: Springer-Verlag, 1982.

HPA: *Helvetica Physica Acta*

HSPS: *Historical Studies of the Physical Sciences*

JETP: *Soviet Physics. Journal of Experimental and Theoretical Physics* (Translation)

LET: Hermann, A., Meyenn, K. V., and Weisskopf, V. F., *Wolfgang Pauli, Wissenschaftlicher Briefwechsel mit Bohr, Einstein, Heisenberg, u.a.* New York: Springer-Verlag, 1979 (vol. 1), 1985 (vol. 2), quotation translated by the authors where necessary.

NC: *Nuovo Cimento*

NP: *Nuclear Physics*

PL: *Physics Letters*

PM: *London, Dublin, and Glasgow Philosophical Magazine and Journal of Science*

PR: *Physical Review*

PRL: *Physical Review Letters*

PRpts: *Physics Reports*

PRSA: *Proceedings of the Royal Society* (London), section *A*

PT: *Physics Today*

PTP: *Progress in Theoretical Physics*

PZ: *Physikalische Zeitschrift*

RMP: *Reviews of Modern Physics*

SIL: Pais, A., *"Subtle Is the Lord . . .": The Science and Life of Albert Einstein*. New York: Oxford University Press, 1982.

TP: Kevles, D. J., *The Physicists*. New York: Knopf, 1978.

ZfP: *Zeitschrift für Physik*

CHAPTER 1

1. Interview, Larry Sulak, IMB mine, 6 June 1985. We thank the officers of Morton-Thiokol, Frederick Reines, Dan Sinclair, and Jack van der Velde for allowing us into the experimental area. Some of the details come from an earlier visit to the mine on 3 May 1985.

2. Cited in Guillemin, V., *The Story of Quantum Mechanics* (New York: Scribners, 1968): 19.

3. Quoted in Badash, L., *Isis* 63: no. 1 (January 1972): 52. The "decimals" is given as the opinion of an "eminent physicist." See also, Galison, P., in Graham, L., et al., eds., *Functions and Uses of Disciplinary Histories*, Vol. VII (Amsterdam: D. Reidel, 1983): 35.

4. This story has been recounted in many places, including Romer, A., *The Restless Atom.* Rev. ed. (New York: Dover, 1982), pp. 15–29; and Shamos, M., *Great Experiments in Physics* (New York: Dryden, 1959): 213ff. Becquerel's memoirs are in *Memoires de l'Académie des Sciences* 46 (1903): 1. Translations of Becquerel's original papers are in Romer, A., ed., *The Discovery of Radioactivity and Transmutation* (New York: Dover, 1964).

5. Cited in Glasser, O., *Wilhelm Conrad Röntgen und die Geschichte der Röntgenstrahlen* (Berlin: Springer–Verlag, 1959): 26.

6. *CR* 122: 150, séance of 20 January 1896.

7. Becquerel, H., *CR* 122: 420, séance of 24 February 1896.

8. Becquerel, H., *CR* 122: 502, séance of 2 March 1896.

9. Becquerel's memoirs note scornfully that one had only to reread the publications of one G. Le Bon "to realize that at the moment he did them, the author didn't have the slightest understanding about the phenomenon of radioactivity" (trans. by authors), Becquerel, *Memoires, op. cit.*, p. 6.

10. *Ibid.*, p. 3.

11. Standard books on Thomson include Strutt, J. W., Lord Rayleigh, *The Life of J. J. Thomson, O. M.* (New York: Cambridge University Press, 1943); Thomson, J. J., *Recollections and Reflections* (London: G. Bell and Sons, 1936); and the odd, amusing Crowther, J. G., *British Scientists of the 20th Century* (New York: Routledge and Kegan Paul, 1952).

12. See, for example, Moralee, D., et al., eds., *A Hundred Years of Cambridge Physics* (2d ed.), Cambridge University Physics Society, 1980.

13. According to Pais, A., *RMP* 49, no. 4 (October 1977): 925, the same announcement was made three months earlier by the Prussian physicist Johann Weichert, who played no role in subsequent events. Walter Kaufman of Berlin also found *elm.*

14. Thomson, J. J., *PM*, series 5, 44, no. 269 (1897): 293. (For a good general history, see Anderson, D. L., *The Discovery of the Electron* [Princeton: Van Nostrand, 1964]). The title of this journal is worthy of attention, as it gives some insight into the origins of physics, and its relation to philosophy. Once upon a time, physics and philosophy were allied disciplines, and a physicist was equally as likely to be referred to as a "natural philosopher," someone who sought truth by rational argument in the natural instead of the human world. This alliance began to break up at the end of the last century; we now place them in separate sections of the curriculum and suppose that different talents are required to pursue each. In current practice, physicists often use *philosophy* to mean "poor physics." Some relics of the old terminology have persisted into this century; Einstein, for instance, taught in the Philosophical Faculty at the University of Zürich. Another is the persistence of the word *philosophical* in the title of this journal. In Rutherford's time, articles treating most branches of science could be found in the *Philosophical Magazine.* Today, with the same title, it is exclusively a journal of solid state physics.

15. The influential German chemist Wilhelm Ostwald went further, arguing that "In fact, energy is the unique real material in the world, and matter is not a carrier, but a manifestation of it" (*Zeitschrift für physikalische Chemie* 9 [1892]: 771, trans. by authors).

16. Thomson, *Recollections, op. cit.*, p. 341.

17. Becquerel, H., *CR* 130: 106, séance of 26 March 1900.

18. Cited in Eve, A. S., *Rutherford* (New York: Macmillan, 1939): 15. Rutherford's original letters to his mother and fiancée seem to have disappeared. Eve is the standard biography, but see also: Badash, L., ed., *Rutherford and Boltwood: Letters on Radioactivity* (New Haven: Yale University Press, 1969); Birks, J. B., ed., *Rutherford at Manchester* (New York: W. A. Benjamin, 1963); Oliphant, M., *Rutherford: Recollections of the Cambridge Days* (New York: Elsevier, 1972); and Wilson, D., *Rutherford: Simple Genius* (Cambridge, Massachusetts: MIT Press, 1983). For the discovery of the nucleus, we have greatly relied on the excellent Heilbron, J. L., *AHES* 4 (1967): 247.

19. Quoted in Wilson, *Rutherford, op. cit.*, p. 90.

20. Oliphant, *Rutherford, op. cit.*, p. 29.

21. Rutherford, E., *PM*, series 5, 47, no. 284 (January 1899): 116.

22. Eve, *Rutherford, op. cit.*, pp. 54–55.

23. The figure is from Champlin, ed., *The Young Folks Cyclopedia* (New York: Holt, 1916). Typical of the awe with which radium was regarded is the 1921 edition of *The American Educator* encyclopedia (Chicago: Ralph Durham), which unhesitatingly defines radium as "the most valuable and possibly the most wonderful substance in the world. . . . In proportion to its weight radium is a hundred times more valuable than diamonds, being worth $3,200,000 an ounce."

24. Rutherford, E., *PM*, series 6, 5, no. 49 (February 1903): 177; *Nature* 79, no. 2036 (5 November 1908): 12.

25. Becquerel, H., *CR* 136: 1517, séance of 22 June 1903; *CR* 141: 485, séance of 11 September 1905.

26. This conclusion was too much for Becquerel as well. Because a particle's momentum is the product of its mass and its velocity, Becquerel argued that it was not the velocity that was increasing but the *mass*; the alpha particle somehow must be collecting other bits of matter in flight, like a snowball rolling downhill.

27. Rutherford, E., *PM*, series 6, 11, no. 61 (January 1906): 166.

28. *Ibid.*, p. 174.

29. Cf. Eve, *Rutherford, op. cit.*, p. 183.

30. Letter, W. H. Bragg to E. Rutherford, 1 October 1908, quoted in Heilbron, *op. cit.*, p. 262; see Wilson, *Rutherford, op. cit.*, p. 286.

31. Marsden, E., in Birks, J. B., ed., *Rutherford at Manchester, op. cit.*, p. 1.

32. Geiger, H., and Marsden, E., *PRSA* 82, no. 557 (31 July 1909): 495.

33. Probably under Rutherford's guidance, Geiger and Marsden had written, "If the high velocity and mass of the α particle be taken into account, it seems surprising that some of the α particles, as the experiment shows, can be turned within a layer of 6×10^{-5} cm. of gold through an angle of 90°, and even more." To produce the same effect with a magnet, they noted, an "enormous field . . . would be required" (*Ibid.*, p. 498).

34. Crowther, J. A., *PRSA* 84, no. 570 (15 September 1910): 226.

35. On 9 February 1911, for instance, Rutherford wrote to his friend William Henry Bragg, another expert in radioactivity: "I have looked into Crowther's scattering paper carefully and the more I examine it the more I marvel at the way he made it fit (or thought he made it fit) JJ's theory. . . . I believe it was only by the use of imagination and failure to grasp where the theory was inapplicable that led him to give numbers showing such an apparent agreement." Bragg was delighted, writing back three days later: "Your opinion of Crowther's paper agrees with mine; do you know, I think it is quite an immoral paper, very nearly dishonest, though I am sure the dishonesty was not intentional." With cheerful unfairness, Bragg continued, "I must say, I have often found myself wrong when I have felt inclined to condemn utterly. But I think the censure is just, this time. . . . [Crowther's article] has the worst fault an experimental paper can have, because it unctuously brings round a lot of facts to suit a theory backed by a great name, and it is so jolly cock-sure. . . ." Quoted in Heilbron, *op. cit.*, pp. 294–5.

36. Rutherford, E., *Proceedings of the Manchester Literary and Philosophical Society*, series 4, 55, no. 1 (March 1912): 18.

37. The term *nucleus* came into use quickly. It may have been suggested by Bohr (Interview, G. Hevesy, *AHQP* [25 May 1962]: 2). Rutherford first used it six months later in *PM*, series 6, 24, no. 142, (October 1912): 453.

38. A Japanese physicist, Hantaro Nagaoka, had previously proposed a somewhat similar model (*Proceedings of the Tokyo Physico–Mathematical Society*, series 2, no. 2 [February 1904]: 92; *PM*, series 6, 7, no. 42, [May 1904]: 445). But G. A. Schott pointed out that the model envisioned by Nagaoka was unstable (*PM*, series 6, 8, no. 45 [September 1904]: 384); after some argument, Nagaoka accepted. For a discussion, see Yagi, E., *Japanese Studies in the History of Science* 3 (1964): 29.

39. Rutherford, E., *Radioactive Substances and their Radiations* (Cambridge, England: Cambridge University Press, 1913).

40. Many histories have attributed the problems of the Bohr atom to radiative instability, but mechanical instability was the more fundamental issue, scientifically and historically speaking. See Heilbron, J. L., and Kuhn, T. S., *HSPS* 1 (1969): 211.

CHAPTER 2

1. Interview, J. Franck, *AHQP* (7 December 1962): 11. The frustrated collaborator was Dirk Coster.

2. The following account is based on the articles in Rozental, S., ed., *Niels Bohr* (New York: Wiley, 1967), particularly Rosenfeld, L., and Rüdinger, E., p. 40; Heilbron, J. L., and Kuhn, T. S., *HSPS* 1 (1969): 211; Bohr's letters in Rosenfeld, L., ed., *Niels Bohr Collected Works* (New York: Elsevier, 1972); autobiographical writings in Bohr, N., *Essays 1958–1962 on Atomic Physics and Human Knowledge* (New York: Wiley, 1963); interviews in the *AHQP*; and the sources cited below.

3. Interview, Niels Bohr, *AHQP* (1 November 1962): 1.

4. Letter, Niels Bohr to Margrethe Nørlund, quoted in Rosenfeld and Rüdinger, *op. cit.*, p. 40. Heilbron and Kuhn (*op. cit.*, p. 223), give the date of the letter as 26 September 1911.

5. Interview, Niels Bohr, *AHQP* (1 November 1962): 6. In the interests of readability, we have switched the first two and last two sentences and slightly changed the punctuation in the transcript.

6. Letter, Niels Bohr to Harald Bohr, 23 October 1911, in Rosenfeld, *op. cit.*, p. 529. Our view here differs from Heilbron and Kuhn, *op. cit.*, who argue that Bohr enjoyed himself at Cambridge. In her *AHQP* interview, Margrethe Bohr discusses her husband's discouragement (Interview, Margrethe Bohr, *AHQP* (30 January 1963): 2–3. Wheeler, J., in Steuwer, R., ed., *Nuclear Physics in Retrospect: Proceedings of a Symposium on the 1930s* (Minneapolis: University of Minnesota Press, 1977): 236ff, also describes Bohr's unhappiness.

7. Bohr, N., *Essays, op. cit.*, p. 31.

8. Planck, M., *AdP* 4, no. 3 (1 March 1901): 553.

9. Planck, M., *Scientific Autobiography and Other Papers*, trans. by Gaynor, F. (New York: Philosophical Library, 1949). His hostility to the atomic theory is described on pp. 32–33. Also useful is the introduction by M. von Laue, which quotes Planck's teacher's advice on p. 8. Much of the following material is covered in Planck, M., *Physikalische Abhandlungen and Vorträge*, (Braunschweig: Vieweg, 1958), especially vol. 3, which has historical articles. The same story is given in Meissner, W., *Science* 113, no. 2926 (26 January 1951): 75. A good popular account is in Segré, E., *From Atoms to Quarks* (San Francisco: W. H. Freeman, 1980): 61ff.

10. In his fine biography of Einstein, the physicist Abraham Pais writes that Planck's mathematical reasoning was unjustifiable "by any stretch of the classical imagination" (*SIL*, p. 371). Kuhn, T. S., *Blackbody Theory and the Quantum Discontinuity, 1894–1912* (New York: Oxford University Press, 1978), chap. 4, argues the opposite.

11. Hermann, A., *Frühgeschichte der Quantentheorie 1899–1913* (Baden: Mosbach, 1969): 32. Planck said, "I had to obtain a positive result, under any circumstances and at whatever cost."

12. A very small number, h is equal to 6.6262×10^{-27} erg–sec.

13. Interview, J. Franck, *AHQP* (7 September 1962): 6.

14. Einstein, E., *AdP* 17, no. 1 (9 June 1905): 132. Einstein hypothesized that freely moving electromagnetic radiation of some types behaves "as though it consisted of distinct independent energy quanta of magnitude $[h\nu]$." (He used other, then-current symbols instead of $h\nu$.) He noted that if light "behaves like a discontinuous medium consisting of energy quanta of magnitude $[h\nu]$, it is reasonable to inquire if the laws of emission and transformation of light are so constituted as though the light were composed of these same energy quanta." (p. 143).

15. Quoted in *SIL*, 382 and 357, respectively.

16. The invention of the term *photon* is in Lewis, G. N., *Nature* 118, no. 2981 (18 December 1926): 874.

17. Letter, E. Rutherford to W. H. Bragg, 20 December 1911, quoted in Eve, A. S., *Rutherford* (New York: Macmillan, 1939): 208.

18. Interview, Niels Bohr, *AHQP* (31 October 1962): 4; (1 November 1962): 13. Interestingly, as shown in Heilbron and Kuhn, *op. cit.*, p. 246, Bohr's decisive rejection of Thomson's model, which took place at this time, was to some extent based on a mathematical error.

19. Letter, Niels Bohr to Harald Bohr, 19 June 1912, Rosenfeld, *op. cit.*, p. 559.

20. Letter, Niels Bohr to Margrethe Nørlund, quoted in Rosenfeld and Rüdinger in Rozental, *op. cit.*, p. 49. There dated "the beginning of July."

21. Most of this memorandum is reprinted in L. Rosenfeld's introduction to Bohr, N., *On the Constitution of Atoms and Molecules* (Copenhagen: Niels Bohr Institute, 1963).

22. In his note to Rutherford, Bohr admitted that he wouldn't even try to give "a mechanical foundation (as it seems hopeless)" to this hypothesis. He added, "This seems to be nothing else than what was to be expected, as it seems to be rigorously proved that [the usual set of physical assumptions] is not able to explain the experimental facts. . . ." (*Ibid.*, pp. xxvi–xxvii).

23. Interview, Niels Bohr, *AHQP*, no. 2 (1 November 1962): 13.

24. During this time, Bohr also worked on the absorption of alpha particles, work which helped stimulate his more well-known studies of the atom. At the honeymoon's end, Bohr showed his bride the pleasures of smoggy Manchester when he took the alpha-ray paper to Rutherford.

25. Balmer, J., *AdP*, series 3, 25, no. 1 (15 April 1895): 80.

26. Rosenfeld in Bohr, *Constitution, op. cit.*, p. xxxix.

27. Hamiltonians were invented by the eponymous William Rowan Hamilton, a nine-

teenth–century Irish mathematical prodigy who is said to have mastered a dozen languages by the time he was nine. Appointed to the astronomy faculty of Trinity College, Dublin, while still a twenty–two–year–old student at the school, Hamilton married unhappily, struggled with alcoholism, and wrote reams of bad poetry, of which he was extremely proud. He was a poor astronomer, maintaining the observatory only with the reluctant assistance of three of his sisters. Shortly after his appointment, in 1827, he recast the science of optics into a more general form, inventing the function that now bears his name. Soon after, in 1833, he used the methods of his optics to reformulate Newtonian mechanics as a whole.

28. Letter, Niels Bohr to Ernest Rutherford, 6 March 1913, quoted in Eve, A. S., *Rutherford* (New York: Macmillan, 1939): 220–21. Bohr wanted Rutherford's opinion, but also lacked standing to submit it himself. Until the 1930s, young scientists had their papers "communicated" to journals by their seniors.

29. Letter, Ernest Rutherford to Niels Bohr, 20 March 1913, quoted partially, *ibid.*, pp. 220–21, and Rosenfeld and Rüdinger, *op. cit.*, p. 54.

30. Bohr, N., PM, series 6, 26; no. 151 (July 1913): 1; no. 153, (September 1913): 476; no. 155, (November 1913): 857.

31. Millikan, R., in *Nobel Lectures 1922–1941* (New York: Elsevier, 1965): 54.

32. Radon, in 1902.

33. Rutherford, E., *PM*, 37, no. 222 (June 1919): 536. The story goes that Rutherford was reprimanded for missing a war meeting and that he responded, "I have been engaged in experiments which suggest that the atom can be artificially disintegrated. If it is true, it is of far greater significance than a war." The statement is sufficiently prescient as to make one skeptical of its veracity. (Jungk, R., *Brighter Than A Thousand Suns*, trans. by J. Cleugh [New York: Harcourt Brace Jovanovich, 1958]: 1).

34. Speech before 1920 Cardiff meeting of British Association for the Advancement of Science in Rutherford's *Collected Papers*, vol. 2 (New York: Interscience, 1964).

35. Curie, I., *CR*, 193, séance of 21 December 1931, p. 1412; Joliot, F., *ibid.*, p. 1415; Curie, I., and Joliot, F., *CR*, 194, séance of 11 January 1932, p. 273.

36. Chadwick, J., *Nature* 129, no. 3252 (27 February 1932): 312. See also, Chadwick, J., in *Proceedings of the Tenth International Congress of the History of Science*, Ithaca, N.Y., 28 August–2 September 1982 (Paris: Hermann, 1964): 159.

37. Quoted in Wilson, D., *Rutherford: Simple Genius* (Cambridge, Massachusetts: MIT Press, 1983): 391. Punctuation and capitalization slightly altered for readability.

38. Interview, Samuel Devons, his office, 23 May 1984. See also, Devons, S., *PT* 24, no. 12 (December 1971): 38.

CHAPTER 3

1. Interview, Howard M. Georgi III, his office, Harvard University, 14 June 1983.

2. Georgi, H., and Glashow, S., *PRL*, 12, no. 8 (25 February 1974): 438.

3. Interview, Howard M. Georgi III, restaurant in New York, 29 January 1985.

4. Interview, Howard M. Georgi III, his office, Harvard University, 14 June 1983.

5. The first use of the term *quantum mechanics* seems to have occurred around 1919 (Interview, P. A. M. Dirac, *AHQP* [10 May 1963]: 6–9). Max Born claimed the first usage occurred in Born, M., *ZfP* 26, no. 6 (20 August 1924): 379 (Born, M., *My Life: Recollections of a Nobel Laureate* [New York: Scribners, 1975]: 215).

6. The literature about Einstein and relativity is vast, and cannot be more than alluded to here. Two authoritative biographies are *SIL* and Hoffman, B., and Dukas, H., *Albert Einstein, Creator and Rebel* (New York: Viking, 1972). See also Schilpp, P., ed., *Albert Einstein: Philosopher–Scientist* (New York: Tudor, 1949). Many popular explanations of relativity exist; one of Einstein's own is *Relativity*, trans. by Lawson, R. W. (New York: Crown, 1961).

7. See *SIL*. Special relativity, which Einstein developed in 1905, is the branch most germane to particle physics.

8. Lorentz, H., *Versuch einer Theorie der Electrischen und Optischen Erscheinungen in Bewegten Körpern*, reprinted in his Collected Papers, *op cit.*, vol. 5, 1.

9. See the discussion in *SIL*, chap. 6.

10. Thomson, J. J., *PRSA* 96: no. 678 (15 December 1919): 311.

11. *Times* (London), 6 November 1919, p. 12.

12. *New York Times*, 9 November 1919, p. 1.

13. *Ibid.*, 10 November 1919, p. 17. There were articles on Einstein daily for the next week.

14. Quoted in interview, Walter Heitler, *AHQP* (18 March 1963): 4. As Einstein complained, "In the principle of relativity, everything is so clear. But in quantum theory it is

horrible. What a mess it is in!" (In interview, James Franck, *AHQP* (9 July 1962): 12). Franck was unsure of the date, but the substance of Einstein's remark suggests that it must have been made after 1915.

15. Strictly speaking, the charge e of an electron is 4.8×10^{-10} absolute electrostatic units; the charge of the proton is the same.

16. More technically, these quantum numbers are n, l, and m, which are, respectively, the number of nodes in the electron wave equation (which is directly related to the allowed energies); the angular momentum; and one of its three space components, the Z-axis projection of the angular momentum.

17. Pauli, W., *ZfP* 31: nos. 5/6 (19 February 1925): 373. See also letter, Wolfgang Pauli to Alfred Landé, 24 November 1924, *LET*. At the time, it was thought these quantum numbers might be properties of the nucleus, rather than the electrons.

18. This account is based on the *AHQP* interviews with both men; Goudsmit, S. A., *PT* 29: no. 6 (June 1976): 40; Uhlenbeck, G., *ibid.*, p. 46; van der Waerden, B. L., in *Theoretical Physics in the Twentieth Century* (New York: Interscience Publishing, 1960), 199; and letters among van der Waerden, Goudsmit, and Uhlenbeck on *AHQP* microfilm 66, section 6.

19. Uhlenbeck, *op. cit.*, p. 43.

20. Interview, George Uhlenbeck, *AHQP* (31 March 1962): 11.

21. Interview, George Uhlenbeck, his office, Rockefeller University, 10 May 1984.

22. Kronig, R. d. L., in *Theoretical Physics, op. cit.*, 21; Van der Waerden, *op. cit.*, pp. 211–12.

23. Uhlenbeck, *op. cit.*, p. 43.

24. There is, curiously enough, no full English-language biography of Heisenberg, although many writers have treated his life and work as part of other studies. In German, there is Hermann, A., *Heisenberg* (Hamburg: Rowolt, 1976). Heisenberg's own autobiographical and philosophical writings include *Physics and Beyond: Encounters and Conversations* (New York: Harper & Row, 1971); *Physics and Philosophy* (New York: Harper & Row, 1962); *Philosophical Problems of Quantum Physics*, trans. by Hayes, F. C. (Woodbridge, Connecticut: Ox Bow Press, 1979); "Theory, Criticism, and a Philosophy," in International Center for Theoretical Physics, ed., *From A Life of Physics*, Supplement to the IAEA Bulletin (Vienna: International Atomic Energy Agency, n.d.); the lengthy romanticized *AHQP* interviews; and *HDQ*.

25. Heisenberg, *Physics and Beyond, op. cit.*, pp. 8–9.

26. Interview, Werner Heisenberg, *AHQP* (30 November 1962): 3.

27. Heisenberg, *Physics and Beyond, op. cit.*, pp. 8–9.

28. Born, M., *My Life, op. cit.*, pp. 212–13.

29. See Paulsen, F., *The German Universities*, trans. by Perry, E. D. (New York: Macmillan, 1895) for an institutional description.

30. Heisenberg, "Theory," *op. cit.*, pp. 34–35.

31. Letter, Heisenberg to Pauli, 9 July 1925, *LET*.

32. This account follows Heisenberg's own account of events. But see Beller, M., *AHES* 33, no. 4, winter 1985, p. 337. For a more skeptical view, see Mackinnon, E., *HSPS* 8 (1976): 137.

33. Heisenberg, W., *Physics and Beyond, op. cit.*, p. 60.

34. *Ibid.*, p. 61 (both quotes).

35. The term *commute* was introduced a year later by a number of physicists, including Paul Dirac (see Dirac, P. A. M., in *History of Twentieth Century Physics* [New York: Academic Press, p. 129]). The term first appears in print in Born, M., Heisenberg, W., and Jordan, P., *ZfP* 35, no. 8/9 (4 February 1926): 557.

36. Heisenberg, W., *ZfP* 33, no. 12 (18 September 1925): 879. Heisenberg concludes, "Whereas in classical theory [*ab*] is always equal to [*ba*], this is not necessarily the case in quantum theory" (p. 884). The bracketed letters are used in lieu of some notation otherwise not used in this account. Mackinnon, *op. cit.*, pp. 178–81, notes that Heisenberg's mathematical manipulations were not entirely correct.

37. Interview, Werner Heisenberg, *AHQP* (22 February 1963).

38. It is not clear whether Heisenberg mentioned his new work in the public lecture, which was on aspects of line spectra. Cf., interview, Werner Heisenberg, *AHQP* (5 July 1963): 1–2.

39. A fine popular exposition of matrices can be found in Kramer, E., *The Nature and Growth of Modern Mathematics*, 2d ed. (New York: Princeton University Press, 1982).

40. *HDQ*, vol. 3, 43–44; see also Born, *My Life, op. cit.*, pp. 218–20.

41. This breakdown was relatively minor; Born had a more complete breakdown in 1928–29. The recounting of these developments in Born, *My Life, op. cit.*, p. 218ff (from which

the quotation is taken), should be used with care. More reliable chronologies if nothing else are provided in the massive, exasperating *HDQ*, vol. 3, especially chap. 2.

42. Born, M., and Jordan, P., *ZfP* 34: no. 11/12 (28 November 1925): 858, received on 27 September.

43. Interview, Werner Heisenberg, *AHQP* (27 February 1963): 22.

44. Letter, Heisenberg to Pauli, 23 October 1925, *LET*.

45. Born, Heisenberg, and Jordan, *ZfP* 35, no. 819, 4 February 1926, p. 557. The article was received on 16 November 1925.

CHAPTER 4

1. Letter, Werner Heisenberg to Wolfgang Pauli, 16 November 1925, *LET*.

2. This is a necessarily truncated summary of a complex series of arguments carefully put in Beller, M., *The Genesis of Interpretations of Quantum Physics, 1925–1927* (Ph.D. dissertation, University of Maryland, 1983). It would be difficult to overstate the extent to which we have been assisted by Dr. Beller's thesis, portions of which were published, in somewhat different form, in *Isis* 74, no. 274 (December 1983): 469 and *AHES*, 33, no. 4 (Winter 1985): 337. Other useful secondary sources are MacKinnon, E., *HSPS* 8 (1977): 137; MacKinnon, E., in Suppes, P., ed., *Studies in the Foundations of Quantum Mechanics*, PSA, 1980; van der Waerden, B. L., in Mehra, J., ed., *The Physicist's Conception of Nature* (Boston: D. Reidel, 1973): 276; Hendry, J., *The Creation of Quantum Mechanics and the Bohr-Pauli Dialogue* (Dordrecht: D. Reidel, 1984), and the sources cited below.

3. Pauli, W., *ZfP* 36, no. 5 (27 March 1926): 336, received on 17 January.

4. For Pauli's description of his solution, see *Naturwissenschaften* 18, no. 26 (27 June 1930): 602. Heisenberg's delight is in his letter to Pauli, 3 November 1925, *LET*, vol. 1.

5. Dirac, P. A. M., *PRSA* 110, no. 755 (1 March 1926): 561. A little later, almost identical reasoning was published in Wentzel, G., *ZfP* 37, nos. 1/2 (22 May 1926): 80.

6. Strictly speaking, they could not calculate the intensity of the hydrogen spectrum lines, and this information was necessary to go on to more complicated systems.

7. A good biography of Louis de Broglie is *Louis de Broglie, physicien et penseur* (Paris: Albin Michel, 1953). See also Segrè, E., *From X-Rays to Quarks* (San Francisco: W. H. Freeman, 1980): 149–53; the thesis itself is de Broglie, L., *Recherche sur la théorie des quanta*, University of Paris, defended 25 November 1924 (Paris: Masson et Cie, 1924).

8. Schrödinger, E., *AdP*, series 4, 79, no. 4 (13 March 1926): 361. Curiously, this initial formulation of Schrödinger's work was nonrelativistic, that is, it did not take into account the Einsteinian effects of the very rapid motions of the electron. Schrödinger had initially worked on a relativistic model, but had gotten what he thought were wrong answers; in fact, the apparent discrepancy was caused because he did not know about the spin of the electron. Thus, he published a nonrelativistic paper.

9. The first instance of this term appears to be in a letter from Schrödinger to Albert Einstein, 28 April 1926, quoted in the "Schrödinger" entry in the *DSB*.

10. Bernstein, J., *Science Observed* (New York: Basic Books, 1982): 149.

11. Schrödinger, E., *AdP*, series 4, 79, no. 6 (6 April 1926): 489.

12. Typical positive reactions are quoted in Jammer, M., *The Conceptual Development of Quantum Physics* (New York: McGraw-Hill, 1966): 271.

13. Letter, Max Born to Erwin Schrödinger, 6 November 1926, *AHQP*.

14. Schrödinger, E., *AdP*, series 4, 79, no. 4, *op. cit.*, p. 375. English versions are in Schrödinger, E., *Collected Papers on Wave Mechanics*, trans. by Shearer, J. F., and Deans, W. M. (London: Blackie & Son, 1928). We use these translations, but amend awkward diction. The original papers are photographically reprinted in Schrödinger, E., *Die Wellenmechanik* (Stuttgart: Ernst Battenberg Verlag, 1963).

15. *Ibid.*, p. 506 (both quotes trans. by authors; emphasis in original).

16. *Ibid.*, p. 514.

17. Heisenberg's letter to Born is described in Born, M., *My Life: Recollections of a Nobel Laureate* (New York: Scribners, 1975): 233; the letter to Pauli was on 8 June 1926 (*LET*, vol. 1). "And now for an unofficial remark about physics. The more I think about the physical side of Schrödinger's theory, the more loathsome I find it. . . . When Schrödinger writes about the visualizability of his theory, I find it crap [*Mist*]."

18. Heisenberg, W., *Physics and Beyond: Encounters and Conversations* (New York: Harper & Row, 1971): 72; Mehra, J., *The Birth of Quantum Mechanics* (Geneva: CERN service d'information scientifique, 1976): 39.

19. Letter, Wolfgang Pauli to Pascual Jordan, 12 Apr. 1926, *LET*, vol. 1. There is an English translation in van der Waerden, B. L., *op cit.*, p. 282ff. As van der Waerden makes clear, the letter also contains what is now called the Klein-Gordon equation.

20. Schrödinger, E., *AdP*, series *4*, 79: no. 8 (4 May 1926): 734. Two weeks after Schrödinger submitted his paper, an American physicist, Carl Eckart, of the California Institute of Technology, independently sent in a third equivalence proof. (Eckart, C., *Proceedings of the National Academy of Sciences* 12, no. 7 [15 July 1926]: 473).

21. See, for example, his letter to Lorentz, 6 June 1926, in Przibaum, K., ed., *Letters on Wave Mechanics*, trans. by Klein, M. J. (New York: Philosophical Library, 1967): especially 64–65. For the attitude toward older physicists, see *HDQ*, vol. 3, p. 167.

22. Schrödinger, *AdP*, *op cit.*, p. 736.

23. Letter, Erwin Schrödinger to Hendrik Lorentz, 6 June 1926, in Przibram, *Letters, op cit.*, pp. 63–65. Schrödinger writes, "But what I had not clearly recognized yet at that time [a few weeks before] was and is that the rules set up for this purpose by Born, Jordan, and Heisenberg are actually false if one applies them to generalized coordinates. . . ." Heisenberg later confessed that throughout his career he had used "rather dirty mathematics," but said that this forced him "always to think of the experimental situation . . . [and] somehow you get closer to reality than by looking for the rigorous methods." (Heisenberg, W., in International Center for Theoretical Physics, ed., *From a Life in Physics*, Supplement to the IAEA Bulletin [Vienna: International Atomic Energy Agency, n.d.]: 39).

24. Van Vleck, J., in Price, Chissick, and Ravensdale, eds., *Wave Mechanics: The First Fifty Years* (New York: Wiley, 1973): 26. Van Vleck dryly remarks, "The last days of the old quantum theory were the golden age of empiricism, where physicists often obtained correct answers by appropriate doctoring of formulas based on questionable theory, and some of this empiricism survived in the early days of quantum mechanics" (p. 31). Heisenberg, Born, and Jordan had developed their methods for a simple, two-dimensional model, then substituted three-dimensional terms in the equations, an erroneous procedure that Schrödinger, at least, had spotted.

25. Schrödinger, E., *AdP*, series *4*, 80: no. 13 (13 July 1926): 437. The steady drumbeat of Schrödinger's papers—one a month for five months—must have seemed particularly alarming to Heisenberg, who spent the same time struggling with his increasingly unwieldy matrices.

26. Heisenberg, *Physics and Beyond*, *op cit.*, p. 73.

27. Schrödinger, E., *AdP* 83, no. 15 (9 August 1927): 956.

28. Heisenberg, W., in Rozental, S., ed., *Niels Bohr* (New York: Wiley, 1967): p. 103. We have slightly amended the unlovely translation. Another version of the same story can be found in Heisenberg, *Physics and Beyond*, *op cit.*, pp. 73–76.

29. Interview, I. I. Rabi, his apartment, New York City, 9 May 1984. See also, *TP*, pp. 213–14. (All quotes from the same interview.)

30. See, for instance Born, *My Life*, *op cit.*, p. 226, and his *AHQP* interview.

31. Born, M., *ZfP* 38, no. 11/12 (14 September 1926): 803.

32. Letter, Pauli to Heisenberg, 19 October 1926, *LET*.

33. *Ibid.*

34. Letter, Heisenberg to Pauli, 28 October 1926, *ibid.*

35. There is amazingly little biographical writing about Jordan. We have relied on his *AHQP* interview.

36. Jordan, P., *ZfP* 30, no. 4/5 (29 December 1924): 297; Einstein, A., *ZfP* 31, no. 10 (21 March 1925): 784.

37. Jordan, P., and Kronig, R. d. L., *Nature* 120: (3 December 1927): 807.

38. This point is developed at length in Jammer, *Conceptual Development*, *op cit.*, p. 331. As Beller has pointed out, however, Jammer's conclusions are outdated.

39. Jordan, P., *Naturwissenschaften* 15, no. 5 (4 February 1927): 105, especially 108. See also, Beller, *AHES*, *op. cit.*

40. Letter, Heisenberg to Pauli, 5 February 1927, *LET*.

41. This is a reconstruction based on the several somewhat contradictory accounts Heisenberg wrote of his train of thought. See, for instance, interviews, Heisenberg, W., *AHQP* (22 February 1963): 26 (5 July 1963): 10, and (11 February 1963); Heisenberg, *Physics and Beyond*, *op cit.*, pp. 77–78; and Heisenberg in Rozental, *Niels Bohr*, *op cit.*, p. 105ff. The inconsistencies in these reports are due in part to the difficulty of conveying in ordinary language the creative jumble of a mind at work. A good synthetic account is in Robertson, P., *The Early Years, The Niels Bohr Institute 1921–1930* (Copenhagen: Akademisk Forlag, 1979): 117–23.

42. Letter, Heisenberg to Pauli, 23 February 1927, *LET*. We have slightly amended the letter by using modern notation.

43. Interview, Heisenberg, W., *AHQP* (25 February 1963): 17.

44. This point has been brought out very clearly by Beller, *Genesis, op cit.*, p. 245ff. The sources here are those in note 41.

45. Interview, Heisenberg, W., *AHQP* (25 February 1963): 17.

46. Heisenberg, W., *ZfP* 43: no. 3/4 (31 May 1927): 172.

47. Letter, Erwin Schrödinger to Max Planck, 4 July 1927, reprinted in Przibaum, *Letters, op cit.*, pp. 17–18.

48. Beller, *Isis, op cit.*, pp. 490–91. The synthesis of Heisenberg's and Schrödinger's approaches was acknowledged at the fifth Solvay Conference on Physics in October 1927, when many physicists viewed quantum mechanics as essentially complete (*Proceedings of the Fifth Solvay Conference* [Paris: Gauthiers-Villars, 1928]). Einstein, however, violently objected; in a now famous series of discussions each day at breakfast he proposed a hypothetical experiment designed to "trick" a particle in some Rube Goldberg–like manner into revealing its exact position and momentum at the same time. Bohr, Pauli, and Heisenberg anxiously tried to answer Einstein's question at the table; by dinner, they had succeeded, and Einstein retired to dream up another experiment. When the conference closed, Einstein's every objection had been satisfied, except for his gut feeling that something was wrong with quantum mechanics. See *SIL*, pp. 444–57, for an account of Einstein's opposition.

49. Wigner, E. P., in Wigner, E. P., ed., *Symmetries and Reflections* (Bloomington, Indiana: Indiana University Press, 1967): 172.

50. Well-known exemplars of this genre are Zukav, G., *The Dancing Wu Li Masters* (New York: William Morrow, 1979); and Capra, F., *The Tao of Physics* (New York: Random House, 1975).

51. Zukav, *op cit.*, p. 53.

52. Interviews, Robert Serber, his office, Columbia University, 1 and 8 April 1985.

53. Bridgman, P. W., *Harper's Bazaar*, March 1929, p. 451.

CHAPTER 5

1. Schweber, S., "Some Chapters for a History of Quantum Field Theory: 1938–1952," 1983 Les Houches lectures, unpublished ms., p. 19. We are grateful to Dr. Schweber for allowing us to see a draft of his excellent survey. This account is based primarily on Dirac's *AHQP* interviews and autobiographical writings, and *HDQ*, vol. 4.

2. Interview, P. A. M. Dirac, *AHQP* (4 January 1962): 5–6.

3. Interview, P. A. M. Dirac, *AHQP* (7 May 1963): 15.

4. The story is told that Peter Kapitza, a Russian physicist who passed the 1920s at the Cavendish, told Dirac to read *Crime and Punishment*, and later asked about his reaction. "It is nice," was the diplomatic reply, "but in one of the chapters the author made a mistake. He describes the Sun as rising twice on the same day." Cited in the anecdotal Gamow, G., *Thirty Years That Shook Physics* (New York: Doubleday, 1966): 121–22.

5. Quoted in *HDQ*, vol. 4: 11.

6. Dirac, P. A. M., *BPP*, p. 46.

7. Interview, Samuel Devons, his office, Barnard College, 23 May 1984. See also, interview, H. B. G. Casimir, *AHQP* (5 July 1963): 8.

8. Interview, I. I. Rabi, his apartment, New York, 9 May 1984.

9. Interview, H. B. G. Casimir, *AHQP* (5 July 1963): 8.

10. Heisenberg, W., *ZfP* 33, no. 12 (18 September 1925): 880.

11. Dirac, P. A. M., *The Development of Quantum Theory* (Oppenheimer Memorial Lecture; New York: Gordon and Breach, 1971): 23–24.

12. Dirac, P. A. M., in *Rendiconti della Scuola Internazionale di Fisica "Enrico Fermi,"* 57th Course (Rome: Università degli Studi, 1974): 134.

13. Letter, Heisenberg to Dirac, 20 November 1925, quoted in *HDQ*, vol. 4, pp. 159–60.

14. Dirac. P. A. M., in C. Weiner, ed., *History of Twentieth Century Physics* (New York: Academic Press, 1977): 125.

15. *Ibid.*, p. 128.

16. Schrödinger, E., *AdP*, series 4, 79, no. 4 (13 March 1926): 361.

17. Dirac, P. A. M., *BPP*, p. 44.

18. Interview, P. A. M. Dirac, *AHQP* (10 May 1963): 28.

19. Gamow, *Thirty Years, op cit.*, chap. 6; Dirac's *AHQP* interviews; interview, Werner Heisenberg, *AHQP* (27 February 1963): 17.

20. Williams, L. P., *The Origins of Field Theory* (Lanham, Maryland: University Press of

America, 1980); Maxwell, J. C., "On Physical Lines of Force, III," in Niven, W. D., ed., *The Scientific Papers of James Clerk Maxwell* (New York: Dover, 1966): 489.

21. Einstein, A., *AdP* 17: no. 1 (June 1905): 132.

22. As far back as 1910, when Paul Debye tried it, physicists had thought of trying to quantize the field, but they had not possessed the necessary quantum mechanical techniques (Debye, P., *AdP*, series *4*, 33, no. 16 [20 December 1910]: 1427).

23. He also derived the creation and annihilation operators first described in Einstein, E., *Verhandlungen der Deutschen Physikalischen Gesellschaft*, series 2, 18, no. 13/14 (30 July 1916): 318.

24. Dirac, P. A. M., *PRSA* 114, no. 747 (1 March 1927): 243. (This and other important articles on quantum electrodynamics are reprinted in Schwinger, J., *Selected Papers on Quantum Electrodynamics* [New York: Dover, 1958].)

25. *Ibid.*, 243.

26. This was pointed out in similar language by Weisskopf, V. F., *BPP*, pp. 56–57.

27. Dirac, P. A. M., *PRSA* 114, no. 747 (1 April 1927): 243 (both quotes).

28. Schweber (*op. cit.*, p. 36) makes this point with elegance and care.

29. It is worth nothing that the concept of empty space was itself relatively new. Until the turn of the century, the Universe was supposed to be permeated by a substance known as the ether, intangible stuff which was thought to be the very fabric of the cosmos. When Dirac banished the vacuum, he got rid of a concept some older physicists had just learned to accept.

30. Dirac, P. A. M., *BPP*, pp. 50–52.

31. Klein, O., and Gordon, W., *ZfP* 45, no. 11–12 (18 November 1927): 751. Strictly speaking, Klein and Gordon recovered relativistic invariance for the Schrödinger equation, but changed the wave equation to an equation of motion.

32. Dirac, P. A. M., *BPP*, p. 51.

33. Dirac, P. A. M., *PRSA* 117, no. 766 (1 February 1928): 610; Dirac, P. A. M., *PRSA* 118: no. 779 (1 March 1928): 351.

34. The Klein-Gordon equation also had two energy values, which Dirac knew full well (see *ibid.*, p. 612).

35. Cf. Weisskopf, V. F., *BPP*, p. 63.

36. Møller, C., *AdP* 14, no. 5 (15 August 1932): 531; Klein, O., and Nishina, Y., *ZfP* 52, no. 11–12 (9 January 1929): 853; and Dirac's two electron papers.

37. Interview, Sir Rudolf Peierls, *AHQP* (18 June 1963): 5–6; confirmed in interview, Sir Rudolf Peierls, his office, Oxford University, 9 October 1984.

38. See *TP*, chap. 14.

39. The first major contribution by an American to quantum theory came in 1926, when Carl Eckart, of the California Institute of Technology, demonstrated the equivalence of Schrödinger's wave mechanics and Heisenberg's matrices. See chap. 4, note 20, above.

40. Interview, I. I. Rabi, his apartment, New York City, 9 May 1984.

41. Heisenberg, W., and Pauli, W., *ZfP* 56: no. 1–2 (8 July 1929): 1; *ibid.*, 59, no. 3–4 (2 January 1930): 168.

42. Rosenfeld, L., *ZfP* 76, no. 11–12 (12 July 1932): 729.

43. Dirac, P. A. M., *PRSA* 126, no. 801 (1 January 1930): 360; P. A. M. Dirac, *BPP*, p. 50.

44. Halpern L., and Thirring, W., *The Elements of the New Quantum Mechanics*, trans. H. Brose (London: Methuen, 1930–31): 101.

45. Weisskopf, V. F., *BPP*, p. 63.

46. Dirac, P. A. M., in *ibid.*, p. 51.

47. *Ibid.*

48. Dirac, P. A. M., *PRSA* 126 (*op cit.*): 360.

49. Weyl, H., *The Theory of Groups and Quantum Mechanics*, trans. by Robertson, H. P. (New York: Dutton, 1931, reprinted New York: Dover, 1930): 234; see also Oppenheimer, J. R. *PR* 35, no. 5 (1 March 1930): 562.

50. He did. Dirac, P. A. M., *PRSA* 133, no. 1 (1 September 1931): 60.

51. Interview, P. A. M. Dirac, *AHQP* (14 May 1963): 30. We have reversed the order of the two sentences.

52. Wilson, C. T. R., Nobel Prize lecture, 12 December 1927, in *Nobel Lectures in Physics 1922–1941*. (New York: Elsevier, 1965): 194.

53. Wilson came to his realizations over a period of years. The publications are listed in Blackett, P. M. S., "Charles Thomson Rees Wilson 1869–1959," in *Biographical Memoirs of the Royal Society*, vol. 6, 1960, p. 294ff.

54. The critical factor is the expansion ratio, that is, the volume of the chamber before and after the pistol is pulled out. With dust in the air, clouds form at a very small expansion ratio; without dust, the ratio rises to at least 1.25 (cf. Janossy, L., *Cosmic Rays* [Oxford: Clarendon Press, 1948]: 56–58).

55. Anderson, C. D., with Anderson, H. L., *BPP*, p. 136. Anderson discovered that adding ethyl alcohol to the vapor made the tracks brighter. The first person to use cloud chambers to study cosmic rays was the Russian scientist Dmitry Skobeltzyn, in the late 1920s. See Skobeltzyn, D., *BPP*, p. 111.

56. Interview, G. D. Rochester, his office, Durham University, 8 October 1984.

57. Anderson, C. D., *American Journal of Physics* 29, no. 12 (December 1961): 825.

58. Anderson, C. D., *PR* 43, no. 6 (15 March 1933): 491.

59. Anderson, C. D., *Science* 76, (9 September 1932): 238.

60. Hanson, N. R., *The Concept of the Positron* (Cambridge, England: Cambridge University Press, 1963): 139, note 2.

61. Blackett, P. M. S., and Occhialini, G. P. S., *PRSA* 139, no. 839 (3 March 1933): 699; Dirac, *Development, op cit.*, pp. 59–60.

62. Oppenheimer, J. R., and Plesset, M. S., *PR* 44, no. 1 (1 July 1933): 53; Fermi, E., and Uhlenbeck, G., *PR* 44: no. 6 (15 September 1933): 510.

63. Kevles, D. J., *The Physics Teacher* 10, no. 4 (April 1972): 175, provides some discussion of the excitement of the time.

67. Rutherford, E., comment after discussion, in *Structure et Propriétés des Noyaux Atomiques*, Rapports et Discussions du Septième Conseil de Physique tenu à Bruxelles du 22 au 29 Oct. 1933 (Paris: Gauthiers-Villars, 1934): 177–78 (trans. by authors).

CHAPTER 6

1. Letter, Pauli to Dirac, 17 February 1928, *LET*. Italics in original; "What do you think about this?" (Was meinen Sie dazu?) was a favorite Pauli phrase. These notions are discussed in a somewhat different fashion in Weinberg, S., *Daedalus* 106 (Spring 1977): 17, especially pp. 24–30.

2. We are leaving out factors of pi and the like.

3. Thomson, J. J., *PM*, series 5, 11, no. 68 (April 1881): 229.

4. See *SIL*, p. 157.

5. Lorentz, H., *Collected Papers*, Vol. 5. (The Hague: Nijhoff, 1937): 127. See also, Lorentz, H., *The Theory of the Electron* (Leiden: Teubner, 1916). This discussion is presented in considerably more technical form in Pais, A., *Developments in the Theory of the Electron* (Princeton: Institute for Advanced Study, 1947); and Pais, A., in Salam, S., and Wigner, E. P., eds., *Aspects of Quantum Theory* (New York: Cambridge University Press, 1972): 79.

6. Interview, Sir Rudolf Peierls, his office, Oxford University, 9 October 1984.

7. In a letter to Oskar Klein on 16 March 1929, Pauli still wanted to "discuss with you this perpetually obscure question: can the self-energy be eliminated by a simple permutation of the factors [in the equations] . . ."? *LET*.

8. The colleague was Victor Weisskopf; cf., Pauli's entry in *DSB*. We are indebted to Prof. Abdus Salam for the story about Pauli's assistant, Ralph Kronig. There is no full biography of Pauli.

9. Pauli, W., *Encyklopädie der mathematischen Wissenschaften*, vol. 5, part 2 (Leipzig: Teubner Verlag, 1921).

10. Pauli, W., *ZfP* 31, no. 10 (21 March 1925): 765.

11. Cited in Darrigol, O., "Les débuts de la théorie quantique du champs (1925–1948)," Thèse pour le Doctorat de Troisième Cycle, Université de Paris-Panthéon-Sorbonne, unpublished, 1982, (catalog no. I 8211–4): 6.

12. The literature on Oppenheimer is both large and curiously incomplete; his physics contributions have received scant attention compared to his military work and his trial. See, for example, Rabi, I. I., Serber, R., Weisskopf, V. F., Pais, A., and Seaborg, G., *Oppenheimer* (New York: Scribners, 1967); Oppenheimer, J. R., *Letters and Recollections*, Smith, A. K., and Weiner, C., eds. (Cambridge, Massachusetts: Harvard University Press, 1980); and the strongly criticized dual biography, Davis, N. P., *Lawrence and Oppenheimer* (New York: Simon and Schuster, 1968).

13. According to *HDQ* 1A (Preface, p. xxv), Oppenheimer did not, as is commonly alleged, speak Sanskrit well.

14. Letter, P. Ehrenfest to Pauli, 26 November 1928, *LET*. See also, Born, *My Life* (New York: Scribners, 1975): 210ff.

15. Letter, Pauli to P. Ehrenfest, 15 February 1929, *LET*.

16. Interview, George Uhlenbeck, his office, Rockefeller University, 10 May 1984.

17. It was thought at first that the paper might be signed by Heisenberg, Pauli, and Oppenheimer; it would be a sequel to the Heisenberg–Pauli formulation of quantum electrodynamics. Circumstances seem to have intervened, and although Oppenheimer used the ideas of the other two men, he wrote and published the paper himself. Interview, J. Robert Oppenheimer, *AHQP* (20 November 1963): 22. (We are grateful to Dr. Abraham Pais for drawing this to our attention.) The interaction was not entirely ignored—only what physicists call the "off-shell" part of it (see chap. 16).

18. Oppenheimer, J. R., *PR* 35, no. 5 (1 March 1930): 461. Concurrent calculations were made by Ivar Waller, another young theorist influenced by Pauli. Waller, however, explored the self-energy of an electron floating through space, not bound in a hydrogen atom (Waller, I., *ZfP* 62, no. 9/10 [15 June 1930]: 673).

19. Conversation, Abraham Pais, a cafeteria in Rockefeller University, 31 October 1984. Others have said that Oppenheimer may have been too influenced by the fatalistic world view of the Hindu religion (Rabi, I. I., in Rabi, et al., *Oppenheimer, op cit.*, p. 7; Elsasser, W. M., *Memoirs of a Physicist in the Atomic Age* [New York: Science History Publications, 1978]: 52–53).

20. The names for the orbitals come from the old spectroscopists' approximate description of the lines they produce. "Sharp," "principal," "diffuse," and "fundamental" have become S, P, D, and F. See chapter 7.

21. Oppenheimer, J. R., and Furry, W., *PR* 45, no. 4 (15 February 1934): 245. "A simple change in notation" is, of course, hindsight.

22. *Ibid.*, p. 253, italics added.

23. *Ibid.*, p. 260.

24. *Ibid.*, p. 260–61.

25. Dirac, P. A. M., in *Structure et propriétés des noyaux atomiques, Conseil de Physique Solvay, VII, rapports et discussions* (Paris: Gauthiers–Villars, 1934): 203. The Solvay Conference was held in late October, just as Oppenheimer and Furry were working on their paper.

26. The quotation is from Dirac's original English-language manuscript, "Theory of the Positron," Churchill College Archives, Section 2, File 8, p. 4.

27. *Ibid.*, p. 6.

28. Letter, Pauli to Heisenberg, 14 June 1934. Pauli said Dirac's "subtraction physics" filled him with "disgust." See also, letter, Pauli to Dirac, 1 May 1933 (both *LET*).

29. Weisskopf, V. F., *Physics in the Twentieth Century* (Cambridge, Massachusetts: MIT Press, 1972): 10–12.

30. Weisskopf, V. F., *ZfP* 89, no. 1/2 (15 May 1934): 27–39.

31. Weisskopf, V. F., *ZfP* 90, no. 11/12, (17 September 1934): 817–18.

32. Interview, Victor F. Weisskopf, his office, CERN, Geneva, 24 September 1984.

33. Interview, Werner Heisenberg, *AHQP* (12 July 1963): 11.

34. Heisenberg, W., *ZfP* 90, no. 3/4 (10 August 1934): 211.

35. They must also be gauge invariant, as will be discussed in chapter 10. One of the first to point this out directly was Peierls, R., *PRSA* 146, no. 857 (1 September 1934): 420.

36. This was first noticed by Mott, N. F., *Proceedings of the Cambridge Philosophical Society* 27, no. 2 (April 1931): 255 (read 26 January 1931); Sommerfeld, A., *AdP*, series 5, 11, no. 3 (29 September 1931): 257; and Bethe, H., and Heitler, W., *PRSA* 146, no. 856 (1 August 1934): 83.

37. Letter, Pauli to Ralph Kronig, 20 November 1935, *LET*.

38. Interview, Victor F. Weisskopf, his office, CERN, Geneva, 24 September 1984. Dr. Weisskopf has kindly allowed us to put together here a number of statements he made at different points during our conversation.

39. Interview, Robert Serber, his office, Columbia, 6 May 1985.

40. The National Research Council was established during World War I as an attempt by the American Academy of Sciences to help in the war effort. It was to promote scientific research that would benefit and protect "the national security and welfare." In 1919, the NRC received money to begin the first national program of postdoctoral fellowships in U.S. history. (The history of this program is recounted in the excellent book *TP*, especially pp. 110–13 and 149–51.)

41. Interviews with Robert Serber, Edward Uehling, and Wendell Furry; Franken, P., in ter Haar, D., and Scully, M., eds., *Willis E. Lamb, Jr.: A Festschrift on the Occasion of his 65th Birthday* (New York: North-Holland Publishing, 1978): VII; Rabi, I.I., in *Oppenheimer, op cit.*, p. 6; Lamb, W., *BPP*, p. 311, especially pp. 313–14.

42. Serber, R., *PR* 48, no. 1 (1 July 1935): 49; Uehling, E. A., *PR* 48, no. 1 (1 July 1935): 55. The Caltech spectroscopic results are described in the next chapter.

43. Uehling, E. A., *ibid.*, 61.

44. Interview, Robert Serber, his office, Columbia, 26 November 1984.

45. Weisskopf, V. F., *Det Kongelige Danske Videnskabernes Selskab, Mathematisk-fysiske Meddelelser* 14, no. 6 (1936). Serber and Uehling had considered only unchanging fields; Weisskopf treated varying external fields as well.

46. Serber, R., *PR* 49, no. 7 (1 April 1936): 545.

47. Interview, Robert Serber, his office, Columbia, 6 May 1985.

48. Bloch, F., and Nordsieck, A., *PR* 52, no. 2 (July 1937): 54.

49. Cf., *BPP*, p. 212, 270; Serber interviews cited above.

50. Serber, R., *PR* 49 (*op. cit.*): 546.

51. Interview, Robert Serber, a restaurant in Manhattan, 22 November 1983 (all quotes).

52. Weisskopf, V. F., in *Physics in the Twentieth Century* (Cambridge, Massachusetts: MIT Press, 1972): 13.

53. Heisenberg, W., *AdP*, series 5, 32, no. 1/2 (8 April 1938): 20–33.

54. This attempt is abundantly documented in *LET*, vol. 2.

55. Dirac, P. A. M., *PRSA* 167, no. 929 (5 August 1938): 148 (here he also had to assume that signals could travel faster than light within the electron); Dirac, P. A. M., *Annales de l'Institut Henri Poincaré* 9: no. 2 (1939): 13; P. A. M. Dirac, *Communications of the Dublin Institute for Advanced Studies*, series A, vol. 1, 1943 (see p. 35).

56. Sakata, S., and Hara, O., *PTP* (Kyoto) 2, no. 1 (January–February 1947): 30; Pais, A., *Verhandlungen Koninklijke Akademie van Wetenschappen te Amsterdam*, 19, 1947.

57. Weisskopf, V. F., *PR* 55, no. 7 (1 April 1939): 678 (minutes of the American Physical Society meeting in New York, 23–25 February 1939); Weisskopf, V. F., *PR* 56, no. 1 (1 July 1939): 72.

58. Interview, Victor Weisskopf, his office, CERN, 24 September 1984.

CHAPTER 7

1. Telephone interview, Markus Fierz, 21 September 1984. For the hydrogen atom, alpha is the velocity of the electron in the ground state divided by the speed of light; it also determines the position of every other energy level in the atom, making it of crucial importance to spectroscopists.

2. For example, at the conclusion of the paper in which he formulated both the full version of quantum electrodynamics and the basic subtraction method, Heisenberg asserted that "a contradiction-free union of the conditions of quantum theory with the corresponding predictions of field theory is only possible in a [theory] that provides a particular value for Sommerfeld's constant $e^2/\hbar c$." (Heisenberg, W., *ZfP* 90, no. 3/4 [10 August 1934]: 231, trans. by authors.) See also, letters, Heisenberg to Pauli, 8 June 1934 and 25 March 1935, *LET*; Born, M., *Proceedings of the Indian Academy of Sciences*, 2A (1935): 533.

3. We are grateful to Richard Learner for providing us with this information.

4. The story is recounted in Chandrasekhar, S., *Eddington: The Most Distinguished Astrophysicist of His Time* (New York: Columbia University Press, 1985). Spoofs of Eddington abounded in the literature of the period. See, for example, Beck, G., Bethe, H., and Riezler, W., *Naturwissenschaften* 19 (1931): 39, and Born, M., *Experiment and Theory in Physics*. (Cambridge: Cambridge University Press, 1944): 37. The latter connects 137 to the Book of Revelations.

5. A major proponent of the importance of studying the physical constants was Raymond T. Birge (1887–1980), who went to Berkeley in 1918 and became chairman of its department in 1932.

6. Spedding, F. H., Shane, C. D., and Grace, N. S., *PR* 44, no. 1 (1 July 1933): 58.

7. Houston, W. M., and Hsieh, Y. M., *Bulletin of the American Physical Society* 8, no. 6, (24 November 1933): 5; Houston, W. V., and Hsieh, Y. M., *PR* 45, no. 2 (15 January 1934): 130 (meetings of Stanford American Physical Society meeting, 15–16 December 1933). Quotation from *PR*. Their long paper (see below) was received September 16, although not published until five months later. Houston's name is pronounced "Hoo-stun," not "Hew-stun."

8. Kemble, E. C., and Present, R. D., *PR* 44, no. 2 (15 December 1933): 1031.

9. Kent, N. A., Taylor, L. B., and Pearson, N. H., *PR* 30, no. 3 (September 1927): 266.

10. Telephone interviews, Robley Williams, 20–21 November 1984.

11. Williams, R. C., and Gibbs, R. C., *PR* 44, no. 5 (15 August 1933): 325; also *PR* 44, no. 12 (15 December 1933).

12. Confusingly, this line is called the alpha line for entirely unrelated reasons.

13. Gibbs, R. C., and Williams, R. C., *PR* 45, no. 3 (1 February 1934): 221.

14. Houston, W. V., and Hsieh, Y. M., *PR* 45, no. 4 (15 February 1934): 263.

15. *Ibid.*, p. 263.

16. *Ibid.*, p. 272.

17. Telephone interview, Edward A. Uehling, 20 July 1984.

18. Uehling, E. A., *PR* 48, no. 1 (1 July 1935): 55, especially p. 61.

19. Williams, R. C., and Gibbs, R. C., *PR* 45, no. 7 (1 April 1934): 475.

20. Spedding, F. H., Shane, C. D., and Grace, N. S., *PR* 47, no. 1 (1 January 1935): 38. The quote below is from p. 38; see also pp. 43–44.

21. Telephone interviews, Robley Williams, 20–21 November 1984.

22. Williams, R. C., and Gibbs, R. C., *PR* 49, no. 5 (1 March 1936): 416 (minutes of Saint Louis American Physical Society meeting, 31 December 1935–2 January 1936).

23. Houston, W. V., *PR* 51, no. 6 (15 March 1937): 446. The distinguished European experimenter, Hans Kopfermann, had also weighed in on the side of the discrepancy–finders (Kopfermann, H., *Naturwissenschaften* 22, no. 14 [6 April 1934]: 218). His initial results fed the desire of Pauli and Weisskopf to keep at the self-energy problem. Then Kopfermann gave the problem to his student, Maria Heyden, who found no discrepancy, perhaps because of the poor quality photographic materials she used (Heyden, M., *ZfP* 106, nos. 7/8 [3 August 1937]: 499).

24. The atomic velocity causes a change in frequency known as a Doppler effect, after Johann Christian Doppler (1803–53), the Austrian physicist and astronomer who first predicted its existence in 1842.

25. The apparatus is sufficiently clever to warrant description. The hydrogen is kept in a 20-centimeter U–shaped tube (a Wood's tube, after R. F. Wood, who invented it in 1928) plunged into a closed tank of liquid air. Physicists connect a 2000–volt electrical source to terminals sealed into the tube. This excitation causes the hydrogen to radiate—that is, to emit light. The light goes into the interferometer, whose heart consists of two unbelievably flat glass plates about the size of a pocket watch, clamped face to face in a small, doughnut–shaped frame. The two inside faces of the plates are coated with a ghost–thin film of mirror silvering and are perhaps a centimeter apart. The light from the tube of hydrogen passes through the first plate, hits the mirror side of the second, bounces back, hits the mirror side of the first, bounces forward, and continues back and forth as many as thirty times before dissipating. At each reflection, however, a small amount of light makes its way through the silvering and passes through the forward lens. The thirty or so successively reflected beams emerge out of the interferometer out of alignment with each other, crests and waves ever so slightly out of step. By coating the interferometer just right and adjusting the spacing, an experimenter can "tune" the equipment so that in some directions the various crests reinforce each other. The reinforced crests can be focused to form a series of glowing, photographable concentric rings.

26. Bethe, H., in *Handbuch der Physik*, 2d ed., H. Geiger and K. Scheel, eds., vol. 24, pt. 1 (Berlin: Julius Springer, 1933): 273, especially sections 10, 43, and 44.

27. The Stark effect and the Zeeman effect, as it happens. Franck was just passing through. Bethe joined the Cornell faculty, and was in more contact with the experimenters (Letter, Hans Bethe to authors, 21 October 1985).

28. Telephone interview, Hans Bethe, 4 December 1984.

29. Telephone interview, Robley Williams, 20 November 1984.

30. Williams, R. C., *PR* 54, no. 8 (15 October 1938): 558.

31. Telephone interview, Robley Williams, 21 November 1984.

32. Pasternack, S., *PR* 54, no. 12 (15 December 1938): 1113. The quote below is from p. 1113. Houston was still working on the problem, as can be seen in the work of his student C. F. Robinson, *PR* 55, no. 4 (15 February 1939): 423 (minutes of the Los Angeles American Physical Society meeting on 19 December 1938). The presentation seems not to have been elaborated into a formal paper, however.

33. Telephone interview, Robley Williams, 20 November 1984. Pasternack died in 1982.

34. Telephone interview, J. W. Drinkwater, 19 October 1984.

35. Drinkwater, J. W., Richardson, O., and Williams, W. E., *PRSA* 174, no. 957 (1 February 1940): 164.

36. *Ibid.*, p. 184.

37. *Ibid.*, p. 187.

38. Interview, Dick Learner, his office, Imperial College, University of London, London, 5 October 1984.

39. Biographical information from Lamb, W. E., *BPP*, p. 311; Lamb, W. E., in *A Festschrift for I. I. Rabi*, Transactions of the New York Academy of Sciences, series 2, 38 (4 November 1977): 82; Lamb, W. E., in Marshak, R., ed., *Perspectives in Modern Physics* (New York: Wiley, 1966): 261.

40. Telephone interview, Willis Lamb, 9 August 1984.

41. The proposal is reprinted in Lamb, *Festschrift, op cit.*

42. The technical description can be found in Lamb, W. E., *PR* 79, no. 4 (15 August 1950): TK, and Trigg, G. L., *Landmark Experiments in Twentieth Century Physics* (New York: Crane, Russak, 1975): 99.

43. Telephone interview, J. Bruce French, 11 July 1984; interview, Victor Weisskopf, his office, CERN, 24 September 1984.

44. We are here relying heavily on the account in Schweber, S., "Some Chapters for a History of Quantum Field Theory: 1938–1952," 1983 Les Houches Lectures, unpublished ms.

45. See, for example, letter, J. R. Oppenheimer to H. A. Kramers, 14 April 1947, National Archives, Oppenheimer Collection, Case File, Box 44, Kramers file. Oppenheimer writes that he was "planning to go to this strange meeting on Eastern Long Island in early June, though I am as much in the dark as you as to what it is all about."

46. Letter, V. F. Weisskopf to K. K. Darrow, 18 February 1947, cited in Schweber, *op cit.*, p. 131.

47. Weisskopf, V. F., "Foundations of Quantum Mechanics, Outline of Topics for Discussion," unpublished ms., April (?) 1947, National Archives, Oppenheimer Collection, Case File, Box 72, File "Theor. Physics Conf.—Corres.—Shelter Isl. '47."

48. Kramers, H., *NC* 15, (1938): 108; see also, Kramers, H., *Nederlandsch Tijdschrift voor Natuurkunde* 11 (1944): 134, and his Shelter Island outline (same as note 47).

49. Interview, Julian Schwinger, a restaurant in Los Angeles, 4 March 1983.

50. Interview, Robert Serber, his office, Columbia, 18 July 1984.

51. Telephone interview, Hans Bethe, 8 August 1984.

52. Interview, Richard Feynman, *AHQP*, 1966, p. 454.

53. Letters, J. Robert Oppenheimer to Alfred N. Richards, 1 December 1947; Alfred N. Richards to J. Robert Oppenheimer, 2 December 1947, both in Library of Congress, Manuscript Division, Oppenheimer Collection, General Case File, Box 72, folder labeled "Theor. Physics Conf.—Corres.—Poconos 1948."

54. Interview, I. I. Rabi, his apartment, New York City, 9 May 1984.

55. Nafe, J. E., Nelson, E. B., and Rabi I. I., *PR* 71, no. 12 (15 June 1947): 914; see also Kusch, P., and Foley, H. M., *PR* 72, no. 12 (15 December 1947): 1256.

56. The reconstruction is based on interviews and the surprisingly good account by Steven White in the *Herald Tribune*, 3 June 1947.

57. Interview, Julian Schwinger, a restaurant in Los Angeles, 4 March 1983.

58. See chap. 9 and Marshak, R., *BPP*, p. 381.

59. Schwinger, J., *BPP*, p. 332.

60. Interview, Hans Bethe, *AHQP* (9 May 1966): 168.

61. Telephone interview, Hans Bethe, 8 August 1984. Bethe was directly inspired by Kramers, whose idea of comparing the self-energies of a free electron and a $^2S_{1/2}$ electron was simpler than that of Weisskopf and French, who compared the $^2S_{1/2}$ and $^2P_{1/2}$ states (Letter, Hans Bethe to authors, 21 October 1985).

62. Oppenheimer Collection, National Archives, General Case File, Box 20, Bethe folder. After further numerical calculations by Bethe's students, the paper was submitted two weeks later.

63. Lamb, W. E., *BPP*, p. 324. Weisskopf's chagrin was touched with exasperation; he believed Bethe had used but not acknowledged the ideas Weisskopf had developed (Interview, Victor Weisskopf, his office, CERN, 24 September 1984). Indeed, Bethe's rough draft in the Oppenheimer ms. collection does not mention Weisskopf.

64. Telephone interview, Willis Lamb, 9 August 1984.

CHAPTER 8

1. Telephone interview, Lloyd Motz, 10 December 1984; biographical information also from Schwinger, J., *BPP*, p. 329.

2. Interview, I. I. Rabi, his apartment, New York City, 9 May 1984.

3. Interview, Julian Schwinger, a restaurant in Los Angeles, 4 March 1983.

4. Interview, George Uhlenbeck, his office, Rockefeller, 10 May 1984.

5. Telephone interview, Lloyd Motz, 10 December 1984.

6. Letters, Victor Weisskopf to J. Robert Oppenheimer, 22 and 29 August and 22 and 14 December 1948, Library of Congress, Manuscript Division, Oppenheimer Papers, General Case File, Box 72, folder labeled "Theor. Physics Conf.—Corres.—Poconos 1948"; letter, Victor Weisskopf to authors, 5 August 1985; telephone interview, J. Bruce French, 11 July 1984; Weisskopf, V., *Physics in the Twentieth Century* (Cambridge, MIT Press, 1972): 17–18.

7. Schwinger, J., *PR* 73, no. 4 (15 February 1948): 416 (received on 30 December 1947); French, J. B., and Weisskopf, V., *PR* 75, no. 8 (15 April 1949): 1240; Kroll, N., and Lamb, W., *PR* 75, no. 3 (1 February 1949): 388.

8. The New York City American Physical Society meeting was held at Columbia on 29–31 January 1948.

9. Interview, Julian Schwinger, a restaurant in Los Angeles, 4 May 1983.

10. Dyson, F., *Disturbing the Universe* (New York: Harper & Row, 1979): 55.

11. Letter, Sin-itiro Tomonaga to J. Robert Oppenheimer, 2 April 1948, Library of Congress, Manuscript Division, Oppenheimer papers, General Case File, Box 73, Tomonaga folder. See also, Tomonaga, S., *PTP* 1, no. 2 (August–September 1946): 27 (a Japanese version appeared in 1943).

12. Takabayasi, T., *BPP*, p. 280.

13. Night letter, J. Robert Oppenheimer to Sin-itiro Tomonaga, 13 April 1948, found in Tomonaga folder (note 11, above); Oppenheimer's letter is in General Case File, Box 72, folder labeled "Theor. Physics. Conf.—Corres.—Pocono—1948."

14. Letter, Sin-itiro Tomonaga to J. Robert Oppenheimer, 14 May 1948; *Physical Review* note of receipt, 1 June 1948, both in Box 73, Tomonaga folder (see above).

15. Schwinger, J., *PR* 73, no. 4 (15 February 1949): 416.

16. Weisskopf, V. F., *BPP*, p. 76.

17. Pauli's reaction can be seen in the letters to Oppenheimer of 6 January 1948, 22 February 1949, and (28?) February 1949 in the Oppenheimer Collection, Library of Congress, Manuscript Division, General Case File, Box 56, Pauli file. For Dirac's attitude, see, for example, Dirac, P. A. M., *BPP*, p. 39.

18. Feynman's high school recollections can be found in Feynman, R. P., "*Surely You're Joking, Mr. Feynman!*" (New York: W. W. Norton, 1985): 15–30.

19. Feynman, R. P., in Badash, L., *Reminiscences of Los Alamos* (Dordrecht, Holland: D. Reidel, 1980): 105. Feynman, "*Surely,*" *op. cit.*, pp. 107–55, has a somewhat similar account.

20. Feynman, R. P., in "The Pleasure of Finding Things Out," *Nova* broadcast, WGBH, Boston, 25 January 1983.

21. Introduced to quantum physics in Dirac, P. A. M., *Physikalische Zeitschrift der Sowjetunion* 3, no. 1 (January 1933): 64.

22. Interview, Robert Serber, his office, 22 October 1984; interview, John Wheeler, Columbia University Faculty Club, 14 December 1983; interview, Richard Feynman, *AHQP*, vol. VII, pp. 450–86.

23. Interview, Richard Feynman, his office, Caltech, 22 February 1985 (all direct quotes).

24. Dyson, F. J., *PR* 75, no. 3 (1 February 1949): 486.

25. Interview, Julian Schwinger, a restaurant in Los Angeles, 4 March 1983; interview, Victor Weisskopf, his office, CERN, 24 September 1984.

26. Cited in Chandrasekhar, S., *Eddington, the Most Distinguished Astrophysicist of His Time* (New York: Cambridge University Press, 1985): 3.

27. Weisskopf, V. F., *BPP*, p. 78; confirmed during interview cited above.

28. His father's name was Stückelberg, with an umlaut; in the United States, he changed his name to Stueckelberg for convenience.

29. Interviews, Valentine Telegdi, CERN cafeteria, 24 and 25 September 1984; interview, Charles Ruegg, his office, Geneva Institute of Physics, 21 September 1984; interview, John Iliopoulos, his office, Rockefeller University, 8 May 1984; telephone interview, Jean Rivier, 2 October 1984; interview, André Petermann, his home, Geneva, 10 February 1984.

30. Stueckelberg, E. C. G., *HPA* 11, no. 3 (1938): 221, especially 242–43; pts. II, III (1938): 298. On p. 317, incidentally, he introduces the concept of baryon conservation, the violation of which is currently a subject of considerable theoretical interest. See also *AdP* 21 (1934): 367.

31. The classical approach is found in Stueckelberg, E. C. G., and Patry, J. F. C., *HPA* 13

(1940): 167; Stueckelberg, E. C. G., *HPA* 14 (1941): 51.

32. Stueckelberg, E. C. G., *HPA* 28 (1945): 21; Stueckelberg, *HPA* 19 (1946): 241.

33. Rivier, D., *HPA* 22, no. 3 (1949): 265.

34. Stueckelberg, E. C. G., and Petermann, A., *HPA* 24 (1951): 317; Stueckelberg, E. C. G., and Petermann, A., *HPA* 26 (1953): 499.

35. Interview, E. C. G. Stueckelberg, his home, Geneva, 10 February 1984.

CHAPTER 9

1. Wulf, T., *PZ* 11, no. 18 (15 September 1910): 812. The paper gives the time, place, and temperature of the observations.

2. Becquerel, H., *CR* 122: 559, séance of 9 March 1896.

3. Elster, J., and Geitel, H., *PZ* 2, no. 38 (22 June 1901): 560. Elster and Geitel, two physicists at Wolfenbüttel, near Brunswick, Germany, were inseparable colleagues known as the Castor and Pollux of physics. Their collaboration was so extensive that the story is told that a man who looked like Geitel was once greeted, "Herr Elster!" He said, "I'm not Elster, I'm Geitel. And I'm not Geitel, either." Elster was the prototypical absentminded professor: Upon being informed he had put a stamp of greater denomination than necessary on a letter, he crossed it out and put a correct stamp on. Because he had a large and a small dog, he cut a big hole and a little hole in his apartment door. And so on.

4. Wilson, C. T. R., *Proceedings of the Cambridge Philosophical Society* 11 (read 26 November 1900): 32.

5. McLenna, J. C., and Burton, E. F., *PR* 16, no. 3 (March 1903): 184.

6. Rutherford, E., and Cooke, H. L., *PR* 16, no. 3 (March 1903): 183.

7. Piel, C., *Der Mathematische und Naturwissenschaftliche Unterricht* 1, no. 3 (February 1949): 105.

8. Simpson, G. C., and Wright, C. S., *PRSA* 85, no. 577 (10 May 1911): 175; Simpson, G. C., *Meteorology. British Antarctic Expedition 1910–1913*, vol. 1 (Calcutta: Thacker, Spink, 1919).

9. See, for example, Wulf, T., *PZ* 10, no. 5 (1 March 1909): 152; Kurz, K., *PZ* 10, no. 22 (10 November 1909): 834. For a contrary view, however, see Richardson, O., *Nature* 73, no. 1904 (26 April 1906): 607, and sources cited therein.

10. Wulf, T., *PZ* 11, no. 18 (15 September 1910): 811.

11. Hess entry, *DSB*.

12. Hess, V., *Thought* 15, no. 57 (June 1940): 229.

13. Gockel, A., *PZ* 12, no. 14 (15 July 1911): 595.

14. Hess, V. F., *PZ* 12, nos. 22/23 (15 November 1911): 998.

15. The photograph is reproduced in Kraus, J., *Our Cosmic Universe* (Powell, Ohio: Cygnus-Quasar Books, 1980): 196–200.

16. Hess, V., *PZ* 13, nos. '21/22 (1 November 1912): 1090.

17. King, L. V., *PM*, series 6, 23, no. 134 (February 1912): 248.

18. Kolhörster, W., *Verhandlungen der Deutschen Physikalischen Gesellschaft* 16, no. 14 (30 July 1914): 719.

19. For an account of the interactions between these groups, see Cassidy, D., *HSPS* 12, no. 1 (Winter 1981): 1.

20. Darrow, K. K., *Bell System Technical Journal* 11, no. 1 (January 1932): 148.

21. There are many sources of information about Millikan. See, for example, Millikan, R. A., *The Autobiography of Robert A. Millikan* (New York: Prentice–Hall, 1950); Kevles, D. J., *Scientific American* 240, no. 1 (January 1979): 142; *TP*, pp. 88–89, 231–42, and *passim*. Millikan's own writing should be enjoyed, but read with suspicion.

22. The first was Albert Michelson, for his experimental proof of the nonexistence of the ether.

23. Millikan, R., *Science* 57, no. 1483 (1 June 1923): 630.

24. Cited in *TP*, p. 180.

25. Millikan, R. A., and Bowen, I. S., *PR* 22, no. 2 (August 1923): 198; subsequently rewritten and reinterpreted in *PR* 27, no. 4 (April 1926): 360.

26. Millikan, R. A., and Otis, R. M., *PR*, series 2, 23, no. 6: 778; subsequently rewritten and reinterpreted in *PR* 27, no. 6 (June 1926): 645.

27. Hess, V. F., *PZ* 27, no. 12 (15 June 1926): 405; Kolhörster, W., *Naturwissenschaften* 15, no. 5 (4 February 1927): 126.

28. Millikan, R. A., *Science* 62, no. 1612 (20 November 1925): 445; a revised version of this

paper is Millikan, R. A., *Proceedings of the National Academy of Sciences* 12, no. 1 (January 1926): 48. Millikan's full account is Millikan, R. A., and Cameron, G. H., *PR*, series 2, 28, no. 5 (November 1926): 851. It is interesting to compare Millikan's original description of his first conclusions with those given a year later.

29. Quoted in *TP*, p. 179.

30. Among the angry papers written in the dispute: Hess, V. F., *PZ* 27, no. 6 (15 March 1926): 126; Bergwitz, K., Hess, V. F., Kolhörster, W., and Schweidler, E., *PZ* 29, no. 19 (1 October 1928): 705; Millikan, R. A., *Nature* 126, no. 3166 (5 July 1930): 14. See also *TP*, p. 179.

31. Lemon, H. B., *Cosmic Rays Thus Far* (New York: Norton, 1936): 56–57.

32. Bothe entry, *DSB*.

33. Bothe, W., and Kolhörster, W., *ZfP* 56, nos. 11/12 (16 August 1929): 751.

34. Quoted in Kevles, *Scientific American, op. cit.*, p. 145.

35. Millikan, R. A., *Annual Report of the Smithsonian Institution*, 1931, p. 270, especially 282–83.

36. The results are summarized in Störmer, F., in *Zeitschrift für Astrophysik* 1, no. 4 (1930): 237.

37. Clay, J., *Verh. Koninklijke Akademie van Wetenschappen te Amsterdam*, 30, no. 9/10 (1927): 711.

38. Millikan, R.A., and Cameron, G.H., *Nature*, 121 no. 3036 (supp.) (7 January 1928): 20.

39. Bothe, W., and Kolhörster, W., *Sitzungsberichte der Preussischen Akademie der Wissenschaften zu Berlin* 24 (1930): 450. The real reason for their failure to discover the latitude effect is that cosmic rays are sufficiently powerful to be affected by the earth's magnetic field only near the equator, where it is strongest. The calculations were carried out successfully in Rossi, B., *NC* 8, no. 3 (March 1931): 85.

40. Hess, *Thought, op. cit.*, p. 234.

41. Jaffe, B., *Outposts of Science* (New York: Simon and Schuster, 1935): 399.

42. The letter is reproduced in Auguste Piccard, *A 16000 Metri*, trans. by D. S. Suardo (Milan: Mondadori, 1933) facing p. 300.

43. Jaffe, *op. cit.*, p. 400.

44. Millikan, R. A., *Cosmic Rays* (New York: Macmillan, 1939): 58.

45. Quoted in *New York Times*, 15 September 1932, p. 23.

46. Neher, H. V., *BPP*, p. 127.

47. *New York Times*, 31 December 1932, p. 1; 1 January 1933, p. 16; 5 February 1933, p. 81; see also the reworked American Association for the Advancement of Science speech in Millikan, R. A., *PR* 43, no. 8 (15 April 1933): 661.

48. Auger, P., and Skobeltzyn, D., *CR* 189, séance of 1 July 1929, p. 55; Blackett, P. M. S., and Occhialini, G. P. S., *PRSA* 139, no. 839 (3 March 1933): 699.

49. Interview, Pierre Auger, his apartment, Paris, 3 October 1984.

50. Rossi, B.; Auger, P., and Leprince-Ringuet, L., both in *International Conference on Physics, London 1934* (Cambridge: University Press, 1935): 233, 195.

51. Anderson, C. D., *BPP*, p. 146.

52. Bethe, H., and Heitler, W., *PRSA* 146, no. 856 (1 August 1934): 83.

53. W. Heitler, *The Quantum Theory of Radiation*, 1st ed. (Oxford: Oxford University Press, 1936 [Preface dated November 1935]). Heitler's remark was typical; see also Oppenheimer, J. R., *PR* 47, no. 1 (1 January 1935): 44, especially p. 47.

54. Anderson, C., *BPP*, p. 147.

55. Anderson, C. D., and Neddermeyer, S. H., *PR* 50, no. 4 (15 August 1936): 263.

56. Yukawa, H., *Tabibito*, trans. by L. Brown and R. Yoshida. (World Scientific, 1982): 79–80.

57. *Ibid.*, p. 119.

58. Schwinger, J., *BPP*, pp. 354–55.

59. Yukawa, *op. cit.*, p. 174.

60. Heisenberg, W., *Zfp* 77, no. 1/2 (19 July 1932): 1; 78, no. 3/4 (21 September 1932): 156; 80, no. 9/10 (16 February 1933): 587.

61. Yukawa, H., *Tabibito, op. cit.*, p. 195.

62. Yukawa, *ibid.*, pp. 194–195.

63. Fermi, E., *NC* 11, no. 1 (January 1934): 1.

64. Letter, Wolfgang Pauli to Lise Meitner, Hans Geiger, et al., 4 December 1930, *LET*; the hypothesis first printed in a comment by Pauli in *Structure et Propriétés des Noyaux Atomiques*,

Rapports et discussions du Septième Conseil de Physique, Institut International de Physique (Paris: Gauthier-Villars, 1934): 324.

65. Rasetti, F., comment in Fermi, E., *Collected Papers* (Chicago: University of Chicago Press, 1962): 538.

66. Tamm, I., *Nature* 133, no. 3374 (30 June 1934): 981; Iwanenko, *ibid.*, 981.

67. Yukawa, *Tabibito, op. cit.*, p. 201. The argument below is simpler than Yukawa's original thoughts.

68. Yukawa, H., *Proceedings of the Physico-Mathematical Society of Japan* 17, no. 2 (February 1935): 48.

69. *Ibid.*, p. 48.

70. *Ibid.*, p. 53.

71. Brown, L., Konuma, M., and Maki, Z., *Particle Physics in Japan, 1930–1950*, Research Institute for Fundamental Physics preprint 408, vol. 2 (September 1980): 31.

72. Tanikawa, Y., in Yukawa, H., *Scientific Works* (Tokyo: Iwanami Shoten, 1979): xvi.

73. Yukawa, *Tabibito, op. cit.*, p. 203.

74. Hayakawa, S., *BPP*, p. 88; the work culminated in Yukawa, H., and Sakata, S., *Proceedings of the Physico-Mathematical Society of Japan* 19, no. 12 (December 1937): 1084; sections III–V, published with various collaborators in the same journal the next year.

75. Brown, Konuma, and Maki, vol. 2, *op. cit.*, p. 29.

76. Tomonaga, S., et al., in Brown, Konuma, and Maki, *op. cit.*, pp. 14–15.

77. Neddermeyer, S., and Anderson, C. D., *PR* 51, no. 10 (15 May 1937): 884–86. The ms. was received March 30, 1937.

78. Street and Stevenson reported their results to the American Physical Society meeting on 29 April 1937 (an abstract appeared in *PR* 51, no. 11 [1 June 1937]: 1005; the information that it was delivered on April 29 is from the final note to the Anderson and Neddermeyer paper). They sent a two-page Letter to the Editor to the *Physical Review* on 6 October (Street, J. C., and Stevenson, E. C., *PR* 52, no. 9 (1 November 1937): 1003. Later it was discovered that experimenters had been photographing mesons for years without knowing their significance. See, for instance, Kunze, P., *ZfP* 83, no. 1–2 (6 June 1933): 1.

79. Oppenheimer, J. R., and Serber, R., *PR* 51, no. 12 (15 June 1937): 1113.

80. Stueckelberg, E. C. G., *PR* 52, no. 1 (1 July 1937): 41.

81. Stueckelberg, E. C. G., *Nature* 137, no. 3477 (20 June 1936): 1032; telephone interview, Jean Rivier, 2 October 1984; interviews, Valentine Telegdi, CERN cafeteria, 24 and 25 September 1984.

82. Anderson, C. D., *BPP*, p. 148. The letter itself is in Anderson, C. D., and Neddermeyer, S., *Nature* 142, no. 3602 (12 November 1938): 878; see also, Kemmer, N., *PTP* (Supp. 35 ext.), November 1965, 605.

83. *E.g.*, Millikan, R. A., *Electrons (+ and −), Protons, Photons, Neutrons, Mesotrons, and Cosmic Rays*, rev. ed. (Chicago: University of Chicago Press, 1947): 508ff.

84. Compton, A., *RMP* 11, no. 3 (July–October 1939): 122.

85. Euler, H., and Heisenberg, W., *Ergebnisse der Exakten Naturwissenschaften* 17 (1938): 1.

86. Oppenheimer, J.R., *PT* 19, no. 11 (November 1966): 58.

87. Interview, Bruno Rossi, his office, MIT, 1 November 1984.

88. Amaldi, E., *Scientia* 114, no. 1 (1979): 51.

89. Piccioni, O., *BPP*, p. 225. We have relied greatly on this and Conversi, M., *ibid.*, p. 242. We are grateful for interviews with Marcello Conversi (his office, University of Rome, 16 February 1984) and Oreste Piccioni (Fermilab, 3 May 1985).

90. Comment by Piccioni, O., *BPP*, p. 272.

91. Piccioni, *BPP*, p. 226.

92. Interview, Oreste Piccioni, Fermilab, 3 May 1985.

93. Conversi, M., and Piccioni, O., *NC* 2, no. 1 (1 April 1944): 40. After the war, the two men learned that Rossi had made an earlier determination by a different method (Rossi, B., and Nereson, N., *PR* 62, nos. 9–10 [1 and 15 November 1942]: 417; Rossi, B., Hilberry, H., and Hoag, J., *PR*, 56, 8 [15 October 1939]: 837–38). A group in occupied France was also working on the meson lifetime, *i.e.*, R. Chaminade, André Fréon, and Roland Maze, *CR* 218, no. 10 (6 March 1944): 402.

94. Tomonaga, S., and Araki, G. *PR* 58, no. 1 (1 July 1940): 90. As Piccioni put it, "Because at that time previous work had well established that 55% of the mesotrons were positive and 45%

negative, we planned to show that only 55% of the stopped particles decayed." (Piccioni, *BPP*, p. 229.)

95. Conversi, M., Pancini, E., and Piccioni, O., *PR* 71, no. 3 (1 February 1947): 209.

96. The full explanation was given in Fermi, E., Teller, E., and Weisskopf, V., *PR* 71, no. 5 (1 March 1947): 314.

97. Rochester, G. D., *CI*, p. 169.

98. Interview, George Rochester, his office, Durham University, 8 October 1984.

99. There is considerable literature on the Japanese "Manhattan Project," much of which is discussed in Phillip S. Hughes, *Social Studies of Science* 10, no. 3 (August 1980): 345 and Wilcox, R.K., *Japan's Secret War* (New York: William Morrow, 1985). The material shortages were exacerbated when a German *U*-boat bound for Japan with two tons of uranium was sunk by Allied vessels. See for example, Shapley, D., *Science*, 13 January 1978, p. 152.

100. Interview, Michiji Konuma, History of the Weak Interaction Conference, Wingspread, Racine, Wisconsin, 30 May 1984.

101. Brown, Konuma, and Maki, *op. cit.*, p. 32.

102. Much of this story is recounted in *BPP*. esp. p. 285.

103. Nishina, Y., Sekido, Y., Miyazaki, Y., and Masuda, T., *PR* 59, no. 4 (15 February 1941): 401; Brown, Konuma, and Maki, vol. 1, *op. cit.*, p. 55.

104. In Japanese, *Mesio-kai*, a pun on "illusion meeting."

105. Sakata, S., *Bulletin Physico–Mathematical Society of Japan* 23, no. 4 (April 1941): 283; *ibid.*, p. 291.

106. U.S. Strategic Bombing Survey, Records, Section 2, Japanese records, 13 b.(4), USSBS report from Yoshio Nishina, National Archives.

107. Sakata, S. and Inoue, T., *PTP* 1, no. 1 (November-December 1946): 143.

108. Marshak, R., *BPP*, p. 385.

109. Weisskopf, V. F., *PR* 72, no. 6 (15 September 1947): 510.

110. Lattes, C. M. G., Muirhead, H., Occhialini, G. P. S., and Powell, C. F., *Nature* 159, no. 4047 (24 May 1947): 694.

111. Marshak, R. E., and Bethe, H. A., *PR* 72, no. 6 (15 September 1947): 506. Interestingly, a similar hypothesis had been made by Christian Møller, summarizing work by himself and Abraham Pais at an English conference (Møller, C., in *Fundamental Particles and Low Temperature Physics*, vol. 1, Cavendish Laboratory, Cambridge, 22–27 July 1946 [London: Taylor & Franci, 1947]: 184). The paper also postulates something like mu–electron universality.

112. Neither Rabi nor anyone else can recall where this now–famous comment was first made, but Rabi thinks that it was at an American Physical Society meeting in New York City.

113. Rochester, G. D. and Butler, C. C., *Nature* 160, no. 4077 (20 December 1947): 855.

114. Marshak, R., *BPP*, p. 376.

115. This and the following quotation are from a lecture by G. D. Rochester at the History of the Weak Interaction conference, Wingspread, Racine, Wisconsin, 31 May 1984.

116. Interview, G. D. Rochester, his office, Durham, 8 October 1984.

117. Leprince-Ringuet, L., *Les rayons cosmiques: les mesotrons* (Paris, Editions Albin Michel, 1945): 137–38, trans. by authors.

118. The history of the Jungfraujoch laboratory is recounted in Debrunner, H., ed., *50 Jahre Hochalpine Forschungsstation Jungfraujoch* (Bern: Wirtschaftsbulletin no. 23 of the Kantonalbank of Bern, October 1981).

119. The description of the Jungfraujoch is from our visit to the site in February 1984.

120. Interview, Antonino Zichichi, his office, CERN, 3 February 1984.

121. Cf. Marshak, R., *Meson Physics* (New York: McGraw-Hill, 1952): 359.

122. The term was coined in Wigner, E., *PR* 51, no. 2 (15 January 1937): 106.

123. A slightly more technical description is to say that isotopic spin is a vector orientated by the presence of an abstract charge space in a way analogous to the magnetic field's orientation of the ordinary spin. Heisenberg thus proposed that different projections in this abstract charge space of the Z axis of this imaginary isotopic spin would be a mathematically convenient way of representing the charge states within a family of particles. Just as the different components of ordinary spin are separated in energy by the magnetic field, causing the fine structure in spectra, so the various components of isotopic spin in a particle family are separated in mass by the effects of the electromagnetic force. This is the origin, by the way, of the slight mass differences between the proton and neutron.

124. The papers are, respectively, Tuve, M. A., Heydenberg, N., and Hafstad, L. R., *PR* 50, no. 9 (1 November 1936): 806; Breit, G., Condon, E. U., and Present, R. D., *ibid.*, p. 825; Cassen, B., and Condon, E. U., *ibid.*, p. 846.

125. Kemmer, N., *Proceedings of the Cambridge Philosophical Society* 34 (1938): 354.

126. Oppenheimer, J. R., and Serber, R., *PR* 53, no. 8 (15 April 1938): 636. They credited the idea to conversations with Gregory Breit, and used it to construct a theory for nuclear transitions that is beyond the focus of this work. Experimental demonstration of isotopic spin conservation in the realm of particle physics first occurred when Enrico Fermi and his collaborators showed it held true for pion–nucleon interactions (Fermi, E., "Pion Scattering in Hydrogen," Third Rochester Conference, 1952).

127. Pais, A., *Physica* 19, no. 9 (July 1953): 869.

128. Telephone interview, Abraham Pais, 5 February 1985.

129. We are here skipping over an important step for the sake of narrative clarity. The Japanese and, later, Pais further hypothesized something called "associated production," which is to say that the V particles could be created only in pairs. At the time, the evidence for associated production was not very good: By the spring of 1951, when the Japanese started their work, less than 150 V particles had been observed, and some of the pictures were poor (Butler, C. C., *Progress in Cosmic Ray Physics* [Boston: North–Holland, 1952] p. 60). Four Japanese articles on associated production were published back to back in *PTP*, a journal already known as a graveyard of good ideas. These were duly interred (Nambu, Y., Nishijima, K., and Yamaguchi, Y., *PTP* 6, no. 4 [July–August 1951]: 615; 619; Aizu, K., and Kinoshita, T., *ibid.*, p. 630; Miyazawa, H., *ibid.*, p. 631; Oneda, S., *ibid.*, p. 633; Pais, A., *PR*, 86, 5 [1 June 1952]: 672).

130. Pais, *Physics, op. cit.*, pp. 869–70.

131. Telephone interview, Murray Gell–Mann, 5 February 1985; see also Gell–Mann, M., *CI*, p. 395.

132. A word about the origins of the names. Hadron is credited to L. Okun; lepton from Møller, *op. cit.*, p. 184; baryon from Pais, A., *Proceedings of the International Conference of Theoretical Physics* (Kyoto and Tokyo, September 1953): 157.

133. Peaslee, D. C., *PR* 86, no. 1 (1 April 1952): 127.

134. Gell-Mann, M., "On the Classification of Particles," typescript, his files.

135. Nishijima, K., *PTP* 9, no. 4 (April 1953): 414 presents similar but much less clearly formulated work. Implicit but unexplained here is the notion of baryon conservation, which will be treated later.

136. Telephone interview, Abraham Pais, 5 February 1985.

137. The first publication of the rules was Amaldi, E., et al., *PT* 6, no. 12 (December 1953): 24. The identical text was published in *NC*, *CR*, *Nature*, and *Naturwissenschaft* in early 1954, and the proceedings of *Congrès International sur le Rayonnement Cosmique*, Bagnères de Bigorre, July 1953.

138. Leprince-Ringuet, L., "Discours de clôture," in proceedings of the *Congrès International sur le Rayonnement Cosmique*, Bagnères de Bigorre, July 1953, typescript, pp. 289–90.

CHAPTER 10

1. Interview, Sheldon Glashow, his office, Harvard, 2 December 1982.

2. The description of the ceremony is based on interviews cited below with the three winners and personal visits to Stockholm.

3. Weinberg, S., *RMP* 52, no. 3 (July 1980): 515.

4. Weinberg, S., *PR* 118, no. 3 (1 May 1960): 838.

5. See, for instance, the discussion of gauge invariance in Wigner, E., "Invariance in Physical Theory," *Proceedings of the American Philosophical Society* 93, no. 7 (December 1949): 521, especially 524–26. It is worth noting that a striking exception to this statement is the study of crystal formation.

6. According to *SIL* (p. 140), the terms *Galilean invariance* and *Galilean transformations* were coined in 1909.

7. For more, see *ibid.*, pp. 121–26 and 140–44.

8. Weinberg, S., *Gravitation and Cosmology* (New York: Wiley, 1972).

9. Interview, Steven Weinberg, his office, University of Texas, 28 November 1984.

10. Literature on Emmy Noether includes Weyl, H., *Scripta Mathematica* 3, no. 3 (July 1935): 201, which is the printed version of Weyl's Memorial Address on Noether; Brewer, J. W., and Smith, M. K., eds., *Emmy Noether, a Tribute to Her Life and Work* (New York: Marcel Dekker, 1981); Dick, A., *Emmy Noether, 1882–1935* (Basel: Birkhäuser Verlag, 1970 [Beihefte zur Zeitschrift Elemente der Mathematik no. 13]); Kramer, E., *The Nature and Growth of Modern Mathematics* (Princeton: Princeton University Press, 1982): 656–79; and her *DSB* entry.

11. Kimberling, C., in Brewer, and Smith, *op. cit.*, pp. 10–12.

12. Quoted in Reid, C., *Hilbert* (New York: Springer Verlag, 1970): 127.

13. See, for instance, Mehra, J., *Einstein, Hilbert, and the Theory of Gravitation* (Boston: D. Reidel, 1974).

14. Quoted in Weyl, *Scripta Mathematica, op. cit.*, p. 207.

15. Quoted in Kimberling, *op. cit.*, p. 18 (trans. by authors).

16. It is said that following the memorial address Weyl sent an obituary notice to the *New York Times*, upon which an editor commented, "Who is Weyl—have Einstein wrote something, as he is the mathematician recognized by the world" (Kimberling, *op. cit.*, p. 52). Einstein did so, and his letter to the *Times* was printed on May 4, 1935. He wrote, "In the judgment of the most competent living mathematicians, Fräulein Noether was the most significant creative mathematical genius since the higher education of women began."

17. Weyl, *Scripta Mathematica, op. cit.*, p. 219.

18. The photograph is reproduced in Brewer and Smith, *op. cit.*, between pp. 17 and 18.

19. Noether, E., "Invarianten beliebiger Differentialausdrücke" and "Invariante Variationsprobleme," *Nachrichten von der Gesellschaft der Wissenschaften zu Göttingen*, 1918, pp. 37–44 and 235–57.

20. Strictly speaking, Noether's theorem applies only to continuous symmetries, such as those for space and time, and not discrete symmetries, such as the similarity between an object and its mirror image.

21. The argument is most fully developed in Weyl, H., *Raum, Zeit, und Materie*, 3rd ed. (Berlin: Springer Verlag, 1920).

22. The German terms are *Masstab invarianz* and *Eich invarianz*, respectively.

23. Weyl's most full description of gauge invariance was Weyl, H., *The Theory of Groups and Quantum Mechanics* (London: Methuen, 1931).

24. Weyl, H., *Mathematische Zeitschrift* 2, nos. 3–4 (30 October 1918): 384; Weyl, H., *AdP*, series 4, 59, no. 10 (20 June 1919): 101. We have not seen a somewhat earlier paper in the *Sitzungsberichte der Koniglichen Preussen Akademie der Wissenschaften* (1918, p. 465).

25. The history of this incident is covered in more detail in *SIL*, p. 341; Yang, C. N., *Annals of the New York Academy of Sciences* 294, no. 1 (8 November 1977): 86. Einstein was exactly right. Indeed, if you take a clock around a room, its wave function phase *does* change (Yang, C.N., *Proceedings of the International Symposium on the Foundations of Quantum Mechanics* [Physical Society of Japan, 1984]: 5).

26. London, F., *Naturwissenschaften* 15, no. 8 (15 February 1927): 187; London, F., *ZfP* 42, nos. 5–6 (14 April 1927): 375. Gauge invariance was independently rediscovered in Klein, O., *ZfP* 37, no. 12 (10 July 1926): 895; Fock, V., *ZfP* 39, nos. 2–3 (2 October 1926): 226. See also Gordon, W., *ZfP* 40 (1927): 119.

27. Weyl, H., *ZfP* 56, nos. 5–6 (19 July 1929): 330.

28. Among the exceptions to this generalization is Peierls, R., *PRSA* 146, no. 856 (1 August 1934): 420.

29. Yang's recollections of Fermi and his graduate school days can be found in Segrè, E., ed., *The Collected Papers of Enrico Fermi*, vol. 2 (Chicago: University of Chicago Press, 1965): 673. It is reprinted in Yang, C. N., *Selected Papers, 1945–1980* (San Francisco: W. H. Freeman, 1983): 305.

30. Telephone interview, Chen Ning Yang, 3 April 1983.

31. Telephone interview, Robert Mills, 7 April 1983. According to Mills's modest assessment, his main contribution was to suggest means of handling the self-interactions of the gauge field quanta.

32. Yang, C. N., and Mills, R. L., *PR* 96, no. 1 (1 October 1954): 191–92, 195. We have slightly amended the notation by ignoring the difference between the terms for the field and four-vectors. Moreover, we have replaced "$\pm e$" by "± 1" for reader convenience.

33. We are grateful to Dr. Abraham Pais for allowing us to see some of the manuscript of his *Inward Bound* (Oxford University Press, 1986), which documents this point.

34. Yang, C. N., *Selected Papers 1945–1980, With Commentary* (San Francisco: W. H. Freeman, 1983): 20.

35. Telephone interview, Chen Ning Yang, 3 April 1983.

CHAPTER 11

1. Goldhaber, M., and Goldhaber, G., *PR* 73, no. 12 (15 June 1948): 1472.

2. Chadwick, J., *Verhandlungen der Deutschen Physikalische Gesellschaft* 16 (1914): 383.

3. For example, Ellis, C. D., *PRSA* 99, no. 698 (1 June 1921): 261; Ellis, C. D., *PRSA* 101, no. 708 (1 April 1922): 1.

4. Ellis, C. D., and Wooster, W. A., *PRSA* 117, no. 776 (1 December 1927): 109.

5. Meitner, L., and Orthmann, W., *ZfP* 60, no. 3–4 (14 February 1930): 143.

6. Among the places Bohr urged this are, Bohr, N., *Journal of the Chemical Society* (1932): 382–88; Bohr, N., in *Convegno di Fisica Nucleare della Reale Accademia d'Italia* (Rome: Reale Academia, 1932): 119.

7. Letter, Wolfgang Pauli to Hans Geiger, Lise Meitner, et al., 4 December 1930, *LET*.

8. Interview, V. F. Weisskopf, *AHQP*, 10 July 1963, p. 12.

9. See citations below.

10. We owe this metaphor to Steven Weinberg, who attributes it to George Gamow.

11. Interview, Maurice Goldhaber, his office, Brookhaven National Laboratory, 21 December 1983.

12. This scene has been described by two of the three listeners: Amaldi, E., *CI*, 261ff; Segrè, E., *Enrico Fermi, Physicist* (Chicago: University of Chicago Press, 1970): 72.

13. Rasetti, F., in Fermi, E., *Collected Papers*, vol. 1 (Italy 1921–38) (Chicago: University of Chicago Press, 1962): 539.

14. The first note is E. Fermi, *Ricerca Scientifica* no. 4 (December 1933): 491. The follow-up papers are *NC* 11, no. 1 (January 1934): 1; and *ZfP* 88, no. 3/4 (19 March 1934): 161.

15. Letter, Ernest Rutherford to Enrico Fermi, 23 April 1934, quoted in Fermi, *Collected Papers*, vol. 1, *op cit.*, p. 641.

16. Fermi, *NC*, *op. cit.*, sec. 9.

17. Gamov, G., and Teller, E., *PR* 49, no. 12 (15 June 1936): 895.

18. Fermi, *NC*, *op. cit.*, sec. 10. The curve actually used by later experimenters, however, was more complicated. It was first described in Kurie, N. D., Richardson, J. R., and Paxton, H. C., *PR* 49, no. 5 (1 March 1936): 368.

19. Ellis, C. D., and Henderson, W. J., *PRSA* 146, no. 856 (1 August 1934): 213ff (phosphorus); Alichanow, A. I., Alichanian, A. I., and Dzelepow, B. S., *ZfP* 93, no. 5/6 (19 January 1935): 350.

20. Konopinski, E. J., and Uhlenbeck, G. E., *PR* 48, no. 1 (1 July 1935): 7; Part 2, *ibid.*, p. 60, no. 4 (15 August 1941): 308.

21. A remarkably straightforward review article of developments until 1943 is Konopinski, E. J., *RMP* 15, no. 4 (October 1943): 209. Particularly illustrating the erroneous results produced by thick sources is Tyler, A. W., *PR* 56, no. 2 (15 July 1939): 125.

22. Wu, C. S., and Albert, R. D., *PR* 75, no. 2 (15 January 1949): 315. See also Wu, *RMP* 22, no. 4 (October 1950): 386.

23. The attempts are reviewed in Crane, H. R., *RMP* 20, no. 1 (January 1948): 278.

24. Bethe, H., and Peierls, R., *Nature* 133, no. 3366 (5 May 1934): 689.

25. See, for instance, Dancoff, S., *Bulletin of the Atomic Scientists* 8, no. 5, June 1952, p. 139.

26. Reines, F., *CI*, p. 238.

27. Letter, Enrico Fermi to Frederick Reines, 8 October 1952, quoted in *ibid.*, p. 241.

28. Reines, F., and Cowan, C. L., Jr., *PR* 92, no. 3 (1 November 1953): 830.

29. The telegram is reproduced in Reines, *CI*, *op. cit.*, p. 249. The discovery is reported in Cowan, Reines, et al., *Science* 124, no. 3212, (20 July 1956): 103.

30. Danby, G., et al., *PRL* 9, no. 1 (1 July 1961): 36.

31. Reines, F., Sobel, H., and Gurr, H. *PRL* 37, no. 6 (9 August 1976): 315.

32. Pontecorvo, B., *PR* 72, no. 3 (August 1947): 246.

33. Lee, T. D., Rosenbluth, M., and Yang, C. N., *PR* 75, no. 5 (1 March 1949): 905; Klein, O., *Nature* 161, no. 4101 (5 June 1948): 897; Puppi, G., *NC* 5, no. 6 (1 December 1948): 587; *ibid.*, 6, no. 3 (3 May 1949): 194; Tiomno, J., and Wheeler, J. A., *RMP* 21, no. 1 (January 1949): 153.

34. Dalitz, R. H., *PR* 94, no. 4 (15 May 1954): 1046. The history of parity is evoked in Yang, C.N., *CI*, p. 439.

35. Discussion comment, Martin Block, Weak Interaction Conference, Wingspread, Wisconsin, 30 May 1984. Separately confirmed in an interview the same day.

36. Yang, C. N., in Ballam, et al., eds., *High Energy Nuclear Physics*, Proceedings of the Sixth Annual Rochester Conference, April 3–7, 1956 (New York: Interscience, 1956): VIII–1.

37. *Ibid.*, VIII–22.

38. *Ibid.*, VIII–27; a misprint has been corrected. It is worth noting that Martin Block recalls Yang's reaction as being considerably more negative (Interview, Martin Block, Weak Interaction Conference, Wingspread, Wisconsin, 30 May 1984).

39. Quoted in Gardner, M., *The Ambidextrous Universe* (New York: Basic Books, 1964): 240.

40. This account is primarily based on Lee, T. D., in Zichichi, A., ed., *Elementary Processes at High Energy*, Proceedings of the 1970 Majorana School (New York: Academic Press, 1971): 830; Lee, T. D., "Broken Parity," in Lee, T. D., *Collected Works*, forthcoming; Yang, C. N., in Yang, C. N., *Selected Papers, 1945–1980* with commentary (San Francisco: W. H. Freeman, 1983): 26–31; Wu, C. S., *Adventures in Experimental Physics*, Vol. Gamma, p. 101; Franklin, A., *Studies in the History and Philosophy of Science* 10, no. 3 (Fall 1979); 201.

41. Siegbahn, K., ed., *Beta- and Gamma-Ray Spectroscopy* (Amsterdam: North-Holland, 1955).

42. Lee, T. D., and Yang, C. N., *PR* 104, no. 1 (1 October 1956): 254–55. The order of the quotes has been changed.

43. Dyson, F., *Scientific American* 199 (September 1958): 74.

44. Interview, Gerald Feinberg, his office, Columbia University, 1 March 1985.

45. Wu, C. S., Ambler, E., Hayward, R. W., Hoppes, D. D., and Hudson, R. P., *PR* 105, no. 4 (15 February 1957): 1413.

46. Garwin, R. L., Lederman, L. M., Weinrich, M., *PR* 105, no. 4 (15 February 1957): 1415. See also, Weinrich, M., Ph.D. thesis, Columbia University, 1958, issued as Nevis Report NEVIS–56 (February 1958); Garwin, R., *Adventures in Experimental Physics*, Vol. Gamma, p. 124.

47. Friedman, J. I. and Telegdi, V. L., *PR* 106, no. 5 (1 March 1957): 1290.

48. Letter, Wolfgang Pauli to Victor Weisskopf, 17 January 1957, quoted in Yang, *Selected Papers, op. cit.*, p. 30.

49. Letter, Wolfgang Pauli to Victor Weisskopf, 27 January 1957, in Kronig, R. d. L., and Weisskopf, V. F., eds., *Collected Scientific Papers of Wolfgang Pauli*, vol. 1 (New York: Wiley-Interscience, 1964): xvii–xviii.

50. Yang, C. N., *Selected Papers. op. cit.*, p. 35; Bernstein, J., *New Yorker*, 12 May 1962, p. 58–59.

51. In the volume of *Adventures in Experimental Physics* devoted to the discovery of parity nonconservation, Goudsmit replied to the criticism, offering reasons to support his insistence that the original be rewritten (*op. cit.*, p. 137). All review papers on the subject, however, have treated all three experiments on an equal basis.

52. The history is covered in Franklin, op. cit.; Cox, R., *Adventures in Experimental Physics*, vol. Gamma, 1973, p. 145; and Grodzins, L., *ibid.*, p. 154. Physicists who raised the general question of parity before Yang and Lee include Purcell, E. M. and Ramsey, N. F., *PR* 78, no. 66 (15 June 1950): 807; Wick, G. C., Wightman, A. D., and Wigner, E. P., *PR* 88, no. 1 (1 October 1952): 101.

53. Interview, Martin Block, Weak Interaction Conference, Wingspread, Wisconsin, 30 May 1984.

54. E.g., letter, Victor Weisskopf to J. R. Oppenheimer, 3 February 1964, in Oppenheimer Papers, Library of Congress, General Case File, Box 20, Weisskopf file; and the historical accounts by Lee and Yang cited above.

55. Lee, T. D., in "High Energy Nuclear Physics," *Proceedings of the Seventh Annual Rochester Conference*, 15–19 April 1957, VII–1, VII–7.

56. The neutron beta decay experiment was Robson, J. M., *PR* 100, no. 3 (1 November 1955): 933; the neon experiment was Maxson, D. R., Allen, J. S., and Jentschke, W. K., *PR* 97, no. 1 (1 January 1955): 109; the helium was Rustad, B. M., and Ruby, S. L., *PR* 89, no. 4 (15 February 1953): 880; *ibid.*, 97, no. 4 (15 February 1955): 991.

57. Wu, C. S., in *Proceedings of the Seventh Annual Rochester Conference*, *op. cit.*, VII–20.

58. Sudarshan, E.C.G. and Marshak, R. E., "Origin of the Universal V-A Theory," Virginia Tech preprint, VPI-HEP 84/8, pp. 7–10.

59. Sudarshan, E. C. G. and Marshak, R. E., *Proceedings of the Padua Conference on Mesons and Recently Discovered Particles* (1957): V–14.

60. Telephone interview, Robert Marshak, 14 January 1985.

61. Gell–Mann, M., Caltech preprint CALT–68–1214, pp. 15–16.

62. Feynman, R. P., *"Surely You're Joking, Mr. Feynman!"* (New York: W. W. Norton, 1985): 250–51. The paper was Feynman, R. P., and Gell–Mann, M., *PR* 109, no. 1 (January 1958): 193. See also the later Sakurai, J. J., *NC* 7, no. 5 (1 March 1958); and Sudarshan, E. C. G., and Marshak, R. E., *PR* 109, no. 5 (1 March 1958): 1860.

63. See, for instance, the experiments referred to in Sudarshan and Marshak, *ibid.*, pp. 1860–62.

64. Goldhaber, M., Grodzins, L., and Sunyar, A., *PR* 109, no. 3 (1 February 1958): 1015.
65. Telephone interview, Murray Gell–Mann, 18 May 1984. He was not alone. Cf., Lee, T. D., in Zichichi, *Elementary Processes, op. cit.*, pp. 837–38.

CHAPTER 12

1. Schwinger's impression on his students—Conversation, Herb Goldstein, 3 November 1983; interview, I. I. Rabi, his apartment, New York City, 9 May 1984; interview, Sheldon Glashow, his office, Harvard University, 2 December 1982.
2. Martin, P. C., *Physica*, 96A, nos. 1/2 (April 1979, Schwinger *Festschrift* issue): 70.
3. Interview, Julian Schwinger, a restaurant in Los Angeles, 4 March 1983.
4. Quoted in Bernstein, J., *New Yorker*, 12 May 1962, p. 82.
5. Schwinger, J., *Annals of Physics* 2, no. 5 (November 1957). See also, Schwinger, J., "Théorie des Particules Elementaires," mimeograph, June 1957 Paris lectures, Glashow's files.
6. The scientist asked for anonymity.
7. Schwinger's predecessors apparently include the Japanese physicists Shoichi Sakata and Takeshi Inoue in 1942 and Mituo Taketani, Seitaro Nakamura, and two others after the war (according to Brown, L., et al., in Brown, Konuma, and Maki, eds., *Particle Physics in Japan, 1930–1950*, vol. 1 (Kyoto: Research Institute for Fundamental Physics, September 1980): 58–60.
8. Schwinger, *Annals of Physics, op. cit.*, p. 425.
9. Flato, M., Fronsdal, C., and Milton, K. A., eds., *Selected Papers (1937–1976) of Julian Schwinger* (Boston: D. Reidel, 1979): 82.
10. His attendance is recorded in "High Energy Nuclear Physics," *Proceedings of the Seventh Annual Rochester Conference*, 15–19 April 1957 (New York: Interscience, 1957).
11. Interview, Sheldon Glashow, his office, Harvard University, 2 December 1982. The discussion here has been slightly condensed with Professor Glashow's kind permission.
12. Biographical information and all quotes from interview, Sheldon Glashow, his office, Harvard University, 2 December 1982.
13. The description of Bronx Science is based on telephone interviews with Charles Hellman, 13 February 1983; Herman Gewirtz, 13 February 1983; and Daniel Greenberger, 16 February 1983, in addition to the interviews cited below.
14. Interview, Gerald Feinberg, a restaurant in New York, 27 January 1983 (this and subsequent quotations).
15. We are grateful to the librarians of the Bronx High School of Science for showing us the yearbooks and course descriptions.
16. Interview, Sheldon Glashow, his office, Harvard University, 2 December 1982.
17. Telephone interview, Sheldon Glashow, 8 April 1983.
18. Telephone interview, Silvan Schweber, 2 January 1984.
19. Telephone interview, Sheldon Glashow, 14 February 1983.
20. Letter, Sheldon Glashow to Gary Feinberg, 16 November 1954, Feinberg's files.
21. Telephone interview, Sheldon Glashow, 8 April 1983.
22. Interview, Sheldon Glashow, his office, Harvard University, 2 December 1982.
23. Interview, Sheldon Glashow, his office, Harvard University, 25 April 1983. The thesis quote is from Glashow, S. L., "The Vector Meson in Elementary Particle Decays," Harvard University Ph.D. Thesis, July 1958 (typescript), p. 75.
24. Glashow, S., *NP* 10, no. 1 (1 January 1959): 107.
25. Interview, Sheldon Glashow, his office, Harvard University, 2 December 1982.
26. Interview, Abdus Salam, his office, International Center for Theoretical Physics (*ICTP*), Trieste, 23 February 1984.
27. Salam, A., *NP* 18, no. 4 (September 1960): 681; Salam, A., and Komar, A., *NP* 21, no. 4 (December 1960): 624; Salam, A., *PR* 127, no. 1 (1 July 1962): 331.
28. Interview, Sheldon Glashow, his office, Harvard University, 2 December 1982.
29. Interview, Murray Gell–Mann, aboard a plane, 3 March 1983.
30. Glashow, S., *NP* 22, no. 4 (February 1961): 579.
31. *Ibid.*, pp. 579, 583, 584.
32. After mixing with the photon, Glashow's Z_s is known as the B, which is today's Z^0.
33. Glashow used (*NP* 22, op. cit., 585) $1/\sin \theta$, whereas Weinberg used $\sin \theta$.
34. Gell–Mann, M., in *Proceedings of the 1960 Annual International Conference on High Energy Physics at Rochester*, 25 August–1 September 1960 (New York: Interscience, 1960): 510.
35. Telephone interviews, Murray Gell–Mann, 18 May 1984 and 13 July 1983.
36. Glashow, *NP* 22, *op. cit.*, p. 586.

37. Glashow, S., and Gell-Mann, M., *Annals of Physics* 15, no. 3 (September 1961): 437.

38. This and the following quote from interview, Murray Gell-Mann, aboard a plane, 3 March 1983.

39. Interview, Abdus Salam, a hotel near Fermilab, 3 May 1985.

CHAPTER 13

1. Salam's childhood is described in an interview with the *Illustrated Weekly of India*, 1 February 1981, p. 10. See also Salam, S., *Ideals and Realities* (Singapore: World Scientific, 1984).

2. Interview, Abdus Salam, an office at CERN, 19 September 1983.

3. Interview, Abdus Salam, his office, ICTP, 23 February 1984.

4. Salam, A., *PR* 82, no. 2 (15 April 1951): 217; Matthews, P. T., and Salam, A., *RMP* 23, no. 4 (October 1951): 311.

5. Interview, Abdus Salam, *Illustrated Weekly of India*, 1 February 1981, p. 12.

6. Shaw, R., "The Problem of Particle Types and Other Contributions to the Theory of Elementary Particles," Cambridge University Ph.D. thesis, unpublished, 1954.

7. Salam, A., *NC* 5, no. 1 (1 January 1957): 299. Salam showed that, just as the photon's masslessness is a manifestation of gauge symmetry, parity violation is a manifestation of chiral symmetry.

8. Letter, Wolfgang Pauli to Abdus Salam, 11 March 1957, courtesy of Professor Salam.

9. Interview, Abdus Salam, his office, ICTP, 23 February 1984.

10. Discussion comment, Abdus Salam, Weak Interaction Conference, Wingspread, Wisconsin, 30 May 1983. The manuscript is Salam, A., "On Fermi Interactions," unpublished ms., courtesy Professor Salam.

11. Salam, A., and Ward, J. C., *NC* 9, no. 4 (16 February 1959).

12. Interview, Abdus Salam, his office, ICTP, 23 February 1984. The paper is Salam, A., and Ward, J. C., *PL* 13, no. 2 (15 November 1964): 168. Of some interest is Salam and Ward, *NC* 19, no. 1 (1 January 1961): 165.

13. Interview, Abdus Salam, New York Hilton, 27 May 1984.

14. Salam, A., "On Fermi Interactions," unpublished ms., courtesy Professor Salam.

15. Interview, Abdus Salam, an office at CERN, 19 September 1984.

16. Superconductivity was first discovered by Heike Kamerlingh Onnes, a Dutch physicist who was fascinated by the behavior of matter at low temperatures, and the electrical resistance of metals. Onnes was the first to liquefy many gases, such as helium, which he then used as a coolant to study low–temperature electrical resistance. In 1911, Onnes was surprised to discover that somewhere near absolute zero the resistance of mercury abruptly dropped so low that he could not measure it—the first evidence of the phenomenon later called superconductivity. In 1913 he was awarded the Nobel Prize for his low temperature studies. Onnes was fanatic about precision; he once said that every physics laboratory should have the motto: "Door meten tot weten" ("Through measuring to knowing"). Onnes entry, *DSB*.

17. Bardeen, J., Cooper, L. N., and Schrieffer, J. R., *PR* 106, no. 1 (1 April 1956) 162; *Ibid.*, 108, no. 5 (1 December 1957): 1175. Equivalent work was done independently by Bogoliubov, N., *JETP* 34, no. 1 (July 1958): 41, and following articles.

18. Among the articles which contributed to this realization are Anderson, P. W., *PR* 112, no. 6 (15 December 1958): 1900; Nambu, Y., *PR* 117, no. 3 (1 February 1960): 648. Nambu's paper contains a more complete list of references in note 3.

19. Telephone interview, Jeffrey Goldstone, 23 January 1984.

20. Goldstone, J., *NC* 19, no. 1 (1 January 1961): 154.

21. Interview, Jeffrey Goldstone, an office at Columbia, 26 November 1984.

22. Interview, Steven Weinberg, his office, University of Texas, 28 November 1984; Nambu, Y., and Jona-Lasinio, G., *PR* 122, no. 1 (1 April 1961): 345. See also, Nambu, Y., *PRL* 4, no. 7 (1 April 1960): 380. Here we give rather short shrift to Nambu; his important work will be discussed further in the following section on strong interactions.

23. Interview, Steven Weinberg, outside Harvard University, 7 May 1985.

24. Interview, Abdus Salam, his office, ICTP, 23 February 1984.

25. Interview, Jeffrey Goldstone, an office at Columbia, 26 November 1984.

26. Goldstone, J., Salam, A., and Weinberg, S., *PR* 127, no. 3 (1 August 1962): 965–70. This paper related the Goldstone boson to relativistic effects, a subject to be discussed below.

27. Weinberg, S., *RMP* 52, no. 3 (July 1980): 516.

28. Interview, Sheldon Glashow, his office, Harvard University, 10 November 1982.

29. Schwinger, J., *PR* 125, no. 1 (1 January 1962): 397.

30. Anderson, P. W., *PR* 130, no. 1 (1 April 1963): 439.

31. *Ibid.*, 441, 442.

32. Telephone interview, Philip Anderson, 28 January 1985.

33. Klein, A. and Lee, B. W., *PRL* 12, no. 10 (9 March 1964): 266–68; Gilbert, W., *PRL* 12, no. 25 (22 June 1964): 713–14. The article is the last Gilbert—who was Salam's graduate student at Cambridge University in England—wrote before switching to biology; he was honored for his biology work with a Nobel Prize in 1980.

34. Priority should be given to the Belgians, if given to anyone (Englert, F., and Brout, R., *PRL* 13, no. 9 [31 August 1964]: 321). They did not show the "eating" mechanisms explicitly, however.

35. Interview, Jeffrey Goldstone, an office at Columbia, 26 November 1984; letter, R. Brout to the authors, 22 October 1984; Higgs, P., "SBGT and all that," Weak Interaction Conference, Racine, Wisconsin, 30 May 1984. Higgs' papers are: Higgs, P. W., *PRL* 13, no. 16 (19 October 1964): 508; Higgs, P. W., *PR* 145, no. 4 (27 May 1966): 1156–63. Brout's quote below is from letter, R. Brout to authors, 13 September 1985.

36. Interview, Steven Weinberg, outside Harvard, 7 May 1985.

37. Interview, Steven Weinberg, his office, University of Texas, 28 November 1984.

38. Telephone interview, Steven Weinberg, 30 September 1985.

39. Weinberg, S., *PRL* 19, no. 21 (20 November 1967): 1264.

40. *Ibid.*, p. 1266. We have replaced "meson" by "boson" and dropped subscripts.

41. Guralnick, G. S., Hagen, C. R., and Kibble, T. W. B., *PRL* 13, no. 20 (16 November 1964): 585; Kibble, T. W. B., *Oxford Conference on Elementary Particles*, 19–25 October 1965 (Oxford: Rutherford High Energy Laboratory): 19; Kibble, T. W. B., *PR* 155, no. 5 (25 March 1967): 1554.

42. Interview, Abdus Salam, his office, ICTP, 23 February 1984.

43. Salam, A., in "Elementary Particle Theory," *Proceedings of the Eighth Nobel Symposium*, 19–25 May 1968 (New York: John Wiley, 1968): 367.

44. Gell–Mann's conference summary does not even mention the idea and mentions Salam only once in passing (Gell–Mann, M., in *Proceedings of the Eighth Nobel Symposium, op. cit.*, p. 395).

45. Interview, Thomas Kibble, his office, Imperial College, 5 October 1984.

46. Coleman, S., *Science*, 206, 14 December 1979, p. 1290; Koester, D., Sullivan, D., and White, D., *Social Studies of Science* 12, no. 1 (February 1982): 73. Weinberg worked a little on trying to renormalize SU(2) × U(1), but got nowhere (Stuller, R.L., "Are Symmetry Broken Models of Weak Interactions Renormalizable?" Ph.D. thesis, MIT, February 1971, typescript).

47. Weinberg, S., *Coral Gables Conference on Fundamental Interactions at High Energy* (3rd) (New York: Gordon and Breach, 1971): 182.

48. Interview, Steven Weinberg, faculty club, University of Texas, 28 November 1984.

CHAPTER 14

1. We have been greatly assisted in writing chapters 14–18 by the history of particle physics by Pickering, A., *Constructing Quarks* (Chicago: University of Chicago Press, 1984). For information about early English laboratories, see Sviedras, R., *HSPS* 7 (1976): 405–36. See also *ibid.*, 2 (1970): 127–45.

2. Glashow, *RMP* 52, no. 3, p. 1319.

3. Letter, Leland Hayworth to T. H. Johnson, 9 September 1953, courtesy Brookhaven National Laboratory. Samios is paraphrasing part of p. 1.

4. Wideröe, R., *Wissenschaftliche Zeitschrift der Friedrich-Schiller-Universität Jena* 13, no. 4 (Winter 1964): 431; Wideröe, R., *Archiv für Elektrodynamik* 21, no. 4 (1928): 400; Wideröe, R., "Some Memories and Dreams from the Childhood of Particle Accelerators," unpublished ms., January 1983, AIP Archives, 2. He subsequently built a linear machine.

5. Lawrence, E. O., and Edlefson, N. E., *Science* 72 (10 October 1930): 376. Livingston, M., *PT* 12, no. 10 (October 1959): 18; for a description of Lawrence, see Childs, H., *An American Genius: The Life of Ernest Lawrence* (New York: E. P. Dutton, 1968); Davis, N. P., *Lawrence and Oppenheimer* (New York: Simon and Schuster 1968).

6. Rutherford's attitude from interview, Norman Feather, *AHQP*, 25 February 1971, p. 17. Results of accelerator, Cockroft, J., and Walton, E. T. S., *PRSA* 136, no. 830 (1 June 1932): 619; ibid. 137, no. 831 (1 July 1932): 229.

7. Livingston, *op. cit.*, p. 21.

8. The precise definition is the energy acquired when an electron is accelerated through a

potential difference of one volt. A kilocalorie, which is the unit typically used, is roughly equivalent to 10^{23} electron volts.

9. Wilson, R. R. in Marshak, R., ed., *Perspectives in Modern Physics* (New York: Wiley, 1966); McMillan, E. M., *PT* 12, no. 10 (October 1959): 24; a vivid picture of this era is in Hilts, P., *Scientific Temperaments* (New York: Simon and Schuster, 1982): 46–55.

10. Interestingly, Lawrence's original plan for the 100 MeV machine would not have worked because he didn't take into account the relativistic effects of accelerating particles to such high energies.

11. Some of this process is described in Needell, A. A., *HSPS* 14, no. 1 (Winter 1983): 99.

12. Talk by Robert Seidel at the International Symposium on Particle Physics in the 1950s, Fermilab, 3 May 1985 (proceedings forthcoming).

13. Roos, M., *RMP* 35, no. 1, (April 1963): 314; the errors were pointed out by Zweig, G., in Isgur, N., *Baryon 1980*, Proceedings of Fourth International Conference on Baryon Resonances (Toronto: University of Toronto, 1980): 454–55.

14. Barkas, W. H., and Rosenfeld, A. H., Lawrence Berkeley Laboratory publication UCRL–8030, 1958; Particle Data Group, *RMP* 56, no. 2 (April 1984).

15. Stueckelberg, E. C. G., *HPA* 11 (1938), 317.

16. Fermi, E., and Yang, C. N., *PR* 76, no. 12 (15 December 1949): 1739. It is worth recalling that the antiproton had not yet been discovered, although its existence was widely assumed.

17. Examples are Goldhaber, M., *PR* 92, no. 5 (December 1953): 1279: *ibid.*, 101, no. 1 (1 January 1956): 433; Miyazawa, H., *PTP* 6, no. 4 (July–August 1951): 631.

18. Engels, F., *Dialectics of Nature*, trans. C. Dutt (New York: International Publishers, 1979). Engels wrote another book on natural science which was published during his lifetime: *Herr Eugen Dühring's Revolution in Science* (now known as *Anti–Dühring*) (New York: International Publishers, 1939). V. I. Lenin also wrote an influential work on natural science, *Materialism and Empirio Criticism* (New York: International Publishers, 1927). For a typical contemporary exposition of Marxist natural philosophy and its connection to Marxist social philosophy see Sheptulin, A. P., *Marxist–Leninist Philosophy* (Moscow: Progress Publishers, 1978).

19. Sakata's recollections are in two essays in Hara, O., et al., eds., *Shoichi Sakata: Scientific Works* (Tokyo: Horei Printing, 1977). Quotations from pp. 393 and 370, respectively.

20. Sakata, S., *PTP* 16, no. 6 (December 1956): 686.

21. Ikeda, M., Ogawa, S., and Ohnuki, Y., *PTP* 22, no. 5 (November 1959): 715; Yamaguchi, Y., *PTP* (supp.) 11 (1959): 1; *ibid.*, p. 37.

22. Pevsner, A., et al., *PRL* 7, no. 11 (1 December 1961): 421. As a sign of the lack of favor met by Sakata's school, the paper contains no mention of the prediction.

23. Sakata, S., in Hara et al., *op. cit.*, pp. 360–61. The paper has an epigram from *Dialectics of Nature*: "Natural scientists may adopt whatever attitude they please, they are still under the domination of philosophy."

24. Gell–Mann, M., Caltech report CALT–68–1214, n.d., p. 3.

25. The story is described in an interview with Gell–Mann by Schultz, R., *Omni*, May 1985, p. 54.

26. Interview, Murray Gell–Mann, aboard a plane, 3 March 1983.

27. Gell–Mann, M., *PR* 106, no. 6 (15 June 1957): 1296. Also important here is Sakurai, J. J., *Annals of Physics* 11, no. 1 (September 1960): 1.

28. Interview, Murray Gell–Mann, his office, Caltech, 21 February 1985.

29. Gell–Mann, M., Caltech report CTSL–20 (unpublished), 15 March 1961, p. 25.

30. Gell–Mann, Caltech report CALT–68–1214, n.d., pp. 22–23.

31. Interview, Murray Gell–Mann, his office, Caltech, 21 February 1985. The last three sentences of the quotation come from a telephone interview on the same subject on 24 April 1985.

32. More exactly, the first two generators are the third axis of isotopic spin, which indicates the particle's electric charge, and "hypercharge," which is the strangeness plus the baryon number. The remaining six are all associated with special vector bosons that change the values of isotopic spin and hypercharge. SU(3) is a "special" unitary group because two of the members are identical; an ordinary unitary group would have different members. A simple and clear exposition of the work can be found in Chew, G. F., Gell–Mann, M., and Rosenfeld, A. H., *Scientific American* 210, no. 2 (February 1964): 74.

33. Gell–Mann, M., Caltech report CTSL–20, *op. cit.*; reprinted in Gell–Mann, M., and Ne'eman, Y., eds., *The Eightfold Way* (New York: W. A. Benjamin, 1964). Originally written as a

forty-six–page report in January 1961, the eightfold way was first published formally as a three–page section at the end of a much longer paper in *Physical Review*: Gell–Mann, M., *PR* 125, no. 3 (1 February 1962): 1067.

34. Ne'eman, Y., *NP* 26, no. 2. (July 1961): 222; interview, Yuval Ne'eman, Fermilab cafeteria, 3 May 1985 (all subsequent quotes); letter, Yuval Ne'eman to authors, 6 August 1985; interview, Abdus Salam, ICTP, 23 February 1984. Some of this story is available in Ne'eman, Y., *Proceedings of the Israel Academy of Sciences and Humanities*, Section of Sciences, no. 21 (Jerusalem: 1983), and Ne'eman, Y., in *Symmetries in Physics, First International Symposium on the History of Scientific Ideas*, Universitat Autonomia de Barcelona, Catalonia, Spain, September 1983, forthcoming.

35. When Gell–Mann first put together SU(3), he visited Berkeley, where some experimenter friends horrified him by presenting him with a coffee cup that had SIGMA-LAMBDA-ODD written around its circumference. They had found evidence that the sigma and lambda did not have the same parity, and hence were not in the same multiplet. Gell–Mann sweated for months before deciding the rumors must be wrong. Eventually he was proven right. (Tripp, R. D., Watson, M. B., and Ferro-Luzzi, M., *PRL* 8, no. 4 [15 February 1962]: 175.)

36. Prentki, J., ed., *Proceedings of the 1962 International Conference on High Energy Physics at CERN* (Geneva: CERN, 1962): 795–805.

37. Within SU(3) systematics, the pyramid-shaped decimet was not the only possible place for the deltas, xi-stars and sigma-stars. Another scheme, with twenty-seven particles, started with an upper layer of strangeness plus-one particles, above the deltas. Such particles could be produced by scattering K + or K⁰ mesons (with one positive unit of strangeness) on protons or neutrons. The husband-wife team of Gerson and Sulamith Goldhaber, working at Berkeley, had just reported negative results from such an experiment. For Ne'eman, this was conclusive. It pointed to the decimet as the unique SU(3) fitting solution. (Letter, Yuval Ne'eman to authors, *op. cit.*)

38. This story compiled from interviews: Gerson Goldhaber, a room at Fermilab, 2 May 1985; Yuval Ne'eman, a hotel near Fermilab, 3 May 1985; Murray Gell–Mann, telephone interview, 5 February 1985. Gell–Mann's prediction is Gell–Mann, M., comment in Prentki, *op. cit.*, p. 805.

39. Interview, Nicholas Samios, his office, Brookhaven, 12 February 1985 (this and following quotes).

40. Interview, Robert Palmer, an office at Brookhaven, 18 May 1983.

41. Gaston, J., *Originality and Competition in Science* (Chicago: University of Chicago Press, 1973): 83–88.

42. Ne'eman, *Symmetries, op. cit.*, p. 42. Later, when the particle was found, Samios forgot to call Ne'eman. He later sent some photographs and a note reading, "Please excuse the oversight, but you knew it existed before we did!"

43. Barnes, V. E., et al., *PRL* 12, no. 8 (24 February 1964): 204; Fowler, W. B. and Samios, N. P., *Scientific American* 211, no. 4 (October 1964): 36.

CHAPTER 15

1. Telephone interview, Robert Serber, 4 June 1983.

2. Interview, Murray Gell–Mann, aboard a plane, 3 March 1983 (all quotes in this section).

3. Gell–Mann, M., *PL* 8, no. 3 (1 February 1964): 214.

4. Well, actually, not so easy. When Gell–Mann assigned strangeness values to particles back in 1953, he knew nothing about strange quarks. As a consequence, he awarded strangeness +1 to mesons with strange *antiquarks*, which means that a particle with strangeness −1 has a strange quark inside it. The principle of one unit of strangeness per strange quark is correct; only the signs are reversed.

5. Telephone interview, Sheldon Glashow, 12 May 1983.

6. Gell–Mann, M., Caltech report CALT–68–1214, p. 30 (unpublished).

7. Interview, Jacques Prentki, his office, CERN, 7 February 1984.

8. Interview, Yuval Ne'eman, a room at Fermilab, 3 May 1985.

9. Ne'eman, *Proceedings of the Israel Academy of Sciences and Humanities*, Sections of Sciences, no. 21 (Jerusalem: 1983): 7–8.

10. Goldberg, H., and Ne'eman, Y., *NC* 27, no. 1 (1 January 1963): 1.

11. Letter, Victor Weisskopf to J. R. Oppenheimer, 13 February 1964, in Oppenheimer Collection, Manuscript Division, Library of Congress, General Case File, Box 77, Weisskopf folder; Schwinger, J., *PR* 135B, no. 3 (10 August 1964): 817 (received 23 March).

12. Zweig, G., CERN preprint 8182/TH401, 17 January 1964 (unpublished); Zweig, G., CERN preprint 8419/TH412, 21 February 1964 (unpublished); some of it was eventually published as Zweig, G., in Zichichi, A., ed., *Symmetries in Elementary Particle Physics*, Proceedings of the "Ettore Majorana" International School of Physics, Erice, Italy, August 1964 (New York: Academic Press): 192.

13. Gell–Mann, M., Caltech report CALT–68–1214, p. 33.

14. Zweig, G., in Isgur, N., ed., *Baryon '80: Proceedings of the Fourth International Conference on Baryon Resonances*, 14–16 July 1980 (Toronto: University of Toronto, 1981): 457 (both quotes).

15. The stimuli were the $V - A$ theory of the weak interactions, which Gell–Mann and Feynman showed was compatible with an intermediate vector boson, and a demonstration by Nambu that the Hofstadter nucleon structure experiments entailed the existence of a spin–one boson (Nambu, Y., *PR* 106, no. 6 [15 June 1957]: 1366).

16. Feinberg, G., *PR* 110, no. 6 (15 June 1958): 1482. Japanese physicists had first raised the possibility of two types of neutrinos immediately after the war (cf., Brown, Konuma, and Maki, *Particle Physics in Japan* [Kyoto: Research Institute for Fundamental Physics] mimeograph, vol. 1, p. 59). Many Western physicists came up with the same idea independently in the next decade; including Feynman and Gell–Mann at the December 1957 meeting at Stanford of the American Physical Society. For some references, see footnotes 4 and 5 in Danby, G., et al., *PRL* 9, no. 1 (1 July 1962): 95. Feinberg's article was by far the most influential.

17. The opinion is held so commonly that it would be unkind to cite specific sources.

18. Interview, Mel Schwartz, his home, 19 February 1985 (all quotes); Schwartz, Mel, *PRL* 4, no. 6 (15 March 1960): 306; Lee, T. D., and Yang, C. N., *ibid.*, p. 307. The Italian–Soviet physicist Bruno Pontecorvo presented a slightly different, more impractical version of the idea in Pontecorvo, B., *Soviet Physics–JETP* 37, no. 6 (June 1960): 1751.

19. Lederman, L., Schwartz, M., and Gaillard, J.–M., in *Proceedings of the International Conference on Instrumentation*, Berkeley 1960 (New York: Wiley-Interscience, 1960): sec. V.1d. This paper and several others are included in Schwartz, M., *Adventures in Experimental Physics*, Vol. Alpha (1972): 82.

20. Lederman, L., "Neutrino Physics," Brookhaven Lecture Series, no. 23, BNL 787(T–300), (9 January 1963): 5–6; story confirmed in interview, Leon Lederman, his office, Fermilab, 14 February 1984.

21. Telephone interview, Murray Gell–Mann, 3 June 1983; interview, Sheldon Glashow, his office, Harvard University, 7 May 1985. See, for example, Glashow, S. L., and Coleman, S., *PRL* 6, no. 8 (15 April 1961): 423. This paper and others are reprinted in Gell–Mann, M., and Ne'eman, Y., *The Eightfold Way* (New York: W. A. Benjamin, 1964).

22. Telephone interview, James Bjorken, 10 May 1983; other fourth quark papers include Hara, Y., *PR* 134B, no. 3 (11 May 1964): 701; and Amati, D., Bacry, J. Nuyts, and J. Prentki, *NC* 34, no. 6 (16 December 1964): 1732, which contains a list of further references in notes 4 and 5.

23. Bjørken, B. J., and Glashow, S. L., *PL* 11, no. 3 (1 August 1964): 255.

24. Glashow, S., *Scientific American*, 233, no. 4 (October 1975): 47. The charmed quark suppresses a second–order process, something more extensively discussed in the following chapter.

25. Interview, Sheldon Glashow, his office, 2 December 1982.

26. Interview, Steven Weinberg, outside Harvard, 7 May 1985.

27. Weinberg, S., *PR*, sec. B., 138, no. 4 (24 May 1965): 990.

28. The chain of papers that created SU(6) began with Gursey, F., and Radicati, L., *PRL* 13, no. 5 (3 August 1964): 173; Pais, A., *PRL* 13, 5 (3 August 1964): 175; and Gursey, F., Pais, A., and Radicati, L., *PRL* 13, no. 8 (24 August 1964): 299. The defects of the model are well summarized in the introduction to Dyson, F., ed., *Symmetry Groups in Nuclear and Particle Physics: A Lecture Note and Reprint Volume* (New York: W. A. Benjamin, 1966).

29. Gell–Mann, M., *Physics* 1, no. 1 (July–August 1964): 63. *CQ*, p. 122, note 63, cites data that Regge pole work and quark ideas constituted about 75 percent of the theoretical papers in *Physics Letters* from 1967 to 1969.

30. Regge, T., *NC* 14, no. 5 (1 December 1959: 951. See also the exhaustive review in Collins, P. D. B., *PRpts* 1, no. 2 (January 1971): 103, and Chew, G., *The Analytic S–Matrix: A Basis for Nuclear Democracy* (New York: W. A. Benjamin, 1966). These ideas were not wrong; indeed, on the mass shell, where quarks are not seen, field theory does manifest itself as nuclear democracy and Regge poles. But the language of quarks was ultimately seen as having explanatory power over a greater range.

31. Biographical material from interviews, Yoichiro Nambu, Weak Interaction Conference, Racine, Wisconsin, 31 May 1984; his office, University of Chicago, 13 February 1985 (all quotes from latter).

32. Nambu, Y., *PR* 117, no. 3 (1 February 1960): 648. References to other workers are in notes 3 and 4. Curiously, one reason for the scorn given to the Heisenberg-Pauli unification (see chap. 20) was that it relied upon a "degenerate vacuum" similar to that posited by Nambu a few years later.

33. Greenberg, O. W., *PRL* 13, no. 20 (16 November 1964): 598. Another way out involved trying to award quarks radial quantum numbers, like electrons in an atom.

34. Nambu, Y., in Perlmutter, A., Kursunaǧlu, B., and Sakmar, I., eds., *Symmetry Principles at High Energy*, Second Coral Gables Conference, 20–22 January 1965 (San Francisco: W. H. Freeman, 1965): 274. Similar ideas were presented somewhat earlier by Bacry, H., Nuyts, and van Hove, L., *PL* 9, no. 3 (15 April 1964): 279.

35. Han, M. Y., Syracuse University preprint NYO–3399–21/1206—SU–21 (unpublished); Han, M. Y. and Nambu, Y., *PR* 139B, no. 4 (23 August 1965): 1006.

36. The term *color* was first suggested by Lichtenberg, D. B. (*Unitary Symmetry and Elementary Particles*, first ed. [New York: Academic Press, 1970]). The colors were first called red, white, and blue, then changed to red, blue, and green, apparently in the mistaken belief that these are the primary colors (see the second 1978 edition of Lichtenberg, *op. cit.*, p. 221).

37. Quote above from Nambu, Y., in de–Shalit, A., Feshbach, H., and Van Hove, L., eds., *Preludes in Theoretical Physics* (New York: Wiley, 1966): 133. The article arose when, quite unexpectedly, Nambu received a letter inviting him to contribute to a *Festschrift*, a book of papers in honor of the sixtieth birthday of Victor Weisskopf, then the head of CERN. Although Nambu had not been a student of Weisskopf and did not know him well, he was happy to be asked. The paper, "A Systematics of Hadrons in Subnuclear Physics," contains all but two of the basic statements about the strong interaction that later became gospel. The exceptions are the Yang–Mills formulation, which Nambu almost but not quite embraced, and the fractional quark charge, which had few strong adherents except Murray Gell–Mann. Because he avoided fractional charge, Nambu's color was not confined. Gell–Mann, too, had been invited to contribute to the *Festschrift*, but his article never arrived. "I was terribly embarrassed about not contributing," he said. "Of all the people who should have contributed to that, I was certainly the—I was someone who *should* have contributed to that, being one of his students. I was asked, of course I was asked, but I have such trouble writing things. I was worried and confused about something, as usual, and I didn't produce anything in time for publication. So I never read it." The coincidence still rankles, and he returned to it a number of times in different conversations. "If I'd seen Nambu's paper, I probably would have pieced everything together. But who knows? Maybe I wouldn't have. But I was so ashamed. I never write articles on time. I always send them in months or years late. We all just sat there, stewing around, and it was so *needless*." (Interviews, Murray Gell–Mann, 3 March 1983, 22 May 1984, 5 February 1985, 21 February 1985.)

38. Interview, Howard M. Georgi III, his office, Harvard University, 14 June 1983. The discussion has been slightly condensed.

39. Interview, James Bjorken, his office, Fermilab, 13 February 1985 (all quotes).

40. Telephone interview, Richard Taylor, 26 February 1985 (all quotes).

41. Panofsky, W., in Prentki, J., and Steinberger, J., eds., *Proceedings of the Fourteenth International Conference on High–Energy Physics*, Vienna, 28 August–5 September 1968 (Geneva: CERN): 23 (quote below, p. 37).

42. E.g., Feynman, R., *Acta Physica Polonica* 24, no. 6 (December 1963): 697; interview, Richard Feynman, his office, Caltech, 22 February 1985.

43. Telephone interview, Henry Kendall, 26 February 1985.

44. Bjorken, J., and Paschos, E. A., *PR* 185, no. 5 (25 September 1969): 1975. Feynman himself did not publish anything on the model until 1972 (Feynman, R., *Photon–Hadron Interactions* [Reading, Massachusetts: W. A. Benjamin, 1972]).

45. Letter, Richard Feynman to authors, 18 September 1985.

46. Interview, Murray Gell–Mann, his office, Caltech, 21 February 1985.

47. *CQ*, pp. 140–52, carefully summarizes the experiments.

48. Interview, Steven Weinberg, outside Harvard, 7 May 1985.

CHAPTER 16

1. Biographical information and all subsequent quotes from interview, J. Iliopoulos, an office at Rockefeller University, 8 May 1984.

2. Interview, Claude Bouchiat, his office, École Normale Supérieure, 4 October 1984.

3. Interview, Jacques Prentki, his office, CERN, 7 February 1984.

4. See, for example, Feinberg, G., and Pais, A., *PR* 131, no. 6 (15 September 1963): 2724. The classification system was introduced in Lee, T. D., *NC*, section A, 59, no. 4 (21 February 1969): 579.

5. Bouchiat, C., Iliopoulos, J., and Prentki, J., *NC*, section A, 56, no. 4 (21 August 1968): 1150; Iliopoulos, J., *NC*, section A, 62, no. 1 (1 July 1969): 209.

6. Cabibbo, N., *PRL* 10, no. 12 (15 June 1963): 531. More exactly, Cabibbo said that the quark coupled in the weak interaction to the up quark is $d_c = \alpha d + \beta s = d \cos \Theta + s \sin \Theta$, where d and s are the quarks in the strong interaction and $\alpha^2 + \beta^2 = 1$. Historically, Cabibbo reintroduced a parameter into the eightfold way that Gell–Mann and Lévy (among others) had invented in 1959.

7. Cabibbo, N., and Maiani, L., *PL*, section B, 28, no. 2 (11 November 1968): 131; Cabibbo, N., and Maiani, L., *PR*, section D, 1, no. 2 (15 January 1970): 707.

8. Interview, Luciano Maiani, his office, University of Rome, 17 February 1984 (all quotes).

9. Interview, Sheldon Glashow, his office, Harvard, 25 April 1983.

10. This had been realized before in different ways by, for example, Feinberg and Pais, *op. cit.*, p. 2728. Many unobserved processes disappeared when the quark model was considered, but not, as Glashow, Iliopoulos, and Maiani realized, this one.

11. Glashow, S., Iliopoulos, J., and Maiani, L., *PR*, section D, 2, no. 7 (1 October 1970): 1285; quotes from p. 1290.

12. The anecdotes above are based on the interviews with Glashow, Iliopoulos, and Maiani cited above, and interview, Samuel Ting, an office at CERN, 9 February 1984.

13. Glashow, S. L., and Iliopoulos, J., *PR*, section D, 3, no. 4 (15 February 1971): 1043.

14. Glashow, S. L., in *Meeting on Renormalization Theory*, 14–18 June 1971 (Marseille: Centre de Physique Théorique, C.N.R.S., 1971): 159.

15. Interview, Sheldon Glashow, his office, Harvard, 2 December 1982.

16. Interview, Martinus Veltman, an office at Columbia, 12 December 1983 (all quotes).

17. Schwinger, J., *PR* 3, no. 6 (15 September 1959): 296. But see also, Gotô, T., and Imamura, T., *PTP* 14, no. 4 (October 1955): 396.

18. Veltman, M., *PRL* 17, no. 10 (5 September 1966): 553.

19. Bell, J. S., *NC*, section A, 50, no. 1 (1 July 1967): 129–30; see also, Bell, J. S., in *Rendiconti della Scuola Internazionale di Fisica "E. Fermi,"* 41st Course (Rome: Università degli Studi, 1966).

20. This is a truncated summary of a longer process. Using current algebra, S. L. Adler and W. I. Weisberger independently discovered a relation between the weak coupling constant and pion–nucleon interactions. Although the Adler–Weisberger relation fit the data tolerably well, one ran into Schwinger terms in its derivation. Veltman removed them by imposing gauge invariance. Bell realized that Veltman's conditions implied that current algebra was really a consequence of Ward identities, a set of equations that describe many of the properties of quantum field theory. Puzzling over Bell's paper, Veltman learned, during a casual encounter with Feynman, that the fields, in the form restated by Bell, were Yang–Mills fields. Feynman however insisted that they were applicable to the strong interactions. (Interview, Martinus Veltman, an office at Columbia, 13 December 1983.)

21. Interview, John Bell, his office, CERN, 6 February 1984. Bell told us his paper had the modest aim of trying to understand current algebra, and that Veltman had found more in the paper than its author had intended. "The illumination to Tini later was a much bigger thing—that gauge invariance is more than a technical device. He underestimated the intelligibility of the paper because he was looking for more than it had, and afterwards, when he found it in his own head, he exaggerated its depth."

22. Veltman, M., *NP*, section B, 7 (1968): 637.

23. Interview, David Politzer, his office, Caltech, 21 February 1985.

24. Fadeev, L. D., and Popov, V. N., *PL*, section B, 25, no. 1 (24 July 1967): 29. For a good review of the Soviet work, see Veltman, M., in Rollnick, H., and Pfeil, W., eds., *International Symposium on Electron and Photon Interactions at High Energies*, Bonn 1973 (London: North Holland, 1974): 429 and especially appendix.

25. Veltman, M., *NP*, section B, 21, no. 1 (1 August 1970): 288.

26. Letter, Gerard 't Hooft to authors, 30 July 1985; subsequent quotes from interview, Gerard 't Hooft, his office, University of Utrecht, 26 September 1984.

27. 't Hooft, G., *NP*, section B, 33, no. 1 (1 October 1971): 173.

28. Bell, J. S., and Jackiw, R., *NC*, section *A*, 60, no. 1 (1 March 1969): 47; Adler, S. L., *PR* 177, no. 5 (25 January 1969): 2426.

29. Cited in *CQ*, p. 178.

30. Interview, David Politzer, his office, Caltech, 21 February 1985 (all subsequent quotes).

31. 't Hooft, G., *NP*, section *B*, 35, no. 1 (1 December 1971): 167.

32. Interview, Martinus Veltman, an office at Columbia, 13 December 1983.

33. Lee, B. W., *PR*, section *D*, 5, no. 4 (15 February 1972): 823; Lee, B. W., and Zinn–Justin, J., *PR*, section *D*, 5, no. 12 (15 June 1972): 3121; *ibid.*, p. 3137; *ibid.*, p. 3155.

34. Interview, Steven Weinberg, his office, University of Texas, 28 November 1984.

35. Adler, S. L., and Bardeen, W. A., *PR* 182, no. 5 (25 June 1965): 1517. The full history of the anomaly is quite interesting, and we no more than touch on it here. It was shown to spoil the renormalizability of the model unless colored quarks were introduced (Bouchiat, C., Iliopoulos, J., and Meyer, Ph., *PL*, sec. *B*, 33, no. 7 [3 April 1972]: 519).

36. Adler, S. L., in Devons, S., ed., *High–Energy Physics and Nuclear Structure.* Proceedings of the Third International Conference on High Energy Physics and Nuclear Structure, New York City, September 1969 (New York: Plenum Press, 1970): 654.

37. Bardeen, W. A., Fritzsch, H., and Gell–Mann, M., in *Proceedings of the Topical Meeting on Conformal Invariance in Hadron Physics* (Frascati, May 1972); Bardeen, W. A., Fritzsch, H., and Gell–Mann, M., in Gatto, R., *Scale and Conformal Symmetry in Hadron Physics* (New York: Wiley, 1973): 139. An ancillary point: Color without colored hadrons is mathematically equivalent to parastatistical quarks without parahadrons. But parastatistics with real parahadrons is not like color with colored hadrons, which is why Gell–Mann et al. did not treat the two ideas similarly. Gell–Mann quote from telephone interview, Murray Gell–Mann, 21 February 1985.

38. Telephone interview, Murray Gell–Mann, 28 May 1985.

39. Fritzsch, H., and Gell–Mann, M., in Jackson, J. D., and Roberts, A., *Proceedings of the XVI International Conference on High Energy Physics*, Fermilab, September 1972 (Batavia, Illinois: National Accelerator Laboratory): 135. See also Gell–Mann's rapporteur talk on p. 333.

40. In their 1972 paper, Fritzsch and Gell–Mann proposed the scheme only tentatively, but it worked so well that the following year they joined with a Swiss physicist in proposing it explicitly: Fritzsch, H., Gell–Mann, M., and Leutwyler, H., *PL*, section *B*, 47, no. 4 (26 November 1973): 365.

41. Telephone interview, Murray Gell–Mann, 28 May 1985.

42. Symanzik, K., *Communications in Mathematical Physics* 18, no. 3 (1970): 227; Callan, C. G., Jr., *PR*, section *D*, 2, no. 8 (15 October 1970): 1541.

43. More exactly, the renormalization group describes such scale transformations in terms of changes of an effective coupling constant in the underlying field theory. (Stueckelberg, E. C. G., and Petermann, A., *HPA* 26, p. 499; Gell–Mann, M., and Low, F., *PR* 95, no. 5 (1 September 1954): 1300.

44. Weinberg, S., in Guth, A. H., Huang, K., and Jaffe, R. L., eds., *Asymptotic Realms of Physics* (Cambridge, Massachusetts: MIT Press, 1985): 1.

45. Wilson, K. G., *PR*, section *B*, 4, no. 9 (1 November 1971): 3174; *ibid.*, p. 3184; Wilson, K. G., *PR*, section *D*, 3, no. 8 (15 April 1971): 1818.

46. Interview, David Gross, an office at Brookhaven, 2 April 1985 (all quotes).

47. Interview, Frank Wilczek, his office, University of California at Santa Barbara, 22 February 1985 (all quotes).

48. Letter, Gerard 't Hooft to authors, 30 July 1985; letter, John Iliopoulos to authors, 22 August 1985.

49. Gross, D., and Wilczek, F., *PRL* 30, no. 26 (25 June 1973): 1343; Politzer, H. D., *PRL* 30, no. 26 (25 June 1973): 1346. See also, Gross, D. J., and Wilczek, F., *PR*, section *D*, 8, no. 10 (15 November 1973): 3633; *ibid.*, 9, no. 4 (15 February 1974): 980; Politzer, H. D., *PRpts C*, 14, no. 4 (1974): 129.

50. Telephone interview, Tony Zee, 5 October 1985; Zee, T., *PR*, sec. *D*, 7, no. 12 (15 June 1973): 3630. See also, Zee, T., *Thy Fearful Symmetry* (New York: Macmillan, 1986).

51. Weinberg, S., *PRL*, 31, no. 7 (13 August 1973): 494; Gross and Wilczek, *op. cit.*, 3633.

52. Interview, Sheldon Glashow, his office, Harvard, 2 December 1983.

CHAPTER 17

1. Interview, Samuel Devons, Barnard, 4 January 1984.

2. See, as a typical example, the discussion in Bernstein, J., *Elementary Particles and Their*

Currents (San Francisco: W. H. Freeman, 1968), p. 43. We have been greatly aided in constructing this account by the writings of Peter Galison and Andy Pickering cited below.

3. Weinberg, S., *PRL* 27, no. 24 (13 December 1971): 1688; Weinberg, S., *PR*, section *D*, 5,
no. 6 (15 March 1972): 1415, quotes from p. 1415. The predictions involved interactions with one
pion, but approximately the same held true for other reactions.

4. Lagarrigue, A., Musset, P., and Rousset, A., "Projet de Chambre à Bulles à Liquides
Lourdes de 17m^3," typescript, 10 February 1964; Allard, J. F., et al., "Proposition de construction d'une grande chambre a bulles liquides lourdes destinées à functioner auprés du synchrotron à protons du CERN," typescript, 1 January 1965, both from Musset's files; other
particulars provided by letter, Paul Musset to authors, 12 July 1984.

5. Interview, Jacques Prentki, his office, CERN, 7 February 1984.

6. On the other hand, Lagarrigue, at least, was aware of the puzzle posed by their nonexistence, saying in a speech about the uses of large bubble chambers: "The lack of neutral currents
is one of the major points [to be studied] of the weak–interaction physics with neutrinos"
(Lagarrigue, A., in Puppi, A., ed., *Old and New Problems in Elementary Particles* [New York:
Academic Press, 1968]: 148).

7. Cundy, D. C., et al., *PL*, section *B*, 31, no. 7 (30 March 1970): 478. The upper limit,
expressed as a ratio of neutral to charged currents, was 0.12 ± .06, the limit cited by Weinberg.
See also, Gargamelle collaboration, "Proposal for a Neutrino Experiment in Gargamelle,"
CERN–TCC/70—12, 16 March 1970, which mentions the use of Gargamelle "to reduce this
limit to 0.05" or even 0.03 through various tests that, in fact, were never performed (p. 6).

8. Interview, Paul Musset, his office, CERN, 2 February 1984.

9. This point is expanded upon in some detail in Pickering, A., "Making Meaning, or
Editing and Epistemology: Three Accounts of the Discovery of the Weak Neutral Current," a
talk before the Joint Seminar for the History and Philosophy of 20th Century Science, 26
October 1984, typescript, p. 15. It is interesting to note that one of Donald Perkins's graduate
students, E. C. M. Young, had conducted a study of the earlier round of neutrino experiments
at CERN with results that could, with hindsight, be interpreted as supporting the existence of
neutral currents (Pickering, A., *Studies in History and Philosophy of Science* 15, no. 2 [June
1984]: note 34). However, Young claimed "reasonably good agreement" between the expected
and observed background events (Young, E. C. M., "High Energy Neutrino Interactions,"
CERN Yellow Report 67–12, 21 April 1967, p. 56).

10. The following description of Fermilab's construction is based on visits to the laboratory
in February and May 1985; interviews with Dick Carrigan, James Bjorken, Leon Lederman,
Carlo Rubbia, Larry Sulak, and Samuel C. C. Ting; and Hilts, P. J., *Scientific Temperaments*
(New York: Simon & Schuster, 1982): 17–99.

11. Quoted in Hilts, *op. cit.*, pp. 98–99.

12. Interview, Larry Sulak, an office at New York University, 28 November 1983. Harvard–Pennsylvania–Wisconsin Collaboration, "NAL Neutrino Proposal," Proposal 1, Fermilab
Archives, Proposal Shelf, 15 April 1970; addendum, July 1970. Hereinafter the group will be
referred to as "HPW Collaboration."

13. Interview, Carlo Rubbia, his office, CERN, 11 February 1984.

14. Interview, Larry Sulak, a restaurant in Ohio, 5 June 1985.

15. Palmer, B., "Very Preliminary Results of Neutral Current Search in the Neutrino–Freon Experiment," handwritten ms. in Dr. Palmer's files, May 1972; notes to talk at
Brookhaven, handwritten ms. in Dr. Palmer's files, July 1972.

16. Letter, Robert Palmer to Peter Galison, 8 June 1983, Palmer's files.

17. Lee, W., *PL*, section *B*, 40, no. 3 (10 July 1972): 423. See also, Lee, B. W., *PL*, section
B, 40, no. 3 (10 July 1972): 420.

18. Prentki, J., and Zumino, B., *NP*, section *B*, 47, no. 1 (September 1972): 99; Lee, B. W.,
PR, section *D*, 6, no. 4 (15 August 1972): 1188.

19. Glashow, S. L., and Georgi, H., *PRL* 28, no. 22 (29 May 1972): 1494.

20. Letter, André Lagarrigue to Willibald Jentschke, 12 April 1972, quoted in Galison, P.,
RMP 55, no. 2 (April 1983): 484. However, Lagarrigue's letter is pessimistic on whether the
question of their existence could be resolved on CERN's low energy accelerators (Letter, Paul
Musset to authors, 12 July 1984).

21. Baltay, C., et al., CERN technical memorandum TC–L/PA/UC/WF–/fv, 14 July 1972;
ibid., TC–L/WFF/ju, 14 July 1972. (All CERN memoranda from Musset's files unless otherwise
specified.)

22. Galison, *op. cit.*, pp. 494–95; interview, Larry Sulak, a restaurant in Ohio, 5 June 1985;
interview, Carlo Rubbia, his office, CERN, 11 February 1984.

23. Galison, *op. cit.*, pp. 486–87; we have rendered Faissner's "Bilderbuch–example" as "picture–book example."

24. Musset, P., *Bulletin of the American Physical Society* 18, no. 1 (winter 1973): 73.

25. Musset, P., CERN technical memorandum TC–L/BC/PM/fv, 19 March 1973; Musset, addendum, CERN technical memorandum TC–L/BC/PM/fv, 26 March 1973.

26. Interview, Paul Musset, his office, CERN, 2 February 1984.

27. Musset recalls meeting Rubbia at a French Physical Society conference in Dijon in early 1973. "I explained what kind of events we had, and that we were beginning to see a signal. I did not give any final number because it was too early. Rubbia said, 'Okay, we can do the same thing with the Fermilab apparatus.' So I knew that they certainly were working on that. But I had the attitude not to try to be informed about what their result was, because really what we had to understand was ours." (Interview, Paul Musset, his office, CERN, 2 February 1984).

28. Interview, Paul Musset, CERN bubble chamber photo storage room, 6 February 1984.

29. Letter, Robert Palmer to P. Galison, *op. cit.* Palmer also tried to get Brookhaven to do their own experiment; Brookhaven National Laboratory Memorandum to R. R. Rau from R. B. Palmer and N. P. Samios, 21 May 1973, Palmer's files. The publication is Palmer, R. B., *PL*, section *B*, 46, no. 2 (17 September 1973): 240.

30. Musset, P., CERN technical memorandum TC–L/PA/PM/ju, 17 April 1973.

31. Musset, P., *Journal de Physique*, Colloque C3, 34, nos. 11/12 (supp.) (November–December 1973): C3–1, at C3–6 (trans. by authors).

32. Interview, Larry Sulak, an office at New York University, 28 November 1983; interview, Carlo Rubbia, his office, CERN, 11 February 1984; interview, David Cline, Weak Interaction Conference, Wingspread, Wisconsin, 31 May 1984; Benvenuti, A., et al., *PRL* 30, no. 21 (21 May 1973): 1084; Galison, *op. cit.*, p. 495.

33. Hasert, F. J., et al., *PL*, section *B*, 46, no. 1 (3 September 1973): 121.

34. Letter, Ettore Fiorini to Paul Musset, 11 July 1973; letter, J. Sacton to Paul Musset, 11 August 1973 (so dated, but internal evidence suggests the actual date is 11 July).

35. Both letters quoted in Galison, *op. cit.*, p. 495.

36. Hasert et al., *op. cit.*, p. 138.

37. Interview, Paul Musset, his office, CERN, 2 February 1984.

38. Letter, Patrick Coomey to Carlo Rubbia, 24 July 1973, Rubbia's files.

39. Benvenuti, A., et al., typescript. The published paper (see below) was "received 3 August 1973." See also, Rubbia, C., and Sulak, L., Harvard technical memorandum, 18 August 1973.

40. Interview, David Cline, Weak Interaction Conference, Wingspread, Wisconsin, 31 May 1984; Galison, *op. cit.*, pp. 497–98.

41. Moreover, the novel manner of presentation of the HPW results drew shouts of protest from other experimenters. "People didn't even understand the statistics," Sulak said later. "It took us a while to learn how to state the argument. Now it's really very common. But we just had to hammer away, and people sort of refused to understand that if you ask—take old physics. Old physics is a neutrino comes in and a muon goes out. Given old physics, how many would you expect with a neutrino in and nothing out, just because you miss [the muon]? And for that type of event, something being missed, we had way too many muonless events. It was like a five or six standard deviation effect, which is a really big effect in terms of statistical power." Most physicists were then used to the question, how well is the value of the ratio known? The certainty was then much less, which distressed some experimenters. They did not understand the significance, Sulak said, "of saying, how many muonless events do you *expect*—it turns out to be eight—and how many do you *see*—thirty-four. Well the probability that if you expect eight to all of a sudden see thirty-four, that's roughly the numbers, is just a far more significant thing. Somehow we could not convince people to understand that" (interview, Larry Sulak, a restaurant in Ohio, 6 June 1985). All quotes below by Sulak from same interview.

42. Interview, David Cline, Weak Interaction Conference, Wingspread, Wisconsin, 31 May 1984.

43. The actual angular difference between the first and second setup was small, as can be deduced from the description above. The difference between a "wide–angle" and "narrow–angle" muon was a matter of a few degrees, because the particles were traveling with sufficient forward momentum to swamp the effects of the sideways motion imparted by the interaction.

44. Actually, he provided others with descriptions of the HPW results. See Myatt, G., in Rollnik, H., and Pfeil, W., *Proceedings of the Sixth International Symposium on Electron and Photon Interactions at High Energies*, Physikalisches Institut, University of Bonn (27–31 August

1973): 389, 395, and 405; Musset, P., *Proceedings of the II International Conference on Elementary Particles*, Aix–en–Provence, 6–12 September 1973. Musset was informed by L. Sulak.

45. Lubkin, G., *PT* (November 1973): 17.

46. Barish, B. C., and Sciulli, F., "Neutral Current Investigations at NAL," Proposal 262, Fermilab Archives, Proposal Shelf, 24 October 1973. The original team proposal (Sciulli, F., et al., "Neutrino Physics at Very High Energies," Proposal 21, Fermilab Archives, Proposal Shelf, 10 June 1970) did not mention neutral currents.

47. Cline, D., Wisconsin technical memoranda, 1 and 11 October 1973, quoted in Galison, *op. cit.*, pp. 497–98.

48. The referees also criticized the way the statistics were handled, not understanding the arguments cited in note 41 above.

49. A copy of the first page of the typescript is in Galison, *op. cit.*, p. 500.

50. Musset, P., "Les reactions de courant neutre dans Gargamelle et le schema des particules elementaires," photocopied typescript, 6 November 1973, his files; interview, Paul Musset, his office, CERN, 2 February 1984.

51. The letter is reproduced in Galison, *op. cit.*, p. 501; account confirmed in interviews with Musset, Rubbia, and Sulak.

52. Imlay, R., Wisconsin technical memorandum, 29 November 1973, Imlay's files. The full handwritten calculation by Aubert, Imlay, and Ling was done by December 18.

53. Telephone interviews, Richard Imlay, 26 June 1985, Larry Sulak, 7 October 1985; Cline, D., Wisconsin technical memorandum, 13 December 1973, Imlay's files, includes the transparencies and Cline's quote below.

54. A follow-up HPW paper appeared eight weeks later (Aubert B., et al., *PRL* 32, no. 25 [24 June 1974]: 1454); Hasert, F. J., et al., *NP*, section *B*, 73, 1 (24 June 1974): 1. The Caltech group didn't publish until the next year; Barish, B. C., et al., *PRL* 34, no. 9 (3 March 1975): 538.

55. Telephone interview, Sheldon Glashow, 29 August 1983.

56. Glashow, S. L., in Garelick, D. A., ed., *Experimental Meson Spectroscopy—1974*, AIP Conference Proceedings, No. 21 (New York: American Institute of Physics, 1974): 392.

57. Iliopoulos, J., in Smith, J. R., ed., *Proceedings of the Seventeenth International Conference on High Energy Physics*, London, 1–10 July 1974 (London: Science Research Council, 1974): section III, pp. 97–100.

CHAPTER 18

1. The earliest piece of evidence for charm turned up earlier, when Kiyoshi Niu, a cosmic ray experimenter from the University of Tokyo, found what is likely a charmed particle while conducting experiments in the cargo hold of a Japan Air Lines flight with a new type of emulsion chamber designed to study high–energy showers. Niu and company had heard nothing of charm, but they realized that they had made an important discovery. They sent their result to *Physical Review Letters*, which rejected it because the journal editors had no confidence in cosmic ray results (Telephone interview, Kiyoshi Niu). It eventually appeared in a Japanese journal (Niu, K., Mikumo, E., and Naeda, Y., *PTP* 46, no. 5 [November 1971]: 1644). What might have been the last great cosmic ray particle discovery remained buried. Julian Schwinger used it as a bolster to his prediction of the J/psi (Schwinger, J., *PR*, section *D*, 8, no. 3 [1 August 1973]: 960–64).

2. Interview, Nicholas Samios and Robert Palmer, Brookhaven National Laboratory, 18 May 1983.

3. Anonymous, *Brookhaven Bulletin*, 25 January 1974, p. 1. The details of the run given below were provided for us from the Brookhaven archives by Neil Baggett and Michael Murtagh, whom we take this opportunity to thank.

4. Interview, Milda Vitols, Brookhaven scanning room, 3 May 1983.

5. Interview, Michael Murtagh, 17 May 1985.

6. We thank Milda Vitols for providing us with a copy of the sketch.

7. Interview, Helen LaSauce, May 1983.

8. Interview, Nicholas Samios and Robert Palmer, 18 May 1983.

9. Brookhaven Nuclear Laboratories (BNL) memorandum, R. B. Palmer and N. P. Samios to R. R. Rau, 21 May 1973; BNL memorandum, R. H. Phillips to R. B. Palmer, 32 December 1974, both from Samios's files.

10. Telephone interview, Samuel C. C. Ting, 6 July 1983.

11. The various pieces of biographical information were collected from prefatory information

in Ting's Nobel lecture and the interviews with Becker, Chen, Deutsch, Lederman, Pipkin, and Ting cited below.

12. See, for instance, Kroll, N. M., *NC*, section *A*, 45, no. 1 (1 September 1966): 65.

13. The CEA was completed in 1962.

14. All quotes from interview, Francis Pipkin, his office, Harvard, 7 May 1985.

15. Blumenthal, R. B., et al., *PR*, 144, no. 4 (29 April 1966): 1199.

16. Interviews, Samuel C. C. Ting, an office at CERN, 9 February 1984; his office, MIT, 15 October 1985.

17. Interview, Samuel C. C. Ting, an office at CERN, 9 February 1984.

18. The Cornell group had presented their first results at an American Physical Society meeting some months before; Talman, R., *Bulletin of the American Physical Society* 11, no. 3 (26 April 1966): 380. We were aided by a telephone interview with Richard Talman, 19 June 1985.

19. Another controversy involved something called the "Omega-Rho Interference Effect." A review of early experimental work on the effect is G. Goldhaber, in Baltay, C., and Rosenfeld, A. H., eds., *Experimental Meson Spectroscopy* (New York: Columbia University Press, 1970): 59–128. The Ting group's final results were presented in Alvensleben, H., et al., *PRL* 27, no. 13 (27 September 1971): 888. Quotes here from interview, Samuel Ting, his office, MIT, 15 October 1985.

20. Letter, Samuel Ting to Min Chen, 8 December 1971, Chen's files.

21. Becker, U. J., et al., "AGS Proposal," no. 598, Brookhaven archives, 11 January 1972; letter, Robert Phillips to Sam Ting, 29 February 1972, Brookhaven archives.

22. Letters, Robert Phillips to Sam Ting, 3 and 26 May 1972, Brookhaven archives.

23. Dates of AGS operation taken from the AGS operations journal, June to December 1974, AGS control room, Brookhaven. Dates confirmed by examination of Ting's experimental log books, Ting's files.

24. Interviews, Min Chen and Ulrich Becker, their offices, MIT, 1 March 1984.

25. Reproduction of printout taken from Becker's files; to fit the page, we have changed the number of asterisks representing events from four to three. A copy of this graph recently displayed in an MIT exhibit is incorrectly dated 2 September.

26. Account and quotes from interview, Ulrich Becker, his office, MIT, 2 March 1984; telephone interview, Wit Busza, May 1985.

27. Interview, Min Chen, his office, MIT, 1 March 1984.

28. Letter, Samuel Ting to Ronald Rau, 20 September 1974; letter, Ronald Rau to Samuel Ting, 26 September 1974, both in Brookhaven archives.

29. An amusing account of this and other experiments that did not observe charm can be found in Lederman, L., in Gaillard, M. K., and Stora, R., eds., *Théories de jauge en physique des hautes énergies*, Les Houches summer school, session 37, 1981 (New York: North Holland, 1983): especially 837–47.

30. SPEAR history from Richter, B., *Slac Beam Line*, Special Issue No. 7, November 1984; quotes below from interview, Burton Richter, his office, SLAC, 20 February 1985. Description of ADA from visit to Frascati in February 1984.

31. This, and much of what follows, is from G. Goldhaber, in *Adventures in Experimental Physics* 5: 131–40. Richter's quote from letter, Burton Richter to authors, 12 August 1985.

32. The magnet, used to bend the beam path 8°, was a landmark in accelerator engineering, the first superconducting magnet used as a standard operating part of an accelerator. Its liquid helium coolant had malfunctioned.

33. Interview, Martin Deutsch, his home, Cambridge, 7 May 1985.

34. Interview, Samuel Ting, an office at CERN, 9 February 1984.

35. Interviews, Bill Taylor and AGS control staff, AGS control room, 20 May 1985. Ting also visited the office of *Physical Review Letters* to learn about the rules for unrefereed publication.

36. We thank Dr. Schwartz for checking his travel records to provide us with exact dates.

37. Telephone interview, Jayashree Toraskar, 12 April 1985.

38. This account and subsequent quote from interview, Mel Schwartz, his home, California, 19 February 1985; see also, Ting, S. C. C., *Adventures in Experimental Physics* Vol. Epsilon (1978): 115.

39. Interviews cited above with Deutsch and Chen.

40. Interview, Gerson Goldhaber, his office, Berkeley, 19 February 1985 (all quotes).

41. Interview, Min Chen, his office, MIT, 1 March 1984.

42. The shoulder is described in Christenson, J. H., et al., *PRL* 25, no. 21 (23 November 1970): 1523. There was another, more theoretical reason: The ratio of the number of muons to

pions was a mysterious number that theorists thought should be explained by the number of heavy photons, but wasn't by the three known. Ting redid the calculations adding in the J, and found that that didn't explain it, either. This led him to hope that there might be others just around the corner (Ting, S., *RMP* 49, no. 2 [April 1977]: 244).

43. Interviews with Richter and Goldhaber cited above, supplemented by notes and photographs from Goldhaber's files.

44. AGS Utilization Daily Report, Brookhaven Archives; interview, Austin McGeary, 20 May 1985.

45. Telephone interview, Burton Richter, 5 July 1983.

46. Aubert, J., et al., *PRL* 33, no. 23 (2 December 1974): 1404; Augustin, J., et al., *PRL* 33:23:1406; Bacci, C., et al., *PRL* 33:23:1408.

47. Abrams, et al., *PRL* 33, no. 24 (17 December 1974): 1453.

48. The two physicists quoted here, like many of their colleagues, talked freely on the subject but requested anonymity.

49. Interview, Martin Deutsch, his home, Cambridge, 7 May 1985.

50. By the time of the J/psi discovery, David Politzer and Thomas Appelquist had written a paper predicting a narrow peak for "charmonium" the charm–anticharm meson; Politzer's conservatism delayed the appearance of the paper until after the announcement; D. Politzer and T. Appelquist, *PRL* 34, no. 1 (6 January 1975): p. 43, and letter, David Politzer to authors, 25 July 1985.

51. Telephone interview, Burton Richter, 5 July 1983.

52. Cazzoli, E. G., et al., *PRL* 34, no. 17 (28 April 1975): 1125. Nguyen–Khac, L., in *"La Physique du Neutrino à Haute Énergie," Proceedings,* (Paris: 18–20 March, 1975): 173.

53. Interview, Sheldon Glashow, 6 June 1983.

54. Interview, Gerson Goldhaber, Summer 1983.

55. Telephone interview, Roy Weinstein, Summer 1983.

56. The standard model faced two other experimental obstacles in these years, both erected by the HPW group. In 1974 and again in 1976 the HPW group reported an inexplicably high value of a number called Y, while in March 1977 they announced six events with three muons; these findings, they said, could not be explained by the standard model. Both the "High-Y" and "trimuon" anomalies, however, were spurious, as a CERN team showed in July and August 1977 to the considerable embarrassment of the Fermilab group. The first HPW papers were: Aubert, B., et al., *PRL* 33, no. 16 (14 October 1974): 984; Benvenuti, A., et al., *PRL* 38, no. 20 (16 May 1977): 1110. The CERN group papers were Holder, M., et al., *PRL* 39, no. 8 (22 August 1977): 433–36; see also Barish, B., et al., *PRL* 38, no. 11 (14 March 1977): 577–80.

57. Bouchiat, M. A., and Bouchiat, C., *PL*, section *B*, 48, no. 2 (21 January 1974): 111.

58. Telephone interview, Edward Fortson, 24 June 1985.

59. *Ibid.*; Feinberg, G., *Nature* 271, no. 5645 (9 February 1978): 509.

60. Interview, Patrick Sandars, his office, Oxford, 9 October 1984.

61. Baird, P. E. G. et al., *Nature* 264, no. 5585 (9 December 1976): 528.

62. Close, F. E., *Nature* 264, 5585 (9 December 1976): 505–6.

63. Interview, Steven Weinberg, outside Harvard, 7 May 1985.

64. Walgate, R., *New Scientist*, 31 March 1977, p. 766.

65. Barkov, L. M. and Zolotorev, M. S., *JETP Letters* 27, no. 6 (20 March 1978): 357.

66. Fortson and Sandars had performed calculations based on an experimentally determined number of 0.35 for $\sin^2 \theta$. In the summer of 1977, this was revised to 0.25 and some months later it fell to 0.21. Moreover, atomic physicists had discovered other factors that affected the way the Weinberg–Salam model was applied to atoms. The effect of the two changes was that the number predicted for the line studied by the Seattle experiment was now 10.5×10^{-8}— within the margin of error of the original experiment! The team had then released more results in contradiction with the new prediction. The Oxford experiment, in contradiction from the start, remained so. The Russian result, originally seeming to confirm the $SU(2) \times U(1)$ prediction, did so no longer.

67. Interview, Charles Prescott, SLAC, 20 February 1985.

68. Proposals E94, E122, SLAC proposal file, SLAC archives. Interview, Richard Taylor, his office, SLAC, 20 February 1985. We have been assisted further by letter, Richard Taylor to authors, 6 August 1985.

69. The results were first published as Prescott, C. Y. et al., *PL*, section *B*, 77, no. 3 (14 August 1978): 347–52.

CHAPTER 19

1. Interview, Frank Wilczek, his office, University of Santa Barbara, 22 February 1985.

2. The W and Z were discovered by two CERN collaborations with a new accelerator that collided protons and antiprotons. For their role in the construction of the machine and the subsequent discoveries, Carlo Rubbia and Simon van der Meer, both of CERN, were awarded the 1984 Nobel Prize for physics. A lengthy popular account of the finding is given in the final chapters of Sutton, C., *The Particle Connection* (New York: Simon and Schuster, 1984); considerably shorter and less technical is Crease, R., and Mann, C., *Science Digest*, September 1984, 51; see also the forthcoming Taube, G., *Nobel Dreams* (New York: Simon & Schuster, 1986). References to original papers are in Sutton, *op. cit.*, p. 174.

3. Bjorken, J., in *Neutrino '79: Proceedings of the International Conference on Neutrinos, Weak Interactions, and Cosmology*, 18–22 June 1979, Bergen, Norway (Bergen: University of Bergen and Nordita, 1980): 9. Bjorken's talk carried the very good point to his fellow theorists that most physical theories have been disproven, and all are at the least incomplete; he warned against the "dangers common to any orthodoxy."

4. These came from an SU(4) model done by Wigner, which treated spin and isotopic spin identically. The fundamental representation consisted of four particles: neutron, spin up and spin down; and proton, spin up and spin down. When the SU(3) eightfold way was invented, that was combined with the SU(2) spin to produce SU(6); quarks were incorporated to produce another version. Salam and two other physicists produced a relativistic version, which was called U(6,6) as a relativistic theory of Zerseth order Lagrangian vertices. It was mistaken for an S-matrix theory. Then a variety of "no–go" theorems were produced that said no fundamental theory could accommodate particles of different spins. Nowadays the arrival of supersymmetry has pointed out a flaw in the various no–go theorems. Salam "demolished"—interview, Abdus Salam, an office at CERN, 19 September 1983.

5. Interview, Abdus Salam, a hotel in New York, 27 May 1984.

6. Berlin, I., *Russian Thinkers*, Hardy, H., and Kelly, A., eds. (New York: Penguin Books, 1978): 22.

7. Lecture, Pati, J., Conference on the History of the Weak Interactions, Wingspread, Racine, Wisconsin, 31 May 1984. More technically, why must the proton and electron in beta decay have the same spin? It would be perfectly possible to give the neutron and electron the same spin in the theory by rewriting the currents.

8. Bjorken, J. D., in *Proceedings of the XVI International Conference on High Energy Physics*, vol. 2 (Batavia, Illinois: Fermilab, 1972): 304. Pati and Salam came up with four models, of which this is the one they most extensively discussed. We have used somewhat more modern notation.

9. Pati, J., and Salam, A., *PR*, section D, 8, no. 4 (15 August 1973): 1249. Worth mentioning here is an earlier, more confused attempt at unification: Bars, I., Halpern, M. B., and Yoshimura, M., *PR* 7, no. 4 (15 February 1973): 1233.

10. Pati, J. C. and Salam, A., *PRL* 31, no. 10 (3 September 1973): 661–64.

11. Salam, A., *RMP* 52, no. 3 (July 1980): 530.

12. Weinberg, S., comment in *Second International Conference on Elementary Particles*, Aix–en–Provence, 6–12 September 1973, *Journal de Physique* 34, no. 10 (supp.), Colloque C1, p. 47.

13. Interview, Abdus Salam, an office at CERN, 19 September 1983.

14. Interview, Howard M. Georgi III, his office, Harvard, 14 June 1983.

15. Telephone interview, Murray Gell-Mann, 22 May 1984.

16. Interview, Howard M. Georgi III, a restaurant in Manhattan, 29 January 1985.

17. The proton decays in Georgi and Glashow's SU(5) for entirely different reasons than in Pati and Salam's model. The proton decay of the Pati and Salam model does not arise from the single coupling constant, but stems instead from their special Higgs bosons. Sometimes the same phenomenon can be predicted for different theoretical reasons.

18. Quotes below from interview, Howard M. Georgi III, his office, Harvard, 14 June 1983; interview, Sheldon Glashow, his office, Harvard, 2 December 1982.

19. In the same experiment that discovered the neutrino, Reines, Cowan, and Maurice Goldhaber used the data to set a limit on proton lifetime (Reines, F., Cowan, C., Jr., Goldhaber, M., *PR* 96, no. 4 [15 November 1954]: 1157). Georgi and Glashow used the limit from Gurr, H. S., et al., *PR*, 158, no. 5 (25 June 1967): 1321. For an historical review, see Goldhaber, M., Langacker, P., and Slansky, R., *Science* 210 (21 November 1980): 851.

20. Georgi, H., and Glashow, S. L., *PRL* 32, no. 8 (25 February 1974): 438.

21. Telephone interview, James Bjorken, 5 July 1983.

22. Telephone interview, Howard M. Georgi III, 11 May 1984.

23. Interview, Steven Weinberg, outside Harvard, 7 May 1985. We have condensed the discussion slightly. It is worth noting here that the paper contains a prediction of the weak mixing angle, $0.2°$, that was initially quite low. During the 1970s, the value of the angle drifted down, to where it is now in good agreement.

24. A good nontechnical introduction to cosmology with some historical overtones is Weinberg, S., *The First Three Minutes* (New York: Basic Books, 1977).

25. Penzias, A., in Reines, F., ed., *Cosmology, Fusion, and Other Matters: George Gamow Memorial Volume* (Boulder, Co.: Colorado Associated University Press, 1972).

26. Interview, Sheldon Glashow, his office, Harvard, 13 June 1983.

27. If the proton does decay, it may solve a long-standing puzzle about the relative proportions of matter and antimatter in the Universe, namely the preponderance of matter. The exact mechanism is rather complicated, involving parity–violating effects. See Yoshimura, M., *PRL* 41, no. 5 (31 July 1978): 281; errata, 42, no. 11 (12 March 1978): 746; and Sakharov, A., *JETP Letters* 5, no. 1 (1 January 1967): 24.

28. Biographical information from interview, M. G. K. Menon, his office, Planning Commission, New Delhi, 21 March 1985. A history of the Tata Institute can be found in Lala, R. M., *The Heartbeat of a Trust* (New Delhi: Tata-MacGraw Hill, 1984), chaps. 8–10.

29. Krishnaswamy, M., et al., *PL*, section *B*, 106, no. 4 (12 November 1981): 339; interview, V. S. Narasimham, Kolar Gold Fields, 14 March 1985 (all quotes).

CHAPTER 20

1. Hawking, S., "Is the End in Sight for Theoretical Physics?" in Boslough, J., *Stephen Hawking's Universe* (New York: Morrow, 1985).

2. The exact source of this often-quoted remark is difficult to discover (telephone interview, Freeman Dyson, 1 December 1985).

3. Einstein, A., in Lorentz, H. A., et al., *The Principle of Relativity* (New York: Dover Publications 1952): 191.

4. Wolfgang Pauli to J. R. Oppenheimer, no date (but certainly in early 1958), Library of Congress, Manuscript Division, General Case File, Box 56, Pauli file. George Gamow reproduces a similar letter in Gamow, G., *Thirty Years That Shook Physics* (New York: Doubleday, 1966): 162.

5. Interview, Gerard 't Hooft, his office, University of Utrecht, 26 September 1984.

6. Telephone interview, Murray Gell–Mann, 5 February 1985.

7. Kaluza, T. F. E., *Sitzungsberichte der Berliner Akademie* 54 (1921): 966. The article was communicated by Einstein on 8 December. At first Einstein had said that he would be pleased to submit the article. (Letter, Albert Einstein to Theodor Kaluza, 21 April 1919, quoted in Freedman, D., and van Nieuwenhuizen, P., *Scientific American*, March 1985, p. 78.) Then, a week later, Einstein wrote to Kaluza with the thought that although he had found nothing wrong in the paper, "the arguments brought forward so far do not appear convincing enough," and asked for more detailed calculations (letter, Albert Einstein to Theodor Kaluza, 28 April 1919). Kaluza apparently did. Einstein didn't respond. Two and a half years later he unexpectedly relented (Letter, Albert Einstein to Theodor Kaluza, 14 October 1921).

8. Klein, O., *ZfP* 37, no. 12 (10 July 1926): 895.

9. Veneziano, G., *NC*, section A, 57, no. 1 (1 September 1968): 190; interview, Yoichiro Nambu, his office, University of Chicago, 13 February 1985. Nambu's first string article is Nambu, Y., in Chand, R., ed., *Symmetries and Quark Models*, Proceedings of an International Conference at Wayne State University, 18–20 June 1969 (New York: Gordon & Breach, 1970): 269. Later, independent discoveries of the string are listed in Scherk, J., *RMP* 47, no. 1 (January 1975): 123.

10. First described by Lovelace, C., *PL*, section B, 34, no. 6 (29 March 1971): 500 (which on p. 502 dismisses the twenty-six dimensions in passing as "obviously unphysical"); proven by Brower, R.C., *PR*, section D, 6, no. 6, 15 September 1972.

11. Neveu, A., and Schwarz, J., *NP*, section B, 31, no. 1 (1971): 56. A related model for spin-½ particles was proposed subsequently by Pierre Ramond, then incorporated in the following: Neveu, A., and Schwarz, J., *PR*, section D, 4, no. 4 (15 August 1971): 1109; Thorn, C.B., *ibid.*, p. 1112; Schwarz, J., *NP*, section B, 46, no. 1 (1972): 61.

12. Schwarz, J., and Scherk, J., *NP*, section B, 81, no. 1 (1974): 118.

13. The superstring theories in note 11 are historically the first example of supersymmetry.

14. Green, M., and Schwarz, J., *PL*, section B, 148, no. 1 (1984): 117; *ibid.*, 151, no. 1 (1985): 21.

15. As of the beginning of 1986, the most favored candidate was proposed by Gross, D., Harvey, J., Martinec, E., and Rohm, R., *PRL* 54 (1985): 502.

16. Princeton theorist Ed Witten has flatly predicted that superstring theory "will dominate the next half century, just as quantum field theory has dominated the previous half century" (quoted in *PT*, July 1985, p. 20).

17. Interview, Howard Georgi, a restaurant in New York City, 29 January 1985.

18. Bridgman, P., *The Logic of Modern Physics* (New York: Macmillan, 1946): 207.

19. Interview, Sheldon Glashow, his office, Harvard, 10 November 1982.

20. Interview, Julian Schwinger, a restaurant in Los Angeles, 4 March 1983.

21. Interview, Steven Weinberg, faculty club, Harvard, 7 May 1985; faculty club, University of Texas, Austin, 28 November 1984.

INTERVIEWS

Although we have based our work upon traditional historical sources, this book would not have been possible if many physicists had not been willing to set aside their own work for a while and talk to us. Most sessions were in person, at length, and taped, although there are exceptions; telephone interviews are indicated below with an asterisk (*). The list below contains only those interviews whose length or direct relevance to the material in this book merited their listing. We hope the others will not mind, and thank all of them for their indulgence.

STEPHEN ADLER   30 May 1985*

P. W. ANDERSON   28 January 1985*

PIERRE AUGER   3 October 1984

ULRICH BECKER   1 and 2 March 1984

JOHN BELL   6 February 1984

HANS BETHE   8 August 1984*, 4 December 1984*

JAMES BJORKEN   10 May 1983*, 5 July 1983*, 13 February 1985, 3 May 1985, 23 November 1985*

MARTIN BLOCK   30 May 1984

CLAUDE BOUCHIAT   4 October 1984

MARIE-ANNE BOUCHIAT   4 October 1984

GIORGIO BRIANTI   7 February 1984

WIT BUSZA   5 June 1985*

NICOLA CABIBBO   21 February 1984

MIN CHEN   1 and 2 March 1984, 6 September 1985*

DAVID CLINE   31 May 1984

MARCELLO CONVERSI   16 February 1984

RICHARD DALITZ   11 October 1984

ALVARO DE RUJULA   9 February 1984

MARTIN DEUTSCH   31 May 1985, 7 May 1985

SAMUEL DEVONS   4 January 1984*, 23 May 1984, 4 October 1985

J. W. DRINKWATER   19 October 1984*

GERALD FEINBERG   27 January 1983, 11 May 1984, 9 January 1985, 1 May 1985, 25 June 1985*

MARKUS FIERZ   21 September 1984*

ETTORE FIORINI   8 February 1984

RICHARD FEYNMAN   23 February 1985

E. NORVAL FORTSON   1 July 1985*

J. BRUCE FRENCH   11 July 1984*

RAOUL GATTO   29 May 1984

MURRAY GELL-MANN   3 March 1983, 3* and 27* June 1983, 13 July 1983*, 10 November 1983*, 18* and 22* May 1984, 5* and 21 February 1985, 24 April 1985, 28* and 29* May 1985

HOWARD GEORGI   14 June 1983, 11 May 1984*, 29 January 1985, 4 September 1985*

SHELDON GLASHOW   9 and 10 November 1982, 2 December 1982, 5*, 8* and 25 April 1983, 12 May 1983*, 6*, 8* and 13 June 1983, 1 July 1983*, 16 January 1984, 8 and 15 May 1984*, 21 May 1984*, 8 May 1985

WOODY GLENN   20 May 1985

GERSON GOLDHABER   11 July 1983*, 1 June 1984, 19 February 1985

MAURICE GOLDHABER   4 February 1983, 2 April 1983, 18 May 1983, 21 December 1983

JEFFREY GOLDSTONE   26 November 1984, 23 January 1985*, 7 May 1985
OSCAR GREENBURG   4 February 1984*
DANIEL GREENBERGER   16 February 1983*
DAVID GROSS   2 April 1985
PETER HIGGS   29 and 30 May 1984
CHARLES HILL   7 February 1984
GERARD 't HOOFT   26 September 1984
JOHN ILIOPOULOS   8 May 1984
RICHARD IMLAY   25 July 1985*, 16 October 1985*
HENRY KENDALL   26 February 1985*
THOMAS KIBBLE   5 October 1984
MICHIJI KONUMA   31 May 1984
MASATOSHI KOSHIBA   6 March 1985
NORMAN KROLL   11 July 1984*
HELEN LASAUCE   May 1983
WILLIS LAMB   9 August 1984*
RICHARD LEARNER   5 October 1984
LEON LEDERMAN   14 February 1984, 24 May 1984*
Y. Y. LEE   7 June 1983*, 11 January 1984
LUCIANO MAIANI   17 February 1984
WILLIAM MARCIANO   2 April 1983
ROBERT MARSHAK   30 May 1984, 14 January 1985*
AUSTIN MCGEARY   20 May 1985
SIMON VAN DER MEER   7 February 1984
M. G. K. MENON   21 and 22 March 1985
PHILIP MORRISON   19 June 1984
LLOYD MOTZ   19 December 1984*
MICHAEL MURTAGH   3 May 1983, 17 May 1985*
PAUL MUSSET   2 and 6 February 1984
YOICHIRO NAMBU   30 and 31 May 1984, 13 February 1985, 30 May 1985*, 5 June 1985*
V. S. NARASIMHAM   15 and 16 March 1985
YUVAL NE'EMAN   3 May 1985
ABRAHAM PAIS   5 February 1985*, 30 March 1985
ROBERT PALMER   4 February 1983, 18 May 1983
RUDOLF PEIERLS   9 October 1984
ANDRÉ PETERMANN   10 February 1984
ORESTE PICCIONI   3 May 1985
FRANCIS PIPKIN   7 May 1985
DAVID POLITZER   21 February 1985
JACQUES PRENTKI   7 February 1984
CHARLES PRESCOTT   20 February 1985
I. I. RABI   9 May 1984
BURTON RICHTER   5 July 1983*, 24 February 1985
ALAN RITTENBERG   7 June 1983*
GEORGE ROCHESTER   30 May 1984, 8 October 1984
BRUNO ROSSI   1 November 1984
CARLO RUBBIA   11 February 1984, 17 May 1984*
ABDUS SALAM   19 September 1983, 23 February 1984, 27 May 1984, 2 May 1985
NICHOLAS SAMIOS   4 February 1983, 18 May 1983, 12 February 1985
PATRICK SANDARS   9 October 1984
HERWIG SCHOPPER   1 February 1984
MEL SCHWARTZ   16 June 1983*, 19 February 1985
JOHN SCHWARZ   22 February 1985
SILVAN SCHWEBER   19 March 1984
JULIAN SCHWINGER   4 March 1983
HERWIG SCHOPPER   1 February 1984
ROBERT SERBER   22 November 1983, 18 July 1984, 22 and 29 October 1984, 19 and 26 November 1984, 28 March 1985, 2, 8 and 26 April 1985, 6, 15, 23, and 29 May 1985, 5 June 1985
N. F. SHUTT   May 1983*
B. V. SREEKANTAN   14 March 1985

ERNST STUECKELBERG   10 February 1984

LAWRENCE SULAK   14 July 1983*, 28 October 1983, 14 February 1985*, 3 March 1985*, 5 and 6 June 1985

RICHARD TAYLOR   20 February 1985, 26 February 1985*

SAMUEL TING   6 July 1983*, 9 February 1984, 15 October 1985

ED UEHLING   20 July 1984*, 25 October 1984*

GEORGE UHLENBECK   10 May 1984

MILDA VITOLS   May 1985

MARTINUS VELTMAN   20 July 1983*, 13 December 1983, 26 February 1985*

STEVEN WEINBERG   28 November 1984, 7 May 1985, 30 September 1985*

ROY WEINSTEIN   Summer 1983*

VICTOR WEISSKOPF   24 September 1984

JOHN WHEELER   14 December 1983

FRANK WILCZEK   22 February 1985

ROBLEY WILLIAMS   20 and 21 November 1984*

C. N. YANG   3 April 1983*, 10 October 1985*

GAURANG YODH   11 January 1984*

TONY ZEE   5 October 1985*

ANTONINO ZICHICHI   3 and 15 February 1984

Glossary

THE FOLLOWING ENTRIES are not intended as precise scientific definitions, but rather as jogs to memories temporarily overloaded by scientific jargon.

Accelerator: Device used to push subatomic particles to high velocities and drive them into each other or into a small target. The resultant collisions give clues to particle behavior. See pp. 253–58 .

Alpha: See fine structure constant.

Alpha decay/particle/ray: Form of radioactivity in which certain atomic nuclei spontaneously eject a particle composed of two protons and two neutrons—a helium atom without its electrons. See pp. 16–18.

Antimatter/antiparticle: Almost every type of particle has an "antiparticle" equal in mass and most other properties but opposite in charge; the positively charged positron is the antiparticle of the negative electron, for instance. When particles and antiparticles meet, they annihilate each other, producing light. See pp. 86–91.

Balmer lines: A portion of the hydrogen spectrum described by a simple mathematical pattern. Niels Bohr used this pattern as a clue to the placement of the electrons around the nucleus. See p. 27.

Baryon: Originally, the proton and any heavier particle; now used for any particle made up of three smaller entities called quarks. See pp. 176, 282–83.

Baryon conservation: A law saying that the total number of baryons minus the total number of antibaryons in an interaction—or in the cosmos as a whole—never changes. Baryon conservation is questioned by grand unified theories.

Beta decay/ray/particle: Form of radioactivity in which nuclei eject electrons, a process that long seemed mysterious because the nucleus does not contain electrons. See pp. 16, 196–214 .

Boson: One of two basic classes of particles defined by the value of their spin, in this case, integral values. Bosons include photons, W and Z particles, pions, and gluons. See pp. 194–95.

Bubble chamber: Device used to create a vat of liquid on the verge of boiling that is unstable enough to register the passage of particles from collisions produced in an accelerator. See pp. 274–79.

Charm or charmed quark: One of the six known types or "flavors" of quark. See **GIM mechanism**, and pp. 290–92.

Cloud chamber: Device used to create a small, rapidly cooling cloud unstable enough to register the passage of incoming particles, whose tracks are then photographed. See pp. 88–90.

Color: Quality somewhat like electric charge that distinguishes otherwise identical quarks. Color is the source of the attraction among quarks. Has nothing to do with ordinary color. See pp. 298–99.

Commutative Law: The mathematical principle a × b = b × a. Often violated in quantum systems, e.g., *a* and *b* do not commute. See p. 48.

Conservation law: A rule that the total value of some quantity does not change in any interaction. Famous examples are the laws of conservation of energy and momentum, which state that the total energy and momentum in any interaction never changes.

Cosmic ray: Charged particles—primarily protons—that bombard the earth from space, some at huge energies. See pp. 147–50.

Coupling constant: A number indicating the strength or likelihood of interaction caused by a given force. See p. 111.

Decay: The transmutation of a particle into one of more other particles; when the particles are in atomic nuclei, these, too, are said to decay.

Eightfold way: A method of classifying particles based on a mathematical system called SU(3) that was invented in 1961 by Murray Gell-Mann and Yuval Ne'eman. See pp. 265–68.

Elastic scattering: Interactions in which no particles change their properties. See p. 303.

465

Electromagnetic radiation: Light. Can be regarded as either consisting of streams of particles called photons, or of alternating electric and magnetic fields that ripple outward in waves.

Electron: Negatively charged particle which orbits about the nucleus of every atom. See pp. 12–14.

Electron volt: Small unit of energy used by particle physicists. Defined as energy acquired by an electron after going through a potential difference of 1 volt. 1 eV = $1.6 \times 10-19$ joule. See p. 257.

Electroscope: Mechanism for measuring small charges of static electricity by observing the separation of two pieces of gold foil. The impossibility of keeping the pieces charged for long intervals—"residual discharge"—first alerted researchers to the existence of cosmic rays. See pp. 147–49.

Electroweak theory: Theory that links together the weak and electromagnetic forces. See pp. 217–28.

Element: A substance that cannot be decomposed into more simple substances. The atoms of an element all have the same number of protons in their nuclei.

Event: Interaction among subatomic particles.

Excited state: State in which an atom or particle temporarily possesses more energy than in its ground state. Usually produced in the lab by bombardment with particles or radiation.

Fermion: One of two basic classes of particles defined by the value of their spin, in this case half-integer values ($-\frac{1}{2}$, $+\frac{3}{2}$, etc.). Fermions include electrons, protons, neutrons, and quarks.

Field: A region in space in which a quantity is precisely defined at every point. See p. 81.

Fine structure constant: An important number in physics that links together relativity, classical theory, and quantum theory, and is the coupling constant for electromagnetism. Although it is still unexplained, its value—about 1/137, a number called alpha—is known with great precision. See pp. 110–12.

Flavor: Name given to the different varieties of quark or lepton. See pp. 282–83.

Frequency: The number of wavelengths that go by a point in a given time. Low-frequency waves have long wavelengths. See p. 27.

Gamma decay/ray: Form of radioactivity in which the nucleus emits a highly energetic photon or particle of light—a gamma ray.

Gauge invariance/theory/symmetry: Gauge theories are theories that depend on gauge symmetries, which are a type of symmetry in which the value of a quantity is changed from point to point in space-time. In quantum electrodynamics, the phase of the wave function can be changed arbitrarily without affecting the electric charge. A feature of gauge theories is that the symmetry cannot be maintained without the introduction of intermediate particles, in this case the photon. There are two types of gauge invariance, global and local. Global gauge invariance occurs when the same change is made everywhere; local gauge invariance, in which changes are made randomly from place to place, can occur in Yang-Mills theories. The gauge theories in the standard model are locally invariant. See pp. 189–95.

GeV: Thousand million electron volts.

GIM mechanism: Theoretical stratagem that explains how the existence of a fourth quark, charm, drastically reduces neutral currents in strange particles. See pp. 314–17.

Global symmetry: See **gauge invariance/theory/symmetry.**

Goldstone boson: Particles almost inevitably produced as a consequence of the spontaneous breaking of a symmetry. See pp. 239–42.

Grand unified theories: Theories attempting to unify the weak, strong, and electromagnetic interactions. See pp. 393–404.

Ground state: The energy level an atom or particle goes to when left undisturbed.

Group theory: Mathematics of symmetry. A group is a collection of elements with properties defined by an operation such as rotation which transforms one element into another. Elementary particle forces are described in terms of "Lie groups" with names like SU(2) and SU(3). See pp. 228–29.

h, $\hbar$: See **Planck's constant.**

Hadron: Particles affected by the strong interaction. There are two classes, baryons and mesons. Both are made from quarks.

Hamiltonian: A type of equation used in quantum physics but invented in the nineteenth century as part of a reformulation of Newtonian mechanics. See pp. 27–28.

Heisenberg uncertainty principle: See **uncertainty principle.**

Higgs boson: A special feature of Yang-Mills theories. The Goldstone bosons in spontaneously broken Yang-Mills theories are "eaten up" by Higgs bosons, which have the property of endowing the intermediate particles with mass. In the process, the Higgs boson is thought to give mass to the W and Z particles. See pp. 242–46.

Inelastic scattering: Particle interactions in which particles change their identities. Generally of higher energy than elastic scattering. Used to study scaling. See p. 303.

Invariance: The property of maintaining a symmetry. See pp. 185–87.

Ion: Atom with more or fewer than its normal complement of electrons. Ordinary atoms are neutral; ions have a net charge. See p. 147.

Lagrangian: A type of equation used in quantum physics but invented in the nineteenth century as part of a reformulation of Newtonian mechanics. See pp. 136–37.

Lamb shift: A slight shift in the energy level of the $2S_{1/2}$ orbital of the hydrogen atom caused by self-energy effects. Its discovery in 1947 was a crucial vindication of quantum electrodynamics. See pp. 110–22.

Latitude effect: Shift in distribution of incoming cosmic rays across the globe caused by their interaction with the earth's magnetic field. See pp. 153–55.

Lepton: Particles unaffected by strong interactions. The leptons are the electron, muon, tau, and at least two and probably three distinct accompanying neutrinos. Leptons have spin 1/2 and are pointlike objects. They are not made of quarks. See p. 176.

Meson: Originally, any subatomic particle heavier than an electron but lighter than a proton. Now means a quark-antiquark pair. See p. 163.

MeV: Million electron volts.

Muon: Heavier version of electron whose similarity in mass to the pion caused a decade of confusion from 1937 to 1947. See pp. 168–69.

Neutral current: Generally used for weak neutral currents, i.e., weak interactions mediated by Z_0 particles. Based on data from strange particles, neutral currents were for years thought not to exist. See GIM mechanism. See pp. 227–28.

Neutrino: Massless, electrically neutral particle that experiences only weak interactions and thus is extremely difficult to stop or detect. See pp. 196–97.

Neutron: Neutral constituent of atomic nuclei. See pp. 31–32.

Noether's theorem: Briefly, for every symmetry in a system, there is a conservation law. See pp. 187–89.

Nuclear force: Force that clamps together protons and neutrons in the nucleus. Now regarded as a side effect of strong interactions. See pp. 158–61.

Nucleon: Generic name for proton and neutron. See p. 174.

Orbital: A region around an atomic nucleus in which there is a high probability of finding a particular electron. Each orbital is characterized by a specific energy and set of quantum numbers, which are used to label it. See pp. 110–11.

Parity violation: Strong and electromagnetic forces have no preference for left or right and are said to conserve parity. The weak force, on the other hand, only affects left spinning particles, ignoring right spinning particles—it violates parity. See pp. 203–10.

Photon: A particle of light. Massless, chargeless. See p. 25.

Pion: Particle whose exchange among protons and neutrons is responsible for the nuclear force. Its similarity to the muon long created confusion. See pp. 168–69.

Planck's constant: The ratio h of a photon's energy E to its frequency v, as given by $E = hv$. Planck's constant divided by twice pi is written as $\hbar$. See pp. 23–27.

Proton: Positively charged particle in the nucleus of every atom now known to consist of three quarks: two ups and a down. See pp. 30–31.

Quantum: A specific amount of energy released or absorbed in a process. Subatomic interactions occur with the exchange of particular packets of energy—quanta—that have certain given values, and no others. The quantum of electromagnetic energy is the photon. See pp. 23–25.

Quantum chromodynamics: Quantum field theory describing the interactions of quarks and gluons in terms of color. See pp. 327–36.

Quantum electrodynamics: Quantum field theory describing the interactions of electrically charged particles and fields. See pp. 82–85.

Quantum field theory: Theory describing creation and destruction of particles in elementary particle interactions. Matter and radiation are regarded as the quanta of particular fields; in a sense. these quantum fields thus become the ultimate reality. See pp. 34–37.

Quantum number: A number in the series . . . $-1\frac{1}{2}$, -1, $-\frac{1}{2}$, 0, $+\frac{1}{2}$, $+1$, $+1\frac{1}{2}$ . . . that specifies the value of a quantized physical quantity such as spin, momentum, or orbital energy. See pp. 41–42.

Quark: Entities that comprise hadrons and have fractional electric charges of $+\frac{2}{3}$ and $-\frac{1}{3}$.

They come in six "flavors" (up, down, strange, charm, top, and bottom) each of which can come in three "colors" (red, blue, and green). See pp. 280–84.

Radioactivity: The spontaneous emission, by atomic nuclei, of one or more subatomic particles. Such a process is a radioactive decay. Three principal types of radioactivity exist: alpha, beta, and gamma. Beta and gamma radioactivity also affect individual particles. See pp. 10–12.

Radium: A very rare, powerfully radioactive element. See p. 16.

Renormalization: Equations in quantum field theory often seem to give nonsensical infinite predictions. Renormalization is the complex procedure necessary to show that they in fact do not. See p. 108.

Residual discharge: See **electroscope.**

Scalar: A type of wave function and its associated spin 0 particle. See pp. 210–11.

Scaling: A type of behavior that occurs when high-energy electrons are directed at protons, revealing the existence of constituents in the latter. See pp. 304–308.

Schrödinger equation: An equation characterizing subatomic particles in terms of abstract waves. See pp. 53–56.

Self-energy: Generally, the name for the way a subatomic particle can interact with itself, which can be likened to the way a jet can be buffeted by its own sonic boom. The calculations of self-energy seemed to hopelessly complicate many theories and discouraged theorists for years. See pp. 92–94.

Spectroscopy: Originally, the study of atomic spectra; now used to describe the study of the exact properties of subatomic particles.

Spectrum: The array of light frequencies given off by atoms; each element has a unique spectrum.

Spin: A property of certain elementary particles whereby they act as if they were spinning about an axis. Particles spin with fixed values of momentum, e.g., electrons and protons always have a spin of one-half. See pp. 42–44.

Spontaneous symmetry breaking: A peculiarity of some theories in which the lowest energy state does not show a symmetry that is apparent at higher energies; as the system moves from higher to lower energies, it is said to "break" the symmetry. See pp. 238–47.

Strangeness or strange quark: One of the six known types or "flavors" of quark. Strange particles, such as the K or lambda, have strange quarks in them. See pp. 172–77.

String theory: A new theory describing subatomic particles as consisting not of points in space but of tiny, one-dimensional strings. See pp. 414–16.

Strong force/interaction: The strongest force in the subatomic world, it has great power but miniscule range. The theory of the strong interaction is called quantum chromodynamics. See pp. 175–76.

SU(2), SU(3), etc.: Mathematical ways of characterizing elementary particle theories. U(1) is associated with quantum electrodynamics, SU(2) × U(1) with the electroweak theory, and SU(3) with quantum chromodynamics. Another use of SU(3) occurred in the eightfold way. See pp. 228–29.

Symmetry: Used by physicists to denote quantities which remain the same despite a change in the system affecting them. A circle has one kind of symmetry because rotating it or flipping it over leaves it unaffected; the Eiffel Tower has another kind because one half mirrors the other. See pp. 184–87.

Uncertainty principle: A key consequence of quantum mechanics. The principle states that it is impossible simultaneously to determine the position and momentum of a given subatomic particle. Another version of the principle states the same thing for the exact energy of the particle and an exact time. See pp. 61–71.

Vacuum polarization: A change in the physical properties of the space about an electrically charged particle that causes complications in the theory used to describe such particles. See pp. 102–108.

Vector: A quantity with a direction, such as momentum; also, a type of wave function associated with a particle with a spin of 1. All elementary particle interactions are vector interactions transmitted by spin − 1 virtual particles. See pp. 210–11.

Vector boson: A particle that transmits one of the elementary particle forces. Photons, gluons, Ws and Zs are vector bosons. See pp. 194–95.

Virtual particle: A particle created *ex nihilo* for a fraction of an instant as allowed by the uncertainty principle. It is believed that all forces are transmitted by virtual particles generally known as vector bosons. See p. 83.

W particle: One of the particles carrying the weak force; the other is the Z_0. The W can be either positive (W^+) or negative (W^-). See pp. 217–18.

Wave equation/function: An equation describing the properties of a particle in terms of a wave. The wave refers not to something rippling but rather to a rise and fall in the probability of finding the particle in a particular place with particular properties. See pp. 53–56.

Wavelength: The distance between two successive crests or troughs of a wave. Light of short wavelengths has a high frequency.

Weak interaction/force: Feeble, short-range interaction responsible for beta radiation and some other types of particle decay. Now incorporated into the electroweak theory and described in terms of W and Z particles. See pp. 175–76.

Z (Z_0) particle: One of the particles carrying the weak force; the other is the W. See pp. 225–28.

Index

About the Authors

Robert P. Crease teaches the history of science at Columbia University. **Charles C. Mann** writes for a variety of publications here and abroad. They both live in New York City.

NATIONAL
GEOGRAPHIC

ILLUSTRATED GUIDE
TO
Nature

NATIONAL GEOGRAPHIC

ILLUSTRATED GUIDE
TO
Nature

FROM YOUR BACK DOOR
TO THE GREAT OUTDOORS

Wildflowers | Trees & Shrubs | Rocks & Minerals
Weather | Night Sky

NATIONAL GEOGRAPHIC

WASHINGTON, D.C.

Lightning illuminates the sky at Three Lakes Wildlife
Management Area in Florida.

PRECEDING PAGES: *The moon rises over Canyonlands National
Park from Green River Overlook, Utah.*

CONTENTS

Introduction 6

‖‖

‖‖

Fall colors surround oak trees in the
Shawangunk Mountains of New York.

Knowledge, Pleasure & Wonder
Exploring the World of Nature

"The world is too much with us," wrote the poet William Wordsworth, and by that, he meant the world of bustle and busyness, deadlines and expectations, noise and hubbub. That world has only grown louder and more predominant in the intervening two centuries, built up with trucks and highways, skyscrapers and traffic, television and smartphones. Our ears and minds and days are crammed full of information. We yearn for forest silence, for the lapping of ocean waves on beach sand, for early morning birdsong.

Studies conducted at the University of Kansas show the benefit of spending more time in nature. Testing the theory that "nature has specific restorative effects on the prefrontal cortex"—the part of the brain responsible for creativity and planning—the researchers gave a standardized problem-solving test to backpackers ages 18 to 60 before and after a four-day sojourn into the wilderness. Campers returning scored 50 percent higher.

Why? "Nature is a place where our mind can rest, relax, and let down those threat responses," says Ruth Ann Atchley, lead investigator. When we spend time in nature, Atchley proposes, "we have resources left over—to be creative, to be imaginative, to problem-solve—that allow us to be better, happier people who engage in a more productive way with others."

This book is designed to invite people, young and old, to step with greater pleasure into the out-of-doors. We intend the book to both educate and entertain. Selecting one realm of nature—the world of wildflowers, trees and shrubs, rocks and minerals, the weather, or the night sky—and getting to know

it more intimately can bring lifelong satisfaction. Or just open up this book occasionally to learn the names of clouds above or flowers at your feet.

Working with experts in the various fields of natural history, we have selected 781 species and phenomena from five of the great realms of nature and offer images, identifying features, and brief descriptions of each for you to learn and enjoy. This is an arbitrary number, and relatively low, given that the book's geographic range covers the entire continental United States and Canada. This illustrated guide is thus a starting point from which every nature enthusiast will proceed to fuller, longer, and more detailed topic-specific field guides as well as a perennial treasury to which you and others in your household will return again and again, weaving the natural world more intimately into your daily life.

"The more high-tech we become, the more nature we need," writes Richard Louv, author of *The Nature Principle* and *Last Child in the Woods*. May this volume bring enjoyment, relaxation, productivity, and wonder as you step out into the world of nature that shares this amazing planet of ours.

Native wildflowers blanket the crest of Rowena Plateau along the Columbia River in Oregon.

1 Wildflowers

Nature's Exterior Decoration

Well before the trees have leafed out, wildflowers signal the new season. Woodland spring ephemerals poke through the moist leaf litter, reaching for the sun before the tree canopy blots it out. The wildflower parade continues throughout the growing year—long or short depending on the species and its location—filling fields, woods, roadsides, desert washes, and myriad other habitats throughout the continent with a kaleidoscopic flowering array.

◼ What Is a Wildflower?

Basically, a wildflower is a noncultivated flowering plant that is not a shrub or tree. Most wildflowers are classified as herbaceous plants—flowering plants with nonwoody stems. (We tend to think of herbs as plants with flavorful leaves that enhance our culinary efforts, but that is too narrow a definition.) Under certain conditions, some wildflower species can grow into shrubs or even small trees, so the distinction is not absolute.

In botanical terms, a wildflower is an angiosperm, a seed-producing flowering plant. All wildflowers bear flowers, although they may be small, inconspicuous, or so short-lived that we may know the plant mainly by its leaves, or foliage. Wildflowers can be annuals, going through a flowering and seeding period in a single year; biennials, setting a rosette of leaves one year and flowers and seeds the next; or perennials, living and flowering for three or more years.

Wildflowers come in all shapes and sizes, meeting the basic needs of survival and reproduction in a mind-boggling variety of flower configurations,

leaf shapes, and fruits (the part of the flower that houses the seeds). Equally intriguing are their numerous strategies for pollination (fertilization) and seed dispersal. Flowers can have the tidy ring of petals in a buttercup, the banners and folds of the Pea family, or the elegant, tapered tepals (lookalike petals and sepals) of many lilies. The leaves also take a multitude of shapes—simple and compound, with variations upon variations. The leaves at the base of a wildflower plant may differ greatly from those on the stem. Fruits are likewise diverse, in forms such as seed heads, pods, capsules, and berries. Learning all these distinctions heightens appreciation of nature's wildflower bounty.

◼ Wildflower Versus Weed

Technically, every weed deserves to be called a wildflower, but whether some wildflowers deserve to be called weeds is a judgment call. One person's wildflower is often another person's weed. Wildflowers that thrive where they're often not welcome—in a turf lawn, a flower bed, or a vegetable plot—tend to be called weeds. The dandelion is a classic example. Viewed against the standard of neatly groomed, cultivated flower species, many wildflowers look unkempt and unruly, so they often end up in the weed category. Of course, there are wildflower species of indisputable beauty by any standard. But all wildflowers deserve a look, and on their own terms. Every wildflower, no matter how homely or unwelcome in the eye of a viewer, has a story to tell about its life, survival strategies, and place in the world of living organisms.

■ Exotics & Invasives

Humans have a long history with wildflowers. Aesthetics figure into this relationship, but so does practicality: potential for use as food, beverage, medicine, cosmetics, or decoration. Wildflowers have often been exploited for these purposes to the point of extermination, such as some wild orchids for their beauty and rarity.

Over time, humans have become very attached to certain wildflowers, and as people take up roots and settle in different regions or on different continents, they want their familiar plants to accompany them. Nostalgia for vine-covered cottages and buildings,

for instance, helped bring English Ivy to North America. Highly successful in its new environment, the species can outcompete native species and establish itself as a monoculture, thwarting biodiversity. Introduction of nonnative species can occur unintentionally as well. For example, unwanted seeds can enter crevices in footwear and vehicle tires, and can contaminate grain seed shipments.

An introduced, nonnative plant is called an exotic. When it runs rampant and impinges on other species (which not all introduced species do), it is designated as invasive. Invasive exotics have significantly changed North American plant communities. Certain native plants can become invasive, but they tend to have a lesser impact. Many local conservation groups are leading efforts to remove exotic invasives and restore native plant communities. Most offer opportunities for volunteers to aid these efforts.

■ Identifying Wildflowers

The wildflower species included in this chapter are just a small sampling of the thousands of flowering plants in North America north of Mexico.

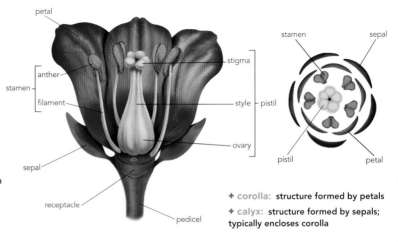

PARTS OF A FLOWER

+ **petal:** individual segment of corolla, typically colored
+ **stamen:** male part of a flower
+ **anther:** part of stamen that makes and stores pollen
+ **filament:** slender support for anther
+ **sepal:** individual segment of calyx, usually green and leaflike
+ **receptacle:** enlarged area bearing flower organs
+ **pedicel:** stalk bearing flower
+ **pistil:** female part of flower
+ **stigma:** receives pollen during pollination
+ **style:** connecting stalk between stigma and ovary
+ **ovary:** base of pistil containing egg or eggs

+ **corolla:** structure formed by petals
+ **calyx:** structure formed by sepals; typically encloses corolla

LEAF/LEAFLET SHAPE	LEAF TYPE	LEAF ARRANGEMENT	LEAF/LEAFLET EDGE
oval	simple	opposite	smooth
round	compound pinnate	alternate	toothed
palmate	compound palmate	whorled	wavy

The 160 species represent different types of common wildflowers from different regions of the continent, showing a range of flower shapes and sizes, plant structures, and distribution. Each entry offers characteristics that will help you identify a species and compare it to similar flowers; additional information about habitat, range, and bloom period; and other details useful for understanding and appreciating the species. The species are presented by color groups—initially the most useful distinguishing characteristic for identification—and within those groups, by similarities of flower shape, foliage, and general appearance. Though some entries mention culinary or medical uses of various wildflowers, readers should not eat any part of any plant unless they are 100 percent certain that it is 100 percent safe. Other species are described as being toxic; if a species is not so described, it should not be deduced that any part is safe to eat. Many books and online references offer detailed information on which plants are unsafe.

A weed is no more than
a flower in disguise,
Which is seen through at once,
if love give a man eyes.
—JAMES RUSSELL LOWELL

■ Wildflowers at Home

When you make native perennial wildflowers part of your home landscape, everyone wins: you, the wildlife, and the local ecosystem. Native wildflowers add beauty and feed birds, butterflies, and other animals in ways that most imported annuals cannot begin to match, and they allow other native plant species to gain a foothold. Where can you find native species for home planting? Though you should never collect plants on public or private lands without permission, you might find specimens for sale through local nature centers, botanical gardens, and native plant societies. These organizations often have demonstration gardens that will give you ideas about plant combinations and light and moisture requirements. Friends with extensive plantings are another source, as are growing numbers of commercial suppliers. Always try to find local ecotypes of the plants you want; they're best suited to your growing conditions and the surrounding ecosystem.

You can take this endeavor one step further by making native wildflowers the foundation of a wildlife habitat that provides the basic necessities for all manner of creatures, from insects to birds to mammals. Wildflowers provide nectar and pollen for insects; nectar, berries, and seeds for birds; and those things plus flowers and foliage for a wide variety of mammals. Vining and shrubby plants provide shelter and places for nesting; and a birdbath, fountain, or small pond provide crucial water. Your yard can satisfy both you and the wildlife.

OFFICIAL FLOWERS OF THE U.S. STATES AND CANADIAN PROVINCES

+ **Alabama: Camellia** (*Camellia japonica*)
+ **Alaska: (Asian) Forget-me-not** (*Myosotis alpestris asiatica*)
+ **Arizona: Saguaro Cactus blossom** (*Carnegiea gigantea*)
+ **Arkansas: Apple blossom** (*Malus pumila*)
+ **California: California Poppy** (*Eschscholzia californica*)
+ **Colorado: Rocky Mountain (Colorado Blue) Columbine** (*Aquilegia coerulea*)
+ **Connecticut: Mountain Laurel** (*Kalmia latifolia*)
+ **Delaware: Peach blossom** (*Prunus persica*)
+ **District of Columbia: American Beauty Rose** (*Rosa "American Beauty"*)
+ **Florida: Orange blossom** (*Citrus × sinensis*)
+ **Georgia: Cherokee Rose** (*Rosa laevigata*)
+ **Hawaii: Pua Aloalo** (*Hibiscus brackenridgei*)
+ **Idaho: Syringa Mock Orange** (*Philadelphus lewisii*)
+ **Illinois: Violet** (*Viola* spp.)
+ **Indiana: Peony** (*Paeonia lactiflora*)
+ **Iowa: Wild Prairie Rose** (*Rosa arkansana*)
+ **Kansas: Sunflower** (*Helianthus annuus*)
+ **Kentucky: Goldenrod** (*Solidago* spp.)
+ **Louisiana: (Southern) Magnolia** (*Magnolia grandiflora*)
+ **Maine: Eastern White Pine tassel and cone** (*Pinus strobus*)
+ **Maryland: Black-eyed Susan** (*Rudbeckia hirta*)
+ **Massachusetts: Mayflower (Trailing Arbutus)** (*Epigaea repens*)
+ **Michigan: Apple blossom** (*Malus pumila*)
+ **Minnesota: Pink-and-white Lady's Slipper** (*Cypripedium reginae*)
+ **Mississippi: (Southern) Magnolia** (*Magnolia grandiflora*)
+ **Missouri: Hawthorn** (*Crataegus* spp.)
+ **Montana: Bitterroot** (*Lewisia rediviva*)
+ **Nebraska: Goldenrod** (*Solidago gigantea*)
+ **Nevada: Sagebrush** (*Artemisia tridentata*)
+ **New Hampshire: Purple Lilac** (*Syringa vulgaris*)
+ **New Jersey: (Common Blue) Violet** (*Viola sororia*)
+ **New Mexico: (Soapweed) Yucca** (*Yucca glauca*)
+ **New York: Rose** (*Rosa* spp.)
+ **North Carolina: Flowering Dogwood** (*Cornus florida*)
+ **North Dakota: Wild Prairie Rose** (*Rosa arkansana*)
+ **Ohio: Scarlet Carnation** (*Dianthus caryophyllus*)
+ **Oklahoma: Mistletoe** (*Phoradendron leucarpum*)
+ **Oregon: Oregon Grape** (*Berberis aquifolium*)
+ **Pennsylvania: Mountain Laurel** (*Kalmia latifolia*)
+ **Rhode Island: Violet** (*Viola* spp.)
+ **South Carolina: Yellow Jessamine** (*Gelsemium sempervirens*)
+ **South Dakota: Pasqueflower** (*Anemone patens* var. *multifida*)
+ **Tennessee: (German) Iris** (*Iris germanica*)
+ **Texas: Bluebonnet** (*Lupinus* spp.)
+ **Utah: Sego Lily** (*Calochortus* spp.)
+ **Vermont: Red Clover** (*Trifolium pratense*)
+ **Virginia: Flowering Dogwood** (*Cornus florida*)
+ **Washington: Coast Rhododendron** (*Rhododendron macrophyllum*)
+ **West Virginia: (Big) Rhododendron (Great Laurel)** (*Rhododendron maximum*)
+ **Wisconsin: Wood Violet** (*Viola papilionacea*)
+ **Wyoming: Indian Paintbrush** (*Castilleja linariifolia*)

Canadian Provinces and Territories
+ **Alberta: Wild Rose** (*Rosa acicularis*)
+ **British Columbia: Pacific Dogwood** (*Cornus nuttallii*)
+ **Manitoba: Prairie Crocus** (*Anemone patens*)
+ **New Brunswick: Purple (Marsh Blue) Violet** (*Viola cucullata*)
+ **Newfoundland and Labrador: (Purple) Pitcher Plant** (*Sarracenia purpurea*)
+ **Northwest Territories: Mountain Avens** (*Dryas integrifolia*)
+ **Nova Scotia: Mayflower (Trailing Arbutus)** (*Epigaea repens*)
+ **Nunavut Territory: Purple Saxifrage** (*Saxifraga oppositifolia*)
+ **Ontario: (Large-flowered) White Trillium** (*Trillium grandiflorum*)
+ **Prince Edward Island: (Pink) Lady's Slipper** (*Cypripedium acaule*)
+ **Quebec: Blue Flag** (*Iris versicolor*)
+ **Saskatchewan: Western Red Lily** (*Lilium philadelphicum*)
+ **Yukon Territory: Fireweed** (*Epilobium/Chamerion angustifolium*)

WILDFLOWERS

||||||||||||||||||||||||||||||||||||||||||||||||||||||||

Yellow Trout Lily

Erythronium americanum H 5–10 in (13–25 cm)

A scaly bulb gives rise to a single stem bearing the blossom of the woodland Yellow Trout Lily. Bloom and bulb shape once gave it a misleading common name of Dogtooth Violet.

> **KEY FACTS**
>
> The 6-part pendent yellow flower has tepals (look-alike petals and sepals) that are bent back; 2 mottled leaves clasp the flower stem.
>
> **+ habitat:** Moist woods, swamps, and meadows
>
> **+ range:** Eastern U.S. and Canada, except for Florida
>
> **+ bloom period:** March–June

From a basal pair of long, elliptical, brown-mottled leaves, the Yellow Trout Lily blossom tops a single stalk, although there may be many such blooms in the species' large colonies. The leaf markings resemble those of a Brown or Brook Trout and give the native perennial its common name. The bent-back petals and sepals provide a signature look for this lily, a spring ephemeral. The flowers are yellow on the outside and brownish within, and display six brownish stamens with prominent anthers. Yellow Trout Lily often occurs in large colonies that include many nonflowering plants with solitary leaves.

Greater Yellow Lady's Slipper

Cypripedium parviflorum var. *pubescens* H 6–28 in (15–70 cm)

A fat, bright-yellow lip petal gives the Greater Yellow Lady's Slipper its name. Usually, only one of the striking flowers crowns the top of a leafy stem, but at times there are two or three.

> **KEY FACTS**
>
> The blossom is yellow and brownish. The stem has up to 6 pointed leaves.
>
> **+ habitat:** Bogs, marshes, and moist woods
>
> **+ range:** Throughout much of the U.S. and Canada, except for far northern areas, and parts of southern and western U.S.
>
> **+ bloom period:** May–July

The Greater Yellow Lady's Slipper, a member of the Orchid family, shows a typical orchid structure with some unique features. The bulbous lower petal of the native species has a rounded opening at the base. The two lateral petals are long, thin, and very twisted. Broad, bright-green leaves have pointed ends and prominent lengthwise ribs. The leaves sheathe the stem at their bases and grow alternately along it. There are both large and small varieties of Yellow Lady's Slippers, and some are quite fragrant. Like most orchid species, their current populations struggle as a result of indiscriminant and often illegal collecting.

Swamp Buttercup

Ranunculus hispidus var. *nitidus* H 12–36 in (30–90 cm)

Despite its name, this native buttercup emerges in both wet and dry habitats. Once classified as a separate species *(R. septentrionalis)*, it now stands as a variety of the Hispid Buttercup.

KEY FACTS

The flower, up to 1 in (2.5 cm), has 5 yellow petals; 3-lobed, hairless, deeply divided leaves are usually as wide as they are long.

+ habitat: Swamps, marshes, other wet habitats, woods, and grasslands

+ range: Throughout much of central and eastern U.S. and Canada, except in northern areas

+ bloom period: April–July

The Swamp Buttercup produces leaves that serve as a base for its flowers, and also alternate leaves on stems that spread along the ground. The yellow flowers rise on stalks, with petals that are bright and shiny most of their length, but become a pale greenish yellow toward the base. Fine lines on the petals serve as guides to the nectar. Bees, flies, butterflies, and beetles visit the flowers to feed on both nectar and pollen. The Swamp Buttercup resembles the Early Buttercup *(R. fascicularis)*, which true to its name gets an earlier start and has more pointed leaves longer than they are wide. It is also significantly hairier than the Swamp Buttercup.

Black Mustard

Brassica nigra H 12–96 in (30–240 cm)

A poultice of Black Mustard leaves placed on the chest was once the go-to remedy for colds, bronchitis, and similar ailments. The seeds are used to make the iconic yellow condiment.

KEY FACTS

Narrow flower clusters have 4-petaled yellow flowers. Upper leaves are toothed and narrow; lower leaves are deeply lobed.

+ habitat: Grassy hills, fields, roadsides, and wasteland

+ range: Naturalized throughout the U.S. and Canada, except in far northern areas

+ bloom period: June–October

The Black Mustard plant is capable of growing very tall and lanky to the point that it flops sideways. At its tallest, the lower leaves can measure up to 10 inches (25 cm) long and may wilt in dry heat. Each flower turns into a slender seedpod that grows close to the flower stalk. A native of Eurasia, the prolific, weedy Black Mustard has adapted well to most of North America. It shares a genus with vegetables such as Brussels sprouts, broccoli, cauliflower, and cabbage and has its own extensive list of culinary and medicinal uses. Many songbird species favor the plant's tiny black seeds.

Silverweed

Argentina anserina L runners to 36 in (90 cm); H to 8 in (20 cm)

The low-growing Silverweed gets its name from the silvery, hairy undersides of its compound basal leaves, which grow on hairy leaf stalks.

KEY FACTS

Solitary, 5-petaled flowers grow on long, hairy stalks; pinnate compound leaves are composed of mostly alternating toothed leaflets.

+ habitat: Pond and stream banks, shores, marshes, wet meadows, and roadsides

+ range: Much of northern and southwestern U.S. and far northern areas into Canada

+ bloom period: June–August

The Silverweed's bright yellow blossoms with blunt petals and a cluster of short yellow stamens strongly suggest the flower's place in the Rose family. The flowers may appear to be at a distance from the plant's foliage, but they are connected; both the flower stalks and the compound leaves spring from nodes on often inconspicuous runners. Long used for food and for medicinal purposes, the Silverweed's cooked root has the taste of parsnips or sweet potatoes. A coastal variety is less hairy or even hairless and is found hugging the Pacific coast and the Atlantic coast from New England northward.

Puncture Vine

Tribulus terrestris L creeper to 36 in (90 cm)

Similar to the Silverweed but far less benign, the Puncture Vine can injure grazing animals with the sharp spines on its fruits as well as a toxic substance that causes sun sensitivity.

KEY FACTS

The 5-petaled yellow flowers grow on short stalks that emerge from the axils of the opposite, compound leaves.

+ habitat: Fields, open ground, and also disturbed areas; often in sandy or gravelly areas

+ range: Naturalized throughout most of the U.S. and parts of Canada

+ bloom period: June–September

The Puncture Vine, a European import, sprawls along the ground on creeping stems that send up compound pinnate leaves with leaflets that are opposite but often of unequal lengths. The seed case breaks into five sections, each with a pair of strong, sharp spines. These can cause great damage to livestock when gathered into a bale of hay used for fodder, and they make the species a noxious weed in the eyes of farmers and ranchers. They also give the plant its common names, including Goathead and Caltrop—the latter a reference to the spiked weapon deployed on the ground to impede horses, troops, and vehicle tires.

Common St. John's Wort
Hypericum perforatum H 12–30 in (30–75 cm)

A prolific European import, Common St. John's Wort wasted no time getting established in North America. It now outcompetes many native wildflower species.

KEY FACTS

The small dark glands edge the bright yellow, 5-petaled flowers with many long and yellow stamens.

+ **habitat:** Open woods, meadows, and along roadsides

+ **range:** Introduced and naturalized throughout much of the U.S. and Canada

+ **bloom period:** June–September

The Common St. John's Wort, a hardy perennial, is named for St. John the Baptist. When its flower petals are pinched, they turn red from the oil in their glands; this unusual property associated them with the saint's beheading. Tiny, stiff, linear-to-oval-shaped leaves grow opposite each other on numerous stems. Oil glands also cover the leaves, visible as tiny, translucent dots. The upper stems culminate in multiple flower clusters. The plant as a whole gives off a faint odor of turpentine or balsam. Part of nature's pharmacopoeia for millennia, Common St. John's Wort has pronounced antiviral and antidepressive effects among its many benefits.

Downy Yellow Violet
Viola pubescens H 6–18 in (15–45 cm)

The Downy Yellow Violet is named for its soft, hairy leaves and stems. Leaves near the bottom of the stem are heart-shaped and have scalloped edges.

KEY FACTS

The petals are bright yellow in color; the lateral ones are bearded, and the lower petal is heavily veined with purple.

+ **habitat:** Woods, woodland edges, and thickets

+ **range:** Eastern U.S. and Canada

+ **bloom period:** May–June

The large flowers of the Downy Yellow Violet appear on slender stalks from the axils of the plant's upper leaves. In addition to the notable veins on the lower petal, other petals may have one or two thin lines. A beard of hairs on the lateral petals encourages visiting insects to brush up against the stigma and anthers, where they pick up pollen on their way to sipping nectar from the spur. In summer, most Downy Yellow Violets grow specialized flowers without petals on short stems growing from the plant's root. These never open, but nonetheless become self-fertilized inside the bud.

Common Evening Primrose
Oenothera biennis H 12–72 in (30–180 cm)

The Common Evening Primrose bides its time during the day, waiting until dusk to unfurl its yellow flowers—each destined to last only one night—and allow moths to enter for pollination.

KEY FACTS

The 2-in (5-cm) flower has 4 rounded and lemon yellow petals.

+ habitat: Open areas with sandy soil, such as embankments, meadows, and along roadsides

+ range: Central and eastern U.S. and Canada

+ bloom period: June–September

After a night in bloom, the Common Evening Primrose flowers wither by morning—at least before noon. The plant reveals its blossoms gradually during the blooming period: Lower ones open first and by the time upper flowers bloom, the lower ones have become seedpods. The plant's tall stem is stout and covered with soft hairs. The long, lance-shaped leaves grow alternately on the stem. They and the stem share a coarse texture. Among its many medicinal applications, the Common Evening Primrose is harvested for an oil that is successfully used to treat hormonal imbalances and eczema.

Common Bladderwort
Utricularia macrorhiza H 4–16 in (10–40 cm)

The Common Bladderwort is a rootless, native, carnivorous aquatic plant. It captures prey by means of trigger hairs at the mouth of underwater "bladders," or inflated segments.

KEY FACTS

The 5-lobed flowers are yellow, or rarely pink or purple; the lobes are fused to form 2 lips, the lower lip with a sickle-like spur.

+ habitat: Lakes, ponds, slow-moving water, and ditches

+ range: Throughout U.S. and Canada

+ bloom period: June–August

Common Bladderwort flowers are distinctive, forming loose clusters on a stalk that clears the water; however, much of the interest lies underwater, where fine branches that fork between three and seven times display numerous bladders. The bladders are a transparent green when young, turning darker and brownish as they age. Once a bladder has captured its prey, including water fleas and mosquito larvae—an action that may take as little as $1/460$ of a second—enzymes within it begin to digest the prey. Some experts consider the scientific name of the Common Bladderwort to be *U. vulgaris macrorhiza.*

Common Mullein

Verbascum thapsus H 24–72 in (60–180 cm)

Many people don't realize that the Common Mullein, a denizen of empty lots and roadsides, grows flowers in its second year. They know the Eurasian import only by its grayish leaves.

KEY FACTS

The spikes of dense 5-petaled yellow flowers grow on the woolly stems.

+ habitat: Fields, roadsides, disturbed areas, and waste ground

+ range: Naturalized throughout U.S. and Canada, except north-western Canada

+ bloom period: June–September

Common Mullein, a member of the Figwort family, is a biennial wildflower that takes off in its second season. A long stem covered with downy hairs emerges from a rosette of large basal leaves, and as the bright yellow flowers start to bloom, the erect spikes take on the appearance of candles or torches—an apt association on several counts. The leaves, when dried, are highly flammable and tradition-ally were used as candlewicks; the flower spikes were dipped into melted fat and set alight as torches. Even today, the flowers are brewed into a tea that treats respi-ratory congestion.

Common Monkeyflower

Mimulus guttatus H 6–36 in (15–90 cm)

Clearly defined hairy ridges on the lower lip of the distinctive blossom are a characteristic of the Common Monkeyflower, also known as Yellow Monkeyflower, a member of the Lopseed family.

KEY FACTS

The yellow flowers are sometimes soli-tary or in clusters, on long stalks, strongly 2-lipped, with red to maroon dots.

+ habitat: Wet places such as stream banks and other wet areas

+ range: Native to the western U.S. and in Canada; highly natu-ralized in the East

+ bloom period: March–August

The Common Monkeyflower can vary sig-nificantly in height and structure, appearing dwarflike with small leaves at times, and tall and imposing at others. The flowers are similar in size and appearance to their relative, the Common Snapdragon. The leaves of Common Monkey-flower also vary significantly in size and shape: Some are oval, some are roundish, some are kid-ney shaped, and some are heart shaped. Leaves at the bottom of the plant gener-ally have stalks; those at the top often are stalkless and clasp the stem. The species can reproduce by means of a creeping stem, or stolon, that sends down new roots at intervals to form new plants.

Coastal Sand Verbena

Abronia latifolia H 1–3 in (2.5–8 cm)

The Coastal Sand Verbena sprawls along Pacific beaches and dunes, well adapted to the drying effects of wind and sun with its moisture-retaining stems and leaves.

KEY FACTS

The small, yellow, trumpet-shaped flowers form a globular head that grows on a succulent stem.

+ **habitat:** Sandy soils of beaches and dunes

+ **range:** Western coast of the U.S. and Canada to British Columbia

+ **bloom period:** May–August

The perennial native Coastal Sand Verbena forms low, extensive mats of succulent stems that often are buried in the sand and anchored by deep roots. The stems bear many smooth-edged fleshy leaves that are oval to kidney shaped and have slightly undulating surfaces. The flower cluster rises gracefully—*Abronia* comes from the Greek for "graceful"—on a stem that grows from a leaf axil. Two other Sand Verbenas grow in the same habitat along the Pacific coast. The Red Sand Verbena or Beach Pancake (*A. maritima*) is similar but has reddish purple flowers, while the flowers of the Beach Sand Verbena (*A. umbellata*) range from white to deep pink.

Greater Celandine

Chelidonium majus H 12–30 in (30–76 cm)

An introduced member of the Poppy family, the Greater Celandine is noted for a yellowish orange sap traditionally used to treat warts, source of another common name, Wartwort.

KEY FACTS

The bright yellow, 4-petaled flower less than an inch (2.5 cm) wide forms loose clusters.

+ **habitat:** Woods, fields, roadsides, waste ground, and disturbed areas

+ **range:** Native to Eurasia; naturalized throughout northern and central U.S. and Canada in disparate regions

+ **bloom period:** April–August

The individual flowers of the Greater Celandine grow on short stems and form a cluster at the end of a slender stalk up to 4 inches (10 cm) tall. The plant's ribbed leaves alternate on their branching stems and are deeply lobed into rounded segments with irregularly toothed margins. Leaves can be up to 14 inches (35 cm) long and typically are composed of five to nine lobes. In addition to serving as a wart remover, the plant's caustic yellow sap was also used to make eyedrops. A very aggressive import, Greater Celandine pushes out native wildflowers, especially woodland species.

Curlycup Gumweed

Grindelia squarrosa H 12–36 in (30–90 cm)

The flower head of this species is clothed by a distinctive cluster of overlapping bracts with recurved tips—the curly cup. The cup also is very resinous—the gumweed reference.

KEY FACTS

The flower head typically has both disk and ray yellow flowers, although the rays sometimes are absent.

+ **habitat:** Fields, prairies, and waste areas

+ **range:** Throughout the U.S. and Canada, except in far northern areas and southeastern U.S.

+ **bloom period:** July–September

The native daisy-like Curlycup Gumweed, a member of the Aster family, rises on a stalk with 2 to about 20 companions. The flower head, up to 1.5 inches (3.75 cm) long, bears many rays and disks, which are darker than the rays. Waxy, oblong, stemless leaves, often with toothed, curled edges and a pointy tip, alternately clasp the smooth, erect stalk that often is branched near the top. With roots up to 6 feet (180 cm) long, the plant does well under drought conditions. American Indians used the species extensively to treat respiratory ailments and stomach and liver problems, and even as a topical treatment for saddle sores.

Eastern Prickly Pear

Opuntia humifusa H to 18 in (45 cm) in clumping form

Sparse, slender spines and short, barbed bristles on fleshy stems make the Eastern Prickly Pear—eastern North America's main claim to flowering cactus fame—require careful handling.

KEY FACTS

The yellow- to orangish-tinged flowers are 2–3 in (5–7.5 cm) wide; multiple flowers grow on the fleshy segments.

+ **habitat:** Sandy, rocky, and hilly areas; sometimes planted in gardens

+ **range:** Central and eastern U.S. and Canada, excluding the upper Northeast

+ **bloom period:** May–August

The Eastern Prickly Pear takes several forms. In most parts of its range, it is a low clumping plant consisting of numerous flattened, ovoid, shiny green, and fleshy segments. These are covered with clusters of reddish brown barbed bristles called glochids and less numerous long, thin spines. The glochids can be as injurious as the cactus's spines, as they have barbed tips and lodge stubbornly in the skin. Low-growing plants can spread some 3 feet (90 cm) in diameter. In some parts of its range, notably in Florida, the Eastern Prickly Pear grows into an erect shrub, more than 6 feet (180 cm) tall.

Creeping Wood Sorrel

Oxalis corniculata H 4–8 in (10–20 cm)

The Creeping Wood Sorrel is a member of the genus *Oxalis,* so named for the Greek word for "sour," a reference to the sour-tasting oxalic acid that occurs throughout the plant.

KEY FACTS

The small yellow flowers with 5 petals and 5 stamens form an umbrella-like cluster; the petals are separate down to the base.

+ habitat: Lawns, flower beds, fields, nursery grounds, and disturbed areas

+ range: Naturalized throughout most of the U.S. and Canada

+ bloom period: June–September

A low, creeping wildflower with heart-shaped, green to purple leaves, the Creeping Wood Sorrel sends down roots and forms stems from nodes as it grows along the ground. It also spreads by very small brown seeds that explode from a ripe capsule. The angular seeds adhere to many objects and easily transport themselves to new areas. The name sorrel itself comes from the German for "sour," and sorrels were an old-time salad ingredient, rich in vitamin C, though we now know that ingestion of too much oxalic acid can block the absorption of calcium. *Oxalis* plants are often sold commercially as "shamrocks."

Black-eyed Susan

Rudbeckia hirta H 12–36 in (30–90 cm)

The Black-eyed Susan flower head rises singly on a long, erect stem and gets its name from a dark purple center composed of tubular florets that appear almost black from a distance.

KEY FACTS

The flower head is large and daisy-like with yellow rays and a dark, cone-shaped disk.

+ habitat: Fields, prairies, and the open woods

+ range: Throughout much of the U.S. and Canada, except in parts of the Southwest and the far northern areas

+ bloom period: June–October

In the Black-eyed Susan, a member of the Aster family, radiating green bracts surround the emerging blossom. As the flower head opens, its long, orange-yellow rays extend to meet the collar, some 2 to 3 inches (5–7.5 cm) across. The rays sometimes bear reddish brown splotches. In its first year, the plant forms a rosette of lance-shaped or oval hairy leaves; in the second year, tall hairy stems produce the conspicuous flower heads. The species flourishes in challenging conditions and provides nectar for bees, butterflies, and other insects, as well as seeds favored by a variety of birds.

Common Sunflower

Helianthus annuus H 3–10 ft (1–3 m)

We mostly know the Common Sunflower from the mammoth-headed, heavily seeded cultivars that nod from their weight. The wild flower heads are smaller, with smaller centers.

KEY FACTS

The flower heads have dense yellow rays and a brownish red central disk. Oval to heart-shaped leaves are sometimes toothed.

+ habitat: Fields, plains, roadsides, and waste areas

+ range: Throughout much of the U.S. and Canada, except in far northern areas

+ bloom period: June–September

A member of the Aster family, the Common Sunflower grows tall on an erect, hairy stem with many rough, hairy leaves. The stem branches at the top and produces multiple flower heads. The flower heads follow the sun during the day as it moves across the sky. American Indians utilized many parts of the species: They made dyes from the flower rays and seeds of its flat central disk; a tea from the flowers to treat lung ailments; and poultices from the leaves for snake bites. They also ground the seeds into flour and used seed oil for cooking and to dress hair.

Common Dandelion

Taraxacum officinale H 2–16 in (5–40 cm)

An introduced species from Eurasia that has spread from coast to coast, the Common Dandelion is easily recognized with its yellow flower heads, jagged leaves, and wispy seed heads.

KEY FACTS

The single yellow flower head composed of many ray-like toothed ligules is cupped by green bracts on a solitary stem with leaves all in a basal rosette.

+ habitat: Lawns, fields, roadsides, and disturbed areas

+ range: Naturalized throughout the U.S. and Canada

+ bloom period: Mainly March–September

The bane of lawn warriors, the Common Dandelion would figure on many top ten weeds lists. The name derives from the French for "lion's tooth," a reference to the distinctive edges of the leaves. This hardy wildflower, a member of the Aster family, blooms much of the year—a few stragglers even show up in the dead of winter. All parts of the plant exude a milky juice when broken. The species is highly edible and valued in traditional medicine for its support of liver and kidney function and diuretic properties. The roots are fermented into beer, and the flowers are used to make a potent wine.

Tickseed

Coreopsis species H 6 in–8 ft (15–240 cm)

Members of the large Aster family, tickseeds in North America encompass more than two dozen species that share mostly yellow, daisy-like flower heads of varying size and delicate leaves.

KEY FACTS

Flower heads are largely daisy-like, yellow, mostly with both disk and ray flowers; some are sunflower-like; a few have pink flowers.

+ **habitat:** Old fields, prairies, roadsides, open woods, swamps, and sandy areas

+ **range:** Throughout U.S. and Canada, except far northern areas; widely cultivated

+ **bloom period:** May–September

Just as there is wide variation among the flowers, tickseed leaves and leaflets can appear at the base of the plant or on the stem, they can be opposite or alternating or appear in whorls, and can be lance-shaped, threadlike, or may be lacking entirely on bare leaf stems. Tickseeds often form large colonies whose member plants bloom over a long period of time. The genus and common names for tickseeds refer to the bedbug, an image related to the appearance of the plants' flattened, seedlike fruits. Many tickseed species have escaped cultivation and are widespread outside their customary native ranges.

Goldenrod

Solidago species H 12–96 in (30–240 cm)

Goldenrods unfairly take the fall for another common wildflower—Ragweed—as a major allergen, but the genus traditionally is a potent healer, as its name—Greek for "make well"—testifies.

KEY FACTS

The goldenrod species have both disk and ray flowers, which are relatively small and are usually yellow.

+ **habitat:** Fields, prairies, open woods, thickets, along roadsides, and in salt marshes

+ **range:** Throughout the U.S. and Canada

+ **bloom period:** July–November

The many members of the genus *Solidago* bear flower heads that vary in shape, size, and arrangement on the stem, but typically are a distinctive golden yellow, although *S. bicolor* has white ray flowers, and *S. ptarmicoides* has white disks and rays. Flower head arrangement can vary from flat-topped rounds to loosely branching, arched clusters that sometimes are one-sided. Leaves are basal or alternate on the stem; they are succulent in salt marsh species. Goldenrods are late summer flowers that rely on insects for pollination. The real culprits in the seasonal sneezefest are Ragweeds and other wind-pollinated species. A number of goldenrod species will hybridize.

Common Jewelweed/Spotted Touch-me-not

Impatiens capensis H 24–60 in (60–150 cm)

Pendent yellow-and-orange flowers ornament this leafy annual with a preference for shade and moisture. Touching the swollen, seed-filled, ripe fruits causes them to burst explosively.

KEY FACTS

Drooping yellow and orange tubular flowers have brown, red, and orange spotting; 2 grow on each flower stalk, flowering one at a time.

+ **habitat:** Shady wetlands, woods, stream and riverbanks, and pond edges

+ **range:** Throughout much of northern, central, and eastern U.S. and Canada

+ **bloom period:** June–October

If not for the graceful, nodding flower—the species' obvious jewel—this plant would seem more of a weed: long, slender, succulent stems laden with oval leaves growing mostly alternately along the stem. The nectar-rich flowers are a favorite of hummingbirds, bees, and butterflies. Sap from the plant's leaves and stems thwarts rash development and soothes itching from Poison Ivy exposure, and often is made into soap for this purpose. The sap also serves as an effective fungicide and treatment for athlete's foot. The similar yellow Pale Jewelweed (*I. pallida*), found in eastern North America, shares these properties.

Western Wallflower/Prairie Rocket

Erysimum capitatum H 6–36 in (15–90 cm)

Western Wallflowers get their common name from a similar Old World relative in the Mustard family that often is seen embedded in drystone walls, a common landscape feature in Europe.

KEY FACTS

Orange to yellow flowers with 4 petals and 4 sepals grow in loose terminal clusters on short stalks.

+ **habitat:** Plains, foothills, deserts, open woods, cliffs, and coniferous forests

+ **range:** Throughout much of western and central U.S. and Canada

+ **bloom period:** March–July

Western Wallflowers often stand out in their habitats in dense clusters of usually erect stems bearing showy flowers with knobby stigmas. The plants are quite leafy—and hairy—with lance-shaped hairy lower leaves and smaller upper leaves, both growing alternately on a hairy stem that is sometimes branched. Sandy and rocky environments suit this hardy plant. It is also known as Prairie Rocket, not for its shape but for its relationship to Salad Rocket, another species in the Mustard family that also is known as Arugula. The Western Wallflower is highly variable, appearing in a number of other color palettes, including yellow, orange-brown, and maroon.

California Poppy

Eschscholzia californica H 6–24 in (15–60 cm)

Fields of California Poppies grace the landscape in the Golden State, where they have been the state flower since 1903. Bright sun opens the flowers; they close at night and on cloudy days.

KEY FACTS

Yellow to brilliant orange flowers have 4 satiny petals that often have an orange spot at the base.

+ **habitat:** Coastal dunes, open forests, meadows, plains, and desert edges

+ **range:** West Coast of the U.S. from southern California into southern Washington

+ **bloom period:** February–September

You can distinguish a California Poppy from other poppy species in its genus by its disk-like collar that sits at the base of the blossom. The flower has many stamens; often a dozen or more crowd the center. A closed flower takes the shape of a nightcap. The California Poppy's leaves have three deeply divided, feathery lobes and typically are blue-green with a grayish cast. Depending on the location, the species can be low and spreading or tall and erect. Also depending on environmental circumstances, it can be either an annual or perennial. The poppy contains several compounds with potential for treating cancer.

Common Orange Daylily

Hemerocallis fulva H 24–72 in (60–180 cm)

This Asian native has long been cultivated in North America as an ornamental plant. Escapees have naturalized widely, making the ephemeral flowers a common roadside attraction.

KEY FACTS

Orange flower is composed of 3 petals with wavy edges and 3 sepals with smooth ones.

+ **habitat:** Roadsides, meadows, woodland edges, stream banks, and disturbed areas

+ **range:** Naturalized in much of U.S. and Canada, except southwestern U.S. and western Canada

+ **bloom period:** May–July

True to its name, the flower of the Common Orange Daylily lasts only one day. But a number of the funnel-shaped blossoms form on each leafless stalk to prolong blooming. The flowers face out, not downward like other lilies, and the petals curve backward. The multibranched stalks, or scapes, rise directly from the root and are taller than the profuse strap-shaped leaves. The blossoms feature yellow centers and radiating thin stripes as well as pronounced veins. The species very rarely forms seeds; most Common Orange Daylily plants produce by cloning from fleshy roots or rhizomes. Another escaped cultivar produces doubled flowers.

Blackberry Lily/Leopard Lily

Iris domestica (Belamcanda chinensis) H 18–48 in (45–120 cm)

The Blackberry Lily spreads wide its distinctly spotted tepals (look-alike petals and sepals) as if to draw attention to its short-lived beauty, as each blossom lasts only one day.

KEY FACTS

The flower has 6 orange tepals that are heavily spotted with red; the flowers grow in multiples on naked, branched stems.

+ **habitat:** Roadsides, grasslands, meadows, open woods, rocky outcrops, and disturbed areas

+ **range:** Naturalized throughout central and eastern U.S.

+ **bloom period:** June–August

A native of China, the Blackberry Lily has escaped cultivation to become widely established in North America. Showy flower sprays appear in the midst of fan-shaped clusters of long, narrow, flat, medium-green leaves. Pear-shaped seedpods form in late summer. When ripe, they split to reveal a cluster of shiny blackberry-like seeds, the source of the plant's common name; the spots, of course, lend another name—Leopard Lily. A species of a different genus also goes by the name Leopard Lily; *Lilium pardalinum,* native to California, has somewhat similarly spotted tepals that curl. Its range does not overlap with that of *Iris domestica.*

Wood Lily

Lilium philadelphicum H 12–36 in (30–90 cm)

Despite its name, the Wood Lily flourishes equally in sandy and brush-covered habitats, with its striking blooms standing erect at the top of their stalks and not drooping like other lilies.

KEY FACTS

Orange-red blooms have tepals that lighten toward the tapered base and are spotted with brown.

+ **habitat:** Prairies, open woods, roadsides, power line cuts, dunes, barrens, and mountain meadows

+ **range:** In U.S. and Canada, except far northern areas and parts of western and southeastern U.S.

+ **bloom period:** June–August

From one to four funnel-shaped blossoms crown the summit of the Wood Lily plant. Narrow, lance-shaped leaves, pointed on the ends, are in whorls of three to six, and usually alternate on each tall, erect stem. American Indians ate the flavorful bulbs and sprinkled the nutritious pollen on numerous dishes. A medicinal tea treated stomach ailments and fever, and aided childbirth. The bulbs and leaves were used in poultices on wounds, bruises, and spider bites. The Wood Lily is losing ground within portions of its range due to development, overpicking, grazing, and browsing by increasing numbers of White-tailed Deer.

Butterfly Weed/Orange Milkweed

Asclepias tuberosa H 12–36 in (30–90 cm)

Like its cousin the Common Milkweed, the Butterfly Weed attracts insects with an abundance of tempting nectar, but this species serves it up in brilliantly colored orange flowers.

KEY FACTS

Clusters of small, orange flowers top erect, hairy stems; long, oblong leaves alternate on the stem.

+ habitat: Prairies, open woods, roadsides, and disturbed areas

+ range: Throughout most of the U.S., except the Northwest, and eastern Canada; highly cultivated

+ bloom period: May–September

The individual flowers of the showy, flat-topped clusters have five bent-down petals and a crown of five erect hoods. The clusters themselves are about 2 inches (5 cm) across. Slits in the flower's central column contain pollen sacs, and the plant relies on visiting bees and butterflies getting a leg tangled in a slit and picking up and carrying off a load. Unlike the Common Milkweed, this species exudes a clear, not milky, sap when the leaves are bruised. If trying to attract butterflies to your garden, instead of planting imported purple *Buddleia* species, choose native Butterfly Weed, which supports species such as the Monarch and the Queen.

Fiddleneck

Amsinckia species H 4–30 in (10–75 cm)

Fiddleneck flowers develop along curled stems in a characteristic fiddlehead shape, unfolding as they grow into hairy, one-sided clusters of yellow trumpet-like flowers.

KEY FACTS

Small, 5-lobed trumpets ranging from pale yellow to yellow-orange populate one side of the flower stem.

+ habitat: Dry areas: meadows, open woods, shrubby and disturbed areas

+ range: Native to western U.S. and Canada but were introduced east of the Rockies in scattered locations

+ bloom period: February–June

Fiddlenecks are a hairy lot. The flowers develop at the tips of bristly, green stems and branches, with bristly, lance-shaped leaves. As the stem uncoils, the flowers appear first from the bottom of the cluster. Later, each flower produces a fruit comprising four triangular nutlets that start out green before turning gray to black. Ten different species of the native annual fiddleneck populate western North America and locations to the east where they have been introduced. Differences among them are often very subtle, making identification difficult. Some species have red splotches in the throats of their flower tubes.

Hoary Puccoon

Lithospermum canescens H 4–20 in (10–50 cm)

The word "Puccoon" in this wildflower's common name is of American Indian origin and refers to plants that yield pigment for dye or paint, in this case usually a deep yellow.

KEY FACTS

Flat-topped, 5-lobed, yellow-orange flowers grow in clusters. The leaves are long and narrow, and largely stalkless.

+ habitat: Fields, open woods, grassland edges, dry and sandy areas

+ range: Throughout much of central and eastern U.S. and Canada

+ bloom period: April–June

The leaves and stems of the Hoary Puccoon are densely covered with soft, grayish white hairs, both color and fuzziness suggesting the adjective "hoary." The half-inch-long (1.25 cm) flowers grow in flat or curled clusters at the ends of stems with alternating, lance-shaped leaves. A similar species, the Hairy Puccoon (*L. caroliniense*), sports hairier leaves with longer hairs. In addition to making dye and paint from the roots of the plant, American Indians made an herbal tea from its leaves and used it to wash feverish and convulsive patients. Related species of puccoon produce red or purple pigments.

Orange Milkwort

Polygala lutea H 4–16 in (10–40 cm)

Like other species in the Milkwort family, the Orange Milkwort seems misnamed. A milky juice does not ooze from leaves, stems, or elsewhere when crushed.

KEY FACTS

Bright orange flower has 5 sepals, 2 of which form "wings," and 3 fused petals that form a tube with a fringe. Flowers grow in compact clusters.

+ habitat: Moist habitats, including pine barrens, damp sandy or peaty soil, meadows, and ditches

+ range: Coastal states from mid-Atlantic to Gulf Coast

+ bloom period: June–October

The dense, brilliant orange flower heads of the Orange Milkwort grow on single or branched stems heavily populated with alternating, spoon-shaped leaves. Leaves at the base of the plant are the same shape but significantly broader and form a rosette. Milkworts get their name from the traditional belief that consuming the plants increased milk production in cows and nursing human mothers. The species' scientific name also seems to contain a misnomer. *Lutea* means "yellow"—clearly not the case with plants in bloom, but appropriate when the blossoms dry out. The Milkwort flower's complex structure makes pollination a tricky business for insects.

Nodding Onion
Allium cernuum H 6–24 in (15–60 cm)

A pronounced bend in the stem just under the flower head gives the Nodding Onion its distinctive shape. This perennial species carries the typical onion odor and taste.

KEY FACTS

Tiny, bell-shaped flowers form an umbel, or cluster, at the top of a leafless stem.

+ habitat: Prairies, glades, rocky embankments, meadows, stream banks, and moist, cool soils at higher elevations

+ range: Throughout most of U.S. and Canada, except much of Southeast and far northern areas

+ bloom period: June–October

Up to several dozen tiny bells composed of six pointed white or pale pink tepals (look-alike petals and sepals) attach by slender stalks to the tip of the scape, supported by green bracts. The umbel nods on a bent scape, a stem that is leafless and arises directly from the plant's underground bulb. Three to five thin, flat, bladelike leaves up to 16 inches (40 cm) long emerge at the base, and sheathe the scape. American Indians, trappers, and settlers found food and medicinal value in this species, the most widely distributed wild onion, and the similar, erect Wild Autumn Onion (*A. stellatum*).

Lesser Purple Fringed Orchid
Platanthera psycodes H 6–40 in (15–100 cm)

The pronounced beauty of the Lesser Purple Fringed Orchid has been its undoing. This native orchid species is listed as threatened or endangered throughout much of its range.

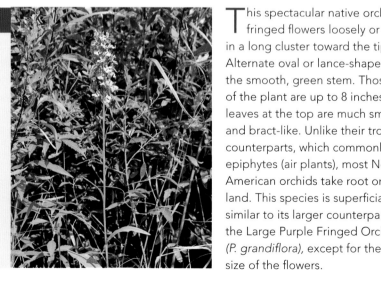

KEY FACTS

White, pink, or purple blooms have a highly fringed, 3-lobed lip, and a long, tubular spur at base.

+ habitat: Moist woods, stream banks, meadows, slopes, marshes, old fields, and roadsides

+ range: Throughout much of eastern U.S. and Canada, except Southeast and far northern areas

+ bloom period: June–September

This spectacular native orchid bears many fringed flowers loosely or densely arranged in a long cluster toward the tip of the stem. Alternate oval or lance-shaped leaves sheathe the smooth, green stem. Those at the bottom of the plant are up to 8 inches (20 cm) long; leaves at the top are much smaller and bract-like. Unlike their tropical counterparts, which commonly are epiphytes (air plants), most North American orchids take root on land. This species is superficially similar to its larger counterpart, the Large Purple Fringed Orchid (*P. grandiflora*), except for the size of the flowers.

Lady's Thumb

Persicaria maculosa H 8–36 in (20–90 cm)

A prolific European import, Lady's Thumb is generally considered a weed. The name comes from a dark spot in the middle of the leaf that was seen to resemble a woman's thumbprint.

KEY FACTS

Tiny pink to purple flowers, composed of 5 tepals joined at the base, form slender, elongated spikes at the end of pinkish stems.

+ **habitat:** Moist areas including roadsides and disturbed areas

+ **range:** Throughout the U.S. and Canada, except in some far northern areas

+ **bloom period:** March–November

The delicate, pink flower spikes of Lady's Thumb make subtle arches on the ends of jointed stems. The stems support lance-shaped leaves, arranged alternately, that can be smooth or covered with short, stiff hairs, and have a papery sheath around the stem at their bases. Lady's Thumb can grow straight up or it can sprawl over the ground. It is opportunistic, often taking up space at the expense of native wildflower species. The plant requires very little substrate to gain a foothold; with enough moisture, it will sprout easily in a thin layer of organic debris in a gutter. The species is often referred to as Smartweed.

Virginia Spring Beauty

Claytonia virginica H 6–12 in (15–30 cm)

An early herald of spring, the Virginia Spring Beauty's delicate blossoms pop up on forest floors, bearing the name of 18th-century American botanist John Clayton.

KEY FACTS

The blossoms have 5 whitish to pink petals veined with deeper pink and 5 stamens; they grow in loose clusters.

+ **habitat:** Moist woods, clearings, seeps, bluffs, lawns, and roadsides

+ **range:** Through-out much of central and eastern U.S. and Canada

+ **bloom period:** January–May

The low-growing, perennial Virginia Spring Beauty often makes an appearance while it is still winter, pushing up from underground potato-like tubers. The plant has two levels of leaves, a basal pair and a pair of opposite stem leaves that are pointed and sometimes fused where they meet. The loose clusters of small, candy-striped flowers make a striking display among the emerging green of the forest, especially in a large stand. But by the time its glossy seeds have ripened, the rest of the plant has all but disappeared above ground. Both American Indians and colonists relished the tuber, which tastes like a chestnut.

One-seeded Pussy Paws

Cistanthe monosperma H 4–20 in (10–50 cm)

The top-heavy, upturned "cat feet" of the One-seeded Pussy Paws may lie on the ground or rise above the plant's basal rosette. This perennial wildflower thrives in a dry, rocky environment.

KEY FACTS

Small flowers ranging from white to pink have 4 petals and uneven sepals, giving the flower a ruffled appearance.

+ **habitat:** Sandy, rocky, and gravelly areas in mid to high elevations

+ **range:** Western U.S. to British Columbia

+ **bloom period:** April–September

The blossoms of the One-seeded Pussy Paws often form one-sided, umbrella-shaped clusters that look like the pads of a cat's paw. Fleshy, spatula-shaped leaves create a rosette at the base of the plant. Two or more stems rise from the basal leaves; these can be erect or prostrate, sometimes varying with the time of day. Leaves higher up are smaller. The perennial One-seeded Pussy Paws are members of the Miner's Lettuce family (previously in the Purslane family), related to Spring Beauty and Bitterroot. The Latin species name, *monosperma*, refers to the seed capsule, which often contains only one shiny black seed.

Bitterroot

Lewisia rediviva H 1–3 in (2.5–8 cm)

In 1806, Lewis and Clark brought back a Bitterroot plant from Montana that received Lewis's name, and the name *rediviva*, meaning "reborn," after the dried specimen was coaxed into bloom.

KEY FACTS

The flower is composed of a variable number of oblong rose-pink petals, usually 10–19; 20–50 stamens crowd the flower's center.

+ **habitat:** Rocky, gravelly areas and wooded and brushy slopes

+ **range:** Western U.S. and Canada

+ **bloom period:** March–June

Bitterroot, Montana's state flower, has large flowers about 2 inches (5 cm) wide that can range from white to light pink to deep pink. Their height is exaggerated by the fact that they grow low to the ground on very short solitary stems. By the time they bloom, the fleshy, cylindrical green leaves that pushed up earlier have all but disappeared, leaving the flowers to grace patches of rock or gravel with moisture and good drainage. Northwestern American Indians routinely ate the nutritious fleshy taproots after boiling them long enough to remove the bitterness, which also causes them to become jellylike.

New England Aster
Symphyotrichum novae-angliae H 24–72 in (60–180 cm)

One of the largest and most imposing asters, with flower heads measuring more than an inch (2.5 cm) across, the New England Aster has a much wider range than its name suggests.

KEY FACTS

Dense petal-like rays that range from blue to rose-purple and rarely white surround an orange-yellow disk.

+ habitat: Wet meadows, moist fields, stream banks, road-sides, and thickets

+ range: Much of U.S. and Canada, except southernmost areas, desert lands, and far northern areas

+ bloom period: July–October

The striking flower heads of the New England Aster have a base of hairy bracts with curled tips and grow on hairy and sticky stalks that form a cluster at the top of the plant, making a dramatic statement. The plant's mainly smooth-edged leaves measure up to 4 inches (10 cm) long and alternate along the stout plant stems, clasping them at their bases. The species is also known as the Michaelmas Daisy because its blooms linger into the fall and often are still present at the time of the feast of St. Michael on September 29. This aster species is pollinated by long-tongued bees and butterflies.

Deptford Pink
Dianthus armeria H 6–24 in (15–60 cm)

Named for the London suburb where it once flourished, the Deptford Pink is an introduced member of the Carnation family admired for its handsome pink flowers.

KEY FACTS

Deep pink flower has 5 spreading, jagged-edged petals dotted with white; sepals form a tube.

+ habitat: Pastures, old fields, woodland edges, roadsides, disturbed areas, and waste areas

+ range: Introduced; naturalized through-out most of U.S. and Canada, except far northern areas

+ bloom period: May–September

The Deptford Pink resembles a simplified carnation, its highly saturated pink blooms highlighted by the profusion of light green, leaf-like bracts that stand erect around the loose, flat-topped flower clusters. Narrow, erect leaves appear in opposite pairs at wide intervals on slender, branched stems. This Eurasian native was introduced and has escaped cultivation. It self-seeds prolifically, giving it a well-deserved reputation as a weedy invasive. The Deptford Pink is somewhat similar to another introduced species, the Sweet William *(D. barbatus)*, which is bulkier with fatter petals and is widely cultivated, but less prone to escape.

Soapwort/Bouncing Bet
Saponaria officinalis H 12–30 in (30–75 cm)

When crushed, the leaves of the Soapwort exude a gluey juice that lathers in water, a trait that gives this member of the Carnation family its common name.

KEY FACTS

Pinkish or white 5-petaled, tubular flowers grow in a flat-topped cluster; the petals are notched or deeply cleft.

+ **habitat:** Waste ground, such as along roadsides and railroad beds

+ **range:** Naturalized throughout most of the U.S. and Canada, except in far northern areas

+ **bloom period:** July–September

The Soapwort stem is wide, smooth, and straight and rarely branches. Long, oval, veined leaves grow opposite each other on the stem. The Soapwort has two methods to aid its rapid spread: The hardy plant spreads by means of seeds and also by underground runners that send up new plants. Soapwort was one of the first flowers introduced by colonial settlers, for its useful cleaning properties as well as a host of medicinal uses. Its alternative common name, Bouncing Bet, may refer to the activity of barmaids, known as Bets, who cleaned ale bottles by filling them with water and a plant sprig and shaking them up.

Pacific Bleeding Heart
Dicentra formosa H 8–18 in (20–45 cm)

Clusters of pendent pink blossoms dangle from arched stalks to create a delightful natural valentine, an alluring benefit of a walk in damp western habitats.

KEY FACTS

The flowers are heart-shaped, puffy, and range from creamy white to rose-purple; tips of outer petals curve outward.

+ **habitat:** Damp woods, clearings, meadows, and stream banks

+ **range:** Native from British Columbia through California; widely cultivated in the U.S. and Canada

+ **bloom period:** March–July

Pacific Bleeding Heart blossoms, about three-quarters of an inch (2 cm) long, appear in small, branched clusters on a stalk that rises directly from the crown of the plant's root. The deeply cut, fern-like green basal leaves, often with a bluish cast, develop in the same manner. After flowering, this native perennial produces kidney-shaped seeds in a long conical pod. Some Northwest Indian tribes chewed the raw, succulent taproots of the species to alleviate toothache and also made a decoction of the roots that was taken to purge worms. The shade-loving plant is widely cultivated as a garden ornamental.

Spotted Geranium/Wild Cranesbill

Geranium maculatum H 12–24 in (30–60 cm)

The Spotted Geranium builds on the principle of five: sepals, petals, leaf lobes, and often blooms in a cluster. The alternate common name refers to the shape of the beaked fruit capsule.

KEY FACTS

The pink, violet, or white flower has 5 fine-lined petals and 5 pointed green sepals underneath.

+ habitat: Moist woods, meadows, thickets, and woodland edges

+ range: Throughout most of central and eastern U.S. and Canada

+ bloom period: March–July

Spotted Geranium flowers appear in small clusters of three to five above green palmate leaves with five deep lobes. The lobes themselves are lobed or toothed on the margins. The species reproduces by means of seeds and also by rhizomes that generate new growth, but it does so in moderation and does not overtake other plants in its environment. The genus name, *Geranium*, derives from the Greek for "crane." American Indians commonly used tea made from boiled plants to treat diarrhea, and they also bathed mouth sores with an infusion that capitalized on the plant's astringent properties.

Swamp Rose

Rosa palustris H 36–96 in (91–244 cm)

A member of the sizeable Rose family, the Swamp Rose usually grows into a medium-size upright shrub, seeking out moist habitats with rich, loamy soil.

KEY FACTS

The flower—up to 3 in (7.5 cm)—has 5 pink petals, a ring of yellow stamens, and a central flattened mass of stigmas.

+ habitat: Swamps, marshes, ditches, stream banks, and pond and lake edges

+ range: Native throughout most of the eastern U.S. and Canada; widely cultivated

+ bloom period: May–June

Swamp Rose blossoms punctuate the shrub's bushy, leaf-filled branches and prickly stems for only a month or two each year. The leaves are odd-pinnate, usually composed of seven leaflets, and the prickles are curved. Red hips, or fruits, replace the flowers; as in other rose species, they are rich in vitamin C. A Swamp Rose bush is like a wildlife grocery, supplying food for scores of different species of insects, birds, and mammals including White-tailed Deer and American Beaver. It is also like a condo, providing sheltered nest sites for many bird species, including warblers and the Northern Cardinal.

Purple Loosestrife

Lythrum salicaria H 12–60 in (30–150 cm)

Introduced from Europe, Purple Loosestrife makes a dramatic statement in large stands, at a cost to native aquatic plants and wildlife. As such, it is classified as a noxious weed in many states.

KEY FACTS

Lavender or rarely magenta to red flowers with 5–7 wavy petals occur in small clumps on long flower spikes.

+ habitat: Wet areas, including meadows, ditches, and floodplains

+ range: Throughout much of the U.S. and Canada, except in far northern areas and parts of southeastern U.S.

+ bloom period: June–September

The terminal flower spike of the Purple Loosestrife takes up about the last foot (30 cm) or more of the plant's long, stout, angled stem. Lance-shaped leaves without stalks appear opposite each other or in whorls along the stem, and rounded or heart-shaped ones grow at its base. Dozens of stems can rise from a single rootstock, and the flowers on these can produce millions of tiny seeds; abundant nectar encourages pollination. The genus name derives from the Greek for "blood," a reference to the plant's color or possibly to the styptic (blood-staunching) quality of some loosestrife species.

Large Beardtongue

Penstemon grandiflorus H 18–48 in (45–120 cm)

The Large Beardtongue is a *Penstemon*, a genus of more than 200 species that is endemic to North America. These attractive wildflowers have been developed into popular cultivars.

KEY FACTS

A large broadly tubular pink to lavender corolla flares into lobes, forming an upper lip with 2 lobes and lower lip with 3 lobes.

+ habitat: Dry prairies, plains, woods, and thickets

+ range: Throughout most of the central U.S.; escaped from cultivation elsewhere

+ bloom period: May–June

The Large Beardtongue bears large, inflated flowers, up to 2 inches (5 cm) long, that grow horizontally from the axils of leafy bracts. The bracts are smaller but similar in appearance to the upper stem leaves. Fine purple lines in the flower's throat serve as nectar guides for pollinators such as long-tongued bees. Leaves at the base form a rosette in the plant's first year; in the second and subsequent years, flower stalks emerge. *Penstemon* means "five stamens," which all the species in this genus have. One of them typically is sterile and hairy—the source of the common name Beardtongue.

Virginia Meadow Beauty
Rhexia virginica H 6–36 in (15–90 cm)

The Virginia Meadow Beauty mostly confines itself to wet places where its classic looks stand out. The attractive native perennial also goes by the name Handsome Harry.

KEY FACTS

The flower has 4 widely spreading, pink to purple petals and 8 showy stamens with long, slender anthers.

+ habitat: Wetlands, wet meadows, pond edges, and ditches

+ range: Much of central and eastern U.S. and Canada

+ bloom period: May–October

The flowers of the Virginia Meadow Beauty form loose clusters at the ends of hairy, squared, slightly winged, erect stems. The blossoms have petals nearly as long as they are wide and long stamens with large, bent, bright yellow anthers. Oval leaves clasp the stem opposite to each other and have rounded bases and three prominent veins. The fruit of this species is a reddish, distinctively urn-shaped, four-pointed capsule that holds many tiny seeds. Members of the Melastome family typically are tropical, but a number of different *Rhexia* species are native to temperate North America.

Fireweed
Chamerion angustifolium H 24–96 in (60–240 cm)

The aptly named Fireweed often makes an appearance after a wildfire, when it moves into a burned-out forest area, taking advantage of sunny conditions to form large, impressive stands.

KEY FACTS

The flower is composed of 4 pink, rose-purple, or rarely white petals; it has 8 even-length stamens, and a 4-lobed stigma.

+ habitat: Open areas, slopes, and roadsides

+ range: Mainly western and northern U.S. and throughout most of Canada

+ bloom period: June–September

A member of the Evening Primrose family, the Fireweed sends up reddish stalks with loose clusters of blossoms in terminal spikes. The blossom, though four-petaled, is less symmetrical than that of the Virginia Meadow Beauty. Its leaves are long, narrow, alternate, and willowlike, and its seeds form in slender pods that peel back from the top when ripe. The hair-tufted seeds disperse efficiently—and far—in the wind, helping this opportunistic species establish itself rapidly. Following the eruption of Mount St. Helens, it was one of the first plant species to populate the wasteland, adding an early burst of color.

Tongue Clarkia/Diamond Clarkia

Clarkia rhomboidea H 12–36 in (30–90 cm)

The Tongue Clarkia's unusual diamond-shaped petals with toothed bases help identify this annual member of the Evening Primrose family. It also is noted for its blue-gray pollen.

KEY FACTS

The flower is rose or lavender-pink, sometimes spotted, with 4 diamond-shaped petals each with 2 small teeth at base.

+ habitat: Open woods and pine forests; slopes to about 10,000 ft (3,000 m)

+ range: Western U.S. and northward into British Columbia

+ bloom period: June–July

Tongue Clarkia keeps a fairly narrow profile. It sends up a thin, erect, usually unbranched stem with sparse numbers of short, smooth-edged, lance-shaped leaves. The leaves attach alternately to the stem by means of short stalks. Few flowers adorn the upper portions of the stems. The distinctive flower buds are nodding, and often have pointed tips. This genus is named for William Clark, the explorer who accompanied Meriwether Lewis, and with him served as co-leader of the Corps of Discovery, on a transcontinental expedition to the Pacific Northwest from 1804 to 1806. The expedition collected many floral and faunal specimens.

Trailing Arbutus/Mayflower

Epigaea repens H 2–3 in (5–8 cm) L to 16 in (40 cm)

For those patient and lucky enough, searching among woodland leaf litter in early spring may reward with a glimpse of the Trailing Arbutus with its fragrant flowers among evergreen foliage.

KEY FACTS

The half-inch (1.25 cm) corollas are pink to white, more or less trumpet-shaped, and have 5 lobes and hairy throats.

+ habitat: Moist and dry woods and in rocky places

+ range: Eastern U.S. and Canada

+ bloom period: February–May

Creeping along the ground, the tangled, hairy stems of Trailing Arbutus, a low-growing member of the Heath family, are replete with leathery leaves and flowers that grow on stem ends or emerge from the leaf axils (the crook where the leaf joins the stem). Fleshy capsules replace the flowers. The species serves as a bellwether for environmental disturbance, being sensitive to changes in climate and habitat; it has become more rare over time. Eastern Indian groups, such as the Cherokee and the Iroquois, made decoctions of the plant to treat mainly stomach and kidney ailments.

Dark-throated Shooting Star

Dodecatheon pulchellum H 2–16 in (5–40 cm)

Less is more for the Dark-throated Shooting Star, also known as the Western Shooting Star. Usually only a few flowers emerge in a cluster from the tip of a single scape (a long leafless stalk).

KEY FACTS

The lobes of the 5-lobed magenta flower sweep back, allowing the stamens and style to protrude.

+ **habitat:** Moist, wet meadows, prairies, bogs, and stream banks from sea level to the timberline

+ **range:** Throughout the western U.S. and Canada

+ **bloom period:** April–August

The flowers of the Dark-throated Shooting Star point downward until they are pollinated, and then turn heads up, all the while impersonating their celestial namesake. Leaves that are mostly oblong grow in profusion at the base of the plant. The leaves can be toothed or have smooth edges, and can have smooth or hairy surfaces. There is also significant variation in the number of flowers on a scape of this native perennial. Some western variations seem to inter-breed and produce intergrades. An eastern species, *D. meadia*, has paler flowers that range from white to pale pink to lavender.

Pink Turtlehead/Lyon's Turtlehead

Chelone lyonii H 12–36 in (30–90 cm)

This member of the Plantain family sports a distinct flower that is unmistakably turtle-like. Its genus name is derived from the Greek for "tortoise."

KEY FACTS

The 2-lipped flowers are deep pink to purple with yellow bearded interior of lower lip.

+ **habitat:** Moist woods, stream banks, and coves

+ **range:** Native to southeastern U.S.; escaped cultivars naturalized in the Northeast

+ **bloom period:** July–October

Compact clusters of striking pink blossoms against deep green foliage make the Pink Turtlehead an unusually attractive wildflower. The shape of its flower adds a comical element. The flowers are borne mainly at the tops of erect stems with many opposite, oval or lance-shaped, toothed leaves. The species prefers a moist footing. Pollinating bees enter the concave, upper, shallowly two-lobed lip of the turtlehead's "mouth," attracted by the ridged, hairy, slightly three-lobed lower lip with its yellow "signals." The species is named for John Lyon, a transplanted Scot who made important contributions to early 19th-century American botany.

Rose Pink

Sabatia angularis H 12–36 in (30–90 cm)

At the center of the Rose Pink flower, a member of the Gentian family, is a star formed by yellow marks at the base of each petal. Dark pink outlines the star.

KEY FACTS

Rose pink to white flower has 5 petals that are slightly joined at the base, and a greenish yellow center with yellow stamens and anthers.

+ **habitat:** Prairies, marshes, open woods, thickets, old fields, and roadsides

+ **range:** Throughout much of eastern and southern U.S. and Canada

+ **bloom period:** July–September

The biennial Rose Pink starts in the first year as a basal rosette of leaves. Its four-angled main stem shoots up the following year, branching near the top. Smooth oval leaves more than an inch (2.5 cm) long closely attach opposite to each other on the stems. A single blossom about an inch wide forms at the tip of the main stem, and others join it on side stems to form a cluster called a cyme. The fragrant flowers open during the day and close at night. After flowering, a single-chambered capsule develops that contains many light, tiny seeds. These can easily disperse on the wind.

Spreading Dogbane

Apocynum androsaemifolium H 6–48 in (15–120 cm)

The poisonous nature of this milkweed relative gives the species its common and genus names: *Apocynum* derives from Greek words meaning "away dog," referring to its toxicity to dogs.

KEY FACTS

Small, pink or white, 5-lobed bell-shaped flowers grow in a terminal cluster.

+ **habitat:** Woods, woodland edges, prairies, meadows, fields, and roadsides

+ **range:** Throughout much of the U.S. and Canada, except in far northern areas and some southeastern states

+ **bloom period:** June–August

Spreading Dogbane produces many forked, branching stems that are green to reddish and bear loose clusters of slightly nodding flowers at the ends. The interior of the fragrant flower has darker pink stripes that lead to the nectar sources. The opposite, lance-shaped leaves either spread out from the stem or droop noticeably; they tend to have smooth upper surfaces and slightly hairy lower ones. Like milkweeds, Spreading Dogbane exudes a milky sap when its stems or leaves are crushed. A related species, *A. cannabinum* or Indian Hemp, has smaller greenish white flowers. American Indians used its long stem fibers to make rope.

Rose Vervain

Glandularia canadensis H 5–10 in (13–25 cm)

A member of the Vervain family, the Rose Vervain either creeps horizontally, forming a mat, or grows upright. In either case, it sends up flat clusters of small, trumpet-shaped flowers.

KEY FACTS

The pink to lavender, trumpet-shaped flower has 5-lobed corollas, the lobes are sometimes notched at the tips.

+ **habitat:** Prairies, open woods, pastures, fields, lake edges, and rocky or gravelly areas

+ **range:** Central and eastern U.S. and Canada

+ **bloom period:** February–September

Small flowers, about a half inch (1.25 cm) wide, crowd the Rose Vervain's terminal flower spikes on erect or sprawling hairy stems. The leaves of this plant are lance-shaped overall, though deeply lobed with roughly serrated edges; they are positioned opposite each other on the stem. Stems often branch. As the plant ages, its flower clusters elongate. This native perennial is often confused with escaped cultivars of the genus *Verbena* that are introduced in many areas but are generally less hardy, especially in the winter. Flowers of purple or white might indicate a cultivar.

Common Milkweed

Asclepias syriaca H 24–72 in (60–180 cm)

The most abundant milkweed, the Common Milkweed yields a thick, bitter, milky liquid when its leaves or stem are bruised. The plant contains compounds used to treat heart disease.

KEY FACTS

Pink to purplish flowers growing in rounded clusters are slightly pendent and formed of 5 bent-back corolla lobes and a central crown of 5 curved lobes.

+ **habitat:** Roadsides, old fields, and disturbed areas

+ **range:** Native throughout most of central and eastern U.S. and Canada

+ **bloom period:** June–August

Common Milkweed blossoms grow from the axils of the upper leaves. The green leaves in general are broadly oval and hairy—sparsely above and fuzzier below, creating a grayish cast. In autumn, the flower heads give way to large, coarse seedpods filled with seeds tipped by silky hairs that eventually burst, dispersing black seeds on the breeze. The caterpillar of the monarch butterfly feeds exclusively on the foliage of milkweeds, whose toxic properties transfer to the bodies of both larva and butterfly and offer protection from predators. Milkweeds attract other butterfly species and other insects, including bugs, beetles, and bees.

Wild Bergamot

Monarda fistulosa H 12–48 in (30–120 cm)

A showy perennial in the Mint family, Wild Bergamot and its relatives attract butterflies and bees with dense globes of pink, lavender, white, or scarlet flowers.

KEY FACTS

Pink to lavender flowers have 2 lips; upper lip has 2 lobes, lower lip is broad with 3 lobes.

+ **habitat:** Fields, prairies, meadows, open woods, thickets, and woodland edges

+ **range:** Native throughout most of the U.S. and Canada east of the Rockies, naturalized throughout wider area

+ **bloom period:** May–September

A single Wild Bergamot flower cluster with conspicuous projecting stamens appears atop a whorl of leaflike bracts at the summit of a square, hairy stem. The blossoms attract butterflies, bees, and hummingbirds. The stem bears opposite, toothed, lance-shaped, and fragrant leaves often used to make a tea. Along with a close relative, *M. didyma* or Bee Balm, Wild Bergamot has been used by American Indians and others to treat a wide range of ailments, including stomach and intestinal troubles, respiratory problems, fevers, and skin eruptions. Wild Bergamot is not related to the orange subspecies *Citrus aurantium bergamia,* which is commonly used to flavor tea.

Rocky Mountain Beeplant

Cleome serrulata H 12–60 in (30–150 cm)

True to its name, the Rocky Mountain Beeplant attracts numerous bees. Many American Indian groups used the foliage and seeds for food and medicinal purposes.

KEY FACTS

Pink to purplish flowers have 4 petals and 6 long, projecting stamens attached along a terminal spike.

+ **habitat:** Prairies, open woods, washes, disturbed areas, and scrub; occurs to 5,000 ft (1,500 m)

+ **range:** Native to western U.S. and Canada; introduced eastward, mostly absent in the Southeast

+ **bloom period:** July–September

The showy pink clusters of the Rocky Mountain Beeplant, a member of the Caper family, grow on the ends of erect, branched stems with alternate palmate leaves, usually composed of three to seven leaflets. The flower spikes continue to elongate as additional flowers open, and it is not unusual for a single spike to display buds, flowers, and long, slender seedpods all at the same time. The species is also called Stinkweed in recognition of the unpleasant odor given off by the leaves when crushed. The similar Yellow Spiderflower (*C. lutea*) has yellow flowers and a more limited distribution.

Purple Passionflower/Maypop

Passiflora incarnata L vine to 30 ft (9 m)

This vining plant with showy blooms owes its name to features resembling symbols of Christ's passion. It grows from slender stems along vines that trail or climb by means of curly tendrils.

KEY FACTS

The purple-and-white blossom is 2–3 in (5–7.5 cm) across; 3-lobed leaves are toothed.

+ habitat: Fields, fencerows, open woods, sandy thickets, stream and riverbanks

+ range: Native to southeastern U.S. and parts of the Midwest; introduced elsewhere

+ bloom period: April–September

Missionaries named the intricate flower for the Christian symbolism they read into its parts: Three central stigmas signify crucifixion nails; five stamens correlate to Christ's wounds; the fringe of purple-and-white filaments resembles the crown of thorns; and the ten combined petals and sepals under the fringe represent the faithful apostles (minus Judas and Peter). A yellowish orange to yellowish green egg-shaped fruit, known as a Maypop, is filled with abundant small seeds embedded in yellow pulp. American Indians used Purple Passionflower to heal wounds and bruises. Today, it is valued for its mildly sedative properties and is used in teas.

Eastern Purple Coneflower

Echinacea purpurea H 12–36 in (30–90 cm)

Its large flower head pointed skyward, the instantly recognizable Eastern Purple Coneflower, also known as Echinacea, is one of the most widely used and studied medicinal plants.

KEY FACTS

The stiff, brownish central disk is surrounded by densely packed, notched, and drooping magenta rays.

+ habitat: Prairies, open woodlands, and dry forest margins; widely cultivated as an ornamental and for herbal medicine

+ range: Central and eastern U.S. and Canada

+ bloom period: June–October

The Eastern Purple Coneflower blossom rises on a stiff, hairy stem. Leaves that are also stiff and hairy alternate along the stem; the upper ones are smooth edged, but the lower ones are toothed, and each bears five distinct ribs. The scientific name for this perennial plant comes from the Greek *echinos*, which means "hedgehog," a reference to the numerous sharply pointed scales in the central disk of the flower head. For centuries, Echinacea was taken internally and applied externally by American Indians for a wide range of ailments. Today, its perceived properties as an immune system booster make it a popular remedy for fighting colds and flu.

Sweet Joe-Pye Weed

Eutrochium purpureum H 24–84 in (60–210 cm)

Projecting styles on the purplish florets of its rounded terminal flower clusters give the Sweet Joe-Pye Weed a fuzzy, out-of-focus look. Despite its name, the species is a popular garden addition.

KEY FACTS

The small pink to purplish 5-lobed disk flowers with long styles make up each of the many flower heads that form a cluster.

+ **habitat:** Open woods, woodland edges, wet meadows, and ravines

+ **range:** Throughout much of the eastern U.S. and Canada

+ **bloom period:** July–September

A member of the Aster family, the Sweet Joe-Pye Weed lacks the rays on its flower heads that are characteristic of many other species in this family. The plant has distinctive stems that are mostly green but become dark purple where leaves attach; the stems bear whorls of three to six coarse, toothed leaves with short stalks, broad bases, and sharply pointed ends. The leaves are up to 10 inches (25 cm) and may give off the scent of vanilla when crushed. After flowering, the clusters convert to balls of slender, five-sided brown seeds that are tufted with brownish hairs to help them disperse in the wind.

Common Ironweed/Prairie Ironweed

Vernonia fasciculata H 24–48 in (60–120 cm)

Ironweeds received their name from the toughness of their stems. The genus name, *Vernonia*, honors William Vernon, an English botanist who collected in Maryland in the late 17th century.

KEY FACTS

The tightly compressed, 5-lobed magenta disk florets with prominent divided styles form a compound flower.

+ **habitat:** Wet prairies, marshes, field edges, and sloughs along railroads

+ **range:** Native throughout much of central U.S. and Canada; introduced elsewhere

+ **bloom period:** July–September

Rugged round central stems support the dense clusters of Common Ironweed flower heads. Narrow, lance-shaped leaves with serrated edges grow alternately on the stem, which can be white, light green, or reddish purple. In addition to its strong constitution, the plant also tends to resist pests and disease and tolerates short bouts of flooding. Common Ironweed blooms attract many bees and butterflies, and its foliage hosts the caterpillar of the American Painted Lady. But its bitter foliage is not a favorite of mammalian herbivores, who tend to shun Ironweed until it is one of the last plants left in a heavily grazed pasture.

Dense Blazing Star/Gayfeather

Liatris spicata H 12–72 in (30–180 cm)

A stand of Dense Blazing Star rising above a damp meadow or prairie is a wondrous sight. The tall plants bear spikes of plumy purple flowers, the source of another common name, Gayfeather.

KEY FACTS

The flower is a rayless, rose-purple disk with long projecting styles that emerges above a base of overlapping bracts.

+ habitat: Prairies, damp meadows, open woods, and marsh edges

+ range: Eastern U.S. and Canada; cultivated widely elsewhere

+ bloom period: July–September

The feathery flower heads of the Dense Blazing Star attach without stalks and crowd a long spike, a foot (30 cm) or longer; they bloom from the top down. The plant stem grows from a base of grasslike leaves, also a foot long. Dense stem leaves appear alternately, becoming progressively smaller toward the top. The flower heads of the conspicuous terminal spike have no scent, though they are highly attractive to bees, butterflies, and moths; two moth caterpillar species feed specifically on the foliage, stems, and developing fruits. The perennial Dense Blazing Star is a member of the Aster family.

Bull Thistle

Cirsium vulgare H 24–72 in (60–180 cm)

The large and distinctive Bull Thistle, a Eurasian import, mounts many defenses to protect itself: Blossoms, stems, and leaves all have spines or prickles in various combinations.

KEY FACTS

The flower head up to 3 in (8 cm) wide is composed of slender purple disk flowers that are supported by spiny green bracts.

+ habitat: Roadsides, fields, pastures, disturbed areas, waste places

+ range: Introduced and naturalized in most of the U.S. and Canada

+ bloom period: June–September; year-round in mild climates

Bull Thistle flower heads emerge at the tips of thick, angular stems and branches armed with sharp, hairy, and prickly leaves and sticky bracts that discourage marauding insects. The spiny leaves are deeply lobed and toothed and the stout stems bear spiny "wings." The biennial plant's defenses are combined with a very efficient seed-dispersal system. The downy bristles of its fruit act like parachutes to transport the light, dry seed. The Bull Thistle is similar but much larger than its cousin the Canada Thistle (*C. arvense*), which also has a spineless stem but is classified as a noxious weed in many states.

Fuller's Teasel

Dipsacus fullonum H 18–72 in (45–180 cm)

The prickly Fuller's Teasel name derives from the use of the dried flower heads by fullers, or cloth cleaners, to tease up the nap of washed cloth, particularly wool.

KEY FACTS

Dense bands of 4-lobed purplish flowers encircle a prickly, ovoid flower head with spiny, upwardly angled bracts.

+ **habitat:** Old fields, roadsides, ditches, and disturbed areas

+ **range:** Naturalized throughout much of the U.S. and Canada, except far northern areas and the extreme Southeast

+ **bloom period:** July–October

The Fuller's Teasel comes armed with multiple defenses. In addition to the prickles and sharply pointed scales of the flower head, short prickles project from the erect stem and from the midrib of the lance-shaped leaves that clasp the stem and form a joined pair from opposite sides. The basal leaves are wider with wavy margins; these leaves die off in the biennial plant's second season. A native of Eurasia, Fuller's Teasel forms large expanses that crowd out many native species. The genus name, *Dipsacus*, means "thirsty," and is thought to refer to the water-catching cup formed by the joined leaf pairs.

Skunk Cabbage

Symplocarpus foetidus H 12–24 in (30–60 cm)

The stinky Skunk Cabbage is often the first wildflower to bloom each year. As early as February, the purplish spathes push up, their internal heat melting surrounding snow.

KEY FACTS

Tiny yellow to purplish flowers form on a spadix (dense spike) that is inside a hoodlike spathe.

+ **habitat:** Wet woodlands, stream banks, open swamps and marshes, and other wet areas

+ **range:** Mid-Atlantic and northeastern U.S. and Canada

+ **bloom period:** February–May

The spathe precedes the leaves of the Skunk Cabbage in seeking the light of day. It changes color as it ages, starting out pale green and barely streaked with brown, and ending up darker and heavily stained with purplish brown. The spathe shelters a thick fleshy spadix, on which large numbers of small flowers form. The almost hidden flowers give off a rank odor, intensified by the plant's internal heat, which can rise more than 30°F (17°C) higher than the outside temperature. When the flowers wither, the leaves emerge from the ground. They are large and green, resembling cabbage leaves.

Jack-in-the-pulpit

Arisaema triphyllum H 12–36 in (30–90 cm)

Tiny "Jack," the spadix, peeps out of his "pulpit," the spathe, in this native woodland perennial. The Jack-in-the-pulpit prefers rich soil and grows well in cultivation.

KEY FACTS

A curved, green spathe, striped with white or purplish brown, folds over a tube with a club-shaped spadix (spike) inside.

+ **habitat:** Damp woods, swamps, bogs, and marshes

+ **range:** Throughout central and eastern U.S. and Canada

+ **bloom period:** March–July

The hooded spathe, or sheath, of the Jack-in-the-pulpit rises from a tube on its own stalk and hovers over the small spadix, or spike, covered with tiny male or female flowers. The spathe emerges near a canopy provided by two (sometimes only one) large, palmate leaves rising on separate stalks. The leaves usually divide into three lobes. At maturity, the spadix produces shiny red berries that provide food for birds and mammals. American Indians collected the short, broad tubers of the plant to dry and cook; this practice supplied the alternate common name Indian Turnip. The roots and other plant parts can blister and irritate the skin.

Checker Lily/Mission Bells

Fritillaria affinis H 6–40 in (15–100 cm)

The Checker Lily may or may not be "checkered" (that is, mottled with brown, red, or purple). Its range of appearance made botanists wonder whether some variants were separate species.

KEY FACTS

A pendent yellow-green flower, composed of 6 lance-shaped tepals (look-alike petals and sepals), is often heavily mottled to appear dark purple.

+ **habitat:** Prairies, open woods, grassy and brushy slopes to higher elevations

+ **range:** Western U.S. and Canada

+ **bloom period:** February–June

Two to five bell-shaped flowers nod from the top of a solitary, erect, unbranched stem of this member of the Lily family. Narrow, lance-shaped leaves grow alternately on the upper portion of the stem and in whorls of three to five lower down. The Checker Lily sprouts from unusual, scaly bulbs that are surrounded by a mass of rice-size bulblets. The genus name comes from the Latin for "fritillary," which means "dice box," a reference to the checker pattern of these lilies and the butterflies known as fritillaries. Another common name for this species is Mission Bells.

Scarlet Rose Mallow

Hibiscus coccineus H 36–96 in (90–240 cm)

The Scarlet Rose Mallow, a type of hibiscus, beguiles with its large, showy red blossoms. Although the perennial species has a small native range, it is cultivated widely in climate-friendly zones.

KEY FACTS

The flower has 5 separated crimson petals and anthers; the stamens form a long tube around the styles and the stigmas.

+ **habitat:** Swamps, marshes, and ditches

+ **range:** Native to southeastern U.S.; also widely cultivated

+ **bloom period:** June–September

Scarlet Rose Mallow sets its blossoms from the upper axils of its palmated, highly divided, pointed, and toothed deep green leaves, which bear a strong resemblance to the foliage of the Hemp plant, *Cannabis sativa*. The large flowers, up to 8 inches (20 cm) wide, are star shaped and appear above undersized green bracts. The stamen column formed around the styles and stigmas is a distinctive Mallow feature. The species forms clumps several feet wide and blooms later in the season than many other wildflowers. Some *Hibiscus* cultivars have more rounded, overlapping petals.

Common Blanketflower

Gaillardia aristata H 6–30 in (15–75 cm)

This member of the Aster family gets its common name from similarities between the cheerful flower and the colorful patterns woven into American Indian blankets.

KEY FACTS

Flower head is formed of yellow rays, usually 3-forked at the tip, with dark red bases around a center of dark red disk flowers.

+ **habitat:** Prairies, woodlands, meadows, and disturbed areas

+ **range:** Native in much of northwestern U.S. and Canada; highly cultivated and escaping farther south and east

+ **bloom period:** July–September

Blanketflowers often grow as single flower heads on stalks rising from a base of rough, hairy, gray-green leaves that alternate on the stem. The leaves, up to 6 inches (15 cm) long, may be lance shaped or lobed, with or without toothed edges. Occasionally, rays will be absent, leaving only the dark central disk flowers. Though not much favored by grazing animals, Blanketflower's presence in a pasture indicates that the preferred plants there have reached their grazing readiness. The species has had many uses in traditional medicine and is being investigated for potential anticancer and antibacterial properties.

Wild Ginger

Asarum canadense H 6–12 in (15–30 cm)

Wild Ginger's pair of large, heart-shaped leaves tower over the small, low-growing and solitary brownish maroon flower that is often overlooked, obscured by the woodland leaf litter.

KEY FACTS

The bell-shaped brownish maroon flower has 3 pointy lobes, about 1 in (2.5 cm) long, and it grows on a short stalk.

+ **habitat:** Moist, rich woods and wooded slopes

+ **range:** Central and eastern U.S. and Canada

+ **bloom period:** April–May

Unrelated to the zesty tropical spice, Wild Ginger gets its name from its similar underground stems and sharp taste. The brownish maroon flower shares its color and fetid smell with rotting meat, making it very attractive to its chief pollinators—various fly species. The plant's distinctive leaves rise on long stalks from the creeping stems, and the pair will often sway together in a gentle breeze. They have dark, woolly, and attractively veined upper surfaces and lighter undersides. The flowers emerge near the ground on their short stalks in the fork between the two leaf stalks.

Red Columbine/Wild Columbine

Aquilegia canadensis H 12–24 in (30–60 cm)

The genus name for columbines, members of the Buttercup family, derives from *aquila*, the Latin for "eagle," and is thought to refer to the talon-like appearance of the flower's distinctive spurs.

KEY FACTS

The flower's 5 yellow-faced petals form red tubular spurs and its red sepals project forward; the yellow stamens protrude.

+ **habitat:** Woods, hillsides, cliffs, bogs, and roadsides

+ **range:** Native to central and eastern U.S. and Canada; also highly cultivated

+ **bloom period:** March–July

The Red Columbine is a woodland wildflower with the whole package: colorful, distinctive, bell-like, pendent blossoms that grow gracefully on branched stems adorned with delicate, three-lobed compound leaves. The spurred red flowers present a challenge for pollinators and rely on hummingbirds and long-tongued insects to get the job done. The perennial plant spreads by prolific self-seeding. A similar species, the Western Columbine (*A. formosa*), has petal-like sepals that spread out. Various native columbines hybridize among themselves. As cultivars, columbines are much admired for their hardiness, persistence, and shade tolerance.

Fringed Redmaids
Calandrinia ciliata H 2–16 in (5–40 cm)

Low-growing Fringed Redmaids got their common name from botanists who encountered the little red flowers on an 18th-century expedition to Peru and Chile.

KEY FACTS

The flower has 5 pale pink to rose-red petals, yellow-tipped stamens, and sepals with a hairy fringe.

+ **habitat:** Grasslands, fields, gravelly washes, desert, rocky slopes, and disturbed areas

+ **range:** Western U.S. to British Columbia; Mississippi; Massachusetts

+ **bloom period:** February–May

Fringed Redmaid flowers grow in loose clusters from the axils of leaflike bracts on erect or prostrate stems that bear lance-shaped, alternately arranged, fleshy leaves. The foliage varies somewhat, but the flower appearance remains consistent. This member of the Miner's Lettuce family (previously in the Purslane family) often keeps low to the ground and spreads prolifically. Fringed Redmaids yield huge quantities of black oil-rich seeds that are a favorite food of many songbirds and rodents. California Indians also collected the seeds, which they parched and ground into meal. They sometimes set grasslands on fire to encourage this fire-following species.

Red Clover
Trifolium pratense H 6–24 in (15–60 cm)

Red Clover, a Eurasian import, is often planted in pastures and as a cover crop. Like its cousin White Clover *(T. repens)*, it can quickly become a widespread landscape feature.

KEY FACTS

Numerous pink to red tubular flowers composed of 5 narrow petals form dense egg-shaped masses to globes.

+ **habitat:** Meadows, pastures, old fields, lawns, and disturbed areas

+ **range:** Throughout the U.S. and Canada, except in far northern areas

+ **bloom period:** May–September

Red Clover, a member of the Pea family, sometimes has red blossoms, but more often they are pinkish, or rarely, white. This species has a compound leaf structure of three oval leaflets with smooth edges, with a V-shaped pattern in the middle of each. Flowering branches are leafier than those of White Clover, carrying one or several leaves just below the flower head. Red Clover flowers have a honey-like fragrance, and the foliage, especially in a dense patch, smells, well . . . like clover. Farmers and home gardeners often improve soil quality by planting this nitrogen-fixing species as a winter cover crop or in rotation.

Scarlet Gilia

Ipomopsis aggregata H 12–36 in (30–90 cm)

You can dress a Scarlet Gilia up, but you cannot hide its less attractive feature: Glandular foliage gives this member of the Phlox family a skunky odor.

KEY FACTS

The trumpet-shaped flower is composed of 5 fused petals with red, pink, or white spreading and pointed lobes, which are sometimes speckled.

+ **habitat:** Meadows, open woods, and rocky slopes

+ **range:** Western U.S. into British Columbia

+ **bloom period:** August–October

Many wildflowers deserve to be called striking, but for the Scarlet Gilia, it is part of the scientific name: *Ipomopsis* derives from the Greek for "striking appearance." The species' leaves don't compete with its flowers; deeply and narrowly lobed, they are sparsely scattered on the stem. Stems often spring from a basal rosette of leaves that may persist for several years after flowering has ended. This common western wildflower was previously placed in the genus *Gilia* and still retains that name in some horticultural literature. Its odor gives it another moniker: Skunkflower. Lewis and Clark brought back a specimen collected in Idaho in 1806.

Indian Warrior

Pedicularis densiflora H 6–24 in (15–60 cm)

This western species, a member of the Broomrape family (previously in the Figwort family), gets its common name from the shape of its flower cluster, which resembles a feathered plume.

KEY FACTS

The red or fuchsia, 2-lipped tubular flower has a straight, hooded upper lip and a smaller lower one.

+ **habitat:** Open woods, chaparral, and slopes to 6,000 ft (1,828 m)

+ **range:** Native to California and Oregon; introduced elsewhere

+ **bloom period:** March–May

The Indian Warrior's clublike flower stalks rise above toothed bracts at the end of a hairy stem. The fernlike leaves, up to 6 inches (15 cm) long, occur mainly at the base; those farther up the stem are much smaller. The species is part of a group of plants once thought to transmit lice to cattle, the reason for the genus name *Pedicularis*, derived from the Latin for "louse." Indian Warrior will opportunistically parasitize the roots of other plants, especially Manzanita shrubs. The species has many traditional medicinal uses, including as a muscle relaxant and nerve-pain reliever.

Scarlet Indian Paintbrush

Castilleja coccinea H 6–24 in (15–60 cm)

Brilliant red-tipped bracts surrounding tubular greenish yellow flowers in a long spike give the Scarlet Indian Paintbrush the appearance of a quick dip in a pot of scarlet paint.

KEY FACTS

The fan-shaped, 3-lobed, red-tipped to all-yellow leafy bracts appear with the long, tubular greenish yellow flowers.

+ **habitat:** Moist places such as prairies, meadows, and roadsides

+ **range:** Central and eastern U.S. and Canada

+ **bloom period:** May–July

Like the Poinsettia, the blaze of color in the Scarlet Indian Paintbrush belongs to the leafy bracts, not the flower petals. Flowers appear in the axils of the bracts, which form a dense spike at the end of a hairy stem. Often, the stigma-tipped style projects past the corollas. Oval leaves at the base of this native plant create a rosette, while those on the stem are narrowly divided. The Scarlet Indian Paintbrush leads a partially parasitic life, appropriating nourishment from other plants by fastening its roots onto theirs. Other *Castilleja* species are very common in western North America.

Seaside Petunia

Calibrachoa parviflora H to 12 in (30 cm)

Seaside Petunias come in a variety of colors, red being one of them. Despite the name, the species is also found inland in habitats with sandy soil.

KEY FACTS

The small 5-lobed trumpet-shaped flower appears in red, pink, purple, and other colors.

+ **habitat:** Dry streambeds, stream banks, washes, and wetlands

+ **range:** Native to southwestern North America; introduced elsewhere; also highly cultivated

+ **bloom period:** April–September

A member of the Nightshade family, the Seaside Petunia is a sprawling, low-growing species that forms a dense mat, sending down new roots at the leaf nodes as it creeps along the ground. The plentiful evergreen leaves are lance shaped, fleshy, and very sticky; the flowers grow from their axils. The species was first classified as a true petunia, but based on DNA evidence, it was put into the genus *Calibrachoa*, named for the 19th-century Mexican botanist Antonio de la Cal y Bracho, a switch that is not always recognized. Both *Petunia* and *Calibrachoa* are considered basically South American genera.

Cardinal Flower
Lobelia cardinalis H 12–60 in (30–150 cm)

Growing tall on erect stems, the bright red blossoms of the Cardinal Flower signal a bounty of nectar to the Ruby-throated Hummingbird, which is one of the plant's chief pollinators.

KEY FACTS

The bright red, tubular flowers have a 2-lobed upper lip and a 3-lobed bottom one.

+ habitat: Moist areas, especially along streams and ponds

+ range: Throughout much of the U.S. and Canada, except in the Northwest and in far northern areas

+ bloom period: July–September

The flowers of the Cardinal Flower, a member of the Bellflower family, grow in loose to dense, elongated spikes. A flower's stamens are fused together and protrude above the tube. Alternate, lance-shaped leaves line the tall stalk at close intervals. The lower leaves have stalks and are toothed; the upper leaves are smoother and clasp the stem. A perennial that prefers a moist footing, the Cardinal Flower often sends up new plants from its creeping, underground rootstalk. The closely related Great Blue Lobelia (*L. syphilitica*) is similar in almost all respects except the color of its flowers, and the two hybridize occasionally.

Blood Sage/Scarlet Sage
Salvia coccinea H 12–24 in (30–60 cm)

Square stems with showy whorls of flaming scarlet blossoms make Blood Sage impossible to overlook. A long blooming season adds to its prominence.

KEY FACTS

The red tubular flower has 2 lips: a small upper one with 2 lobes and a larger bottom one with 3.

+ habitat: Sandy sites in open woods, woodland edges, thickets, and disturbed areas

+ range: Mainly on Coastal Plain of southeastern U.S.; highly cultivated and sometimes escaping farther northward

+ bloom period: February–October

The striking flowers of Blood Sage, also commonly known as Salvia, appear in well-separated whorls, forming an elongate, loose cluster atop a straight stem. This member of the Mint family has bright green, roughly triangular leaves that grow opposite one another. Of the many native *Salvia* species, Blood Sage is uncommon in having red flowers, a feature that of course makes it very popular with hummingbirds. Its foliage is pungent, however, which discourages some herbivores such as the White-tailed Deer. *Salvia coccinea* has been developed into many flashy cultivars.

Virginia Spiderwort
Tradescantia virginiana H 8–24 in (20–60 cm)

The deep purplish blue blossoms of the Virginia Spiderwort live very short lives, blooming only for a day before the petals begin to contract and form gummy blobs by evening.

KEY FACTS

Flower has 3 broad oval petals and 3 hairy sepals above leaflike bracts; 6 orange-tipped stamens with long, hairy filaments provide contrast.

+ habitat: Woods, woodland edges, prairies, meadows, stream banks, roadsides, and disturbed areas

+ range: Eastern North America and California

+ bloom period: April–August

Clusters of Virginia Spiderwort flowers grow at the end of rounded stems that can be solitary or branched at the base of the plant. Long, narrow leaves with a pronounced midrib form an arched clump. The plants multiply mainly by underground runners that root and then send up new plants. There are a number of proposed explanations for this wildflower's common name. One is that the leaf growth pattern resembles a crouched spider; another suggests that the fine filaments of the stamens look like a spider's web. Yet a third attributes the name to a threadlike secretion that flows from the cut stem.

Eastern Bluestar/Blue Dogbane
Amsonia tabernaemontana H 12–36 in (30–90 cm)

The rather unwieldy scientific name of the Eastern Bluestar honors the 16th-century botanist Jacobus Theodorus Tabernaemontanus, considered by many the father of German botany.

KEY FACTS

Star-shaped flower has 5 blue, spreading lobes above a slender tube, with a white or yellow patch at the base of each lobe.

+ habitat: Open woods, thickets, meadows, low prairies, marshes, stream banks, and roadsides

+ range: South-central and eastern U.S., except in some northeastern areas

+ bloom period: March–July

Eastern Bluestar's easily recognized sky blue flowers appear in loose clusters on branching stems. The stems are dense with alternate, willowlike, lance-shaped, green leaves that turn yellow in the fall. After the flowers bloom, two cylindrical seedpods up to 5 inches (12.5 cm) long replace each of the pollinated flowers. At maturity, the pods split along the side to reveal a row of cylindrical seeds. The plant reproduces by self-seeding. The species, which contains a toxic, milky sap in its leaves and stems, also goes by the name Blue Dogbane and is a member of the Dogbane family.

Round-lobed Hepatica

Anemone americana H 2–6 in (5–15 cm)

What look like colorful petals on the Round-lobed Hepatica, a member of the Buttercup family, are actually its sepals. They can vary in number, but six is a frequent count.

KEY FACTS

The lavender, blue, pink, or white flower is formed of 5–12 sepals with many white stamens and is supported by 3 green bracts.

+ **habitat:** Woods, rocky slopes, and bluffs

+ **range:** Eastern North America, except in far northern areas

+ **bloom period:** February–April

The solitary, hairy flower stalk of the Round-lobed Hepatica rises from a base of leaves with three rounded lobes that inspired the name Hepatica, which refers to the liver. These leaves can have mottling above and a purplish cast underneath. They grow along the ground on their own stalks, appear mostly after the flowers have bloomed, and persist through the winter before withering. A similar species, the Sharp-lobed Hepatica (*A. acutiloba*), has both pointier bracts and leaves and occurs within roughly the same range. These two species historically were assigned to the genus *Hepatica*.

Pasqueflower/Prairie Smoke

Anemone patens var. *multifida* H 4–16 in (10–40 cm)

The Pasqueflower receives its common name from the fact that its blossoms often emerge during the celebration of Easter, which was known in an old French form as Pasque.

KEY FACTS

The blue to purple flower has 5–8 petal-like, pointed sepals and numerous yellow stamens.

+ **habitat:** Prairies, meadows, and rocky areas

+ **range:** Northern and central U.S. and Canada into western Canada

+ **bloom period:** March–June

The hairy, leafless flower stalk of the Pasqueflower rises from under the ground in early spring. Its large, solitary crocus-like flower—reaching 3 inches (7.5 cm) across when open—begins to bloom as the foliage starts to form. The basal leaves of the plant are silvery, deeply divided, and fernlike, and the bracts beneath the flower are hairy and whorled with linear lobes. The foliage and flower stalks continue to grow after the flower has bloomed. Attractive silky and feathery seed heads replace the blossoms, giving it the alternate name Prairie Smoke. A cultivated variety from Eurasia (*A. patens patens*) has a similar appearance but a wider range of flower colors.

Colorado Blue Columbine

Aquilegia coerulea H 12–24 in (30–60 cm)

This perennial member of the Buttercup family serves as Colorado's state flower and has been taken to the max in cultivation, including forms with double blossoms.

KEY FACTS

The flower has 5 deep blue petal-like sepals and 5 white petals, each with a backward-pointing blue spur.

+ **habitat:** Moist woods and meadows

+ **range:** Mid- to higher elevations of Rocky Mountain states in the U.S.; highly cultivated

+ **bloom period:** June–August

The showy blossoms of the Colorado Blue Columbine often face proudly upward. They appear in bushy clumps with fernlike foliage in the form of compound leaves that are deeply lobed and cleft. There is considerable color variation in the species. Flowers to the north and west of the native range tend to be paler, going even into white forms. Also, native flowers at higher altitudes tend to be a deeper blue than those at lower altitudes. The genus name derives from the Latin for "eagle," and refers to the flower's long spurs that are thought to resemble an eagle's talons. The species name refers to the flower's blue color.

Tall Blue Larkspur

Delphinium exaltatum H 48–72 in (120–180 cm)

The "spur" in the name Larkspur refers to the fact that two of the blossom's petals and one of its petal-like sepals form a single, cone-shaped prong, or spur, that projects backward.

KEY FACTS

Complex purplish blue flower has 4 petals that form a nectar tube at their base, and 5 hairy, petal-like sepals.

+ **habitat:** Open woods, rocky slopes, and limestone bluffs

+ **range:** Scattered native distribution in parts of eastern U.S.; highly cultivated

+ **bloom period:** July–September

Tall Blue Larkspur flowers, which form long clusters, have a lot going on. In addition to their distinctive spurs, they have notched upper petals and bearded, two-lobed lower ones. Lateral sepals have a greenish spot in the center, and lower and upper sepals have a greenish spot at the tips. The three-lobed leaves sometimes have divided basal lobes and are widely dispersed on the stem. This perennial member of the Buttercup family ranks as the tallest of the native larkspurs and typically presents more luxurious foliage than other species. The toxic alkaloids it contains present a danger to grazing cattle.

Great Blue Lobelia
Lobelia siphilitica H 12–48 in (30–120 cm)

Blooming late in the season, the striking flower clusters of the Great Blue Lobelia stand out in their rich, moist habitats.

KEY FACTS

The bright blue or lavender blue tubular flower has a 2-lobed upper lip and a 3-lobed lower lip, with 2 white bumps on the lower lip.

+ **habitat:** Woods, meadows, stream banks, swamps, and ditches

+ **range:** Throughout much of central and eastern U.S. and Canada

+ **bloom period:** July–October

The flowers of the perennial Great Blue Lobelia form long, slender clusters at the ends of tall, erect, leafy stems. The oval to lance-shaped leaves grow up to 5 inches (12.5 cm) long; they lack stalks and are alternate on the stems, which seldom are branched. This member of the Bellflower family is similar to the red Cardinal Flower (*L. cardinalis*) in most aspects except color and some details of flower structure, and sometimes hybridizes with it. Many bee species seek out its nectar and pollen. The plant's species name comes from the erroneous belief that it could cure syphilis; in fact, it contains several toxic alkaloids.

Wild Lupine/Sundial Lupine
Lupinus perennis H 8–24 in (20–60 cm)

Lupines get their name from the Latin for "wolf," a misnomer bestowed when it was believed these members of the Pea family devoured minerals from the soil.

KEY FACTS

The blue flower has 5 petals: a large upper banner petal, 2 forward-projecting side wing petals, and 2 lower ones called a keel that are fused at the base.

+ **habitat:** Dry open woods, woodland edges, and fields

+ **range:** Much of eastern U.S. and Canada; also highly cultivated

+ **bloom period:** April–July

Wild Lupine forms elongated flower clusters up to 10 inches (25 cm) long at the top of erect, branching stems. The clusters tend to be looser than those of the Texas Lupine (*L. texensis*) with its crowded flower stalks. The flowers are typically blue and less commonly white or pink. The Wild Lupine's leaves are palmately divided into seven to eleven leaflets that are hairy on the undersides, as are the stems. This species also is known as the Sundial Lupine. Far from depleting the fertility of the soil, lupines enhance it by fixing nitrogen in their roots.

Texas Lupine/Texas Bluebonnet
Lupinus texensis H 6–18 in (15–45 cm)

The Texas Lupine, better known as the Texas Bluebonnet, received its traditional common name because its flowers resemble the sunbonnets that pioneer women wore.

KEY FACTS

The 5-petaled pea-like blue flower has a white spot on its upper petal that turns dark purplish as the flower ages.

+ **habitat:** Prairies, open fields, and roadsides

+ **range:** Texas, Oklahoma, Louisiana, and Florida; also highly cultivated

+ **bloom period:** March–May

Large, elongated flower clusters dense with pealike flowers top the branching stems of the Texas Lupine, which rise from a basal rosette of leaves. The topmost flowers (not just the central spots) on the dense clusters are conspicuously white. The leaves of the plant are light green and velvety, divided palmately into five pointed leaflets. Long, green pods replace the flowers. When the seeds inside these mature, the plant will die back. The Bluebonnet is the state flower of Texas, but the designation represents all the native *Lupinus* species there, resulting in multiple state flowers, *L. texensis* being just one of the honored plants.

Hookedspur Violet/Western Dog Violet
Viola adunca H 3–6 in (7.5–15 cm)

Hookedspur Violet flowers do not grow from their stems or roots, like other violet species. Instead, they appear from the plant's leaf stalks.

KEY FACTS

Blue-violet flower has 5 petals; the side petals have bearded tufts and the lower petals have dark nectar guides and a long, curved spur.

+ **habitat:** Meadows, open woods, slopes, and stream banks

+ **range:** Throughout northern and western U.S. and Canada

+ **bloom period:** April–August

The flowers of the Hookedspur Violet grow on stalks from the axils of the plant's leaves. The leaves themselves are heart shaped to oval and finely but bluntly toothed and are alternate on the stems. The lower flower petal's long nectar spur may curve and appear above the top flower petals. A number of fritillary and other butterfly species and their larvae favor this violet for its nectar and foliage. Like some other Violet species, the Hookedspur Violet produces small, inconspicuous cleistogamous flowers that never open, but produce viable seeds by self-pollinating. The name Dog Violet apparently arose to separate it from the sweet-scented violets.

Birdfoot Violet

Viola pedata H 3–10 in (8–25 cm)

Birdfoot Violets are named for their three-lobed, deeply divided leaves with secondary lobes that widen toward the tips, a structure that resembles a bird's foot.

KEY FACTS

The blue, purple, or lavender (sometimes two-toned) beardless flower is composed of 5 petals and 5 stamens.

+ **habitat:** Open woods, prairies, rocky slopes, and along roadsides

+ **range:** Throughout much of central and eastern North America

+ **bloom period:** March–June

The flowers of the Birdfoot Violet rank as one of the largest among the blue violets—up to 1.5 inches (3.75 cm) across—their blueness offset by a bright orange center composed of a cluster of conspicuous stamens. The plant has no aerial stems; instead, its leaves and flower stalk emerge directly from the rootstock. A striking variation of the species has dark purple upper petals and pale violet lower ones. The Birdfoot Violet's coppery seeds produce a sugary, ant-attracting gel that encourages the insects to carry them off and disperse them. The species sometimes continues to flower into the fall.

Common Morning Glory

Ipomoea purpurea L vining to 15 ft (4.5 m)

The genus name of the Common Morning Glory derives from Greek words meaning "like a worm," referring to the twining growth of the plant.

KEY FACTS

The funnel-shaped blue, purple, pink, white, or variegated flower has 5 united petals and a style with a 3-parted stigma.

+ **habitat:** Fields, fence lines, along roadsides, and in disturbed areas

+ **range:** Introduced and naturalized throughout much of the U.S. and eastern Canada

+ **bloom period:** June–September

Common Morning Glory plants with their large, trumpet-shaped flowers climb over anything natural or built that provides support, or they otherwise sprawl over open ground. The plants have slender stems that are somewhat hairy, with large, usually heart-shaped, bright green alternate leaves on long slender stalks. Small flower clusters appear from the axils of some of the leaves. The blossoms open in the morning and last only one day, and are replaced by round seed capsules. Introduced from South America, the species has proven very adaptable and prolific, though not as aggressive in its spread as some other morning glories.

Distant Phacelia
Phacelia distans H 6–36 in (15–90 cm)

These pretty blue flowers of arid regions carry a toxic secret: Handling the plants without gloves can cause a severe case of contact dermatitis.

KEY FACTS

The pale blue, funnel- or bell-shaped corollas have 5 rounded lobes and 5 delicate pro- truding stamens.

+ habitat: Desert, washes, slopes, and sandy roadsides

+ range: Southwest- ern U.S.

+ bloom period: February–June

Distant Phacelia flowers emerge in a hairy, coiled terminal cluster, on only one side of the stem. The curl tends to straighten out as the plant transitions from flower to fruit. The leaves of the species are hairy, pinnate, and fernlike, and they grow on hairy, reddish stems that can be erect or lie along the ground. Distant Phacelia is one of about a hundred *Phacelia* species in the western United States, many of them quite similar. Due to the shape of its curled flower coil, a distinguishing feature, it also goes by the name Scorpionweed.

Wild Blue Phlox
Phlox divaricata H 10–20 in (25–50 cm)

Native perennial Wild Blue Phlox spreads on creeping stems that send up delicate, fragrant flowers and dense, green foliage on rich woodland floors.

KEY FACTS

The trumpet-shaped flower of blue, laven- der, or white has 5 flat lobes that can be notched, and stamens and pistils contained within the slender tube.

+ habitat: Woods, fields, and stream banks

+ range: Throughout central and eastern U.S. and Canada; highly cultivated

+ bloom period: March–May

The hairy and sticky stems of Wild Blue Phlox have narrow, lance-shaped leaves that send down roots at their nodes and send up stems with loose, flat-topped terminal clusters of sweetly fragrant flowers. They can reproduce by cloning, forming extensive mats over time. The plants also give rise to infertile shoots that are usually shorter and have more rounded leaves, but do not pro- duce flowers. After flowering, ovoid seed capsules containing small seeds form on the fertile stems, which die back. Infertile shoots persist through the winter, storing energy in their roots that fuels next spring's crop of fertile shoots.

Heal-all/Self-heal

Prunella vulgaris H 12–36 in (30–90 cm)

The two-lipped flowers of the Heal-all mature a few at a time, beginning at the bottom of the dense cylindrical flower head. This prolongs the bloom time of this member of the Mint family.

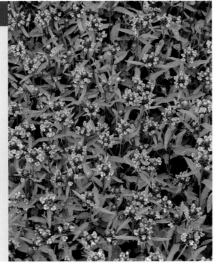

KEY FACTS

The small, 2-lipped purple- to blue-and-white flowers have hairy bracts that form a compact spike at the end of stems.

+ **habitat:** Fields, roadsides, and along woodland edges

+ **range:** Throughout much of the U.S. and Canada, except in far northern areas

+ **bloom period:** June–September

Heal-all flowers appear in profusion on erect flower heads. The leaves are lance shaped, untoothed or lightly toothed, and grow opposite each other; those near the top of the plant often lack stalks. The spikes continue to grow longer after flowering. Heal-all spreads from rhizomes that send up square, red-tinged stems. Introduced from Europe, with a native variety also present, Heal-all has a long tradition of use in treating wounds and afflictions of the throat; it has additional common names of Self-heal, Carpenter's Weed, and Woundwort. In traditional Chinese medicine, Heal-all was commonly used to treat liver and kidney ailments.

Virginia Bluebell

Mertensia virginica H 8–24 in (20–60 cm)

The Virginia Bluebell is beautiful in and of itself, with its nodding clusters of light blue, trumpet-shaped flowers, but to add to the delight, it tends to grow in local profusion.

KEY FACTS

The inch-long (2.5 cm), funnel-shaped flowers have 5 lobes, and the edges of the sepals are fringed.

+ **habitat:** Moist woods, stream banks, floodplains, and wet meadows

+ **range:** Throughout much of eastern U.S. and Canada; also cultivated

+ **bloom period:** March–June

The Virginia Bluebell's striking blue blossoms unfold from pink buds amid thick gray-green foliage, making a plant in transition a very colorful sight. The stem leaves are oval and have smooth edges, and those at the base are longer, up to 8 inches (20 cm). Although the species is particular about the moistness of its habitat, it rewards a suitable venue with an extensive display. *Mertensia* species are often called Lungwort for their similarity to a European species used to treat lung disease. In North America, it is also known as Virginia Cowslip. Butterflies favor this species and aid in its pollination.

Allegheny Monkeyflower

Mimulus ringens H 12–36 in (30–90 cm)

The Allegheny Monkeyflower's comical appearance inspired the genus name *Mimulus,* derived from the Latin *mimus,* referring to a mime, a comic actor, or a buffoon.

KEY FACTS

The light purple to blue flower has a 2-lobed top lip and a 3-lobed bottom lip; both lips are broad and wavy.

+ **habitat:** Wet meadows, stream banks, and pond edges

+ **range:** Throughout much of the U.S. and Canada, except southwestern U.S. and in far northern areas

+ **bloom period:** June–September

Leaves greatly outnumber flowers on the Allegheny Monkeyflower plant. They are lance shaped, toothed, and grow opposite each other, clasping the smooth, square, hollow stems. Sparse blossoms arise from slender stalks in the axils of upper leaves. The flower displays two bright yellow spots in its throat that may serve as a signal to bees. The species keeps small insects from its limited supply of nectar by closing off access to the throat unless the heftier weight of a pollinating bee on the blossom's lower lip creates an opening. A relative is the Common Monkeyflower (*M. guttatus*), described on page 19.

Fringeleaf Wild Petunia/Hairy Ruellia

Ruellia humilis H 12–24 in (30–60 cm)

The Fringeleaf Wild Petunia, an endangered native species, resembles a petunia but belongs to the Acanthus family, and the genus is named for 16th-century French herbalist Jean de la Ruelle.

KEY FACTS

The funnel-shaped flowers about 2.5 in (6 cm) long have 5 lobes and range in color from lavender to blue to purple.

+ **habitat:** Prairies, woodland edges and openings, and thickets

+ **range:** Most of the eastern U.S., except the Northeast

+ **bloom period:** May–October

The blossoms of the Fringeleaf Wild Petunia have long tubes and flared lobes often finely lined with purple to guide hummingbirds and other pollinators to the nectar inside. The flowers develop without stalks on angled, green, hairy stems. The stems bear opposite, broadly oval, smooth-edged and stalkless leaves, which are covered with white hairs above and below. The stalkless flowers and leaves give the plant a bushy appearance. All told, the species really earns the common name Hairy Ruellia. The flowers commonly open during the morning, only to fall off the plant by the end of the day.

Harebell
Campanula rotundifolia H 6–24 in (15–60 cm)

The Harebell is a hardy wildflower that belies the delicate appearance of its slender stems and drooping blossoms, thriving in such challenging habitats as high altitudes in the mountains.

KEY FACTS

The delicate 5-lobed blue-violet bell-shaped flowers nod on thread-like stems.

+ **habitat:** Meadows, woods, rocky slopes, sand shorelines

+ **range:** Throughout nearly all of the U.S. and Canada, except the southeastern U.S.

+ **bloom period:** June–September

The Harebell is a member of the Bellflower family, and the genus has many members with bell-shaped or conical flowers that character-istically nod from slender stalks. In the Harebell's case, the stalks are very slender and fragile look-ing; the flowers grow on them singly or in clusters. Slender, straight leaves on most of the plant stem give way to rounded ones at the base, the source of the species name *rotun-difolia*. As the plant matures, the basal leaves may wither and disappear. This perennial Harebell forms small patches, not large stands. The plant's genus name is from the Latin, meaning "little bell."

Prairie Flax/Lewis's Flax/Wild Blue Flax
Linum lewisii H 12–24 in (30–60 cm)

A close relative of the cultivated flax species from which linen and linseed oil are made, Prairie Flax provides nutritious forage and seeds for livestock, birds, and other wildlife.

KEY FACTS

The pale blue flowers about 1.5 inches (4 cm) across with darker blue veins have 5 pet-als, 5 sepals, and 5 stamens; they grow on multiflowered stems.

+ **habitat:** Grass-lands, meadows, open woods, and mountain brush

+ **range:** Throughout Canada and western U.S.; also a few popu-lations in eastern U.S.

+ **bloom period:** March–September

Attractive and highly symmetrical Prairie Flax flowers grow in profusion in loose clusters. The plants have numerous long, narrow leaves that mostly grow alternate on the stem, but some-times appear opposite one another. As the flower clusters develop and bloom, the stems remain green but begin to shed their leaves. The top-heavy flower stems rarely stand erect but instead lean. The Prairie Flax is used widely to control erosion, to beautify land-scapes, and to restore native plant biodiversity. This species is able to withstand fires and resprout afterward. It is named for Meriwether Lewis, who collected a specimen in 1806.

Common Chicory

Cichorium intybus H 24–72 in (60–180 cm)

Roadsides and other disturbed places are replete with the erect-stemmed Common Chicory, punctuated at intervals with striking blue radiating flowers. It is a member of the Aster family.

KEY FACTS

The stalkless blue flower has numerous strap-shaped florets, the inner ones shorter, with toothed ends.

+ habitat: Roadsides, fields, and disturbed places

+ range: Naturalized throughout much of the U.S. and Canada, except in far northern areas

+ bloom period: June–October

Common Chicory's sky blue flower heads open several at a time on a stem and are ephemeral, lasting only one day. Small, inconspicuous oblong leaves clasp the stem; leaves at the base are jagged and dandelion-like. The plant exudes a milky white sap. The species has a long history as human food and animal fodder, and as a remedy for ailments of the liver, kidneys, and stomach. The French popularized the addition of the dried and ground root to coffee to enhance flavor and to stretch a quantity of the expensive beans. Today, prolific Common Chicory often is regarded as an undesirable weed.

Water Hyacinth

Eichhornia crassipes H 3–36 in (7.5–90 cm)

The floating Water Hyacinth, with its attractive flowers and dense foliage, forms immense rafts, clogging lakes and waterways and impeding water traffic.

KEY FACTS

Six-lobed purplish blue flowers have a distinctive yellow eye on the upper central lobe.

+ habitat: Fresh water, including marshes, bayous, lakes, streams, rivers, and ditches

+ range: Native to South America; naturalized mainly in southeastern U.S. and eastern Canada

+ bloom period: June–September or all year, depending on climate

Water Hyacinth blossoms, some 2 inches (5 cm) wide, appear on stalks amid shiny green, fleshy leaves. The leaf blades are round to kidney shaped and grow on stalks up to 3 feet (90 cm) tall from spongy, inflated bulbous bases. Roots dangle beneath the water's surface. At one time, the Water Hyacinth, a relative of the native Pickerelweed, was considered a desirable ornamental plant. But the species reproduces not only by seeds, but also rapidly by means of offsets—a method that can double the size of a plant mass in as few as six days. This has made it one of the most reviled invasive plants.

Greater Fringed Gentian

Gentianopsis crinita H 12–36 in (30–90 cm)

One of the last wildflowers to bloom in the year, the Greater Fringed Gentian's arresting deep blue flowers sometimes witness the first snowfalls in northern and mountainous parts of its range.

KEY FACTS

The bright blue corollas with 4 heavily fringed, flaring lobes above a tube grow at the end of an erect, often branching stem.

+ **habitat:** Moist meadows, forest edges, and stream banks

+ **range:** Mainly north-central and northeastern U.S. and Canada

+ **bloom period:** August–November

The Greater Fringed Gentian flower follows the sun: It opens in the morning on sunny days and closes in the evening, the buds appearing spirally twisted. Four folded sepals, two shorter than the others, clasp the corolla tube. Leaves are mostly lance shaped and rounded at the base. A rosette of leaves forms the first year; the flowering stalks appear in the second. The plant's common and scientific names tie it to King Gentius, a second-century B.C. king of Illyria on the Balkan Peninsula who championed its medicinal properties that centered on treating digestive ailments. The species' exquisite beauty has led to its decline through collecting and overpicking.

Virginia Iris/Southern Blue Flag

Iris virginica H 12–24 in (30–60 cm)

After the Virginia Iris flowers, its weak stems sometimes collapse, leaving a jumble of stems, leaves, and spent flowers on the ground. This species is a smaller version of *I. versicolor.*

KEY FACTS

The flower has 3 purple-blue petals and 3 petal-like sepals that are fused below; sepals have a greenish yellow nectar guide at the base.

+ **habitat:** Marshes, swamps, wet meadows, edges of ponds, lakes, streams, and ditches

+ **range:** Found mainly in southeastern and south-central U.S.

+ **bloom period:** May–June

The leaves of the Virginia Iris usually grow higher than the species' showy blooms on their sometimes branched stems. Long, narrow, and folded, the stems fan out from the plant base to form a V in cross section. The leaf shape accounts for the plant's common name; it derives from the Middle English *flagge,* which means "rush" or "reed." The conspicuous signal patches on the blossoms attract pollinators to the nectar. The species reproduces by rhizomes and by seeds. A variety of this species, Shreve's Iris (*I. virginica* var. *shrevei),* has more branches, and the stems usually do not collapse after flowering.

Blue-eyed Grass

Sisyrinchium angustifolium H 4–18 in (10–45 cm)

Examined closely, the Blue-eyed Grass's grasslike foliage looks more like a member of the Iris family than a grass. Unlike many other irises, these leaves may be shorter than the flower stalk.

KEY FACTS

The flowers have 3 each of pointed violet-blue petals and sepals, and yellow centers.

+ **habitat:** Damp meadows, low road-sides, prairies, old fields, damp fields, open woods, and ditches

+ **range:** Throughout much of central and eastern U.S. and Canada

+ **bloom period:** March–July

The star-shaped flowers of the Blue-eyed Grass grow at the top of long, flat, twisted, and often branched stems. The flower stalks appear amid a tight clump of bright green, linear leaf-like bracts that are about a quarter inch (0.6 cm) wide. Usually only one flower is in bloom on a stalk at a time. When the plant withers, its foliage turns black. The Cherokee, Iroquois, and other American Indian groups used the leaves and roots of the plant to treat both stomach and intestinal difficulties, including bouts of worms. A distinct yellow "eye" gives the unassuming flower a bit of pizzazz. Although its leaves are very narrow, this is not a true grass.

Old-field Toadflax/Blue Toadflax

Nuttallanthus canadensis H 6–24 in (15–60 cm)

Old-field Toadflax is partial to sandy soil and disturbed areas, seeming to choose sites that present little competition from other plants. The species is closely related to Snapdragons.

KEY FACTS

Half-inch-long (1.3 cm) blue to blue-violet 2-lipped flowers have 2 white ridges on the lower lip; the flower base shows a slender, curved spur.

+ **habitat:** Open, dry, rocky sites, sandy fields, roadsides, and open woods

+ **range:** Most of central and eastern U.S. and Canada and U.S. West Coast

+ **bloom period:** March–September

The blossoms of the Old-field Toadflax grow from short stalks in an open spikelike cluster in the upper portion of a stem that rarely branches. The leaves of this species are more than an inch (2.5 cm) long and are fairly linear and smooth. They have two growing strategies: Very narrow leaves appear alternately on erect plant stems, and others grow opposite each other on prostrate stems. Both the flowering stalks and the leaf stalks are green to reddish green and smooth. Only a few flowers bloom at one time, and a small colony of the plants may bloom continuously for two or three months.

American Brooklime
Veronica americana H 6–36 in (15–90 cm)

An exceptionally pretty member of the group known as Speedwells or Veronicas, the American Brooklime has stout but weak stems that take root at leaf joints when they touch the ground.

A delicate-looking plant, the American Brooklime bears loose, elongated 2- to 6-inch (5–15 cm) clusters of tiny flowers on branches rising upright from the leaf joints of its upper leaves. The leaves are oval to lance shaped and usually slightly toothed; those on the upper flowering stems clasp the stems at their base. This species is in a group called Speedwell for a reason: Much of its growth sprawls rapidly along the ground, punctuated at intervals by the smooth and hollow flower-bearing branches. The American Brooklime frequently grows partly in and partly out of the water.

One-flowered Cancer Root/One-flowered Broomrape
Orobanche uniflora H 3–10 in (7.5–25 cm)

Each stalk of One-flowered Cancer Root sends up one blossom. The parasitic species does not photosynthesize and obtains nourishment by tapping into the roots of its host plants.

One-flowered Cancer Root has scaly, short, and mostly subterranean stems that send up one to four flower stalks, each bearing a single flower. The stem may measure no more than 1.5 inches (4 cm). The plant does not have true leaves; instead, it has small brown scales at the base of the flower stalk. The species hijacks nutrients from a host plant and it parasitizes a wide variety of other plant species. A look around at the other plants surrounding a One-flowered Cancer Root creates a mystery as to which of the species might be hosting the boldly flourishing parasite.

Azure Bluet

Houstonia caerulea H 3–8 in (7.5–20 cm)

Perfection in miniature, slender stems, tiny leaves, and dainty, delicately tinted blossoms distinguish Azure Bluet, a member of the Madder family.

KEY FACTS

The pale blue, trumpet-shaped flowers have 4 spreading lobes and deep yellow centers, and grow singly on unbranched stems.

+ habitat: Fields, meadows, woods, and disturbed areas

+ range: Throughout much of eastern U.S. and Canada, excluding Florida

+ bloom period: April–July

Azure Bluet flowers are mostly less than a half inch (1.25 cm) wide, but they make a big impression, especially as they tend to grow in large patches. They are also named Quaker Ladies for their demure appearance. The plant's genus name refers William Houstoun, an early 18th-century Scottish botanist who collected plants mainly in the American tropics. Although the leaves on most Bluet stems are tiny and sparse, those at the base of the plant are larger and spatula shaped, and they grow in tufts. The Cherokee made an infusion from the plant, which they administered to prevent bed-wetting.

Woodland Forget-me-not

Myosotis sylvatica H 5–12 in (12.5–30 cm)

The Woodland Forget-me-not was certainly not forgotten: The introduced Eurasian native escaped from cultivation and now is naturalized in a number of areas of North America.

KEY FACTS

Small blue to white flowers have 5 spreading lobes that emerge from short tubes; the flowers grow in a rounded cluster.

+ habitat: Moist woods, stream banks, swamps, ditches, and roadsides

+ range: Naturalized in disparate areas mostly in northern and central U.S. and Canada

+ bloom period: April–September

The bractless flower clusters of the Woodland Forget-me-not grow on erect stems that branch from the base of the plant. The leaves near the base are longer and wider than those higher up the stem. The genus name *Myosotis* comes from the Greek for "mouse ear," referring to the various species' furry leaves. In the language of flowers, a presentation of Forget-me-nots signifies true love and remembrance through a legendary association with a tragic lost love. Despite this endearing allusion, the species perpetuates itself by aggressively self-seeding, which increases its potential as an undesirable invasive in some regions.

Common Arrowhead

Sagittaria latifolia H to 36 in (90 cm)

These aquatic perennials form large colonies in watery environments. Their genus name comes from the Latin for "arrow," a reference to the distinctive leaf shape.

KEY FACTS

The white flower is formed of 3 white to pinkish petals and 6 or more stamens; pistils are on separate flowers.

+ habitat: Aquatic habitats such as marshes, swamps, ponds, mud banks, wet sand, and ditches

+ range: Throughout most of U.S. and Canada, except far northern areas

+ bloom period: July–September

Leaves are all basal, with long stalks and large, mostly arrowhead-shaped green blades that vary in width and emerge from the water. Underwater, the leaves may be long and narrow. The emergent leaves are followed by taller stalks with whorls of three white flowers. Some variants display short hairs over much of the plant. The Common Arrowhead's starchy tubers, developing under the mud, are favored by ducks and muskrats and give the plant one of its alternate common names: Duck-potato. The tubers were also an important source of food and medicines for American Indians, who called the plant Wapato.

Soapweed Yucca

Yucca glauca H 24–48 in (60–120 cm)

Soapweed Yucca played many roles in traditional American Indian hygiene and health. Soap and dandruff shampoo were made from the roots, and a solution was used to kill lice in hair.

KEY FACTS

The flower is greenish white and bell shaped and has 6 pointed tepals (look-alike petals and sepals).

+ habitat: Prairies and waste areas at low to middle elevations

+ range: Central U.S. into Alberta; escapes from cultivation

+ bloom period: June–August

A member of the Asparagus family (previously in the Agave family), Soapweed Yucca often forms small colonies. First, a crowded clump of daggerlike, fibrous, evergreen leaves with hairy edges is established. Later, a fast-growing flower stalk to more than 4 feet (120 cm) tall rises from the center of the basal rosette. It is densely but loosely populated with pendent, bell-like blooms on short stalks. Each species of Yucca relies on a symbiotic partnership with its corresponding species of Yucca Moth for reproduction. In exchange for transfer of pollen, the Yucca flower and seedpods offer shelter and food for the moth's eggs and larvae.

Smooth Solomon's Seal
Polygonatum biflorum H 8–84 in (20–210 cm)

"Smooth" is an apt word for identifying this species, referring to its smooth leaves. The almost identical Hairy Solomon's Seal (*P. pubescens*) has minute hairs on the underside of its leaves.

KEY FACTS

The pendent 6-lobed tube, whitish to greenish yellow, is composed of 6 tepals, which are united at their bases.

+ habitat: Rich woods, old fields, thickets, and along roadsides

+ range: Central and eastern U.S. and Canada

+ bloom period: March–June

Flowers usually dangle in pairs from leaf axils on an erect or arching stem that has many smooth, stalkless, parallel-veined, lance-shaped leaves. Thick, fleshy, and gnarly rootstocks send up new stems each year. When these die back, they leave noticeable scars on the roots. These are the "seals," resembling the signet ring of King Solomon, and counting them gives a good estimate of a plant's age. A variety, *P. biflorum* var. *commutatum*, or Great Solomon's Seal, has proportionately larger flowers and leaves, and occasionally may grow to more than 6 feet (180 cm) tall.

Large-flowered White Trillium
Trillium grandiflorum H 8–18 in (20–45 cm)

This trillium boasts the largest flower of all *Trillium* species. It measures up to 4 inches (10 cm) wide and turns from white to light pink as it ages.

KEY FACTS

The flower has 3 large, oval, white petals with wavy edges; 3 green, lance-shaped sepals; and 6 yellow stamens.

+ habitat: Moist woods, swamps, floodplains, and roadsides

+ range: Throughout much of eastern North America; also highly cultivated

+ bloom period: April–June

The showy blossom of the Large-flowered White Trillium, a member of the Bunchflower family (previously in the Lily family), rises on a central stalk above a whorl of three terminal leaves. The leaves—in reality, leaflike bracts—measure up to 6 inches (15 cm) long and have smooth margins and prominent parallel veins. There are variations of this species involving flower color or markings, including some that may be related to the presence of an infection by bacterial microorganisms. American Indians used the roots of the plant to treat a number of ailments. Trilliums of any species should be left undisturbed in their natural habitat.

Gunnison's Mariposa Lily/Gunnison's Sego Lily

Calochortus gunnisonii H 6–18 in (15–45 cm)

The genus name for this wildflower, *Calochortus*, derives from the Greek for "beautiful grass," referring to the plant's grasslike leaves, which wither as the flowers bloom.

KEY FACTS

The flower has 3 white, or sometimes purple, wide petals, each with a hairy, yellow glandular base bordered above by a purple line.

+ **habitat:** Open woods, prairies, dry gulches, and mountain meadows

+ **range:** Western U.S. in the Rocky Mountain and adjacent states

+ **bloom period:** May–July

The tulip-like Gunnison's Mariposa Lily, also known as Gunnison's Sego Lily, emerges from a bulb along with its linear basal leaves. One to three flowers may appear on each unbranched stem. The species occurs as scattered plants or in expansive colonies that make a dramatic impression. Meriwether Lewis collected the first *Calochortus* lily specimen for scientific study in Idaho in 1806. Various American Indian groups used the bulb of *C. gunnisonii* for food and medicine. The namesake of the species, Captain J. W. Gunnison, led an ill-fated surveying expedition to Utah in 1853, during which he and eight colleagues were murdered.

Common Chickweed

Stellaria media H 4–8 in (10–20 cm)

The common name of this Eurasian import refers to the fact that chickens favor its small, delicious flowers and leaves, which humans historically also have enjoyed.

KEY FACTS

The small white flower has 5 radiating petals, which are so deeply cleft that they often appear to be 10 petals.

+ **habitat:** Fields, open woods, lawns, and disturbed areas

+ **range:** Naturalized throughout most of the U.S. and Canada

+ **bloom period:** Nearly year-round under favorable conditions

The flowers of the Common Chickweed have five white petals arranged in a star shape, and five green, hairy sepals. The flowers form a cluster at the end of a stem or can emerge singly from a leaf axil. Leaves are smooth edged and oval, and grow opposite each other. The species forms low, sprawling, dense mats of stems that reach 16 inches (40 cm) long and have a single line of hairs on one side. It often germinates in fall and winter, and can be in bloom at any time of the year. Common Chickweed is often considered an undesirable weed in lawns and crop fields.

Wood Anemone
Anemone quinquefolia H 4–10 in (10–25 cm)

The delicate Wood Anemone, a member of the Buttercup family, graces woodland settings in early spring. Its habit of trembling in the breeze gives it another common name: Windflower.

KEY FACTS

The small white or sometimes pink flower typically has 5 petal-like sepals and a greenish center filled with 30–50 stamens.

+ habitat: Woods, woodland edges, and stream banks

+ range: Throughout much of central and eastern Canada and the eastern U.S.

+ bloom period: April–June

Wood Anemone tends to develop in large stands. Individual plants may start out with one long-stalked basal leaf with three to five coarsely toothed lobes. A single flower, about an inch (2.5 cm) wide, arises on a stalk above a whorl of three stem leaves. The flower may have four to nine petal-like sepals, but most commonly has five. Stem leaves are deeply divided into three, and sometimes five, leaflets. Even the leaves with three leaflets may appear to have five because the side leaflets are deeply cleft. By summer, the plant has died back and all but disappeared.

Early Meadow Rue
Thalictrum dioicum H 8–36 in (20–90 cm)

Male and female flowers of Early Meadow Rue occupy separate plants, giving rise to the species name, *dioicum,* derived from the Greek for "two households."

KEY FACTS

The male flower has elongated yellow stamens dangling from 4 to 5 greenish petal-like sepals; the female flower has a cluster of pistils.

+ habitat: Woods, slopes, and ravines

+ range: Throughout much of central and eastern North America

+ bloom period: April–June

Early Meadow Rue gets a head start on other plants and even on itself in the spring. The small, tasseled flowers of the perennial plants form large clusters from stem ends and leaf axils. The clusters start to bloom before the leaves themselves have fully developed. The leaves grow on long stems and have from 3 to as many as 12 roundish leaflets that may have rounded teeth. Flowers on male plants droop; those on female plants stand erect. Despite gender separation, flowering time is coordinated. The petal-like sepals often drop off, leaving the stamens and pistils.

Eastern Virgin's Bower
Clematis virginiana L 6–20 ft (2–6 m)

This native Virgin's Bower climbs over supporting structures with leaf stalks that act like tendrils, or it sprawls horizontally as a jumbled ground cover.

KEY FACTS

The flower has 4 white petal-like sepals and a center of long stamens in the male flower and long pistils in the female flower.

+ **habitat:** Moist woods, thickets, stream banks, pond edges, and fencerows

+ **range:** Much of central and eastern U.S. and Canada

+ **bloom period:** July–October

Eastern Virgin's Bower, a vigorous vining member of the Buttercup family, rapidly creates a mass of foliage and fragrant flowers that begin to blossom in late summer. The foliage takes the form of compound green leaves with three to five oval toothed leaflets. Male and female flowers grow in clusters from leaf axils on separate plants, relying on bees and butterflies to accomplish pollination. After fertilization, the female flower transforms into a seed head with a showy plume-like tail attached to each fruit. Repeated in large numbers, the plumes give the plant a grayish, bearded look.

Mayapple
Podophyllum peltatum H 12–18 in (30–45 cm)

Large colonies of green umbrella-like leaves shelter the nodding white flowers of the Mayapple, a shade-loving plant that often spreads abundantly across rich woodland soils in spring.

KEY FACTS

The flower has 6 to 9 white petals and a yellow center with double the number of stamens as petals.

+ **habitat:** Woods, shady fields, along roadsides, and stream banks

+ **range:** Central and eastern U.S. and Canada

+ **bloom period:** March–June

The apple blossom–like flower of the Mayapple dangles from its stalk at the junction of a pair of large, palmately lobed and deeply cleft leaves that are up to a foot (30 cm) around. The stalked leaves appear at the tip of a single stem that rises from spreading underground rhizomes. The plant also produces solitary leaves directly from the rhizomes that are not associated with flowers. The petals fall from the flower, and an egg-shaped yellowish fruit develops containing several seeds. All parts of the prolific Mayapple are poisonous, except for the ripe fruits.

Bloodroot

Sanguinaria canadensis H 4–8 in (10–20 cm)

American Indians used the reddish sap produced by the Bloodroot as a dye for clothing and crafts, as well as facial decoration.

KEY FACTS

The flower has 8–16 white petals surrounding a center with yellow-anthered stamens and a green pistil with a yellow stigma.

+ habitat: Woods and stream banks

+ range: Central and eastern U.S. and Canada

+ bloom period: March–April

In early spring, a large, grayish green leaf emerges from the ground that enfolds the stalk of a single flower. Before the leaf can fully unfold to reveal its multiple major and minor lobes and prominent veins, the flower has pushed past to unfurl like a small water lily. Both the flower and the leaf stalks continue to grow. The blossom opens during the day and closes at night for the few days of its existence. The foliage and rhizomes of the Bloodroot produce an acrid reddish juice, which gives the species its namesake "blood."

Twinleaf

Jeffersonia diphylla H 6–18 in (15–45 cm)

The genus name of this delicate and unusual plant is an honor bestowed on Thomas Jefferson, an avid horticulturalist, by American botanist William Bartram.

KEY FACTS

The bowl-shaped flower has 8 oval white petals and a center with 8 yellow stamens.

+ habitat: Moist woods and rocky slopes, often in limestone areas

+ range: Eastern U.S. and Canada, excluding the Northeast and parts of the Southeast

+ bloom period: March–May

The namesake basal leaves of the clump-forming Twinleaf, a member of the Barberry family, rise on long stalks and give a doubled appearance due to their distinct lobes. The cup-shaped white flower tops a separate leafless stalk. The plant bears a passing resemblance to the unrelated Bloodroot, a major difference being the latter's large, five- to nine-lobed leaf. Like the Bloodroot, the Twinleaf leaf and flower stalks continue to grow after the flower has blossomed and as the fruit ripens. The Twinleaf fruit is an unusual pear-shaped brownish pod with a hinged lid.

Annual Prickly Poppy

Argemone polyanthemos H 12–36 in (30–90 cm)

Annual Prickly Poppy is a tad less prickly than some others in its genus. Its name comes from *argema*, Greek for "cataract." *Argemone* species were once used to treat eye ailments.

KEY FACTS

The flower has 6 thin, white, crinkled petals and a mound of 20 or more yellow stamens and a maroon stigma in the middle.

+ **habitat:** Prairies, meadows, roadsides, and disturbed areas

+ **range:** Native to the central U.S.; introduced farther east and northwest

+ **bloom period:** April–October

Delicate large blossoms resembling crepe paper, up to 4 inches (10 cm) wide, top the branched stems of the Annual Prickly Poppy, a prominent annual or biennial wildflower species native to the central plains and prairies. The rest of the plant is covered with prickles (hence, the common name Prickly Poppy). The blue-green, waxy, and spiny lobed leaves are dense on the stiff stems, which are somewhat spiny. The pointed oval seedpod, with its protruding stigma, is also prickly. Disturbances such as highway construction probably aided the spread of this species, particularly to the West. All parts of the plant are poisonous to eat.

Carolina Horse Nettle

Solanum carolinense H 12–36 in (30–90 cm)

Not a true nettle but a member of the Nightshade family, the Carolina Horse Nettle sports prickly stems and leaves that discourage grazing animals.

KEY FACTS

The white or pale violet star-shaped flower is composed of 5 lobed, spreading corollas and 5 stamens with long yellow anthers erect in a group in the center.

+ **habitat:** Fields, prairies, roadsides, and disturbed areas

+ **range:** Much of western U.S. and eastern U.S. and Canada

+ **bloom period:** May–October

The Carolina Horse Nettle's attractive star-shaped flowers with their distinctive banana-like anthers share the plant with hairy and prickly leaves, stems, and stalks. The flowers grow in clusters at the top of erect stems that bear lance-shaped, angular, lobed leaves. The plant's prickly surfaces and Nightshade family toxicity are off-putting to mammalian herbivores; however, the mature yellow fruits that resemble little tomatoes are tolerated and eaten by birds, skunks, rodents, and other animals. For those who consider this species an undesirable weed, gloves should always be worn when attempting to pull it.

Dutchman's Breeches
Dicentra cucullaria H 4–12 in (10–30 cm)

The wide pantaloons Dutch men traditionally wore inspired the common name for this woodland perennial, a relative of Bleeding Hearts.

KEY FACTS

The flower has 2 outer white petals that form stout spurs and 2 inner yellow petals that curve up at the base.

+ **habitat:** Woods, ravines, ledges, and stream banks

+ **range:** Mainly in the eastern U.S. and Canada and the Pacific Northwest

+ **bloom period:** March–May

In the early spring, the double-spurred flowers of Dutchman's Breeches dangle upside down in pairs from a curved, leafless scape, or flower stem, that rises from the plant's rhizome. Their triangular nectar spurs, formed by the outer white petals, require the services of long-tongued bees for efficient pollination. The scape overhangs a profusion of deeply cut, fernlike, grayish green leaves that form a rosette on the forest floor. The compound leaves attach to the rhizome by a long, brownish stalk and have three leaflets, which are divided into three more leaflets that are further divided into linear lobes.

Wild Strawberry
Fragaria virginiana H 3–6 in (7.5–15 cm)

The large and becoming larger cultivated strawberries we enjoy had their start as a hybrid of the Wild Strawberry and a South American species.

KEY FACTS

The small flower is composed of 5 white petals around many bright yellow stamens and numerous separate pistils on a central cone.

+ **habitat:** Wood edges, fields, meadows, open ground, roadsides, and disturbed areas

+ **range:** Throughout the U.S. and Canada, except some far northern areas

+ **bloom period:** April–June

The native perennial Wild Strawberry, a member of the Rose family, hugs the ground, sending up new plants by means of sprawling, hairy runners. The compound leaves have three hairy, oval, toothed leaflets. The small, white flowers appear in clusters on short stems or emerge directly from the underground root. They turn into the small red, iconic strawberry with a multitude of seedlike fruits embedded in depressions on its surface. The plant sometimes goes dormant during the heat of the summer. This species feeds a wide range of animals, from insects to birds to mammals—including, of course, humans.

White Wild Indigo
Baptisia lactea/B. alba H up to 6 ft (180 cm)

This *Baptisia* species produces no blue dye, unlike some other members of the genus. When new growth pushes through the ground, it somewhat resembles asparagus.

KEY FACTS

The pealike white or cream flower is composed of 5 petals: a top banner petal, 2 side wing petals, and 2 lower ones that form a keel.

+ **habitat:** Open woods, marsh and lake edges, and prairies

+ **range:** Much of the central and eastern U.S.

+ **bloom period:** April–July

The White Wild Indigo's long, erect spikes of pealike flowers seem to go on forever at the top of stiff stems that form the bushy plant and can grow to 6 feet (180 cm) tall. At that height, the flowers frequently tower over nearby vegetation. The stems branch near the top and are covered with compound, three-parted green leaves. The leaves turn black in the fall, and the flowers are replaced by hairless, inflated seedpods, about 2 inches (5 cm) long that are green at first and then turn black. The pods often stay on the plant through the winter. Larvae of some butterflies use *Baptisia* as a food plant.

Flowering Spurge
Euphorbia corollata H 12–36 in (30–90 cm)

A relative of the Poinsettia, Flowering Spurge shares some features with the iconic holiday plant, such as bracts—modified leaves—that masquerade as petals.

KEY FACTS

What appears to be a flower incorporates 5 rounded, white glandular appendages that surround a small cup of tiny yellow flowers.

+ **habitat:** Open woods, old fields, prairies, sand dunes, roadsides, and disturbed areas

+ **range:** Central and eastern U.S. and Canada, except far northern areas

+ **bloom period:** June–October

The tiny real flowers of the Flowering Spurge are part of an elaborate architecture. The flowering forms appear in clusters in a multitiered branching structure that occurs at the top of a long, central stem. Linear to oblong leaves climb the stem alternately, ending in whorls of three just below the flower clusters. Typical of spurges, the species has milky sap in all tissues that can cause skin irritation in susceptible individuals. The name "spurge" comes from the Latin for "to purge," and the plant was used for that purpose, although it has other, mildly toxic properties.

Tufted Evening Primrose

Oenothera caespitosa H 6–12 in (15–30 cm)

The Tufted Evening Primrose displays its beauty by night, when the white blossom pops open and emits a sweet fragrance that lets pollinators know it is open for business.

KEY FACTS

The flower has 4 large white petals that are notched and appear heart shaped, and 8 yellow stamens.

+ **habitat:** Open woods, clearings, open desert, arroyos, dry slopes, and roadsides

+ **range:** Much of the western U.S. and Canada, except in far northern areas

+ **bloom period:** April–August

The Tufted Evening Primrose flower is stemless, rising from a substantial basal rosette, but the flowers have a long tube that holds the petals above the base. The rosette is formed of many long, narrowish, lance-shaped leaves that often have scalloped, toothed, or lobed edges. The 4-inch-wide (10 cm) bloom is a one-night wonder, fading to pink as the corolla withers. In warm climates, this perennial species often is evergreen, but it can be deciduous in colder regions. Despite the name, evening primroses are not members of the genus *Primula*, which is in the Primrose family.

Bunchberry Dogwood

Cornus canadensis H 3–9 in (7.5–22.5 cm)

Dogwood as ground cover, Bunchberry Dogwood looks much like its tree relatives and is a rare representative of that family in herb form.

KEY FACTS

The four petal-like white bracts surround a cluster of tiny, white flowers with purple centers.

+ **habitat:** Coniferous and mixed woods, thickets, and swamps, sometimes in mountainous areas

+ **range:** Northern U.S. and Canada and in the southern Rocky Mountain states

+ **bloom period:** May–September

The Bunchberry Dogwood's white bract-and-flower combination stands out in the midst of its lush, green foliage. This low-growing plant has pointed, oval, crowded pairs of opposite leaves that misleadingly appear to grow as whorls of four to six on slender stems. Whorls with four leaves typically do not flower, whereas those with six do. The plant forms extensive colonies by means of rhizomes, underground stems that put out roots and shoots, an important method of reproduction given that fruit production by the species is spotty. When it does fruit, clumps of dark red, two-seeded berrylike drupes replace the flowers.

Indian Pipe

Monotropa uniflora H 3–10 in (7.5–25 cm)

A ghostly denizen of dimly lit woods, the pale Indian Pipe contains no chlorophyll. Instead, it takes in nourishment by parasitizing soilborne fungi that are associated with the roots of trees.

KEY FACTS

The single white to pinkish to reddish flower is composed of 3 to 6 petals and about twice as many stamens.

+ **habitat:** Moist, shady woods and slopes

+ **range:** Throughout most of the U.S. and Canada, except in the central Rockies, Southwest, and far northern areas

+ **bloom period:** June–September

The Indian Pipe has no green foliage to distinguish flower from stem; the bend marks the nodding flower's beginning in the younger plant. The thick stem is fleshy, translucent, and covered in scales. As the flower and then the seed capsule mature, the entire plant begins to turn black, as it also does if it is picked and dried. After flowering, the flower base begins to unbend until the mature fruit is erect. Ripe seeds emerge from slits that open in the five-segmented capsule. This species parasitizes mycorrhizal fungi often associated with the roots of oaks and conifers.

Elliptical-leaved Shinleaf

Pyrola elliptica H 6–12 in (15–30 cm)

The leaves of this native wildflower traditionally were used as a poultice on wounds and bruises to reduce pain and inflammation—that is, a shin plaster—the source of the common name.

KEY FACTS

The white to greenish, cup-shaped flower has 5 petals, often green veined, and a curved, protruding style.

+ **habitat:** Dry or moist woods

+ **range:** Northern U.S. and Canada, except in far northern areas

+ **bloom period:** June–August

In May, a long scape, or leafless flowering stem, grows from the nearly basal cluster of bright green, broadly oval evergreen leaves of the Elliptical-leaved Shinleaf. It will grow up to 12 inches (30 cm) long and develops an elongated floral cluster. It bears up to about 20 fragrant, nodding blossoms. By the time the blooms of spring woodland wildflowers are long gone, the shinleaf flower comes to life in deep shade. The genus name is the Latin diminutive for "pear," referring to the shape of the leaves of some species.

Field Bindweed

Convolvulus arvensis L vining to 72 in (180 cm)

A member of the Morning Glory family, Field Bindweed is one of the more aggressive of the introduced Eurasian species, deep rooted and climbing or trailing its way across the landscape.

KEY FACTS

The funnel-shaped white or pink flower has a 5-petaled, shallowly lobed corolla.

+ habitat: Fields, open areas, roadsides, gardens, and disturbed sites

+ range: Introduced and naturalized throughout the U.S. and Canada

+ bloom period: April–October

The genus name of the Field Bindweed derives from the Latin for "to twine around," and that certainly is the method of operation for this perennial. Its stems are long and hairy, with triangular leaves arranged alternately along them. The flowers emerge singly or in pairs from the leaf axils. The plant adapts easily in many climatic zones and plant communities. Field Bindweed arrived here by the mid-18th century, probably first entering as a contaminant in crop or garden seeds. Many U.S. states and a number of Canadian provinces place this species on their noxious weeds lists.

Jimsonweed/Thorn Apple

Datura stramonium H 12–48 in (30–120 cm)

The name Jimsonweed is a corruption of Jamestown Weed. In 1679, soldiers in the Virginia colony allegedly ate the plant in a salad and suffered extreme psychological effects.

KEY FACTS

The white or purple funnel-shaped flower, up to 3 in (7.5 cm) long, has 5 lobes and a violet throat.

+ habitat: Sandy soils, fields, barnyards, and disturbed areas

+ range: Introduced and naturalized throughout the U.S. and Canada, except in far northern areas

+ bloom period: July–October

The large Jimsonweed blossoms with their pointed lobes appear in the axils of branch forks along the strong purple or green stems. Long, stalked leaves grow alternately from the stem and are usually jaggedly lobed. After blooming, the corolla falls off; in its place a green, oval capsule develops, about 2 inches (5 cm) long and covered in prickles, source of the alternate name Thorn Apple. Jimsonweed, introduced in temperate North America from the tropics, is a poisonous member of the Nightshade family. It contains alkaloid compounds with many toxic and psycho-active properties, making it harmful to humans and livestock.

American Ginseng

Panax quinquefolius H 6–24 in (15–60 cm)

This all-purpose healing plant has been overharvested due to worldwide demand for its roots, which began in the 1700s. Its genus name, *Panax,* derives from a Greek word meaning "cure-all."

KEY FACTS

The tiny, greenish white flowers form a small umbellate cluster on a separate stalk rising from a group of leaves.

+ **habitat:** Rich woods

+ **range:** Native to the eastern U.S. and Canada; widely and commercially cultivated

+ **bloom period:** June–July

The tiny, white umbel of the American Ginseng flower is dwarfed by the trio of palmately compound leaves that surround it. Three of the usually five, toothed leaflets are large; the other two are much smaller. Red berrylike fruits with two seeds each develop from the flowers in the fall. The common name of this species, Ginseng, derives from a Chinese term meaning "manlike," a reference to the root shape. The Chinese and others turned to the North American forms of this important medicinal plant when local species were overharvested.

Fragrant Water Lily

Nymphaea odorata H 2–6 in (5–15 cm)

The Fragrant Water Lily might appear to be tropical—and many people know it as a fish pond ornamental—but it is a widespread native perennial in temperate North America.

KEY FACTS

The white flower, up to 8 in (20 cm) across, has 4 sepals, 20–30 pointed white petals, and dozens of bright yellow stamens.

+ **habitat:** Aquatic habitats such as ponds, lakes, and slow-moving streams

+ **range:** Throughout U.S. and Canada, except some northern areas; highly cultivated

+ **bloom period:** March–October

Floating effortlessly on the surface of a pond or lake, the Fragrant Water Lily is strongly tethered to its rootstock in the muddy bottom. Long, separate stalks link the blossom and leaves to their source. The circular leaves develop first; up to a foot (30 cm) across, they are green on top, purplish below, and deeply cleft on one side. The flowers float lightly on their sepals, opening in the morning and closing at night. They last only a few days, and after withering are replaced by dark, round fruits that mature underwater to release their seeds.

Milk Vetch
Astragalus species H 4–48 in (10–122 cm)

Some milk vetches are the bane of ranchers and livestock farmers. Many species contain a powerful and addictive toxin that slowly debilitates and can eventually kill grazing animals.

KEY FACTS

The pealike flowers in colors from white to yellowish to pink to red to purple appear in loose clusters to dense spikes.

+ **habitat:** A variety, including prairies, desert, open woods, shores, and mountain slopes

+ **range:** Nearly 400 species occur in the U.S. and Canada

+ **bloom period:** March–August

Many of the hundreds of species of native milk vetches tend to sprawl. Some, such as the Canada Milk Vetch (*A. canadensis)*, are more upright than others. Milk vetch leaves arise at the base of the plant and are pinnate, also alternating along the stem in many species. Milk vetches are closely related to Locoweeds (genus *Oxytropis*), which produce many of the same toxic effects. To confuse matters, milk vetches also are commonly called Locoweed, and the typical toxicity of most plant parts of species in both genera, along with overlapping habitats and range, tends to perpetuate the association.

White Sweet Clover
Melilotus albus H 12–96 in (30–240 cm)

Beekeepers often plant White Sweet Clover, a Eurasian import and member of the Pea family, as a fragrant and flavorful nectar source for their hives.

KEY FACTS

The small white flowers have 5 petals: a banner petal at top; two side petals, or wings; and 2 bottom petals joined as a keel.

+ **habitat:** Open areas, roadsides, and disturbed sites

+ **range:** Throughout most of the U.S. and Canada, except in some far northern areas

+ **bloom period:** May–September

The White Sweet Clover's pealike flowers grow as long, slender spikes from the leaf axils of erect stems. The compound leaves have three oblong leaflets and they emit a vanilla-like fragrance when crushed. The same chemical compound responsible for the fragrance is a source of the active ingredient in well-known anticoagulant medications that prevent blood clots. The species is frequently planted as a nitrogen-fixing cover crop to improve agricultural land. White and Yellow Sweet Clover (*M. officinalis*) are virtually indistinguishable when in flower or fruit.

Red Baneberry
Actaea rubra H 12–36 in (30–90 cm)

Beware of plants with "bane" in the name, including the Red Baneberry, a member of the Buttercup family with bright red fruits, which is highly poisonous.

KEY FACTS

The flowers have 4–10 spoon-shaped white petals and numerous long white stamens.

+ **habitat:** Moist woods, thickets, and stream banks

+ **range:** Throughout much of the northern and western U.S. and Canada, except in some far northern areas

+ **bloom period:** April–July

Red Baneberry flowers form dense clusters on the leafy stems of bushy plants. The large, divided leaves are composed of numerous oval leaflets with sharply toothed edges. The flower petals tend to drop off, leaving long stamens that give the clusters a feathery look. Red berries with black dots develop that attach to the clusters with slender stalks. A close relative, the White Baneberry (*A. pachypoda*) has white berries with prominent black dots that suggest its other common name, Doll's Eyes. To confuse matters, the Red Baneberry, which differs from the white in flower stalk, flower, and fruit features, sometimes produces white berries.

Pokeweed
Phytolacca americana H to 10 ft (3 m)

An alternative common name for Pokeweed is Inkberry, which speaks to the practice of using the berry's juice as ink, as was commonly done by Civil War soldiers in letters home.

KEY FACTS

The tiny stalked flower, composed of 5 white petal-like sepals and up to 25 stamens, grows in an elongate cluster.

+ **habitat:** Open woods, thickets, old fields, roadsides, and disturbed areas

+ **range:** Much of the eastern U.S. and Canada, southern U.S., and in parts of the West

+ **bloom period:** July–October

Pokeweed flower clusters are terminal on the stem branches. The plant's leaves are very long, lance shaped, and heavily veined. As the berries form, the heavy axis supporting them starts to droop. Berries start out green and eventually become a blackish purple. Although the species is considered mildly toxic, parts of the Pokeweed have long been used in traditional medicine, and the developing greens are cooked as a spring green in the Southeast. Its documented antiviral properties are being studied as a potential treatment for HIV. "Poke" may derive from "puccoon," an American Indian term for a plant that yields dye.

Solomon's Plume/Feathery False Solomon's Seal
Maianthemum racemosum H 12–36 in (30–90 cm)

Erect or arched stems with branched clusters of small white flowers hold their own against imposing foliage in the Solomon's Plume, a native perennial.

KEY FACTS

The tiny white flowers are composed of 6 tepals (look-alike sepals and petals) that are shorter than the blossoms' 6 stamens.

+ **habitat:** Moist woods, floodplains, and stream banks

+ **range:** Throughout most of the U.S. and Canada, except in some far northern areas

+ **bloom period:** April–July

The flowers of the Solomon's Plume appear on short stalks at even intervals on short branches. The branches form a pyramid-shaped terminal cluster on an unbranching stem. The species name, *racemosum*, references the raceme, which is the term for a kind of flower cluster. Large, oval, veiny green leaves up to 8 inches (20 cm) long attach alternately to the stem. After blooming, the flowers are replaced at first by round green berries with coppery or purple spots that mature to a deep red. Western plants of this species tend to have erect stems; eastern plants often have arched ones.

American Elderberry
Sambucus canadensis H 5–12 ft (1.5–4 m)

Growing often into a stately, multibranched shrub, the American Elderberry has a long history of medicinal and other practical uses. It has immune system–stimulating properties.

KEY FACTS

The small, 5-lobed white flowers form large clusters, up to 10 in (25 cm) across.

+ **habitat:** Stream banks, woods, field edges, thickets, and roadsides

+ **range:** Native to the eastern U.S. and Canada; cultivated and escaped elsewhere

+ **bloom period:** May–July

The foliage and flower-laden American Elderberry often arches, its blossoms emitting a musty smell. The leaves are compound, with lance-shaped, toothed leaflets up to 7 inches (17.5 cm) long. After flowering, purplish black berrylike fruits appear, a favorite of wildlife. Both the flowers and the berries have many medicinal uses, and they are used to make cordials and wine; the juice also is used to flavor commercial beverages. The genus name derives from the Greek for a musical instrument and relates to the soft pith in the twigs, which are fashioned into flutes and whistles.

Field Pussytoes

Antennaria neglecta H 2–12 in (5–30 cm)

Small clusters of fluffy flowers produce a flower head with the shape that gives the Field Pussytoes its name. The genus *Antennaria* includes more than 30 native species and many variations.

KEY FACTS

The rounded white disk flowers grow in small, dense, fuzzy clusters at the top of the stems.

+ habitat: Prairies, fields, pastures, and roadsides

+ range: Throughout the U.S. and Canada

+ bloom period: April–July

Multiple fuzzy and fluffy parts reinforce the feline imagery of the Field Pussytoes. The fluffy, rayless flower heads grow atop fuzzy stems from a fanned-out base of spoon-shaped woolly leaves, each with a single prominent vein. The flowers morph into even fluffier seed heads. Stem leaves are smaller and lance shaped. Some species of the genus *Antennaria* do not have to rely on fertilization to produce seed; the female flower can manufacture fertile seed alone. The plant can also create clones by means of runners. These two reproductive methods lead to many variations in cloned colonies, making identification of Pussytoes species a challenge.

Wild Carrot/Queen Anne's Lace

Daucus carota H 12–48 in (30–120 cm)

The Wild Carrot, or Queen Anne's Lace, a Eurasian import, over time was selectively bred into a subspecies we know as a delicious and nutritious orange vegetable.

KEY FACTS

The flat, lacy white umbels (umbrella-like clusters) grow on hairy stems with deeply dissected leaves.

+ habitat: Fields, roadsides, and in disturbed areas

+ range: Introduced from Eurasia and naturalized throughout the U.S. and Canada

+ bloom period: May–October

A single purplish floret rests in the center of each of the Wild Carrot's exquisite open, umbrella-shaped flower clusters, which in turn are formed of smaller white umbrellas. The mature flower cluster takes a curved, hollowed-out shape that suggests a bird's nest, which is another of its common names. The fernlike leaves give off a distinct "carroty" odor when crushed. The long, white taproot of the Wild Carrot is not something to be munched raw, as is done with its cultivated cousin, but it can become edible through cooking. The species has long been used as an herbal remedy, especially in the treatment of bladder and kidney ailments.

Cow Parsnip

Heracleum maximum H 2–10 ft (1–3 cm)

The Cow Parsnip, largest native member of the Carrot family in North America, has a flower head similar to that of the Wild Carrot, but its leaves and stems differ greatly.

KEY FACTS

The small 5-petaled white or yellowish flowers form small umbels (umbrella-like clusters) that are part of large compound umbels.

+ **habitat:** Meadows, fields, pastures, woods, and marshes

+ **range:** Throughout U.S. and Canada, except far northern areas and the extreme South

+ **bloom period:** February–September

Characteristic of the Carrot family, the Cow Parsnip displays flat-topped umbels composed of small white flowers. This native plant generally has large leaves—up to 20 inches (50 cm) wide with an inflated basal sheath and a blade divided into three maple-like leaflets with serrated and lobed margins. The terminal leaflet is often the largest. The leaves appear alternately on the ribbed, woolly, and hollow stem. Flower buds in waiting form large growths on the stem, often the size of an orange. The genus name may refer to the great size of the plant or suggest that consumption of the Cow Parsnip aided Hercules in his tasks.

Oxeye Daisy

Leucanthemum vulgare H 6–30 in (15–75 cm)

A Eurasian import that is better known than many native wildflowers, the Oxeye Daisy was first planted here as a garden ornamental and now is naturalized nearly everywhere.

KEY FACTS

The flower head has 15–25 white ray florets and an indented center of many yellow disk florets.

+ **habitat:** Fields, pastures, and disturbed areas

+ **range:** Introduced and naturalized throughout the U.S. and Canada, except in some far northern areas

+ **bloom period:** June–August

The Oxeye's iconic daisy flower, up to 2 inches (5 cm) wide, tops an erect, often hairy stem that rises from a base of dark green leaves on short, slender stalks. Leaves farther up the stem are smaller, oblong, and attach directly to the stem. This perennial species is so hardy that its rootstock can survive bulldozing and removal, only to resurrect in another location. The word "daisy" is a corruption of the Old English "day's eye," a reference to the flower head's habit of opening in the morning and closing in the evening. A number of states consider the Oxeye a noxious weed.

White Wood Aster
Eurybia divaricata H 12–36 in (30–90 cm)

The flower heads of the White Wood Aster grow in such profusion on zigzagging stems that they make a big impact wherever they grow.

KEY FACTS

The flower head is composed of 5 to 10 narrow, white ray florets and a dozen or so yellow disk florets.

+ **habitat:** Open woods, woodland edges, and roadsides

+ **range:** Throughout much of the eastern U.S. and Canada

+ **bloom period:** August–October

White Wood Aster flowers grow on flat-topped clusters at the top of very leafy stems. The leaves are heart shaped to oval to lance shaped, pointed, and toothed. The leaves near the base of the plant start to die back as the flowers bloom in late summer and early fall, and the flower heads' yellow centers turn purplish as they age. This native perennial species tolerates dry, shady conditions and spreads prolifically by means of underground rhizomes and seeds. The genus *Eurybia* is named for a Greek sea goddess who was the mother of Astraeus, Titan god of the stars and planets.

Philadelphia Fleabane
Erigeron philadelphicus H 6–36 in (15–90 cm)

The Philadelphia Fleabane is a little daisy on steroids, boasting up to 300 ray flowers a head. The genus name derives from the Greek for "old man," which may refer to the stem's beardlike down.

KEY FACTS

A small flower head has up to 300 white or sometimes pinkish threadlike ray flowers, surrounding dense yellow disk flowers.

+ **habitat:** Fields, meadows, open woods, lake edges, roadsides, and disturbed areas

+ **range:** Throughout U.S. and Canada, except parts of Southwest and far north

+ **bloom period:** March–June

The stem of Philadelphia Fleabane bears a stalked cluster of the small, composite flower heads with their distinguishing abundance of rays. Additional clusters often grow from the axils of the upper leaves, which are sparse near the flower. The leaves are generally lance shaped or oval, may be slightly toothed, and grow alternately, clasping the stem. They are larger and more numerous at the bottom of the plant. This early-flowering, basically biennial species has more or less played out by midsummer, another feature that may contribute to the "old man" name.

White Crownbeard/Frostweed

Verbesina virginica H 36–84 in (91–213 cm)

During frigid weather, fleshy green stems of the White Crownbeard split open and exude sap that freezes into often elaborate ribbonlike shapes called frost flowers.

KEY FACTS

The composite flower head typically has 5 white rays and a center of 5-lobed white disk flowers with white stamens and purple anthers.

+ **habitat:** Open woods, fields, stream banks, and roadsides

+ **range:** Central and southern portions of the eastern U.S. and Canada

+ **bloom period:** August–November

Flower heads of the White Crownbeard appear in large terminal clusters; frequently, individual heads may lack many or most of their rays. Large, oval to lance-shaped, pointed leaves alternate on the plant stems. The bases of the leaves extend downward on the stem, contributing to its distinctive longitudinal flanges, or wings. Below-freezing temperatures cause water and water vapor inside the stems to freeze and split the stems, leading to the formation of the icy shapes. White Crownbeard is an important nectar source for the Monarch Butterfly. American Indians used it to treat a number of medical ailments, including gastrointestinal problems.

Common Yarrow

Achillea millefolium H 12–36 in (30–91 cm)

The Common Yarrow often survives along sides of heavily trafficked roads where dust and pollution have killed off other plants.

KEY FACTS

The compact clusters of small flower heads have yellowish disk flowers that are surrounded by 5 rounded white rays.

+ **habitat:** Roadsides, old fields, and scrublands; also highly cultivated

+ **range:** Throughout the U.S. and Canada

+ **bloom period:** May–September

Common Yarrow flower heads grow on the ends of tough, hairy, gray-green stems. As the blossoms age, the florets at the center of the head turn brown or grayish. Recognized as a valuable medicinal plant for millennia, Common Yarrow contains more than 100 biologically active compounds, including some with strong anti-inflammatory properties. One explanation attributes its genus name, *Achillea*, to the legend that Achilles used the herb to stanch the bleeding of soldiers' wounds during the Trojan War. Another version links the name to the plant's discoverer, also called Achilles. Handling Common Yarrow can sometimes cause a rash.

Green False Hellebore
Veratrum viride H 18–96 in (45–240 cm)

A highly toxic plant, the Green False Hellebore sends up large, ribbed, bright green leaves and long flower clusters in wet meadows, posing a threat to grazing animals and their offspring.

KEY FACTS

The small, hairy, green, star-shaped flowers grow on long spikes.

+ **habitat:** Swamps, wet forests, wet meadows, seeps, and other wet areas, often in montane areas

+ **range:** Separate populations in the western and eastern U.S. and Canada

+ **bloom period:** May–July

A member of the Bunchflower family (previously in the Lily family), the Green False Hellebore emerges in early spring, developing large leaves that are up to a foot long (30 cm) and 6 inches (15 cm) wide, are heavily ribbed, and clasp the stem. The plant occurs in western and eastern North America but is absent from the continent's center, likely the result of glacial advance. Nevertheless, the two populations share common features, and many botanists consider them a single species. Green False Hellebore contains strong alkaloids that are used homeopathically in small doses to treat hypertension and heart disease, and have been studied in mainstream medical trials.

Common Ragweed
Ambrosia artemisiifolia H 12–60 in (30–150 cm)

Despite its genus name, *Ambrosia,* the Common Ragweed is a plain and much maligned wildflower, being a major instigator of the seasonal allergy known as hay fever.

KEY FACTS

The long central flower spikes have stalked drooping yellow-green male flower heads above stalkless, smaller green female flower heads.

+ **habitat:** Fields, roadsides, and disturbed areas

+ **range:** Throughout the U.S. and Canada, except in far northern areas

+ **bloom period:** July–October

The Common Ragweed gives rise to numerous small flower heads, source of the fine yellow pollen that spreads misery to allergy sufferers in late summer and fall. Its highly dissected leaves are more conspicuous; up to 6 inches (15 cm) long, they grow on green to pinkish hairy stems and create a profusion of foliage. The female flowers mature to produce hardy, burr-like fruits that can stay viable for more than five years and take hold in disturbed and nutrient-depleted soils. Although honeybees collect pollen from ragweed, the plant relies on wind pollination. Seed-eating bird species are attracted to the oil-rich seeds.

Eastern Poison Ivy

Toxicodendron radicans H ground cover or shrub to 3 ft (1 m); L vining to 150 ft (50 m)

Learning all the different appearances of Eastern Poison Ivy through the seasons is a good idea. It can save the discomfort of an inflamed rash, caused by a toxic resinous substance, urushiol.

KEY FACTS

The small, yellowish white or yellowish green flowers grow in elongated clusters from leaf axils.

+ **habitat:** Woods, thickets, roadsides, and disturbed areas

+ **range:** Throughout mainly central U.S. and Canada

+ **bloom period:** April–June

The genus name *Toxicodendron*, meaning "poison tree," explains it all. Urushiol, Poison Ivy's irritating oil, is toxic in all seasons and can remain potent on unwashed clothing. The old adage "Leaves of three, let it be" is your best guide, because species recognition is tricky. Poison Ivy leaves are compound, with three leaflets that can be small or large, flat green, shiny, or reddish. Notably, the central leaflet tapers to a stalk. The species can take the form of a trailing ground cover, an upright shrub, or a woody climber. Its stems can be slender or thick, smooth or hairy. Another adage—"Only a dope swings from a hairy rope"—is good advice.

Atlantic Poison Oak

Toxicodendron pubescens H to 10 ft (3 m)

With three-part leaves like its close relative Eastern Poison Ivy, Atlantic Poison Oak most often takes the form of a shrub. The urushiol-laden plant lacks Poison Ivy's climbing capabilities.

KEY FACTS

The small yellowish flowers grow in elongated clusters from the leaf axils.

+ **habitat:** Woods, thickets, glades, old fields, and sand hills

+ **range:** South-central and southeastern U.S.

+ **bloom period:** March–June

Atlantic Poison Oak has the potential to grow to 10 feet (3 m), but more often it stays in the 2- to 4-foot (60 to 120 cm) range as a small-ish upright shrub. Its leaves are compound, with three oak-shaped leaflets that have downy stalks. After blooming, the inconspicuous flowers turn into small, tan, velvety, pumpkin-like fruits, and the foliage turns an alluring reddish brown in the fall. The similar western species, Pacific Poison Oak (*T. diversilobum*), grows as a shrub or vine, and its leaves can have three or five leaflets that are toothed or lobed.

Poison Sumac
Toxicodendron vernix H 6–30 ft (2–9 m)

Poison Sumac prefers wetter habitats than Poison Ivy, so the two rarely appear together. It does share some habitats with nonpoisonous sumacs (*Rhus* species).

KEY FACTS

The 5-petaled greenish flowers grow in loose clusters from leaf axils.

+ habitat: Wet areas, including bogs, fens, swamps, marshes, and stream banks

+ range: Eastern U.S. and Canada, including Texas

+ bloom period: June–July

Poison Sumac is a flowering shrub with a proper trunk, branches, and twigs. The twigs bear alternate, compound leaves composed of an odd number of pinnate oval leaflets, usually between 7 and 13, attached to their stems by reddish stalks. The leaves turn orange or red in the fall. The bark of the trunk is gray and mostly smooth, except for the small bumps of the lenticels, pores allowing gas exchange. Urushiol-containing Poison Sumac resembles *Rhus* species (p. 118), but has smooth, not toothed, leaf margins and its drupes (fruits) are smooth and white, instead of red and hairy.

Multiflora Rose
Rosa multiflora H 6–15 ft (1.8–4.5 m)

An East Asian import, Multiflora Rose grows prolifically as a many-blossomed shrub or climber that is known for its ability to form an impenetrable hedge.

KEY FACTS

The flowers have 5 lance-shaped sepals and 5 white or pink petals with many stamens and pistils.

+ habitat: Fields, roadsides, woodland edges, prairies, and some wetlands

+ range: Introduced but naturalized in the eastern U.S. and Canada and also along the west coast of U.S. and Canada

+ bloom period: May–July

Large clusters of small white flowers grow on long, arching stems full of thorns. Leaves are divided into 5 to 11 toothed leaflets and have fringed stipules—paired, winglike appendages. The fragrant flowers produce bright red hips, or fruits, that are a favorite of birds. The plant reproduces by seeds or by developing new plants where stems root on touching the ground. The species was introduced as a rootstock for ornamental roses and then deployed for erosion control and as barriers. It invades pastures and crowds out native flora, especially in the woodland understory; cultivation is prohibited in numerous states.

Porcelain Berry
Ampelopsis brevipedunculata L vine to 15 ft (4.5 m)

Looking similar to native grape species, the Porcelain Berry vine sends tendril-bearing branches up, out, and over other plants. Its dense foliage robs light and space from native species.

KEY FACTS

Inconspicuous greenish white flowers occur in flat-topped to dome-shaped, large, stalked clusters opposite the shallowly to deeply trilobed leaves.

+ **habitat:** Forest edges, thickets, and disturbed land; along ponds and streams

+ **range:** Introduced; now naturalized in the eastern and midwestern U.S.

+ **bloom period:** June–August

The perennial Porcelain Berry is known more for its multicolored fruits than for its small flowers. Attractive clusters of berries ranging from green to lavender to porcelain blue often occur together on the same cluster. These berries and the lush, quite variable foliage made it a popular ornamental plant when brought here from Asia in the 1900s. Birds and other animals feed on the berries and disperse the seeds, although the species self-sows efficiently. Porcelain Berry proliferates very rapidly and is difficult to eradicate once it has taken hold. The plant resists pests, and it tolerates poor soil if there is enough light.

English Ivy
Hedera helix L vine to 100 ft (30 m)

Brought to North America by nostalgic European settlers hoping to create a familiar presence, English Ivy wasted no time climbing buildings and trees, and crawling across landscapes.

KEY FACTS

Tiny, inconspicuous, yellow-green flowers grow in clusters at ends of mature climbing plants that have reached open sunlight.

+ **habitat:** Woodlands, forest edges, fields, and coastal areas with not too much moisture

+ **range:** Native to Eurasia; widely naturalized in western, central, and eastern U.S.

+ **bloom period:** September–October

Most people never see English Ivy in bloom. The flowers appear in the fall only near the tops of mature vines. In its creeping and early climbing phase, English Ivy produces evergreen, three- to five-lobed leaves on woody stems that attach to surfaces by small rootlets aided by a sticky substance. The leaves of mature flowering stems are unlobed and oval. The vines block light needed for photosynthesis—often killing the tree from the bottom and leaving only a "broccoli top"— and potentially causing it to uproot or break during storms. English Ivy along the ground also crowds out native species, creating a monoculture.

Japanese Honeysuckle

Lonicera japonica L vine to 30 ft (9 m)

For many, summer nostalgia includes plucking and sipping the sweet nectar from a fragrant Japanese Honeysuckle blossom. Its luxuriant vines climb over shrubs and small trees.

KEY FACTS

A strongly 2-lipped, tubular flower with long, projecting stamens; ovate, hairy, untoothed leaves grow opposite each other.

+ **habitat:** Woods, wood edges, thickets, and roadsides

+ **range:** Naturalized in central, eastern, and southwestern U.S., and in California

+ **bloom period:** April–September

The blossom of the Japanese Honeysuckle starts out white, as one of a pair at leaf axils on a hairy, woody stem that climbs over bushes or trails on the ground; over time, the flower ages into a golden yellow. Often overlooked are the black berries that form later. One of several introduced Asian honeysuckles, Japanese Honeysuckle can easily overtake and choke out woodland trees. As with many exotic species, there is a less aggressive native equivalent that can provide a similar look in the garden. Among the possibilities, Trumpet Honeysuckle (*L. sempervirens*) has long, red blossoms and produces nectar favored by hummingbirds.

Japanese Wisteria

Wisteria floribunda L vine to 70 ft (21 m)

In some areas, Japanese Wisteria is a common sight, draped over a porch front or arbor. But this aggressive, woody vine escapes cultivation and spreads rapidly, thwarting native flora.

KEY FACTS

Violet-blue, lilac, white, or pink pealike flowers grow in large, pendulous clusters from 6–18 in (15–45 cm) long; compound, pinnate leaves have 7–19 leaflets.

+ **habitat:** Various in cultivation; escapees thrive in forests.

+ **range:** Native to Japan; escaped cultivars widely established in central and eastern U.S. and Canada

+ **bloom period:** April–June

The strong, climbing vines and fragrant, showy flowers of the Japanese Wisteria, a member of the Pea family, were introduced to North America in the mid-19th century. Seeds form in slender, fuzzy, flattened pods about 6 inches (15 cm) long. Escaped Japanese Wisteria, often from abandoned homesteads and nurseries, infiltrates forests, where it can smother trees and crowd out native species. Gardeners who want the same look should seek out cultivars of American Wisteria (*W. frutescens*), a native species, with narrower and more delicate flower clusters. Its native range is the southeastern and midwestern United States.

Morning sun silhouettes the towering trees of Redwood National Park, California.

2 | Trees & Shrubs
The Living Landscape

What is a tree? What is a shrub? Most of us would say we knew, and that we could tell you the difference. But because this chapter deals with 160 species of trees and shrubs likely to be encountered in the United States and Canada, it would be worth a few minutes to have a closer look at this issue. Debating how to define all species as either a shrub or a tree, especially considering how variable many species are, seems less valuable than having a look at how these plants are able to exist in the first place. The word is *wood.*

Herbaceous annuals, biennials, and perennials grow mostly at their tip, or apex, which contains the apical meristem, where the growing cells divide. So do trees and shrubs, but they also grow outward, laterally. Lateral growth occurs in lateral meristems. Growth patterns and aging result in several kinds of tissues that in concert allow these sometimes behemoth plants to arise. Woody plants include small trees, bushes, shrubs, and many vines.

■ Anatomy of a Tree Trunk

Let us imagine that we are looking at the stump of a just-felled tree that had been evolved just so naturalists could learn the basic anatomy of wood. We are looking at a cross section of the tree, taken near its base, in which we can see perfectly its bark and other layers. Let us consider those layers, from the outside in, remembering that the same layers may be found on a smaller scale on the tree's branches, twigs, and roots. There are three main layers—the bark, the cambium, and the wood, some of them divided further.

Bark, of course, is a tree's external covering. This layer of dead cork-like cells protects the interior from injury and disease and conserves water.

The **inner bark,** or **phloem,** is a transport system that moves food and other substances made during photosynthesis (usually taking place in the leaves) to the rest of the tree. As new bark is produced, it pushes older cells outward. Eventually the older bark cannot retain all this newer bark, and its smooth complexion cracks, flakes, or peels.

Between the inner bark and the wood lies the **cambium,** the lateral meristems, where the cells that make the tree grow wider multiply. New cells form to its outside and its inside.

Toward the center of the trunk, inside the cambium, is the **wood,** or **xylem.** Wood contains strengthening fibers and other conducting vessels. There are two kinds of wood. The sapwood, adjacent to the cambium, transports water and dissolved minerals and nutrients from the soil (taken up by roots) to the rest of the tree.

When the cambium creates new wood cells, their size and shape depend on the conditions in the area at the time, so in the generally wetter spring, new cells are bigger, and in the drier summer, they are smaller. These differences make visible annual rings (each actually representing a layer of wood), which of course are an index of a tree's age.

With time, some xylem layers closer to the center of the trunk die, no longer needed for transport. This older wood, called *heartwood,* fills with substances, like resins, that protect it from rot. It is often darker than the surrounding wood.

Without the development of this miraculous system, we would not have trees and shrubs. The wood varies with the kind of tree or shrub, and properties of the different parts of the trunk also vary. People have exploited these differences for millennia, using different species in making canoes, clothing, medicines, food, shelter, charcoal, decorations, or fire.

■ Chapter Organization

The chapter is organized in several ways. The overarching scheme is the most major. First discussed are the gymnosperms. The word means "naked seed," and for our purposes, it is another way of saying conifers. The conifers include the cedars, the cypresses, the pines, and the spruces, whose seeds are bare in the cone and not surrounded by a fruit. (The Ginkgo, which does make fruit, and a nasty one at that, is considered a conifer.) The other major group is the angiosperms, whose seeds are enclosed by fruits (the mature ovary). The angiosperms are divided into the dicots (which have two cotyledons, or seed leaves), which include oaks, the Mountain Laurel, birches, locusts, and the Jumping Cholla, and the monocots (which have but one cotyledon), which include the Joshua Tree, the palms, and the palmettos.

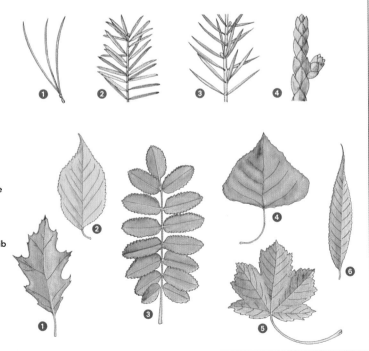

KEY IDENTIFYING FEATURES

Conifers have four types of leaf arrangement:

❶ **Fascicles** are bunches of needles growing from one bud.

❷ **Pectinate rows** are arranged in ranks along opposite sides of the shoot.

❸ **Awl-shaped leaves** are set radially around the shoot.

❹ **Scalelike leaves** are set densely around the shoot.

Broadleaf trees have more varied foliage than the conifers. Some examples of types of foliage shape include:

❶ **Pinnately lobed:** many lobes radiating from the leaf's midrib

❷ **Ovate:** egg-shaped with broad end at the base

❸ **Pinnately compound:** many small leaflets arranged in pairs on either side of a central midrib

❹ **Deltoid:** triangular with stem attached to the middle of the base

❺ **Palmately lobed:** hand-shaped structure with more than three lobes branching from a single point at the base

❻ **Lanceolate:** general shape of a lance—longer than wide and a pointed apex

Snow dusts aspen and spruce trees in Colorado's Rocky Mountain National Park.

■ Principles Used in the Entries

As you use the entries, keep in mind a few principles that are followed. *General* is probably the most important trait for these descriptions. Multivolume sets have been published on wood morphology, reproductive strategies, taxonomy, ranges, and ecology of trees and shrubs. Detailed, torturous keys get to the nitty-gritty of identification. All these books are available at libraries and bookstores. This chapter is meant to whet readers' interest and to truly help them identify trees and shrubs that they may see on field trips and travels, and to tell them something about the species' ecology, value to wildlife and humans, and status—including whether or not they are native.

Geographic ranges are general and refer to a plant's native range. It is not noted that conifers are evergreen, which most are; some, however, are deciduous (the Baldcypress, for example), and this is

*Why are there trees
I never walk under
but large and melodious thoughts
descend upon me?*
—WALT WHITMAN

pointed out. Likewise, it is not stated when an angiosperm tree is deciduous, because most of them are; but some are evergreen or mostly so (the Wax Myrtle, for example), and this too is pointed out.

Though some entries mention culinary uses of various tree or shrub parts, readers should not eat any part of any plant unless they are 100 percent certain that it is 100 percent safe. Other species are described as being toxic; if a species is not so described, it should not be deduced that any part is safe to eat.

The descriptions of the trees and shrubs are designed to convey the species' unique features and those that are most helpful in identification, most important the height (following the scientific name) and the information under Key Facts; the photographs and illustrations complement each other. Features of twigs are rarely described. Floras often describe twigs, as do books that teach users to identify deciduous trees in winter. Depending on latitude, altitude, soil characteristics, or sunlight, two individuals of a species can look quite different. The species entries take into account much of the variation by a descriptive phrase, or a range of the numbers.

▪ Trees & You

Not only will this chapter help you get to know and more deeply appreciate this starter set of 160 species, it will help you see the forest in addition to the trees—and something larger than that, albeit not definable or quantifiable. The spirit of North America, and of its peoples, indigenous and not, is inseparable from these amazing woody plants.

OFFICIAL TREES OF THE U.S. STATES AND CANADIAN PROVINCES

+ **Alabama: Longleaf Pine** (*Pinus palustris*)
+ **Alaska: Sitka Spruce** (*Picea sitchensis*)
+ **Arizona: Palo Verde** (*Parkinsonia florida*)
+ **Arkansas: Pine Tree** (*Pinus*)
+ **California: Coast Redwood** (*Sequoia sempervirens*), **Giant Sequoia** (*Sequoiadendron giganteum*)
+ **Colorado: Blue Spruce** (*Picea pungens*)
+ **Connecticut: White Oak** (*Quercus alba*)
+ **Delaware: American Holly** (*Ilex opaca*)
+ **Florida: Cabbage Palmetto** (*Sabal palmetto*)
+ **Georgia: Live Oak** (*Quercus virginiana*)
+ **Hawaii: Kukui or Candlenut** (*Aleurites moluccana*)
+ **Idaho: Western White Pine** (*Pinus monticola*)
+ **Illinois: White Oak** (*Quercus alba*)
+ **Indiana: Tuliptree** (*Liriodendron tulipifera*)
+ **Iowa: Oak** (*Quercus*)
+ **Kansas: Eastern Cottonwood** (*Populus deltoides* ssp. *deltoides*)
+ **Kentucky: Tuliptree** (*Liriodendron tulipifera*)
+ **Louisiana: Baldcypress** (*Taxodium distichum*)
+ **Maine: Eastern White Pine** (*Pinus strobus*)
+ **Maryland: White Oak** (*Quercus alba*)
+ **Massachusetts: American Elm** (*Ulmus americana*)
+ **Michigan: Eastern White Pine** (*Pinus strobus*)
+ **Minnesota: Red Pine** (*Pinus resinosa*)
+ **Mississippi: Magnolia** (*Magnolia*)
+ **Missouri: Flowering Dogwood** (*Cornus florida*)
+ **Montana: Ponderosa Pine** (*Pinus ponderosa*)
+ **Nebraska: Eastern Cottonwood** (*Populus deltoides* ssp. *deltoides*)
+ **Nevada: Singleleaf Piñon** (*Pinus monophylla*)
+ **New Hampshire: Paper Birch** (*Betula papyrifera*)
+ **New Jersey: Northern Red Oak** (*Quercus rubra*)
+ **New Mexico: Piñon Pine** (*Pinus edulis*)

+ **New York: Sugar Maple** (*Acer saccharum*)
+ **North Carolina: Longleaf Pine** (*Pinus palustris*)
+ **North Dakota: American Elm** (*Ulmus americana*)
+ **Ohio: Ohio Buckeye** (*Aesculus glabra*)
+ **Oklahoma: Eastern Redbud** (*Cercis canadensis*)
+ **Oregon: Douglas-fir** (*Pseudotsuga menziesii*)
+ **Pennsylvania: Eastern Hemlock** (*Tsuga canadensis*)
+ **Rhode Island: Red Maple** (*Acer rubrum*)
+ **South Carolina: Cabbage Palmetto** (*Sabal palmetto*)
+ **South Dakota: Black Hills Spruce** (*Picea glauca* var. *densata*)
+ **Tennessee: Tuliptree** (*Liriodendron tulipifera*)
+ **Texas: Pecan** (*Carya illinoensis*)
+ **Utah: Blue Spruce** (*Picea pungens*)
+ **Vermont: Sugar Maple** (*Acer saccharum*)
+ **Virginia: Flowering Dogwood** (*Cornus florida*)
+ **Washington: Western Hemlock** (*Tsuga heterophylla*)
+ **West Virginia: Sugar Maple** (*Acer saccharum*)
+ **Wisconsin: Sugar Maple** (*Acer saccharum*)
+ **Wyoming: Plains Cottonwood** (*Populus deltoides* ssp. *monilifera*)

Canadian Provinces and Territories
+ **Alberta: Lodgepole Pine** (*Pinus contorta*)
+ **British Columbia: Western Redcedar** (*Thuja plicata*)
+ **Manitoba: White Spruce** (*Picea glauca*)
+ **New Brunswick: Balsam Fir** (*Abies balsamea*)
+ **Newfoundland and Labrador: Black Spruce** (*Picea mariana*)
+ **Northwest Territories: Jack Pine** (*Pinus banksiana*)
+ **Nova Scotia: Red Spruce** (*Picea rubens*)
+ **Nunavut Territory: none**
+ **Ontario: White Pine** (*Pinus strobus*)
+ **Prince Edward Island: Northern Red Oak** (*Quercus rubra*)
+ **Quebec: Yellow Birch** (*Betula alleghaniensis*)
+ **Saskatchewan: White Birch** (*Betula papyrifera*)
+ **Yukon Territory: Subalpine Fir** (*Abies lasiocarpa*)

TREES & SHRUBS

Ginkgo/Maidenhair Tree

Ginkgo biloba H to 130 ft (40 m)

The Ginkgo is considered a conifer, though it does not produce cones. It is deciduous, quickly losing its distinctive leaves after a bright yellow autumn phase.

KEY FACTS

This tree fares well in gardens and cities; *biloba* refers to its leaves, usually 2-lobed.

+ leaves: Fan-shaped, recalling maidenhair fern

+ flowers/fruits: Sexes on different trees, male in catkin-like clusters, female 1 or 2 on short stalk

+ range: Survives mainly in cultivation

Ginkgo biloba is the only species in a genus that millions of years ago had nearly world-wide distribution. The native of China is now prized around the world for its beauty and its ability to withstand pollution and other adverse conditions—including atomic bomb blasts. (A group of trees just one mile/1.6 km from ground zero at Hiroshima survived, regained their health, and still grow there.) The fruit mature on the female flower stalks and recall small plums—until they fall to the ground and decay, earning the tree another alias, "stinkbomb tree."

Alaska Cedar/Yellow Cypress/Nootka Cypress

Chamaecyparis nootkatensis H 49–125 ft (15–38 m), but shrub-like at highest altitudes

Typical of false cypresses (often incorrectly called cedars), the Alaska Cedar is aromatic and resinous. Its cones mature in two years (one year in its southern extent, like other false cypresses).

KEY FACTS

Branches droop in flat sprays.

+ leaves: Scales pressed together, bluish green; unpleasant odor

+ cones: Male and female on same tree at branchlet tips; mature seed cones of 4–6 short-pointed scales, reddish brown

+ range: Coastal mountains, southeastern Alaska to northern California

Prized for its durable, fine-grained wood, the Alaska Cedar is slow growing and long lived, so that a tree can be much older than the diameter of its trunk would suggest. One individual is 3,500 years old. The tree was important to indigenous peoples of its range, including the Nuu-chah-nulth (formerly called Nootka). It is used in landscaping in the coastal Northwest. This species was originally assigned to the genus *Cupressus* (the true cypresses), then to *Chamaecyparis*, but it may be assigned to a new genus, *Xanthocyparis*, because it is felt to differ strongly from the other two genera.

Atlantic White Cedar
Chamaecyparis thyoides H to 82 ft (25 m)

A medium to large tree reminiscent of the junipers, the Atlantic White Cedar grows in pure stands or, less often, with wet-woods trees like Baldcypress and Blackgum.

KEY FACTS

Branches are in fan-shaped sprays; bark is not shredded.

+ **leaves:** Scales tightly overlapping

+ **cones:** Male and female on same tree but separate branches; mature cones of 6 pointed, woody scales

+ **range:** Maine to Florida to Mississippi, mainly in swamps and wet sands of Coastal Plain

The wood of the Atlantic White Cedar is so appealing that the species has been overharvested—since the time of the American Revolution. Its stands are much reduced, though it is still commercially important in parts of its range. Its wood is light, strong, and easily worked, ideal for use in posts and telephone poles, barrels, shipbuilding, and decoy carving. Aromatic, it repels insects, retards decay, and resists disease, such as cedar-apple rust (unlike the Eastern Redcedar, for which the tree can be mistaken if not in fruit). Shade tolerant and attractive, the tree is often planted in yards and gardens.

Eastern Redcedar
Juniperus virginiana H to 100 ft (30 m)

The Eastern Redcedar is our most widespread conifer and one of the most common, found in fields, fencerows (thanks to seed dispersal by birds on fences), dry, open woods, and along roads.

KEY FACTS

Branchlets often droop.

+ **leaves:** Mature leaves scalelike, overlapping; foliage needle-like

+ **cones:** Male and female cones are mostly on separate trees; female cones mature into bluish berries.

+ **range:** East of Great Plains, from southeastern Canada to the Gulf of Mexico

The aromatic wood of this slow-growing juniper (it is not a true cedar) is perhaps best known for its use in lining cedar chests and closets; its aroma repels moths and other insects. It is also used for posts, paneling, and carving. Its reddish brown bark peels off in strips. Many cultivars have been developed for landscaping. The tree is the primary host to the fungus that causes cedar-apple rust, which produces masses of orange, gelatinous spore tubes among its needles. The fungus's secondary hosts are members of the rose family, such as pear and apple trees, where it is more damaging, affecting fruit crops.

Redwood/California Redwood

Sequoia sempervirens H to 375 ft (115 m)

The giant Redwood is locally common in protected areas along rivers experiencing extensive rainfall and constant fogs off the ocean.

KEY FACTS

The world's tallest tree, named for its thick, reddish bark, can live longer than 2,000 years.

+ needles: Dark green

+ cones: Male and female in separate clusters on same tree; seed cones oblong, red-brown

+ range: Pacific coast, southwestern Oregon to central California

Redwoods have limbs to a height of 50–100 feet (16–30 m), above which the trunk has many vertical stems. This unique habitat accumulates water, develops soils, and supports vertebrates, insects, and epiphytic plants. Not only is the tree's habitat characterized by fog, the tree induces fog drip, condensing water and supplying organisms in its crown and the ground below. Its wood is smooth, straight grained, and strong, with great appeal for building and woodworking. Since Spanish settlement, more than 90 percent of the largest Redwoods have been logged. The tree is now well managed.

Giant Sequoia

Sequoiadendron giganteum H 250–290 ft (76–88 m)

This massive tree was named for Sequoyah, a 19th-century Cherokee silversmith, who devised a character set for writing in the tribe's language and published papers and books in Cherokee.

KEY FACTS

Columnar trunk is fluted; bark is reddish brown, as thick as 2 ft (0.6 m) at the base.

+ leaves: Scalelike, pointed, blue-green

+ cones: Egg-shaped, woody seed cones, similar to a Redwood's, but larger

+ range: Restricted to 75 groves in California, on western slopes of the Sierra Nevada

The sole surviving *Sequoiadendron* species, the Giant Sequoia includes some of the world's largest and oldest organisms. The crown can develop secondary trunks and hollow spaces, providing cover, nesting, and foraging opportunities for wildlife. The wood resists decay, but it is brittle, and harvest difficulties have also limited its commercial importance. Fire suppression threatens the Sequoia, as species naturally limited by fire can thrive in the understory. The Sequoia's thick bark can normally resist fires, but the understory trees act as fire ladders, allowing the flames to reach the giants' highly flammable crowns.

Baldcypress
Taxodium distichum H 65–130 ft (20–40 m)

A majestic, deciduous conifer of wet areas, the Baldcypress in the deeper South is commonly home to the epiphytic, grayish bromeliad Spanish moss, hanging from its limbs.

KEY FACTS

The trunk base is enlarged, fluted, and surrounded by knobby, woody "knees."

+ **leaves:** Flat, 2-ranked, feather-like, turning russet in autumn

+ **cones:** Rounded, wrinkled seed cones

+ **range:** Southeast; wet areas in Coastal Plain, inland to Texas and up the Mississippi River to Illinois and Indiana

An array of cultivars of the Baldcypress have been developed, and the tree is grown coast to coast, and it can survive in areas that are drier and decidedly less gothic than its familiar swampy habitat—including on city streets. The species' most menacing disease is "pecky cypress," a sometimes deadly brown pocket rot caused by a fungus that attacks the heartwood. Insects damage the leaves, cone, and bark. Humans, too, pose a threat, by draining wetlands and overharvesting for timber. The attractive wood is easily worked and decay resistant, used in shingles, flooring, cabinetry, trim work, beams, and barrels.

Arborvitae/Eastern White Cedar
Thuja occidentalis H to 65 ft (20 m)

The Eastern White Cedar is at home in moist or wet soil, where it outcompetes other trees. It is most populous in the Great Lakes region and farther north and is somewhat shade tolerant.

KEY FACTS

Short branches to the ground form flat sprays; reddish, shredding bark.

+ **leaves:** Scales sometimes long-pointed

+ **cones:** Male and female cones on same tree; female greenish, turning brown, with 8–12 pointed scales

+ **range:** Manitoba to Hudson Bay, south to Iowa and South Carolina

This native evergreen can live more than 800 years, and individuals growing in Ontario are the oldest trees in eastern North America. The Ojibwe discovered its value in construction and medicine. After they taught 16th-century French explorer Jacques Cartier to use its vitamin C–rich foliage to treat scurvy, it earned the name Arborvitae ("tree of life" in Latin). The decay-resistant, fragrant wood is used as posts and in boats, cabins, and shingles. Its oils find purpose in disinfectants and insecticides. The Eastern White Cedar is popular as an ornamental, and numerous cultivars have been developed.

Pacific Silver Fir
Abies amabilis H to 151 ft (46 m)

The Pacific Silver Fir is commonly found in moist coastal coniferous forests, to the tree line in the mountains but to sea level from Vancouver Island, British Columbia, northward.

KEY FACTS

Spire-like crown rounds with age.

+ **needles:** Attached spirally but twisted at base, lying flat on and above the stem; dark green above, silvery below

+ **cones:** Male and female on same tree; cylindrical, upright seed cones, not stalked, green, becoming purple to brown

+ **range:** Alaska to northwestern California

Soft and weak, the Pacific Silver Fir's wood is an important source of pulp used in plywood, crates, and poles. But the tree is attractive, and cultivars have been developed, though they are planted mostly in its native range and in parts of New Zealand and Scotland that can supply the requisite cool, humid summers. Indigenous peoples in its range used its sap as chewing gum, made bedding of the branches, and burned the wood. The trees are associated with hemlocks and other firs, among others, assemblages that provide food and habitat for sooty and spruce grouse, squirrels, and a bird named Clark's Nutcracker.

Balsam Fir
Abies balsamea H to 66 ft (20 m)

The Balsam Fir, whose shape is a slim pyramid, is common in moist places and swamps, as well as on well-drained hillsides. Growth is strongest and fastest in full sun.

KEY FACTS

Seed bracts extend slightly if at all beyond cone scales.

+ **needles:** Two-ranked, flattened, grooved above

+ **cones:** Erect, rounded cylinders; purple, browning with age

+ **range:** From Alberta to Labrador south to the Great Lakes and northeastern states, south to Virginia in higher mountains

The Balsam Fir is common, often dominant in its moist habitats, occurring with White Spruce, Black Spruce, Trembling Aspen, or Paper Birch. It has been heavily damaged by the balsam woolly adelgid, an invasive insect introduced from Europe. The foliage and seeds provide food and cover for wildlife, including birds, squirrels, moose, and porcupines. In addition to providing pulpwood, this handsome tree is planted as an ornamental and favored as a Christmas tree. Its fragrant resin (balsam) is used in incense, rodent repellent, and to make Canada balsam, which has served as a cement for eyeglass lenses and been used to treat colds.

Fraser Fir/She-balsam

Abies fraseri H mostly to 80 ft (25 m)

The population of the Fraser Fir has been reduced as much as 95 percent by the Balsam Woolly Adelgid, an insect introduced into North America from Europe in the early 1900s.

KEY FACTS

The seed bracts protrude between cone scales.

+ needles: As in Balsam Fir, with a pine-like scent

+ cones: Male and female on same tree; seed cones purple, browning with age

+ range: Rare in high Appalachian forests of southwestern Virginia and adjacent North Carolina and Tennessee

In the wild, the Fraser Fir grows in pure stands or with other evergreens and with birches and other hardwood. It is popular as a Christmas tree, for which it is farmed, but it is intolerant of warm climate, so its culture is being reconsidered in some areas where temperatures are increasing. For better or for worse, it is sometimes called She-balsam because resin can be "milked" from its bark blisters (not true of the balsam-bearing Red Spruce *Picea rubens,* called He-balsam). "Balsam" describes fragrant oily or resinous substances found in a number of trees and shrubs that are not necessarily close botanical relatives.

Grand Fir

Abies grandis H to 260 ft (80 m)

The majestic Grand Fir reaches its greatest size in the rain forests of Washington's Olympic Peninsula and inhabits moist forests from sea level to 4,900 feet (1,500 m).

KEY FACTS

Branches are short and drooping.

+ needles: Shorter at stem tips; 2-ranked, dark green above, 2 white bands beneath

+ cones: Seed cones barrel-shaped, upright on crown branches; seed bracts not visible

+ range: Southern British Columbia to northern California

This stately, long-lived fir is common in its range, growing in mixed coniferous and hardwood stands, often with Douglas-fir and Larch. The Grand Fir's resin has been used in wood finishing, and its wood is valuable in the paper and building industries. The tree has also been used medicinally. Commercially, it is most important in Idaho, although in general it has not proved to be a good species for landscaping outside its native range. It is an important source of food and cover for small mammals, seed-eating birds, and game birds, and it is often a host to the parasitic fir Dwarf Mistletoe.

California Red Fir
Abies magnifica H to 120 ft (37 m)

A large evergreen conifer that often dominates the montane forests of California and Oregon, the California Red Fir is distinguished by its blue-gray needles and reddish brown bark.

KEY FACTS

Short, horizontal limbs make the tree slim.

+ **needles:** Sharp, curving atop shoot; with whitish coating when young, aging to bluish green, with white bands beneath

+ **cones:** Seed cones rounded, cylindrical, borne at crown; bracts not visible beyond scales

+ **range:** Mountains of Oregon and California

At home in altitudes from 5,250 to 9,350 feet (1,600 to 2,850 m), the long-lived California Red Fir often grows in dense, pure stands, but is sometimes associated with other conifers when near the tree line. Its branches have a camphor-like aroma. As in many firs, the fatty seeds provide energy-rich food for rodents and other small mammals, and the shoots are fodder for deer. Its wood is stronger than that of other firs, and it is an important source of wood for the pulp industry and construction and as fuel and Christmas trees. The tree is plagued by heart rot and Dwarf Mistletoe.

Tamarack/Eastern Larch
Larix laricina H to 80 ft (25 m)

A slow-growing, conic, deciduous conifer, the Tamarack is heat intolerant, which explains its greatly northern distribution and why it is seldom sought in landscaping far from its native range.

KEY FACTS

Bark is scaly, pinkish to red-brown.

+ **needles:** Soft, 3-sided, in tufts on spurs, turning yellow before falling

+ **cones:** Male and female in separate clusters on same plant; seed cones upright, unstalked, egg-shaped, persisting

+ **range:** All Canadian provinces, Great Lakes and northeastern states

A common tree of swampy forests and moist uplands in its boreal zone, the Tamarack is one of our most northerly trees and has a large natural range. It grows in pure stands or is found in conjunction with Balsam Firs and other conifers. Browsed but not seriously harmed by deer, a tree can fall victim to porcupines that girdle it while feeding on its bark. The larch sawfly can damage or even kill Tamaracks by defoliating them. The wood is hard and resinous and provides lumber used for railroad ties, construction, poles, and pulp. Wild animals use its seeds.

White Spruce
Picea glauca H to 131 ft (40 m)

Our northernmost tree, the White Spruce is widespread in bogs, on bodies of water, and on rocky hills, often found with such trees as Balsam Fir, Eastern Hemlock, and Red Maple.

KEY FACTS

This species is dense, with a conical crown, becoming cylindrical.

+ **needles:** Four-sided, individual, on peg-like stalk

+ **cones:** Unisexual, on same tree; seed cones pendulous, slender, maturing to brown; scales flexible

+ **range:** Alaska through all Canadian provinces, Great Lakes states, to New England

The White Spruce is valued in Canada for lumber and paper pulp, and it is popular for Christmas trees, landscaping, and as a windbreak. Its needles are aromatic—but they are not exactly mountain fresh and have earned it the labels "skunk spruce" and "cat spruce." *Picea* is Latin for "pitch," a reference to resins in the bark. Spruce Beetles have destroyed 2.3 million acres in Alaska; it is susceptible to Spruce Budworm and Spruce Sawfly, and rusts make it shed its needles early. South Dakota's state tree is a variety known as the Black Hills Spruce (*P. glauca* var. *densata),* but not all authorities recognize the variety.

Black Spruce
Picea mariana H to 98 ft (30 m)

A widely distributed, small, slow-growing conifer of the continent's coldest regions, the Black Spruce inhabits moist flatlands and lake margins and, in its southern extent, sphagnum bogs.

KEY FACTS

This spruce is spire-like; its branches droop with ends upturned.

+ **needles:** Four-sided, blue-green above, paler, powdery below

+ **cones:** Unisexual, on same tree; seed cones clustered in crown

+ **range:** Alaska through all Canadian provinces to Great Lakes states and New England

In addition to reproducing by seeds, the Black Spruce can propagate asexually; the lower limbs can touch the ground, often under the weight of snow, and sometimes take root, creating a circle of smaller trees around the main trunk of the parent. The tree is similar to the Red Spruce but is usually found in more extreme conditions and is more northern. Its form varies; for example, at the tree line, it is often prostrate. Because the tree is small, the wood is useful for little more than pulp and fuel. It is of low value for wildlife food but provides cover for small mammals and important nesting habitat for birds.

Blue Spruce/Colorado Spruce

Picea pungens H 65–82 ft (20–25 m)

The slow-growing Blue Spruce is a medium-size evergreen probably best known from its widespread planting in Canada, the U.S., and Europe. At least 38 cultivars have been developed.

KEY FACTS

In aging, this spruce becomes more layered.

+ **needles:** Bluish green, sharp, 4-sided, on all sides of stem

+ **cones:** Unisexual, on same tree; seed cones slender, cylindrical, hanging from upper branches

+ **range:** High in Rocky Mountains from Idaho and Wyoming across Utah and Colorado to Arizona and New Mexico

The Blue Spruce has been called the "most beautiful species of conifer" for its stately form and "vibrant" blue to silvery hue (though this is more common in cultivars than in natural trees). Its primary value is in its looks, which horticulturists have exploited. It usually has a sporadic distribution or is a component of mixed-conifer forests, with Douglas-fir, Engelmann Spruce, and others. Not a favorite forage for wildlife, it does provide seeds for birds and cover for deer. It is plagued by the Spruce Bark Beetle, Spruce Gall Aphid, Spruce Budworm, and Spruce Spider Mite.

Engelmann Spruce

Picea engelmannii H to 164 ft (50 m)

With the Subalpine Fir *(Abies lasiocarpa)*, Engelmann Spruce forms one of the most frequent forest types in the Rocky Mountains. At high elevations, it is one of the largest trees.

KEY FACTS

A spire-like tree, its trunk has few small limbs between branches.

+ **needles:** Linear, 4-sided, blue-green

+ **cones:** Unisexual on same tree; seed cones yellow to purple-brown; bracts hidden by thin scales notched at tip

+ **range:** Yukon, British Columbia, and Alberta south to Arizona and New Mexico

Soft and knotty, Engelmann Spruce wood is not prized for lumber, restricted to framing and other unseen construction uses and to paper manufacturing. Instead, it enters the limelight in construction of guitars and piano soundboards. The bark provides tannin, and cord has been made from its branches and roots. Young male cones are eaten raw or cooked, used as flavoring, and added to breads and cereals; the seeds are small but edible. A vitamin C–rich infusion is made from the young shoots. Another, made from the bark, is used to treat respiratory ailments. Various resins have been used to treat eczema.

Red Spruce/He-balsam

Picea rubens H to 148 ft (45 m)

A handsome, medium-size evergreen, the Red Spruce remains one of the most important forest trees in the Northeast, despite having been overexploited for myriad purposes.

KEY FACTS

Crown is conical but broader than in other eastern spruces.

+ **needles:** Four-sided, curved, on all sides of twig, bright green

+ **cones:** Male and female on same tree; seed cones long, egg-shaped, red-brown; scales untoothed

+ **range:** Cape Breton to Ontario, south in mountains to North Carolina

Red Spruce grows in pure stands or with other conifers such as Eastern White Pine, Balsam Fir, and Black Spruce. Light, straight-grained, and resilient, its wood is used in construction and paper-making; its resonance makes it sought after for building guitars, violins, piano soundboards, even organ pipes. Its resin was used into the 20th century in chewing gum (replaced now by a substance from a tropical plant). The buds, foliage, and seeds provide food for small mammals and birds (up to half the diet of a White-winged Crossbill). The Spruce Bud-worm is especially damaging where this spruce grows with Balsam Fir.

Sitka Spruce

Picea sitchensis H to 197 ft (60 m)

The largest spruce, the tall, grand Sitka, can be found in its long band of Pacific habitat, from 980 feet (300 m) to sea level, its range inland determined by that of ocean fogs.

KEY FACTS

The trunk base is buttressed.

+ **needles:** Yellow- to blue-green, sharp, flat, with 2 white bands beneath

+ **cones:** Male and female on same plant; seed cones cylindrical, hanging from upper shoots; scales thin, irregularly toothed

+ **range:** Narrow strip along coast from south-central Alaska to northern California

Different parts of the Sitka Spruce have many uses in food, varnish, medicine, and for making rope and cord. The wood is exceptionally strong, thanks to its long, straight grain, especially when its light weight is considered. That combination caught the imagination of airplane designers in both world wars. So did its cost, then less than that of steel or aluminum. It is still used in aerobatic craft. (The largest plane ever built was an immense seaplane, made entirely of wood, called the "Spruce Goose"—even though it was made of birch.) Sitka Spruce is a food source for birds and small mammals. It is rarely cultivated.

Jack Pine

Pinus banksiana H to 89 ft (27 m)

Our most northerly ranging pine, the Jack Pine is usually a small to medium-size tree that soon loses the pyramidal form of its youth to assume a more gnarled look.

KEY FACTS

Fire is often needed to open cones.

+ needles: In 2s, olive to gray-green, stiff, curved to slightly twisted; margins rough

+ cones: Crooked toward twig tip; scales thick, stiff, with fragile spine; resinous

+ range: Much of Canada, Great Lakes states, New England, south to Missouri, West Virginia

Large stands of Jack Pine are requisite as breeding habitat for the endangered Kirtland's Warbler *(Setophaga kirtlandii)*, which nests almost exclusively in the Lower Peninsula of Michigan, near the southern extent of the tree's range. Historically, fires maintained this open habitat, but now it must be maintained by controlled burning. The tree supplies wood for use as pulp, posts, and firewood. Young trees are a host for the sweet fern blister rust, which causes orange cankers to grow on the trunks, and galls to form on the lower branches. Mature trees are defoliated by the Jack Pine Budworm.

Lodgepole Pine

Pinus contorta H to 98 ft (30 m)

Often found in expansive pure stands, the slender Lodgepole Pine is expected to decline drastically as a result of warming in its cool range, drought, and epidemic levels of Pine Sawyers.

KEY FACTS

Inland, this pine is tall, slender, and straight; in coastal or wet areas, it is shrubby and twisted.

+ needles: In 2s, curved, thick, stiff

+ cones: Broadly oval; scales with curved spine

+ range: Alaska to Baja California; most abundant in northern Rockies and Pacific coast area

American Indians used the slim, flexible limbs to build tepees (thus its common name), the soft inner bark for food, and the sap in medicines. The wood is used in framing, posts, railroad cross ties, and paneling, and to make pulp. It is a close relative of the Jack Pine, with which it hybridizes where the two occur together. Like its relative, it is plagued by Pine Sawyers; the insects spread fungi, and, girdling the tree, they kill it. But in an odd turn of events, the dead trees provide fuel for fires, enhancing germination of Lodgepole seeds. Seedlings outperform those of other species.

Shortleaf Pine
Pinus echinata H 80–100 ft (24–30 m)

The Shortleaf Pine is second in commercial importance (behind Loblolly) in the Southeast, where it is used for turpentine production, plywood, flooring, beams, and (taproot included!) pulp.

KEY FACTS

The crown is pyramidal to rounded; the trunk is clear of scrubby limbs.

+ needles: In 2s and 3s, long, slender, dark blue-green

+ cones: Cylindrical, conical, or egg-shaped; scales thin, prickle-tipped

+ range: Southeast; also New York and New Jersey to Missouri, Oklahoma, and Texas

The most wide-ranging yellow pine of the Southeast, the Shortleaf has needles that could be deemed short only in comparison with those of other native southern pines. The branches are irregular, compared with the straighter ones of the Loblolly, with which it hybridizes where their ranges overlap (as it also does with the Pitch Pine). Its seeds provide cover and forage for birds, squirrels, chipmunks, and mice. The Shortleaf's plated bark has resin ducts, appearing as small holes. In addition to Pine Sawyers, it is threatened by littleleaf disease, caused by a root fungus in poorly drained soils.

Piñon Pine/Two-needle Piñon
Pinus edulis H to 20 ft (6 m)

A handsome, small, gnarled tree, the Piñon has a dense, rounded, conic crown. Increased droughts have recently killed trees whose habitat is the less dry portion of the species' range.

KEY FACTS

Branches persist almost to the trunk base; the limbless portion is taller, and the silhouette is more irregular with age.

+ needles: In 2s, sometimes 1s or 3s, dark green, curving

+ cones: Spherical; scales lack prickle

+ range: Utah, Arizona, Colorado, New Mexico; small numbers in adjacent states

The seeds of the slow-growing Piñons do not fall when the cone opens. Their dispersal depends on the foraging of Pinyon Jays (named for the tree), which pluck the seeds from the cone scales and then stash them for future eating. Some seeds inevitably go forgotten and eventually germinate. The jays have mammalian competitors. Humans also collect the seeds but do not lose track of them so often. Indigenous peoples have harvesting rights to these traditional foods—called pine nuts in English, *pignoli* in Italian—and there is a commercial harvest. The wood is used as fuel, for posts, and in incenses.

Slash Pine

Pinus elliottii H 59–100 ft (18–30 m)

Usually inhabiting slashes, or overgrown swampland, Slash Pines grow fast and so are favorites in reforestation programs.

KEY FACTS

These have a rounded crown and reddish brown bark, becoming plated.

+ needles: In 2s or 3s; long, straight, stout

+ cones: Long, conical to ovoid, glossy, stalked; scales with outcurved prickle

+ range: Coastal Plain of South Carolina to Florida Keys and Louisiana; naturalized in nearby states

A tree of warm, humid flatwoods in poor, sandy Coastal Plain soils, the Slash Pine is grown in suitable areas worldwide for its wood, which is heavy, strong, durable, and compares favorably with that of the Longleaf Pine; its resin is used to make turpentine and rosin. (The species has become invasive in Hawaii and Australia and is under scrutiny in the entire Pacific Rim.) It is like the related Loblolly, except the Slash Pine often has needles in twos (mostly threes in the Loblolly) and larger cones. Large stands are especially hard-hit by the fusiform rust fungus, the most serious disease of pines in the Southeast.

Sugar Pine

Pinus lambertiana H 98–164 ft (30–50 m)

Our largest, most majestic member of the genus *Pinus,* the long-lived Sugar Pine inhabits mainly mixed-conifer forests. It can take 100 years or more to begin producing seeds.

KEY FACTS

This pine grows mostly on damper northern slopes. It has heavy seed cones (largest of any tree), which can weigh down branch tips.

+ needles: In 5s, blue-green, straight, slender

+ cones: Long, brown (purple when young), long-stalked, hanging

+ range: Mountains of Oregon and California

The Sugar Pine takes its name from the sweet-smelling and sweet-tasting resin that oozes from injured wood and forms candy-like beads. It is said to taste better than maple sugar, a trait native peoples took advantage of—though in moderation, because it is a laxative. They also used the resin to affix points and feathers to arrow shafts. The timber is appreciated for its workability, lightness, and strength and is used in framing and molding. Despite the tree's rapid growth rate, it cannot keep pace with harvests. Although it is planted in its native range, it has not been successful in culture elsewhere.

Singleleaf Piñon
Pinus monophylla H 16–66 ft (5–20 m)

The Singleleaf Piñon is often the dominant tree, a key element in the piñon–juniper woodlands of the mountains. As do other piñons, it depends on the Pinyon Jay for dispersal of its seeds.

KEY FACTS

Bark is gray and scaly, developing reddish furrows; the trunk branches to the base.

+ **needles:** Gray-green to blue-green, mostly solitary; stout, curved, sharp-pointed

+ **cones:** Broadly egg-shaped; a few woody scales

+ **range:** Idaho south to California, Nevada, Arizona, and New Mexico

Humans have long eaten the seeds of the Singleleaf Piñon. The seeds are known as pine nuts, but American Indians still roast the cones and grind the seeds to make traditional soups and cakes. This practice was disrupted in the late 1800s, when the resin-rich trees were harvested for use as a fuel in smelting silver. More recently, deemed low-quality forage for livestock, the trees have been destroyed in favor of other species. Other uses are as fence posts and Christmas trees. The parasitic Dwarf Mistletoe affects the Singleleaf, and a number of insects and fungal diseases especially harm piñons.

Western White Pine
Pinus monticola H 45–160 ft (14–49 m)

Closely resembling its relative the Eastern White Pine, this species grows fast and is quick to become established following a disturbance, such as fire.

KEY FACTS

This tall and narrow pine's trunk is clear of scrubby branches.

+ **needles:** In 5s, long, sometimes twisted, blue-green, whitish waxy wash

+ **cones:** Long, narrow cylinders hang from a long stalk; scales thin, curved back

+ **range:** British Columbia to California; Alberta to Nevada and Utah

Primarily a mountain tree, this species ranges to sea level at its northern extent. The wood is light, soft, straight-grained, and nonresinous. Easily worked and excellent for finer products than that of many conifers, it is used in interior details, trims, floors, and for matches and toothpicks. White Pine blister rust, caused by a fungus native to Asia and introduced into North America via Europe around 1900, attacks the five-needle pines and is especially grave in this species, having killed as much as 90 percent of trees in some regions. Bears, tearing into trunks for the sweet sapwood, also cause irreparable harm.

Longleaf Pine

Pinus palustris H 80–130 ft (24–40 m)

The grand Longleaf Pine was once widespread, the keystone of unique habitats of the southeastern Coastal Plain.

KEY FACTS

The needles are longest of our pines; the branches are sparse, relatively short, and stout; the crown is open and asymmetrical.

+ **needles:** In 3s, slender, flexible, usually hanging

+ **cones:** Largest of eastern pines; woody, hanging

+ **range:** Coastal Plain sandhills and flats, Virginia to Texas

As a seedling, the Longleaf Pine passes through a "grass stage" during which it grows thicker instead of taller, and its taproot becomes established; the long needles at this unbranching stage recall tufts of bunchgrasses. Historically, fires created a savanna characterized by the enormous pines and other fire-adapted species. Overharvest (for lumber, pulp, turpentine, and resin, especially for naval purposes), elimination of fires, and management for other species wiped out up to 95 percent of this habitat, but the species and the savannas are being reestablished in much of its former range by extensive plantings maintained by prescribed burning.

Ponderosa Pine

Pinus ponderosa H 60–140 ft (18–39 m)

Our most widely distributed and most abundant pine, the stately Ponderosa gets its name from the Spanish adjective meaning heavy, or ponderous.

KEY FACTS

Mature bark is cinnamon and plated.

+ **needles:** Mostly in 3s, with tiny teeth on edges, sharp, with turpentine odor

+ **cones:** Male and female on same tree; seed cones reddish, egg-shaped, not stalked; scales with stout prickles

+ **range:** Pacific coastal mountains and Rockies

The Ponderosa Pine is found in pure stands and in mixed-conifer forests, where it towers over other trees. One of the most important sources of timber in the West, its wood is used for construction and cabinetmaking, as well as for pulp and firewood. Though sometimes used for Christmas trees, it is not farmed. Larvae of the moth *Chionodes retiniella* feed exclusively on the Ponderosa's needles. Government fire suppression since the early 1900s threatened the Ponderosa's native parklike habitat, because its competitors were no longer killed by the flames. It is now often managed with the aid of controlled burning.

Red Pine/Norway Pine

Pinus resinosa H 50–100 ft (15–30 m)

Norway Pine is a misnomer for this North American native. It was probably so labeled by European explorers and settlers for whom it recalled the Norway Spruce.

KEY FACTS

This pine of cold regions usually towers over other trees; the crown is dense and rounded.

+ **needles:** In 2s, sometimes twisted, snapping when bent

+ **cones:** Seed cones near branch tips, small, egg-shaped, not prickly

+ **range:** Cape Breton Island to Manitoba, around the Great Lakes, south to Virginia

Formerly one of the main timber trees in much of its range, the Red Pine has light, close-grained wood, which is well suited for construction and the manufacture of pulp. As in many pines, the resin is made into turpentine. Vanillin, used in flavorings, is isolated during pulp processing. The tree is often cultivated as an ornamental or as a shade tree, and several cultivars, including a dwarf version, have been developed. Fire is needed to create the condition the seeds need for germination, namely bare mineral soil, and when there are no fires, other species will eventually replace it.

Eastern White Pine

Pinus strobus H 50–220 ft (15–67 m)

The Iroquois tree of peace, the Eastern White Pine once had extensive stands in the Northeast but has been so exploited that only one percent of those stands survive.

KEY FACTS

Branches are tiered, long, and horizontal.

+ **needles:** In 5s, light green to bluish, edges rough

+ **cones:** Seed cones slim, hanging, stalked; resinous, and somewhat curved

+ **range:** Newfoundland to Manitoba, around Great Lakes; New England to West Virginia; mountains south to Georgia

The White Pine was already being cut by colonists by the 1650s, and aggressive logging for the next two centuries did irreparable damage. The tree provided resins, pitch, and turpentine for use on ships, and its straight, strong trunks were ideal for ships' masts. Today, the wood is used in construction and for trim, furniture, and pulpwood, the demand met by cultivation. Many cultivars are grown, including dwarf and weeping varieties. In the United Kingdom, it is called the Weymouth Pine, for British explorer George Weymouth, who returned in 1605 with seeds he had collected in what is now Maine.

Loblolly Pine
Pinus taeda H 60–140 ft (18–43 m)

Cultivated varieties of the large, fragrant, and resin-rich Loblolly Pine include a 20-foot version often planted in windscreens; the wild type is not suitable because it sheds its lower branches.

KEY FACTS

This fragrant tree is pyramid-shaped when young, and eventually drops lower branches.

+ **needles:** In 3s, slender, stiff, sometimes slightly twisted

+ **cones:** Seed cones woody; scales thin with short prickle

+ **range:** Mainly in Coastal Plain and Piedmont region from southern New Jersey to eastern Texas

The Loblolly's rapid growth habit is a mixed blessing. The tree is one of the first to be naturally established in abandoned fields and is a good soil stabilizer. It is the most commercially important southern pine, but often at a price. Raised in dense monocultures, it is productive in lumber and forest products and renewable for those purposes, but these tree farms never will revert to the mixed-hardwood stands that they often replaced and that otherwise would develop. Mixed forests sustain greater biodiversity, and their wholesale substitution by depauperate Loblolly stands is receiving attention in restoration projects.

Virginia Pine
Pinus virginiana H 25–49 ft (10–15 m)

Common in the Piedmont and foothills of the Appalachians, the Virginia Pine has been called the Oldfield Pine because it is one of the first trees to grow on abandoned agricultural fields.

KEY FACTS

This smallish pine's limbs grow irregularly from the trunk, which often bears stubs of old branches.

+ **needles:** In 2s, short, pointed, twisted, spreading, serrate

+ **cones:** Seed cones small, conical to egg-shaped, woody; scales with slim, curved prickle

+ **range:** New York to Ohio, south to Mississippi

Thanks to its scrubby look, the Virginia Pine is also known as Scrub Pine, but its low branches remain on the tree, which helps make it the most popular species in the South for use as a Christmas tree. It grows in pure stands or mixed with various other trees, including hardwoods, which, in nature, will eventually shade it out. The Virginia Pine pioneers on old fields and is planted on abandoned fields and cutover lands. The wood is often used for pulp or rough lumber. In older trees, woodpeckers nest in the trunk, creating cavities where fungi have softened the wood.

Douglas-fir

Pseudotsuga menziesii H 80–150 ft (24–46 m)

The grand, long-lived Douglas-fir is among the most important of our timber trees. Unlike the upright cones of the true firs, those of the Douglas-fir hang from the branches.

KEY FACTS

Pyramidal when young, the tree loses its lower branches later for a cylindrical appearance.

+ **needles:** Flattened, on slim twigs

+ **cones:** Seed cones with 3-pointed seed bracts projecting beyond the rounded scales

+ **range:** Coast to mountains from British Columbia to California, and to tree line in Rockies

One of the largest trees in the world (unusually can reach 300 feet/92 m), the Douglas-fir produces a great volume of timber, which is used in support beams, interior woodworking, and plywood veneer. The species is used in landscaping, and numerous cultivars have been developed. In fact, the tree is planted as an ornamental in cool regions worldwide, and can be invasive outside its natural range. A variety called the Rocky Mountain Douglas-fir (*P. menziesii* var. *glauca*) is smaller, its foliage with a bluish or grayish cast (i.e., glaucous). Its seed bracts curve back on themselves, pointing toward the cone base.

Eastern Hemlock

Tsuga canadensis H 66–100 ft (20–30 m)

In the southern reaches of its range, the Eastern Hemlock is under attack by the Hemlock Woolly Adelgid, an insect from Asia that feeds on a tree's sap, killing the tree in only a few years.

KEY FACTS

Branches are drooping, with pubescent twigs and fissured bark.

+ **needles:** Flattened, in 2 ranks

+ **cones:** Seed cones to 0.8 in (2 cm), egg-shaped, stalked, borne at branch tips

+ **range:** Ontario to Nova Scotia, Minnesota through northeastern states, to mountains of Georgia and Alabama

The wood is of poor quality, sometimes used in general construction and as pulp. Many cultivars have been developed, including dwarf, shrubby, and weeping forms. The related but geographically distinct Western Hemlock—*T. heterophylla* (H 130–230 ft/40–70 m)—is the largest species of the genus, common and widespread in its range from Alaska to California (especially Oregon and Washington), as well as Montana. Its seed cones are 0.75–1 inch (2–2.5 cm) long, unstalked, its twigs finely pubescent. The hard, durable, and light wood makes it commercially superior to that of other hemlocks. It is rarely cultivated.

Pacific Yew
Taxus brevifolia H 15–50 ft (5–15 m)

A slow-growing evergreen, the Pacific Yew has lithe branches that often appear drooping or weeping.

KEY FACTS

The Yew is nonresinous, with thin, reddish brown bark.

+ **leaves:** Flat, pointed; borne spirally but appear in 2 ranks

+ **cones:** Berrylike, one seed surrounded by red, fleshy aril

+ **range:** Alaska to central California in mountains; isolated population from southeastern British Columbia to Idaho

A chemical compound isolated from the Pacific Yew (and found in other yews) is used in chemotherapy for cancer. The compound, known as the generic drug paclitaxel, from the name of the genus, is now manufactured with semisynthetic and cell culture procedures that do not harm the trees. The attractive wood is hard, heavy, strong yet flexible, fine grained, and easily worked. But the supply is limited, and the small size of the trunk and limbs limits their use to such things as canoe paddles, carvings, musical instruments, harpoon and spear handles, archery bows, and firewood. Grown ornamentally, the tree is especially good in hedges.

Common Sumac/Smooth Sumac
Rhus glabra H to 23 ft (7 m)

Common Sumac is a familiar shrub or small tree along roadsides, field edges, and train tracks. In the fall, red leaves and, on female plants, clusters of dark red berries, are its trademark.

KEY FACTS

This fast-growing sumac is nonpoisonous, but is related to Poison Ivy.

+ **leaves:** Pinnately compound; 7–31 narrow, toothed leaflets

+ **flowers/fruits:** Male and female inflorescences are on separate plants, at branch ends.

+ **range:** All 48 contiguous U.S. states and across Canada; most common in East

Common Sumac is popular in gardens for its striking red autumn foliage—giving it another name, Scarlet Sumac. But it can be weedy, spreading via underground stems. It may be confused with its look-alike cousin Poison Sumac (p. 91, *Toxicodendron vernix*), but the latter is rarely encountered, restricted to moist, even wet areas. Also, in *T. vernix*, male and female flowers grow on the same plant, the leaves have fewer leaflets, and the fruit clusters grow in the axils where two branches diverge. Sumac seeds provide food for birds and insects into the winter, and those of Common Sumac can be made into a lemonade-like beverage.

Pawpaw

Asimina triloba H to 40 ft (12 m)

The Pawpaw has few insect pests, possibly because its leaves contain unsavory chemical compounds. Compounds in the seeds show promise for use in cancer chemotherapy.

KEY FACTS

Colonies are often created by suckering.

+ **leaves:** Alternate, to 1 ft (30 cm) long; oval, narrowing basally

+ **flowers/fruits:** Flowers bell-shaped, 6-petaled, purple; berries fleshy, large, roughly cylindrical, ripening to yellow, then brown

+ **range:** Southern Ontario, eastern states west to Nebraska, Texas

The Pawpaw is a northern representative of the tropical custard apple family. Pawpaws were once harvested, but their supply has dwindled as forests have been cut. The fruit is delicious, but fierce competition from wildlife makes it a prize indeed. Contact with the fruit and leaves can occasionally cause dermatitis. The species is cultivated in its native range, but its wood has no value. The flowers' putrid scent attracts small flies and beetles, which effect pollination. *Asimina* comes from an American Indian name for the tree; the common name probably echoes the Spanish *papayo*, a word of Arawak origin in the West Indies describing a similar but unrelated fruit, the papaya.

American Holly

Ilex opaca H 20–49 ft (6–15 m)

An evergreen broadleaf, the American Holly is a common shrub or small tree of the understory, most common in the humid southeastern United States.

KEY FACTS

This holly sometimes has multiple trunks; bark is smooth and gray.

+ **leaves:** Alternate, oval, stiff, sparsely but strongly toothed; tooth tips sharp

+ **flowers/fruits:** Flowers unisexual, on separate trees, tiny, inconspicuous; fruits orange or red berrylike drupes, often dense

+ **range:** Mainly eastern and southern U.S.

The American Holly is the favorite of the evergreen hollies, with its handsome foliage, pyramidal form, and beautiful berries, ubiquitous in holiday and winter arrangements wherever it grows or is grown. More then 300 cultivars have been developed (but a female tree is needed, with a male nearby, if berries are expected). The berries also provide food for many birds and small mammals, which disperse the seeds. The wood is white and not important as a source of lumber. It is used in veneers, as pulpwood, and in creation of small wooden objects. The slow-growing tree tolerates a range of soil types.

Common Winterberry
Ilex verticillata H 8–15 ft (2.4–4.5 m)

The deciduous Common Winterberry is a popular ornamental and is striking in winter with its red berries borne on bare branches with dark, smooth bark.

KEY FACTS

This small tree is multi-trunked.

+ **leaves:** Alternate, elliptic, pointed; margin toothed, lacking spines

+ **flowers/fruits:** Flowers unisexual, on separate trees; greenish white; fruits bright-red berrylike drupes, often dense

+ **range:** From Quebec to Newfoundland south to Florida and Louisiana

The Common Winterberry is most abundant in wet areas, along streams, and in moist wooded habitats of the Coastal Plain. The red berries lend color to the winter landscape until the birds make off with them. A number of cultivars have been developed, and male and female cultivars must be paired up to ensure that they flower at the same time. The bark of the Winterberry was listed in the U.S. Pharmacopeia for most of the 19th century. Concoctions made from it have been used as an astringent, a bitter tonic, or a febrifuge (thus another of its common names, Feverbush).

Yaupon
Ilex vomitoria H 12–25 ft (3.7–7.6 m)

As in other hollies, the fruits of the Yaupon are an important winter food for birds and small mammals, which distribute the seeds.

KEY FACTS

This evergreen shrub branches heavily inside; can form dense thickets.

+ **leaves:** Alternate, oval, shiny, thick, small

+ **flowers/fruits:** Flowers unisexual, on separate plants, inconspicuous, greenish white; fruits bright-red drupes, numerous

+ **range:** Virginia to Florida, west to Oklahoma and Texas

The Yaupon's species label, *vomitoria*, is a misnomer. Certain American Indian tribes drank a ceremonial infusion that caused purging, and Yaupon was erroneously thought to be the emetic agent. Its twigs and leaves contain caffeine and are used to make a tealike drink not unlike maté, a beverage made from another holly, *I. paraguariensis*, and enjoyed in parts of South America. The name "Yaupon" is from a Catawban word related to the word for tree. This holly, too, is a popular ornamental, used for winter color in the landscape and in holiday decorations; cultivars include dwarf and weeping forms.

Devil's Walkingstick
Aralia spinosa H 15–23 ft (4.6–7 m)

The Devil's Walkingstick is in the Ginseng family, along with the invasive English Ivy, the *Panax* ginsengs, and the popular houseplant Schefflera.

KEY FACTS

This plant bears short, strong spines.

+ **leaves:** Alternate, to 32 in (81 cm) long, compound to doubly so; leaflets opposite, oval, pointed

+ **flowers/fruits:** Flowers unisexual, on same plant; tiny, white, in broad clusters; fruits fleshy, black berries

+ **range:** New Jersey to Florida, west to Texas and Missouri

Though its description may sound menacing, the Devil's Walkingstick is appreciated and even grown ornamentally for its intricate foliage, appealing inflorescences and fruits, and fall color. Its leaves turn bright yellow, and the black fruit are borne in clusters on bright pink stalks. The berries were eaten by the Iroquois, and also provide food for birds and small mammals, as well as black bear. The plant, a shrub or small tree of the understory, often forms dense clusters and has an open, irregular, flat crown. It is aromatic and deciduous, and its bark, roots, and berries have been used medicinally.

Big Sagebrush
Artemisia tridentata H 2–13 ft (0.6–4 m)

This evergreen member of the Aster family is one of the most wide-ranging shrubs in western North America, a dominant species to the tree line in much of the Great Basin.

KEY FACTS

The shrub is covered with fine gray hairs.

+ **leaves:** Alternate, wedge-shaped, 3-lobed at tip

+ **flowers/fruits:** Flowers perfect, tiny, tubular, yellow, in heads of 3–12; fruits seedlike, dry, hard, flat, broadest toward tip

+ **range:** British Columbia to Baja California, east to the Dakotas

The coarse, aromatic Big Sagebrush is a plant of dry, rocky soils that can live 100 years or longer. While not related to the sages, which are in the mint family, it has an aroma and a growth form that are not unlike those of some sages. Protein-rich (more nutritious than Alfalfa), it serves as forage for many large herbivores, especially in winter. Today it is used mainly as firewood and in smudges and incenses—its aromatic oils burn strongly and fragrantly. Native peoples, too, used it for those purposes but also made ropes and baskets from it and used it medicinally to treat a range of symptoms, as a tea and as a disinfectant.

Red Alder
Alnus rubra H to 60 ft (18 m)

Alnus rubra is our tallest alder and, unlike the others, a commercially viable source of timber. In fact, it is the Pacific Northwest's most important hardwood tree.

KEY FACTS

Inner bark and heart-wood are red, thus the common name.

+ **leaves:** Alternate, oval, pointed, with rounded teeth

+ **flowers/fruits:** Catkins unisexual, on same plant, in clumps, male drooping, female erect; fruits small cones

+ **range:** southeastern Alaska to southern California, within 125 miles of Pacific

Native peoples made a red dye from the inner bark and used it to color their fishing nets, which made them less visible to fish. The tree also had medicinal uses, possibly because it contains, as do the willows, aspirin-like chemical compounds that show antitumor properties. Its rapid growth makes it a good pioneer on disturbed lands, which it stabilizes. It is short-lived because when other trees invade, it cannot survive in the resulting shade. Associated with its roots are nitrogen-fixing bacteria, which provide the trees nitrogenous nutrients. The wood is used in furniture, pallets, plywood, spools, and boxes, and in pulp and as firewood.

Hazel Alder
Alnus serrulata H 12–20 ft (3.6–6 m)

Primarily an Atlantic coastal species, the Hazel Alder is a common large shrub or small tree of moist lowlands and stream banks, where it is mostly found in mixed stands.

KEY FACTS

This alder can be spindly with crooked trunks.

+ **leaves:** Alternate, oval, edge wavy, finely serrate

+ **flowers/fruits:** Catkins unisexual, on same plant, male pendent, female erect; fruits small, woody cones, lasting into winter

+ **range:** Nova Scotia and New Brunswick to Florida, west to Oklahoma and Texas

The Hazel Alder's growth habit makes the tree a superior colonizer and ideal in restoring wetlands, stabilizing stream banks, and mitigating storm-water runoff. When it is sold, it is generally for such purposes. The plant has an extensive root system and forms broad thickets by suckering. The roots are extensive and have symbiotic bacteria that fix nitrogen, providing the tree with nitrogenous nutrients. It is an important component of the American Woodcock's habitat and provides food and shelter for many other birds and small mammals. White-tailed Deer browse on the plants. The tree has had a range of medicinal uses, including as a pain reliever during childbirth.

Yellow Birch

Betula alleghaniensis H 60–75 ft (18–23 m)

The largest and most important birch, this species provides 75 percent of the birch wood used in the United States, and half of the species' entire stock is in Quebec.

KEY FACTS

Yellow-bronze bark (hence the common name) has horizontal lenticels.

+ **leaves:** Alternate, sharp-tipped, toothed

+ **flowers/fruits:** Catkins unisexual, on same tree, male pendent, female upright; fruits cone-like

+ **range:** Southeastern Canada; northeastern U.S., Great Lakes states, south in mountains to Georgia

A species of lower elevations, the Yellow Birch is a slow-growing, long-lived, single-trunked tree. When solitary, it develops a broad, open shape; when crowded, the trees grow tall and slender. The leaves turn a brilliant yellow in the fall. Deer and moose browse on the plants, and birds eat the seeds. A syrup is sometimes derived from the sap, and a tea is made from the twigs or inner bark, which have the aroma and flavor of wintergreen. The dark reddish brown to creamy-white wood is durable and heavy and is used widely in interior finishes, veneers, flooring, furniture, and cabinets.

River Birch

Betula nigra H 50–70 ft (15–21 m)

The fast-growing, shade-intolerant River Birch is a vigorous pioneer species that stabilizes soil on stream banks. The largest specimens grow in the Mississippi Valley.

KEY FACTS

The bark of young trees sheds in papery coils.

+ **leaves:** Alternate, broad, strongly serrate, pointed, wedge-shaped at base

+ **flowers/fruits:** Catkins unisexual, on same tree, male pendent, female upright; fruits cone-like

+ **range:** New Hampshire west to Minnesota, south to northern Florida and west to Texas

The lithe River Birch is the only spring-fruiting birch and the only birch whose native range includes the southeastern Coastal Plain. When young, it has a pyramidal shape, which becomes rounder with age. It often has multiple and highly branching trunks. Most common in floodplains and wet areas, it also grows in drier locations. Though the wood is close grained and sturdy, it is too knotty to be of much commercial value. It is used for furniture, baskets, small objects, and pulpwood. But it is an attractive tree, used as an ornamental; a number of cultivars have been developed. The sap has been concentrated for use as a sweetener.

Water Birch
Betula occidentalis H 20–35 ft (6–11 m)

The Water Birch is a shrub or small tree of wet, wooded sites, often growing near waterways or on banks, occurring sporadically throughout its range.

KEY FACTS

Bark is not exfoliating.

+ leaves: Alternate, rounded to wedge-shaped, with small marginal serrations on the larger teeth

+ flowers/fruits: Catkins unisexual, on same tree, male pendent, female erect; fruits are cones

+ range: In the west, especially in the Rocky Mountains, east to northwestern Ontario

The crown of the Water Birch is irregular and open; the branches are slender and drooping; and it usually has multiple trunks, forming thickets. The tree is too small to have much timber value, but its hard and heavy wood does find utility as posts and firewood. It is seldom cultivated but is planted as a stabilizer and buffer along stream banks; it is rather shade tolerant. Water Birch is an important component in wildlife habitat, and large mammals browse on the plants. Native peoples used a tea prepared from the tree as a diuretic and as a treatment for kidney stones.

Paper Birch/White Birch
Betula papyrifera H 50–70 ft (15–21 m)

The bark of the Paper Birch eventually becomes chalky white and papery, with prominent black lenticels, peeling off in horizontal strips to reveal orange beneath.

KEY FACTS

Pyramidal when young, but shape becomes less regular with age.

+ leaves: Alternate, oval, pointed; margins doubly toothed

+ flowers/fruits: Catkins unisexual, on same tree; fruits are cones

+ range: Alaska to Labrador, south into Rockies, Plains states, and Pennsylvania to North Carolina in mountains

For most of us, the Paper Birch is the tree that probably comes to mind when we think of birch trees, because of its beauty and because it is the most widespread of our birches. It can have one or multiple trunks. A number of cultivars have been developed. The wood is soft but moderately heavy and so is used for fuel, as well as in veneers, plywood, cabinets, furniture, and pulp. The tree's buds, catkins, and seeds provide food for many small mammals and birds, and a range of larger mammals find food in the bark and stems, including moose and porcupines.

Virginia Roundleaf Birch/Ashe's Birch

Betula uber H to 40 ft (12 m)

The first tree species listed as endangered by the U.S., the Virginia Roundleaf Birch is now listed federally as threatened (though still deemed endangered in Virginia, where it is endemic).

KEY FACTS

When crushed, the twigs smell of wintergreen; the bark eventually splits into ragged plates.

+ **leaves:** Alternate, round to oval, with heart-shaped base, serrate

+ **flowers/fruits:** Catkins unisexual, on same tree, male pendent, female erect; cones erect

+ **range:** Endemic to a site in Smyth County, Virginia

The majestic Virginia Roundleaf Birch was first described in 1918, but it was not seen again and was believed extirpated until 1975, when a stand of 41 mature trees was found on a creek bank in Smyth County, Virginia. Thanks to propagation programs, in 2006, the count stood at 961, including 8 trees from the original 1975 group. In 1918, the species was described as a botanical variety of the Sweet Birch *(B. lenta)* but was elevated to its own species in 1945, a decision that is still debated. The 1975 rediscovery went mainstream with a feature in the *New Yorker* the following January.

Southern Catalpa/Indian Bean Tree

Catalpa bignonioides H 25–45 ft (7–14 m)

One of the South's trademark trees, the Southern Catalpa is the host plant of a Sphinx Moth larva, the Catawba Worm, a popular bait for bream; the tree is also called Catawba and Fish Bait tree.

KEY FACTS

This tree can be wider than tall.

+ **leaves:** Opposite or whorled, large, rounded or heart-shaped

+ **flowers/fruits:** Flowers perfect, tubular, 2-lipped, 5-lobed, white, in showy clusters, throat with yellow and purple; fruits slender, beanlike pods

+ **range:** Georgia to Florida, west to Mississippi

The Southern Catalpa has been planted beyond its natural range for shade, beauty, and interest almost since its initial discovery by colonists, but drawbacks to some people are its falling flowers, unpleasant smell, large leaves, seedpods, and the tendency to sucker. It has become naturalized in those areas. The Northern Catalpa, *C. speciosa*, is the most north-ranging member of its family, with a limited natural range that includes southwestern Indiana and southern Illinois, Tennessee, and Arkansas. It is a larger tree, 75–100 feet (23–30 m), with larger leaves; smaller, fewer, and somewhat less attractive flowers; and thicker and longer pods.

Saguaro
Carnegiea gigantea H 40–60 ft (12–18 m)

The unforgettable shape of this tree-size cactus instantly brings to mind the Old West, an association that is exploited in movies and advertisements.

KEY FACTS

The trunk and branches are ribbed, with dense clusters of spines.

+ leaves: None

+ flowers/fruits: Flowers open at night, crown of white petals with yellow center, 3 in (7.5 cm) across, near branch ends; fruits oval, red, sweet; up to 4,000 seeds

+ range: Sonoran Desert in Arizona and into California

Our largest cactus does not develop its trademark upcurved arms until its 75th year; until then, the columnar plant is called a "spear." It can live 200 years. When it rains, the plant absorbs water and physically expands, then uses the water as needed. The cactus grows faster in areas with more rainfall. Its waxy coating helps conserve water. The Saguaro offers nesting habitat to birds including the Gila Woodpecker, Cactus Wren, and Pygmy Owl, and its seeds and fruit provide food. Arizona law makes it a crime to harm a Saguaro, and a permit is required to move one in order to build.

Jumping Cholla
Cylindropuntia fulgida H to 12 ft (3.6 m)

The name "Jumping Cholla" reflects the ease with which the spiny joints separate from the plant when bumped, sticking to one's clothes or skin, seemingly having jumped from the plant.

KEY FACTS

This shrubby cactus has one low-branching trunk.

+ leaves: Reduced to spines

+ flowers/fruits: Flowers pink to magenta, at branch and old fruit tips; 5–8 petals; fruits green, pear-shaped, many-seeded berries

+ range: Sonoran Desert in California and Arizona into New Mexico, Nevada, and Utah

The drooping branches of the Jumping Cholla are jointed, with cylindrical segments. When a fruit persists into the following year, a flower and another fruit can form at its tip, creating a chain of 25 years' worth of end-to-end fruits. This habit gives the plant another name, the Hanging Chain Cholla. The segments are armed with sharp, strong spines in crowded tufts. Because the segments disconnect easily, the ground at the base of the plant is often littered with plant parts, which can take root and form new plants. The fruits and seeds are food sources, especially for rodents, and the flesh of the plant provides water for desert animals, especially during droughts.

Roughleaf Dogwood
Cornus drummondii H 15–25 ft (4.6–7.6 m)

The Roughleaf is a dogwood whose flower clusters are not surrounded by large showy bracts, which may surprise those of us who think of the bracts as the symbol of dogwoods.

KEY FACTS

This dogwood can be a small tree or a multiple-trunked shrub.

+ **leaves:** Opposite, simple, entire; rough above

+ **flowers/fruits:** Flowers perfect, with 4 short, off-white petals, in clusters at branch tips; fruits hard berrylike drupes, white, in clusters

+ **range:** Great Plains, Midwest, Mississippi Valley

The Roughleaf Dogwood is a common component in the understory of rich woodlands. Though not showy, it is nonetheless attractive and has found uses in hedges and borders, around patios and in foundation plantings, and in parking lots and medians. As with many of its relatives, its fruits are eaten and its seeds are dispersed by many birds. And its suckering habit, resulting in those multiple trunks, is helpful in controlling erosion and in buffers. In the leaves, the veins come off the midrib, then arc toward the tip, and eventually nearly parallel with the margin, a dogwood trait.

Flowering Dogwood
Cornus florida H 15–35 ft (4.5–11 m)

In 2012, the U.S. sent 3,000 saplings of the Flowering Dogwood to Japan to mark the 100th anniversary of the famous cherry trees in Washington, D.C., a gift from Japan in 1912.

KEY FACTS

The branching is often tiered, with limbs to the ground.

+ **leaves:** Opposite, red in fall

+ **flowers/fruits:** Flowers perfect, tiny, in heads surrounded by 4 large white or pink petal-like bracts notched at the tip; fruits red, ovoid drupes

+ **range:** Mainly the eastern half of the U.S.

An eye-catching tree of the understory, the Flowering Dogwood is one of our most beautiful trees. But dogwood anthracnose, a fungal disease discovered in the U.S. in 1978, has spread throughout its range, causing significant losses of wild and planted trees, especially at higher altitudes and in shadier or moister sites. It first attacks leaves and stems, which die back, then it causes cankers, and the trees die. Guidelines are available for mediating the disease, and the many cultivars include a number of anthracnose-resistant varieties. The hard, dense wood has been used for golf club heads, mallets, tool handles, jewelry boxes, and spools.

Pacific Dogwood
Cornus nuttallii H 30–45 ft (9-14 m)

The drupes of the Pacific Dogwood are eaten and their seeds dispersed by birds and small mammals, as in other dogwoods. Large herbivores browse on the young plants.

KEY FACTS

This dogwood can bloom a second time, in early fall.

+ **leaves:** Opposite, simple, oval, hairy

+ **flowers/fruits:** Perfect, tiny flowers, in crowded heads with 4–8 showy, white to pinkish, petal-like bracts; fruits berrylike drupes, pink to red or orange

+ **range:** Pacific coast, British Columbia to California

The graceful Pacific Dogwood is a striking tree of the Pacific coast, where it is found inland to about 200 miles. It is much like the Flowering Dogwood, with its "flowers" of a small inflorescence and showy bracts. The Pacific Dogwood's flowers are larger, their bracts lack a terminal notch, and the two species' ranges are widely distinct. Anthracnose plagues the Pacific Dogwood; low air circulation and too much water worsen it. Dogwoods are sometimes called cornels, a name referring to their genus. The name is probably derived from the Greek *kerasos*, or "cherry tree," presumably because of the color of the drupes. A few cultivars are available.

Common Persimmon
Diospyros virginiana H 30–40 ft (9-12 m)

When fully ripe (following a solid frost, or if aged beyond technically ripe), a wild persimmon is delicious, tasting somewhat like a date. Earlier, it is so bitter and astringent as to be inedible.

KEY FACTS

Bark forms plates recalling charcoal briquettes.

+ **leaves:** Alternate, simple, entire, oval

+ **flowers/fruits:** Male and female flowers on separate trees; female, white, bell-shaped; male clustered, tubular, smaller; fruits fleshy, round, orange to purplish

+ **range:** East; southern, but found to Connecticut

Persimmons persist on the tree long after the leaves have fallen, which makes them one of the latest available wild fruits. This was nicely timed with the harvest feasts of American Indians and European settlers alike, so persimmons were served. The fruits are eaten fresh and are made into cakes, puddings, breads, candies, pies, jams, and beverages. Humans must vie with raccoons, other mammals, and birds to get the prized fruits. A member of the ebony family, the Persimmon has dark, heavy, close-grained wood, which is used in heads for golf club woods, in lathe work, and in carving. Several cultivars have been developed.

Mountain Laurel
Kalmia latifolia H 3–23 ft (1–7 m)

The blight that decimated the American Chestnut in the early 1900s benefited the Mountain Laurel, which prospered as sunlight gained entry to the once denser reaches of the forest.

KEY FACTS

This evergreen broad-leaf shrub is dense and gnarled.

+ **leaves:** Alternate, elliptic, leathery

+ **flowers/fruits:** Flowers perfect, white to pink clusters; 5 petals, fused into saucer shape; fruits small, round, capsules with withered pistil

+ **range:** East of a line from New Brunswick to Louisiana

The Mountain Laurel grows mostly in rounded stands sprawling on the forest floor. It is one of our most magnificent native shrubs (occasionally reaching tree size), common in the wild, and is planted for its showy beauty in yards and parks. Many cultivars have been developed. The stamens are held under tension by the petals, springing loose when tripped by a bee, peppering the insect with pollen, which it transfers to other flowers. All parts of the plant are toxic. It was used among native peoples as a means of suicide, as well as a source of a yellow-brown dye.

Sourwood
Oxydendrum arboreum H 40–60 ft (12–18 m)

The Sourwood finds many uses in landscaping, thanks especially to its leaves, whose vivid fall color has few rivals and which persist longer than most other leaves.

KEY FACTS

The branches often extend to the ground and droop.

+ **leaves:** Alternate, vivid orange or red in fall

+ **flowers/fruits:** Flowers cup-shaped, small, white, in spikes along one side of stem, showy; fruits are capsules, silver-gray in fall

+ **range:** Southeast; west to Kentucky, south to the Gulf and northern Florida

This narrow, small to medium-size tree of the understory has a graceful, open, and irregular shape. It was being cultivated as early as the mid-1700s, and cultivars have been developed. In nature, it is found in mixed stands on drier, upland, wooded sites, often with other heaths (as members of its family are known), such as *Rhododendron* species. It is sensitive to air pollution. The Cherokee and Catawba used the young trees to make arrows, and though the Sourwood's dense, close-grained wood has been used for paneling, fuel, and pulp, it has not had great commercial success.

Sparkleberry/Farkleberry

Vaccinium arboreum H 12–28 ft (3.6–8.5 m)

The Sparkleberry is our only tree-high *Vaccinium*, though some highbush blueberries come close. It is more tolerant of more alkaline soils than are its relatives.

KEY FACTS

The tree is spindly with twisted branches and is mostly evergreen.

+ **leaves:** Mostly evergreen, alternate, rounded, leathery, bright red in fall

+ **flowers/fruits:** Flowers white, bell-shaped, 5-lobed, in drooping clusters; fruits black, shiny

+ **range:** Southeast from Virginia to Missouri, to Florida and Texas

The name "Sparkleberry" probably refers to the fruit's shininess; "Farkleberry" may be a play on that name. The fruits are dry, bitter, and mostly ignored. The name has been tied to two Arkansas governors: Orval Eugene Faubus was nicknamed "Farkleberry" by a cartoonist who found it apt for satirizing Faubus and the state's politics. Frank White was dubbed "Governor Farkleberry" after saying that his family had been so poor they had to eat Farkleberries. This attractive plant provides habitat and food for wildlife, and the bark, peeling in red, gray, and brown patches and often splotched with lichen, adds visual interest.

Blueberries

Vaccinium species H various, to 16 ft (5 m)

Though these plants have provided food for humans and wildlife for centuries, cultivars have been developed that provide plumper, juicier, and sweeter berries.

KEY FACTS

Blueberries are mostly upright, small shrubs.

+ **leaves:** Deciduous or evergreen, oval to tapering at ends

+ **flowers/fruits:** Flowers bell-shaped, 5-lobed, white, pink, or red, in clusters at end of branches; fruits are blue berries

+ **range:** Arctic Circle south, especially in cooler areas

The genus *Vaccinium* includes 450 to 500 species of shrubs, vines, and small trees mainly in cooler areas of the Americas, Europe, and Asia, but also in southern Africa, Madagascar, and Hawaii. The "typical" blueberries are endemic to North America. Their classification is difficult. Genetic and molecular studies are incomplete, there are different schools of thought, and hybridization is a further complication. The genus includes the important Highbush (*V. corymbosum*), Lowbush (*V. angustifolium*), Velvetleaf (*V. myrtilloides*), and Rabbiteye (*V. virgatum*) blueberries and at least nine others, including the Cranberry (*V. macrocarpon*), a low-growing evergreen in acidic bogs in the Northeast.

Mimosa/Silktree
Albizia julibrissin H 20–50 ft (6–15 m)

A native of Asia, from Iran to Japan, the exotic Mimosa was introduced here as an ornamental in the 1700s. It remains popular in gardens, yet it is invasive on roadsides and in disturbed areas.

KEY FACTS

Flowers are borne at branch ends; trunks are often multiple.

+ leaves: Opposite, large, twice compound, feather-like; leaflets opposite

+ flowers/fruits: Flowers pink filamentous pompons, fragrant, clustered; fruits are thin, flat pods

+ range: Nonnative; widely naturalized, invasive, especially in Southeast

The Mimosa is vase-shaped, with a crown of large, feathery leaves. It has bacteria in its roots that can change inert atmospheric nitrogen into forms that can serve as plant nutrients. Wildlife eats its fruit and disperses its many seeds. The roots or an old stump can send up sprouts. As a result, and because it is nonnative, it has no natural enemies to keep it in check, and it is often invasive. The University of Florida Center for Aquatic and Invasive Plants suggests native alternatives, such as Serviceberry (*Amelanchier arborea*), Redbud (*Cercis canadensis*), and Fringe Tree (*Chionanthus virginicus*), but people still plant the Mimosa, and there are many cultivars.

Yellow Paloverde
Parkinsonia microphylla H to 25 ft (7.6 m)

The Yellow Paloverde is a trademark shrub or small tree of the Arizona desert. Its branches are photosynthetic, which gives it a yellow-green cast and its common name.

KEY FACTS

The branches are erect, not flowing.

+ leaves: Alternate, twice compound; 2 leaflets; secondary leaflets tiny, oval, soon dropping

+ flowers/fruits: Flowers perfect, in small clusters; 5 petals—4 yellow, 1 white; fruits are pods constricted between seeds

+ range: Sonoran Desert in California and Arizona

Like its relative the Jerusalem Thorn, the Yellow Paloverde is drought deciduous: It loses its leaves in the hottest and driest part of the year, which lessens water loss. It will even drop some branches if weather is especially severe. It is a "nurse plant" to the Saguaro; the cactus's seeds germinate in its shade, and it protects the seedlings from wind and trampling. Yellow Paloverde wood is used mainly as fuel, and it is cultivated in other arid regions. Buffelgrass (*Cenchrus ciliaris*) was introduced from Africa to grow as food for livestock, but it has become invasive and harms the Yellow Paloverde by robbing it of scarce water from the soil.

Eastern Redbud
Cercis canadensis H 15–28 ft (4.6–8.5 m)

The Eastern Redbud stages a striking display in spring, its pink flowers appearing before most trees have leafed out. Flowers are borne on all parts of the tree, even the trunk.

KEY FACTS

The flower, as in many legumes, recalls that of the familiar garden pea.

+ **leaves:** Alternate, simple, entire, heart-shaped, papery, pendent; yellow in fall

+ **flowers/fruits:** Flowers perfect, pink, stalked, in clusters of 4–8; fruits flat, hanging pods

+ **range:** Eastern half of the U.S.

The Eastern Redbud is not native to what is now Canada; *canadensis* in the scientific name refers to a rather different Canada, a French colony that extended into what is now the U.S., including areas where the Redbud is native. This graceful, often multistemmed shrub or small tree grows on a range of soils and in different light regimes, which, with its beauty at flowering time, have made it a successful ornamental. Cultivars include white-flowered varieties. The tree is too small to be a commercial lumber source. Like many leguminous plants, the Redbud can fix nitrogen.

Honeylocust
Gleditsia triacanthos H 30–80 ft (9–24 m)

The large, forked thorns on the trunk and lower branches of the Honeylocust have been used as pins or even nails. Thornless cultivars are popular in landscaping.

KEY FACTS

The bark becomes ridged, its edges curling.

+ **leaves:** Once or twice compound; leaflets oval

+ **flowers/fruits:** Flowers in hanging unisexual clusters on same tree (some flowers perfect), yellow-green, fragrant; fruits twisting, leathery pods, turning brownish

+ **range:** Midwestern and south-central U.S.

The wood of the Honeylocust is of good quality and durable when in contact with the soil. It is used to make furniture, interior finishings, and utilitarian products such as posts and firewood. The seedpods provide a sweet pulp (thus the common name) that was eaten by native peoples and has been fermented into a beer. The inconspicuous flowers are a source of pollen and nectar for honey. Pharmacognosists have isolated chemical compounds from the tree that have anticancer properties or that show promise in treating rheumatoid arthritis. The pods, bark, and young shoots are often eaten by mammals, the seeds by birds.

Kentucky Coffeetree

Gymnocladus dioicus H 60–80 ft (18–24 m)

The stout branches of the Kentucky Coffeetree lose their leaves early and so can go leafless for half the year. *Gymnocladus* means "naked branch."

KEY FACTS

The bark is fissured with scaly ridges.

+ **leaves:** Huge, twice compound; leaflets oval, pointed

+ **flowers/fruits:** Flowers fragrant, in upright clusters; 5 petals and calyx lobes, alternating, greenish white, sexes on same tree; fruits broad, leathery

+ **range:** Midwest to Appalachians, south to Louisiana

Though the Kentucky Coffeetree is widespread and tolerates a range of conditions, it is not common. Animals that might have dispersed its large seeds—Mastodon, Mammoth, and Giant Sloth—are long extinct, which may have curbed the tree's population. Colonists roasted and milled the seeds to make a coffee-like beverage, but the unroasted seeds are toxic. The drink contains no caffeine and was forgotten when coffee beans became available. In landscaping, male trees are usually used because females drop one part after another, which must be cleaned up. The males, however, lack the interesting pods in winter.

Honey Mesquite/Glandular Mesquite

Prosopis glandulosa H 20–30 ft (6–9 m)

Though only a large shrub or small tree (at its biggest in areas of highest moisture), the Honey Mesquite is still the largest tree in much of its range.

KEY FACTS

It has feathery foliage and long, paired spines.

+ **leaves:** Alternate, twice compound, smooth, with 2 leaflets, each with small leaflets

+ **flowers/fruits:** Flowers perfect, white to yellow, fragrant, in spikes; fruits are long cylindrical pods, constricted between seeds

+ **range:** Central Texas to California

Before Europeans introduced livestock, mesquites may have had more clearly defined ranges. Animals that eat the pods disperse the seed, and have thus blurred the boundaries. In addition to seeds, the Honey Mesquite reproduces by sprouting from the roots. It has found many uses and is being eyed as a promising species for the world's food supply (as well as for animal feed and building materials). Native peoples used the pods for food and to make alcoholic beverages and flour. Mesquite honey is prized for its flavor. Yet where stock is grazed, mesquites are considered weeds and are often destroyed.

Velvet Mesquite
Prosopis velutina H to 30 ft (9 m)

Overgrazing removed range plants that served as fuels in natural fires. Coupled with dispersal of its seeds by livestock, mesquite expanded into grazing lands, where it is considered a pest.

KEY FACTS

Larger than other mesquites, it is spiny, with feathery foliage.

+ **leaves:** Alternate, twice compound, velvety, 2 leaflets, each with smaller leaflets

+ **flowers/fruits:** Flowers perfect, creamy, fragrant, in spikes; fruits are long cylindrical pods, constricted between seeds

+ **range:** Central and southern Arizona

Mesquites such as the Velvet were essential to native peoples of the Southwest. The trees often grew in bosques (wooded assemblages that extended for miles along rivers), allowing the fruit to be gathered almost as if it had been planted. The beans were dried and ground into flour that was used to make bread, a dietary staple; the pods are about 25 percent sugar, and the seeds are high in protein. The trees were also important for articles of everyday life, such as tools, rope, baskets, and weapons. The original extent of the mesquite bosques is diminishing as the mesquites are cut for wood and cleared for agricultural and other development.

Black Locust
Robinia pseudoacacia H 30–60 ft (9–18 m)

The roots of the Black Locust have nodules that contain bacteria capable of fixing nitrogen, that is, of transforming inert nitrogen gas in the air into chemical forms that plants can use.

KEY FACTS

Most leaves have a pair of thorns at the base.

+ **leaves:** Alternate, compound, feathery; entire, large, oval leaflets

+ **flowers/fruits:** White, fragrant flowers, typical pea flower, in large, showy, hanging clusters; fruits are flat pods, sometimes persisting

+ **range:** Central Appalachians and Ozarks

The Black Locust's many seeds and ability to sprout from its roots enable it to establish quickly, even on poor soils—ideal in reclamation projects and erosion control, but not so much in a garden. Nevertheless, it has been widely planted and is naturalized in the 48 contiguous states and much of Canada. It grows (and can be invasive) in many parts of the world, sometimes farmed as an alternative to taking tropical woods. It produces a strong, hard wood used in boats (instead of teak), furniture, paneling, and flooring, in addition to more utilitarian uses including the manufacture of charcoal.

American Chestnut

Castanea dentata H to 100 ft (30 m) before the blight

"Not only was baby's crib likely made of chestnut, but chances were, so was the old man's coffin," wrote George H. Hepting in "Death of the American Chestnut" (*Journal of Forest History*, 1974).

KEY FACTS

The tree is spreading and dense.

+ leaves: Alternate, 5–8 in (13–20 cm), pointed; base rounded; margin sharp-toothed

+ flowers/fruits: Flowers unisexual, both sexes on same plant—male in catkins, female near catkin bases; fruits spiny capsules with 3 nuts

+ range: Eastern United States, Ontario

The American Chestnut has been called the ideal tree. Majestic, massive, it made up 25 to 50 percent of the hardwoods in its range. Its wood was unequalled, it was a key source of tannins used in tanning leather, and the nuts had a ready market in the East. But a fungus, brought from Asia around 1900, caused a blight that destroyed nearly all mature chestnuts in 40 years. The species survives in unaffected areas to which it had been transplanted. Also, the root systems did not succumb and send up sprouts. Studies are under way to exploit what natural resistance exists, and to explore hybridization with a resistant chestnut from China.

American Beech

Fagus grandifolia H to 66 ft (20 m)

No matter how remote the tree, it seems initials will be found carved into the trunk of an American Beech, the carving sealed by the bark and remaining for posterity.

KEY FACTS

The smooth bark is light gray.

+ leaves: Alternate, to 5.5 in (14 cm), pointed; margin toothed

+ flowers/fruits: Flowers unisexual, male in round heads, female 2–4 in spikes, on same tree; fruits spiny husks with 2 3-faced, winged nuts

+ range: New Brunswick to Florida to Mississippi Valley

The American Beech is a large, imposing, slow-growing, and long-lived tree usually found in mixed hardwood forests, where it is the dominant or a codominant tree, sturdy and with many branches. The tree is deciduous, but many leaves persist through the winter, their light reddish brown somehow almost showy against the gray bark. The nuts are delicious, but competitors are many, including squirrels, chipmunks, mice, raccoons, porcupines, opossums, rabbits, deer, black bear, and foxes, not to mention ruffed grouse, turkeys, bobwhite, and pheasants. Beech wood is used in furniture, veneers, and flooring, and for fence posts and fuel.

White Oak
Quercus alba H to 82 ft (25 m)

The White Oak is possibly our most abundant tree, usually dominant when present, and our most commercially important oak. There are 58 species of oak in North America north of Mexico.

KEY FACTS

The leaves are red-purple in fall.

+ leaves: Alternate, to 7 in (18 cm); 7–10 rounded lobes

+ flowers/fruits: Flowers in unisexual catkins on same tree; male lax, green-yellow; female reddish, in leaf axils; acorn caps cover a third of nut

+ range: Southern Ontario and Quebec to eastern and central U.S.

The White Oak is usually found in mixed hardwood forests. Half of U.S. hardwood production is White Oak, because of its quality and abundance. It is used in furnishings, interiors, shipbuilding, and wine and whiskey barrels. This tree is the flagship of the White Oak group, whose acorns mature the first year on new branches. The inside of the acorn shell is not hairy, and the cap's scales are brown and flat. The meat is sweet to somewhat bitter. The leaf lobes are rounded and do not have bristles at the tip. The bark is often pale gray and blocky. Species of the White Oak group will hybridize with one another.

Arizona White Oak
Quercus arizonica H to 60 ft (18 m)

The Arizona White Oak is found in arid and semiarid areas in oak and piñon woodlands, growing larger in areas with more moisture, such as canyons.

KEY FACTS

The trunk is short, and the branches are twisted.

+ leaves: Evergreen or nearly so, alternate, to 3.5 in (9 cm) thick

+ flowers/fruits: Catkins typically oak; acorns oblong, to 1 in (2.5 cm); caps cover a third of nut

+ range: Central Arizona to southwestern New Mexico into Texas

The long-lived Arizona White Oak (White Oak group) is a shrub to a medium-size tree, yet one of the largest of the southwestern oaks. It has stout, spreading branches and an irregular crown, and its light gray bark can be an inch (2.5 cm) thick on a mature tree. The acorns are eaten by cattle and deer, but they are not the preferred acorn. Birds forage in the stands, and deer browse on the sprouts. The wood is neither straight enough nor large enough to provide commercial timber, and, though difficult to cut and work, it is used for fuel and sometimes in furniture.

Scarlet Oak
Quercus coccinea H to 100 ft (30 m)

The Scarlet Oak is widely planted in the U.S., Canada, and Europe as an ornamental and a shade tree, and cultivars have been developed. The leaves turn brilliant red in autumn.

KEY FACTS

The trunk of this oak swells at the base.

+ **leaves:** Alternate, to 6 in (15 cm); 5- to 9-lobed, each with a bristled tip; sinuses are deep and C-shaped

+ **flowers/fruits:** Catkins typically oak; acorns to 1.2 in (3 cm); caps cover half of nut

+ **range:** Eastern and central U.S.

The Scarlet Oak (a member of the Red Oak group; see Northern Red Oak, p. 143) is a large, fast-growing tree with stout, upright, spreading branches and an open, rounded crown, the bark thin and gray-black. It is common in dry upland forests, where it grows in large pure stands and mixed stands. The wood is reddish brown, coarse-grained, and strong, inferior to that of the Northern Red Oak but, as wood of many other species of the Red Oak group, is marketed as Red Oak. It is used as lumber. Small mammals and birds use the acorns and seedlings as food, and the trees provide nesting sites, including cavities.

Coastal Sage Scrub Oak
Quercus dumosa H 3–7 ft (1–2 m)

The Coastal Sage Scrub Oak usually grows within sight of the Pacific Ocean—that is, on prime real estate. The tree, which has never been common, is at great risk from human encroachment.

KEY FACTS

The tree has a scraggly look.

+ **leaves:** Evergreen or nearly so, to 1 in (2.5 cm), irregularly toothed or shallowly lobed, with erect curly hairs on underside

+ **flowers/fruits:** Catkins typically oak; acorns narrow, pointed; reddish caps cover a third of the nut

+ **range:** Southern California

Most Coastal Sage Scrub Oaks of the White Oak group (see White Oak) have at one time been considered *Quercus dumosa*, but the species concept is now much narrower. It refers to a plant growing in a limited part of southern California and restricted to Coastal Sage Scrub habitats. Most trees once considered this species are now included as the more widespread and common California Scrub Oak, *Q. berberidifolia*, which ranges to the high north coast mountains and to the foothills of the Sierra Nevada. Its acorns are more rounded, and its leaves lack hairs on the underside. Its range does not overlap with that of *Q. dumosa*.

Southern Red Oak

Quercus falcata H to 100 ft (30 m)

A medium-size to large tree, the Southern Red Oak is also known as Spanish Oak, probably because it grows in some of the former Spanish colonies.

KEY FACTS

Its trunk is straight, its crown rounded.

+ **leaves:** Alternate, to 9 in (23 cm); 3–7 lobes, terminal longest, sharp, often curved; sinuses deep, U-shaped

+ **flowers/fruits:** Catkins typically oak; acorns round, orangish, 0.5 in (1.3 cm); caps cover a third of nut

+ **range:** Southeast to southern Missouri, eastern Texas

The Southern Red Oak (a member of the Red Oak group; see Northern Red Oak, p. 143) is a handsome tree that is rather fast growing and fairly long lived. It grows on dry, poor, sandy, upland soils. The bark is dark and thick with deep furrows between broad, scaly ridges. It is an important timber tree, providing light-red wood that is coarse grained, hard, and strong. It is used in furniture, cabinets, veneer, and for pulp and fuel. Native peoples used oaks for myriad medicinal purposes. The Southern Red Oak was used to treat such conditions as indigestion, chapped skin, and fever, and as an antiseptic and tonic.

Gambel Oak

Quercus gambelii H 10–30 ft (3–9 m)

For the caterpillar (larva) of the Colorado Hairstreak Butterfly to undergo metamorphosis, it must feed on the leaves of the Gambel Oak.

KEY FACTS

The tree has crooked branches.

+ **leaves:** Alternate, 3–6 in (8–15 cm), yellow-green, broadest beyond middle to uniformly wide; 5–9 lobes; deep sinuses

+ **flowers/fruits:** Catkins typically oak; acorns to 1 in (2.5 cm), mostly round; caps cover to a third of nut

+ **range:** Utah, Wyoming, Arizona, New Mexico

A tall shrub or small tree, the Gambel Oak (White Oak group) is the most common deciduous oak in most of the Rocky Mountains, widespread and abundant in the foothills and lower elevations on dry slopes and in canyons. The slopes can be covered in clonal thickets of the tree, the result of sprouting from tuber-like parts of the root system. Farther south, it exists in mixed stands. The tree's small size prevents the fine wood from being commercially significant, but it is used for fires and to make fence posts. Deer and livestock browse the foliage, and acorns provide food for turkeys and squirrels, as well as domestic stock.

Oregon White Oak/Garry Oak
Quercus garryana H to 50 ft (15 m)

The only oak native to British Columbia, Washington, and northern Oregon, the Oregon White Oak is the most commercially important oak in the West.

KEY FACTS

The tree often bears mistletoe.

+ **leaves:** Alternate, 3–4 in (8–10 cm), broadest beyond middle to uniformly wide; 5–9 lobes; teeth variable, sinuses deep

+ **flowers/fruits:** Catkins typically oak; acorns to 1.5 in (3.8 cm); caps cover to a third of nut

+ **range:** Central California to southwestern British Columbia

The Oregon White Oak is a shrub to medium-size tree with a dense, spreading crown. It exists in pure stands or mixed with other hardwoods or conifers. Its ecological foes include fire suppression. Natural grass fires did not harm the oaks but burned out young Douglas-firs beneath them. In the absence of fires, the Douglas-firs become established, and the oaks cannot survive in their shade. Other threats are development and invasive nonnatives. In Oregon White Oak woods in British Columbia, more than 80 percent of the understory is nonnative, and in Oak Bay on Vancouver Island, a permit is required to cut even a branch from established trees.

Shingle Oak
Quercus imbricaria H to 65 ft (20 m)

A stately, medium-size tree, the Shingle Oak is planted for shade or as a windbreak, a hedge, an ornamental, or a street tree. It is our most cold-hardy oak.

KEY FACTS

The lower branches may droop.

+ **leaves:** Alternate, 4–6 in (10–15 cm), oval, unlobed, broadly pointed, with one bristle; margins somewhat wavy; yellow-brown to russet in fall

+ **flowers/fruits:** Catkins typically oak; acorns to 0.7 in (1.8 cm); caps cover about half of nut

+ **range:** Midwest and upper South

The Shingle Oak (a member of the Red Oak group; see Northern Red Oak, p. 143) is never a dominant species in a mixed woods, probably because of its intolerance of shade. It is not a commercially important timber tree, although one of its main practical uses gives it its common name. The wood is pale reddish brown, heavy, hard, and coarse grained, and is used to make split shingles. This use is probably related as well to its scientific name, *imbricaria*, derived from the Latin word for "tile." It also may refer to the overlapping scales of the winter bud. The acorns are especially bitter.

Valley Oak/California White Oak

Quercus lobata H to 82 ft (25 m)

The largest oak tree in the U.S. and Canada is a Valley Oak in California's Round Valley called the Henley Oak, which is 151 feet (46 m) tall and more than 500 years old.

KEY FACTS

The twigs often weep; salt in the air makes the tree scrubby to 4 mi (6.4 km) inland.

+ **leaves:** Alternate, to 4 in (10 cm); 9–11 deep, rounded lobes

+ **flowers/fruits:** Catkins typically oak; acorns 1.2–2.5 in (3–5.6 cm), narrow, acutely conical or bullet-shaped; caps covering nut base

+ **range:** Endemic to California

A California icon, the Valley Oak (White Oak group) is a tree of valleys in the state's inner and middle coastal ranges. When English explorer George Vancouver visited the Santa Clara Valley in 1792, the expanses of Valley Oak savanna reminded him of a closely planted stand with the understory removed, and he called the trees the "stately lords of the forest." More than 90 percent of the Valley Oak stands were gone by World War II—victims of orchard and vineyard agriculture. The San Francisco Estuary Institute is investigating "re-oaking," hoping to reintegrate Valley Oaks and other natives into California's highly developed landscape.

Overcup Oak

Quercus lyrata H to 65 ft (20 m)

The Overcup Oak's scientific name refers to the overall shape of the leaf, which is said to recall the outline of a lyre.

KEY FACTS

The tree has a short trunk.

+ **leaves:** Alternate, 6–10 in (15–25 cm); 5–9 lobes, irregular, outer pair often making cross shape; deep sinuses

+ **flowers/fruits:** Catkins typically oak; acorns about 1 in (2.5 cm), round, often almost covered by cap; very long stalk

+ **range:** Deep South and Mississippi Valley

The Overcup Oak (White Oak group) is a slow-growing, small to medium-size tree of the warm, humid Southeast. Also called Swamp Post Oak, it grows among the Baldcypress and Water Tupelo common in many southern swamps. Its wood warps easily and is generally inferior to that of most of the White Oak group; nonetheless, it is marketed as White Oak and used for general construction and barrels. The Overcup Oak is planted to enhance land for wildlife, and its acorns are eaten by deer, turkey, and squirrels; ducks also eat the acorns, though because these nuts are so large, they are less useful than those of other oaks.

Bur Oak/Mossycup Oak

Quercus macrocarpa H to 98 ft (30 m)

The Bur Oak has the largest acorn of all our oaks, and its scientific name reflects that: *Macrocarpa* means "big seed."

KEY FACTS

The bark forms rectangular blocks.

+ **leaves:** Alternate, to 12 in (30 cm); widest beyond middle; 5–9 lobes; deep sinuses

+ **flowers/fruits:** Catkins typically oak; acorns to 2 in (5 cm); caps cover three-quarters of nut, with mossy or bur-like fringe

+ **range:** Midwest and eastern U.S., south-central Canada

A medium-size to large, spreading tree, the Bur Oak (White Oak group) is cold tolerant and, because of its long taproot, drought resistant. Its large acorns are important forage for wildlife, including black bears. It is planted as an ornamental and is the most forgiving White Oak of the urban environment. The Bur Oak savannas of the eastern prairie were vital to settlers, providing wood and grazing land. In the early 1900s, these savannas covered 32 million acres. By 1985, as a result of development, including agriculture, and fire suppression, high-quality savanna had declined to 6,400 acres (2,592 ha)—a loss of more than 99 percent.

Cherrybark Oak/Swamp Red Oak

Quercus pagoda H to 130 ft (40 m)

The acorns of the Cherrybark Oak provide food for domestic hogs and for wildlife, including larger birds and a number of mammals.

KEY FACTS

The leaf shape recalls a pagoda (thus its scientific name).

+ **leaves:** Alternate, 5–8 in (13–20 cm), broadest at base; 7–11 lobes, regularly shaped

+ **flowers/fruits:** Catkins typically oak; acorns 0.6 in (1.6 cm); caps cover a third to half of nut

+ **range:** Southeastern United States, Mississippi Valley

One of the largest oaks in the South, the Cherrybark Oak (Red Oak group; see Northern Red Oak, p. 143) is similar to the Southern Red Oak, of which it was formerly considered a botanical variety. It prefers moist areas, in riverbanks and in floodplains, especially in the Coastal Plain. The bark recalls that of the Black Cherry, smooth, with flaky ridges, and with a red tint. The trunk is straight, with relatively few limbs, and the wood is strong, making this a good source of timber. The light red-brown wood is heavy, hard, and coarse grained and is used in fine work, including interiors, furniture, cabinets, and floors.

Pin Oak
Quercus palustris H to 82 ft (25 m)

"Pin Oak" is said to refer to persisting dead branches that resemble pins driven into the tree's trunk; another theory is that its branches were used as dowels, or "pins," in barn construction.

KEY FACTS

The tree's interior is dense.

+ leaves: Alternate, to 6 in (15 cm), broadest beyond middle; 5–9 lobes, bristle-tipped; very deep sinuses

+ flowers/fruits: Catkins typically oak; acorns round, with short beak at tip; caps cover only nut base

+ range: Mid-Atlantic and central states and extreme southern Ontario

The fast-growing Pin Oak (Red Oak group; see Northern Red Oak) thrives in wet, soggy flood-plains and flatlands and tolerates some flooding, but it also inhabits some well-drained and upland areas. The branches spread in age, and the lower ones often droop toward the ground. The wood is weaker than that of the Red Oak and is used in general building, as posts, and for fuel. The Pin Oak is planted extensively as an ornamental and shade tree (even in parts of Australia and Argentina), and cultivars have been developed. Birds that eat its acorns include mallards, wood ducks, blue jays, turkeys, and woodpeckers.

Willow Oak
Quercus phellos H to 98 ft (30 m)

The Willow Oak is so named because its leaves, narrow and tapering at both base and tip, recall the leaves of willows. It is not, however, related to the willows.

KEY FACTS

The leaves are borne stiffly on the twigs.

+ leaves: Alternate, to 5 in (13 cm), tipped by tiny awn; unlobed

+ flowers/fruits: Catkins typically oak; broadly rounded acorns, to 0.5 in (1.3 cm); caps cover a third of thin, saucer-like nut

+ range: Atlantic and Gulf Coastal Plains, Mississippi Valley

The Willow Oak (Red Oak group; see Northern Red Oak) inhabits bottomlands and drier areas and is often planted in landscapes and on streets. It consistently has good acorn crops and is important to wildlife. While acorns feed larger animals, the trees host thousands of insect species, few of them pests. For many butterflies and moths, an egg-laying stopover on an oak (for some, a particular species) is required. The larvae need oak leaves to eat, or they cannot metamorphose into adults. Leafhoppers, wasps, and beetles also use oaks, and all these insects provide food for birds and other animals.

Northern Red Oak
Quercus rubra H to 98 ft (30 m)

One of our top timber trees, the Northern Red Oak produces hard, strong, coarse-grained wood, often used in interiors. It is widely planted in Europe for such purposes.

KEY FACTS

The dark gray-brown bark is ridged.

+ **leaves:** Alternate, 4–9 in (10–23 cm), often widest beyond middle; 7–9 lobes; sinuses rounded halfway to midrib

+ **flowers/fruits:** Catkins typically oak; acorns to 1 in (2.5 cm); caps cover a quarter of saucerlike nut

+ **range:** Extreme southeastern Canada, eastern U.S.

The attractive and moderately fast-growing Northern Red Oak is the banner tree of the Red Oak group, which also includes Black Oaks. These oaks' acorns mature in two years, on the previous year's branches. The husk is thin, clinging, and papery, and woolly on the inside. The acorn meat is typically very bitter due to the high levels of tannins, which had to be leached out before native peoples could use them as food. The leaf lobes are often toothed and bear bristles at their tip. The bark is dark, and the texture varies. Species of the Red Oak group will hybridize with one another.

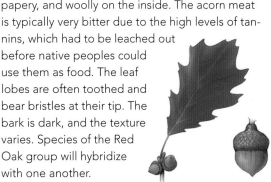

Shumard Oak
Quercus shumardii H to 115 ft (35 m)

A large, stately southern oak, the Shumard Oak is long lived, the crown opening with age, and the trunk buttressed in older trees.

KEY FACTS

This deep-rooted oak is often planted near paved areas.

+ **leaves:** Alternate, to 8 in (20 cm), broad; 5–9 paired lobes; tips toothed, bristled; sinuses deep

+ **flowers/fruits:** Catkins typically oak; acorns broadly ovate, to 1 in (2.5 cm); caps cover a third of nut

+ **range:** Eastern, central United States, southwestern Ontario

Acorns of the Shumard Oak (Red Oak group) feed many wildlife species. Humans, too, ate acorns, and there is now renewed interest in the nuts' dietary potential. But acorns, especially in the Red Oak group, contain bitter tannins, which make unprocessed acorns inedible. The Menominee in Wisconsin removed the tannins in a process that was probably not unique to them. They toasted the acorns and removed the husks. They then boiled the acorns in water, which was discarded. They boiled them again, adding alkaline wood ash to accelerate leaching. A final boil removed the ash. The processed acorns were dried and ground into meal or stored.

Post Oak

Quercus stellata H to 65 ft (20 m)

Not a very important timber species, the Post Oak does produce lumber, sold as White Oak, that is used in railroad ties and flooring and, of course, for fence posts.

KEY FACTS

The trunk is gray to light red-brown, becoming ridged.

+ **leaves:** Alternate, oblong, to 6 in (15 cm); 5 lobes, the center 2 largest, lending a cross-like shape

+ **flowers/fruits:** Catkins typically oak; acorns to 1 in (2.5 cm); caps cover half of nut, not fringed

+ **range:** Eastern and central U.S.

Identification of the Post Oak (White Oak group) is difficult because it varies in plant habit, leaf shape, and bark; aberrant populations have been incorrectly designated as a separate species. A character sometimes deemed certain for identification is its cross-shaped leaf, created by two main lobes, which are square and large and project opposite each other from the leaf axis. But other species can have the same arrangement. A better way, though a bit technical, involves use of a hand lens to examine the twigs and the underside of the leaves for stellate hairs, which resemble tufts branching in a starlike (stellate) pattern.

Black Oak

Quercus velutina H to 82 ft (25 m)

The bark of the Black Oak is so rich in tannins that their extraction was once a commercial pursuit. Tannins, used to tan leather, are now synthesized.

KEY FACTS

The bark is gray-black, in vertical plates.

+ **leaves:** Alternate, to 10 in (25 cm); 5–9 lobes, pointed, bristle-tipped; sinuses deep, U-shaped

+ **flowers/fruits:** Catkins typically oak; acorns 0.5 to 0.75 in (1–2 cm); caps cover half of nut

+ **range:** Extreme southwestern Ontario, eastern and central U.S.

The Black Oak (Red Oak group) is common in dry, upland deciduous forests, as well as in savannas, where the eastern forests cede to the prairie. Oak flowers are similar among species, unisexual, and borne on separate catkins, which appear on the same tree before or with the leaves. They usually go unnoticed until they fall onto sidewalks and cars. The male catkins are yellow to greenish, long, lax, and pendulous, bearing many flowers that make pollen. The stiffer, smaller female catkins bear one to several cupules, each of which contains a flower and will become an acorn cap. Oaks, like most trees bearing catkins, are wind pollinated.

Live Oak

Quercus virginiana H to 115 ft (35 m)

The young U.S. Navy bought extensive Live Oak stands for the use of its shipbuilders. The strong, curved limbs were ideal for fashioning the ships' ribs and other supports.

KEY FACTS

The trunk becomes buttressed with age.

+ **leaves:** Nearly evergreen, alternate, to 5 in (12 cm); oblong, widest beyond middle, stiff, waxy; sometimes toothed, sometimes curled under

+ **flowers/fruits:** Catkins typically oak; acorns long, slim; caps scaly, deep

+ **range:** Coastal Plain from Virginia to Texas

The long-lived and fast-growing Live Oak (White Oak group) has been described, without exaggeration, as majestic, noble, and picturesque. The tree is wide spreading with an extensive, rounded crown. Its massive branches can sprawl so far from the trunk that they bend to touch the ground, sometimes even curving up again. The dark gray bark recalls an alligator's skin. Spanish moss, which often graces its limbs, completes the picture of this icon of the Deep South. The Live Oak can also be shrubby, and it sprouts vigorously from stumps. Native peoples extracted an oil similar to olive oil from its slim acorns.

Ocotillo/Coach Whip

Fouquieria splendens H 25 ft (7.6 m)

The Ocotillo is unique and unmistakable. As many as 100 unbranching, whiplike canes grow from the root crown, all angling out slightly for a narrow, vaselike outline.

KEY FACTS

Spiny canes arise from the root crown.

+ **leaves:** Alternate, 2 in (5 cm), oval, fleshy, in bunches

+ **flowers/fruits:** Flowers perfect, scarlet, tubular, clustered at stem tips; fruit capsules with winged seeds

+ **range:** Southern tip of Nevada through Mojave and Sonoran deserts from California to Arizona

The Ocotillo is called drought deciduous, but it is fairer to say that it leafs out after a rain several times a year. When leafless, the tortuous, gray-green canes handle the photosynthesis. The shrub is even more striking when flowering, the tips of the stems bearing bright red flowers that attract hummingbirds and bees, which pollinate them. *Ocotillo* means "little ocoto," the ocoto being the Montezuma Pine, *Pinus montezumae,* of Mexico and Central America. But the plants don't look alike; they are both just extremely resinous. The stems of the Ocotillo are sometimes removed at the root, and then sunk in the ground to make a living fence.

Witch Hazel

Hamamelis virginiana H to 20 ft (6 m)

Cutting a Y-shaped twig from a Witch Hazel is the first step in making a dowsing rod (several other species are also favored), which "water witches" use to locate, or dowse, water.

KEY FACTS

The Witch Hazel flowers in the fall.

+ leaves: Alternate; somewhat scalloped; base rounded to wedge shaped; yellow in fall

+ flowers/fruits: Flowers perfect, small but showy, 4 petals, narrow, curly, spreading ribbonlike; fruit capsules 2-beaked, becoming woody

+ range: Eastern U.S., Nova Scotia

The Witch Hazel is a shrub or small tree, common in the understory of many eastern woods. The biggest specimens are seen in the mountains of the Carolinas. It has upright branches and an irregular crown, often leaning strongly. Its wood and fruits are of minor value, but it has been used medicinally for centuries. The native peoples boiled many parts of the tree, then used the astringent liquid to treat inflammations. Settlers followed suit, and commercial production eventually began and continues today. Witch Hazel is used as an aftershave lotion, as an eyewash, to treat insect bites, and to soothe hemorrhoids, and may act as a UV protectant.

Sweetgum

Liquidambar styraciflua H to 150 ft (47 m)

Many people may decry the spiny gumballs the Sweetgum produces, but it is widely cultivated for its attractively shaped leaves and its brilliant red, yellow, and orange fall foliage.

KEY FACTS

Maple-like, but the leaves are alternate (opposite in maples) and star shaped.

+ leaves: 5 lobes, pointed, deep, wide

+ flowers/fruits: Flowers in unisexual clusters on same tree; fruits spiky balls, to 1.5 in (3.8 cm)

+ range: Southeast, to Mississippi and Ohio valleys and on the coast to New York

A medium-size to large tree, the Sweetgum is one of the most common hardwoods in the eastern United States, and one of the most commercially important, second only to the oaks in production. The wood, marketed as satin walnut, takes a finish well and is used in furniture, plywood, cigar boxes, barrels, and pulp. Many cultivars have been developed. The genus name refers to a juice that oozes from the bark, a balsam similar to turpentine, which is extracted from the bark by boiling and made into resin. The juice also will dry on the tree, and dried bits were picked off and used as chewing gum by Native Americans.

Black Walnut

Juglans nigra H to 125 ft (38 m)

The Black Walnut is allelopathic, that is, most plants cannot grow beneath it. It produces a compound, juglone, that inhibits metabolism in many plant species.

KEY FACTS

The trunk is often straight for half its height.

+ **leaves:** Alternate, compound, to 2 ft (0.6 m); 15–23 leaflets.

+ **flowers/fruits:** Male flowers in catkins, female on new growth, on same tree; fruits single or paired; husk thick, green; nut black, grooved

+ **range:** Eastern U.S., southern Ontario

The Black Walnut was once so common that furniture was solid walnut, a beautifully grained, brownish wood. The large walnuts are gone, and the wood is used in gun stocks and some furniture and cabinets. Most walnut furniture is only walnut veneer. Walnuts for ice cream, candy, and cakes support a commercial harvest. These nuts have a stronger, richer flavor than English walnuts, but extracting the meat is real work. The hull is thick and hard, and the interior is a catacomb from which bits must be removed with a nut pick. Anyone trying to remove a nut from its husk will soon have brown fingers: No surprise that dyes and inks are made from the husks.

Bitternut Hickory

Carya cordiformis H to 115 ft (35 m)

The Bitternut Hickory is so named because the nuts are too bitter to eat—even for squirrels. But colonists extracted their oil to burn in lamps, and the husks yield a yellow-brown dye.

KEY FACTS

Leaf buds are yellow.

+ **leaves:** Alternate, compound, to 8 in (20 cm); 7–11 leaflets

+ **flowers/fruits:** Male flowers in catkins, female 1 or 2 on spikes, on same tree; fruits clustered, husk thin, opening along winged sutures; nut shell bony, tip sharp

+ **range:** Eastern U.S., southern Ontario and Quebec

The Bitternut Hickory is one of the largest hickories and the most abundant, found in a range of soil conditions and temperatures. Despite its bitter nuts, it is nonetheless important to our palates, though indirectly. Pecans are grafted onto the rootstock of this wide-ranging hickory so that pecans can be produced north of the Pecan tree's usual, rather southerly range. The Bitternut's dense, strong wood is used in furniture, tool handles, ladders, in woodworking, pulp, and as fuel. Native peoples made bows from the wood and used it in making birch-bark canoes. As with other hickory woods, it is used for smoking meat and to make charcoal.

Pignut Hickory

Carya glabra H 82–132 ft (25–40 m)

The Pignut Hickory, it is thought, got its common name from colonists who noticed that their free-range pigs, rooting in the woods for tree nuts, were especially fond of those of this species.

KEY FACTS

The bark is ridged but not shaggy.

+ **leaves:** Alternate, compound, to 10 in (25 cm), smooth; 5–7 leaflets

+ **flowers/fruits:** Male flowers in catkins, female few, in clusters, on same tree; fruits pear-shaped, husk thin, splitting partly; nut unwinged; unribbed

+ **range:** Eastern U.S., southwestern Ontario

A slow-growing, short-branched, medium-size tree, the Pignut Hickory is the most common hickory in the Appalachians and an important species in oak–hickory forests of the East. The nuts make up as much as 25 percent of the diet of squirrels and are important food items for chipmunks, raccoons, crows, and wood ducks. But for the record, the fruit of a hickory is not a nut but a *drupe:* Fleshy tissue surrounds a pit, which contains the seed. In the hickories, the fleshy tissue is the husk, the pit is the nut, and the seed is the nut meat. A peach is constructed in much the same way.

Pecan

Carya illinoinensis H 70–100 ft (21–30 m)

The word "pecan" is descended from an Algonquian word *pakan,* which meant a nut so hard that it had to be cracked with a rock. The word is similar in other native languages.

KEY FACTS

Huge limbs support an oval crown.

+ **leaves:** Alternate, compound, 12–20 in (30–51 cm); 9–15 leaflets

+ **flowers/fruits:** Male flowers in catkins, female 1 to few on spike, on same tree; clustered; fruits thin-husked; nut to 2 in (5 cm); shell thin, brown, often with black

+ **range:** Centered in lower Mississippi Valley.

It is hard to imagine anyone in the world who is not acquainted with the pecan, the soul of pralines, used in ice creams, salads, and scones, and more recently encrusting everything from salmon to seitan. Pecan trees, the largest hickory, are grown far and wide. Spanish and Portuguese explorers took the first trees to Europe in the 16th century. Nevertheless, the United States still supplies about 80 percent of the world's demand. Total U.S. production reached 294 million pounds (133 million kg) (unshelled) in 2010. The long-lived Pecan is also appreciated for its shade, and many cultivars are grown extensively in North America and abroad.

Big Shellbark Hickory/King Nut Hickory

Carya laciniosa H 60–80 ft (18–24 m)

The Big Shellbark Hickory is also called the King Nut Hickory, and deservedly so, because its fruit can be as big as 2.5 inches (6.5 cm) across—just slightly smaller than a tennis ball.

KEY FACTS

The twigs are orange.

+ leaves: Alternate, compound, to 24 in (61 cm); 7 leaflets

+ flowers/fruits: Male flowers in catkins, female in clusters, on same tree; fruit to 2.5 in (6.5 cm); husk splitting to base; nut large, flattened; shell hard, thick

+ range: Midwest, Ohio and upper Mississippi Valleys

The Big Shellbark Hickory is very slow growing, and its fruits are so large that dispersal is hampered; thus, it has never been common. The medium to large tree grows mostly in mixed stands, especially on floodplains and in river swamps. Its nuts are prized for their flavor. The species name, *laciniosa*, from the Latin for "flap," means "shaggy," and the Big Shellbark is similar to the Shagbark Hickory. Key differences distinguish them. The Big Shellbark's bark curls up in strips, though less than the Shagbark's; its leaves have 7 (vs. 5) leaflets, the leaflets have no hairs at the tip of the teeth, the husk of the fruit is thicker, and it prefers wetter habitats.

Shagbark Hickory

Carya ovata H to 150 ft (46 m)

Hickory bark syrup is made from shed Shagbark bark by oven-toasting the bark, then boiling, reducing the liquid's volume, and adding sugar. It is used on meats, in drinks, and on pancakes.

KEY FACTS

Bark on older trees is shaggy.

+ leaves: Alternate, compound, to 14 in (36 cm); 5 leaflets

+ flowers/fruits: Male flowers in catkins, female on short spike, on same tree; fruit to 1.6 in (4 cm); nut usually ridged, shell thick, meat sweet

+ range: Most of eastern U.S., southern Ontario and Quebec

The Shagbark Hickory is similar to the big shellbark but usually has 5 (vs. 7) leaflets, and the leaves have hairs at the tooth tips. The bark is in long, loose plates that curve out, shaggier than in the Big Shellbark. Shagbark nuts are the best of the hickory nuts and were a staple in the diet of native peoples in the tree's range. A food prepared from Shagbark nuts was called *pawcohiccora* in an Algonquian language and altered by settlers to "hickory." Pawcohiccora was a sort of nut milk made from nut meal added to boiling water, and the resulting rich, oily gruel was collected for use in breads and stews.

Mockernut Hickory

Carya tomentosa H to 100 ft (30 m)

The name "Mockernut Hickory" suggests that the nut directs ridicule at the would-be nut eater who, having finally cracked its extremely thick shell, is rewarded by just a small kernel of meat.

KEY FACTS

The bark is not shaggy.

+ **leaves:** Alternate, compound, 9–14 in (23–36 cm); 5–9 leaf-lets, toothed, hairy beneath

+ **flowers/fruits:** Male flowers in cat-kins, female 2–5 in short spike, on same tree; fruit to 2 in (5 cm); nut 4-angled with pointed tip

+ **range:** Southern Ontario, most of eastern U.S.

The slow-growing, tall, straight Mockernut is the most common hickory in the South, and it is a key species in oak–hickory forests of the East. Its wood is among the best of the hickories, used to make tool handles, wood splints, furniture, and charcoal. The species label *tomentosa* refers to the hairiness of the underside of the spicy-smelling leaves. The nut meat, though there is not much of it, is sweet, and competition for the nuts from birds and wild mammals is strong. Many cavity dwellers inhabit the tree, including wood-peckers, Chickadees, and Black Rat Snakes.

Sassafras

Sassafras albidum H 32–60 ft (10–18 m)

Safrole, which makes Sassafras aromatic, is used to manufacture the drug ecstasy. Sassafras is not a good source, but an Asian relative is, and safrole is illegal in Canada and the U.S.

KEY FACTS

The roots and bark smell spicy.

+ **leaves:** Alternate, simple, to 7 in (18 cm); 0, 2, or 3 lobes

+ **flowers/fruits:** Male, female flowers on different trees, yellow-green, small, showy in clusters on bare tree; fruit a ber-rylike drupe, blue, on red stalk

+ **range:** Eastern U.S., Ontario

Sassafras is planted for its fragrance and for its foliage, which turns yellow to red in the fall. All parts are aromatic. The bark of the root was used to make a tea and to perfume soaps; the shoots were used to make root beer. But the aromatic element, safrole, turned out to be carcinogenic, and these products are now artificially flavored. The Choc-taw, native from Florida to Louisiana, have long dried the mucilaginous leaves and ground them to a powder for flavoring and thickening dishes. Named *filé* by French-speaking settlers, the powder is used in Creole and Cajun kitchens to thicken gumbo, soups, and gravy. The leaves are safrole-free.

Northern Spicebush
Lindera benzoin H 6–12 ft (1.8–3.7 m)

The species label *benzoin* refers to the spicy, fragrant leaves, recalling benzoin, a resin obtained from unrelated plants and used in incense (devoid of the compound also called benzoin).

KEY FACTS

The shrub often has several stems.

+ **leaves:** Alternate, simple, entire, 3–6 in (7.6–15 cm), oval, aromatic

+ **flowers/fruits:** Male, female flowers on different trees, yellow, showy in clusters on bare tree; fruit a red, spicy berrylike drupe

+ **range:** Texas, Oklahoma, eastern U.S., south Ontario

The flowers of the Spicebush are among the first to appear in the spring before the trees leaf out. Though small, the flowers are spectacular if there are enough of this small, graceful tree in the understory. The common name derives from its fruit, which, dried and ground, is used like allspice. The fruit, leaves, and twigs are made into an herbal tea. The Spicebush is the main host (the related Sassafras is another) of the Spicebush Swallowtail, a butterfly that lays its eggs on the plant, in whose foliage the eventual caterpillars will undergo metamorphosis. Spicebush is also a host of the Eastern Tiger Swallowtail.

California Laurel/Oregon Myrtle
Umbellularia californica H 23–100 ft (7–30 m)

When the bank in North Bend, Oregon, closed in 1933, coins worth up to $10 were made of myrtlewood, letting commerce continue. The coins are still accepted but most are in collections.

KEY FACTS

Plant is shrubby in exposed situations.

+ **leaves:** Evergreen, alternate, to 4 in (10 cm), lance shaped

+ **flowers/fruits:** Flowers perfect, to 0.6 in (1.4 cm), yellow, in umbrella-like clusters (thus the genus name); fruit a berry, to 1 in (2.5 cm), ripening to purplish

+ **range:** Southern California to Oregon

The California Laurel is an attractive, medium-size tree with a short trunk that soon divides. This is the only species in the genus, but it has many common names, including California Bay. Its peppery leaves (it is also called Pepperwood) are used as the familiar bay leaf, but they have a stronger flavor. For native tribes, this plant treated many maladies, including headaches, though it can also cause headaches and so is called Headache Tree. The olive-like fruit, called the California Bay nut, is like a tiny avocado (the plants are close relatives). It has a relatively large pit surrounded by oily flesh, both of which are eaten.

Tuliptree/Tulip Poplar/Yellow Poplar

Liriodendron tulipifera H 98–165 ft (30–50 m)

Liriodendron derives from the Greek words for "lily" and "tree"; *tulipifera* means "tulip bearing." The Tuliptree is in the Magnolia family, as study of the flower suggests. It is not a poplar.

KEY FACTS

The twigs turn upward.

+ leaves: Alternate, uniquely 4-lobed; yellow in fall

+ flowers/fruits: Flowers perfect, to 3 in (7.6 cm), at branch tips, tulip-like; 3 green sepals; 6 upright petals, yellow-green, orange at base; fruit a cone-like cluster of winged seeds on a stalk

+ range: Eastern U.S.

The majestic Tuliptree is one of our biggest hardwoods, with a single, straight trunk, which is often half bare. Native peoples used the trunks to make dugout canoes. It is an important tree in eastern hardwood forests, especially at lower elevations. The tree is valued for its light, straight-grained wood, used in interior detail, furniture, general construction, plywood, and pulp. Many animals depend on the tree for seeds, browse, sap, or cover, but bees especially avail themselves of this important honey plant: The flowers from one 20-year-old Tuliptree can provide them with nectar sufficient for 4 pounds (1.8 kg) of honey.

Cucumber-tree

Magnolia acuminata H 60–95 ft (20–30 m)

The name "Cucumber-tree" describes the unripe fruit, a composite of many follicles with their developing seeds. When green, this "cone" is shaped much like a cucumber.

KEY FACTS

The bark is furrowed.

+ leaves: Alternate, simple, entire, 6–10 in (15–25 cm), oval, pointed; gold to maroon in fall

+ flowers/fruits: Flowers perfect, to 3 in (7.6 cm) across; 9 yellow tepals, outer 3 bent back; fruit to 3 in (7.6 cm), cone-like; seeds red

+ range: Louisiana to New York, southern Ontario

The Cucumber-tree has a wide distribution, but it is never abundant, usually scattered in the eastern oak–hickory forests. It is most often found in the mountains, and the largest specimens grow in the southern Appalachians. The hardiest magnolia, it is the only one native to Canada, found, but rare, in Ontario. Its slightly fragrant flower is smaller and not as showy as that of other magnolias, and it is the only *Magnolia* species that has yellow sepals. A number of cultivars have been developed. The wood is similar to Tuliptree wood and is sold alongside it. It is used in crates, cabinets, and some paneling.

Southern Magnolia
Magnolia grandiflora H 65–95 ft (20–29 m)

The Southern Magnolia had traveled to England by 1726, via Mark Catesby, an English naturalist who published and illustrated the first book on the flora and fauna of North America.

KEY FACTS

The bark is smooth, gray-brown.

+ leaves: Evergreen, alternate, simple, entire, 3–8 in (7.6–20 cm), oval, shiny, stiff, rusty below

+ flowers/fruits: Flowers perfect, to 8 in (20 cm) across, showy, fragrant; 6–12 broad, creamy white tepals; fruit compound, dry; seeds red

+ range: Coastal Plain, North Carolina to Texas

This small to medium-size tree of the lowlands is botanically and culturally identified with the South, growing largest in moist and well-drained soils and popular as an ornamental far from its native range. More than 100 cultivars have been developed. Other than the seed, the Southern Magnolia is not used much by wildlife, and the wood is of limited importance, but all parts of the tree have yielded compounds of potential pharmaceutical value. The genus name (and that of the family, Magnoliaceae) was assigned by botanist Carl Linnaeus in honor of French botanist Pierre Magnol of Montpellier, who conceived of the family as an organizing unit in plant taxonomy.

Sweetbay
Magnolia virginiana H to 90 ft (27 m)

Sweetbay is often called "tardily deciduous," but it varies from deciduous in its more northern extent to evergreen in its southern. The tree also grows larger in the South.

KEY FACTS

The bark is smooth.

+ leaves: Semi-evergreen, alternate, simple, entire, to 6 in (15 cm), blunt-tipped, silvery below

+ flowers/fruits: Flowers perfect, showy, fragrant, to 5.5 in (14 cm) across; 9–12 creamy-white tepals; fruit and seeds red

+ range: Coastal Plain and Piedmont, eastern U.S.

A graceful, small to medium-size tree, the Sweetbay prefers low elevations and wet, sandy, acid soils. Its foliage and twigs are aromatic, and its flower has a lemony scent. When a breeze catches the foliage, briefly flashing the leaves' silvery white undersides, the tree ripples with light. Primarily southeastern, it ranges north to Massachusetts, where the discovery in 1806 of a Sweetbay swamp in Gloucester caused a stir among botanists. Jacob Bigelow made the find public in 1814 in his *Plants of Boston*, calling it "our only species of this superb genus." The stand is now the focus of an annual "Save the Sweetbay" event in Gloucester.

Chinaberry Tree
Melia azedarach H to 49 ft (15 m)

The fruit of the Chinaberry contains very hard, reddish, five-grooved seeds that were widely used to make rosaries and other beaded objects before the advent of plastics.

KEY FACTS

The clustered fruits are persistent and toxic.

+ leaves: Alternate, once or twice compound, to 2.2 in (6 cm); toothed or lobed, smooth leaflets

+ flowers/fruits: Perfect, small, fragrant, in loose clusters; 5 petals, lavender, with purple tubular corona; fruit a yellow-brown drupe

+ range: Nonnative; widespread

Native to the Himalaya and eastern Asia, the Chinaberry (in the Mahogany family) was brought to South Carolina in the late 1700s as an ornamental and a shade tree. It escaped and is now invasive, especially common in thickets in old fields and on disturbed sites. The Chinaberry is ideally suited for invasion. It is fast growing and short lived, and produces huge numbers of seeds, which are dispersed by birds. Its seedpods persist on the tree into winter; it forms colonies by sprouting; and it is drought tolerant and a soil generalist with essentially no pests. The weak, soft wood has been used in furniture and cabinetry, as well as for firewood.

Osage Orange
Maclura pomifera H 40–60 ft (12–18 m)

A female tree will bear normal-looking fruit even if there is no male nearby to provide pollen; the fruit will not, however, contain seeds. The tree is not an orange but is in the Mulberry family.

KEY FACTS

The bark has an orange cast.

+ leaves: Alternate, simple, to 5 in (13 cm), oval, pointed; leaf axils with 1-in (2.5-cm) spine

+ flowers/fruits: Male, female flowers on separate trees; fruits orange-like, bumpy, to 5 in (10–13 cm) across; skin with milky sap

+ range: South-central U.S.

The Osage Orange is a medium-size tree of rich bottomlands but tolerates a range of conditions. Though of a rather narrow natural range, it has been planted and has naturalized through most of the eastern United States and in Ontario. It is planted in hedges and, because of its spines, was used in cattle "fences" before the advent of barbed wire. The bright orange wood is not commercially important, though it is good firewood. Squirrels may excavate the fruit to eat its seeds, but the tree is not important to wildlife. A large, extinct animal, such as the Mammoth or a horselike mammal, may have been the natural disperser of Osage Orange seeds.

White Mulberry
Morus alba H 50 ft (15 m)

The fastest known movement in plants is that of a mechanism in the White Mulberry's male flowers that ejects pollen into the air. The structure triggers at more than half the speed of sound.

KEY FACTS

The leaves are usually smooth above.

+ **leaves:** Alternate, oval, to 4 in (10 cm), serrate, often deeply lobed

+ **flowers/fruits:** Male and female catkins on same or different trees; fruit aggregates of drupes, ¾ in (2 cm), ovoid, red turning white, pink, or purple-black

+ **range:** Eastern U.S. and Ontario

The White Mulberry has been cultivated for several thousand years in China as a host for the silkworm. This spreading tree was introduced to North America for the same purpose, but the attempt failed—by no fault of the trees. They are naturalized across the eastern United States and into Ontario, often seen on roadsides and in abandoned fields. The invasive nonnative also crosses with the native Red Mulberry. Distinguishing the White Mulberry from the Red can be tricky, as the species overlap in leaf shape and hairiness. The White has leaves that are most often glossy above, often curling up.

Red Mulberry
Morus rubra H to 66 ft (20 m)

Choctaw women wore cloaks of cloth woven of fiber cord processed from the inner bark of the young shoots of Red Mulberry. Other garments were likewise made of the cloth or of the cord.

KEY FACTS

The leaves are often rough above.

+ **leaves:** Alternate, to 5.5 in (14 cm), entire or lobed, rounded and long-pointed

+ **flowers/fruits:** Male and female catkins on separate trees; fruits aggregates of drupes, 1.2 in (3 cm), cylindrical, red to purple to almost black

+ **range:** Eastern U.S., southern Ontario

The Red Mulberry is our most common native mulberry. It is also often planted in yards, but in summer the ground under a female tree will be covered with overripe berries, so some people buy only male trees. Those people will not enjoy the berries, which for centuries have been consumed raw and in beverages, preserves, cakes, and dumplings. The fruits are usually used immediately, and are seldom seen for sale because they do not keep for long. Genetic pollution of *Morus rubra* by the nonnative *M. alba*, the White Mulberry, is recognized as a problem. In Canada, the problem is considered serious enough that the Red Mulberry is listed as endangered.

California Wax Myrtle/Pacific Bayberry

Morella californica H to 26 ft (8 m)

Pacific Bayberry fruits are a source of wax and fragrance that have been used to make candles and soap. Dyes were also made from the berries.

KEY FACTS

The leaves are sticky.

+ **leaves:** Evergreen, alternate, to 4 in (10 cm), narrow, broader, sparsely toothed toward tip

+ **flowers/fruits:** Male, female, perfect flowers in catkins on same plant, in various combinations, inconspicuous; fruit a purple drupe with white wax coating

+ **range:** Lower British Columbia south

The aromatic Pacific Bayberry is a shrub or small tree of the coast. It is often planted as an ornamental, for its evergreen foliage, to attract wildlife, to serve as a hedge or screen, or to retard erosion. The roots contain bacteria that fix nitrogen from the air, making it usable as a plant nutrient. In 2007, a leaf blight, an infection by a fungus-like organism that is known also to damage apples and pears, was noted on the Pacific Bayberry in Oregon. The lower plant is affected first, but much of the plant is eventually defoliated. The leaves may regrow, but repeated defoliations could kill affected branches.

Wax Myrtle/Southern Bayberry

Morella cerifera H to 40 ft (12 m)

One of the birds that take cover in and eat the berries of the Wax Myrtle is *Setophaga coronata coronata*—the Myrtle Warbler, named for its association with the shrub.

KEY FACTS

The gray bark can be almost white.

+ **leaves:** Evergreen, alternate, to 4 in (10 cm), narrow, slightly toothed, broader near tip

+ **flowers/fruits:** Male and female catkins on separate plants; drupe green with light-blue wax coating, borne on female plants

+ **range:** Eastern and Gulf Coastal Plains

A large, evergreen shrub or small tree, the Wax Myrtle is a coastal plant of sandy to moist soils, often forming thickets. It is widely planted, and cultivars have been developed. *Cerifera* means "wax bearing," and all bayberries have been used to make candles. The berries are boiled and the melted wax skimmed from the surface of the water and molded or dipped to make fragrant candles. *Bayberry* means *"berryberry"*: In this usage, the word "bay" comes from the French *baie*, or "berry." But the word also hints at the leaf's aroma, invoking the bay leaf, a culinary herb for which aromatic Wax Myrtle leaves have long been substituted by native peoples and Europeans to flavor seafood and other dishes.

Tasmanian Bluegum
Eucalyptus globulus H 263 ft (80 m)

An ingredient of cough drops and chest ointments, aromatic eucalyptus oil from Tasmanian Bluegum also has been used to treat arthritis and in mouthwashes and insect repellents.

KEY FACTS

The bark shreds in long strips.

+ **leaves:** Evergreen, alternate, to 7.1 in (18 cm), narrow, tapering to a long point, hanging

+ **flowers/fruits:** Flowers single, in upper leaf axils; 4 petals, fused into a cap that falls from the bud; fruit a woody capsule

+ **range:** Nonnative in far West

Thousands of acres of Bluegum were planted for timber in California in the late 1800s. A flop wood-wise but still grown as a windbreak or ornamental, it escaped and now thrives north into British Columbia. But Bluegum is most invasive in and poses the greatest threat to California. Its flammable oils and old bark that piles up beneath it make it incendiary. Of the energy behind the 1991 Oakland Hills firestorm, which killed 25 and burned 3,000 homes, 70 percent came from eucalyptus. (More than 100 other "eucs," including the River Redgum, *E. camaldulensis,* and the Forest Redgum, *E. tereticornis,* also grow in California.)

Water Tupelo
Nyssa aquatica H 115 ft (35 m)

As its common and scientific names suggest, the Water Tupelo often grows in wet places, including Baldcypress and other swamps and periodically flooded areas.

KEY FACTS

Its feet are wet to flooded.

+ **leaves:** Alternate, simple, to 8.5 in (22 cm), pointed, with few teeth; long stalk

+ **flowers/fruits:** Male flowers in heads, female solitary, some perfect, on same tree; fruit a purple drupe, 1.5 in (4 cm)

+ **range:** Virginia to Texas (except peninsular Florida), Mississippi Valley

The Water Tupelo is a long-lived, medium-size to large tree of the Coastal Plain. If it grows in very wet places, the trunk is swollen and buttressed, sometimes hugely so. Above that (or if not buttressed), the trunk is long, straight, and clear, supporting a narrow, open crown. Many fruits are produced, and the seeds are dispersed mainly by water. The fruits are also eaten by wildlife, including songbirds, turkeys, and groundhogs. The wood is weak and soft, used in paneling, boxes, and crates. In addition to its use in pulp, the wood of the swollen base is a preferred medium of wood carvers. Honey from this and other tupelos is prized.

Blackgum/Black Tupelo
Nyssa sylvatica H 131 ft (40 m)

The trunk of the Black Tupelo is straight, extending to the top of the tree. The leaves are brilliant yellow, red, orange, and purple in the fall. It is cultivated and popular as an ornamental tree.

KEY FACTS

Its feet are dry to moist.

+ leaves: Alternate, simple, to 6 in (15 cm), rounded or broadly pointed; stalk short

+ flowers/fruits: Male flowers in dense heads, female 2 to several in cluster, some perfect, on same tree; fruit drupe 0.5 in (1.2 cm), usually 3–5, blue

+ range: Eastern United States, south Ontario

The Black Tupelo is a widely distributed, medium-size to large tree of the open woods in uplands and well-drained alluvial bottoms, especially in acidic soils. The tree has dense foliage, and its crown is conic to flat. In old trees, the bark is blocky, resembling the skin of an alligator. Its fruits are eaten by many birds and mammals, and bees collect nectar from its flowers to make the popular tupelo honey. In addition to the usual uses as lumber, veneer, and pulp, Black Tupelo is used to make rollers, bowls, blocks, mallets, gun stocks, wheel hubs, and pistol grips.

White Ash
Fraxinus americana H to 82 ft (25 m)

This tree is called the White Ash because the underside of the leaflets, in contrast to the deep green above, are so light green as to appear white.

KEY FACTS

Its twigs are smooth.

+ leaves: Opposite, compound, to 12 in (30 cm); 5–9 leaflets, to 6 in (15 cm), white-green below; yellow to red to purple in fall

+ flowers/fruits: Unisexual clusters on separate trees; fruit a samara, broadest near tip

+ range: Eastern U.S.; southern Ontario to Cape Breton Island

Our most common native ash, the White Ash is a tree of well-drained, rich soils, usually growing in mixed hardwood stands. It is the most valuable ash for timber, going into furniture and paneling. Also, most baseball bats are made from its wood, as are snowshoes, electric guitar bodies, and lobster pots. All *Fraxinus* species are menaced by the Emerald Ash Borer, a metallic green beetle native to Asia that was first seen in North America in 2002, and has been reported in 18 states as of 2013. The borer destroys a tree's transport system, killing half the branches in a year and most of the crown in two. It has killed 60 million ashes so far.

Green Ash

Fraxinus pennsylvanica H to 66 ft (20 m)

Its broad tolerance for different types and acid levels of soils makes the Green Ash a popular and widely planted ornamental and shade tree, as well as a successful street tree.

KEY FACTS

The leaves are golden in fall.

+ leaves: Opposite, compound, to 12 in (30 cm); 7–9 leaflets, to 6 in (15 cm), on narrowly winged stalks

+ flowers/fruits: Unisexual clusters on separate trees; fruit a samara, broadest at or above middle

+ range: Alberta to Cape Breton, to Texas and northern Florida

The Green Ash is the most widely distributed ash in North America, inhabiting wet uplands, floodplains, and stream banks, growing in mixed-hardwood forests or pure stands. It is most common in the Mississippi Valley. Small to medium-size, it has a tall, slender trunk, and its crown is irregular to rounded. It produces a large seed crop that is important to turkeys, quail, squirrels, and other small mammals. Moose and deer browse on the shoots and leaves. Its pale brown hardwood is used for similar purposes as that of the White Ash, and it, too, is affected by the Emerald Ash Borer.

Empress Tree/Princess Tree/Royal Paulownia

Paulownia tomentosa H to 60 ft (18 m)

Paulownia was named for Anna Pavlovna, queen of the Netherlands, daughter of Emperor Paul I of Russia, and granddaughter of Catherine the Great.

KEY FACTS

The dry pods are persistent.

+ leaves: Opposite, to 12 in (30 cm), sometimes 3-lobed, hairy above, densely so below; long-pointed; base heart-shaped

+ flowers/fruits: Perfect, to 2.5 in (6 cm), tubular, purple, fragrant, in pyramidal clusters to 14 in (35 cm); fruit capsules large

+ range: Widely invasive

The Empress Tree, a native of China, has been cultivated in Japan and Europe for several hundred years. It was introduced to North America in 1834 for its appealing clusters of purple flowers. Now naturalized and invasive, and cold hardy as far north as Massachusetts, it is seen planted in gardens and parks, but also in vacant lots where it has seeded itself. Each capsule contains 2,000 seeds; one tree can make 20 million in a crop. For a different vision of how prolific it is, consider that in the 1800s, when porcelain was shipped out of China, instead of excelsior, they packed it in the tiny, fluffy-winged seeds of the Empress Tree.

American Sycamore

Platanus occidentalis H 98–130 ft (30–40 m)

English settlers named the tree because its leaves recalled the Sycamore Maple *(Acer pseudoplatanus)*, from England. Leaves are similar, but sycamores and maples are not related.

KEY FACTS

The bark falls in plaques.

+ leaves: Alternate, simple, 4–8 in (10–20 cm); 3–5 very shallow lobes

+ flowers/fruits: Male and female heads on same tree; fruit a narrow achene, with tuft aiding dispersal, in dense sphere from long stem, to 1.4 in (3.5 cm), eventually shattering

+ range: Eastern U.S.

The American Sycamore, the most massive tree of eastern North America, can be identified from far away by its thick single trunk, with its distinctive, irregular patches of green, tan, and white resulting from shedding, older bark. Found among hardwoods of the bottomlands in nature, the tree is grown for shade and is planted on city streets. It is sometimes infected with plane anthracnose, which causes leaf dieback and is host to the Eastern Mistletoe. It was under one of these trees on Wall Street that the New York Stock Exchange began in 1792 with 24 parties signing what became known as the Buttonwood Agreement. The Sycamore is also called Buttonwood.

London Plane Tree

Platanus × acerifolia H 66–115 ft (20–35 m)

Half the planted trees in London are the London Plane Tree. First used in England in the mid-1600s, it became *the* tree in the capital because it could withstand the challenges of urban life.

KEY FACTS

Fruiting balls appear threaded.

+ leaves: Alternate, simple, 4–8 in (10–20 cm); 3–7 shallow lobes

+ flowers/fruits: Male and female heads dense, on same tree; fruit a narrow achene with tuft aiding dispersal, in dense spheres (mostly 2) on stem, eventually shattering

+ range: Hybrid; grown widely in North America

The London Plane Tree is easily confused with the American Sycamore, especially because of their similar patchy, multishaded trunks. The leaves and the fruits distinguish the two. This tree is a hybrid between the American Sycamore and the Oriental Plane Tree *(P. orientalis)*. The latter developed resistance to plane anthracnose, a disease from Asia, with which it evolved. Resistance inherited from the Oriental Plane Tree has made the London hybrid popular in eastern North America, where anthracnose has marred many American Sycamores, which have no resistance. But its resistance varies.

California Sycamore
Platanus racemosa H to 115 ft (35 m)

The largest California Sycamores are found in the canyons, though the species also grows on stream banks and in other moist areas in the central and southern areas of the state.

KEY FACTS

The tree often leans.

+ **leaves:** Alternate, simple, to 10 in (25 cm); 3–5 lobes, half as long as leaf

+ **flowers/fruits:** Flowers tiny, male and female heads on same tree; fruit a narrow achene, with tuft aiding dispersal, in dense maroon spheres, 2–7 appear to be strung zigzag on a stem

+ **range:** California

The California Sycamore, native in the foothills and Coast Ranges, is similar to the American Sycamore, but with more strongly lobed leaves. Its bark, too, is shed in plaques, and the trunk is an irregular patchwork of brown, gray, and white, but on older trees, it is furrowed with broad ridges. The species also often has multiple trunks. Native peoples used its larger leaves to wrap baking bread, the inner bark was used medicinally and for food, and the branches were used to build homes. Plane anthracnose can cause leaf dieback, and the tree is a host to the Bigleaf Mistletoe, also known as Sycamore Mistletoe.

California Buckthorn/California Coffeeberry
Rhamnus californica H 20 ft (6 m)

The California Buckthorn is often called Coffeeberry for its fruit, which progresses from green, to red, to black, much as the coffee plant's fruit. It does not, however, provide a coffee substitute.

KEY FACTS

This species often has multiple trunks.

+ **leaves:** Evergreen or nearly so, alternate, simple, to 3 in (7.6 cm), lustrous

+ **flowers/fruits:** Flowers perfect, small, inconspicuous, star-shaped, greenish, in small clusters in leaf axils; fruit is a berry, 0.5 in (1.3 cm)

+ **range:** California

The California Buckthorn grows as a shrub or small tree on coastal chaparral, on hillsides, and in ravines. It also occurs in parts of Oregon, Arizona, and New Mexico. The plant flowers and sets seed profusely and so is valuable in erosion control. Horticulturally, it is planted for its colorful berries and their contrast with the foliage. The berries are eaten by goats, deer, black bear, and livestock, as well as birds. They are sweet but laxative; native peoples in the Buckthorn's range used it medicinally for that reason, as well as to soothe toothaches by holding a warmed root against the gum and to treat various skin ailments.

Red Mangrove
Rhizophora mangle H to 82 ft (25 m)

Red Mangrove seeds germinate when still attached to the tree. Seedlings grow to 12 in (30 cm), resembling hanging pods. They fall and float until reaching a suitable substrate, then take root.

KEY FACTS

Aerial roots arise from the stems.

+ leaves: Evergreen, opposite, simple, to 6 in (15 cm), leathery, shiny

+ flowers/fruits: Perfect, to 1 in (2.5 cm) across, pale yellow, 2 or 3 at leaf axils; fruit egg-shaped, seeds germinating and seedlings elongating before falling

+ range: Southernmost Florida

Some 70 trees and shrubs in more than 20 families are called mangrove, uniquely adapted to the tropical intertidal zone. Only one, the Rhizophoraceae, is the Mangrove family, and only one of its species is native to North America, the Red Mangrove. Its mangrove look is due to its odd seedlings and to its arching aerial roots, which admit oxygen, often lacking in the stagnant sediments. The White Mangrove, *Laguncularia racemosa* (family Combretaceae), inhabiting higher ground, is also native to Florida. It, too, has aerial roots and precocious seedlings. The United Nations estimates that 20 percent of mangrove ecosystems were lost worldwide from 1980 to 2010.

Saskatoon/Pacific Serviceberry
Amelanchier alnifolia H to 40 ft (12 m)

The city of Saskatoon, Saskatchewan, was named after the berry (and not vice versa), in 1882. The word "Saskatoon" comes from a Cree phrase meaning "early berry."

KEY FACTS

Fall leaves are orange to red.

+ leaves: Alternate, to 2.4 in (6 cm), oval to round, toothed at tip

+ flowers/fruits: Flowers perfect, to 1.2 in (3 cm), fragrant, 3–20 at branch ends; 5 white petals; fruit a blue, sweet berry-like pome, to 0.4 in (1.1 cm).

+ range: Alaska to north California, to Wisconsin, Ontario

The word "Saskatoon" refers to both the berry and the plant. This shrub or small tree often forms thickets by sprouting from its extensive root crown and rhizome network, which makes it useful in erosion control. Twigs, foliage, and bark make the plant important to large game species, such as deer and moose, especially in winter, and the fruits provide food for many birds and mammals, including the black bear. The fruit looks like a blueberry. Native peoples have long eaten them, including in fermented form, and added to pemmican for its flavor and as a preservative. It is raised commercially, and growers market it as a superfruit, like blueberries and acai berries.

Hawthorn
Crataegus species H 16–49 ft (5–15 m)

People born on Manitoulin Island, Ontario, in Lake Huron, call themselves haweaters, because the island is home to many Hawthorns.

KEY FACTS

Fruits recall tiny apples.

+ **leaves:** Alternate, to 4 in (10 cm), entire, toothed, or lobed, even in same species

+ **flowers/fruits:** Flowers perfect, to 1 in (2.5 cm) across, single or clustered; 5 petals, white to red; fruit a pome, to 1 in (2.5 cm), yellow, red, or nearly black

+ **range:** All but northernmost Canadian provinces

While a tree may rather easily be identified as a Hawthorn, determining its species is often difficult. Variation within a species can sometimes be greater than that between two species, and they have a complicated breeding biology. As a result, more than 1,700 names have been published worldwide for what are now believed to represent around 200 species. Here, we are considering the genus only. The thorns are modified branches that can be 3 inches (7.6 cm) long. Densely branched, Hawthorns provide excellent cover and nesting sites for songbirds, and many, especially waxwings and thrushes, eat the fruits, called "haws."

American Plum
Prunus americana H to 36 ft (11 m)

The American Plum was likely cultivated in the Great Plains before European contact. Its branches figure in the Cheyenne Sun Dance, and the Navajo made a reddish purple dye with the roots.

KEY FACTS

The tree is spiny.

+ **leaves:** Alternate, to 4 in (10 cm), pointed, tapered at base, finely toothed

+ **flowers/fruits:** Flowers perfect, to 1 in (2.5 cm) across, clustered 2–5, with 5 white petals; fruit drupe to 1 in (2.5 cm), orange to red

+ **range:** Eastern and midwestern North America

The American Plum is the most widely ranging wild plum in North America, a tree of mixed hardwood forests and edge habitats. It is probably not native to Canada, but its range has been expanded around the world by cultivation. Like many woody species of the rose family, this one readily sends up suckers, which makes it valuable in erosion control. Though its small size makes its wood unimportant, its showy, fragrant flowers have made it a popular choice as an ornamental tree in parks and yards, and it is raised in orchards for its fruits, which are made into jellies and pies.

Mexican Plum

Prunus mexicana H to 35 ft (10.6 m)

Although found in northern Mexico, the Mexican Plum is a tree of the south-central states. It was named in 1882 from a specimen collected in the northeastern Mexican state of Coahuila.

KEY FACTS

The bark is furrowed.

+ **leaves:** Alternate, simple, to 4 in (10 cm), twice fine-toothed, hairy below; hairy stalk

+ **flowers/fruits:** Flowers perfect, to 1 in (2.5 cm) across, clustered 2–4; 5 white petals; fruit drupe, to 1.2 in (3 cm), red to purple

+ **range:** Midwest, south-central states

The Mexican Plum is sometimes confused with the American Plum. The former's hairy leaf-stalks and larger fruits can help distinguish it. It is common, though usually scattered, in a variety of habitats, including woodland edges and open fields, as well as richer woodlands. It seldom sends up suckers and is mostly single trunked, which makes it useful for grafting cultivated plum varieties. Its branching pattern and dark, peeling bark make it a favorite in bonsai. Though the juicy flesh is tart, it is improved with sugar and is often made into jelly and jam. It is often planted as an ornamental.

Black Cherry

Prunus serotina H to 110 ft (34 m)

The Black Cherry was likely being grown as an ornamental in Paris as early as 1630. Introductions in Europe continued, and it is now called a forest pest in Denmark, Germany, and the Netherlands.

KEY FACTS

The bark is plated in age.

+ **leaves:** Alternate, to 6 in (15 cm), finely blunt-toothed

+ **flowers/fruits:** Flowers perfect, showy in pendulous spike to 6 in (15 cm); 5 white petals; fruit drupe, to 0.4 in (1 cm), on flower spike, almost black

+ **range:** East, to Nova Scotia; New Mexico, into Arizona

The Black Cherry is our largest and most widely distributed cherry. A fast-growing pioneer in abandoned fields, it is also found in mixed stands of hardwoods and conifers. Its red-brown wood is hard, polishes well, and rivals walnut for use in cabinetry. Many birds and small animals eat its fruits, as do humans. The small cherries are bitter and so are made into sweet jams, but they are also used in soft drinks and ice cream because their flavor is more intense than that of sweet cherries. The tree is often damaged by tent caterpillars and Japanese beetles. The species label, *serotina*, means "late"; this species blooms after other cherries.

Klamath Plum
Prunus subcordata H to 26 ft (8 m)

The first European settlers in Pacific states soon began to cultivate the Klamath Plum, which they considered the region's most useful fruit. It took botanists until the mid-1800s to discover it.

KEY FACTS

The tree is spiny.

+ **leaves:** Alternate, to 3 in (7.6 cm), almost round; margin once or twice toothed

+ **flowers/fruits:** Flowers perfect, small, clustered 2–7; petals white or pinkish; fruit drupe, to 1.2 in (3 cm), dark red or yellow, flesh juicy

+ **range:** Pacific states

The Klamath Plum is the only plum tree native to the Pacific coast, a large, thicket-forming shrub or a small tree growing in pure stands or with other hardwoods or evergreens. It is found on the eastern slopes of the Coastal Ranges and Sierra Nevada of California and in the dry valleys east of the Cascade Mountains of Oregon and Washington. Although the plum is tart, it is eaten both fresh and dried, and a number of cultivars have been developed. The species name, *subcordata*, means that the base of the leaf is almost cordate, or heart shaped. The tree is too small for its wood to have commercial value.

Chokecherry
Prunus virginiana H to 26 ft (8 m)

Chokecherry fruit is unappealing (thus the common name) but is used in jellies, pies, and wine. Some cultivars have nonbitter fruit, and researchers hope to improve the species as a food plant.

KEY FACTS

The fruit is astringent.

+ **leaves:** Alternate, to 4 in (10 cm); finely, sharply toothed

+ **flowers/fruits:** Flowers in loose clusters to 6 in (15 cm), fragrant, 5 white petals; fruit drupe, to 0.4 in (1 cm), red or purple, thick-skinned, juicy

+ **range:** British Columbia to Newfoundland, northern U.S.

Native peoples used the Chokecherry to treat many ailments; the U.S. Pharmacopeia listed the bark as recently as 1970. The very nutritious fruits have always been eaten, appearing in stews, in pemmican, or alongside salmon. Some people believe that the Chokecherry holds more promise even than the Saskatoon for large-scale culture because of the cherry's hardiness and fruit yield. At present, Chokecherries come from wild trees, and not even 5 percent are harvested. The Chokecherry became North Dakota's state fruit, because archaeologists have found its pits at many sites there. Birds are so fond of the fruit that the tree is sometimes called the Virginia Bird Cherry.

Black Cottonwood
Populus trichocarpa H to 164 ft (50 m)

The Black Cottonwood is the first woody species to have had its genome mapped. It was important to map all the genes of a tree because of wood's economic importance.

KEY FACTS

The trunk is clean.

+ **leaves:** Alternate, to 4 in (10 cm), pointed, finely toothed; leafstalk not flat

+ **flowers/fruits:** Male catkins dense, to 2 in (5 cm), female loose, to 3 in (7.6 cm), on separate trees; fruit capsule 3-parted, to 0.5 in (1.3 cm)

+ **range:** Alaska through California, Idaho

The Black Cottonwood is the largest hardwood in the West and our biggest poplar. This majestic tree is especially successful on deep, rich soils built up by river silting, often developing pure stands. It also grows alongside other hardwoods and conifers. The capsule fruits of cottonwoods develop in the pendent female catkin, appearing as though attached to a thread. The seeds are tiny and bear cottony tufts that aid in their dispersal and that give the trees their name. Native tribes used the inner bark medicinally. It contains salicin, a compound related to aspirin that can reduce inflammation and break a fever.

Eastern Cottonwood
Populus deltoides ssp. *deltoides* H 72–100 ft (22–30 m)

The common name refers to the fluffy-tufted seeds. Another name is Necklace Poplar, a reference to the female catkin's bearing mature fruits and resembling a strand of beads.

KEY FACTS

The trunk branches are massive.

+ **leaves:** Alternate, simple, to 7 in (17.8 cm), triangular, tapering to tip; margin toothed, wavy; leafstalk slender, flat

+ **flowers/fruits:** Male and female catkins to 3 in (7.6 cm), on different trees; fruit capsule, to 0.5 in (1.3 cm), in 3–4 parts

+ **range:** Eastern U.S.

The Eastern Cottonwood is a tall tree of moist, rich lowland forests and is therefore largely absent from the higher Appalachians. The trunk soon branches, creating a wide, spreading crown, and the upright limbs arch at their ends for a vase-like form. The Eastern Cottonwood is often planted for quick shade, but any planting should be well thought out because the extensive roots can invade pipes and buckle sidewalks. Many people plant the male tree, which does not produce cotton. The species label, *deltoides*, refers to the triangular shape of the leaf. The flattened and extra-flexible leaf stem makes the leaf shake in even a calm breeze.

Plains Cottonwood

Populus deltoides ssp. *monilifera* H to 88 ft (27 m)

Isolated populations west of the Rocky Mountains in the Pacific Northwest were recently determined to be the Plains Cottonwood, previously believed to grow only east of the Rockies.

KEY FACTS

This is the largest tree of the Plains.

+ **leaves:** Alternate, to 3.5 in (8.9 cm); tapering to pointed tip; margin coarsely toothed, wavy; leafstalk slender, flat

+ **flowers/fruits:** Male and female catkins on different trees; fruit capsule, to 0.4 in (1.1 cm), conical

+ **range:** Manitoba to Texas Panhandle

The Plains Cottonwood has been considered a separate species, *P. sargentii,* but it is now usually thought of as a subspecies of *P. deltoides;* the subspecies has smaller leaves that have fewer teeth. It is the largest tree and the most abundant cottonwood of the Great Plains, found in pure stands in rich riverside habitats but otherwise mixed with other hardwoods. As in most catkin-bearing trees, cottonwood catkins form before the trees leaf out, the hallmark of pollination by wind. The Dakota ate the inner bark of the spring sprouts and also fed them to their horses. The subspecific label, *monilifera,* means "resembling a string of beads."

Quaking Aspen

Populus tremuloides H 39–60 ft (12–18 m)

The reason the trees quake has long been pondered. In a recent study, leaf stems were splinted to prevent quaking. Insects damaged nonquaking leaves 27 percent more than unsplinted leaves.

KEY FACTS

The white bark has black scars.

+ **leaves:** Alternate, simple, to 3.2 in (8 cm); round, finely toothed, whitish beneath

+ **flowers/fruits:** Male and female catkins on separate trees; fruit capsule, 70–100 in catkin 4 in (10 cm) long

+ **range:** Alaska to Labrador, south to Mexico, Nebraska, Missouri, Virginia

In the Quaking Aspen, suckering from roots and rhizomes produces more than just a thicket; it establishes dense clonal stands. In fact, the world's heaviest (and oldest) "organism" is a clone of aspens in Utah that is 80,000 years old, weighs 13.2 million pounds (6 million kg), and covers more than 106 acres (43 ha). Named Pando (Latin for "I spread"), the stand of 43,000 male trees is in decline, possibly stressed from disease, insect damage, and drought. The aspen is remarkable visually. Its leaves have a flattened stem, which is flexible, allowing the leaves to quake in the slightest breeze. They turn brilliant yellow in fall.

Pussy Willow

Salix discolor H to 30 ft (9.1 m)

Many of us first saw the Pussy Willow not outdoors but in an arrangement of its bare spring branches with their pearl-gray male catkins, which are said to look like cats' feet.

KEY FACTS

The catkins are unstalked.

+ **leaves:** Alternate, to 4.8 in (12 cm), oval, irregularly toothed, whitish green below

+ **flowers/fruits:** Male and female catkins on different trees, dense, fuzzy; fruit capsule beaked; seeds with cottony down

+ **range:** British Columbia to Newfoundland, south into Smoky Mountains

The Pussy Willow is a small tree or a shrub with many tall stems found mostly in pure stands, and especially in wet areas. Because it tends to be shrubby, it is often planted as a hedge. There are male and female trees, and the trademark catkins are male, emerging very early in the spring; a male tree must be planted if these catkins are to be expected. They are also eaten by birds. The specific label, *discolor,* means "two-colored" (not discolored), referring to the difference in color between the green upper leaf surface and the much lighter, whitish underside.

Bebb Willow

Salix bebbiana H 15–25 ft (4.5–7.5 m)

Bebb Willows are the premier diamond willows that, probably because of a fungus, develop large, diamond-shaped depressions on the trunk—a striking motif when the wood is carved.

KEY FACTS

The bark is gray-maroon.

+ **leaves:** Alternate, to 3.5 in (9 cm), oval, white below; veins netlike; leafstalks to 0.3 in (8 mm)

+ **flowers/fruits:** Catkins unisexual, on different trees, to 3 in (7.5 cm); fruit capsule beaked, seeds with threads

+ **range:** Coast to coast in Canada and northern U.S.

The Bebb Willow is the most common willow tree in Alaska and Canada. As is common among the willows, its buds, shoots, bark, and wood provide food for many mammals. A large shrub or a small, bushy tree, it has a short trunk and a rounded crown and is found most often in moist, rich soils. It is fast growing and sprouts from the roots, creating clonal thickets, and thus is an excellent pioneer species in areas that provide the plants with sufficient moisture. It is often found with other willows. The wood is used to make baseball bats and wicker furniture.

Black Willow
Salix nigra H 30–60 ft (9–18 m)

Wood of the Black Willow is light, does not splinter readily, keeps its shape, glues well, and is easy to work. It was at one time used in the manufacture of artificial limbs, or wooden legs.

KEY FACTS

The trunk is massive.

+ **leaves:** Alternate, to 6 in (15 cm), narrow, finely toothed, green above and below

+ **flowers/fruits:** Catkins unisexual, on separate trees, to 3 in (7.6 cm), upright at branchlet ends; fruit capsule ovoid, seeds tiny and silky

+ **range:** Maine to Minnesota, south to Texas

The Black Willow is our largest native willow and our only commercially important one. A species of wet sites, it is largest in the lower Mississippi River region. It is fast growing and forms dense root networks and so is a good pioneer species, used to reduce erosion on stream banks. Medicinal use of willow bark dates to the time of Hippocrates (400 B.C.), when the bark was chewed to relieve inflammation. An active ingredient is salicin, a chemical compound related to aspirin (acetylsalicylic acid). Both words echo the scientific name of the family, Salicaceae, and genus, *Salix*. Other substances give willow bark antioxidant, fever-reducing, immune-boosting properties.

Weeping Willow
Salix babylonica H to 60 ft (18 m)

The father of taxonomy, Linnaeus, erred in naming the Weeping Willow *babylonica* for the willow of Babylon, mentioned in Psalm 137. That tree was the Euphrates poplar, *Populus euphratica*.

KEY FACTS

The twigs are flexible and greenish gold.

+ **leaves:** Alternate, to 7.1 in (18 cm), narrow, finely toothed

+ **flowers/fruits:** Catkins unisexual on separate trees, to 1 in (2.5 cm), upright; fruit capsule on catkin, to 1 in (2.5 cm)

+ **range:** Nonnative; planted widely

The graceful Weeping Willow is a native to China and one of the world's most loved trees. It was first brought to North America from Europe in 1730 for use as an ornamental. The slender, flexible branches hang vertically to the ground, in curtain-like masses. Like other willows, it grows best in moist sites and is often almost stereotypically seen growing beside a lake. Its hanging twigs are greenish gold, and its leaves turn bright yellow in the fall. It has a short trunk, with pale to dark gray, roughly ridged bark often marred with burls from which grow pale yellow shoots.

Bigleaf Maple

Acer macrophyllum H to 98 ft (30 m)

In the Bigleaf Maple, researchers hunt unique traits to amplify them in cultivars. For example, trees that bear red leaves in northern California, and some in Washington, produce triple samaras.

KEY FACTS

The leaf is the largest of any maple.

+ **leaves:** Opposite, to 12 in (30 cm); 5 lobes, deep, pointed; teeth few, irregular

+ **flowers/fruits:** Flowers unisexual, green-yellow, fragrant, on same hanging clusters; fruits paired samaras, in long clusters

+ **range:** Coastal, British Columbia to southern California

In the Pacific Northwest, the only maple tree that is medium size to large is the Bigleaf Maple. It grows mostly in mixed stands on moist sites but is planted for shade in cities. Its trunk is straight and its branches stout, and it is a source of commercial lumber, the wood used in musical instruments, furniture, and cabinets. Older trees produce burls that lend special beauty to veneers. Because the bark of the Bigleaf Maple holds moisture, in places like the Quinault Rain Forest in Washington's Olympic Mountains, the tree is often densely clothed in epiphytic mosses, ferns, and liverworts.

Box Elder/Manitoba Maple

Acer negundo H to 65.6 ft (20 m)

"Box Elder" refers both to the poor-quality wood, suitable for nothing more elegant than boxes and crates, and to the supposed similarity of the leaves to those of elders *(Sambucus)*.

KEY FACTS

Fruiting is often prolific.

+ **leaves:** Opposite, compound, smooth; 3–9 leaflets, to 4.8 in (12 cm); teeth pointed, irregular

+ **flowers/fruits:** Clusters unisexual on separate trees; paired samara fruits, often in hanging clusters

+ **range:** Eastern United States, northwest to Alberta, to Ontario

The Box Elder, or Manitoba Maple, is our only compound-leaved maple. It is often grown ornamentally, and the cultivars include a seedless variety. It is not universally popular, however, naturalizing easily and sometimes invasive, even in North America. When it has three leaflets, it can recall Poison Ivy, but the latter has alternate leaves (the Box Elder's are opposite) and always has three leaflets (the Box Elder can have as many as nine), and its twigs are not green (the Box Elder's are). The tree is the main host of the Boxelder Bug, a black and red insect, but the bugs are more annoying than harmful.

Red Maple
Acer rubrum H to 92 ft (28 m)

The often showy Red Maple earned its species label, *rubrum,* meaning "red." The flowers, fruit, stems, buds, and fall leaves are all red or reddish.

KEY FACTS

The samaras are red, pink, or yellow.

+ leaves: Opposite, pale below; 3–5 lobes, pointed, once or twice toothed

+ flowers/fruits: Perfect or unisexual (sometimes in unisexual clusters, on same or different trees), on slender stalk; fruits double samaras on slender stalk

+ range: Eastern U.S., Quebec

The Red Maple is one of the most abundant trees in eastern North America, growing on many types of soil, in many tree assemblages, and from sea level well into the Appalachian Mountains. It is often planted for shade in yards and on streets. Cultivars have been developed that are more adapted to the urban environment or that enhance red and orange foliage in the fall. The fruits are eaten by squirrels and birds, and deer and rabbits browse the shoots. The wood is used in flooring, cabinets, and furniture. Syrup is made from the sap, but it is inferior to that of the Sugar Maple.

Silver Maple
Acer saccharinum H to 98 ft (30 m)

The Silver Maple is tapped and its sap used to make syrup and sugar. For commercial success, the sugar level is too low, despite its species name, *saccharinum,* from the Latin for "sugar."

KEY FACTS

The trunk often branches low.

+ leaves: Opposite, to 7.9 in (20 cm), silvery below; 5 lobes; sinuses narrow, margin irregularly toothed

+ flowers/fruits: Unisexual clusters on same or different trees; fruits double samaras on slender stalk, prominently veined

+ range: Eastern U.S., New Brunswick

This abundant tree of wet lowlands is related to the Red Maple but lacks the red. Its leaves are intricately divided, and they seem to flicker in a breeze, their silvery undersides (thus the common name) coming in and out of view. The fruits are the largest produced by a native maple and provide food for squirrels and birds. The tree was once planted widely as an ornamental and street tree, but now people think better of it; it splits, breaks, and sheds twigs, it rots and suckers, the seeds are a nuisance, the roots can ruin pipes, and its pests excrete a sticky sap on the ground, or car, beneath.

Sugar Maple

Acer saccharum H to 98 ft (30 m)

Many maple species grow in Canada, but many Canadians consider the Sugar Maple the national tree. An iconic, five-lobed leaf of a maple is the red focus of the nation's flag.

KEY FACTS

The fall foliage is brilliant.

+ leaves: Opposite, to 8 in (20 cm), paler below; 3–5 lobes, pointed; teeth few, coarse

+ flowers/fruits: Flowers perfect or unisexual, on slim stalk to 3 in (7.5 cm), in clusters on same tree; fruits double samaras

+ range: Nova Scotia to southern Manitoba, south to Tennessee

The magnificent Sugar Maple is one of eastern North America's most famous trees. European colonists learned from the native peoples that the tree's sap could be collected and concentrated to make maple syrup, and concentrated some more to make sugar. In addition to its syrup, throughout its range every fall, the tree attracts thousands of people who want to see the expanse of bright color, blinding bright yellow, orange, and red—sometimes all on the same tree. A key species of the eastern hardwood forests, it is also planted in yards, on streets, and in parks. Many cultivars have been developed.

Norway Maple

Acer platanoides H to 98 ft (30 m)

The Norway Maple is one of North America's most common street trees. The popular European import has been bred for improved color, leaf shape, growth form, and urban performance.

KEY FACTS

The leaf stems exude milky sap when broken.

+ leaves: Opposite, to 6.3 in (16 cm), pointed teeth on margin; 5 lobes, mostly pointed

+ flowers/fruits: Flowers perfect or unisexual, sexes on different clusters; fruits double samaras, wings almost in a straight line

+ range: Nonnative; planted widely

Introduced to the northeast by Philadelphia botanist John Bartram in 1756, the Norway Maple is one of the most popular trees in North America, widely planted, especially in cities and towns, from Ontario to Newfoundland, from Maine to Minnesota, and south to North Carolina and Tennessee, as well as from British Columbia to Oregon, east to Idaho and Montana. Sadly, the popular non-native has jumped its bounds to invade forests in the northeast and the Pacific Northwest, outcompeting native species and overshading understory species. The Norway is nevertheless still widely sold in nurseries.

Ohio Buckeye/Fetid Buckeye

Aesculus glabra H 30–49 ft (9–15 m)

The seed of *Aesculus glabra* is called a buckeye, because it recalls the eye of a male deer. Carried in the pocket as a good luck charm, it was also charged with preventing rheumatism.

KEY FACTS

All parts are rank when crushed.

+ leaves: Opposite, compound; 5–7 leaflets, oval, pointed, to 6 in (15 cm)

+ flowers/fruits: Flowers perfect or unisexual, yellow, in terminal clusters to 7 in (17.8 cm); fruit a spiny capsule, to 2 in (5 cm); seeds brown, shiny, to 1.5 in (3.8 cm)

+ range: East-central United States

A medium-size tree, the Ohio Buckeye fares best and takes on a more pleasing form when it grows in river bottoms or on stream banks, although it grows with oaks and hickories in drier locations. Buckeye trees are not heavily used by wildlife, although squirrels will sometimes eat young buckeyes. All parts of the plant contain a toxic alkaloid that may deter would-be browsers. The Ohio Buckeye is planted often as an ornamental for its fruit and because of its bright orange fall foliage. Its light wood is used in woodcarving and was once used to manufacture artificial limbs.

Tree of Heaven/Ailanthus

Ailanthus altissima H 82 ft (25 m)

Ailanthus occurs in Ontario, Quebec, and 43 states, classed as a noxious or invasive plant in many. It should not be planted. Eliminating trees, especially females, reduces seed production.

KEY FACTS

It damages the native environment severely.

+ leaves: Alternate, compound, to 35.5 in (90 cm); 11–41 leaflets, paired along stem

+ flowers/fruits: Flowers mostly unisexual, sexes on separate trees, tiny, yellow-green, in large, conical clusters at tip of twigs; fruits double, twisted samaras in clusters

+ range: Nonnative; widely invasive

Called "tree from hell" and the "kudzu of trees," this native of China seems biologically engineered to invade. Ailanthus spreads by its profusely produced seeds and by sprouting from its roots. It produces a chemical that causes allelopathy, preventing most other plants from growing beneath it. It may be the continent's fastest growing tree, adding 3 to 6 feet (1–2 m) a year when young. What's more, its value to wildlife is virtually nil, and its wood is useless. First brought to North America in 1784, it was planted extensively as a street tree in the 1800s, faring well under the stresses of city life. But the opportunist soon turns a clearing into a dense, impenetrable mass of stems.

American Basswood

Tilia americana H 65–82 ft (20–25 m)

The American Basswood has been widely used medicinally—for example, to treat headache or digestive ailments. It is now said that overconsumption can cause heart problems.

KEY FACTS

The flower stalk has a long, leafy bract.

+ leaves: Alternate, to 8 in (20 cm), toothed; base unequally heart-shaped

+ flowers/fruits: Flowers perfect, to 0.6 in (1.5 cm), creamy, fragrant, 5 petals, few on stalk; fruit a nutlet, to 0.4 in (1.2 cm) across

+ range: Central, eastern U.S.; southern Ontario, Quebec

This stately tree of the lowland woods is the best known of our native basswoods, although the European species (lindens) are better suited for life in cities. The American Basswood has a straight trunk, unbranched for half its length, and a broad crown. Found mostly in mixed stands in rich soils, it is important to wildlife: It naturally develops cavities that attract cavity nesters like woodpeckers and wood ducks, and squirrels and quail eat its seeds. The timber is important in the Great Lakes region, used in cabinetry, musical instruments, boxes, excelsior, and pulp. Native peoples made thread and cord from its tough inner bark.

Common Hackberry

Celtis occidentalis H 32–60 ft (10–18 m)

Detracting from the Common Hackberry's beauty are "witches' brooms," abnormal but harmless tufts of short twigs often formed in its branches, probably caused by a fungus and a mite.

KEY FACTS

The bark has corky ridges.

+ leaves: Alternate, to 3.5 in (8.9 cm), toothed near pointed tip, base unevenly rounded

+ flowers/fruits: Flowers perfect or unisexual, on same plant; fruit a drupe, to 0.8 in (2 cm), orange turning purple

+ range: Southern Ontario and Quebec; midwestern, northeastern U.S.

The small to medium-size Common Hackberry excels in moist, fertile places but is drought tolerant and thrives in sandy soils, too. It has a straight trunk with thick branches. Its common name refers to the fruit and derives from "hagberry," a British word for a tree similar to the Chokecherry. The fruit is edible and sweet, though usually out of reach. The fruits are somewhat important to wildlife, including raccoons, squirrels, and game birds. The weak, soft wood is nonetheless used in some furniture and in fences, posts, and boxes. The Hackberry is planted in gardens and as a street tree.

American Elm
Ulmus americana H 132 ft (40 m)

American Elms lined many streets, especially in northeastern states, but Dutch Elm disease, a fungus carried by Elm Bark Beetles, has ravaged the trees. All elm species are susceptible.

KEY FACTS

The tree is vase-shaped.

+ **leaves:** Alternate, to 6 in (15 cm), rough above, toothed, pointed, uneven at base

+ **flowers/fruits:** Flowers perfect, small, 3–5 on slim, drooping stalk; fruits flat samaras, to 0.5 in (1.3 cm), deeply notched at tip; seeds with papery wing

+ **range:** East; southern Canada south

Many native peoples selected a towering American Elm as the site for councils and other important events, and the stately, graceful tree held Europeans similarly in awe. In many other ways, this is among the most important trees of eastern North America. In the wild, it tolerates a range of soils, usually growing alongside other hardwoods. Its buds, flowers, and fruits feed mammals and birds, and its twigs provide forage for larger mammals. Some cultivars have been developed that show some resistance to Dutch Elm disease. It is hoped that the same can be done for another deadly threat facing elms—elm yellows (see below).

Slippery Elm
Ulmus rubra H 49–82 ft (15–25 m)

This tree is named for its mucilaginous inner bark. The bark, a cough remedy before European contact, stayed in the U.S. Pharmacopeia until 1960 and is still used in cough preparations.

KEY FACTS

The trunk branches high in the tree.

+ **leaves:** Alternate, to 6.3 in (16 cm), dark green, hairy below

+ **flowers/fruits:** Flowers perfect, small, few, in crowded clusters on short stalk; fruit a samara, to 0.5 in (1.3 cm) across, slightly notched at tip, hairy

+ **range:** Southern Ontario through most of eastern U.S.

This lowland tree usually grows with other hardwoods. Its seeds are not of great importance to wildlife, but it provides some browse for deer and rabbits. The scientific name means "red elm," another of its common names, reflecting the red-brown wood. It is used for boxes, crates, and baskets. Native peoples used its bark for canoe shells when birch was not available. The Slippery Elm (and the American) is threatened by elm yellows, a disease spread by leafhoppers that attacks the trees' transport tissues. The first symptoms are yellowing, drooping, or early loss of leaves, but by the time they appear, it is too late.

Rock Elm

Ulmus thomasii H 82–115 ft (25–35 m)

Even though the Rock Elm's claim to fame is its hard wood, it is often called Cork Elm because its older branches can bear three or four thick, irregular corky wings.

KEY FACTS

The fruits are a key trait.

+ **leaves:** Alternate, to 4 in (10 cm), coarsely toothed, on hairy stalk

+ **flowers/fruits:** Flowers perfect, small, reddish, 2–4 in dangling clusters; fruit samara, flat, broad winged, slightly notched at tip

+ **range:** Northern U.S. Midwest, east to northeastern U.S., southern Canada

The Rock Elm is not a tree of pure stands but is usually found growing alongside such species as American Elm, Basswood, Sugar Maple, and White Ash. While it prefers a moist, well-drained loam, it also grows on drier, upland sites, including those on limestone. Some argue that growth on a rocky substrate earned it its common name, but most say it refers to the wood. The heaviest, hardest elm wood, it has few knots, bends well, and can take shock. That makes it superior when strength is mandatory, so it has seen use in ships' timbers, curved sections of furniture, early refrigerators and car bodies, hockey sticks, and ax handles.

Creosote Bush

Larrea tridentata H 3.3–10 ft (1–3 m)

For native peoples, the resinous, fragrant Creosote Bush served many medical purposes, but the resin also served as a glue for mending broken pottery or cementing a projectile point to its shaft.

KEY FACTS

Nodes make the stem look jointed.

+ **leaves:** Evergreen, opposite, compound; 2 leaflets, to 0.7 in (1.8 cm), resinous, joined basally

+ **flowers/fruits:** Flowers to 1 in (2.5 cm) across, velvety, axillary; 5 yellow petals; capsule 5-parted, fuzzy

+ **range:** Southern California, east to Utah and Texas

The Creosote Bush is a trademark of the Mojave Desert. In full bloom, the flowers lend the plant a yellowish cast, and after a rain, the resin exudes a tarry aroma. Its main value to wildlife is as shelter. Kangaroo Rats and Desert Tortoises are two animals that bed down under the plant. The Creosote Bush Grasshopper eats this plant exclusively, having evolved digestive processes allowing it to process the mostly inedible resin. A Creosote Bush can live 100 years, but when it dies, stems sprout at its edge, creating a Creosote Bush ring. One ring, called King Clone, has been aged with radiocarbon dating at 11,700 years. Its diameter averages 45 feet (13.7 m).

Joshua Tree
Yucca brevifolia H to 50 ft (15 m)

Other members of the genus *Yucca* are probably more familiar to the reader, and it helps to recall their leaves (living and dead) and flowers when contemplating the Joshua Tree.

KEY FACTS

Plants can sprout from rhizomes.

+ **leaves:** Evergreen, to 9 in (23 cm), spear-like, sharply toothed, in rosettes at branch ends

+ **flowers/fruits:** Flowers perfect, white, clustered at branch tips; 6 tepals; fruits to 5 in (13 cm), 3-angled, light red to yellow-brown

+ **range:** Endemic to Mojave Desert

The striking Joshua Tree seems to branch wildly, but closer examination reveals that each branching results in two equivalent, stout branches more or less at a right angle to each other. The branches are armed with horrific spines and are tipped with clusters of bladelike leaves that recall a Common Yucca plant. The tree remains unbranched, though, until it blooms the first time (or until its tip is somehow damaged). Old leaves persist, folded back and shaggy along the branches, which bear a thick, furrowed bark. In a sort of give-and-take, Yucca Moths pollinate the flowers, which open at night, and lay their eggs inside them, where eventually their larvae will undergo metamorphosis.

Cuban Royal Palm/Florida Royal Palm
Roystonea regia H to 131 ft (40 m)

Palms are indispensable to people who need to use them for food, fiber, sugar, wax, and wood. Palms are commercially important, too, as the source of rattan, oils, coconuts, and acai.

KEY FACTS

The crownshaft is conspicuous.

+ **leaves:** To 13 ft (4 m), compound, featherlike; many strap-shaped leaflets

+ **flowers/fruits:** Flowers unisexual, on same tree, to 0.3 in (8 mm) across, white, fragrant, in hanging clusters to 2 ft (0.6 m); drupes, to 0.6 in (1.5 cm), black

+ **range:** Southern Florida

In addition to its long, feathery leaves, the Cuban Royal Palm is distinguished from others by the occasional bulge along its trunk and, just below its canopy, by the bright glossy green "crownshaft." The crownshaft is made up of the bases of the leaves, the older ones closer to the outside. When a leaf has died, its base will eventually separate from the tree, leaving a new ringed scar just below the crownshaft on the light gray trunk. The tree is a tall, unbranched column with a spreading crown of leaves. The flowers are pollinated by bees and bats, and bats, as well as birds, eat the fruit and disperse the seeds.

Cabbage Palmetto

Sabal palmetto H to 82 ft (25 m)

This palmetto is named for its terminal bud, or heart of palm, where new fronds originate. "Swamp Cabbage" may be eaten raw or cooked. Once the bud is taken, the tree stops growing.

KEY FACTS

The trunk is straight and unbranched.

+ leaves: Leafstalk to 7.5 ft (2.3 m), serving as midrib; blade fan-like, to 6.6 ft (2 m)

+ flowers/fruits: Flowers perfect, small, white, in clusters to 6.6 ft (2 m); drupes, to 0.5 in (1.2 cm), black

+ range: Florida Keys, in Coastal Plain into North Carolina

The Cabbage Palmetto is a tree of the marshes, woodlands, and sandy soils of the Coastal Plain and is tolerant of salt spray. It is a straight, slender tree that is often grown in yards and along streets. This plant does not have a crownshaft; its leaves are produced by a terminal meristem. Native peoples harvested the fibers from the leafstalks and used them for scrubbing and to make straps and baskets. The fibers were not just used locally, though. They have been found among artifacts from the Winnebago in Wisconsin and the Iroquois of New York, as far as 700 miles (1,100 km) north of the Cabbage Palmetto's northern limit.

Saw Palmetto

Serenoa repens H to 23 ft (7 m)

Extracts of Saw Palmetto fruit are the top herbal treatment for benign prostatic hyperplasia, and research has shown that the extract can indeed improve urinary symptoms and flow volume.

KEY FACTS

The "saw" is the spiny leafstalk.

+ leaves: Leafstalk to 5 ft (1.5 m), ending in fanlike blade to 3.3 ft (1 m)

+ flowers/fruits: Flowers perfect, small, white, in clusters to 3.3 ft (1 m); drupes, to 0.8 in (2 cm), blue-black

+ range: South Carolina to southern Florida, west to Louisiana

The Saw Palmetto is not an erect palm. Instead, it sprawls along the ground, with fronds growing at the ends of long, branching, horizontal stems running as far as 15 feet (4.6 m) at or just below ground level. The result is a dense understory in pine flatwoods and scrub habitats. Some species associated with Saw Palmetto are the Crested Caracara, Sand Skink, Florida Mouse, and endangered Florida Scrub Jay. When the fruits are ripe, black bears will wander from their home range to eat them. Honey made from Saw Palmetto flowers is of high quality and commercial value.

Key Thatch Palm/Brittle Thatch Palm

Leucothrinax morrisii H to 40 ft (12 m)

Plant names are not static. The Key Thatch Palm, long called *Thrinax morrisii*, was given its own genus, *Leucothrinax*, after taxonomists found that it differed genetically from other *Thrinax* species.

KEY FACTS

The fronds are green and silver.

+ leaves: Leafstalk to 3.3 ft (1 m), smooth; blade fan-shaped, to 32 in (80 cm)

+ flowers/fruits: Flowers perfect, ivory-white, fragrant, tubular, in elongate, branching cluster to 6.6 ft (2 m); drupes, to 0.3 in (7 mm), white

+ range: Southern Florida

The slow-growing Key Thatch Palm has a slender, columnar trunk, with smooth, gray to brownish bark. Capped with a rounded, open crown of magnificent green and silver fronds, it is often planted as a specimen tree in the landscape or in small groups. As in other species, when grown in full sun, it will develop a denser, rounder crown, and in shadier conditions, the canopy is much more open. In nature, this tree is found in a range of habitats including the seashore and pineland sand. The Key Thatch Palm's leaves have long been used to make brooms and mats, though today these are more for decoration than for utilitarian purposes.

California Washingtonia/California Fan Palm

Washingtonia filifera H to 50 ft (15 m)

The iconic palms of Los Angeles were planted for the 1932 Summer Olympics. They will be replaced someday with native broadleafs. The palms are nonnative Mexican Washingtonias.

KEY FACTS

The leaflet edges bear threads.

+ leaves: Leafstalk to 5 ft (1.5 m), with hooked spines, bearing fan to 5 ft (1.5 m)

+ flowers/fruits: Flowers perfect, small, white, fragrant, in clusters to 16.5 ft (5 m); drupes, to 0.4 in (9 mm), almost black

+ range: Southwestern Arizona and southeastern California

The California Washingtonia is a columnar palm that is well known but rare, restricted in nature to streams and canyons that offer more water than the adjacent desert. It is planted in tropical areas around the world, especially along city boulevards like those in south Florida. As beautiful as the California species is, the Mexican Washingtonia (*W. robusta*) is prettier and faster growing. Other traits that make it stand out are its slimmer trunk, its height (to 75 feet/23 m vs. 50 feet/15 m in the California), its bright green leaves (vs. gray-green), and its less thready leaflet edges (*filifera* is Latin for "making threads" or "thread-bearing").

An Arizona slot canyon features Navajo sandstone, formed from windblown sands of an ancient desert.

3 Rocks & Minerals

Earth's Beauty and Power

With all the bustle and show of living things, it can be easy to overlook the ground beneath our feet. But the Earth's rocks and minerals are among the most fascinating aspects of the natural world. They are the record of billions of years of Earth's activity. Within their structures, from the smallest crystal to the largest mountain face, the beauty and dynamism of our planet are revealed.

■ What are Minerals?

Minerals are the solid substances that make up rocks. There are thousands of different minerals on Earth, some rare and valuable like diamond and gold, others, such as quartz, as common as beach sand. Minerals are defined by a few characteristics: They are solids. They are naturally occurring. They have an ordered atomic arrangement. They have a well-defined chemical composition.

Most minerals exist as small, irregular grains within rocks. Occasionally, isolated minerals have beautiful geometric shapes and smooth planar surfaces. These shapes and surfaces are crystals, and they are the outward expressions of the mineral's internal atomic arrangement. Crystals form when a mineral has sufficient time and space in which to grow.

Gems, or gemstones, are materials cut and polished, used for jewelry or other decorative items. Most gems, unless they are synthetically produced, are minerals with gem-specific names. For example, amethyst is a purple variety of quartz, and emerald is a green variety of beryl.

Identifying minerals can be both fun and challenging. It takes close observation, a few basic tests, and the process of elimination—and it definitely gets easier as you gain experience. You don't need many tools: A hand lens and a small hammer will get you started. Mineral identification kits containing streak plates, hardness tools, and a small bottle of diluted acid are helpful.

■ Three Types of Rocks

Rocks are the solid materials that make up the Earth. There are three basic types: igneous, sedimentary, and metamorphic.

Igneous rocks form when molten material, called magma, cools and solidifies. Igneous rocks can be either intrusive, meaning the magma solidified underground, or extrusive, meaning the magma erupted onto the Earth's surface before solidification. (Erupting magma is called lava.) Intrusive igneous rocks cool slowly, allowing minerals to grow into visible grains that form an interlocking, crystalline texture. Extrusive rocks cool quickly, resulting in a fine-grained volcanic texture. Some igneous rocks have two grain sizes with large, well-formed crystals surrounded by a groundmass of finer grains. This is called a porphyritic texture, and the large grains are known as phenocrysts.

In addition to texture, igneous rocks are categorized by chemistry. Mafic rocks are rich in iron and magnesium (minerals such as biotite and hornblende) and poor in silica (such as quartz and feldspar), and they tend to be dark in color. Felsic rocks are light in color, containing more silica-rich minerals and fewer that are rich in iron and magnesium.

Sedimentary rocks form from an accumulation of sediments. The sediments can be preexisting minerals or rock fragments—collectively called clasts—that have been transported by wind, water, or ice; or they can be chemically precipitated or biologically generated in place, as with many types of limestones.

During accumulation, sediments become compacted by the weight of material above. Finally, groundwater or other fluids passing through the formation precipitates a cement between the sediment grains—commonly silica or calcium carbonate—hardening the sediment pile into stone. Clastic sedimentary rocks such as conglomerates, sandstones, and mudstones are classified by grain size.

Metamorphic rocks form when preexisting rocks are transformed by heat and pressure, a process called metamorphism. This process involves physical and chemical changes that occur in the solid state—changes such as the growth of new minerals, and the physical rotation or recrystallization of existing minerals. The parent rock that becomes a metamorphic rock is called a protolith. Limestone, for example, is the protolith for the

---

KEY MINERAL PROPERTIES

+ **Hardness:** See page 184.

+ **Streak:** Streak is the color of the mineral when it is ground into a fine powder. Many mineral identification kits contain small white porcelain plates called streak plates. Scratching an edge of a specimen against this plate will produce a streak. If the mineral is clear, or if it is harder than the porcelain plate, the streak will be white.

+ **Crystal habit or aggregation:** The form of a mineral specimen can be helpful for identification. If the specimen has well-developed crystal faces, the shape of the crystal, known as its habit, can be diagnostic. Many specimens, however, do not have well-developed crystal faces and are instead aggregates of small mineral grains. A variety of terms describe crystal habit and state of aggregation:
  Prismatic means that the crystal is long in one direction with well-developed faces. Some prismatic crystals have blunt ends; they can also be topped with pyramid shapes.
  Columnar means that the crystal is long in one direction and resembles the shape of a rounded column.
  Tabular and **platy** habits refer to crystals that have two dimensions relatively equal in length and one that is short.
  Bladed habit refers to crystals that are both elongate and flat, like a blade.
  Fibrous means that the crystal grows in the shape of threads or filaments.

Massive means that the mineral specimen does not have crystal faces.
Granular describes aggregates of mineral grains that are all approximately the same size and dimension.
Compact specimens are very fine grained so individual grains cannot be distinguished.
Mammillary, from the Latin *mamma*, meaning "breast," describes specimens that are smooth and rounded.
Botryoidal is the term for specimens that are smooth and globular, with smaller spherical shapes than mammillary specimens.

+ **Luster:** The way light is reflected from the surface of a mineral is called luster. Terms describing luster include metallic, glassy or vitreous, pearly, waxy, earthy, and dull. One mineral can display different types of luster depending on its crystal faces or habit.

+ **Color:** Color is an important property used in mineral identification, but it can be misleading because most minerals come in a variety of hues.

+ **Cleavage & fracture:** Some crystals tend to break along one or more smooth, flat surfaces known as cleavage planes. The orientation of these planes is helpful for identifying minerals. Breaks along irregular surfaces instead of cleavage planes are called fractures. Conchoidal fracture is a type of fracture with a smooth, scooped, or curved pattern, common in quartz and obsidian.

In a microscopic view of granite, polarized light interacts with mineral grains, revealing brilliant colors.

metamorphic rock marble, and granite is a pro-tolith for some gneisses. Metamorphism occurs when rocks are subjected to high pressures and/or temperatures—conditions that happen during movements of the Earth's plates. When plates collide and push up large mountain ranges, for example, rocks are squeezed and heated in the thickening crust. Metamorphism also occurs in shallow regions of the crust by the heat given off of large magmatic intrusions.

Landforms & Structures

The study of the Earth is not just about identifying rocks and minerals. In fact, what most people notice first are not individual rocks, but rather the shapes and textures of the land—the natural features that make up the landscape. Why does a valley have a certain profile, for example, or why is one coastline rimmed with cliffs and another

|||||||||||||||||||||||||||||||||||||||||||||||||||||||||||||||||

To him was given
Full many a glimpse
(but sparingly bestowed
On timid man) of Nature's processes
Upon the exalted hills.
—WILLIAM WORDSWORTH

|||||||||||||||||||||||||||||||||||||||||||||||||||||||||||||||||

gentle beaches? These are landforms, and they can tell us a great deal about how our planet works. Most landforms are the result of a combination of features and processes—the structures of the rocks combined with the effects of weathering and erosion. Weathering is the breakdown of rocks by either physical or chemical means. Erosion is the removal and transport of rock materials by the energy of wind, water, or ice.

▣ Fossils

Fossils are the preserved remains or impressions of ancient living things. Fossils form when organisms are buried in sediment and preserved via processes like compaction, chemical alteration, or replacement with minerals. Fossils are extremely important because they are our primary clues about past life-forms and previous environments that existed on our planet—the key to our understanding of evolution. The best places to look for fossils are areas where sedimentary rocks are exposed along road cuts, hillsides, or along streams and rivers. Fossils are most common in fine-grained sedimentary rocks such as mudstones and shales, and marine sedimentary rocks such as limestones.

▣ Use Caution

Collecting rocks, minerals, and fossils is an exciting hobby, but one that carries significant risks and responsibilities. Be sure to check the status of a prospective site beforehand. Gain permission if it is private land, or conform to the rules and regulations regarding prospecting and collecting if the land is public. Information can be obtained from state geological surveys, state parks, the U.S. Forest Service, and the Bureau of Land Management. Always put your safety first. Wear protective gear, especially when using hammers, chisels, or other digging equipment, and be aware of risks like rockfall and unstable slopes. Never enter closed or abandoned mines.

MINERAL HARDNESS SCALE

+ **Hardness:** The resistance of a mineral to scratching or abrasion is called its hardness. It is measured on a relative scale from 1 to 10 in increasing hardness. A mineral with a lower number can be scratched by a mineral with a higher number. Minerals with similar numbers can scratch each other. Each level of hardness is designated by a mineral standard.

HARDNESS	DESCRIPTION	MINERAL STANDARD
1	Very soft, falls apart in fingers	Talc
2	Easily scratched by fingernails	Gypsum
3	Can be scratched by a copper penny	Calcite
4	Can be scratched by a knife	Fluorite
5	Can be scratched by a knife with difficulty; can be scratched by glass	Apatite
6	Cannot be scratched by a knife; will scratch glass	Orthoclase
7	Scratches glass easily	Quartz
8	Scratches glass very easily	Topaz
9	Can cut glass; can be scratched by diamond	Corundum
10	Will cut glass; no other minerals will scratch it	Diamond

ROCKS & MINERALS

Quartz

Class: Silicate Chemical formula: SiO_2

Quartz is one of the most common minerals in the Earth's crust, found in many sedimentary, igneous, and metamorphic rocks. Collectible varieties include amethyst and smoky quartz.

KEY FACTS

Well-formed crystals are hexagonal prisms with pyramid-shaped tops; quartz comes in various colors, usually clear, gray, or milky white; transparent to translucent.

+ hardness: **7**

+ streak: **White**

+ locations: **This is a common rock-forming mineral; gem-quality crystals are found in veins and granitic pegmatites.**

Quartz, an important building block of the Earth's crust, consists of silica and oxygen atoms linked in a strong, three-dimensional framework. Quartz is distinguished by its hardness and, in well-formed crystals, its hexagonal-shaped prisms with pyramidal ends. Traces of other elements substituting for silicon create different varieties. Iron in the structure forms yellow quartz crystals known as citrine, or purple crystals known as amethyst. Smoky quartz contains trace amounts of aluminum. Quartz has a vitreous or greasy luster and no cleavage. Microcrystalline forms and aggregates have conchoidal fracture.

Chalcedony

Class: Silicate Chemical formula: SiO_2

Chalcedony is a variety of quartz made up of crystals so small they cannot be seen with the naked eye. It comes in beautiful colors and patterns, some valued as semiprecious stones.

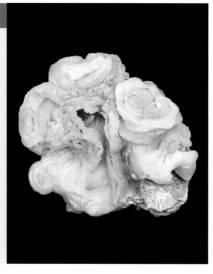

KEY FACTS

This attractive mineral is hard and compact, translucent to opaque, with a waxy or vitreous luster and a conchoidal or uneven fracture.

+ hardness: **6–7**

+ streak: **White**

+ locations: **Cavities in volcanic rocks and as nodules in sedimentary rocks; clasts of chalcedony can be found in some ocean beach and river gravels.**

Chalcedony is microcrystalline or cryptocrystalline quartz. Microcrystalline crystals are so small that they can be identified only under a microscope; cryptocrystalline crystals, from the Greek *krypto*, meaning hidden, are so small that an optical microscope cannot resolve them. Chalcedony can form in rock cavities, most often in fine-grained volcanic rocks. It is commonly found in globular forms—botryoidal or mammillary shapes—and as the outer shell of geodes and the material replacing plant parts in petrified wood. Collectible varieties include agate, bloodstone, carnelian, onyx, and petrified wood.

Agate

Class: Silicate Chemical formula: SiO_2

A concentrically banded form of chalcedony, agate is commonly found filling rounded cavities in volcanic rocks. Cut and polished agates are sold as ornaments and used in jewelry.

KEY FACTS

This microcrystalline to cryptocrystalline quartz is banded typically parallel to its cavity; wide range of natural colors; many bright colors of commercial specimens are artificial.

+ hardness: 6–7

+ streak: White

+ locations: Rounded nodules in various rock types; cavities in volcanic rocks; eroded agates in river deposits

Agate is known for its beautiful colors and concentric bands, which are often enhanced by dyes and polish in commercial samples. Most agates form from silica-rich fluids that have filled open pockets or seams in volcanic rocks. The silica is deposited in bands of tiny, parallel fibrous crystals around the cavity's inside wall. During crystallization, impurities collect along the bands, creating alternating colors and rings. Look for well-developed crystals (often amethyst or smoky quartz) at centers of some agate nodules. Collectors use the term "agate" for nonbanded chalcedony such as moss agate and flame agate.

Opal

Class: Silicate Chemical formula: $SiO_2 \cdot nH_2O$

Opal is a form of cryptocrystalline quartz with water present. It is made of tiny spheres packed together that interact with light to create a distinctive "opalescent" rainbow shimmer.

KEY FACTS

Variable in color, opal exists as compact masses, crusts, and veins with vitreous or pearly luster.

+ hardness: 5.5–6

+ streak: White

+ locations: Near geysers and hot springs; also as pseudomorphs of marine shells or wood, meaning it has replaced the original material while retaining the item's shape and form

Opal is hardened silica gel in the form of tiny, tightly packed spheres with a significant amount of water (5 to 10 percent) in its pores. It is fragile and can easily crack and lose its water content over time, which diminishes its opalescence. Opal is usually white, but can also be green, brown, black, yellow, or gray due to various impurities such as iron and manganese. Opal is deposited by silica-rich waters at relatively low temperatures. It forms in small cavities in basalt, in cracks and small veins, and as pseudomorphs of wood and shells. The word "opal" comes from Greek *opallios*, meaning "precious stone."

Jasper

Class: Silicate Chemical formula: SiO_2

Jasper is a granular form of microcrystalline quartz, distinguished by its opacity, rich colors, and abundant impurities. It is typically deep red, yellow, or brown due to traces of iron oxide.

KEY FACTS

With hues of rich red, yellow, or brown, jasper is opaque and can have an angular, broken appearance or fine color banding.

+ hardness: 6–7

+ streak: White

+ locations: In association with various sedimentary rocks, banded iron formations, and fault zones; jasper pebbles and cobbles also occur in stream deposits.

Jasper is evenly colored or has banding or brecciation, an amalgamation of angular, broken shards. Polished jasper is traded as an ornamental semiprecious stone. Banded jasper differs from agate by its opacity (agate is typically translucent or semitranslucent). Classification of chalcedony, agate, jasper, chert, and flint is inconsistent. Agate and jasper are commonly considered to be varieties of chalcedony, which is cryptocrystalline or microcrystalline quartz. Chert and flint are often classified as sedimentary rocks composed of microcrystalline quartz. Jasper can also be classified as a colorful form of chert.

Chert/Flint

Class: Silicate Chemical formula: SiO_2

Chert, also known as flint, is made of cryptocrystalline or microcrystalline quartz. Chert occurs as layered formations or as nodules in marine sedimentary rocks.

KEY FACTS

Nodules and beds are opaque, hard, and dense masses with conchoidal or irregular fracture; white and gray varieties are often called flint.

+ hardness: 7

+ streak: White

+ locations: Found as nodules in chalk and other marine sedimentary rocks; also in association with Precambrian banded iron formations

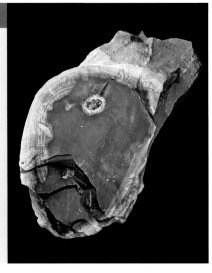

Chert, also known as flint, is a hard, compact form of cryptocrystalline or microcrystalline quartz primarily marine in origin. Chert is commonly gray, white, yellow, black, or brown. It forms irregular sedimentary bands or layers, or nodular masses, in fine-grained limestones and chalk. It forms by accumulation of silica, sometimes from organic sources such as needlelike structures in marine sponges. Some cherts contain small fossils. Flint is the name commonly used to describe nodular chert. Chert is usually classified as a rock, but it is included here with other forms of microcrystalline quartz.

Feldspars

Class: Silicate Chemical formula: $XAl_{(1-2)}Si_{(3-2)}O_8$

Feldspars are a group of minerals made of silicon, oxygen, and aluminum, plus various amounts of sodium, calcium, and/or potassium. They are common constituents of the Earth's crust.

KEY FACTS

These common rock-forming minerals are light in color and are often blocky. They display two cleavage planes that intersect at an angle of approximately 90 degrees.

+ hardness: 6–7

+ streak: Most feldspars have a white streak.

+ locations: Widespread across all rock types; common in intrusive igneous rocks

Feldspars are building blocks of many rocks, especially granites and other intrusive igneous rocks. They are also found in clastic sedimentary rocks and as grains in metamorphic rocks. A key component of the Earth's crust, feldspars are also found in meteorites and rocks from the moon. Feldspars come in various colors, commonly white, pink, gray, or other light shades. They are distinguished from other minerals by their hardness and cleavage planes. Feldspar is mined for use in glassmaking and ceramics, and is used in products such as fiberglass, floor tiles, tableware, sinks, toilets, and paints.

Potassium Feldspars

Class: Silicate Chemical formula: $KAlSi_3O_8$

The potassium feldspars are a subgroup of feldspar minerals that contain potassium in their structures. They are important components of igneous rocks.

KEY FACTS

These feldspars with salmon pink grains are commonly found in granite; however, they can be white, gray, or yellow, and are often difficult to distinguish from plagioclase.

+ hardness: 6–7

+ streak: White

+ locations: Granite and other intrusive igneous rocks; well-formed crystals in granitic pegmatites

Potassium feldspars (also known as K-feldspar or K-spar) include microcline, orthoclase, and sanidine. They form short tabular or prismatic crystals. Microcline is milky white, pink, or blue-green and found in metamorphic and igneous rocks. Well-formed microcline crystals occur in granitic pegmatites. Orthoclase is common in granite and other intrusive igneous rocks; sanidine is a high-temperature form of potassium feldspar common in rhyolite and other extrusive igneous rocks. Orthoclase with fine intergrowths of sodium feldspar (albite) is known as moonstone and displays a unique iridescent luster.

Plagioclase

Class: Silicate Chemical formula: $NaAlSi_3O_8$ (albite)–$CaAl_2Si_2O_8$ (anorthite)

Plagioclase is a common rock-forming mineral and a primary component of many igneous rocks including granite. It includes a feldspar subgroup with sodium and/or calcium in the structure.

KEY FACTS

Commonly white or gray, the crystals have two cleavage planes that intersect at approximately 90 degrees; fine parallel striations on one cleavage face distinguish it from potassium feldspar.

+ hardness: 6–7

+ streak: White

+ locations: Metamorphic rocks including gneisses and amphibolites; common in igneous rocks

Plagioclase is the name for a group of feldspars that have a range of compositions from a pure sodium variety (albite) to a pure calcium variety (anorthite). The plagioclase feldspars include albite, oligoclase, andesine, labradorite, and anorthite. It can be difficult to distinguish between the different types of plagioclase in the field. Plagioclase is usually white or light gray, but can also be dark gray, bluish white, or greenish white. Smooth cleavage planes have a vitreous luster. Like other feldspars, plagioclase is distinguished by its hardness and cleavage planes. Single crystals are rare.

Labradorite

Class: Silicate Chemical formula: $(Ca,Na)(Al,Si)_4O_8$

Labradorite is one of the plagioclase feldspars, containing both calcium and sodium in its structure. Labradorite is known for its large crystal masses with distinct iridescence.

KEY FACTS

Although it is commonly dark blue or gray, some varieties of labradorite can be light in color. The cleavage planes display a diagnostic iridescent color play, called schiller.

+ hardness: 6–7

+ streak: White

+ locations: Found in large crystalline masses in anorthosite; also occurs as a component of gabbro

Labradorite crystals display a striking iridescent color on their cleavage surfaces, known as labradorescence, making it one of the most beautiful of the feldspars. Labradorescence is caused by the interaction of visible light with microscopic layers within the labradorite crystal structure. Labradorite commonly forms large aggregate masses; individual crystals are tabular. Slabs of cut and polished labradorite are used as a decorative facing material of many buildings. Labradorite is named after its type area in northeastern Labrador. It is also common in the anorthosites of the Adirondacks in New York.

Nepheline

Class: Silicate Chemical formula: $(Na,K)AlSiO_4$

Nepheline is the most common feldspathoid, a group of minerals similar to feldspars but undersaturated in silica. It can be found in silica-poor igneous and metamorphic rocks.

KEY FACTS

This silicate is distinguished from feldspars by lack of cleavage, and is distinguished from quartz by hardness (nepheline is softer); quartz is also more resistant to weathering.

+ hardness: 5.5–6

+ streak: White

+ locations: Alkalic volcanic rocks and syenites; cannot coexist in a rock with quartz

Nepheline is a member of the feldspathoids, with chemistries similar to potassium feldspars but with less silica. It is an important constituent of silica-poor igneous rocks such as phonolites and syenites (that is, nepheline syenite). Nepheline is usually white or gray with a greasy luster. It is susceptible to weathering and, thus, can appear pitted or partially dissolved. Well-developed crystals are rare, and often have a prismatic habit with a hexagonal cross section. Nepheline can be difficult to distinguish from quartz but is slightly softer and cannot coexist with quartz in the same rock.

Olivine

Class: Silicate Chemical formula: $(Mg,Fe)_2SiO_4$

Olivine is a green mineral that is part of many iron- and magnesium-rich igneous rocks such as basalt and peridotite. It is the primary component of the Earth's upper mantle.

KEY FACTS

The name refers to its olive green color. It has a vitreous luster and hardness; well-formed crystals are rare.

+ hardness: 6.5–7

+ streak: White or colorless

+ locations: Found in basalts and other mafic volcanic rocks as isolated inclusions; a primary constituent of ultramafic rocks like peridotites and dunites

Olivine grains, with a vitreous luster, translucence, and green or yellow-green color, can look like fragments of green sea glass. Transparent or translucent green crystals are known as the gem peridot. Olivine has variable amounts of iron and magnesium. Rare, pure iron varieties are called fayalite; pure magnesium varieties are called forsterite. Only fayalite coexists with quartz. Olivine is a major constituent of rocks from the upper mantle and oceanic crust. It is found in mafic and ultramafic igneous rocks, and less commonly in metamorphic rocks including marbles. Hawaii's green beaches consist of olivine.

Garnet

Class: Silicate Chemical formula: $X_3Y_2(SiO_4)_3$

Garnets are colorful minerals common in igneous and metamorphic rocks. Individual types of garnet are defined by their chemistry—how different elements fit into their crystal structure.

KEY FACTS

Although usually deep red, garnets can be other colors. The mineral can be distinguished by its hardness and its well-formed 12-sided crystals.

+ **hardness:** 6.5–7.5

+ **streak:** White or colorless

+ **locations:** Most often occurs as small crystals in schists; also found in granitic rocks, pegmatites, and marbles

Garnet types include almandine, pyrope, spessartine, and grossular. Specimens commonly have well-developed crystal faces. In schists and other rocks, the symmetrical, colorful crystals are distinctive, especially when surrounded by flattened grains of other minerals. The most recognizable garnets are deep red with a vitreous luster and high hardness, but they can be orange, yellow, green, blue, purple, pink, brown, and black. Garnets are used as gems and industrial abrasives. Their presence can give geologists a clue to where the rock came from; for example, some garnets indicate origins in the Earth's mantle.

Andalusite

Class: Silicate Chemical formula: Al_2SiO_5

Andalusite is one of three aluminosilicate minerals with the same chemistry but different structures: andalusite, kyanite, and sillimanite. It is found in metamorphic rocks.

KEY FACTS

Prismatic pink, violet, green, or gray crystals have nearly square cross sections; the crystals can be twinned. Andalusite has one perfect cleavage plane and an uneven or conchoidal fracture.

+ **hardness:** 6.5–7.5

+ **streak:** White to colorless

+ **locations:** Can be found in schists and other aluminous metamorphic rocks

Three aluminosilicate minerals have the same chemistry but different crystal structures, known as polymorphs: andalusite, kyanite, and sillimanite. Each polymorph forms under different pressure and temperature conditions, so finding one of these minerals is an important clue about the history of the rock. Andalusite forms under high temperatures at relatively low pressures. Nearly square cross sections of prismatic andalusite crystals are distinctive, but andalusite can also have a massive or compact form. A variety of andalusite called chiastolite contains dark inclusions that create the shape of a cross.

Kyanite

Class: Silicate Chemical formula: Al_2SiO_5

Kyanite is an aluminosilicate mineral commonly forming long tabular or bladed crystals. It is an indicator of medium- to high-pressure metamorphism.

KEY FACTS

Although it is best known for its cyan-blue color, other colors exist in kyanite, including clear, white, gray, and yellow.

+ hardness: 6.5–7 perpendicular to long axis; 4–5 parallel to long axis

+ streak: White

+ locations: Found in mica schists and gneisses; also occurs in some hydrothermal quartz veins

Kyanite is the medium- to high-pressure polymorph of the three aluminosilicates: andalusite, kyanite, and sillimanite. Kyanite specimens commonly display elongate, bladed, or tabular crystals. Kyanite also displays two different levels of hardness corresponding to different regions of the crystal. They are hard (hardness of 6.5 to 7) in the direction perpendicular to the long axis of the crystal and soft (hardness of 4 to 5) parallel to that axis. Kyanite is often found in association with staurolite and garnet in metamorphic rocks. Kyanite is used in a variety of refractory and abrasive products.

Sillimanite

Class: Silicate Chemical formula: Al_2SiO_5

Sillimanite is one of the three aluminosilicates, along with andalusite and kyanite, that have the same chemistry but different structures. Sillimanite grains are often fibrous or needlelike.

KEY FACTS

Characteristic forms of sillimanite include colorless, white, or gray needles as well as fibrous masses. Variations are brownish or blue-green.

+ hardness: 7

+ streak: White

+ locations: Just as its close chemical relatives andalusite and kyanite, this aluminosilicate polymorph occurs in mica schists and gneisses.

Sillimanite is the high-temperature polymorph of the three aluminosilicates: andalusite, kyanite, and sillimanite. It is commonly colorless or white, but can also be brown, yellowish brown, or blue-green. Sillimanite crystals are usually prismatic and fibrous with vitreous luster. Fibrous varieties are also known as fibrolite. It is found in aluminous metamorphic rocks, commonly associated with corundum, cordierite, andalusite, and kyanite. It is the state mineral of Delaware, where it is common in the schists of the Delaware Piedmont, in places as large masses and rounded boulders.

Staurolite

Class: Silicate Chemical formula: $Fe_2Al_9Si_4O_{22}(OH)_2$

Staurolite is a hard, medium- to dark-colored mineral found in some metamorphic rocks. Specimens often have well-developed crystal faces creating prismatic shapes.

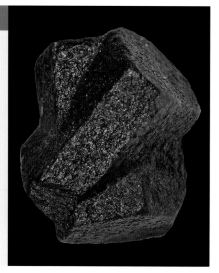

KEY FACTS

Two crystals of staurolite are often found grown together, some at right angles. Right-angled "fairy crosses" or "fairy stones" are thought by some people to bring good luck.

+ hardness: 7–7.5

+ streak: White to gray

+ locations: Like andalusite and kyanite, it can be found in mica schists and gneisses.

Staurolite is distinguished by its hardness and its propensity for growing two "twin" crystals at the same time. The twins are commonly oriented perpendicularly to each other, forming a cross. Some collectors call these "fairy crosses," and they are considered good-luck charms. Other twins intersect at about 60-degree angles. Staurolite crystals are medium to dark in color, usually reddish brown, tan, brown, or black. Prismatic crystals have a vitreous luster and are hexagonal or diamond shaped in cross section. Crystals with well-developed faces have a glassy luster and are hexagonal or diamond shaped in cross section.

Epidote

Class: Silicate Chemical formula: $Ca_2(Al,Fe)_3(SiO_4)_3(OH)$

Epidote is a common mineral with a characteristic pistachio green color. It is usually a secondary mineral, grown during metamorphism or alteration of a rock.

KEY FACTS

The green colors of epidote are distinctive. This mineral has one direction of perfect cleavage—a characteristic distinguishing it from green amphiboles, which have two cleavage planes.

+ hardness: 6–7

+ streak: Colorless to gray

+ locations: Found in metamorphic rocks, pegmatites, and small cavities in basalt

Epidote is a common mineral found in rocks that have undergone metamorphism or alteration, often by the heat of a nearby magma chamber. Epidote forms well-developed columnar prisms or thick, tabular crystals with grooved faces. It also forms thin crusts and seams, filling vesicles (small cavities) and fractures in basalts. Epidote has a characteristic green color, often pistachio green, yellow-green, or greenish black. It has a vitreous luster, and it is transparent to translucent. Epidote crystals are pleochroic, meaning that different colors appear through the crystal prism as the prism is rotated.

Beryl

Class: Silicate Chemical formula: $Be_3Al_2Si_6O_{18}$

Beryl is best known for its gemstone varieties emerald and aquamarine. It is an aluminosilicate mineral with beryllium in its crystal structure. It is most often found in granitic pegmatites.

KEY FACTS

The beautiful colors of highly prized gem-quality beryl include red, pink, green, and blue. The crystals form hexagonal prisms, which add to the mineral's popular appeal.

+ hardness: 7.5–8

+ streak: White

+ locations: Although usually found in granitic pegmatites, beryl also occurs in some metamorphic rocks.

Beryl is often recognized by its hexagonal crystals. Beryls form beautiful prisms in granitic pegmatites. The beryl crystal structure contains relatively large spaces that can accommodate additional atoms of different elements, especially manganese, iron, and chromium, resulting in a wide variety of colors. Beryl with iron in the beryllium site forms the gem aquamarine. Manganese in the aluminum site forms the pink gem morganite. Chromium beryls are emeralds. Beryl crystals have a vitreous to greasy luster and are harder than quartz. In addition to granitic pegmatites, beryls can also be found in some metamorphic rocks.

Tourmaline

Class: Silicate Chemical formula: $Na(Mg,Fe)_3Al_6(BO_3)_3(Si_6O_{18})(OH,F)_4$

Tourmaline forms beautiful columnar crystals with rounded triangular cross sections. Its colors include black, brown, blue, pink, and green, sometimes with several colors in one crystal.

KEY FACTS

Distinguishing characteristics of tourmaline include well-formed columnar crystals with pyramidal faces on one end. Some are multicolored, including a type popularly called "watermelon tourmaline."

+ hardness: 7–7.5

+ streak: White

+ locations: Found in granites, pegmatites, quartz veins, and some metamorphic rocks

Tourmaline crystals have rough, vertical striations and rounded to triangular cross sections. Crystals often have pyramidal faces on one end. Tourmaline is most commonly the black, vitreous variety known as schorl. A popular gem called "watermelon tourmaline" displays concentrically zoned colors with red or pink inside and green outside, formed by chemical changes during crystal growth. Red or pink tourmalines contain manganese; green tourmalines result from iron. The largest crystals are found in pegmatites, though tourmaline is also found in schists and gneisses. Gem-quality tourmalines are collected in Maine, Connecticut, and California.

Pyroxene

Class: Silicate Chemical formula: $XY(Si,Al)_2O_6$

Pyroxenes are a group of important rock-forming minerals found in igneous and metamorphic rocks. The most common pyroxene mineral is augite, a component of some volcanic rocks.

KEY FACTS

These silicates are similar to amphiboles but have cleavage planes intersecting at right angles.

+ hardness: 5–6.5

+ streak: Variable; augite light green to colorless.

+ locations: Mafic and ultramafic igneous and metamorphic rocks; forms large crystals in porphyritic volcanic rocks; also occurs as granular masses

The pyroxene group contains various minerals defined by crystal structures and chemical compositions, including aegirine, augite, diopside, jadeite, and enstatite. These are high-temperature minerals similar to amphiboles but without water content. Pyroxenes form prismatic crystals with vitreous luster, in elongate and short varieties. Most pyroxenes require a hand lens to identify. They differ from amphiboles by their cleavage angles: pyroxene intersecting at right angles; amphibole intersecting in wedge or diamond shapes. Earth's upper mantle is made of olivine and pyroxene. Augite is found in volcanic dikes.

Amphibole

Class: Silicate Chemical formula: $XY_2Z_5(Si,Al,Ti)_8O_{22}(OH,F)_2$

Amphiboles are a group of important rock-forming minerals. They are typically dark in color and are distinguished by their diamond- or wedge-shaped cleavage intersections.

KEY FACTS

Elongate crystals are generally flatter than pyroxenes, with oblique, perfect cleavage planes that intersect at approximately 120 degrees.

+ hardness: 5–6

+ streak: Variable; usually white or colorless

+ locations: Found in many igneous and metamorphic rocks including granite, basalt, gneisses, schists, and amphibolite

Like pyroxenes, amphiboles are minerals rich in iron and magnesium. Unlike pyroxenes, amphiboles are hydrous, meaning they have water in their structure. Geologists sometimes call amphiboles "garbage-can minerals" because many elements fit into the crystal structure. The most common amphibole is hornblende, a constituent of granitic rocks. Others include tremolite and actinolite in schists and metamorphosed impure limestones. Amphiboles form elongate, prismatic crystals, sometimes in radiating aggregates, typically gray (tremolite) to dark green (actinolite). Darker colors indicate increasing iron content.

Hornblende

Class: Silicate Chemical formula: $(Ca,Na,K)_{2-3}(Mg,Fe,Al)_5(Si,Al)_8O_{22}(OH)_2$

Hornblende is the name for various iron- and magnesium-rich amphiboles found in many igneous and metamorphic rocks. The dark-colored mineral in granitic rocks is usually hornblende or biotite.

KEY FACTS

A distinguishing characteristic of hornblende is its wedge-shaped amphibole cleavage. Its most common colors are black and dark brown.

+ hardness: 5–6

+ streak: Colorless to white, gray, or brown

+ locations: Found in many igneous and metamorphic rocks; large, well-formed crystals occur in granitic pegmatites

Hornblende is usually black or dark brown, but can be dark green. It is distinguished from other dark minerals such as biotite, tourmaline, and pyroxenes by its wedge-shaped cleavage angles. Cleavage angles are usually identified under a hand lens. Well-formed crystals are commonly short columnar prisms with six-sided cross sections and a vitreous to dull luster. Hornblende crystals can have a bladed or even fibrous habit and form massive aggregates. Hornblende is a common constituent of many rocks including granite, syenite, diorite, gabbro, basalt, schist, gneiss, and amphibolite.

Talc

Class: Silicate Chemical formula: $Mg_3Si_4O_{10}(OH)_2$

Talc is a soft, sheetlike mineral. It is the main component of soapstone and is also used in lubricants, cosmetics, and in industrial applications as a heat-resistant material.

KEY FACTS

Its greasy touch and low hardness easily identify talc. It is soft enough to be scratched by a fingernail, although talc schist is used as a building material.

+ hardness: 1

+ streak: White

+ locations: Metamorphic rocks including schists, marbles, and metaperidotites; associated with serpentine, pyroxene, and olivine

Talc is an important commercial and industrial material, used for its heat-resistant and lubricant properties. It is found in metamorphic terrains associated with ocean floor and ultramafic rocks such as marbles and peridotites. The common building material soapstone is a talc schist. Talc forms white to light green aggregate masses with a distinctive greasy touch and low hardness. It rarely forms individual crystals, but can replace crystals of other minerals during alteration, and can take on their shape, known as pseudomorphs. Talc is found, among other places, in the Appalachians, California, and Texas.

Muscovite

Class: Silicate Chemical formula: $KAl_3Si_3O_{10}(OH,F)_2$

Muscovite is one of the most common micas, a group of sheetlike minerals called phyllosilicates. It is the clear, light brown, or gray mica found in schists, gneisses, and sometimes in granites.

KEY FACTS

This widespread form of mica has thin, flexible, transparent cleavage sheets and is generally light gray or silvery. It has important uses in electronics.

+ hardness: 2–3

+ streak: Colorless or white

+ locations: Granitic rocks, pegmatites, schists, and gneisses; large crystals found in various areas of the U.S.

Muscovite is a common mica, easily identified by its flaky habit and perfect single cleavage that creates easy-to-peel, thin, flexible sheets. Muscovite is distinguished from biotite by its light colors, usually clear, white, gray, or silvery. It typically forms larger crystal masses than biotite and is more common. Large sheets of muscovite are mined from granitic pegmatites for use in the electronics industry. The name refers to the Muscovy region in Russia where large sheets of muscovite were once used as window glass. Large crystals are common in North Carolina, New England, Colorado, and South Dakota.

Biotite

Class: Silicate Chemical formula: $K(Mg,Fe)_3(Al,Fe)Si_3O_{18}(OH,F)_2$

Biotite, along with muscovite, is part of the mica group of phyllosilicates, or sheetlike minerals. It is a platy black mineral found in some granites and schists.

KEY FACTS

Black or deep brown plates (which are tabular crystals) have a distinctive flaky cleavage. Sometimes a gold tint to the plate gives this mineral a resemblance to pyrite.

+ hardness: 2–3

+ streak: Colorless or white

+ locations: Commonly a component of schists, but also can be found in biotite granites

Biotite is a black or deep brown mica. Gold-tinted biotite can be distinguished from pyrite by its platy cleavage and low hardness. Large biotite "books" can be found in granitic pegmatites. Well-formed crystals can have a hexagonal cross section, and they can easily peel into thin, flexible sheets. Most biotite, however, is found as small, embedded dark-colored grains. Biotite is an important mineral in geological research, used for analyzing the temperature histories of metamorphic rocks and, in some cases, for determining the approximate ages of igneous rocks.

Serpentine

Class: Silicate Chemical formula: $Mg_3Si_2O_5(OH)_4$

Serpentines are a group of minerals that form by the alteration of olivine and pyroxene. Serpentines are associated with rocks that originated in the oceanic crust and upper mantle.

KEY FACTS

This silicate is commonly dark green with a greasy to waxy luster and feel; compact and massive. The name suggests its resemblance to snake skin.

+ hardness: 2–5

+ streak: White

+ locations: Zones of altered rocks, associated with other magnesium-rich minerals. It is commonly found along California's coast ranges.

Serpentine minerals have three main varieties: antigorite, chrysotile, and lizardite. Antigorite is most common, occurring in compact masses that are typically dark green with a greasy luster. Lizardite is fine grained and scaly, and is found associated with altered marbles. Chrysotile is a type of asbestos with a fine fibrous form and very low hardness. Masses of serpentine make up rocks called serpentinites; these can have irregular green patterns and scaly surfaces resembling a snake skin, hence the name "serpent rock." Rocks of serpentine masses and serpentine breccia are used as decorative building stones.

Kaolinite

Class: Silicate Chemical formula: $Al_2Si_2O_5(OH)_4$

Kaolinite is the main component of clay. It forms compact masses that can be layered. It is either a primary mineral, formed in place, or it is the weathering product of minerals such as feldspar.

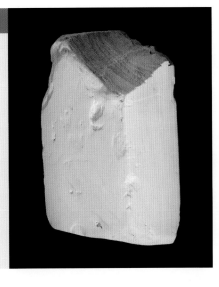

KEY FACTS

Compact, flourlike masses are usually white or earthy red, and they have an earthy odor. This clay has a great variety of uses, ranging from cosmetics to china to construction bricks.

+ hardness: 2

+ streak: White

+ locations: Found in clay beds and in place of altered feldspars in granites and pegmatites

Kaolinite is a type of clay found in areas of chemical weathering. It forms microscopic platy crystals that make up compact, earthy masses. It has a dull luster, is relatively soft, has a distinct earthy odor, powdery feel, and is usually white. Illite and montmorillonite are other common clay minerals in soils. It is difficult to distinguish these without laboratory equipment. Kaolinite is one of the most important nonmetallic minerals, used in many commercial and industrial applications including cosmetics, paints, adhesives, and glossy paper. Kaolinite clay is also used to make bricks, tiles, fine pottery, and china.

Calcite

Class: Carbonates Chemical formula: $CaCO_3$

Calcite is the primary mineral component of limestone and marble. It is the stable form of calcium carbonate at surface temperatures and pressures, and often has a biological origin.

KEY FACTS

Its "fizzing" in dilute acids (for example, hydrochloric acid) is a characteristic feature of calcite. Its hardness is also diagnostic.

+ hardness: 3

+ streak: White

+ location: Component of carbonate rocks including limestone, travertine, and marble; also forms thin veins and the linings of small cavities

Two common forms of calcium carbonate are calcite and aragonite. Calcite is the stable form at the Earth's surface. It is typically clear or white, but can be yellow or gray, with a pearly or vitreous luster. It is found in many different crystal habits and shapes: rhombohedrons, prisms, granular masses, and microcrystalline masses. Crystals display three perfect cleavage planes. Calcite differs from other white or clear minerals by its hardness—unlike quartz, it can be scratched by a knife—and its fizzing reaction in dilute acids. Dolomite, on the other hand, will fizz only under a concentrated hydrochloric acid solution.

Dolomite

Class: Carbonates Chemical formula: $CaMg(Co_3)_2$

Dolomite is an important rock-forming mineral in sedimentary and some metamorphic rocks. It is the primary component of the rock known as dolomite or dolostone.

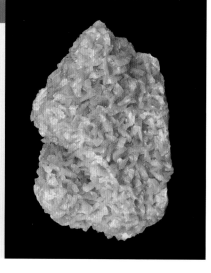

KEY FACTS

Although it is similar to calcite, dolomite reacts only to concentrated hydrochloric acid.

+ hardness: 3.5–4

+ streak: White

+ location: Besides its occurrence in the rock dolomite, the mineral also occurs with calcite in limestone and some marbles. It can be found in hydrothermal veins and small cavities where it forms crystals.

Dolomite is similar to calcite but with magnesium in its structure as well as calcium. Dolomite crystals are commonly rhombohedrons that are curved or slightly saddle shaped; like calcite, it forms granular and microgranular masses. Dolomite is typically white, gray, brown, or yellow. Its hardness, luster, and cleavage are similar to calcite. Calcite, however, will fizz under a diluted hydrochloric acid solution, whereas dolomite requires concentrated acid or a scratched or powdered surface to cause a fizzing reaction.

Malachite

Class: Carbonates Chemical formula: $Cu_2CO_3(OH)_2$

Malachite is a beautiful green, copper carbonate mineral. It is often associated with copper ore deposits and coexists with azurite, a less common copper carbonate that is brilliant blue.

Malachite is an opaque, striking green mineral, often displaying color bands of different shades of green. It is typically found in massive aggregates with a globular or bubbly surface (geologists call this botryoidal or mammillary habit), but can also form thin coatings or stalactites in caverns, or veins/veinlets along fractures in limestone. Well-developed crystals are rarely true malachite crystals; instead, they are pseudomorphs that form by replacing crystals of azurite. Malachite masses take a beautiful high polish. Pure specimens are cut and polished into various forms and collected as semiprecious or decorative stones.

Gypsum & Anhydrite

Class: Sulfates Chemical formula: $CaSO_4 \cdot 2H_2O$

Gypsum and anhydrite are calcium sulfate minerals, gypsum as the hydrous form and anhydrite the form without water ($CaSO_4$). They are typically found in evaporite deposits.

Gypsum and anhydrite occur in deposits formed when lakes and shallow seas evaporate, and from waters circulating through sandstones and clays. Crystals are colorless to white; can be tabular or diamond shaped/rhombic. Radiating blades of gypsum coated in sand are called desert roses. Glassy gypsum crystals are called selenite. Massive gypsum beds are known as alabaster. Gypsum is mined to produce cement, Sheetrock, and fertilizer. It makes up the dramatic dunes of White Sands National Monument in New Mexico. Anhydrite is harder and has three good cleavage planes that create cubic-like fragments.

Galena

Class: Sulfides Chemical formula: PbS

Galena is a gray metallic mineral typically found in hydrothermal ore deposits. Galena is made of atoms of lead and sulfur packed together in a cubic structure and is an important lead ore.

KEY FACTS

This important ore is a distinctive lead-gray color with a metallic luster, and a cubic or octahedral shape. A special characteristic is how heavy it feels in the hand.

+ hardness: 2.5

+ streak: Gray

+ location: Found in hydrothermal deposits, often in association with sphalerite, pyrite, and chalcopyrite

Galena is a major lead ore and a source of silver, bismuth, and thallium. It commonly forms perfect cubes or octahedrons. Galena is opaque gray with a shiny metallic luster when fresh. Specimens easily tarnish with exposure to air, giving some samples a dull luster. Hardness is a key identifying characteristic—it can be scratched with a fingernail—yet it has a high specific gravity, so it feels heavy in the hand. The distinctive gray streak is also diagnostic. Galena is common in silver and lead mining areas of the U.S. and Canada. It is the state mineral of Missouri and Wisconsin.

Chalcopyrite

Class: Sulfides Chemical formula: $CuFeS_2$

Chalcopyrite, a copper iron sulfide, is an important copper ore. It is a brassy and golden mineral found in hydrothermal ore deposits and is a common specimen in rock and mineral collections.

KEY FACTS

Brassy color differs from the bright golden hue of gold and pyrite. Specimens are easily tarnished. Chalcopyrite is softer than pyrite and harder than gold.

+ hardness: 3.5

+ streak: Green-black

+ location: Ore deposits, primarily in hydrothermal veins. An important location is in the greenstone belt of Ontario.

Chalcopyrite is distinguished by its greenish brassy color, distinct from the metallic gold of both pyrite and gold. Its hardness is also distinctive. It is softer than pyrite, whereas gold is softer than both chalcopyrite and pyrite, and is more malleable. Well-formed chalcopyrite crystals are rare and typically have a tetrahedral shape, which distinguishes them from the cubic form of pyrite. Chalcopyrite occurs in sulfide ore deposits formed by deposition of copper during the circulation of medium-temperature fluids in volcanic environments. It occurs as mineral veins, granular masses, and nodules.

Pyrite

Class: Sulfides Chemical formula: FeS_2

Pyrite, commonly known as fool's gold, is made of iron sulfide. It forms distinctive golden, metallic cubic crystals that are popular with collectors.

KEY FACTS

Pyrite's hardness is diagnostic—much harder (but lighter) than gold, and it can scratch glass. Its streak is also a helpful identifier.

+ hardness: 6–6.5

+ streak: Greenish black

+ location: Hydrothermal veins and other sites in sedimentary, metamorphic, and igneous environments. Small crystals can be found in shale.

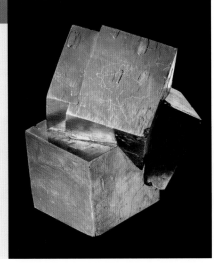

Pyrite is known as fool's gold because of its gold color and high metallic luster. Unlike true gold, however, pyrite is relatively hard—similar in hardness to quartz—and is commonly found in the form of distinctive crystal cubes or crystals with 12 sides in the shape of pentagons. It is also significantly lighter than gold. Crystal faces of pyrite commonly display striations. Another diagnostic feature of pyrite is that it gives off a sulfur smell when broken or hammered. Pyrite is the most common sulfide mineral and is used to produce sulfuric acid. It is sometimes found as a replacement mineral in fossils.

Halite

Class: Halides Chemical formula: NaCl

Halite is the naturally occurring form of table salt and is also known as rock salt. It is found in evaporite deposits and can form thick beds in sedimentary sequences.

KEY FACTS

Distinguished by its salty taste and solubility in water, halite has long been familiar as table salt; however, current grocery store salt is primarily synthetic.

+ hardness: 2.5

+ streak: White

+ location: Evaporite deposits and in salt crusts along shorelines in arid environments

Halite is used as road salt and is important in the chemical industry—for example, to produce hydrochloric acid. Salt crystals are cubic, and are typically clear, white, or sometimes bluish or with blue spots. They have a vitreous or greasy luster. Halite precipitates out of seawater when the water evaporates to less than 10 percent of its original volume. Because of halite's low density, salt beds can rise under pressure through weaknesses in overlying rocks. These disruptions of sedimentary strata create interesting structures in the deserts of the U.S. Southwest.

Fluorite

Class: Halides Chemical formula: CaF_2

Fluorite is a colorful mineral that forms cubic or octahedral crystals. Fluorite fluoresces under ultraviolet light, and the phenomenon is named after this property of fluorite.

KEY FACTS

Transparent to translucent crystals of fluorite come in various colors, including pale purple, pink, green, and yellow. Its fluorescence under ultraviolet light is a famous characteristic.

+ hardness: 4

+ streak: White

+ location: Occurs in hydrothermal veins; also in cavities of granitic pegmatites

Fluorite is usually found as well-formed crystals with cubic and octahedral faces. It can be colorless, but is more often colorful, including dark blue, violet, pink, green, and yellow varieties. The colors are influenced by concentrations of rare earth elements in the crystal structure. Crystals have a vitreous luster. Hardness is a good diagnostic tool. It is softer than quartz but harder than calcite. Some specimens show cleavage planes intersecting at 60 degrees. Fluorite forms in medium- and low-temperature hydrothermal deposits, in crystal pockets of granites and granitic pegmatites, and as a component of some metamorphic rocks.

Silver

Class: Native elements Chemical formula: Ag

Silver is a rare native element that occurs in hydrothermal deposits and in association with other ores. Silver is distinctively malleable and is an important precious metal.

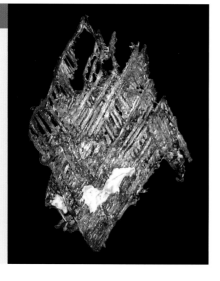

KEY FACTS

This prized metal has a silver-white color and a high metallic luster when it is fresh; however, as is well known, its surface oxidizes or tarnishes easily to dark gray and black.

+ hardness: 2.5

+ streak: Silver-white

+ location: Hydrothermal veins or by alteration of other minerals; also placer deposits

Native silver is rare, found in hydrothermal deposits and alteration zones (where high-temperature fluids have either deposited material or changed existing rocks). It occurs in thin plates, as well as in granular habits and skeletal and wire-like, branching forms. It is distinguished by its opaque, silver-white color and metallic luster that easily tarnishes. It is distinctly malleable and is relatively soft. Most silver is produced by refining other compounds instead of mined in its pure form. It is used as currency and in jewelry, and is important industrially for its conductivity and malleability.

Gold

Class: Native elements Chemical formula: Au

The precious metal gold is used for currency, jewelry, electronics, and more. It occurs in hydrothermal deposits and in quartz veins, but is mostly found as grains in river and beach sands.

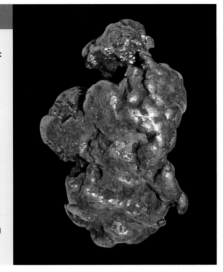

KEY FACTS

Known for its distinc-
tive color and metallic
luster, gold resists
oxidation. Most gold
is discovered in accu-
mulations of particles
in river and beach
sands.

+ hardness: 2.5–3

+ streak: Golden

+ location: Hydro-
thermal deposits,
quartz veins in large
granitic bodies, and
secondary weathering
deposits

Gold rarely forms distinct crystals; instead, it
is found as golden yellow platelets, branching, wiry forms, and as inclusions in quartz. Gold
is soft, but has a very high density. It is malleable,
and it maintains its metallic luster and color without oxidation. Gold differs from pyrite by its hardness (gold is softer) and by a lack of sulfur smell.
Flecks of yellowish mica can resemble gold, but
mica has a vitreous, not metal-
lic, luster and is brittle. Gold
is extremely unreactive and
survives chemical weathering. Most gold is found
as particles in gravels and
sands that have weathered
from gold-bearing veins.

Copper

Class: Native elements Chemical formula: Cu

Native copper occurs as irregular nodules, veins, and wirelike or platelike forms. It is an
important industrial metal because of its high thermal and electrical conductivity.

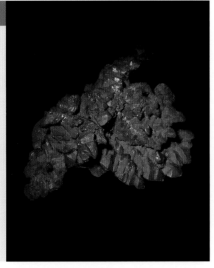

KEY FACTS

The original copper-
red color is often oxi-
dized to black, blue,
and green. Native cop-
per is usually found in
irregular masses, often
with strange shapes.

+ hardness: 2.5–3

+ streak: Copper-red

+ location: Hydro-
thermal veins; also
in rare lava flows
including along the
mid-continent rift in
northern Michigan

Most industrial copper is extracted from other
minerals such as chalcopyrite, but native
copper can be found. Copper rarely exists as well-
formed crystals and is more often found as irregular masses, sometimes with branching or wirelike
shapes, and as the matrix surrounding volcanic
clasts. A key to identifying copper is its copper-red
streak and low hardness—it can be scratched by
a knife. Copper specimens commonly display thin
black, blue, or green oxidation coatings. One of
the world's largest deposits of
native copper is in Michigan's Upper Peninsula,
where it is associated
with a thick series of
lava flows.

Diamond

Class: Native elements Chemical formula: C

Diamond, the hardest mineral, consists of pure carbon, and in nature, forms only at high pressures and temperatures deep in the Earth. Diamonds come to the Earth's surface via volcanic vents.

KEY FACTS

A diamond's famed hardness is diagnostic—nothing will scratch it. The various colors of diamonds are results of impurities such as the elements boron and nitrogen.

+ hardness: **10**

+ streak: **None**

+ location: **Rare; found in association with ultramafic (iron- and magnesium-rich) igneous rocks in continental shields**

Diamonds are a high-pressure crystalline form of carbon with individual atoms packed together by strong chemical bonds. Graphite is also pure carbon, but is a low-pressure mineral in the form of carbon sheets with weak bonds. Diamonds are unstable at the Earth's surface temperatures and pressures, but they cannot convert to graphite on human timescales. Small fragments of quartz can look like diamonds but are distinguished by testing hardness. Diamond crystals are usually octahedrons or cubic forms. Impurities give diamonds different colors: Boron, for example, creates a blue tint; nitrogen casts a yellow hue.

Magnetite

Class: Oxides Chemical formula: Fe_3O_4

Magnetite is a form of iron oxide found in some metamorphic and igneous rocks as well as secondary deposits in black sands. True to its name, its magnetism is diagnostic.

KEY FACTS

This magnetic oxide is black, with a semi-metallic luster. Black sands often found along rivers and on ocean beaches are grains of magnetite that have eroded out of rocks.

+ hardness: **5.5–6**

+ streak: **Dark gray to black**

+ location: **Common accessory mineral in a variety of metamorphic and igneous rocks**

Magnetite is a major iron ore, found as octahedral crystals, granular aggregates, and in tiny veins. It is dark gray to black with metallic luster. Its black streak and magnetism are distinguishing characteristics. Magnetite is important to geologists because it records information about the Earth's magnetic field when it crystallizes. Analyzing magnetite-bearing rocks of different ages allows geologists to understand movement of the Earth's tectonic plates. Large crystals of magnetite occur in pegmatites and in high-temperature veins.

Hematite

Class: Oxides Chemical formula: Fe_2O_3

Hematite is the most common form of iron oxide and an important iron ore. It occurs in various forms including black platy or tabular crystals and irregular masses with a brick red hue.

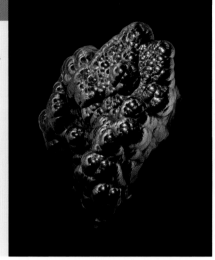

KEY FACTS

A red streak is diagnostic for this major source of iron, found as thick sedimentary beds in some formations. Crystals have a metallic luster, but masses are typically dull.

+ hardness: **5–6**

+ streak: **Bright to earthy red**

+ location: **A common component of many different rocks; large deposits are found in banded iron formations.**

Hematite varies in form and color, appearing steely gray, black, brown, orange, or red. The red streak is a good method to differentiate it from similar minerals such as magnetite and ilmenite. Hematite can form circular aggregates of platy crystals known as iron roses. It also forms tabular and hexagonal crystals, and irregular, rounded masses. Clays with high hematite content are called red ochre, which is used as a pigment. Hematite forms as a deposit in quartz veins, as well as in metamorphic rocks. It is the primary component of banded iron formations.

Corundum

Class: Oxides Chemical formula: Al_2O_3

Corundum is an aluminum oxide known for its high hardness and beautiful gemstone varieties, including pink and red rubies, and blue, green, and yellow sapphires.

KEY FACTS

The hardness of corundum is diagnostic; only diamonds are harder. This mineral forms tabular and prismatic crystals. Impurities provide colored gems such as rubies and sapphires.

+ hardness: **9**

+ streak: **White**

+ location: **Pegmatites, gneisses, schists, and marbles; also in association with ultramafic igneous rocks**

Corundum is an aluminum oxide mineral with a close-packed crystal structure and strong bonds between atoms, giving it high hardness and density. Corundum of pure aluminum oxide is clear, white, or gray. Impurities lead to various colors. Chromium creates pink or red rubies. Iron and titanium create blue sapphires; iron alone creates yellow sapphires. Corundum forms during alteration of volcanic and ultramafic igneous rocks and during metamorphism of aluminum-rich sedimentary and igneous rocks. Corundum is used in industry as an abrasive. Corundum and quartz never coexist in the same formation.

Granite

Type: Igneous

Granite is one of the Earth's most common rocks and an important building block of the continents. It is a light-colored intrusive igneous rock with a distinct granular texture.

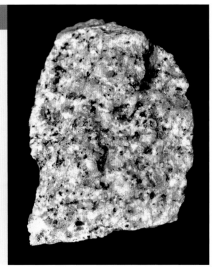

Granite is an intrusive igneous rock containing at least 20 percent quartz and two types of feldspar—plagioclase and alkali feldspar in relatively equal amounts. It can be difficult to distinguish between the feldspars in the field, or to determine the amount of quartz; thus, many medium-grained, light-colored igneous rocks are called granite, even if they are technically something else. Granites form from felsic magma that slowly cools below the Earth's surface. Multiple intrusions of granitic magma build up large rock bodies called batholiths, which form in the cores of mountain belts and are exposed by erosion.

Porphyritic Granite

Type: Igneous

A porphyritic granite has mineral grains of two different sizes. Typically these are large feldspar grains, often with well-formed crystal faces, surrounded by smaller grains of other minerals.

The term "porphyritic" describes igneous rocks that have minerals with two grain sizes. The larger grains are called phenocrysts, from the Greek *phaino*, meaning "visible," and *cryst*, meaning "crystal." Porphyritic granites often have pink or red alkali feldspar phenocrysts surrounded by smaller grains of plagioclase, quartz, biotite, and/or hornblende. These phenocrysts form when the alkali feldspar is the first mineral to crystallize out of the magma. The feldspars grow before the other minerals crystallize and fill the remaining space. Porphyritic granites are used for countertops, wall claddings, and monuments.

Rapakivi Granite
Type: Igneous

Rapakivi granites are granites with a striking texture of large, rounded feldspar grains with different colored rims. The name "rapakivi" is Finnish and means "crumbly stone."

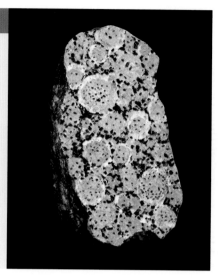

KEY FACTS

This granitic rock is used as a building and decorative interior stone.

+ grain size: Medium to coarse

+ texture: Crystalline, porphyritic

+ composition: Quartz + alkali feldspar + plagioclase feldspar ± hornblende ± biotite ± muscovite

Rapakivi granites are common decorative stones with large pink or brown alkali feldspar grains surrounded by gray or white plagioclase rims. The large grains, called phenocrysts or megacrysts, typically have a rounded shape. Rapakivi granites are found on all continents, usually in the older continental cores. The building stone known as Baltic brown is a rapakivi granite from Finland that is about 1.6 billion years old. It is a popular stone used for kitchen countertops and flooring. The groundmass of this granitic rock is a dark color because it is rich in biotite and hornblende.

Aplite
Type: Igneous

Aplites are light-colored, fine-grained intrusive rocks that commonly form dikes. Most are found associated with crosscutting large granitic bodies and have a composition similar to granite.

KEY FACTS

These rocks have an even, fine grain size, creating a sugary texture.

+ grain size: Fine to medium

+ texture: Crystalline; even grain size, no fabric

+ composition: Quartz + alkali feldspar + plagioclase ± mica

Cutting across many exposures of granitic rocks are relatively narrow dikes that are conspicuously light in color and fine in grain size. These are called aplite dikes and they typically have bulk compositions that are similar to granite. Although the grains are small, quartz, feldspar, and sometimes muscovite are usually visible. Aplites are thought to be formed from the last remnants of the magma that created the surrounding granite. This magma must have cooled quickly to produce the fine texture. Aplites are often more resistant to weathering than the surrounding rock, and can project out as ridges.

Pegmatite

Type: Igneous

A pegmatite is an intrusive igneous body with very coarse-grained, interlocking crystals. Many pegmatites are granitic. Pegmatites are sites for finding rare mineral specimens.

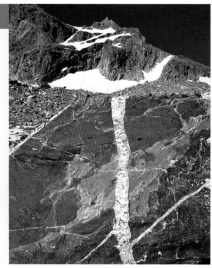

KEY FACTS

Crosscutting dikes, veins, or irregular intrusions with very large grains are features of pegmatite.

+ grain size: **Very coarse**

+ texture: **Crystalline; grain size and mineral distribution can be heterogeneous.**

+ composition: **Variable; most pegmatites are granitic.**

Pegmatites are intrusive bodies of rock notable for very coarse grain size. The crystals vary from a few inches to several yards across. Pegmatite intrusions can be dikes, veins, and lenses that cut across preexistent rock. Geologists differ about how pegmatites form; most agree that they form from the last remaining melt in a magma chamber, which is saturated in water and other fluids. This melt must cool very slowly to grow such large crystals. Its concentration of rare elements leads to crystallization of exotic minerals, including many gems. Pegmatites are important sources for mining feldspar and mica.

Granodiorite

Type: Igneous

Granodiorite is an intrusive igneous rock that resembles granite but contains less alkali feldspar and more plagioclase than a true granite. It is a common building and decorative stone.

KEY FACTS

Usually light to medium gray overall, granodiorite has interlocking white, gray, and black mineral grains.

+ grain size: **Medium to fine**

+ texture: **Crystalline**

+ composition: **Quartz + plagioclase + alkali feldspar ± hornblende ± biotite**

Granodiorite has an intermediate composition between felsic granite and mafic gabbro. It has an overall light- to medium-gray appearance, dominated by white or gray grains of plagioclase and quartz, plus black grains of biotite and hornblende. Granodiorite is often lumped with granite and described as a granitic rock or granitoid. It is a common building and decorative stone, used for countertops, building facades, and structural blocks. The rocks of the Sierra Nevada are primarily granites and granodiorites, with granodiorite making up many iconic formations in Yosemite National Park.

Syenite
Type: Igneous

Syenite is an uncommon intrusive igneous rock similar to granite but with little or no quartz and abundant alkali feldspar. Nepheline syenites are syenites that contain nepheline and no quartz.

> **KEY FACTS**
>
> Syenite is dominated by alkali feldspar grains.
>
> + grain size: Medium to coarse
>
> + texture: Crystalline, sometimes porphyritic
>
> + composition: Alkali feldspar ± quartz or feldspathoid (nepheline) ± plagioclase ± amphibole or pyroxene ± biotite

Syenite is an intrusive igneous rock that is predominantly alkali feldspar. Quartz is either absent or present in small amounts. If the rock contains nepheline, which is never present with quartz, it is known as a nepheline syenite. Syenites are formed in the thick continental crust from magma derived from partially melting granitic rocks, and are associated with failed continental rift valleys. Syenites are mined for use in the glass and ceramics industries to lower the melting temperature of a glass or ceramic mixture. They are also used as decorative stones, especially a blue variety that gains its color from the blue mineral sodalite.

Diorite
Type: Igneous

Diorite is a medium- to dark-gray intrusive igneous rock made of plagioclase feldspar and a dark mineral, usually hornblende. It is commonly found in association with granite and granodiorite.

> **KEY FACTS**
>
> Diorite has a salt-and-pepper appearance with mixed light and dark grains.
>
> + grain size: Medium to coarse
>
> + texture: Crystalline; can be porphyritic
>
> + composition: Plagioclase + hornblende ± biotite

Diorite is an intrusive igneous rock made up of predominantly plagioclase feldspar plus an iron and magnesium mineral component, typically hornblende, sometimes with biotite. It contains less quartz than granodiorite and less quartz and alkali feldspar than granite. Because of its mineral assemblage, diorite is medium to dark gray in overall color, with a salt-and-pepper appearance created by the light plagioclase grains and black hornblende grains. Inclusions of diorite are common in some granitic bodies. Diorite is used as a structural stone, a countertop material, and a decorative stone.

Gabbro

Type: Igneous

Gabbro is a dark-gray, coarse-grained intrusive igneous rock made predominantly of plagioclase and pyroxene. It is an important component of oceanic crust.

> **KEY FACTS**
>
> Gabbro is a plutonic rock that is dark gray to black.
>
> + **grain size:** Coarse
>
> + **texture:** Crystalline
>
> + **composition:** Plagioclase + pyroxene ± olivine

Gabbro is a mafic rock, meaning it is silica-poor and iron- and magnesium-rich. It contains plagioclase feldspar plus pyroxene, usually augite, sometimes with olivine and/or hornblende. Gabbros with predominant orthopyroxene are called norites. Plagioclase in gabbro tends to be darker gray than in granitic rocks. Gabbro is coarse grained because it cools slowly at depth instead of erupting near or on the surface like basalt. It is an important component of oceanic crust. Exposures of gabbro can indicate rock formations that originated as ocean floor but were thrust onto continents by movement of the Earth's tectonic plates.

Anorthosite

Type: Igneous

Anorthosite is more than 90 percent plagioclase feldspar. It is uncommon, found in isolated exposures in some mountain belts and in ancient cores of the continents.

> **KEY FACTS**
>
> Anorthosite has interlocking grains of plagioclase feldspar, some with iridescent labradorite.
>
> + **grain size:** Coarse
>
> + **texture:** Crystalline; some with plagioclase grains in a dark-colored matrix
>
> + **composition:** 90 percent plagioclase, 10 percent ferromagnesian minerals

Anorthosite is an uncommon plutonic rock made almost entirely of plagioclase feldspar. Depending on the feldspar's color, anorthosites can be light gray or bluish. Anorthosite crystals are typically quite large. Some rock samples brought back from the moon are anorthosite. The origins of anorthosite are still debated among geologists. One hypothesis is that it formed from accumulation of plagioclase crystals that floated to the top of a magma chamber and then rose to shallower levels in the crust as a buoyant crystal mush. Anorthosite over 1.1 billion years old is exposed in the Adirondack Mountains of New York.

Peridotite
Type: Igneous

Peridotite is the dominant rock of the Earth's upper mantle. It is a dense, iron- and magnesium-rich intrusive igneous rock made of olivine and pyroxene.

KEY FACTS

Most peridotites have a reddish weathering surface.

+ **grain size:** Medium to coarse

+ **texture:** Crystalline; some layered; olivine often altered to serpentine

+ **composition:** Olivine + pyroxene ± hornblende + accessory minerals, including chromite

Peridotite is an ultramafic rock, meaning it is poor in silica and rich in iron and magnesium, resulting in a dense, often dark-colored rock. Many peridotite outcrops are slabs or pieces of mantle rock thrust onto the continents by large-scale movement of the Earth's plates during the construction of mountains or brought to the surface by volcanic eruptions. Peridotite comes in several subvarieties and many different textures, depending on the relative amounts of olivine and pyroxene, their chemical makeup, and the rock's magmatic history. Most peridotite outcrops contain secondary minerals including serpentine and talc.

Dunite
Type: Igneous

Dunite is a type of peridotite made of more than 90 percent olivine. Fresh exposures are a beautiful green, but most outcrops have a tan or brown weathering surface.

KEY FACTS

Dunite is an intrusive igneous rock of equigranular olivine with black grains of chromite as a common accessory.

+ **grain size:** Medium to coarse

+ **texture:** Crystalline, typically equigranular

+ **composition:** More than 90 percent olivine

Dunite is an uncommon rock that is made almost entirely of olivine. Fresh, unaltered dunites have an equigranular texture of interlocking, glassy green olivine grains. These rocks may include black grains of chromite either dispersed with the olivine or aggregated in chromite layers. A significant supply of commercial chromium comes from chromite concentrations in dunite. Dunite is named after Dun Mountain in New Zealand, where it is part of a group of rocks that were formerly pieces of the ocean floor. Most dunite outcrops have a tan or brown weathering surface. In others, the olivine has been partially or completely altered to serpentine.

Rhyolite

Type: Igneous

Rhyolite is the extrusive equivalent of granite. It is a light-colored, fine-grained volcanic rock that forms from explosive eruptions of a highly viscous lava.

KEY FACTS

Rhyolite is a light to medium gray or pink volcanic rock with occasional flow banding; blocks are brittle with a flinty appearance.

+ grain size: **Fine**

+ texture: **Can be glassy or porphyritic**

+ composition: **Same as granite**

Rhyolite is an extrusive igneous rock that forms from the volcanic eruption of a felsic (silica-rich) magma. Rhyolitic lava is extremely viscous and sticky, so it does not flow very far or accumulate into typical volcanic cones; instead, it builds rounded lava domes. Rhyolitic eruptions are explosive and produce a lot of ash. Rhyolites' appearances include very fine-grained, glassy varieties, and porphyritic varieties with visible quartz. Rhyolite has few dark, iron- and magnesium-rich mineral phases, so it is typically light in color and weight. The countertop and tile stone known as *porfido trentino* is a rhyolite from Italy.

Pumice

Type: Igneous

Pumice is a volcanic rock characterized by many cavities of various sizes called vesicles. Pumice forms when the lava in a volcanic eruption turns frothy from the expulsion of water and gases.

KEY FACTS

A volcanic rock with many holes (vesicles), pumice floats in water. It is typically light in color.

+ grain size: **None to very fine**

+ texture: **Rough surface, vesicular**

+ composition: **Varies; commonly rhyolite or andesite**

Pumice is made of volcanic glass that is extremely porous due to its many gas bubbles, called vesicles. Pumice forms from magma that is saturated in gases and water. During volcanic eruptions, these are released, creating a frothy foam-like lava. This lava cools so quickly that no crystals are able to grow, and the resulting rock is called a glass. Pumice is usually light in color and typically has the same composition as a rhyolite or andesite, depending on the composition of the magma from which it formed. Pumice is used as an abrasive material, including a bath accessory for smoothing rough skin.

Obsidian
Type: Igneous

Obsidian is volcanic glass that forms from viscous lava, typically in rhyolite lava domes. The brittle nature of obsidian and its conchoidal fracture make it an ideal material for sharp tools.

KEY FACTS

This shiny volcanic glass has a conchoidal fracture; banding and inclusions are common; colors are variable.

+ **grain size:** None to very fine

+ **texture:** Glassy

+ **composition:** Can vary; commonly rhyolite

Obsidian is glassy and brittle with a conchoidal fracture, meaning it breaks irregularly along curved surfaces. Paleolithic hunter-gatherers exploited these properties, chipping conchoidal flakes from obsidian pieces to create sharp projectile points and other tools. Obsidian forms in lava domes from lava so sticky and viscous that it is cooled too rapidly for minerals to crystallize, resulting in a glass. Silica-rich lavas like rhyolite or dacite are the most viscous and, thus, most common form of obsidian. Obsidian is frequently opaque black, but it can have various colors and patterns, including intricately banded forms.

Tuff
Type: Igneous

Volcanic eruptions produce massive amounts of ash and other material known as tephra. If enough heat is retained after it falls to the ground, tephra will fuse together, forming tuff.

KEY FACTS

Tuff is a volcanic rock made of consolidated ash and other volcanic fragments; can be porous.

+ **grain size:** Variable

+ **texture:** Visible volcanic clasts; possible layering or bedding

+ **composition:** Can vary; commonly rhyolite or andesite

Material ejected into the atmosphere during a volcanic eruption is classified by size. Particles smaller than 0.08 in (2 mm) are known as ash, particles between 0.08 and 2.5 in (2 and 64 mm) are called lapilli, and anything larger is a volcanic bomb. Collectively, this material is called "tephra," the Greek word for ash. When tephra falls to the Earth and solidifies, it forms a rock called tuff. Tuffs can also form when tephra becomes slowly compacted and cemented into a rock. Tuffs vary in their composition and appearance. Most have visible volcanic fragments of different sizes and shapes, including many that are angular.

Dacite
Type: Igneous

Dacite is a volcanic rock of intermediate composition between rhyolite and andesite. It is the volcanic equivalent of granodiorite. Dacite is associated with highly explosive volcanic eruptions.

KEY FACTS

Dacite often has visible quartz and/or plagioclase; it can be light gray or dark gray to black.

+ **grain size:** Fine to medium

+ **texture:** Commonly porphyritic with phenocrysts of plagioclase, quartz, biotite, or hornblende

+ **composition:** Same as granodiorite

Dacite is a volcanic rock that is slightly less rich in silica than rhyolite, but like rhyolite, comes from a highly viscous lava and produces violently explosive volcanic eruptions. Dacite forms thick rounded lava domes on the surface of volcanoes, including the lava dome at Mount St. Helens. Dacite is named after a locality in the mountains of Romania, part of a region known during the Roman Empire as Dacia. Dacite is typically light gray, but some can be dark gray or even black. Dacite is common in volcanoes that form along continental subduction zones— margins of continents beneath which oceanic slabs sink.

Andesite
Type: Igneous

Andesite is a volcanic rock similar to basalt but with more silica. It is the volcanic equivalent of diorite. Andesitic volcanoes are steep-sided cones, common around the Pacific Rim.

KEY FACTS

A common igneous rock, andesite is typically medium to dark gray; can have greenish or reddish hues.

+ **grain size:** Fine to medium

+ **texture:** Commonly porphyritic with phenocrysts of plagioclase and/or pyroxene or hornblende

+ **composition:** Same as diorite

Andesite is an extrusive igneous rock that forms from the lava flows of stratovolcanoes. Stratovolcanoes are a type of composite volcano built from alternating layers of lava, ash, and cinders, resulting in a steep-sided, symmetrical cone. Some of the world's most beautiful mountains are andesitic stratovolcanoes, including Mount Fuji in Japan and Mount Shasta, Mount Hood, and Mount Adams in the Pacific Northwest. Andesite is named for the Andes Mountains of South America where the rock is also common. Andesite is typically porphyritic, with visible crystals of plagioclase or pyroxene set in a fine-grained groundmass.

Diabase
Type: Igneous

Diabase, also called dolerite, has the same composition as basalt and gabbro. These are distinguished by their grain size: basalt fine, diabase medium, and gabbro coarse.

KEY FACTS

Diabase forms dikes and sills; most outcrops have a light-brown weathered surface; fresh surfaces are dark gray or green overall.

+ grain size: Medium

+ texture: Crystalline; commonly porphyritic

+ composition: Plagioclase + pyroxene + olivine

Diabase, also known as dolerite, is a dark-colored rock that forms dikes, sills, and other relatively small igneous bodies. Diabase is compositionally identical to basalt and gabbro, and all three rock types come from the same type of magma. Basalt, however, forms when the magma erupts onto the Earth's surface, and gabbro forms when the magma cools slowly at great depths. Some outcrops called traprock are made of diabase. Diabase traprock near Washington, D.C., is quarried and crushed for use in concrete and as road base material. The Palisades of the Hudson River is an enormous sill made of diabase.

Basalt
Type: Igneous

Basalt forms extensive lava flows worldwide. The oceanic crust is composed primarily of basalt, and the lunar "maria"—the dark regions visible on the moon—are also basalt.

KEY FACTS

A common dark-colored volcanic rock, basalt is typically black, gray, or brown; surfaces vary from smooth to sharp and cindery.

+ grain size: Fine

+ texture: Can be porphyritic

+ composition: Plagioclase + pyroxene + olivine

Basalt is one of the most common rock types. It is a mafic igneous rock with an aphanitic texture, meaning its grains are so fine that they cannot be distinguished with the naked eye; thus, the rock looks relatively uniform. Porphyritic basalts are exceptions; they contain visible plagioclase or olivine crystals surrounded by a fine-grained groundmass. Basalt lava flows cover much of the Earth's surface, including the ocean floor, volcanoes in Hawaii and Iceland, and large regions of continents known as flood basalts. Columbia River Basalts are flood basalts that extend across Washington, Oregon, and Idaho.

Sandstone
Type: Sedimentary

Sandstone is made of compacted and cemented sand-size particles. Sandstones vary depending on composition of the grains, type and amount of cement, and depositional environment.

KEY FACTS

Sandstone is a sedimentary rock made of sand-sized grains.

+ **grain size:** Medium

+ **texture:** Clastic

+ **composition:** Variable; common components include quartz, feldspar, mica, and rock fragments.

Sandstone is one of the most common sedimentary rocks. It is made of particles or clasts that fall within the size range that classifies sand, 0.0008 to 0.08 in (0.02 to 2 mm). This is a broad definition—the only parameter being grain size—thus, the rocks that are identified as sandstones can vary widely. Depending on how rounded or angular their grains are and the amount of compaction of the sediment, sandstones can be quite porous and thus serve as important reservoirs worldwide for resources such as water and hydrocarbons. Sandstones are used for building facades and unpolished floor tiles. Decorative flagstone is typically sandstone.

Quartz Arenite
Type: Sedimentary

Quartz arenite is a sandstone composed almost entirely of quartz. Quartz arenites form from accumulations of quartz sand, commonly in windblown deserts, beaches, and high-energy rivers.

KEY FACTS

This sedimentary rock is made of sand-size grains of quartz; cement is often quartz but can also be calcite or hematite.

+ **grain size:** Medium

+ **texture:** Clastic

+ **composition:** More than 90 percent quartz

An arenite is a texturally mature sandstone, meaning it is of consistent grain size and shape, and contains few to no clay particles between the grains. Its grains are predominantly quartz. Sediments mature during weathering, transportation, and deposition, as less resistant minerals break down and more resistant minerals become rounded. Quartz arenites form from sediment deposited far from the source, typically in environments such as sand dunes, coastlines, and rivers. These sandstones can take on different colors, commonly cream, orange, or red, depending on small amounts of iron oxide coating the sand grains.

Graywacke
Type: Sedimentary

A wacke is a variety of sandstone with mud or clay in the matrix that surrounds the sand grains. A graywacke is a type in which many of the sand grains are rock fragments.

KEY FACTS

An immature, poorly sorted sandstone, graywacke is typically dark.

+ grain size: At least 50 percent sand size with fine-grained matrix

+ texture: Clastic, poorly sorted

+ composition: Rock fragments, feldspar, quartz, minerals rich in iron and magnesium, clay, and mud

Unlike arenites, wackes are sandstones that are not well sorted—they contain a mixture of grain sizes, from clay to sand, as well as grain shapes from angular to rounded. The term "graywacke" refers to wackes with a composition that includes numerous rock fragments as well as minerals such as micas. The minerals and fragments are fragile, so their preservation in a sandstone indicates that the sediment source cannot be far. Geologists interpret many graywackes as deposits of turbidity currents—strong currents in water moving down a slope. These rocks are some-times called turbidites instead of graywackes.

Arkose
Type: Sedimentary

An arkose is a type of sandstone made up of at least 25 percent feldspar. Arkoses typically also contain quartz and mica grains and rock fragments.

KEY FACTS

A sedimentary rock, arkose is made of sand-size grains of feldspar and quartz; cement is commonly quartz or calcite.

+ grain size: Medium

+ texture: Clastic

+ composition: More than 25 percent feldspar

An arkose or arkosic sandstone contains a mod-erate to high percentage of feldspar grains. Feldspar is susceptible to breakdown via chemical weathering, so its presence in a sandstone indi-cates that the sediment source is not far from its depositional environment. Arkoses thus tend to form relatively close to mountains or other uplands from which feldspar-rich sediment, typically from granitic rocks, is shed. In fact, geologists some-times use arkosic sandstones in the sedimentary record as an indication of previous mountain belts. Arkoses are often pink or red, but can also have gray or even greenish hues.

Limestone
Type: Sedimentary

Limestone is made up mainly of calcite (calcium carbonate). It is primarily a marine sedimentary rock, though some freshwater limestones are known to occur.

KEY FACTS

A light-colored sedimentary rock, limestone is often fine grained; some formations are rich in marine fossils. It fizzes in contact with hydrochloric acid.

+ grain size: **Very fine to coarse**

+ texture: **Variable**

+ composition: **Primarily calcite**

Calcium carbonate can chemically precipitate out of solution in sea or lake water, settling to the bottom, accumulating as a sediment, and forming limestone. Limestone may also form from accumulated animal shells made of calcium carbonate. Thus, fossils are common in limestones; some limestone formations are made entirely of fossil shells. Some limestones contain small, spherical beads of calcite called oolites that develop when calcium carbonate precipitates around a particle such as a sand grain. Limestone forms cliffs in arid environments, but forms caves and other karst landforms in humid environments.

Dolomite/Dolostone
Type: Sedimentary

Dolomite is similar to limestone but with a high percentage of the mineral dolomite—calcium magnesium carbonate—instead of calcite. Dolomite rock is sometimes called dolostone.

KEY FACTS

Although it is similar to limestone, dolomite does not fizz as actively in hydrochloric acid; dolomite's weathering surface is a buff color, whereas limestone tends to be grayer.

+ grain size: **Fine to medium**

+ texture: **Compact, relatively homogeneous**

+ composition: **Primarily dolomite**

Dolomite, also known as dolostone, is similar to limestone and is often found in association with limestone. Most dolomite originated as limestone or lime mud, and became dolomite when calcite (calcium carbonate) was replaced by dolomite (calcium magnesium carbonate), a process called dolomitization. Dolomite and limestone can be difficult to distinguish in the field. Dolomite tends to weather to yellowish beige or brown, whereas limestone tends to be gray. Dolomite is named after French geologist Déodat de Dolomieu, who first described it in the 1700s from cliffs in the Italian Alps now called the Dolomites.

Travertine
Type: Sedimentary

Travertine is a type of limestone formed via direct chemical precipitation of calcite out of solution. This commonly occurs in association with the waters around hot springs.

KEY FACTS

Deposits of travertine are associated with caves, hot springs, and geysers; typically layered with light colors; fossils are common.

+ **grain size:** Very fine

+ **texture:** Massive, nonclastic, porous

+ **composition:** Calcite (calcium carbonate)

Travertine is a beautiful ivory- to peach-colored stone that forms around hot springs when calcium carbonate falls out of solution. This occurs when carbon dioxide is released as heated waters rise from depth or are agitated as in waterfalls and cascades. The results are porous, finely layered deposits of calcium carbonate. Travertine is a popular building stone. Much of Rome, which is surrounded by volcanic thermal springs, was constructed with travertine, including the Colosseum. Travertine is also used to clad modern buildings including the Getty Center in Los Angeles and Lincoln Center in New York.

Chalk
Type: Sedimentary

Chalk is a variety of limestone made of tiny skeletons of minute marine organisms. The skeletons, composed of calcium carbonate, accumulate on the floors of shallow seas.

KEY FACTS

White, pure limestone chalk is relatively soft, made of microscopic shells with a carbonate mud matrix.

+ **grain size:** Fine

+ **texture:** Bioclastic

+ **composition:** Calcite (calcium carbonate)

Chalk is a relatively pure limestone made of accumulated shells of marine microorganisms cemented together with a carbonate mud. Many chalk deposits formed during the Cretaceous period, from about 142 to 65 million years ago, a time when global sea levels were high and parts of the continents were flooded with shallow seas. "Cretaceous" comes from the Latin word *creta*, meaning chalk. Many small, calcareous marine organisms thrived in these environments, and deposits of their skeletons built up on the sea floor to form layers of chalk. Chalk formations commonly contain nodules of chert known as flint.

Conglomerate
Type: Sedimentary

Conglomerates are sedimentary rocks composed of particles larger than sand grains, including pebbles, cobbles, and boulders. They form in various depositional environments.

KEY FACTS

Coarse-grained clasts of conglomerates are surrounded by a fine-grained matrix; relative proportions of clasts and matrix vary; size, shape, and composition of clasts also vary.

+ **grain size:** Larger than 0.08 in (2 mm)

+ **texture:** Clastic

+ **composition:** Variable

Sedimentary rocks made of particles larger than sand grains are known as conglomerates. The particles vary from rounded river pebbles to angular fragments of a debris flow. Conglomerates represent high-energy deposition including river- or beach-deposited gravel beds, alluvial fans, glacial till deposits, and mudslides. Conglomerates are classified by the relative proportion of clasts and matrix—the fine-grained material cementing the rock together. Conglomerates in which the clasts touch are known as clast supported. Those in which the clasts are surrounded by matrix and not touching are called matrix supported.

Breccia
Type: Sedimentary

Breccias contain coarse, angular particles surrounded by finer grains. The term "breccia" is used to describe sedimentary formations and rocks with fault-related or igneous origins.

KEY FACTS

Breccias show angular, broken fragments surrounded by a fine-grained matrix.

+ **grain size:** Larger than 0.08 in (2 mm)

+ **texture:** Clastic

+ **composition:** Variable

Breccias are similar to conglomerates but are distinguished by their angular, broken rock fragments instead of rounded clasts. The angular clasts can come from a single source or from multiple rock types. Like other sedimentary rocks, the cement is commonly calcareous or siliceous. Breccias form in various environments. Sedimentary breccias can form from angular fragments of an underwater debris flow, deposits of a landslide, or other mass-wasting events. Fault breccias form when rocks are fragmented during slip along a brittle fault zone. Volcanic breccias form from angular fragments ejected during an eruption.

Shale

Type: Sedimentary

Shale is a sedimentary rock made of silt- and clay-sized grains—particles smaller than sand. It is distinguished by its fissility, meaning its tendency to split into thin layers parallel to bedding.

KEY FACTS

A clastic sedimentary rock, shale splits easily into layers.

+ **grain size:** Very fine

+ **texture:** Fine grains, fissile

+ **composition:** Variable; common mineral grains include quartz, feldspars, mica, calcite, and clays.

Sedimentary rocks made of grains finer than sand—a mixture of silt and clay—are generally known as mudrocks. Shale is a mudrock characterized by its property of easily breaking in one direction parallel to bedding. This is called fissility, and is formed by alignment of platy minerals such as micas and clays. Shale forms from compaction of fine-grained sediment, a process that typically occurs in still or slow-moving water as in bogs, deltas, and deep regions of the continental shelf. Fossil preservation is enhanced in these sediments and quiet depositional environments; thus, shales are common sources of fossils.

Mudstone

Type: Sedimentary

Mudstone is a fine-grained sedimentary rock made up of a mixture of clay- and silt-size particles. Unlike shale, mudstone is massive and does not easily split into layers.

KEY FACTS

A fine-grained sedimentary rock, mudstone does not easily split into layers.

+ **grain size:** Very fine

+ **texture:** Massive, not laminated

+ **composition:** Variable; common mineral grains include quartz, feldspars, mica, calcite, and clays.

Mudstone is a common sedimentary rock. It is clastic, but its particles are generally too small to see with the naked eye. Both mudstone and shale are made of compacted mud; the difference lies in the way these rocks break. Mudstone has a tendency to break into irregular blocks, whereas shale consistently breaks along a plane that is parallel to bedding. Shales and mudstones come in a variety of colors, most commonly gray, but also red, green, and purple. The colors usually indicate the presence of iron— red when iron is oxidized, green when iron is partially reduced.

Gneiss

Type: Metamorphic

Gneiss (pronounced nice) is a metamorphic rock characterized by compositional bands of separated light and dark minerals. These bands define a coarse layering or foliation in the rock.

KEY FACTS

This foliated metamorphic rock is defined by compositional banding, with few platy or needlelike minerals.

+ grain size: Medium to coarse

+ texture: Crystalline; compositional banding

+ composition: Variable; quartz, feldspar, and hornblende are common components.

Gneisses form from metamorphism of preexisting rocks that are typically rich in quartz and/or feldspar, such as granitic rocks and sandy sedimentary formations. Gneisses with an igneous parent rock are classified as orthogneisses; those with a sedimentary parent are called paragneisses. Gneisses form when the parent rocks are exposed to high temperature and pressure in the Earth's crust, usually by burial during mountain building. Gneisses' compositional bands are layered segregations of light-colored minerals such as quartz and feldspar from dark-colored minerals such as hornblende.

Migmatite

Type: Metamorphic

Migmatites, or migmatitic gneisses, are rocks that have partially melted at high temperatures and pressures, resulting in a hybrid rock with both igneous and metamorphic properties.

KEY FACTS

These rocks are heterogeneous, with segregations of light- and dark-colored minerals usually in irregularly folded layers.

+ grain size: Medium to coarse

+ texture: Crystalline

+ composition: Variable; light layers commonly granitic; dark layers generally contain hornblende and biotite.

Migmatites are named from the Greek *migma*, meaning mixture. They are mixtures of igneous and metamorphic components, formed when metamorphism progresses past the point at which the rock begins to melt. The melting is incomplete, producing segregations of melt that become swirly, light-colored igneous layers surrounded by the dark-colored residual metamorphic rock. Migmatites form under very high temperatures and pressures deep in the Earth's crust, as in the cores of large mountain ranges. These rocks are used as decorative stones, and they are often incorrectly called granites.

Amphibolite

Type: Metamorphic

Amphibolite is a rock rich in amphibole—typically hornblende or actinolite—often with plagioclase. Some amphibolites are massive; others have fine foliation or bands.

KEY FACTS

Colored black or dark green, amphibolite is often associated with gneisses and/or schists.

+ grain size: Medium to coarse

+ texture: Variable; can be massive, foliated, or banded

+ composition: Hornblende or actinolite plus plagioclase, sometimes garnet, and other minerals

Amphibolites are widespread metamorphic rocks made predominantly of amphibole plus plagioclase. Some amphibolites display an easily recognizable salt-and-pepper appearance with grains of black hornblende and white plagioclase. Amphibole grains are typically aligned in a weak to moderate foliation. The protolith, or parent rock, of amphibolites is generally difficult to determine. Most amphibolites form by metamorphism of mafic igneous rocks such as basalt and gabbro, though some originate as sedimentary rocks. Amphibolite is used as a building and decorative stone, which is often called "black granite."

Mylonite

Type: Metamorphic

Mylonites are hard, compact, foliated rocks with very fine grains. They commonly display a strong linear fabric defined by elongate minerals.

KEY FACTS

Strong deformational fabrics, compact texture, and fine grain size distinguish mylonites.

+ grain size: Very fine; some coarse relict grains common

+ texture: Well-developed foliation and lineation

+ composition: Variable

Mylonites form when rocks are deformed by movements along a fault. They typically form in deep zones of faults where rocks deform in a ductile fashion, like warm plastic, rather than breaking brittlely. As a rock is sheared, its minerals are pulverized on a microscopic scale into smaller grains. Harder minerals like quartz and feldspars can resist complete pulverization, resulting in larger, rounded, or oval grains surrounded by the fine-grained material. If there is significant shearing, even the resistant grains will be reduced to the fine matrix. Most mylonites form along discrete shear zones.

Schist

Type: Metamorphic

Schist is a common metamorphic rock characterized by its schistosity, which is a fine layering formed by the preferred orientation of platy and needlelike minerals, typically micas.

KEY FACTS

A finely layered (schistose) rock, schist has more than 50 percent sheetlike minerals, commonly muscovite, biotite, and other micas, and/or chlorite.

+ **grain size:** Medium to coarse

+ **texture:** Well-developed schistosity

+ **composition:** Mica- or chlorite-rich, otherwise variable

The term "schist" comes from the Greek *schistos,* meaning "divided." Schists typically split along parallel planes defined by the alignment of platy minerals such as micas. These planes are a type of foliation called schistosity. Schists form from metamorphism of fine-grained sedimentary rocks such as mudstones and shales. These rocks typically first form slate, then phyllite, an intermediate rock; eventually, as temperatures and pressures increase, they form schist. Schists can also develop from metamorphism of other fine-grained rocks such as volcanic tuffs. Some schists display tight chevron-type folds.

Garnet Schist

Type: Metamorphic

Schists can be classified according to their significant mineral components; for example, a biotite schist or a muscovite–chlorite schist. Garnet schist or garnet–mica schist features garnet.

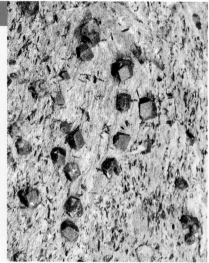

KEY FACTS

This form of schist has rounded garnet crystals, typically deep red, embedded in a scaly matrix of smaller, aligned micas.

+ **grain size:** Medium

+ **texture:** Well-developed schistosity; commonly larger than other minerals

+ **composition** Typically muscovite and garnet; can include biotite, staurolite, kyanite, or sillimanite

If a mudstone or shale is buried at depth by large-scale movements in the Earth's crust (for example, during the building of a mountain belt), the rock undergoes progressively higher pressures and temperatures. These conditions cause the rock to metamorphose into a schist. Types of minerals that grow in the schist depend on the peak pressure and temperature conditions reached by the rock. Thus, the mineral makeup of some schists is an important tool for scientists to understand the geologic history of a region. Garnet schists can indicate that a rock reached relatively high pressures and/or temperatures.

Greenschist

Type: Metamorphic

These fine-grained metamorphic rocks are rich in chlorite, epidote, or actinolite. The minerals give the rock an overall greenish hue.

KEY FACTS

Greenschist is a chlorite schist or other foliated metamorphic rock with chlorite and/or actinolite.

+ **grain size:** Medium

+ **texture:** Foliated

+ **composition:** Chlorite, epidote, actinolite

The term "greenschist" usually describes a chlorite schist, though it can refer to other foliated metamorphic rocks that are rich in green-colored minerals such as chlorite, actinolite, and epidote. Greenschists have a foliation defined by the alignment of chlorite grains. Most greenschists form when a fine-grained mafic volcanic rock such as a basalt is metamorphosed at low temperatures and pressures. The term "greenschist" also describes metamorphic conditions. Greenschist metamorphism occurs during low to medium pressures and temperatures. For example, a slate is not a greenschist, but it develops under greenschist conditions.

Blueschist

Type: Metamorphic

Blueschist is a metamorphic rock that typically has a blue hue from the presence of the mineral glaucophane, a sodium-rich amphibole. The term also describes a set of metamorphic conditions.

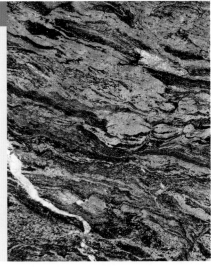

KEY FACTS

Blueschist is a foliated metamorphic rock with a bluish color.

+ **grain size:** Medium

+ **texture:** Foliated, not necessarily schistose

+ **composition:** Variable; blue color comes from glaucophane

A blueschist rock forms under blueschist metamorphic conditions, but not all rocks that experience blueschist metamorphism turn into blueschists. Confused yet? The complication comes from two different uses for the word "blueschist," a similar scenario to the use of the word "greenschist." The rock type called blueschist is a metamorphic rock that contains the bluish amphibole glaucophane. The term "blueschist" can also refer to a set of metamorphic conditions characterized by high pressures and low temperatures. These conditions occur along subduction zones, where one of the Earth's plates is pulled down beneath another.

Slate

Type: Metamorphic

Slate is a fine-grained metamorphic rock that splits easily along a well-developed foliation. Slate is commonly used for roofing tiles as well as for other decorative stones.

KEY FACTS

Typically gray or greenish, slate is a brittle rock with well-developed foliation.

+ **grain size:** Fine

+ **texture:** Compact, well foliated

+ **composition:** Rich in micas and quartz, though often too fine-grained to distinguish individual minerals

Slate is a common metamorphic rock characterized by its ability to split into very thin sheets or plates, some of which are quite large. Slate is actively quarried for use as floor tiles, hearths, and roofing tiles. The original blackboards in classrooms were made of single sheets of slate, though most blackboards today are composed of man-made materials. Slate forms via the metamorphism of mudstones, shales, or tuffs at low pressures and temperatures. If metamorphism progresses, mica crystals grow larger, turning a slate into a schist. Quartz veins and small faults are common in slate outcrops.

Marble

Type: Metamorphic

The metamorphism of limestone or dolomite results in formation of marble, a rock that is composed almost entirely of recrystallized, interlocking grains of calcite or dolomite.

KEY FACTS

A dense stone, marble is usually pure white, gray, or yellow; other varieties are pink, blue, or black.

+ **grain size:** Fine to coarse

+ **texture:** Crystalline; often massive; some show banding

+ **composition:** Calcite and/or dolomite; accessory minerals can include micas, graphite, and serpentine.

Some of the most famous sculptures and buildings are created out of marble, a type of metamorphic rock. Michelangelo's "David" is carved out of an Italian stone known as Carrara marble. The Taj Mahal in India is clad almost entirely in a marble known as White Makrana. In the United States, the Supreme Court Building and the Lincoln Memorial are built with slabs of Alabama marble. Marble is white when it is pure calcite. More commonly, marbles contain impurities and other minerals resulting in different colors and textures. Some coarse-grained marbles can look similar to quartzite, but marble is much softer.

Serpentinite
Type: Metamorphic

Serpentinite is composed of serpentine, often with chromite or magnetite and veins of calcite, dolomite, or talc. It forms by alteration and metamorphism of ultramafic rocks.

KEY FACTS

Light yellow-green, dark green, or black, serpentinite is typically brittle with a greasy or waxy feel.

+ grain size: Fine

+ texture: Variable; brecciated forms, veins common

+ composition: Primarily serpentine; other minerals may include chromite, magnetite, and talc.

Serpentinite is a striking rock, commonly streaked with various colors from yellow-green to black, with a glossy surface that is greasy or even soapy to the touch. Serpentinite forms when olivine-rich rocks like peridotite are altered in the presence of water, a process known as serpentinization. The formation of serpentinites occurs at low metamorphic grades. Serpentinite is a popular stone used for floor tiles and countertops in buildings, and as an ornamental stone. The popular decorative stone known as *verd antique* ("antique green") is a serpentinite with a brecciated or angular, broken texture.

Quartzite
Type: Metamorphic

Quartzite is extremely hard and compact, formed by metamorphism of quartz-rich sandstones or chert at elevated temperatures and pressures.

KEY FACTS

Hard, compact, and quartz-rich, quartzite has a distinctive sugary appearance. It is commonly white or gray.

+ grain size: Medium

+ texture: Compact, interlocking grains

+ composition: Primarily quartz; other minerals may include micas and/or hematite.

Quartzite forms when quartz grains of the protolith—the parent rock before metamorphism—are heated and squeezed together until they form new grains. This process creates a homogeneous mosaic of interlocking grains, resulting in a dense, very hard stone. Textures from the protolith—rounded quartz grains surrounded by fine quartz cement, for example—are destroyed. Quartzites are distinguished by their hardness, conchoidal fracture, and sugary texture in outcrop. The decorative stone aventurine is quartzite. Green aventurine, used in jewelry, is colored by a chromium-rich mica known as fuchsite.

Plants

Plants are among the most diverse groups of organisms. Their fossils help scientists understand the evolution of life as well as changes in the Earth's climate through geologic time.

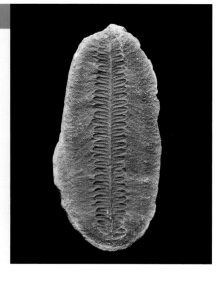

KEY FACTS

Preserved parts of plants include leaves, bark, seeds, spores, and pollen.

+ time period:
About 450 million years ago to present

+ size: Variable

+ where to find:
Fine-grained sedimentary rocks

Plant fossils are the geologic record of ancient plants. The first land plants appeared about 450 million years ago. Complete fossils are rare; most are fragments such as leaves, cones, and stems. Most plants decompose rapidly; to become fossilized, specific conditions have to occur. The plant must be buried by sediment quickly enough to shield it from decomposition, but gently enough to preserve its integrity. Thus, most plant fossils are found in fine-grained sedimentary rocks such as shales and mudstones. Fossil plants, especially leaves, spores, and pollen grains, are important to our understanding of past climates.

Petrified Wood

Petrified wood is a plant fossil preserved by the infiltration of minerals around the organic material, a process called permineralization. Most petrified wood is made of quartz or calcite.

KEY FACTS

Three-dimensional replicas of woody plants, including trunks, branches, and bark; harder and more brittle than unfossilized wood.

+ time period:
About 400 million years ago to present

+ size: Variable

+ where to find:
Eroded out of sedimentary rock formations

Ancient plants fossilized by permineralization are called petrified wood. The pore spaces around and within the plant's organic material are filled by minerals, usually microcrystalline quartz. The organisms retain their original structures, so permineralized fossils are three-dimensional forms, not casts or impressions. Petrified wood is found throughout the U.S. and Canada, especially in western states. In Arizona's Petrified Forest National Park, entire trees over 200 million years old were fossilized in buried logjams of ancient rivers. Petrified wood can be colorful due to the presence of iron, carbon, and manganese.

Graptolites

Graptolites are a group of small, free-floating marine organisms that flourished during the early and middle Paleozoic eras. Graptolite fossils are found worldwide, mostly in black marine shales.

KEY FACTS

These small, twiglike fossils have saw-blade shapes, some in spirals.

+ **time period:** 500–315 million years ago

+ **size:** Typically 1–4 in (2.5–10 cm); some as long as 8–10 in (20–25 cm)

+ **where to find:** Marine shales of Ordovician, Silurian, and Devonian periods

Graptolites are a group of relatively simple marine animals that lived during the early explosion of life in the Earth's oceans beginning about 500 million years ago. Graptolites are wormlike animals that lived in colonies floating through seawater like plankton. Graptolites constructed tubelike shells out of soft collagen, and the tubes were linked to form branching colonies. Many graptolite fossils resemble plant leaves or branches of plants. Others take the shape of tiny saws, tuning forks, and spirals. Graptolites are typically found in black marine shales associated with calm waters of the outer continental shelf.

Bryozoans

Bryozoans are colonial aquatic organisms that resemble small corals. They are made of many small individuals living in compartments of a calcareous skeleton.

KEY FACTS

Lacy or fanlike fossils resemble moss or corals.

+ **time period:** 488 million years ago to present

+ **size:** Individuals are less than 1 mm; colonies range from millimeters to meters.

+ **where to find:** Shallow-marine limestone formations; also shales and mudstones

Bryozoans are colonial organisms that secrete calcareous shells with many compartments. Each compartment contains a body cavity with a retractable food-gathering arm called a lophophore. The lophophore is a group of tentacles armed with beating fibers called cilia that gather food particles from the seawater. Most bryozoans attach themselves to the sea bottom onto a rocky substrate or the discarded shell of another organism. They take many different forms, including sheetlike crusts, netted fans, and small branching trees. They build reefs. Bryozoan fossils are most common in shallow marine limestone formations.

Corals

Corals live in tropical marine settings and are the principal component of reefs. Corals are ancient animals, with fossil specimens dating back more than 500 million years.

KEY FACTS

Skeletons of coral have various shapes including horns, brains, spirals, branches, and disks.

+ **time period:** 540 million years ago to present

+ **size:** Variable; millimeters to meters

+ **where to find:** Limestones

Corals belong to the phylum Cnidaria (organisms with stinging cells), which includes sea anemones and jellyfish—simple, soft-bodied marine organisms. Corals construct a skeleton made of calcium carbonate and are well preserved as fossils. Living corals consist of a soft body cavity called a polyp with a single opening, surrounded by tentacles that gather food from the water. The polyps are attached to a hard skeleton that they construct throughout their life. Coral fossils display many shapes, including disks, cylinders, spirals, and brain-like domes. Because of their limited ecological tolerances, corals are indicators of ancient environmental conditions.

Trilobites

Trilobites are ancient animals, now extinct, that dominated the world's oceans during the Cambrian period. Related to crustaceans, they had a hard outer shell that is well preserved.

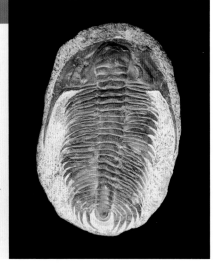

KEY FACTS

Body is segmented, and has an armored shell and crescent-shaped head.

+ **time period:** 540–20 million years ago

+ **size:** Typically 1–4 in (2.5–10 cm)

+ **where to find:** Fine-grained marine sedimentary rocks; famous locales are Burgess Shale in British Columbia, Beecher's Trilobite Bed in New York, and Wheeler Shale in Utah.

Trilobites are among the earliest arthropods—a classification of invertebrate animals with exoskeletons, including insects, spiders, and crustaceans. Most trilobites were 1 to 4 inches (2.5 to 10 cm) long, though one of the largest trilobites, found on the shore of Hudson Bay, measures 28 inches (71 cm). Trilobites were prolific and geographically dispersed, roaming worldwide Paleozoic oceans in deep and shallow water environments. These factors, plus the easy preservation of their shells, resulted in trilobite fossils being both common and widespread around the globe. Trilobite fossils are important time markers for sedimentary formations.

Brachiopods

Brachiopods, also known as lampshells, are marine animals with two shells connected at a hinge. At first glance, they appear similar to bivalves, though these animals are unrelated.

KEY FACTS

The two shells of brachiopods are symmetrical along the midline, perpendicular to the hinge, with the bottom shell typically larger than the top shell.

+ **time period:** 540 million years ago to present

+ **size:** Most are about 0.25–4 in (0.6–10 cm); largest up to 8–12 in (20–30 cm)

+ **where to find:** Limestones and marine shales

Brachiopods are marine organisms with two shells, similar to clams and mussels. Clams and mussels, however, are bivalves, unrelated to brachiopods and different in many ways. In brachiopod shells, the left half is a mirror image of the right; they are commonly asymmetrical top to bottom, with the bottom larger than the top. Though only a few hundred species live today, brachiopods dominated the world's oceans approximately 540 to 252 million years ago, when tens of thousands of species flourished. Many species died off during a mass extinction event approximately 250 million years ago at the end of the Permian period.

Bivalves

Bivalves are marine and freshwater organisms with two shells joined along a hinge. Fossil bivalves are found worldwide from all time periods since the Cambrian.

KEY FACTS

The two shells of bivalves are symmetrical parallel to their hinge; the valves are mirror images of each other.

+ **time period:** 540 million years ago to present

+ **size:** Most are 0.4–4 in (1–10 cm); some are as large as 3.3 ft (1 m) or more.

+ **where to find:** Limestones and shales

Bivalves are aquatic animals with two shells that meet along a hinge. These include living clams, oysters, mussels, and scallops. Bivalves have a muscular foot that emerges from the shell and allows some species to burrow into the sand or mud. Other bivalves anchor themselves to rocks or coral reefs. Others, such as scallops, swim by clapping their shells together, forcing out water to propel them. Fossil bivalves are common, and they come in a variety of shapes and sizes. The symmetry of their two shells is key to their identification. In contrast, brachiopod symmetry is perpendicular to the orientation of the hinge.

Gastropods

Gastropods are snails, most with coiled, hard shells made of calcium carbonate. They are found in a range of habitats, including marine and freshwater environments as well as terrestrial habitats.

KEY FACTS

Snails and sea slugs are gastropods. Snail shells come in various shapes with single chambers. Sea slugs have internal shells, and some lack a shell.

+ time period: About 540 million years ago to present

+ size: about 0.04–35 in (1 mm–90 cm)

+ where to find: Fine-grained clastic sedimentary rocks and limestones

Gastropods first appeared during the early Cambrian period about 540 million years ago. They are a large and diverse group, with species living in marine environments, freshwater, and terrestrial habitats. Most gastropods have a coiled shell with a single chamber. Unlike bivalves, gastropods have heads that are distinguished from their bodies. Like bivalves, they have a single, muscular foot that helps them move. Gastropod and bivalve fossils are abundant and widespread. Because of their excellent preservation and the diversity of their species, they are important to scientists for understanding evolutionary processes.

Ammonites

Ammonites are an extinct group of mollusks best known for their beautiful spiral shells. Ammonite fossils are prolific in the geologic record.

KEY FACTS

The shells of ammonites are divided into chambers.

+ time period: About 400–65 million years ago

+ size: Up to about 6.5 ft (2 m)

+ where to find: Marine sedimentary rock formations; key localities include Badlands National Park in South Dakota and Guadalupe Mountains National Park in Texas.

Ammonites are marine animals related to squid, cuttlefish, and octopuses, but went extinct about 65 million years ago, the same time as dinosaurs' demise. They are best known for their distinctive spiral shells, though a few species evolved with nonspiral forms. Ammonites were prolific and diverse, with many species. Their fossils are among the most abundant on Earth and serve as excellent time markers, allowing geologists to link the rock layers in which they are found to specific time periods. The name "ammonite" refers to the Egyptian god Ammon, who wore ram's horns resembling the spiral shape of the shells.

Echinoderms

Echinoderms are a diverse group of marine animals including sea stars, sea urchins, and sand dollars. Echinoderms' skeletal plates made of calcite are well preserved as fossils.

Echinoderms are a broad group of marine organisms that appeared in the early Cambrian period about 540 million years ago, which has many living members today. The most recognizable are sea stars, sea cucumbers, sea urchins, and sand dollars, which have a unique skeleton made of interlocking plates of calcite. These interior, or endoskeletons, called tests, enclose the organs and are covered with a spiny skin. The skeletons also display a radial symmetry, often with five points, such as the five limbs of a sea star. Echinoderm skeletons are widespread in the fossil record and are an important indicator of past marine conditions.

Crinoids

Crinoids are marine animals that look like flowers, with a flexible stalk topped with a head of waving arms. They are echinoderms, related to sea stars and sea urchins.

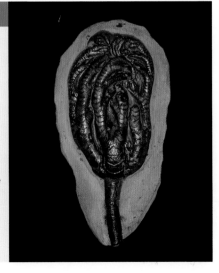

Crinoids, like all echinoderms, have a skeleton interlocking calcite plates that house their vital organs. Crinoids are sessile—meaning that they live attached to some sort of substrate such as a rock, shell, or reef—or they are free-floating, planktonic organisms. Crinoids have three basic body parts: a stem, a cup or calyx, and arms. The cup and arms look like tassels or blossoms atop stems. Crinoid stems are flexible cylinders made of small, circular plates with a star-shaped interior canal. Fragments of stems are common fossils, especially in limestones in the western U.S. and Canada. Some limestone formations consist entirely of crinoid parts.

Fish

Animals classified as fish include modern bony fish, sharks, and rays, as well as many species that are extinct. Some mudstones and shales contain exquisitely well-preserved fish fossils.

KEY FACTS

Most fish have elongate bodies with fins, and are covered with scales or bony plates.

+ **time period:** 500 million years ago to present

+ **size:** Variable; some fossil fish reach over 65 ft (20 m) long.

+ **where to find:** Fine-grained sedimentary rocks; famous locales include Fossil Butte National Monument in Wyoming.

Fish are the first vertebrates in the fossil record, about 500 million years ago. Those fish were jawless, and bony plates covered all or most of their bodies, instead of a hard internal skeleton. Over time, different fish appeared, including sharks and rays about 400 million years ago, and eventually fish with internal bony skeletons about 200 million years ago. The Niobrara Formation in the central and western U.S. contains abundant fish fossils from an interior seaway that existed about 85 million years ago. Wyoming's Green River Formation, about 48 million years old, contains spectacular beds of fossil fish.

Shark Teeth

Shark teeth are relatively common fossils and a favorite of collectors. Sharks shed thousands of replaceable teeth during their lifetime—an abundant source for fossilization in ocean sediments.

KEY FACTS

Individual teeth with various shapes can be found.

+ **time period:** 450 million years ago to present

+ **size:** Up to about 7 in (18 cm)

+ **where to find:** Along beaches and eroding out of marine sedimentary rock formations, even far inland such as in many localities in Wyoming

Sharks are among a group of fish that includes skates and rays. They first appeared about 450 million years ago and are distinguished by an internal skeleton of cartilage, not bone. Without hard parts, sharks are rarely preserved as fossils, except their teeth. Most teeth become fossilized by minerals filling in the pores around the tooth material. This process takes thousands of years and is why many fossilized shark teeth are dark instead of the light whites and yellows of living sharks' teeth. A famous fossilized shark is the giant Megalodon, which lived approximately 20 million to 2 million years ago and whose teeth are up to about 7 inches (18 cm) long.

Insects

Insects are among the most diverse group of animals. Wingless insects first appeared about 400 million years ago, and flying insects appeared about 360 million years ago.

KEY FACTS

Wings with a network of veins are more often fossilized than other body parts; whole insects are found in amber.

+ time period:
About 400 million years ago to present

+ size: Up to about 2 ft (0.6 m)

+ where to find:
Fine-grained sedimentary rocks; famous locales include Florissant Fossil Beds in Colorado.

Insects, one of the most diverse groups of animals on the planet, were the first to develop flight, approximately 360 million years ago. Insects have segmented bodies: a head, a middle section called a thorax, and an abdomen. They have a hard outer skeleton, and their legs are jointed. Insects were particularly abundant about 305 to 270 million years ago, when gigantic dragonflies had wingspans over 2 feet (0.6 m) long. Insects are relatively poorly preserved as fossils because of their fragile bodies. One exception is amber, or fossilized tree resin, which preserves entire insects that became trapped in tree sap.

Amphibians & Reptiles

Amphibians and reptiles are among the earliest animals to inhabit the land. These groups evolved to include a wide diversity of species, both living and extinct.

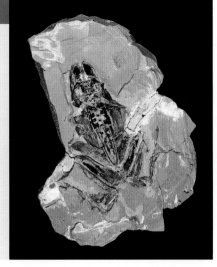

KEY FACTS

Tetrapods, named from the Greek words for "four-footed," are four-limbed vertebrates, and the earliest of these ancient animals able to walk on land were amphibians and reptiles.

+ time period:
368 million years ago to present

+ size: Up to about 130 ft (40 m) long

+ where to find:
Fine-grained sedimentary rocks

The first vertebrates to "migrate" from water to the land were amphibians known as the early tetrapods, or four-legged animals. A famous fossil is a 360-million-year-old skeleton of a large creature called Ichthyostega found in Greenland. It more resembled crocodiles than today's amphibians. Others include the bizarre Diplocaulus, a giant newt-like creature with a boomerang-shaped skull. Fossils resembling modern frogs, toads, newts, and salamanders are found in rocks younger than 200 million years old. Reptiles evolved later, and their abundant fossils include turtle shells, lizard skeletons, dinosaur bones, and reptile eggs.

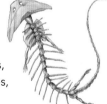

Dinosaurs

Dinosaurs are part of the diapsid group of reptiles. They grew to enormous proportions and dominated most terrestrial habitats until their sudden demise about 65 million years ago.

KEY FACTS

Large bones are found in Triassic to Cretaceous period rocks; dinosaurs had two skull openings behind their eye sockets.

+ **time period:** 225–65 million years ago

+ **size:** Variable; up to about 190 ft (58 m) long

+ **where to find:** Sedimentary rocks; Dinosaur National Monument in Colorado and Utah is a famous site.

Dinosaurs are diverse reptiles that populated the Earth for 160 million years. There are two main groups: Ornithischia, "bird-hipped," with rear-facing pubic bones, and Saurischia, "reptile-hipped," with forward-facing pubic bones. Ornithischia include Triceratops and Stegosaurus. Saurischia include herbivorous sauropods such as Apatosaurus and carnivorous theropods such as Tyrannosaurus and Velociraptor. Paleontologists trace the ancestry of birds to theropods. Recent discoveries indicate that some dinosaurs had extensive feathered plumage. Dinosaur fossils range from footprints to fragments of bone to entire skeletons.

Birds

Bird fossils are of great interest to scientists tracing the evolutionary history from dinosaurs to modern birds. Bird fossils, however, are uncommon because of their fragile hollow bones.

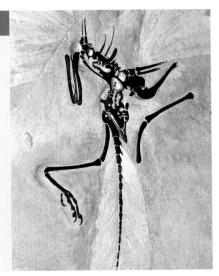

KEY FACTS

Bird fossils are uncommon; evidence for feathers is an important feature.

+ **time period:** About 150 million years ago to present

+ **size:** Variable

+ **where to find:** Sedimentary rock formations; uncommon

One of the oldest and most famous early bird fossils is Archaeopteryx, from a 150-million-year-old limestone quarry in Germany. Archaeopteryx has features of both dinosaurs and birds and has been interpreted as an intermediate evolutionary stage. Like modern birds, Archaeopteryx had wings and feathers, but like dinosaurs, it had teeth, claws, and a long tail. As more discoveries of feathered dinosaurs emerge, however, interpretations of Archaeopteryx and the transition from dinosaurs to modern birds are changing. Birds continued to diversify after non-avian dinosaurs' extinction, and by about 35 million years ago, most groups of modern birds had appeared.

Mammals

The 65-million-year period after dinosaurs' extinction is often considered the Age of Mammals. Their fossils help scientists understand climate changes and movements of the continents.

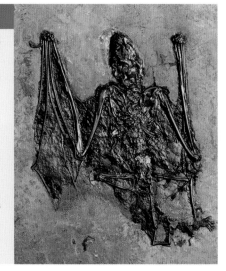

KEY FACTS

Mammal fossils range from bone fragments and teeth to entire skeletons and even hair.

+ time period: About 150 million years ago to present

+ size: Variable

+ where to find: Widespread; famous localities include La Brea Tar Pits in Los Angeles, Mammoth Site in South Dakota, and sites in western Nebraska.

Mammals are warm-blooded, have hair, and feed milk to their young. They appeared around the same time as dinosaurs, but mammals didn't flourish until after dinosaurs' demise about 65 million years ago. Early mammals were relatively small, looked similar to rodents, and fed on insects. As mammals diversified, they became larger and dominated nearly every terrestrial ecosystem. Mammal fossils, especially teeth, are widespread and well preserved. Mammals in famous fossil finds from the Pleistocene, the period of the last ice ages, include skeletons of mammoths, mastodons, saber-toothed cats, native horses, and camels.

Trace Fossils

Trace fossils include footprints, tracks, burrows, bite marks—even fossilized dung. These are traces of living organisms that have been preserved in the geologic record.

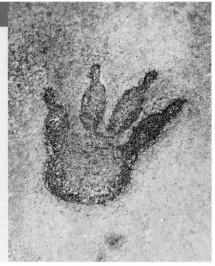

KEY FACTS

Burrow marks, footprints, feeding marks, and excrement are preserved primarily in sedimentary rocks.

+ time period: About 542 million to 10,000 years ago

+ size: Variable

+ where to find: Sedimentary rocks; locales include dinosaur tracks at Dinosaur State Park in Connecticut and Dinosaur Valley State Park in Texas.

Trace fossils preserve organisms' activities rather than their body parts. Invertebrates' trace fossils include impressions of feeding, burrowing, and boring, mostly preserved in fine-grained marine sedimentary rocks. Trace fossils from vertebrates, especially land animals, include bite marks and footprints, even pieces of fossilized dung known as coprolites. Trace fossils provide an understanding of how prehistoric organisms lived, moved, and ate. Sometimes it is difficult to distinguish a trace fossil from a pseudofossil, a geological formation that looks similar to something made by an organism but is inorganic.

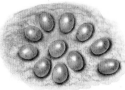

Lineation
Deformed rocks

Lineations are linear structures in deformed rocks. There are several types, including the hinges of folds and the long axes of aligned minerals and stretched clasts.

KEY FACTS

Lineations are structures that can be represented by a line: long in one dimension relative to others.

+ fact: Look for lineation along the surface of a foliation plane.

+ fact: Deformed rocks with strong linear fabrics and without planar fabrics (foliation) are L-tectonites.

+ fact: Common in mountain outcrops and ancient cores of continents

A lineation is a visible texture of linear elements that develops when a rock is deformed. For example, a conglomerate originally made of rounded cobbles develops a lineation if it is squeezed such that the cobbles are stretched out into elongate, cigar-like shapes (it develops a foliation if the clasts are flattened into pancakes). Other common lineations are defined by the alignment of elongate minerals, which can form as metamorphic minerals grow into a preferred orientation, or they can be original minerals that are rotated into a linear fabric during deformation.

Foliation
Deformed rocks

Deformed rocks that exhibit layering or any through-going textures that are planar are said to have foliation. These rocks may split or break along foliation planes, or may be more massive.

KEY FACTS

Foliation is an arrangement of planar or tabular features in a deformed rock.

+ fact: Rocks can have more than one type of foliation in more than one orientation.

+ fact: Layering in schists and gneisses is an easy way to identify foliation.

+ fact: Common in outcrops in mountain belts and the ancient cores of continents

A rock has a foliation if it has a visible texture or fabric created by planar or tabular elements. These form as a rock is squeezed and flattened by forces in the Earth's crust that are typical of mountain-building processes. Slates are metamorphic rocks that cleave or break along foliation planes. The fine layering in schists created by aligned plates of mica is another type. In higher-grade metamorphic rocks such as gneisses, light and dark layers of different minerals define a foliation. Sometimes rocks display more than one foliation.

Augen

Deformed rocks

Augen are large, lens-shaped mineral grains found in deformed rocks. Augen are typically surrounded by fine grains of mica and other minerals aligned in wavy layering or foliation.

KEY FACTS

Eye-shaped minerals, commonly feldspars, are surrounded by platy minerals aligned in a foliation.

+ fact: Augen are larger than other mineral grains in a rock, commonly about 0.2–1.2 in (0.5–3 cm).

+ fact: Found in metamorphic rocks such as schists and gneisses

+ fact: Form when rocks are deformed at depth in the Earth's crust

Augen, German for "eyes," are eye-shaped minerals or clusters of minerals in some deformed rocks. They can be relict crystals from the original rock that were rotated and rounded during high-temperature deformation, or they can be newly grown metamorphic crystals that developed during deformation. Augen are typically harder, compact minerals such as feldspar, quartz, and garnet. They are surrounded by micas and other platy crystals that are aligned in a wavy foliation. Augen form in areas of the Earth's crust called shear zones, where rocks move past each other at relatively high temperatures.

Boudinage

Deformed rocks

Boudinage forms when a relatively rigid rock layer is stretched and pulled apart into sausage shapes known as boudins. Surrounding layers appear to flow around the boudins.

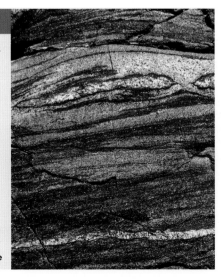

KEY FACTS

A rock layer or quartz vein looks as if it has been pulled apart like taffy.

+ fact: Easiest to see in exposures perpendicular to the rock's dominant foliation or layering

+ fact: Found in deformed veins and foliated rocks

+ fact: Forms when rocks are deformed in the Earth's crust before they are exposed at the surface

Boudinage is a structure in veins or rock layers that are more rigid or competent than the surrounding rock. When these are stretched, the competent vein or layer is pulled apart. The result is an effect that resembles a link of sausages. Boudinage is from the French *boudin,* a type of sausage. Boudinage structures range from thin, undulating ribbons in narrow veins, to larger, blocky, rotated tablets. Boudinage is an example of how different rocks react to the same forces: Rock around boudins behaves like heated plastic, flowing in a solid state to accommodate the strain, while the mechanically stronger boudin layer pulls apart like taffy.

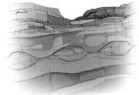

Folds
Deformed rocks

Folds are bends in bodies of rock that form during shortening or contraction. Folds occur at all scales, from deformation of individual crystals to structures the size of mountain ranges.

Folds are formed when rocks shorten or contract due to movements in the Earth's crust and upper mantle. Folds can be found at all scales, from bent individual mineral grains to enormous warps defining mountain ranges. Geologists have many terms to describe different kinds of folds. Folds in the shape of a lowercase *n* are called anticlines; folds in the shape of a lowercase *u* are called synclines. Folds can be upright, asymmetrical, or recumbent. Tight, Z-shaped folds are called crenulations or chevrons. Folded sedimentary rocks can form traps for accumulation of oil and natural gas.

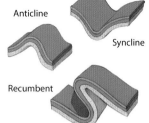

Anticline

Syncline

Recumbent

Faults
Deformed rocks

Faults are brittle fractures that accommodate movement in rocks. The body of rock on one side of a fault moves relative to the body on the other side by sliding along the fault plane.

Three types of faults are thrust or reverse, normal, and strike-slip. Thrust or reverse faults are planes along which deeper rocks move upward relative to shallower rocks. These faults are caused by compressional forces in the crust. Normal faults are planes along which younger, shallower rocks drop down relative to older, deeper rocks. Normal faults accommodate stretching or thinning in the crust and are common in the U.S. Basin and Range province. Strike-slip faults are planes along which rocks slide laterally past each other without significant vertical movement. The San Andreas Fault is a famous strike-slip fault.

Joints
Deformed rocks

Joints are naturally occurring cracks found in almost all rock types. Joints are distinguished from faults by a lack of movement across the fracture plane.

KEY FACTS

To see a joint, look for brittle cracks that cut across rock faces.

+ fact: Joints form from stresses in the Earth's crust.

+ fact: Joints are important conduits for groundwater in aquifers.

+ fact: Joints can have various orientations, appearing in regular sets or as isolated fractures.

Many rock outcrops contain brittle cracks, called joints. There is no movement along joints, however. If there is evidence of sliding along a fracture, the crack is called a fault, not a joint. Joints form by forces that act upon a body of rock. As the Earth's plates move, forces are distributed through the crust, causing some rocks to crack. Rocks can also crack during changes in temperature, from uplift to shallower depths in the Earth's crust, or by cooling of magma. Joints occur as isolated cracks or in widespread fractures. They are affected by weathering and erosion, and are an important control on the evolution of a landscape.

Veins
Deformed rocks

Veins are narrow, sheetlike bodies made up of one or more minerals. Veins form when minerals crystallize out of water-rich fluids.

KEY FACTS

Veins appear as crosscutting stripes or stringers on the surface of an outcrop.

+ fact: The most common veins are white and are filled with quartz or calcite.

+ fact: Veins can also be host to rare minerals and ore deposits.

+ fact: Veins are found in all rock types.

Cutting across many outcrops are narrow, sheetlike bodies made up of one or more minerals. These are called veins, and they are filled with minerals that crystallized out of water-rich fluids. In some cases, veins form when fluids infiltrate existing cracks or joints in a rock formation. Veins following crack systems may form individual tabular bodies or spidery networks of irregular shapes. Some veins are tightly folded, indicating high-temperature deformation of the rock body after the vein crystallized. Veins are important sources of metals and other precious minerals, including gold and copper.

Tafoni
Weathering & erosion

Tafoni, or honeycombing, is a surface feature marked by rounded pits, hollows, and shallow caverns, often clustered in groups and separated by hardened ridges or visors.

KEY FACTS

These features are easiest to identify in cliff faces with a honeycomb appearance.

+ fact: Develop in many different rock types

+ fact: Appear as small pits clustered together or large, rounded hollows

+ fact: Found in sandstone cliff faces and some granitic bodies; common in coastal environments

Clusters of hollows, rounded pits, and shallow caverns on rock faces and outcrops are known as tafoni, honeycombing, or cavernous weathering. Geologists have posed many hypotheses for how cavernous weathering develops. Most agree that salt crystallization is key. Salts crystallize in tiny pits on the surface of a rock, transported by wind or water. The salt expands, breaking off adjacent grains of the host rock and expanding the pit. This process forms larger and larger rounded hollows. In some porous sandstones, cement hardening along internal joints or bedding planes may be important in channeling erosion.

Desert Varnish
Weathering & erosion

Desert varnish is a surface coating common on the sandstone rock walls of the desert Southwest, though it can form on many rock types and in different environments.

KEY FACTS

The varnish is a crust on the surface of a rock; dark colors vary from reddish brown to shiny black.

+ fact: Crust is usually less than 0.02 in (0.5 mm) thick.

+ fact: Common on exposed rock faces in arid environments

+ fact: Many petroglyphs are created by chipping through desert varnish to paler rock beneath.

Desert varnish is a dark, hard coating made of clay minerals, manganese oxide, and iron oxide. It is found on many different rock exposures. Geologists have debated the varnish's formation because manganese and iron oxides could indicate bacterial metabolism. Rocks on Mars appear to have desert varnish; however, most geologists agree that it can develop without the aid of life. Much of the varnish is composed of silica, which reaches the surface via clay minerals in windblown dust or by leaching from the rock's interior. Black, shiny desert varnish is rich in manganese oxide; dull reddish varnish is rich in iron oxide.

Spheroidal Weathering
Weathering & erosion

Spheroidal weathering, also known as onion-skin weathering, is a process that creates rounded outcrop formations and boulders in the shape of spheres.

Spheroidal weathering is a type of weathering that results in rounded outcrop formations and spherical boulders. Spheroidal weathering occurs when a rock formation is broken into blocks by several sets of fractures called joints. Groundwater penetrates the rock formation along the joints, facilitating chemical weathering at a faster rate than where the rock is not fractured. Physical weathering is also accelerated along joints. Weathering progresses, rounding off the edges of two intersecting joints and the corners of three intersecting joints, producing curved rock formations and even perfect spheres.

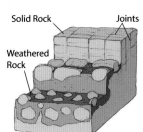

Sheeting or Exfoliation Joints
Weathering & erosion

Sheeting joints, also known as exfoliation joints, are curved fractures that parallel the surface of a body of rock. The joints separate sheets of rock, creating dome-like landforms.

Sheeting joints, also known as exfoliation joints, are common features of granitic terrains. These are sets of curved fractures oriented roughly parallel to the surface of a rock. The joints separate concentric slabs or shells of rock that progressively weather and erode, creating domes and other rounded landforms. One explanation for sheeting joints is that they form during the release of pressure when rocks overlying a body of granite are removed by erosion—a removal called "unroofing." Another explanation is that sheeting joints form from mechanical stresses along the surface of a dome.

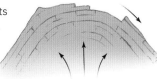

Arches
Weathering & erosion

Natural arches, windows, and bridges are erosional features of some landscapes. They can form by river erosion, wave action along coastlines, and weathering.

Arches, windows, and bridges are slabs of rock overlying natural openings. Many arches are formed by erosion along parallel sets of fractures to create rock fins. Some fins are undermined when there is contact between a permeable sandstone formation and an impermeable formation such as shale. Groundwater percolating through the sandstone stops at the shale and flows laterally, weakening the sandstone. This can undermine a sandstone fin, causing a rockfall or opening an arch. Natural bridges and windows may also form by dissolution of limestone in cave formation.

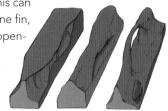

Three stages in arch formation

Towers & Fins
Weathering & erosion

Along with arches, natural stone towers and fins are some of the world's most spectacular landforms. Most of these features form by preferential weathering and erosion along fractures.

Towers and fins are iconic landforms of the Southwest. Most of them have been created by millions of years of weathering and erosion. The key to the creation of towers and fins is in the regional pattern of fractures in the rock, known by geologists as joints. Weathering and erosion become concentrated along these planes, and if a rock contains a set of evenly spaced, parallel fractures, erosion along these planes will leave narrow fins standing. If a rock contains two sets of joints that are oriented roughly perpendicular to each other, weathering and erosion will create freestanding towers.

Hoodoos/Mushroom Rocks/Pedestal Stones
Weathering & erosion

Hoodoos, mushroom rocks, and pedestal stones are names given to rock spires and perched blocks with irregular, sometimes fantastical shapes.

KEY FACTS

The rock spires may or may not have capstones or perched blocks.

+ fact: Common in sedimentary and volcanic rocks

+ fact: Found especially in deserts, badlands, and mountains

+ prime locations: Bryce Canyon National Park in Utah, Badlands National Park in South Dakota, and Tent Rocks National Monument in New Mexico

Thin rock spires with irregular shapes are known as hoodoos, tents, and fairy chimneys. They are found in basins, canyons, and badlands, usually in layered sedimentary or volcanic rocks. Spires supporting perched blocks or wide, rounded caps are known as mushroom rocks or pedestal stones. These are commonly in granitic, volcanic, and sedimentary rocks. Many hoodoos and pedestal stones form when a hard, more resistant rock formation overlies a softer, weaker formation. Weathering attacks these rocks along joints, leaving narrow cones of the softer rock protected by isolated caps of the overlying formation.

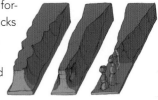

Three stages in formation of hoodoos

Landslides/Slumps/Creep
Weathering & erosion

The movement of rock and soil down a slope creates distinctive land features and gravity-formed deposits. Downslope movement occurs via slides, slumps, flows, or slow creep.

KEY FACTS

Indicators of downslope movement include curved trees, rock scars, scarps, and broken or hummocky land surfaces.

+ fact: Occurs when slopes are destabilized by natural factors and/or human activities

+ fact: Landslides are a major hazard.

+ fact: Slides move as slowly as a few millimeters a year or up to 200 mph (322 kph).

Soil or rock creep is a slow drift of material down a slope. Its indicators can be subtle. Vegetation may continue to grow upward on a creeping slope, resulting in trees with curved trunks. Material in a slide mainly retains its structure and catastrophically fails all at once, usually on underground surfaces lubricated with water. Slumps occur when groundwater disrupts the structure of a body sliding down a curved surface. Flows, including slow creep, occur when material is saturated with water and behaves like a viscous liquid, as in mud and debris flows.

Dikes

Igneous terrain

Dikes are sheets of igneous rock that crosscut surrounding bedrock at a steep angle. Basalt dikes commonly occur in areas of extension in the Earth's crust.

KEY FACTS

This band or sheet of rock crosscuts surrounding rock at a steep angle.

+ fact: Dikes are made of igneous rock, but they can crosscut any rock type.

+ fact: Large numbers of dikes in a region are called dike swarms.

+ fact: Widespread; common in volcanic terrain, granitic regions, and mountain belts

Dikes are igneous intrusions that crosscut bedrock as steeply angled or vertical sheets. Different types of dikes form in different magmatic environments. Dikes form when magma rises in the Earth's crust until reaching neutral buoyancy—the same buoyancy as surrounding rock. Then, the magma spreads laterally, vertically, or horizontally depending on local stresses and weaknesses. Vertical orientations create dikes; lateral orientations create sills. Dikes with finer grain sizes along their margins than in their interior are called chilled margins. These form when magma cools more quickly in contact with surrounding rock.

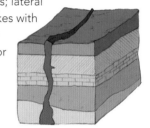

Sills

Igneous terrain

Sills are intrusive sheets of igneous rock oriented horizontally or parallel to the fabric of the surrounding bedrock. They are common in sedimentary basins intruded by magma.

KEY FACTS

This band or sheet of igneous rock parallels the structure of the surrounding bedrock.

+ fact: Sills are commonly thicker than dikes.

+ fact: Sills consist of igneous rock but can intrude any rock type.

+ fact: Widespread; common in volcanic terrain and regions of sedimentary bedrock

Sills are sheetlike intrusive bodies oriented parallel to the layering of surrounding rock, commonly horizontal in flat sedimentary basins. Sills are like dikes, forming from lenses of buoyant magma that spread laterally in the crust. In sills, stresses in the Earth's crust favor emplacement horizontally or parallel to existing structures. These generally form near the Earth's surface and can be thicker than dikes because pressures tend to be lower closer to the surface. The Palisades Sill in New York and New Jersey is a diabase sill that forms spectacular cliffs along the Hudson River.

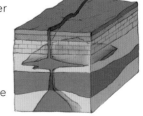

Inclusions
Igneous terrain

Some intrusive igneous rocks contain inclusions of other rocks. Inclusions can result from magma or from blocks of surrounding rock entrained in magma as it cooled.

KEY FACTS

Inclusions are angular blocks or rounded blobs of rock that are compositionally different from the surrounding rock.

+ fact: Inclusions are found in both intrusive and extrusive igneous rocks.

+ fact: Some inclusions are called xenoliths, from the Greek word *xeno*, meaning "foreign."

+ fact: Some xenoliths contain diamonds.

Large exposures of intrusive igneous rocks such as granites are heterogeneous. Granites often show variability in grain size, crystal orientation, and mineralogical makeup. Some granites (and other igneous bodies) contain inclusions of other rock types. Inclusions can be chunks of mantle rock brought to the surface by ascending magma. These inclusions, called xenoliths, can tell us about rocks deep in the Earth's interior. Inclusions of basalt and/or diorite are also common in granitic bodies. These are thought to be blobs of mafic magma that became entrained when the granite was not fully hardened.

Volcanic Bombs
Igneous terrain

Volcanic bombs are pieces of lava ejected during volcanic eruptions. These masses solidify in the air before landing and can have shapes like teardrops, ribbons, or balls.

KEY FACTS

Chunks of volcanic rock have distinctive shapes including teardrops, ribbons, and spindles.

+ fact: Volcanic bombs are 2.5 in (6.4 cm) to many feet in diameter.

+ fact: Can form out of different types of lava; basaltic bombs are common.

+ fact: Occur on flanks of volcanoes such as Mauna Kea in Hawaii

Volcanic bombs are ejected chunks of magma that form streamlined masses as they harden while flying through the air. Volcanic bombs are larger than about 2.5 inches (6.4 cm) in diameter. Smaller particles are called lapilli or ash. Bombs commonly form elongate spheroidal bodies, sometimes with twisted "tails" or spindle shapes. Some are not completely solidified when they hit the surface, and they flatten upon impact. Other shapes include teardrops and ribbons. Volcanic bombs are ejected along vents that liberate gases as large, bursting bubbles and indicate more explosive varieties of volcanoes.

A'a Lava

Igneous terrain

A'a (pronounced ah-ah) is a Hawaiian word for a type of lava flow characterized by sharp, angular rubble. It forms when the top parts of a flow harden and break into chunks.

KEY FACTS

These lava flows produce piles of sharp, rubbly volcanic rock.

+ fact: Develops where lava flows over steep slopes

+ fact: Flows typically advance faster than pahoehoe flows.

+ fact: A'a is common on the Hawaiian Islands.

Basaltic lava flows have three types: a'a, pahoehoe, and pillow lava. A'a creates piles of sharp, jagged rubble. A'a forms when the surface of a flow hardens into a solid and then breaks up into fragments because of the continued movement of lava beneath. The fragments tumble down the front of the flow and are entrained beneath it as molten lava in the interior continues to advance. This movement is similar to the tread of an advancing bulldozer. A'a flows create thick piles of sharp fragments and blocks that are difficult to walk on. Chunks of a'a are black or red volcanic rocks often used in landscaping.

Pahoehoe Lava

Igneous terrain

Pahoehoe (pronounced pa-hoy-hoy) is a Hawaiian word for a type of lava flow characterized by smooth surfaces with wrinkles and ropelike ridges.

KEY FACTS

Pahoehoe forms hardened lava flows with smooth surfaces and curved ridges.

+ fact: Pahoehoe can transition to a'a, but a'a does not turn into pahoehoe.

+ fact: Footpaths on volcanic rock often follow pahoehoe flows.

+ fact: Pahoehoe is common on the Hawaiian Islands.

Pahoehoe, unlike a'a, is a smooth and cohesive lava flow. It forms when the surface of molten lava cools into a thin, smooth skin. The skin is then pushed into folds by faster-moving lava below. Pahoehoe flows can eventually develop into lava tubes. When the flows harden, accumulations of the surface folds create wrinkles and ropelike masses. The surface of pahoehoe is relatively smooth, and footpaths on volcanic rock usually follow exposures of pahoehoe. Basaltic volcanoes often have both pahoehoe and a'a lava flows, depending on the viscosity and temperature of the lava as well as on the shape of the terrain.

Pillow Lava
Igneous terrain

Pillow lava is a type of lava flow that forms underwater. Outcrops look like stacks of elongate lobes and rounded blobs. Exposures of ancient pillow lavas can be highly deformed.

KEY FACTS

Formations consist of stacked lobes and protrusions, circular or elliptical in cross section.

+ **fact:** Outer layer is typically smooth and can be glassy.

+ **fact:** Pillow lavas in ancient rocks indicate the presence of water.

+ **fact:** Active pillow lavas are under Hawaiian waters; ancient pillow lavas are in uplifted oceanic rocks.

Pillow lava is formed when lava erupts underwater, along mid-ocean ridges and submarine flanks of seamounts and other volcanoes. Pillow lava is named for the globular shape of lava as it cools underwater. It develops because the surface of the lava rapidly cools and hardens to form a crust. As the molten lava continues to advance, it breaks through the crust in the shape of rounded blobs or tubes that are hard outside and molten inside. These blobs pinch off from the crust and roll down to the front of the flow. Pillow lavas in exposed ancient rocks are evidence of past oceanic environments.

Lava Tubes
Igneous terrain

Lava tubes are open tunnels made of shells of volcanic rock, typically basalt. They are conduits for molten lava. Some volcanoes have networks of lava tubes along their flanks.

KEY FACTS

Open tunnels are formed by hardened volcanic rock, commonly basalt.

+ **fact:** Lava tubes form in flows of low-viscosity lava.

+ **fact:** Lava tubes can create intricate networks of caves.

+ **fact:** Well-known lava tubes are found in New Mexico, Hawaii, California, Oregon, and Washington.

Lava tubes are large hollow tunnels through which lava travels away from its vent. Lava tubes tend to form in volcanoes that have relatively low eruption rates. High eruption rates can sustain open channels rather than closed tubes. Lava tubes are created when the surface of a moving lava flow cools and hardens into a crust. The crust insulates the lava flow below from the cool atmosphere, allowing it to remain molten. These streams of molten lava typically continue to drain through the tube until the magma supply is cut off. Usually the lava then drains away, leaving an empty tube in the volcanic landscape.

Cinder Cones
Igneous terrain

Cinder cones are small volcanoes that form out of hardened pieces of ejected lava that rain back down from the atmosphere after they are erupted.

KEY FACTS

These cone-shaped hills are made of chunks of hardened lava.

+ fact: Typically less than 1,000 ft (300 m) high

+ fact: They are built up over the course of multiple eruptions over many years.

+ fact: Examples are in Sunset Crater Volcano National Monument in Arizona and Lava Beds National Monument in California.

Cinder cones are a simple type of volcano in which most of the erupted magma is ejected as a fountain-like spray out of a central vent. The magma feeding cinder cones is charged with gasses. These gasses expand as they rise, causing the lava to spray out of the vent. The airborne lava hardens into volcanic bombs and smaller chunks of volcanic rock called cinders, lapilli, and ash. These particles, collectively known as tephra, fall back to Earth and accumulate around the vent, building up a cone-shaped landform. Most cinder cones have a rounded crater at their summit.

Caldera
Igneous terrain

A caldera is a large depression of rock formed in a volcano. Calderas form when volcanic eruptions empty underground magma chambers, causing the rock above to collapse.

KEY FACTS

These large volcanic depressions are often several miles wide.

+ fact: Common features of explosive volcanoes

+ fact: They're sometimes confused with craters, which are smaller, bowl-like formations at the summit of volcanoes.

+ fact: Much of Yellowstone National Park sits within an enormous caldera.

A caldera, named from the Spanish word for cauldron, is a large depression created by volcanic activity. It forms when a magma chamber beneath a volcano is emptied by eruption, causing the rocks above to collapse into a large depression. In very explosive volcanoes, calderas can form by a single eruption. In less explosive volcanoes, magma chambers may take numerous small eruptions to empty; thus, caldera formation is a slow, progressive process. Continued volcanic activity after formation of a caldera may cause the magma chamber to recharge, and then the center of the caldera will rise into a feature called a resurgent dome.

Volcanic Neck or Plug
Igneous terrain

A volcanic neck is the resistant core of a volcano that is exposed by erosion. Volcanic necks, also called plugs, are typically tall, somewhat cylindrical-shaped landforms.

KEY FACTS

A neck is an eroded cylindrical landform made of volcanic rock.

+ fact: Often found with eroded dikes; some surround the neck radially.

+ fact: Volcanic necks sometimes contain pieces of rock pulled from deep in the Earth during eruption.

+ fact: Famous necks include Devils Tower National Monument in Wyoming and Shiprock in New Mexico.

Most volcanoes are fed by a central vent that pipes magma from deep within the Earth. Cone-like outer structures of many volcanoes are created by buildup of lava flows and/or accumulation of ejected ash, cinder, and volcanic bombs. After volcanic activity ends, these are worn away by erosion. Outer flanks of volcanoes are sometimes easily eroded, especially if they consist of unconsolidated cinder and ash. Often the most resistant part of a volcano is the vent that becomes filled with hard volcanic rock at the end of eruption. Remnants of these vents projecting above the landscape are called volcanic necks or plugs.

Columnar Jointing
Igneous terrain

Symmetrical columns of volcanic rock form by a process called columnar jointing. As lava shrinks during cooling, fractures divide the rock into roughly hexagonal columns.

KEY FACTS

Polygonal columns consist of volcanic rock.

+ fact: Form in lava flows, ash flows, and shallow intrusions

+ fact: Columns can be hundreds of feet high.

+ fact: Famous examples are in Devils Postpile National Monument in California and Devils Tower National Monument in Wyoming.

Columnar jointing is a distinctive outcrop feature of some volcanic rocks, especially basalt flows in the West. As thick flows of lava cool, they shrink, causing the hardening lava to crack. In a homogeneous pool of lava without outside stresses, vertical cracks will develop with intersections of approximately 120 degrees, which is the most efficient orientation to accommodate shrinking. This orientation results in vertical columns with hexagonal cross sections. In the real world, however, columnar joints can have variable angles, and the resultant basalt columns take on various polygonal cross sections.

Bedding
Sedimentary terrain

Sedimentary beds are layers of rock with distinguishable tops and bases. Bed boundaries form by changes during deposition of sediment, before the sediment hardens into a rock formation.

KEY FACTS

Layers can be seen in sedimentary rock.

+ fact: Range from the millimeter to tens of meters scale

+ fact: Deposited horizontally with older layers below younger ones

+ fact: Dramatic horizontal patterns in walls of the Grand Canyon are created by bedding planes.

Sedimentary rocks are easily recognized by their layered appearance. The layers, called beds or strata, are made of accumulated sediments compacted and/or cemented into stone. Sediments are transported to their resting place by a mechanism such as wind or water, or they are deposited in place by living organisms such as coral, or by chemical precipitation out of a solution. Changes in this process are caused by changes in the environment or in the sediment source. The changes create natural breaks in the pile of sediment, which later define bed boundaries or bedding planes in the sedimentary rock.

Unconformity
Sedimentary terrain

An unconformity is a break in the sequence of sedimentary beds along which erosion occurred. There are three types: angular unconformities, disconformities, and nonconformities.

KEY FACTS

This feature is caused by disruption in the layering of sedimentary rocks.

+ fact: An unconformity represents a gap in the rock record.

+ fact: Unconformities can indicate deformation, erosion, or changes in sea level.

+ fact: John Wesley Powell described the Great Unconformity in the Grand Canyon in 1869.

In an angular unconformity, sedimentary layers below the boundary are at a different angle from layers above. These occur when the lower rocks are folded and/or faulted so that the layers are no longer horizontal. Later, new sediment is deposited, creating a new horizontal sequence of layers. Disconformities occur when there is a break in sediment deposition and erosion of a surface before the upper layers are deposited. These boundaries often show irregular surfaces between different rock types. A nonconformity is an erosive surface between a nonlayered rock such as granite and a layered sedimentary sequence.

Cross Bedding

Sedimentary terrain

"Cross bedding" is a term for layering that forms within a sedimentary bed and is oriented at an angle to the original horizontal bedding planes of the rock.

KEY FACTS

Angled layers occur within a sedimentary bed.

+ fact: Found in sedimentary rocks, primarily sandstones

+ fact: Some cross beds are a mark of ancient sand dunes.

+ fact: Geologists use cross beds to interpret previous environments and even the direction of water and wind currents.

Cross beds are sublayers in sedimentary rock formations that are oriented at an angle to the bedding planes. Cross beds form from particles moved by water or wind as they are being deposited. The grains accumulate into piles; in windblown sand, the piles can be large dunes. When they grow too high, the grains avalanche down the front and come to rest. Continued movement of particles rebuilds the piles. Repeated cycles create inclined layers or laminations that are preserved in rock. Cross bedding forms dramatic textures in sandstone outcrops in the Southwest, including famous exposures in Zion National Park in Utah.

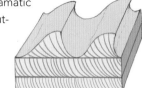

Ripple Marks

Sedimentary terrain

Ripples are low, elongate ridges that form when water or wind move sand particles. Ancient ripples are preserved in sandstones as features known as ripple marks or cross lamination.

KEY FACTS

Planar surfaces of sedimentary rocks contain sets of low, elongate ridges.

+ fact: Found in sedimentary rocks, primarily sandstones

+ fact: Look for ripple marks in outcrops of sedimentary rock in the Southwest.

+ fact: Geologists use ripple marks to interpret prior environments.

Hiking on sedimentary rocks, you may find exposed bedding planes in sandstones that are marked by low, elongate ridges—semiparallel sets of wrinkles protruding from the surface. These are ripple marks, which are essentially the preserved trace of sand ripples that formed during the deposition of the sand particles that make up the rock. Ripples are small, generally about an inch high, and form by the way sand particles aggregate and move as they are transported by currents or waves. Ripples can form from wind in desert environments or by the flow and/ or waves in rivers, lakes, and seas.

Mudcracks
Sedimentary terrain

Mudcracks are the cracks separating polygonal forms in clay and mud beds. Cracks form when wet, fine-grained sediments shrink as they dry.

KEY FACTS

Networks of cracks break dried mud or clay into plates.

+ **fact:** Ancient mudcracks are preserved in claystones and mudstones.

+ **fact:** Dinosaur tracks are commonly found in the same rock formations as mudcracks.

+ **fact:** Look for mudcracks in outcrops in Colorado, New Mexico, Utah, and Arizona.

When a bed of wet, fine-grained sediment such as a layer of clay or mud dries, the surface shrinks and contracts, resulting in a network of cracks. The cracks form plates of sediment with curled edges. The plates can form rough polygonal shapes, commonly hexagons. Environments creating mudcracks include marshes and lake beds that drain and become exposed to the atmosphere. Mudcracks can be preserved in sedimentary rocks if they are filled with sediment and then buried. Geologists use mudcracks in ancient rocks to understand "which way was up" in sedimentary rocks that have been disturbed by folding or faulting.

Karst
Sedimentary terrain

Karst is a distinctive type of terrain formed by the chemical dissolution of rocks. It develops in soluble rocks, primarily carbonates like limestone and dolomite.

KEY FACTS

Karst landforms include caves, springs, towers, rocks in fluted shapes, stalactites, and stalagmites.

+ **fact:** About 10 percent of the Earth's surface is karst.

+ **fact:** Much of the world's life depends on water from karst areas.

+ **fact:** Kentucky's Mammoth Cave is a famous karst, and karst is actively forming in Florida.

Carbonate rocks, especially limestone, dissolve in contact with groundwater or seawater that is undersaturated with calcium carbonate. Rock formations made of evaporite deposits such as gypsum and halite will also dissolve in contact with water. Landforms that develop by dissolution of bedrock are called karst. Karst surfaces are marked by springs, sinkholes, and caves. Dissolution along joints and faults can result in karst towers with fluted surfaces. Formations range from tiny chemical precipitates to entire landscapes. Water in underground conduits and caves, called karst aquifers, are important water resources.

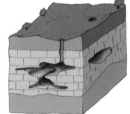

Stalactites & Stalagmites

Sedimentary terrain

Stalactites and stalagmites are features of limestone caves. They are a type of cave deposit called dripstones that are formed by drops of water that precipitate calcium carbonate.

KEY FACTS

Stalactites point down from roofs of caves; stalagmites point up from floors.

+ fact: Stalactites and stalagmites are karst formations.

+ fact: Most are made of calcium carbonate.

+ fact: Spectacular examples include Natural Bridges Caverns in Texas, Carlsbad Caverns National Park in New Mexico, and California Caverns.

Stalactites and stalagmites are cave deposits called speleothems. The word "speleothem" is from the Greek *spelaion*, meaning "cave," and *thema*, meaning "deposit." Stalactites are cone or straw shaped, and protrude downward from the roof. These form from water trickling down from cracks in the cave roof. Stalagmites are cone shaped and protrude upward from the floor; they form from drops of water falling from stalactites. One way to remember the difference is to associate the letter *g* in stalagmite with "ground"—stalagmites point up from the ground. Stalactites and stalagmites can grow together forming columns.

Sinkholes

Sedimentary terrain

Sinkholes are areas where part of the bedrock below the land surface has dissolved and the land surface has collapsed into the space below, sometimes catastrophically.

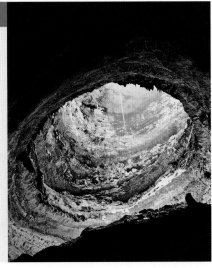

KEY FACTS

Bowl-shaped depressions in the land surface come in various sizes, some as large as hundreds of feet in diameter.

+ fact: Features of karst landscape

+ fact: Can develop rapidly, especially after a heavy rain

+ fact: Common in Florida, Texas, Alabama, Missouri, Kentucky, Tennessee, and Pennsylvania

Sinkholes are landforms that result from a collapse of the land surface. They occur in regions underlain by bedrock that is easily dissolved in groundwater. Examples are carbonate rocks and evaporite deposits such as salt and gypsum. The movement of groundwater causes the rock to dissolve over time, slowly enough that the ground above may stay intact. Eventually, however, dissolution leaves the bedrock unable to support the surface, and the land collapses into the hole. This can happen on various scales and can be catastrophic if buildings or roads are built on the affected surface.

Desert Pavement
Desert environments

Desert pavement is a hard ground surface made of tightly packed pebbles and cobbles overlying a layer of fine sediment. Desert pavements are found in arid and semiarid environments.

KEY FACTS

A thin layer of packed rock fragments and clasts covers the ground.

+ fact: Many clasts are coated with a dark-colored patina called desert varnish.

+ fact: Exposures range from small patches to areas as large as hundreds of square miles.

+ fact: Newspaper Rock in Utah is a famous example.

Desert pavement is a ground surface common in some windblown deserts. It is formed by a thin layer of tightly packed clasts, pebble to cobble size, and generally of various shapes, sizes, and types. Ventifacts—wind-shaped particles—are common among these clasts. Desert pavements are found worldwide, even in Antarctica. The manner of their formation is controversial. Various ideas proposed to explain the features include deposition of the clasts during catastrophic rain events, uplift of the clasts due to wetting and drying or freezing and thawing processes, or creation of the pavement in place by wind erosion.

Ventifacts
Desert environments

Ventifacts are rocks shaped, polished, and etched by windblown sand. They come in all shapes and sizes, from small, polished, aerodynamic pebbles to large mushroom-shaped boulders.

KEY FACTS

These rocks in desert environments have smoothed, polished surfaces and/or pitted or fluted surfaces created by wind erosion.

+ fact: Look for ventifacts in exposures of desert pavement.

+ fact: The shape can indicate wind direction.

+ fact: Ventifacts have been identified on Mars and in Antarctica.

Ventifacts are formed in arid environments by the abrasive power of sand and smaller particles carried by the wind. They have various forms, depending on original characteristics of the rocks and energy and direction of the wind. Some ventifacts have smooth, wind-blasted faces. Others contain surfaces with networks of grooves and pits. Some ventifacts contain several wind-smoothed faces, suggesting that either the rock moved during wind abrasion or the wind direction changed during the ventifact's formation. Ancient ventifacts preserved in beds of sedimentary rock are indications of a stony desert in the past.

Alluvial Fans
Desert environments

Alluvial fans are cones of sediment similar to deltas, which form where there is an abrupt change in topography from steep uplands to a flat plain.

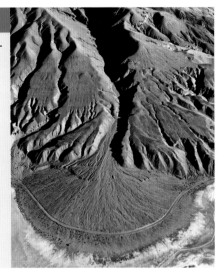

KEY FACTS

These are fan- or cone-shaped accumulations of gravel, sand, and smaller materials.

+ fact: Made of unconsolidated sediment known as alluvium

+ fact: Identified on Mars

+ fact: Spectacular alluvial fans are found in Death Valley National Park.

Alluvial fans are characteristic of arid and semi-arid environments in which mountains or hills abut a wide basin or plain. The high ground is drained along valley systems to the basin, where there is an abrupt change in topography. Here, flows of water and sediment rapidly lose energy and the sediment load is deposited. Repeated deposition builds up cones of sediment at the mouth of drainages. Alluvial fans are similar to deltas but occur above water in arid environments. Alluvial fans from neighboring valleys can converge when built up with enough sediment.

Sand Dunes
Desert environments

Sand dunes are distinctive features of sandy deserts, formed by wind-powered transport and deposition of sand grains. Dune shapes depend on wind direction and sand abundance.

KEY FACTS

Dunes are accumulations of sand grains in piles with straight or curved crests.

+ fact: Some cross beds in sandstones are the mark of ancient sand dunes.

+ fact: Colorado's Great Sand Dunes National Park contains the tallest sand dunes in North America.

+ fact: Sand dunes are features of expansive sandy deserts called ergs.

Arid deserts are characterized by landforms built from wind energy. As wind blows across sand, sand grains are progressively skipped forward by a process known as saltation. Saltating sand grains accumulate until the crest of the pile is unstable and the grains avalanche down the lee side, forming the shape of a dune. Straight-crested dunes called "transverse dunes" form perpendicular to the prevailing wind direction. When there is limited sand supply, individual half-moon shaped dunes called "barchan dunes" form. When there are two or more prominent wind directions, dunes with different orientations and shapes form.

Meandering Rivers
River environments

Rivers that flow in sinuous channels with many bends and loops are called meandering rivers. These rivers tend to occur in low areas with a relatively consistent elevation gradient.

KEY FACTS

Loopy, sinuous channels are signs of meandering rivers.

+ fact: Meandering rivers have wide floodplains.

+ fact: Channels of meandering rivers are deepest on the outside of each bend.

+ fact: The Mississippi River in the southeastern United States is a classic meandering river.

Two primary river patterns account for side-to-side movement of a river course: meandering and braided rivers. Meandering rivers form in areas with a fairly consistent and low topographic gradient. The current occupies a single channel and swings back and forth across bends and loops on each side of its overall flow direction. The river scours the outside of the bends and deposits material on the inside, deepening the loops over time. Eventually, the current can break through adjacent loops of the deepest meanders, causing the channel to shortcut and change course. The cutoff meander becomes an oxbow lake.

Braided Rivers
River environments

Rivers in a network of interconnected channels are called braided rivers or streams. These tend to form with a sudden change in the elevation gradient and/or the sediment load.

KEY FACTS

Rivers with multiple channels converge and diverge in a braided pattern.

+ fact: Found in areas of relatively steep topography

+ fact: Common in regions with glacial outwash

+ fact: Beautiful examples of braided rivers are found in Alaska.

When the elevation gradient of a river changes suddenly, as in a transition from a mountainous region to a broad plain, the river pattern that develops is called braided. In this pattern, the river occupies numerous channels that branch off from one another, converge, and then diverge again irregularly. Braided streams are also characterized by a high sediment load. They leave behind stones and gravels concentrated in high strips that divide channels, and silts and clays deposited in hollows between them. Braided streams are common in outwash plains and alluvial fans at the foot of mountains.

Deltas
River environments

Deltas are broad regions of deposited sediment at river mouths. They form when rivers suddenly lose their gradient and deposit their sediment load at the coast of a sea or lake.

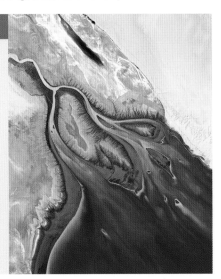

KEY FACTS

Broad accumulations of sediment occur where rivers empty into oceans or lakes.

+ **fact:** Sediment becomes finer grained along the far edges of the delta.

+ **fact:** River deltas create important agricultural lands.

+ **fact:** Some ancient delta sediments are petroleum reservoirs.

Deltas are broad deposits of fine sediment, mostly silt, at the mouth of a river where it meets a lake or sea. The river loses its gradient and rapidly slows, and this decrease in energy causes the river to drop its sediment load. Deltas are similar to alluvial fans, but occur along coastlines, and the deposited materials are finer because they are farther from the source. Deltas are named after the Nile River Delta, whose shape resembles the triangular capital letter delta in the Greek alphabet. Deltas have different shapes. The Mississippi Delta is called a "bird's foot delta" because of its fingerlike projections of sediment.

Potholes
River environments

Potholes are smooth, circular depressions in surfaces of exposed bedrock. They form by the abrasive process of sand and gravel whirled around a depression by flowing water.

KEY FACTS

Circular depressions appear in bedrock, some with polished surfaces.

+ **fact:** Some are called giants' cauldrons or kettles.

+ **fact:** Many in the U.S. and Canada formed by glacial meltwaters during the last ice age.

+ **fact:** Some of the world's deepest are found in basalt at Interstate State Park in Minnesota.

Potholes are rounded depressions in rock, with shapes and sizes ranging from small, cylindrical holes to gigantic, cauldron-shaped bowls. Most potholes are formed by river abrasion: the scouring action of stream-carried sediment against exposed bedrock. Sand and gravel, trapped in small fissures and depressions, are whirled by the current, scouring out potholes. This process is greatest during floods, when energy is high enough to transport heavy, coarse sediment with more erosive power. Potholes can indicate previous glacial activity because meltwater from a receding glacier is a powerful scouring agent.

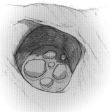

Barrier Islands
Coastal environments

Barrier islands are sandbars or sandspits not connected to a coastline. They are separated from the coast by a sound or a lagoon, and they are high enough for dunes to form.

KEY FACTS

Sandy islands are separated from the continent by a sound, bay, or lagoon.

+ fact: Barrier islands move and change shape with ocean currents and storm events.

+ fact: Vegetation helps stabilize barrier islands.

+ fact: The Outer Banks of North Carolina are barrier islands.

Barrier islands are linear deposits of sand off-shore, separated from the mainland by a sound, bay, or lagoon. They typically have a sandy beach on the seaward side, then an area of dunes, and then a marsh leading to a shallow lagoon on the inland side. Barrier islands are somewhat stabilized by dune vegetation, but they can change dramatically from season to season and storm to storm. Barrier islands "migrate" in two ways: Alongshore ocean currents move sediment from one tip of the island to the other, or storms remove beach sand from the ocean side and deposit it on the marsh side as overwash.

Shoals/Sandspits/Sandbars
Coastal environments

Shoals, sandspits, and sandbars are coastal landforms that develop where streams and alongshore currents create linear deposits of sand and mud.

KEY FACTS

These landforms consist of linear deposits of sand and mud.

+ fact: Shoals create local areas of shallow water and are navigation hazards for boats.

+ fact: A nearly continuous chain of sandspits and barrier islands extends from Long Island, New York, to Florida.

+ fact: Cape Cod, Massachusetts, is a sandspit.

Shoals are elongate deposits of sand and mud that extend into water in coastal environments. Shoals that are attached to a headland but that extend into the water past a headland bend are called spits or sandspits. These form via alongshore drift. When waves meet the beach at an oblique angle but recede perpendicularly, the change in current direction transports sand down the shore. When the angle of a headland changes, alongshore drift transports sand past the bend, extending the sand deposit into the sea until the current loses energy. Sandbars are isolated shoals detached from headlands and surrounded by water.

Marine Terraces or Platforms
Coastal environments

Marine terraces or platforms are bench-like coastal landforms that usually indicate a lowering of sea level or a rise of the land. These terraces are cut by wave erosion before they are exposed.

KEY FACTS

Broad "benches" are found along coasts.

+ fact: Some terraces along the California coast are uplifted by movement along the San Andreas Fault.

+ fact: Dramatic stacked terraces are exposed in the Palos Verdes Hills in Los Angeles County, California.

+ fact: Common on the Pacific coast of North America

Marine terraces are horizontal platforms separated by steep cliff faces on exposed coastlines. Ocean waves are one of nature's most powerful forces. Wave energy becomes concentrated on exposed headlands, cutting into the rock like a horizontal saw, causing overlying rock to collapse and retreat landward. This process cuts a nearly horizontal platform adjacent to a steep scarp. If the land rises with respect to sea level, this platform is exposed above the water. Periodic uplift (or sea level decline) will expose marine terraces in a steplike pattern, as along the coast of California.

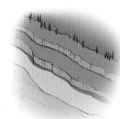

Sea Stacks
Coastal environments

A sea stack is a remnant of a coastal headland, carved by the powerful erosive force of ocean waves. Sea stacks generally form where resistant bedrock is exposed along a high-energy coast.

KEY FACTS

Isolated rock towers rise out of the ocean along a coastline.

+ fact: Can form from many different types of rock, including limestone, sandstone, and volcanic tuff

+ fact: Sea stacks are important bird habitats.

+ fact: Common along the West Coast; famous along Harris Beach, Oregon

Sea stacks are isolated towers or monoliths that rise steeply out of the sea. They are generally clustered along steep coastlines marked by resistant bedrock. Sea stacks often form when waves attack jointed or faulted rocks, preferentially eroding along these weak planes, leaving pillars or towers surrounded by seawater. Eventually, wave attack will continue to erode the stacks, and they will topple into the ocean and become pulverized to smaller and smaller particles. Sea stacks can also form in limestone bedrock by the combined effects of wave attack and limestone dissolution.

Cirques
Glaciated terrain

Cirques are bowl-shaped scoops on mountainsides, which are formed by glaciers. Glaciers carve cirques backward into the mountainside by breaking down and removing rock beneath the ice.

KEY FACTS

Bowl-shaped depressions appear in the sides of mountains.

+ fact: Cirques are named from the French word for "amphitheater."

+ fact: Those found in nonglaciated regions are important evidence for previous ice ages.

+ fact: Cirques filled with lakes are called tarns.

Mountain glaciers form along the flanks of high peaks and are relatively small compared with continental glaciers or ice sheets. The bowl-shaped depressions at the head of a mountain glacier are called cirques. Cirques are easily recognizable by their distinctive rounded, hollowed-out form, like the path of a gigantic ice cream scoop on the side of a mountain. Cirques form as the glacial ice and meltwater below the ice gouge out the rock behind the glacier. When two cirques on different sides of a mountain erode deeply enough to intersect, they form a sharp ridge between them called an arête.

Glacial Erratics
Glaciated terrain

Glacial erratics are rocks that have been deposited by a glacier, usually after being transported great distances from their bedrock source.

KEY FACTS

Erratics are out of place in their current environment due to transport by a glacier.

+ fact: Can range in size; boulders are well known.

+ fact: The largest known erratic in the U.S. and Canada is New Hampshire's Madison Boulder, weighing almost 6,000 tons.

+ fact: Some boulders are perched precariously on top of smaller stones.

Glacial erratics are clasts of rock, including large boulders, transported and deposited by glaciers. They are called erratics because they are incongruous with the surrounding bedrock, having been moved sometimes great distances by a glacier. Scientists use erratics to map the extent of the continental ice sheet that covered much of Canada and the U.S. during the last ice age, which reached its maximum extent approximately 21,000 years ago. Glacial erratics originate when glaciers pluck rocks as they scour over topographic highlands. When the glacier melts, it drops its sediment load into an unconsolidated pile.

Kettle Ponds
Glaciated terrain

Kettle holes or ponds are depressions formed by melting chunks of glacial ice that are no longer connected to the glacier. After the glacier has receded, these holes often fill with water.

KEY FACTS

Kettle ponds are depressions formed in glacial outwash plains that have filled in with water.

+ fact: The ponds are recharged by groundwater and/or rainwater.

+ fact: Thoreau's Walden Pond in Massachusetts is a famous kettle pond.

+ fact: The ponds of Cape Cod, Massachusetts, are kettle ponds.

Kettle ponds are important markers of glaciated terrain. These variably sized ponds are found in regions that were previously glaciated. They form when chunks of ice calve off the front of a receding glacier and are buried by glacial sediment called till. When the chunks of ice melt away, the till collapses into the space, creating a hole. Over time, these holes fill with water and become ponds. Kettle ponds are common landforms across the northern region of North America because of the continental ice sheet that reached its maximum extent approximately 21,000 years ago during the last glacial period.

U-shaped Valleys
Glaciated terrain

Valleys carved by glaciers develop a characteristic U-shaped cross section. U-shaped valleys are key indicators of glacial erosion, whereas V-shaped valleys are carved by rivers.

KEY FACTS

Broad troughs in the landscape have U-shaped cross sections.

+ fact: U-shaped valleys filled with seawater are called fjords.

+ fact: U-shaped valleys are important indicators of past glacial periods.

+ fact: Fantastic examples can be seen in Glacier National Park in Montana.

Distinctive markers of glaciated terrain are steep-walled valleys with broad floors that resemble the shape of a U in cross section. U-shaped valleys are formed by the downhill flow of glaciers. The ice carves out steep walls and a relatively flat base. These valleys are also often straighter in course than river-cut valleys. Some U-shaped valleys can be notched by V-shaped river channels from the flow of glacial meltwater. The floors of tributary glaciers can be significantly higher than the main valley floor, leading to waterfalls and hanging valleys after the retreat of the glaciers.

A mountain lake reflects a rainbow at Mount Robson Provincial Park in British Columbia.

4 Weather
Nature You Can't Ignore

Most of the time, people go about their daily lives without thinking much about the weather. Before they leave home for the day, a brief radio forecast or a glance at a website tells them all they need to know: How should I dress to go out for the day? Many people, however, are fascinated by the ever-changing sky and weather, and they want to keep on learning about it.

■ A Quick Guide to Weather Science

Our day-to-day weather is the result of the sun's unequal heating of Earth, with the tropics—the belt circling Earth 1,600 miles (2,575 km) north and south of the Equator—receiving much more solar energy than the regions around the North and South Poles. Storms and ocean currents along with calmer winds move warm and cold air and water, redistributing heat, balancing Earth's heat budget. These movements of warm and cold air create our day-to-day weather.

While Earth sees relatively great temperature extremes, the planet's temperatures allow some forms of life over almost all of the planet. Earth's monthly average temperatures range from as cold as -80°F (-62°C) on Antarctica's polar plateau to 100°F (38°C) on tropical deserts. The world's single-day lowest temperature ever officially recorded was -129°F (-89.2°C) at the Russian Vostok Station in Antarctica on July 21, 1983, during the coldest part of the Southern Hemisphere winter. Earth's single hottest daily temperature on record is 134°F (56.7°C) at Furnace Creek, California, on July 10, 1913, at what is now the headquarters of Death Valley National Park. By the way, on January 8, 1913, Furnace Creek set its cold temperature record, 15°F (-10°C). Furnace Creek has tied this temperature since then, but has never been colder. This tells us something about desert climates: Temperatures drop rapidly after sunset because such dry climates have little water vapor in the air to absorb heat radiating away from the ground and radiate it back down.

■ Atmosphere & Water

If Earth didn't have an atmosphere and oceans, the planet's daily temperatures would probably be much like those on the moon, which range from daytime highs of approximately 200°F (93°C) and overnight lows of -280°F (-173°C). Copious amounts of water are important for Earth's weather, and not just because water fills the oceans. Water is the only natural substance that exists in all three states of matter at temperatures and air pressures found near Earth's surface. Water is the raw material for clouds, and all kinds of precipitation. Without water, Earth's weather would be like that on Mars: winds blowing dust around and wide temperature changes.

The air's invisible water vapor contains energy to power violent weather such as this summer storm building over Florida's Everglades.

■ Water Supplies Storm Energy

In addition to supplying raw material for clouds and precipitation, water supplies some of the energy that powers the weather. Water is the only natural substance that exists in all three states of matter—solid, liquid, and gas—at temperatures and pressures found near Earth's surface. When it changes among its phases, water either takes in energy as heat from the surrounding air, which cools the air, or adds heat to the surrounding air, warming it.

Heat is taken from the surroundings when water evaporates into vapor, when ice melts into water, or when ice sublimates directly to water vapor without first melting. This is why evaporation of perspiration from our skin cools us in hot weather. If the air is too humid for perspiration to evaporate easily, we become sticky and hot.

Heat taken from the surroundings is called latent heat. When water vapor condenses into liquid water or deposits directly to ice, it returns its latent heat to the surroundings as does water when it freezes into ice. Condensation, deposition, and freezing all add energy to the surrounding air,

BECOMING INVOLVED WITH WEATHER

The books, websites, and magazines in "Further Resources" will help you to learn more about weather and climate. If you want to look into becoming a meteorologist, or otherwise become directly involved with weather, you have various possibilities.

+ **If you like to watch storms** and want to learn more about them and help the U.S. National Weather Service, you should consider becoming one of more than 290,000 trained volunteers who send reports of severe weather to a local NWS office. To become involved, contact the Warning Coordination Meteorologist at your local NWS office and arrange to take the training course these offices offer. To do this, go to *www.stormready.noaa.gov/contact.htm* and follow the directions. You will learn the basics of safety.

You can also check with your local weather office about becoming a volunteer cooperative weather observer. Since 1890, volunteers in this program have been sending data from NWS instruments on their property—mostly temperature and precipitation—to the NWS. These data are used to supplement climate reports from places away from NWS and other official weather stations.

+ **The Community Collaborative Rain, Hail and Snow Network** is looking for volunteer observers to add to its thousands of volunteers. It is now the largest provider of daily precipitation observations in the United States, and is moving into Canada. Information is available at *www.cocorahs.org*. Becoming an observer is a good way to learn about weather.

+ **Environment Canada's Meteorological Service** has a storm spotter program called the Canadian Weather Amateur Radio Network, which began as a program for amateur radio operators but now is open to anyone who wants to be a storm spotter. Contact your nearest Environment Canada weather office for information.

+ **The U.S. National Oceanic and Atmospheric Administration (NOAA)**—the parent agency of the National Weather Service—offers a free smartphone app that enables anyone to report precipitation anonymously. Anyone can use the app to report storms to researchers at NOAA's Severe Storms Laboratory and the University of Oklahoma. NOAA and the university are using the reports to build a database. The reports are not used for warnings.

A thundercloud builds before a storm at Fort Whyte Centre in Winnipeg, Manitoba.

warming it. Energy added to rising air by condensation, deposition, and freezing adds heat to the air, making the heat rise faster and farther. Such latent heat releases the supply of energy for thunderstorms and hurricanes.

As air descends, as when falling rain drags it down, it warms, causing water to evaporate and ice to melt into water or sublimate directly into water vapor. These take energy from the falling air, making it colder and thus heavier. Normally, sinking air warms, but the heat taken away by water's phase changes more than offsets this warming. Under the right conditions, the air grows cool and heavy enough to smash into the ground as a damaging microburst.

▪ Stable & Unstable Atmospheres

Meteorologists use the words "stable" and "unstable" in a technical sense to describe what kind of

‖‖

*Sunshine is delicious,
rain is refreshing, wind braces us up,
snow is exhilarating; there is really
no such thing as bad weather,
only different kinds of good weather.*
—JOHN RUSKIN

‖‖

weather to expect at particular times and places. Nevertheless, the words carry similar connotations when used to describe people. A "stable" person is one who stays calm even under pressure. When the atmosphere is stable, you might have rain or snow, but it will fall over a wide area without thunderstorms. An "unstable" person reacts with road rage when another driver gets in the way. When the atmosphere is unstable, the weather can produce strong thunderstorms, maybe even tornadoes.

When the atmosphere is stable, air that is given an upward shove (as by wind blowing upward on a mountain) will stop rising and begin sinking when the shove stops. In an unstable atmosphere, the air continues rising after the shove ceases. The temperature profile of the atmosphere makes the difference. Rising air cools by 3.5°F for each 1,000 feet (6.4°C/1,000 m) of altitude gained. When water vapor begins evaporating, it releases latent heat, warming the air, which offsets the cooling to some degree. This means humid air can rise farther and faster than dry air.

Here's how this works. If the temperature is 60°F (15.5°C) at the surface, a bubble of air will cool to 54.5°F (12.5°C) when it rises 1,000 feet. If the temperature of the surrounding air here is 53°F (11.7°C), the bubble will be warmer. It will continue rising. The atmosphere here is unstable. When the bubble becomes colder than the surrounding air, it stops rising; the atmosphere at this height is stable. Forecasters need to know the stability to make predictions.

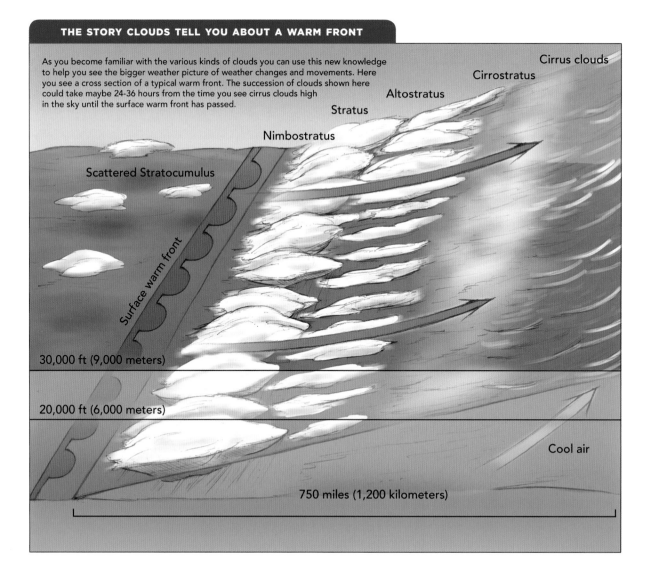

THE STORY CLOUDS TELL YOU ABOUT A WARM FRONT

As you become familiar with the various kinds of clouds you can use this new knowledge to help you see the bigger weather picture of weather changes and movements. Here you see a cross section of a typical warm front. The succession of clouds shown here could take maybe 24-36 hours from the time you see cirrus clouds high in the sky until the surface warm front has passed.

Cirrus clouds

Cirrostratus

Altostratus

Stratus

Nimbostratus

Scattered Stratocumulus

Surface warm front

30,000 ft (9,000 meters)

20,000 ft (6,000 meters)

Cool air

750 miles (1,200 kilometers)

WEATHER

||||||||||||||||||||||||||||||||||||||||||||||||||||||||||||||||||

What Causes the Seasons?

Earth's North Pole–South Pole axis tilts 23.5° in relation to its yearly path around the sun. This tilt causes the seasons, by changing the distribution of sunlight throughout the year.

KEY FACTS

Seasons are formed by the tilt of Earth on its axis, which causes unequal sunshine in various regions of Earth.

+ fact: On the equinoxes—March 21–22, and Sept. 22–23—the sun is directly above the Equator.

+ fact: On the Dec. 20–21 solstice—Northern Hemisphere winter—the sun is directly above the Tropic of Capricorn, latitude 23.5° S.

On September 22 or 23, the sun shines equally on all of Earth. Until December 21 or 22, days grow shorter in the Northern Hemisphere as the sun moves higher in the sky and days grow longer south of the Equator. On December 21 or 22, the winter solstice, the sun never rises north of latitude 66.5° N (the Arctic Circle); it never sets south of latitude 66.5° S (the Antarctic Circle). This process reverses until March 21 or 22, when the sun shines equally on all of Earth. Until June 20 or 21 the Southern Hemisphere receives less sun, turning colder as the Northern Hemisphere receives more sun and warms up.

Climate Zones

The amount of solar energy any part of Earth receives determines whether it will have a tropical, temperate, or polar climate. A location's elevation and nearness to an ocean also affect climate.

KEY FACTS

A region's climate depends on the latitude.

+ fact: Tropical climates have average temperatures above 64°F (18°C).

+ fact: Temperate climates have four seasons without extreme temperatures.

+ fact: Polar climates have no monthly averages higher than 50°F (10°C). On some days the sun never sets; on others it never rises.

Climate is the long-term average weather of an area, including temperatures and precipitation. The tropics, immediately north and south of the Equator, receive the most solar energy: The sun is almost directly overhead all year, and so on average the tropics are Earth's warmest region. The polar regions, centered on the North and South Poles, receive the least amount of solar energy because the sun is always low in the sky, and it doesn't come up at all for part of the year in some regions. On average, these are Earth's coldest regions. In between, the temperate zones have variable climates without widespread tropical or polar extremes.

Heat Balance

To a large degree, atmospheric winds and oceanic currents even out the temperature contrasts caused by day and night, and the seasonal variations caused by Earth's tilted axis.

KEY FACTS

Without oceans and atmosphere, Earth's temperatures would be like the moon's: -387°F (-233°C) at night, 253°F (123°C) during the day.

+ fact: Winds aloft and at Earth's surface move warm air toward the poles, cool air toward the Equator.

+ fact: Ocean-top currents carry warm water toward the poles; underwater currents haul cold water toward the Equator.

In the tropics, warm, humid air rises, forming towering Intertropical Convergence Zone thunderstorms. Air flowing in to replace the rising air creates tropical trade winds above the oceans. Air from the storms moves toward the south or north, but Earth's rotation upsets this simple, theoretical pattern. Earth's complex wind patterns include high-altitude jet streams and air sinking at some places, rising at others. These winds and the water vapor they carry are major drivers of storms, warm and cold outbreaks, and other weather events. We can wonder when we'll start seeing other expected changes.

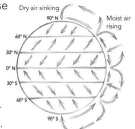

How Humans Affect the Weather

Earth's average overall temperature changes slowly and has warmed and cooled in the past. It is now warming, with greenhouse gases added to the air by humans playing an important role.

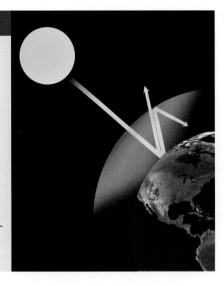

KEY FACTS

Without the natural greenhouse effect, Earth's average temperature would be 0°F (-19°C) instead of today's 60°F (15°C).

+ fact: Many prefer the phrase "climate change" because warming is only one phenomenon now occurring.

+ fact: Greenhouse gas emissions are falling in developed countries but are rising in developing nations.

Climate scientists are unwilling to say that climate change caused any particular event, such as a hurricane or heat wave, but evidence is strong that the Earth is warming faster than it would without human-related causes, especially by adding greenhouse gases, which trap heat that would otherwise escape. The number of local record high temperatures is increasing while record low temperatures are decreasing. Globally, glaciers are shrinking, and ice on lakes and rivers is breaking up earlier in the spring. Some of the biggest changes are happening in the Arctic where sea ice is shrinking. Scientists in the past had frequently predicted that many of the changes we are now experiencing would eventually occur.

Cloud Composition and Colors

Clouds are made of tiny water drops or ice crystals held up by rising air. Rain or snow falls when the drops or crystals grow large enough to overcome the resistance of rising air.

KEY FACTS

The faster air rises into clouds the bigger the drops or crystals can grow before they begin falling.

+ fact: Air rises as slowly as inches an hour in some clouds, faster than 100 mi (160 km) an hour in others.

+ fact: The lowest clouds touch the ground as fog; the highest can reach above 60,000 ft (18,000 m).

The water drops and ice crystals that make up clouds are large enough to scatter all wavelengths of sunlight, which makes the tops and sunny sides of clouds white. Bottoms of thin clouds are also white. Dark cloud bottoms don't necessarily mean rain. At least half of the sunlight hitting a cloud less than 3,000 feet (900 m) deep makes it through the cloud. This causes many cloud bottoms to be gray while the tops and sides facing the sun are white. Little sunlight passes through clouds more than 3,000 feet deep, which makes their bottoms dark. Shadows of other clouds also darken clouds. Sunrise and sunset turn clouds yellow, orange, or red.

Creating Clouds

Clouds form when rising air becomes cold enough for the air's humidity to begin turning into cloud drops or ice crystals.

KEY FACTS

The temperature of the surrounding air does not affect the cooling rate of rising air.

+ fact: When condensation begins, water vapor releases heat, which reduces rising air's cooling rate to less than 5.4°F per 1,000 ft (10°C per 1,000 m).

+ fact: Air is rising in all clouds at speeds ranging from inches an hour in many clouds to 100 mi (160 km) an hour in fierce thunderstorms.

Even when rising air is cool enough for water vapor to begin condensing into cloud drops, the vapor needs help. Condensation begins with water vapor molecules attaching to tiny particles in the air known as cloud condensation nuclei. These nuclei include dust and a variety of other natural substances as well as some kinds of pollution. Satellite images show long bright clouds created by nuclei from ship exhaust gases. These illustrate how added nuclei create many tiny cloud drops that reflect more light than nearby ordinary clouds with larger drops. This is one illustration of the complexities of clouds and why scientists continue working diligently to learn more about these important phenomena.

Cloud Names

Today, with a few additions and enhancements, we still use a system of cloud names devised in the early 1800s.

KEY FACTS

Latin words borrowed for cloud names are *cumulus*, meaning "heap"; *stratus*, "layer"; *cirrus*, "wispy"; and *nimbus*, "rain."

+ fact: Cloud names can be combined into a new word, as in cumulonimbus and cirrostratus, or into a "species" name, such as cumulus congestus.

+ fact: Meteorologists added "alto" for middle-level clouds.

Scientist Luke Howard organized clouds much as the Swedish biologist Carl Nilsson Linnaeus had devised the system of "genus" and "species" for plants and animals. A British chemist whose company manufactured medications, Howard practiced meteorology on his own. He first presented his cloud system during an 1802 lecture in London, which was widely publicized and accepted by those developing the science of meteorology. Howard's decision to incorporate Latin into his naming system, as Linnaeus had done with biology, helped make the names international. At the time, scientists were beginning to figure out how clouds form and their importance in understanding weather. Howard's focus on clouds as visible indications of atmospheric changes helped advance the science.

Describing Cloud Cover

Meteorologists use specific terms for the general public and different ones for pilots to describe how much of the sky clouds cover currently or are forecast to cover.

KEY FACTS

Clouds cover ⅛ to ¼ of sky. For pilots: few clouds. For public: mostly clear at night, mostly sunny in daytime.

+ fact: Clouds cover ⅜ to ½ of sky. For pilots: scattered clouds; for public: partly cloudy.

+ fact: Clouds cover ⅝ to ⅞ of sky. For pilots: broken clouds; for public: mostly cloudy or considerable cloudiness.

Although the terms "scattered" and "partly cloudy" mean the same thing, as shown in Key Facts, the National Weather Service (NWS) uses "scattered" for pilots and "partly cloudy" for the public, and you might hear both. Pilots and controllers also need to know how far above the ground the bottoms of the clouds are. Weather stations have automated ceilometers, which send pulses of infrared light straight up and measure the time needed to reflect back to the instrument to calculate cloud heights. Each minute, the instrument uses the last 30 minutes of data on when clouds were above it to calculate cloud cover. For pilots and the public, "overcast" refers to clouds covering more than seven-eighths of the sky. Gradually increasing cloudiness can show that rain or snow is coming.

Cumuliform Clouds

Puffy cumulus clouds form in an unstable atmosphere humid enough for its water to condense into cloud drops in bubbles of warm air rising as narrow streams called thermals.

Meteorologists say the atmosphere at a particular time and place is unstable when rising air, which always cools at a steady rate, stays warmer—thus lighter—than the surrounding air. It continues in this direction—at times to great heights—as long as the rising air remains warmer than the surrounding air. Relatively warm air at the surface and cold air aloft creates instability. As air is rising to form cumulus clouds, some air aloft is sinking around the clouds, keeping the air between clouds clear. These up-and-down air movements—termed "convection"—form individual cumulus clouds of various sizes instead of a single, widespread cloud covering a large area.

Stratiform Clouds

Stratiform clouds form in a stable atmosphere that doesn't support cumuliform thermals because air doesn't continue rising after the initial shove upward.

Meteorologists say the atmosphere at a particular time and place is stable when rising air, which always cools at a steady rate, grows colder—thus heavier—than the surrounding air. Once the air becomes heavier than the surrounding air, it will stop rising, which means cumulus clouds cannot form. Stratiform clouds form when all of the air in a particular area is pushed up, such as when wind carries warm air up and over heavier cold air at the surface. As stable air is pushed up, it spreads out to form a layer or layers of stratiform clouds that are thinner but wider than cumulus clouds.

Orographic Clouds

Rain or snow falling from orographic clouds are important sources of water for many regions, including much of the western United States.

KEY FACTS

When the atmosphere is stable, air continues rising as it reaches mountaintops and can create thunderstorms.

+ fact: When the atmosphere is unstable, air flowing over mountains forms turbulent rising and sinking that can stretch 100 mi (160 km) or more.

+ fact: Arid regions downstream of large mountains are called rain shadows.

When humid winds blow over hills or mountains, orographic clouds form. In some locations, these clouds bring most of the year's snow and rain. Air flowing up mountains cools enough for its humidity to condense into raindrops or to form snow. This orographic precipitation waters trees and other plants that cover mountains, even while nearby lower elevations remain arid. Spring and summer melting mountain snow feeds rivers, fills reservoirs, and becomes the major source of water for places far from the mountains. In the Sierra and Cascades ranges of the West, for example, flowing air loses moisture over the mountains, creating the dry Great Basin east of the mountains.

Cirrus Clouds

Meteorologists use specific terms for the general public and different ones for pilots to describe how much of the sky clouds cover currently or are forecast to cover.

KEY FACTS

Cirrus clouds block little sunlight but absorb infrared radiation from Earth, contributing to the greenhouse effect.

+ fact: Cirrus clouds are made mostly of ice crystals—they are located in -50°F (-46°C) air.

+ fact: At any one time, cirrus clouds cover approximately 25 percent of the Earth.

Cirrus clouds form when water vapor that has been pumped high into the air, often by a storm, turns directly into ice crystals—a process called deposition. Cirrus clouds can be a sign that rain or snow is on the way, maybe in a day or so. The only precipitation these clouds produce is light snow, which evaporates long before reaching the ground. Meteorologists identify these wisps of snow as "fall streaks," but they are commonly called "mares' tails." Some cirrus clouds are as thin as roughly 300 feet (100 m); others are as thick as 5,000 feet (1,500 m). Cirrus clouds are white—except around sunrise and sunset—and are generally transparent.

Nacreous Clouds

Nacreous clouds form in the stratosphere high above the Arctic and Antarctic, where they help to destroy stratospheric ozone. They are sometimes, but very rarely, seen at lower latitudes.

KEY FACTS

Destruction of stratospheric ozone is a concern because it blocks harmful ultraviolet radiation.

+ fact: Nacreous clouds are 49,000–82,000 ft (5,000–25,000 m) high, where temperatures are below -108°F (-78°C).

+ fact: They are named for iridescent nacre, a material that some mollusks make.

Nacreous clouds can be seen from the northern regions of Europe, Asia, and North America and over Antarctica, but they are very rarely seen elsewhere. Little was known about them until 1982, when NASA scientists using satellite data described them in detail, including the fact that they form regularly over Antarctica in the ozone layer and sometimes over the Arctic. The researchers named them "polar stratospheric clouds." In 1986 and 1987, scientists working in Antarctica found that these clouds are, in effect, natural laboratories that enable man-made substances including chlorofluorocarbons (such as Freon) to destroy stratospheric ozone. Temperatures over Antarctica are low enough for the clouds to form every winter, but they form only occasionally over the Arctic.

Cirrocumulus Clouds

Cirrocumulus clouds are white patches of high cloud without gray shadows and with lumps called cloudlets, often a sign that rain or snow could arrive in a day or sooner.

KEY FACTS

Cirrocumulus clouds are the least common high clouds.

+ fact: Cirrocumulus cloudlets appear to be the size of the tip of your little finger held at arm's length (or smaller).

+ fact: Cirrocumulus clouds are mostly ice crystals but often include liquid drops that have not frozen in well-below-freezing air.

Cirrocumulus clouds are usually found higher than 20,000 feet (6,100 m). If you watch one for a while, you might see it turn into a cirrostratus or cirrus cloud or dissipate because any particular cirrocumulus cloud tends to have a short life. The cloudlets show that convection—up-and-down air movement—is occurring, as it is in any cloud with "cumulus" as part of its name. These clouds are usually in relatively warm air that is moving over cooler, denser air at the surface, maybe 500 miles (800 km) or more ahead of a warm front, which is the surface boundary between warm and cold air.

Cirrostratus Clouds

Cirrostratus clouds are high clouds, above 20,000 feet (6,100 m). They are sometimes too thin to hide the sun or moon, even though they might be as much as 1,000 feet (305 m) thick.

KEY FACTS

Some very thin cirrostratus clouds appear only as a halo around the sun or moon.

+ fact: Forecasters use "hazy sunshine," to describe the milky look of the sky with cirrostratus clouds.

+ fact: Fibrous cirrostratus clouds with no halos are called "cirrostratus fibratus."

Cirrostratus clouds are made of ice crystals and are generally thin and uniform. They form when warm air moving over heavier cold air ahead of a warm front or rising air in the center of a surface low-pressure area lifts humid air into the upper atmosphere. If you see them replace cirrus clouds over a few hours and then grow thick enough to hide the sun or moon, you know snow or rain has a good chance of arriving within 24 hours—and maybe sooner. Just as other high clouds do, cirrostratus clouds frequently reflect red and yellow patterns that create spectacular sunrises and sunsets.

Contrails

Contrails are long, thin clouds that form behind high-flying airplanes as water vapor in the airplane's engine exhaust creates a narrow stream of air humid enough for a cloud to form.

KEY FACTS

Both piston engines and jet engines exhaust water vapor, which can create contrails.

+ fact: During World War II, contrails helped enemy pilots and anti-aircraft gunners spot high-flying bombers and fighters.

+ fact: Ice crystals in contrails fall approximately 6.56 ft (2 m) a second.

"Contrail" is short for "condensation trails," which are narrow clouds of condensed water vapor that become visible behind airplanes flying higher than 26,000 feet (8,000 m), where the temperature is colder than approximately -40°F (-40°C). Water vapor in the airplane's exhaust added to the air's own water vapor transforms into ice crystals around nuclei of material including particles in the exhaust. Contrails grow long when the air at their altitude is humid, a phenomenon that can indicate rain or snow is on the way. Minimal contrails behind a high-flying airplane means the air at its altitude is dry, so no contrails form or they quickly evaporate. When contrails persist for more than a few minutes, winds aloft often push them into wavy paths.

Altocumulus Clouds

These puffy clouds, white and gray with darker patches, generally form about 6,500 to 20,000 feet (2,000 to 6,100 m) above the ground and can cover large areas of the sky.

KEY FACTS

Altocumulus clouds on warm, humid mornings indicate thunderstorms that day.

+ fact: Individual clouds appear as wide as your thumb with your hand at arm's length.

+ fact: A sky with wavy, mixed altocumulus and cirrocumulus clouds and blue-sky gaps is called a "mackerel sky" because it resembles fish scales.

Altocumulus clouds form in various ways. They might develop from cumulus clouds that grow to an altitude where rising air slows and spreads out, or they might be transformations of other clouds including altostratus, stratocumulus, or nimbostratus. The clouds are usually made of water drops, although ice sometimes is found. They are usually less than 3,000 feet (1,000 m) thick and sometimes produce virga—rain that evaporates on the way down—but they rarely produce rain that reaches the ground. They may form distinct layers or parallel bands of clouds, called cloud streets, with air rising into the clouds and sinking between the bands.

Undulatus Asperatus Clouds

Since 2009, news reports in the United States and the United Kingdom have suggested that this is a new variety of cloud. Although such clouds are rare, there is no reason to think they are new.

KEY FACTS

Billow clouds are the undulatus variety you are most likely to see, often with a blue-sky background.

+ fact: Gavin Pretor-Pinney, founder of the Cloud Appreciation Society in the U.K., suggested the cloud's name.

+ fact: Undulatus clouds highlight the atmosphere's many natural wavy motions

An "undulatus" cloud, according to the American Meteorological Society glossary, is composed of long, parallel elements, merged or separate, that look like undulating ocean waves. The term "asperatus" comes from the Latin verb for "rough or difficult." Peggy LeMone, a scientist at the National Center for Atmospheric Research, recalls taking her first photo of such a cloud "on a wintry day in Columbia, Missouri, probably in the 1970s," and says she has seen others since then. Is it a new type? Maybe, instead, this cloud reminds us that if you look at enough clouds, you're bound to see some that are hard to classify. If you continue closely observing clouds, you'll learn that such ambiguity is common. This is a reason why many people find the sky and weather fascinating.

Lenticular Clouds

You're a few miles from a mountain or mountain range and see something like a flying saucer. You are almost surely looking at a lenticular cloud formed by wind blowing over mountains.

KEY FACTS

Lens-shaped clouds atop mountains, sometimes stacked up, are called "mountain wave clouds."

+ fact: Airline pilots avoid mountain wave turbulence, but pilots of sailplanes use the waves to climb above 50,000 ft (15,000 m).

+ fact: Lenticular clouds are most common east of the Sierra Nevada and the Rockies.

When the atmosphere is stable, wind goes down and up in a wavy pattern after it has crossed mountains. (When it is unstable, the air keeps rising.) Unless the air is very dry, a lens-shaped cloud or line of clouds forms atop the wave or waves. Air rising toward the top of a wave cools, and its humidity condenses into cloud drops. Air descending from the top of a wave warms, which evaporates cloud drops. The combined processes create this lens-shaped cloud. Lenticular clouds change shape little and stay the same distance from the mountains instead of traveling downstream with the winds. These clouds remain stationary as air moves through them.

Mammatus Clouds

Mammatus clouds are pouches that hang from the bottoms of clouds, most commonly from thunderstorm anvils, a cloud spreading out from the storm's top.

KEY FACTS

"Mammatus" comes from the Latin word *mamma*—meaning "udder" or "breast."

+ fact: While mammatus clouds are most common on thunderstorm anvils, they also appear rarely under altocumulus, altostratus, and cirrus clouds, and even contrails.

+ fact: Most clouds form in rising air; mammatus clouds form in sinking air.

Mammatus pouches, which can be transparent or opaque, form when blobs of cold air containing water drops, ice crystals, or both begin sinking into clear air below a cloud. They descend into increasing air pressure, which warms the falling drops and crystals. At the same time, the crystals or drops are evaporating into water vapor, which cools them, offsetting the warming to some extent and keeping them cooler, thus heavier, than the surrounding air. The blob of crystals or drops sinks below the bottom of the cloud and looks like a pouch. Individual pouches may last ten minutes, but a cluster can last for hours.

Altocumulus Castellanus Clouds

Altocumulus castellanus clouds look somewhat like the turrets of a castle. If you see these early in the day, it could signal rain or thunderstorms later that day.

KEY FACTS

Tall, narrow castellanus clouds are called turkey necks.

+ fact: These can show that rising air is breaking through a warm layer that has been suppressing thunderstorms.

+ fact: Castellanus towers are easier to see from the side rather than along the narrow cloud's length.

When rolls of altocumulus clouds begin sprouting towers, the atmosphere above the clouds is becoming unstable enough that they might grow into towering cumulus and possibly thunderstorms that day. Such instability means the atmosphere above altocumulus clouds is cold enough for rising air to stay warmer than the surrounding air and continue rising. The towers are not a guarantee of thunderstorms, just a possible predictor. If the air near the ground is humid, thunderstorms are more likely. If you are on an airplane about to take off with castellanus clouds overhead, expect some turbulence as the airplane flies through the clouds.

Stratocumulus Clouds

Like other kinds of stratiform clouds, stratocumulus clouds spread across large areas with small breaks between individual clouds. In addition, they display a rounded cumuliform shape.

KEY FACTS

Stratocumulus clouds cover an average of 23 percent of the oceans and 12 percent of land.

+ fact: Individual stratocumulus clouds appear roughly the size of your fist when you extend your arm full length toward the cloud.

+ fact: Stratocumulus clouds produce little precipitation—mostly drizzle, light rain, or snow.

Stratocumulus clouds are most common over chilly oceans in the subtropics where air is slowly sinking from aloft, warming it. The layer of warm air blocks humid air from rising far from cool oceans. Thus, stratocumulus clouds generally don't rise above 8,000 feet (2,400 m). Stratocumulus clouds also form over land where air is descending into a surface high-pressure area and warming. If cumulus clouds are forming in such an area, they stop growing taller at the warm layer and spread out. Because the tops of stratocumulus clouds reflect a good share of the sunlight that hits them, they tend to cool the Earth.

Stratus Clouds

Stratus clouds are featureless gray layers with generally uniform bases. They sometimes, but rarely produce drizzle, small ice crystals, and snow grains.

KEY FACTS

Instead of rain or snow, drizzle, snow grains, or tiny ice crystals usually fall from stratus clouds.

+ **fact:** A stratus fractus cloud has parts of different sizes and brightness that change rapidly.

+ **fact:** When you can see the sun through the clouds, its outline is usually clearly discernible.

Stratus clouds are common in coastal areas because the air contains abundant humidity at low levels and the atmosphere is likely to be stable, which favors their development. Fog moving in from the ocean at night can set the stage. Stratus sometimes forms when the bottom of a layer of fog evaporates, and the lower part of the fog rises off the ground. Stratocumulus clouds can also form into stratus when the bottom of the stratocumulus descends and spreads out under a layer of warm air aloft. Stratus clouds are most common at night and in the early morning, before the sun begins to evaporate them. Stratus clouds can form in air moving into thunderstorms.

Nimbostratus Clouds

When dark, gloomy nimbostratus clouds move in with their steady rain or snow, they leave no doubt that you are in for a period of wet weather and no sun.

KEY FACTS

An altostratus becomes a nimbostratus cloud if it totally blocks the sun or its precipitation reaches the ground.

+ **fact:** Unattached cloud fragments called "pannus" or "scud" clouds often form below nimbostratus, cumulus, and cumulonimbus clouds.

+ **fact:** Nimbostratus clouds do not bring lightning, thunder, or hail.

Nimbostratus clouds are low clouds that hang below 6,500 feet (2,000 m) above ground level. They can be a few thousand feet thick and are heavy with suspended water drops, possibly ice crystals, and falling rain or snow that is likely to last for several hours or even more than a day. Little or no sunlight makes it through the cloud. The rain or snow from a nimbostratus is steady, and it falls at a light or moderate rate, not the on-and-off but sometimes drenching rain showers that cumulonimbus clouds produce. The bottoms of nimbostratus clouds tend to be ragged instead of well defined.

Altostratus Clouds

Given their name because they are the highest (alto-) of the stratus or sheet-type clouds, the thin gray altostratus sky cover signifies a storm could be on the way.

KEY FACTS

Not enough sunlight passes through altostratus clouds to cast shadows on the ground.

+ fact: Altostratus clouds tend to be translucent; you can often see a watery sun or moon through them.

+ fact: A watery moon appearing and disappearing through altostratus clouds is a classic horror movie scene.

If the sky is covered as far as you can see by gray or blue-gray clouds that are 6,000 to 20,000 feet (1,800 to 6,100 m) above the ground, they are altostratus clouds. If the clouds are white or have areas of white, you are instead looking at cirrostratus clouds. Altostratus clouds are made of ice crystals, usually near the top of the cloud, and water drops, near the bottom. They usually arrive ahead of storms that carry widespread, steady rain or snow. As rain or snow continues falling from the altostratus, the cloud's bottom can sink below 6,000 feet, and it becomes a nimbostratus cloud.

Fair-weather Cumulus Clouds

These small, puffy, white clouds portend calm, dry weather for the immediate future, at least early in the day. As the day goes on, they can grow into thunderstorm clouds.

KEY FACTS

The scientific name for fair-weather cumulus clouds is *cumulus humilis,* from the Latin word for "low, lowly, small, or shallow."

+ fact: If you're on an airplane taking off under fair-weather cumulus clouds, expect mild turbulence until you're above the clouds.

+ fact: Fair-weather cumulus commonly form under cirrostratus clouds.

Fair-weather cumulus develop as the sun warms the ground, creating thermals that rise until water vapor begins condensing. Warm air aloft, or air aloft that cools too slowly with height, blocks air from rising higher than the cloud tops. If this continues, the day will remain calm. If you see one or a few clouds growing higher than the others, you know that either air is rising into them with enough force to break through the warm layer, or that the atmosphere above the clouds has cooled, making it more unstable. When this happens, one or a few of the smaller cumulus clouds can grow into cumulus congestus (opposite) and even thunderstorms.

Cumulus Congestus Clouds

These are impressive, hard-to-ignore clouds that tower high in the sky with solid-looking, cauliflower-like towers. Some of these clouds grow into fierce thunderstorms.

KEY FACTS

Cumulus congestus clouds are also called "towering cumulus" because they are usually taller than wide.

+ fact: Congestus clouds can produce heavy and prolonged rain or snow showers without growing into cumulonimbus clouds.

+ fact: Cumulus congestus clouds can form as individual clouds or as a "wall" of clouds.

Cumulus congestus clouds form on days when the atmosphere up to great heights is unstable. Thermals of warm air begin rising from the ground at about 50 mph (80 kph) or faster. As the rising air cools, water vapor begins condensing into tiny cloud drops. The cauliflower-like parts of the cloud are made of tiny water drops that reflect more light than the softer, fibrous parts of the cloud, which are made of ice crystals. When parts of the cloud begin to take on the softer look, it shows the cloud is "glaciating"—ice crystals are forming, which is the beginning of its transformation into a cumulonimbus or thunderstorm.

Cumulonimbus Clouds

Cumulonimbus, commonly called a thunderstorm, is the only cloud requiring caution when nearby. Its potential dangers include lightning, tornadoes, and dangerous straight-line winds.

KEY FACTS

Thunderstorm tops are typically 20,000 ft (6,000 m) above the ground and at times 75,000 ft (23,000 m) high.

+ fact: Cumulonimbus clouds usually reach their peak strength in late afternoon.

+ fact: Thunderstorms occur as individual storms, in large and small clusters, and in lines 200 mi (300 km) or more long.

Meteorologists consider a cumulus congestus to have become a cumulonimbus when at least the top part of the cloud has taken on the smooth, fibrous, glaciated appearance that comes when this part of the cloud is mostly ice. A sure sign is that the top of the cloud begins to flatten out into a characteristic anvil shape. Thunderstorm precipitation is showery. It starts and stops suddenly, and it can be quite heavy over a relatively small area while nearby areas are dry. The general winds in an area push thunderstorms across the countryside, as they do other clouds, sometimes as fast at 50 mph (80 kph) but usually more slowly. Thunderstorms are quite turbulent inside the cloud.

Pyrocumulus Clouds

Pyrocumulus or "fire cumulus" clouds form over large fires, usually wildfires, under certain atmospheric conditions. They can threaten wildfire-fighters with hard-to-forecast wind shifts.

KEY FACTS

Fires sometimes create tornado-like "fire whirls"; most last a few minutes but some persist for 20 minutes and spread the fire.

+ fact: Pyrocumulus clouds sometimes grow into thunderstorms—pyrocumulonimbus.

+ fact: Volcanic eruptions create pyrocumulus clouds. The mushroom cloud of a nuclear bomb is a pyrocumulus.

Wildfires are dangerous enough, but if they burn during days when the atmosphere is unstable and dry with light upper air winds, they become especially dangerous for firefighters. Under those conditions, a fire's hot air will rise straight up, 20,000 feet (6,000 m) and more, to create a pyrocumulus cloud. Air rushing in from around the fire to replace the rising air fans the flames. These winds can also change direction with little warning, especially on mountains and in hills, sending fire in new directions and potentially trapping firefighters. While the bottom of a pyrocumulus might be brown or gray with smoke, the top is bright white, like the tops of other cumulus clouds. Pyrocumulus clouds don't have enough water to produce the rain that would put out a fire.

Billow Clouds

The lovely wave formations atop billow clouds are spawned by forces that are common in the atmosphere and create other phenomena, such as ocean waves.

KEY FACTS

Lord Kelvin, a Scottish physicist, and Hermann von Helmholtz, a German scientist, mathematically analyzed waves in the 19th century.

+ fact: Such waves form at the boundary between fluids with different densities.

+ fact: The distance between each cloud wave is usually between 3,200 and 6,500 ft (1,000 and 2,000 m).

Billow clouds, also called Kelvin-Helmholtz clouds or waves, usually occur at a layer of warm air high above the ground. Up there, with the warm air above and cold air below, winds are blowing in opposing directions, and the curving pattern develops. To envision what happens, imagine holding a ball between the palms of your hands and moving your hands back and forth in opposite directions. To form billow clouds, the air rolls between two outside layers of air moving in opposite directions, just like the ball rolls between your hands. The rolling air might not complete the circle, though, creating forms that look like breaking ocean waves.

Supercell Thunderstorm Clouds

Supercells are the most dangerous but the least common thunderstorms. They produce almost all of the deadliest tornadoes. Distinct features make supercells stand out.

KEY FACTS

Haze and hills in the U.S. Southeast and East often make it difficult to see enough of a thunderstorm to determine if it is a supercell.

+ fact: Strong supercells have a dome on the top, called an "overshooting top."

+ fact: Low-precipitation supercells on the arid High Plains produce little rain or hail.

A supercell is a long-lasting kind of thunderstorm that produces strong tornadoes and other dangerous weather. A mesocyclone—a one- to ten-mile-wide rotating updraft that carries air from the ground to the storm's top—distinguishes supercells from all other thunderstorms. You can often see its barber-pole striations. Supercells develop a wall cloud between the area of precipitation from the thunderstorm and a precipitation-free base.

The wall cloud forms where wet, cool air is being drawn into the thunderstorm's updraft. A wall cloud that lasts more than ten minutes and moves violently is most likely to produce a tornado.

Shelf and Roll Clouds

Shelf and roll clouds form on top of air that descended in a thunderstorm and moves away from the storm. A shelf cloud is attached to its parent thunderstorm; a roll cloud has broken away.

KEY FACTS

Sometimes a shelf cloud spins out a "gustnado," a small, weak tornado that rarely causes damage.

+ fact: All shelf clouds come from thunderstorms; some roll clouds come from cold fronts and sea breeze fronts.

+ fact: Roll clouds spin around a horizontal axis and are not as common as shelf clouds.

Air descending from a thunderstorm travels away from the storm, becoming a dome of cool air that is called a gust front. As its leading edge meets and plows into warm, humid air, it pushes the air up. The warm air cools, and its humidity condenses to form a shelf cloud atop the cool downdraft air. Such a gust front can last more than a day and travel hundreds of miles to help trigger new thunderstorms by pushing up warm, humid air. If the front of the shelf cloud is ragged, pushing rising, small, ragged clouds in front of it, damaging wind squalls and shifts in wind direction will likely result.

Dew

Dew, the water that you often find on the grass and your car on some early mornings, is one of the many ways water moves out of and back into the atmosphere.

KEY FACTS

The term "dew point" refers to the temperature at which condensation begins on the ground or in the air.

+ fact: Dew on grass mostly evaporated from the grass and stayed in the air nearby until condensing back onto the grass.

+ fact: Dew often forms first on car roofs because they radiate heat directly away.

Dew begins condensing when the air cools to the dew point—a temperature point that varies depending on how much water vapor is in the air. Because warm air can hold more water than cold air, water vapor will begin condensing from very humid air at a higher temperature than it would condense from air with very little water vapor. Dew begins forming on grass and other low-lying plants, because air sinks as it cools and air along the ground is usually colder than air just a little higher up. Calm, clear, still nights encourage dew. These are nights without wind that mixes the air, and with heat radiating directly to space. You can't rely on the old folklore that says a dewy morning means the day will be clear, because a new weather system can move in with clouds and rain.

Frost

Frost forms when water vapor in the air becomes ice without first condensing into water through a process called deposition. It forms on clear, calm nights with temperatures below 32°F (0°C).

KEY FACTS

Frost can form on the ground when the official temperature—measured above ground level—is above freezing.

+ fact: "Black frost" occurs when frigid air kills plants without visible frost.

+ fact: The growing season runs from the average date of spring's last frost to the average date of fall's first frost.

Frost forms overnight as white crystals on grass and other objects. With enough humidity in the air, these begin growing new crystals, which are called hoarfrost. Snowbanks are a good place to look for hoarfrost because some of the snow sublimates directly into water vapor during the day, and the vapor stays near the snow on calm days. At night, the vapor deposits into hoarfrost crystals, which cause the snow to sparkle. Frost forms on the inside of windowpanes that are not well insulated when it is moderately humid inside and very cold outside. Imperfections or scratches help shape the patterns that form. Frost can form on double-pane windows.

Rain

Clouds are made of water in the atmosphere, but they do not always cause precipitation. All rain falls from clouds, but most clouds do not produce rain.

KEY FACTS

Meteorologists have agreed on standard descriptions of falling rain.

+ **fact:** Light rain falls at up to 0.10 in (2.54 mm) an hour; scattered drops are seen.

+ **fact:** Moderate rain falls from 0.11–0.30 in (2.79–7.62 mm) an hour; drops aren't clearly seen.

+ **fact:** Heavy rain falls at more than 0.30 in an hour in sheets rather than drops.

A freezing cold rain and a warm rain both involve precipitation, but clouds that are colder than 32°F (0°C) and those that are warmer produce rain in different ways. In both, roughly a million cloud drops come together to produce a raindrop 0.08 inch (2 mm) in size. But in warm clouds, a few slightly enlarged cloud drops fall and sweep up others to grow into raindrops. In freezing cold, when ice and water mix, water vapor migrates into ice crystals that grow heavy enough to fall. If the air below is warmer, the ice crystals melt into rain; if not, they come down as freezing rain. See shapes of small and large drops at right.

Drizzle

Drizzle consists of water drops less than 0.02 in (0.5 mm) in diameter. Ordinary drizzle causes few problems to everyday life, but freezing drizzle turns to ice on impact and can pose a danger.

KEY FACTS

Visibility determines drizzle's intensity. More than a half mile (0.8 km) is light; between a quarter (0.4 km) and a half mile is moderate; less than a quarter mile is heavy.

+ **fact:** Drizzle drops fall close together and float in air currents.

+ **fact:** Drizzle is most likely in November and least likely in July in North America.

D rizzle falls mostly from stratiform or stratocumulus clouds. Climate scientists are especially interested in the drizzle from the shallow stratocumulus clouds that cover large areas of subtropical oceans—the areas just north and south of the tropics. These clouds are important for cooling the Earth, and researchers are investigating the role oceanic drizzle plays on these clouds and their effects. For most of us, drizzle means nothing more than a gloomy day—except for the ice that freezing drizzle leaves on roads and sidewalks. Drizzle-size drops that freeze on impact are especially dangerous for aircraft encountering them in the clouds or as they fall. Drizzle helps illustrate how even tiny atmospheric phenomena can be important.

Sleet

Sleet generally refers to frozen raindrops less than 0.2 in (5 mm) in diameter that fall during winter storms often with, before, or after freezing rain and snow.

KEY FACTS

The eastern United States and southeastern Canada are the only places where sleet accumulates to a thickness of more than 0.8 in (2 cm).

+ fact: The U.S. National Weather Service issues a "sleet warning" when more than a half inch (12.7 mm) is expected.

+ fact: Sleet bounces and can be heard when it hits.

Sleet forms when a layer of air above freezing temperature (32°F; 0°C) lies 5,000–10,000 feet (1,500–3,000 m) above the ground, sandwiched between layers of colder, below-freezing air above and below. Snow falls from the higher layer of cold air into the middle layer of warm air. It melts, falls into the lower layer of cold air, and refreezes into ice pellets. These pellets can be spherical, conical, or irregular in shape. In North America, sleet can mount up to as high as 2 inches (5 cm). Such accumulations occur when a strong storm pushes warm air over a layer of dense, cold air at the surface for a few hundred miles. Sleet likely comes ahead of an advancing warm front.

Freezing Rain

Freezing rain occurs when raindrops that are supercooled—cooled to below 32°F (0°C) but not frozen—instantly turn into ice when they hit cold objects such as roads and power lines.

KEY FACTS

Warm fronts bring freezing rain as a layer of warm air moves over ground-level frigid air.

+ fact: Freezing rain causes black ice: a slick, transparent layer of solid ice atop a flat surface such as a road.

+ fact: Some estimate that as many as one-fourth of winter accidents are caused by freezing rain.

As with sleet, layers of warmer air sandwiched between layers of cold air during winter storms set the stage for freezing rain. In some circumstances, the layer of below-freezing air at ground level isn't thick enough to freeze raindrops into sleet. Then the drops become supercooled and coat everything cold that they hit with a heavy glaze of solid ice. The weight of this clinging ice can bring down tree limbs and power lines. Plows have a harder time clearing ice from roads than snow. The ice is slicker than snow, causing more cars to spin out or slide off roads. When freezing rain is forecast, road crews often spray chemicals such as calcium magnesium acetate on pavement. The chemicals inhibit ice from forming or keep ice that does form from sticking to roads.

Snow

Snow is frozen precipitation that forms as six-sided crystals by the direct deposition of water vapor as ice on freezing nuclei or on tiny ice crystals in clouds.

KEY FACTS

Visibility determines snowfall intensity: more than a half mile (0.8 km is light, between a quarter (0.4 km) and a half mile is moderate, less than a quarter mile is heavy.

+ fact: Snow flurries are light showers with little accumulation.

+ fact: Blizzards are heavy snowfalls lasting 3 hours or more with winds, of 35 mph (56 kph) or faster.

Temperature and humidity determine the shape of snow crystals. They may start as six-sided crystals, but they often reach the ground in more simple or more complex forms, depending on the conditions they encounter as they fall. When supercooled water drops hit a snow crystal, they stick to it as rime, a coating of ice. This process can create snow pellets—spherical white particles up to 0.2 inch (5 mm) across. In near-freezing temperatures, crystals stick together to form snowflakes, sometimes large agglomerations. If you closely examine snow, you might see various forms of hexagonal snow crystals as well as indistinct pieces of ice broken off from crystals. Snow that falls in air close to freezing is heavier than snow in colder air.

Snow Crystals

When water cools to 32°F (0°C), its molecules align themselves as six-sided ice crystals. When these crystals grow in the air, they retain a hexagonal shape but assume shapes of infinite variety.

KEY FACTS

The shape of snow crystals has long fascinated scientists.

+ fact: In 1611, Johannes Kepler wrote that snow crystals have six sides.

+ fact: In 1665, Robert Hooke sketched six-sided ice crystals he saw through a microscope.

+ fact: In the 1930s, Ukichiro Nakaya found that temperature and humidity determine crystal shapes.

If you collect and examine snowflakes as they fall—a dark piece of cardboard works, or even a dark-colored sleeve—you'll discover few like the lacy dendritic (branching) snowflake designs created for winter-season decorations. In fact, you're more likely to see broken bits of crystal shapes that tumbled against one another as they fell. If you don't see crystals with regular shapes the first time you try, don't give up. Eventually you should happen to collect crystals that are more than broken pieces. You could find that observing snow crystals, as with other aspects of weather, can take you deep into physics and other physical sciences.

Thunderstorms

Cumulus clouds grow and proceed through predictable stages as they turn into thunderstorms. Learning about thunderstorm stages is a first step toward understanding them.

KEY FACTS

A thunderstorm's life cycle has three stages.

+ fact: First is a towering cumulus stage as rising air forms a growing cloud.

+ fact: The mature stage, the longest, begins when rain starts falling, dragging down cold air.

+ fact: The dissipating stage begins when updrafts end, leaving only downdrafts until the storm dies.

As an ordinary cumulus cloud grows into a cumulus congestus and then a cumulonimbus, or thunderstorm, it's an indication of surging updrafts rising to form the cauliflower-like clouds you see as water vapor becomes cloud drops, which grow into raindrops. When parts of the cloud become smooth, ice crystals are forming. This process of glaciation releases more energy and generates more updraft, which can help spur the formation of the telltale anvil shape at the top of a thunderstorm cloud. In the mature stage, updrafts can be faster than 100 mph (160 kph); accompanying downdrafts measure half that speed. In fierce thunderstorms, especially supercells, a storm's mature stage can last for hours.

Multicell Cluster Thunderstorms

A large cumulonimbus cloud that is producing lightning with more than one dome on top, maybe an anvil on one side, is a multicell cluster of thunderstorms.

KEY FACTS

Multicell cluster thunderstorms line up in the direction the winds are blowing.

+ fact: New cells usually form at the side from which the wind is blowing; cells mature in the cluster's center.

+ fact: Each cell of a cluster lasts approximately 20 minutes, but the cluster itself can keep going for hours.

Although many thunderstorms go though three distinct stages as separate, single-cell storms, groups of related storms either in clusters or lines are more common. A multicell cluster forms when the downdrafts from one thunderstorm push air up to trigger an adjacent thunderstorm. This storm in turn can trigger a new one as the original storm is dissipating. A cluster will often have at least one storm in the dissipating stage, one mature storm, and one still in the towering cumulus stage, with the clouds of all of them blending. The wind pushes thunderstorms in a cluster in the same direction.

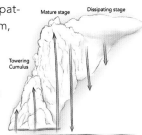

Mature stage   Dissipating stage

Towering Cumulus

Supercells

These thunderstorms are "super" in terms of their size, how long they last, and their potential for causing death and destruction. Supercells produce almost all of the most deadly tornadoes.

KEY FACTS

A supercell lasts for several hours, even though it is producing tornadoes along the way.

+ fact: Tops of Great Plains supercells reach 40,000 ft (12,000 m) above the ground.

+ fact: Landforms influence the size and life span of supercells. Those occurring in the eastern U.S. are generally smaller than those on the Great Plains.

Although most single-cell thunderstorms last less than a half hour, supercells last for hours. Winds from different directions at various altitudes cause the main updraft to lean instead of traveling straight up and down, which means that rain falling from the top of the storm doesn't cool the cloud's rising warm air, and so the updraft keeps going. Many supercells produce only weak tornadoes, but some can be deadly. Researchers are looking for ways to distinguish well in advance which supercells will be the most dangerous and which will bring no major threats. When you see a supercell, you should stay alert for a tornado.

Squall Line Thunderstorms

If you see a long line of approaching thunderstorms preceded by a shelf cloud—a low-hanging, horizontal, wedge-shaped cloud—prepare for strong winds and lightning.

KEY FACTS

A squall line can stretch for hundreds of miles—as far as from Louisiana into Illinois. Airliners detouring around them can cause major delays.

+ fact: Squall line thunderstorms line up at roughly a right angle to the wind direction.

+ fact: Most squall lines last from late morning or early afternoon until after dark.

Steady winds can push a line of thunderstorms together in the same direction. The advancing storms scoop up warm, humid air, which feeds the storms even further. As individual storms die, new ones take their place in the line. Tornadoes sometimes occur, but a squall line's major danger is fierce straight-line winds that blast down in the direction the storms are moving. Individual storms in a line can reach more than 40,000 feet (12,000 m) into the air, too high for airliners to fly over. Squall lines can occur right along cold fronts, but the strongest are usually those several miles ahead of a cold front.

Upper air winds

Derecho

A derecho is an extremely long-lasting, fast-moving thunderstorm squall line that produces winds of at least 57 mph (92 kph) along a path at least 240 miles (386 km) long.

KEY FACTS

The term "derecho" was first used by Gustavus Hinrichs, a University of Iowa professor, in an 1888 scientific journal article.

+ fact: Hinrichs chose *derecho*, Spanish for "straight ahead," to distinguish its straight-line winds from a tornado's rotating winds.

+ fact: The National Weather Service began using "derecho" in 1987.

In the late spring and early summer, especially strong, long-lasting squall lines called derechos move across the Great Plains from the Rocky Mountains and sometimes all the way to the Atlantic coast. As in any squall line, the individual thunderstorms making up a derecho weaken and die, soon replaced by new ones, and so the damage along a derecho's path is not consistent. Winds build up to extreme speeds because the downburst force from individual thunderstorms is added to the speed of the wind pushing the squall line to the east. A derecho dies when it runs into dry air in the upper atmosphere or when the winds pushing it die down. Most derechos occur in the summer, most likely during heat waves, and mostly east of the Rocky Mountains.

Bow Echoes

A bow echo storm is an especially dangerous curved line of thunderstorms. The most dangerous winds occur at the crest or center of the curving bow-shaped formation of clouds.

KEY FACTS

A bow echo can range in size from 12 to 125 mi (20 to 200 km) across and last from 3 to 6 hours.

+ fact: Tornadoes often form on each end of a bow echo, but these are usually weak, doing little damage.

+ fact: On November 2, 1995, a bow echo hit the Hawaiian island of Kauai with winds of 90 mph (145 kph).

At times, winds blast down from part of a squall line or an isolated supercell and race ahead of the line or the supercell right above the ground. This soft, fast-moving wind, called a rear-inflow jet, pushes warm, humid air up and triggers a line of thunderstorms that form in a bow shape. T. Theodore Fujita, the famed 20th-century Japanese American tornado researcher, discovered and named bow echo damage patterns and radar images while investigating a derecho that struck from Michigan to Minnesota on July 4, 1977. Almost all derechos produce bow echoes, which often cause the derecho's most destructive winds.

Mesoscale Convective Complex

A mesoscale convective complex (MCC) is experienced as a long summer night with constant thunderstorm downpours and frequent lightning, usually on the Great Plains.

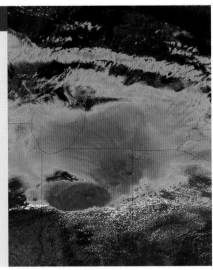

KEY FACTS

Each word in "mesoscale convective complex" describes an element of this sort of storm.

+ fact: Mesoscale: midsize weather phenomena, from a few miles to a few hundred miles across

+ fact: Convective: movement of air up and down

+ fact: As a complex, these storms are interrelated, not just happening to be together.

By definition, a mesoscale convective complex (MCC) is a persistent, nearly circular area of clouds measuring a temperature of -25°F (-32°C). Generally thunderstorm anvils, they cover a huge area—at least 38,500 square miles (99,700 sq km), roughly the size of Iowa. The system begins in the late afternoon and early evening with heavy rain and sometimes strong winds. By early morning, the thunderstorms die and the complex's rotating vortex, taller than 10,000 feet (3,000 m), continues traveling to the east. It can reignite the MCC the next evening. Meteorologists didn't realize MCCs were organized systems until weather satellite images showing cloud-top temperatures, thus their heights, became available in the 1970s. MCCs mostly affect the middle of the U.S.

Lightning

Lightning is a huge electrical spark flashing between areas of opposite electrical charge. It occurs inside clouds, from a cloud to the ground, from a cloud to another cloud, or into empty air.

KEY FACTS

Separate current strokes lasting a few tenths of a second cause the flickering lightning flash you see.

+ fact: Lightning's rapid heating and cooling of the air creates the sound waves we hear as thunder.

+ fact: Thunder rumbles as sounds from different parts of the flash arrive at slightly different times.

The violent churning of mixed ice crystals and water drops within a thunderstorm leaves areas of negative and positive charge in different parts of a cloud, usually with a strong negative charge near the cloud's bottom. Attraction between this and positive charge on the ground causes streams of negative charge to begin working their way down through the air as stepped leaders that zig one way, zag another. These create paths for stronger currents. When one of these connects with something on the ground, such as a tree or a lightning rod, a strong return stroke, which we see as lightning, flashes from the ground to the cloud.

Lightning to Worry About

About 100,000 thunderstorms occur in the United States each year, according to the National Weather Service. Lightning makes each one dangerous, no matter how weak it is.

KEY FACTS

A flash of lightning measures approximately 50,000°F (27,760°C), but it lasts so briefly that a victim does not suffer deep burns.

+ fact: Lightning can cause random neurological damage or stop a victim's heart.

+ fact: Lightning rods, which carry lightning into the ground, have changed little since Benjamin Franklin's invention.

Most lightning flashes are a negative charge attracted to the ground's strong positive charge. To avoid lightning damage, you need to provide lightning a low-resistance path to the ground such as a lightning rod kept in good condition with no breaks in its path into the ground. To prevent injury or death by a lightning strike outside, stay away from lightning's path to the ground by being inside a sturdy building or a hardtop vehicle. Because lightning goes into the ground, it can hit and damage underground utility lines. Water pipes can give lightning a path inside your home. Indoors during a thunderstorm you shouldn't be near plugged-in appliances or computers, take a shower or bath, or talk on a phone with a cord. Cordless and cell phones are safe to use indoors during a thunderstorm.

Upper Atmospheric Lightning

A few thunderstorms put on stunning but hard-to-see shows of phenomena called sprites, elves, and blue jets above the storms as lightning flashes in and below the clouds.

KEY FACTS

Now, thanks to aircraft and imaging technology, we have images of all three phenomena, caused by intense electrostatic currents above dying thunderstorm clouds.

+ fact: Sprites are reddish orange or greenish blue with tendrils hanging down.

+ fact: Blue jets are narrow cones that fan out above the top of a storm cloud.

Going back to at least World War II, airplane pilots reported unusual lights above thunderstorms, causing scientists to wonder what happens above as well as below thunderstorms. Finally, scientists testing a low-light video in 1989 saw the first images of a sprite, a large discharge of energy above a thunderstorm. Unless you're an airline pilot, you will have to make a special effort to see these upper atmospheric lightning phenomena. The best way to try to see them is to be somewhere high, such as the Rocky Mountain Front Range. Choose a clear night and pick a location that is far from city lights. Wait until your eyes become adapted to the darkness, and then look out over the tops of thunderstorms on the Great Plains.

When Lightning Hits Airplanes

On average, lightning hits each airliner flying over North America once a year, but today's aircraft are built to shrug off the charge. Lightning last caused a U.S. airline crash in 1962.

KEY FACTS

Lightning frequently hits aircraft today.

+ fact: Airplanes and rockets can trigger lightning by flying into strong electrical fields.

+ fact: Lightning hit NASA's lightning research plane 714 times.

+ fact: After being hit by lightning 26.5 seconds after blastoff in 1969, Apollo 12 continued on to make the second manned moon landing.

When lightning hits an airplane, the electricity spreads out and flows through the aluminum skin and then back out into the air. The only damage caused includes small burn marks where the lightning entered and left the airplane. Fuel tanks have extra metal around them to keep lightning hits from burning through. Shielding protects electrical systems from the direct flow of lightning currents and from currents lightning can induce in wires. Advanced composite aircraft such as Boeing's 787 Dreamliner have conductive material embedded in the skin. Even if lightning hits, no one aboard a commercial aircraft today is in danger of being shocked. **Lightning's threat to those servicing airplanes on the ground can delay flights.**

Saint Elmo's Fire

Strong electrical fields rip electrons from nitrogen and oxygen molecules in the air causing a harmless blue-violet glow called Saint Elmo's fire, usually on tapered objects.

KEY FACTS

Saint Elmo's fire is named for Saint Erasmus of Formiae—Saint Elmo in English—the patron saint of sailors.

+ fact: Neon gas in neon and mercury vapor in fluorescent tubes creates glow discharges like Saint Elmo's fire.

+ fact: Flying through volcanic ash causes bright Saint Elmo's fire on wings and passenger windows.

Saint Elmo's fire is a natural blue-violet glow caused by a strong electrical field tearing apart the air's molecules of nitrogen and oxygen—a process called ionization. The resulting Saint Elmo's fire is a plasma. Curves and sharp points concentrate electronic fields, making the glow stronger. In past centuries, people saw Saint Elmo's fire at the top of ship masts and church steeples. The glow seemed miraculous because, though it looks like fire, it produces no heat and does not burn ships' wooden masts. Today, you are most likely to see Saint Elmo's fire on the tips of airplane wings. Pilots regularly see it around their windshields. In 1749 Benjamin Franklin became the first scientist to describe Saint Elmo's fire as an atmospheric electrical phenomenon. He sometimes saw it on the tips of lightning rods before lightning hit the rods.

Ball Lightning

Since the time of the ancient Greeks and Romans, people have reported glowing, floating balls in the air, associating them with lightning, yet scientists are still unable to explain what's going on.

KEY FACTS

Most reports of ball lightning indicate that it is harmless. Scientists can't explain it.

+ fact: Ball lightning is roughly the size of a grapefruit and as bright as a 60-watt bulb.

+ fact: Laboratory experiments have created objects with aspects of ball lightning, but have not entirely matched reports of it.

This unusual phenomenon is called ball lightning because it often resembles a floating ball. Almost everyone who has viewed a ball says it appeared either just after a lightning strike or when lightning was striking nearby. Researchers have compiled at least 10,000 reports of ball lightning in recent decades. Some observers report ball lightning passing through windows without causing damage; others report seeing it appear and disappear inside airplanes in flight, again with no damage. Mysteriously, it glows without giving off heat and does not appear to generate power. Individuals and groups of people have reported ball lightning since the time of ancient Greece. Glowing balls approximately six inches in diameter floating in the air have persuaded many scientists that ball lightning is real.

Hail

Hailstones are pieces of ice that form in thunderstorm updrafts, which keep them from falling while more water freezes onto them. Updraft speed determines the size of hailstones.

KEY FACTS

A hailstone ½ inch in diameter needs a 20 mph (32 kph) updraft to form; a ¾-inch stone needs a 64 mph (103 kph) updraft.

+ fact: More hail falls yearly on northeastern Colorado and southeastern Wyoming than elsewhere in the U.S.

+ fact: The National Oceanic and Atmospheric Administration says hail injures 24 people in the U.S. each year, deaths are very rare.

Most hailstones form in multicell, supercell, or cold-front squall line thunderstorms, usually near the center of a storm. Hail begins forming as tiny ice pellets collide with supercooled water droplets that freeze on contact with the ice. As they grow, hailstones may make several up-and-down trips within the center of the storm before the updraft weakens or when they become heavy enough to fall. By definition, a thunderstorm that produces ¾-inch (19 mm) hail is severe because of the strong updrafts and other winds in the system. Very strong updrafts can carry hailstones high into a storm and then sweep to one side, so that they fall outside the storm itself.

Microbursts

A microburst is an intense, concentrated wind that blasts down from a shower or thunderstorm, affecting an area no longer than 2.5 miles (4 km) on a side with strong, gusty winds.

KEY FACTS

Wind damage often shows whether a microburst or a tornado caused it.

+ **fact:** The National Weather Service has 48 Terminal Doppler Weather Radars to detect airport microbursts.

+ **fact:** In 1983, a 138-mph (220 kph) microburst hit Andrews Air Force Base three minutes after *Air Force One* carrying President Ronald Reagan landed.

Microbursts can blow down big trees, so people should avoid standing near trees during a windy thunderstorm. The greater danger of microbursts is to flying aircraft. Meteorologists did not have a name for microbursts until the 1970s, when researcher T. Theodore Fujita coined the term "microburst" to describe the concentrated winds that had caused several airplane crashes. Once the term was in use, pilots and air traffic controllers were better able to observe and avoid them. Through 1985, the U.S. air industry suffered roughly one fatal microburst airliner crash every 18 months, but since then, only one has occurred, thanks to better warnings.

Gust Front

Air coming down from a thunderstorm moves over the ground, as a miniature cold front. Effects include bringing slightly cooler temperatures and triggering new thunderstorms.

KEY FACTS

You may feel a distant thunderstorm's gust front as a cool breeze on a hot day.

+ **fact:** Outflow boundaries interacting with supercell thunderstorms increase the odds that the supercell will produce tornadoes.

+ **fact:** A gust front hitting an airport can endanger takeoffs and landings with unexpected wind shifts.

Meteorologists call gust fronts "outflow boundaries" because they do more than bring a quick shot of refreshing cool air. They can persist longer than 24 hours with small temperature differences in the air across the front. Even small changes can help trigger new thunderstorms and affect existing storms. Air converges along gust fronts, bringing dust, insects, and other small objects that Doppler radar can see long after the gust front forms. Meteorologists use such gust front images to forecast where new thunderstorms are likely to begin. For instance, the meeting point of two gust fronts, each of which may be a day old, is a prime location for where storms may be expected to originate and then develop.

Funnel Cloud

A funnel-shaped cloud full of condensed water stretching down from a towering cumulus cloud or cumulonimbus might be called a tornado that is not touching the ground.

KEY FACTS

When the air is too dry for a condensation funnel to form, the first sign of a tornado can be spinning debris on the ground.

+ fact: A tornado's spinning winds extend wider than the visible funnel.

+ fact: Funnels turn from white to black or other colors when they pick up dirt or dust from the ground.

A funnel dipping down out of a storm cloud is a funnel cloud as long as it neither touches the ground nor kicks up dust or debris from the ground. Once either of those conditions occur, it has become a tornado. This type of condensation funnel forms when falling atmospheric pressure in the vortex, or spinning funnel, cools the air enough for water vapor to condense into a swirling cloud. Almost all tornadoes begin as funnel clouds, but many funnel clouds never become tornadoes. Meteorologists sometimes talk of cold air funnels: They don't form in thunderstorms, they are generally weak, and they don't last long, although a few might touch down briefly as weak tornadoes or waterspouts.

Tornadoes

Tornadoes are rotating columns of air hanging from a cumulonimbus or sometimes a cumulus congestus cloud that have developed long enough that they come in contact with the ground.

KEY FACTS

Tornadoes' strength is ranked from F0 to F5 on the Fujita scale.

+ fact: Weak tornadoes: F0, 65–85 mph (105–137 kph); F1, 86–110 mph (138–177 kph)

+ fact: Strong tornadoes: F2, 111–135 mph (179–218 kph); F3, 135–165 mph (219–266 kph)

+ fact: Violent tornadoes: F4, 166–200 mph (267–322 kph); F5, faster than 200 mph (322 kph)

The wind speed of tornadoes can range from 40 mph (64 kph) to more than 300 mph (483 kph). Tornadoes are the most destructive of all local weather phenomena. Those clocking faster than 166 mph (267 mph) account for more than 70 percent of all tornado deaths and the greatest destruction. Fortunately, about three-fourths of all tornadoes are weaker, with winds no faster than 110 mph (179 kph). Because strong tornadoes destroy wind instruments, meteorologists must often examine damage to determine the wind speed. The U.S. averages more than 1,000 tornadoes a year, more than any other nation; Canada ranks second.

Multiple Vortex Tornadoes

Many tornadoes, especially the larger ones, are packages of twisters with smaller vortices circling around the central tornado's perimeter, sometimes seen but often hidden.

KEY FACTS

Three seems the most common number of subvortices in multiple vortex tornadoes, but spotters have seen as many as seven.

+ fact: Some hurricane eye walls have small vortices that look and act like tornado subvortices.

+ fact: Spotters reported multiple vortices before the May 22, 2011, tornado hit Joplin, Missouri, killing 161 people.

Why does it happen that a tornado can destroy a well-built house while a similar house, right next door, escapes with minor damage? The existence of multiple vortex tornadoes explains this phenomenon. The heavy damage occurs where the winds of the main tornado combine with the winds of a subvortex moving in the same direction, thus multiplying the impact. At times, this combination of forces can add 100 mph (161 kph) to the wind, spotlighting small areas in the tornado's larger path. Tornado spotters most likely see multiple vortices as a tornado is beginning, before dust and debris darken the main funnel. Damage patterns can confirm the vortices.

Waterspouts

A common definition of a waterspout is "a tornado over water," but meteorologists prefer to use the term for weak, non-supercell vortices that form below cumulus congestus clouds.

KEY FACTS

A waterspout's condensation funnel is not water sucked up from below; it is condensed from water vapor in the air above.

+ fact: A waterspout's life cycle is usually less than 20 minutes; winds reach speeds no greater than 85 mph (137 kph).

+ fact: A fall waterspout season often occurs on the Great Lakes.

Waterspouts most often form from the bottoms of dark, growing cumulus clouds. The first sign of a waterspout forming is a dark spot on the water's surface where the still invisible funnel has touched down; it may be kicking up water spray as well. Many waterspouts die out in this early stage, but sometimes they develop to the stage that a condensation funnel is visible, coming down from the cloud to the dark spot. The oceans around southern Florida, especially in the Keys, have more waterspouts than any other part of the U.S. Non-supercell waterspouts are generally weak, which means that they aren't as dangerous as most ordinary tornadoes, but boaters should not ignore them. Waterspouts can be stronger than they appear.

Dust Devils

A dust devil is a small whirlwind that, unlike a tornado, is not attached to a cloud. Dust devils usually occur in hot, dry locations where dust or sand makes them visible.

KEY FACTS

The strongest dust devils occur in the hottest deserts, but small ones can occasionally occur in cities as well.

+ fact: Dust devil winds sometimes build to the speed of an F1 tornado, up to 110 mph (177 kph).

+ fact: The Viking orbiters NASA sent to Mars in the 1970s discovered that dust devils occur on the red planet.

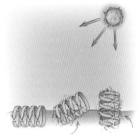

Dust devils form in thermals of rising air and often develop a funnel-like shape as air flows out of the top. The air cools as it rises and then eventually sinks back to the ground, where it can create the light breezes that spin thermals into dust devils. Dust devils can develop to a height of 3,000 feet (900 m) or more on some of the hottest deserts. They are usually benign, but they occasionally cause damage and injuries. On September 14, 2000, for example, a dust devil hit the Coconino County Fairgrounds in Flagstaff, Arizona, causing minor injuries and damaging booths and tents.

Acid Rain

Acid rain describes precipitation made slightly acidic by air pollution often from far away. It wreaks damage on both the natural environment and on limestone and marble buildings.

KEY FACTS

Acid rain's most damaging effects occur in locations where the soils do not naturally neutralize acids.

+ fact: Acid rain leaches calcium from soil, depriving plants of an important nutrient.

+ fact: Robert Angus Smith, a Scottish chemist, coined the term "acid rain" in 1852 and linked it to damage it causes.

Although the phenomenon likely dates back a century or two, only in the mid-20th century did people begin to take responsibility for acid rain and its consequences. Most acid rain in North America develops from sulfur dioxide and nitrogen oxides produced by burning fossil fuels, mostly to generate electricity and run cars and trucks. The acid rain does not come directly from smokestacks or tailpipes. Instead, the emissions react with water in the atmosphere to create damaging compounds that can travel hundreds of miles from their sources. The pollutants can fall as dry materials as well as in rain or snow. The chemicals make soil, rivers, lakes, and ponds slightly acidic, potentially killing smaller animals and plants. Air quality rules are reducing acid rain, and some hard-hit areas in the Northeast are recovering.

Hurricanes

A hurricane is a tropical cyclone in the Atlantic, the eastern Pacific, the Caribbean Sea, or the Gulf of Mexico—in which winds reach a sustained speed of 74 mph (119 kph) for a minute or more.

KEY FACTS

A tropical cyclone becomes a tropical storm when its winds reach 39 mph (63 kph), at which time it is assigned a name from a pre-selected list.

+ fact: Sustained winds of 74 mph (119 kph) make it a hurricane.

+ fact: Hurricanes are classified on a scale from 1 to 5 in order of severity based on wind speed.

Tropical cyclones are born over ocean water that is warmer than 80°F (27°C) and in a condition of high humidity. Because they draw their energy from warm water, these storms weaken and die when they move over cool water or land. A mature hurricane system rotates counterclockwise with a diameter of hundreds of miles and an eye of mostly clear air and light winds in the center. The fastest winds are usually in the eyewall—the ring of thunderstorms encircling the eye. Rain bands of thunderstorms spiral around the storm and into the eyewall. A hurricane pushes water ahead, creating a storm surge when it comes ashore.

Steering current
Air leaving the storm's top
Low-level winds spiraling into the storm

Hurricane Forecasting

Radar, satellites, sea buoys, and reconnaissance flights contribute to improved methods of hurricane forecasting, so that meteorologists can often predict a hurricane's path a week ahead.

KEY FACTS

Hurricane researchers are focusing on ways to improve forecasts for when a storm will strengthen.

+ fact: Weather satellites ensure that no hurricane will hit land without warning.

+ fact: The National Hurricane Center regularly sends airplanes into hurricanes to gather data that are unavailable via satellite.

Hurricane forecasts are fast improving, but they are still far from perfect. From 1970 to 2012, the average error in track forecasts for three days into the future shrank from 518 miles to 138 miles (834 to 222 km). That is a significant improvement, but it means that if a strong hurricane is forecast to hit in three days, anyone located 200 miles (322 km) on each side of the forecast strike zone should be alert for changes in the forecast and ready to evacuate. Forecasters also assess climate variables to predict the number of hurricanes in a given season, still an inexact science as well.

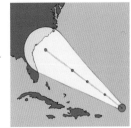

Extratropical Cyclones

In brief, these are large storm systems that form over cold air or cold water away from the tropics. They account for almost all stormy weather in the middle and polar latitudes all year.

KEY FACTS

Fronts are defined by the interaction of air masses at their boundary.

+ fact: At a cold front relatively cold air is advancing to replace warmer air.

+ fact: At a warm front relatively warm air is advancing to replace cooler air.

+ fact: At a stationary front warm air and cold air meet, with neither advancing.

You're not likely to hear a television meteorologist describe the weather system or storm threatening to bring widespread rain or snow as an extratropical cyclone—but chances are it is one. The term applies to any storm that is not a tropical cyclone (or hurricane). Unlike tropical cyclones, these storms contain both warm and cold air masses with fronts as boundaries between them. The temperature contrast between the large masses of cold and warm air supplies the storm's energy. The larger the contrast, the stronger the storm. Extratropical cyclones range in diameter from 600 to 2,500 miles (1,000 to 4,000 km). The fronts usually run like spokes of a wheel from the storm's central area of low pressure.

Winter Storms

Winter storms commonly contain ice, snow, and blizzards, and yet extratropical cyclones that cross North America in the winter can bring thunderstorms, tornadoes, and flooding downpours.

KEY FACTS

Humid winds blowing uphill bring heavy upslope snowstorms to mountains in the West and East.

+ fact: A nor'easter is a winter storm that moves northward along the East Coast of the U.S. and Canada.

+ fact: Alberta clippers are small winter storms that zip from Canada to the East Coast with sharp temperature drops.

From time to time, a strong extratropical cyclone slams into the West Coast from the Pacific Ocean with flooding rain in lower elevations and heavy snow in the mountains. Crossing the Rockies disrupts the storm's surface winds, but as the upper-level winds move over the Great Plains, they stir up surface winds reviving the winter storm with cold air from Canada and warm air from the Gulf of Mexico. Some storms move to the Gulf of Mexico and then northeastward along the Atlantic coast; others travel into the Midwest, bringing blizzard conditions. A winter storm's cold front can stretch into the South to produce severe thunderstorms and sometimes tornadoes. Snow, of course, is also the lifeblood of ski resorts.

Blizzards

A snowstorm becomes a blizzard when winds reach a speed of 35 mph (56 kph) accompanying falling or blowing snow that reduces visibility to less than a quarter mile (400 m).

KEY FACTS

With few trees to slow winds and reduce blowing snow, blizzards are most common on the Great Plains.

+ fact: A severe blizzard has winds faster than 45 mph (72 kph), low visibility, and temperatures of 10°F (-12°C) or lower.

+ fact: Snow already on the ground can cause a "ground blizzard" when wind blows it around.

The combination of low temperatures and poor visibility in blowing snow makes blizzard conditions the most dangerous winter weather. The blowing snow creates a "whiteout" when the horizon disappears, no shadows appear, and objects are hidden. Disoriented victims can become lost in places they know well, even between their house and barn. A more common hazard in today's automobile world is a chain-reaction collision often caused by a driver who stops suddenly when visibility drops to zero. In both cases, low temperatures can be fatal for victims who cannot reach warm shelter soon enough. Many blizzard victims die of carbon monoxide poisoning when they run a snowbound car to keep warm and exhaust gas leaks in, or when they use an unvented heat source such as a charcoal grill indoors.

Lake-Effect Snow

Bitter cold air flowing over much warmer water brings heavy snow to areas downwind of the Great Lakes and a few other bodies of water. The lakes are needed for this weather system.

KEY FACTS

Thundersnow occurs more often with lake-effect snow than with other types of snow.

+ fact: Great Lakes snow bands help to bring as much as 200 in (5 m) of snow a year to West Virginia's mountains.

+ fact: Approximately 500 in (13 m) of lake-effect snow from Utah's Great Salt Lake falls on nearby mountains in a year.

Frigid air blowing across lakes that are at least 20°F (11°C) warmer than the air creates cumulus clouds as the warmer lake water evaporates into the cold air. These clouds dump heavy snow as they move inland and over hills. Places downwind of the Great Lakes, such as Buffalo, New York, often have 100 or more inches (2.5 m) of snow a year—double the amount that falls on places at the same latitude not downwind of the lakes. The greatest amounts of lake-effect snow fall early in the season, before the water in the lakes cools. A lake stops making snow if it freezes over.

River Floods

Rivers flood after prolonged heavy rain over a large area or when deep snow covering a large area melts. Hydrologists can predict river floods in time for those threatened to flee.

KEY FACTS

A floodplain is flat or nearly flat land adjacent to a river that floodwaters naturally cover.

+ fact: Sediments deposited by previous floods make floodplains prime farmland with rich soils.

+ fact: A river flood watch means water is close to rising above flood stage; a river flood warning means flooding may begin soon.

River floods are usually slow-motion disasters, with the highest part of the flood—the crest—moving downstream at less than 10 mph (16 kph). Hydrologists describe actual and forecast flood heights in relation to the "flood stage" at particular gauging stations. Flood stage is the water height at which flooding begins to cause damage at a location. Because flood stage is different at each station, you cannot rely on figures from upstream to indicate what your area should expect. Levees—long, earthen banks that hold back floodwater—sometimes break, which can quickly enlarge the area covered by floodwater. Because forecasting isn't perfect, if flooding is predicted for an area near your home you should be prepared for the flood to be higher than predicted.

Flash Floods

Flash floods, which inundate low-lying areas in less than six hours, are a leading cause of weather deaths in the United States. Intense rainfall or dam failures cause most flash floods.

KEY FACTS

Slowly moving water 2 ft (0.61 m) deep can carry away an SUV-size vehicle.

+ fact: More than half of flash-flood victims were in vehicles driven into water covering a road.

+ fact: A dam failure caused the worst flash flood in U.S. history, killing 2,209 people in Johnstown, Pennsylvania, on May 31, 1889.

Intense rain from thunderstorms or a dying hurricane can create floods that quickly turn usually quiet streams into death traps. In the spring, chunks break off from the ice that has covered a stream. They wash down and pile up into ice dams that can suddenly collapse, creating flash floods. Some of the worst flash floods occur in deserts, where water doesn't soak into the ground. Rain from a thunderstorm too far away to see or hear can race down dry streambeds and catch hikers unaware. Hurricanes and tropical storms that have moved far inland often cause flash floods as well. Those living near streams in hilly or mountainous areas need to be especially aware of the flash-flood danger. A weather radio that turns on and sounds an alarm when a warning is issued can save your life.

Dust Storms

Dust storms, which affect arid regions, fill the air with wind-driven, microscopic dust particles over an extensive area and reduce horizontal visibility to less than $\frac{5}{8}$ of a mile (1 km).

KEY FACTS

The leading edge of a dust storm often looks like a knobby vertical or convex wall.

+ fact: The Arabic word "haboob" is often used for dust storms in the southwestern U.S.

+ fact: Sandstorms do not grow as tall as dust storms because sand particles are larger and heavier. Winds rarely lift sand particles above 50 ft (15 m).

Thunderstorm gust fronts pushing over dry, dusty ground can stir up dust storms. Unlike cold fronts moving into humid areas, which trigger thunderstorms, a cold front in an arid region often lifts dust high into the air. Dust storms can reach a height of roughly 3,000 feet (1,000 m). Cold fronts advancing into dry air create the largest dust storms. In the U.S. Southwest, the dust storm season is May through September. The most severe storms occur when the soil is driest, between April and June, depending on the year's weather. Winds in dust storms are rarely faster than 30 mph (48 kph), but they have been clocked as fast as 62 mph (100 kph). During the 1930s Dust Bowl drought, storms carried dust from the plains to the East Coast.

Volcanic Ash

Volcanic eruptions are not weather events, but can substantially affect the atmosphere. Volcanic ash shot high into the air can stop airplane engines. Some eruptions affect global weather.

KEY FACTS

Between 1982 and 1989, volcanic ash briefly stopped all engines of three Boeing 747s. All three landed safely.

+ fact: Sulfur from the 1991 Mount Pinatubo eruption in the Philippines cooled the Earth for a year.

+ fact: Eruptions in 2010 and 2011 forced *Air Force One* pilots to change President Obama's schedule on three overseas trips.

Both the ash and the gases shot into the air during a volcanic eruption have deleterious effects on weather and daily life. The ash, composed of tiny particles, can block the sun and cool the Earth temporarily. Sulfurous gas shot into the stratosphere during large eruptions forms a haze of sulfuric acid that can block the sunlight and combine with water to form acid rain. When a jet aircraft runs into a cloud of volcanic ash, the tiny particles invade the spaces between moving parts in the engine and drivetrain and can melt and fuse inside the works. This is why volcanic eruptions such as that of Eyjafjallajökull in Iceland in 2010 interrupt air travel until the atmosphere has cleared, which often takes days.

Air Pressure

Even though we hardly notice the air around us, its pressure is one of the most important forces driving the weather. Unequal air pressures in large masses of air cause the winds to blow.

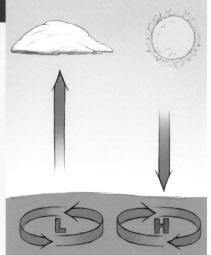

KEY FACTS

At sea level, the air's average pressure is 14.7 lb per in².

+ fact: At 18,000 ft (6,000 m) above sea level, air pressure averages 7.25 lb per in²; half of Earth's air is below that altitude.

+ fact: At 102,000 ft (31,090 m) above sea level, pressure averages 0.147 lb per in²; 99 percent of Earth's air is below.

Dry air consists of roughly 78 percent nitrogen molecules and 20 percent oxygen, with other gases making up the rest. Air is easily compressed, and the pressure at any altitude depends on the weight of all of the air above pressing down. (That is why air pressure decreases rapidly with increasing altitude.) The air's molecules are zipping around at roughly 1,000 mph (1,600 kph)—the higher the temperature, the faster they're going. Fast-moving molecules create pressure pushing in all directions, including up, to oppose the weight of molecules above. We experience the movement of the air as wind. Differences in air pressure at different locations and different altitudes cause winds to blow.

Measuring Air Pressure

Meteorologists measure atmospheric air pressure both at Earth's surface and aloft, because pressure differences between locations determine wind speeds, directions, and weather.

KEY FACTS

The height of mercury in a barometer tube—in inches or millimeters—was the original air pressure measurement.

+ fact: Today the U.S. National Weather Service uses millibars to describe upper air pressures and in surface reports for meteorologists.

+ fact: Canada, like most other nations, uses hectopascals for barometric measurements.

In the late 19th century, as meteorology was becoming a mathematical science, meteorologists began using what are now called hectopascals, a metric unit of pressure, like pounds per square inch, that can easily be used in mathematical formulas. In common parlance, one more likely hears about "inches of mercury," a unit the U.S. National Weather Service uses for surface atmospheric pressure in reports for the public. The phrase harkens back to the mercury barometer, invented by the Italian Evangelista Torricelli in the 1640s. Most weather observers today use electronic devices that sense air pressure rather than mercury barometers. These devices are at the heart of automated barometers, and hikers can easily carry them.

Why Winds Blow

Winds blow as air moves from areas of high atmospheric pressure toward areas of lower pressure. Because the Earth rotates underneath, the wind follows slightly curved paths.

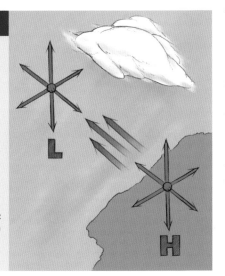

KEY FACTS

The Coriolis force, named for Gaspard G. Coriolis, describes how Earth's rotation causes winds to follow curved paths.

+ fact: It causes counterclockwise winds around large Northern Hemisphere storms and clockwise winds in the Southern Hemisphere.

+ fact: It has no effect on water draining from a sink or down a toilet.

To see why winds blow, let the air out of a balloon or a bicycle tire and feel the escaping air create a mini-wind as it moves from the high-pressure air inside to the lower-pressure outside air. The same phenomenon happens in the atmosphere on a much larger scale as wind blows from an area of high pressure toward an area of lower pressure. Two factors determine the wind's speed: the difference in pressure and the distance between the two areas. Masses of air close together with distinctly different pressure induce strong winds; masses of air far apart with similar pressure induce little to no wind. Because friction with the ground slows winds near the ground, they are normally slower than winds aloft and also low winds above an ocean or large lake.

Measuring Winds

To describe and predict the weather, meteorologists use various anemometers to measure both wind speed and direction. A wind's direction is named for the direction from which it is blowing.

KEY FACTS

Official surface winds are measured by instruments mounted 33 ft (10 m) above the ground.

+ fact: Wind speeds are calculated as 2-minute averages.

+ fact: A squall is a wind 18 mph (30 kph) faster than the sustained (or steady, underlying) wind. A squall happens suddenly and lasts at least 2 minutes.

The observation of wind speed and direction is an ancient art and a modern science. For many years, most weather stations used cup-and-vane anemometers with spinning cups to measure wind speed and with a moveable vane to show the wind direction. Snow and ice could disable these, so measurements during severe weather were disrupted. Today, sonic anemometers are becoming standard. They send ultrasound waves between three arms 4 to 8 inches (10 to 20 cm) apart. Winds slow or speed the sound waves, and a processor inside the instrument determines the time it takes for the sound to travel between the arms, and uses those findings to calculate wind speed and direction. Sonic anemometers work well in turbulence.

Local Winds

Differences in air temperatures over relatively short distances, such as a couple of hundred miles, cause local winds—from gentle sea breezes to the roaring winds that whip up California wildfires.

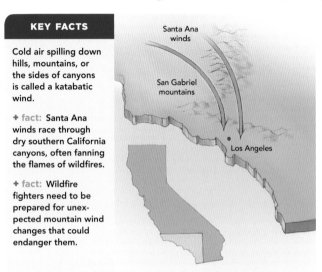

KEY FACTS

Cold air spilling down hills, mountains, or the sides of canyons is called a katabatic wind.

+ fact: Santa Ana winds race through dry southern California canyons, often fanning the flames of wildfires.

+ fact: Wildfire fighters need to be prepared for unexpected mountain wind changes that could endanger them.

Local winds occur apart from large air masses and weather systems, caused by the interaction of winds and geography. California's Santa Ana winds begin when dense, cold air builds up east of the Sierras and the southern coastal range. The cold air spills through the canyons, warming as it falls downhill—a local phenomenon that occurs in many places. Sea breezes begin when land warms faster than water. Air rises over the warmed land and air from over the water flows in to replace it. As mountaintops warm during the day, air begins rising, and air from valleys flows uphill to replace it. At night, as the mountaintop air cools, the air becomes heavier and flows down into valleys, making them colder than nearby elevations.

Regional Winds

Various kinds of winds other than parts of tropical or extratropical cyclones can have major regional effects on the weather. These include monsoon winds and chinook winds.

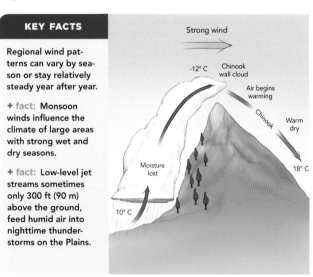

KEY FACTS

Regional wind patterns can vary by season or stay relatively steady year after year.

+ fact: Monsoon winds influence the climate of large areas with strong wet and dry seasons.

+ fact: Low-level jet streams sometimes only 300 ft (90 m) above the ground, feed humid air into nighttime thunderstorms on the Plains.

At the intersection of weather systems, landforms, and major bodies of water, regional weather patterns can be expected. Monsoon climates in Asia and in the southwestern U.S. and adjacent Mexico vary between very dry and very wet seasons. In these regions during the summer, warmed inland air rises and humid winds from the ocean bring humidity, feeding rainstorms. In winter, dry winds blow from inland to the oceans. On the east side of the Rockies, chinook winds warm up as they blow down and melt winter snow. In one case, chinook winds caused temperatures to rise from -54° to 48°F (-48° to 9°C) in 24 hours. Strictly speaking, "monsoon" refers to winds with pronounced seasonal shifts or climates with such shifts. It's also commonly used for heavy rain the humid summer winds bring, or even for any heavy rain.

Global Winds

Global-scale winds blow constantly above the Earth, moving warm air out of the tropics and cold air out of the polar regions, setting the stage for smaller weather events that directly affect us.

In the tropics or the polar regions, the winds blow from the east most of the time. In the middle latitudes, north and south, while the general flow is from the west, storms complicate the surface picture with changing wind directions. The winds high aloft, including jet streams—concentrated horizontal, high-altitude winds—move from west to east with deviations to the north and south. Extratropical cyclones travel generally west to east with diversions like those of the jet streams. When tropical cyclones such as hurricanes move into the middle latitudes, their paths begin curving toward a west-to-east direction. Weather forecasters focus a great deal of their attention on measuring and forecasting global-scale winds because they determine the paths and strengths of storms and their winds and precipitation.

Jet Streams

Jet stream paths follow the locations of cold and warm air at the surface and are intimately linked with the movements of cold air toward the Equator and warm air toward the poles.

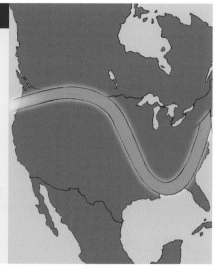

Meteorologists define a jet stream as "a relatively narrow river of very strong horizontal winds embedded in the winds that circle Earth aloft." Jet streams skirt the boundaries between deep layers of warm and cold air—fronts on Earth's surface, and often locations of potentially dangerous weather. A wavy jet stream shows that warm air is moving north and cold air south, possibly destined to mix it up in a new storm. Nevertheless, jet streams and other upper air winds steer storms and determine where areas of high and low pressure form at the surface. Jet streams and surface weather dance with one another, neither one always taking the lead. Fair jet stream winds that dip far over the South are a characteristic of strong winter storms.

The Polar Jet Stream

The northern polar jet stream is an upper-atmosphere band of high-speed winds circling the globe above the ever-shifting boundary between cold, dry polar air and warmer, moist mid-latitude air.

KEY FACTS

The polar jet is fastest and farthest south during the coldest parts of winter, farther north in summer.

+ fact: The polar jet stream helps extra-tropical cyclones form and grow and helps steer them.

+ fact: The Southern Hemisphere's polar jet stream usually circles over the continent of Antarctica all year long.

The polar front separates cold polar air and warm mid-latitude air, but often, polar and mid-latitude air blend with no sharp temperature differences. At such places, the polar jet fades and then forms again where the boundary has larger temperature contrasts. When polar air plunges south, the polar jet turns south and loops around to the north, staying above the warm–cold boundary as a trough. Fast jet streams characterize fierce storms. During the March 12–14, 1963, "superstorm" that paralyzed the East, the polar jet trough dipped all the way over the Gulf of Mexico, and jet stream winds were as fast as 224 mph (360 kph). The extra-tropical storm below this jet stream had surface winds faster than 74 mph (119 kph).

Atmospheric Rivers

Atmospheric rivers are narrow bands of strong low-level winds—5,000–8,000 feet (1,500–2,250 m) above oceans—that feed tropical water vapor to mid-latitude storms.

KEY FACTS

Atmospheric rivers supply between a third and a half of all U.S. West Coast precipitation.

+ fact: An atmospheric river from the Pacific Ocean crossed Central America to feed the February 2010 East Coast "Snowmageddon" blizzard.

+ fact: West Coast meteorologists often call atmospheric rivers the "Pineapple Express."

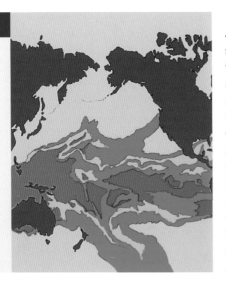

As far back as the 1930s, scientists hypothesized the existence of narrow bands of strong low-level winds that supplied moisture for middle-latitude storms. It took computers and weather satellites to confirm it. In the 1990s, researchers suggested that three to five narrow rivers of air supply 90 percent of the tropical water vapor that reaches the middle latitudes. In 2004, using data collected during airplane flights into these atmospheric rivers and other sources, NOAA scientists confirmed the hypothesis. West Coast forecasters now use these research results to improve forecasts for rain and snow brought by the atmospheric rivers, some of which flow eastward from near Hawaii. Atmospheric rivers also affect Europe and Africa.

Deep Ocean Currents

Water from the Gulf Stream and other currents on the ocean surface are parts of a global conveyor belt that includes underwater currents and transports carbon dioxide and nutrients.

KEY FACTS

Oceanographers estimate that water takes 1,000 years to travel the complete circuit of global currents on and below the surface.

+ fact: Deep ocean water lies more than 6,000 ft (1,800 m) below the surface, where little light penetrates.

+ fact: Deep ocean water is very cold, usually from 32°F to 37°F (0°C to 3°C).

The Gulf Stream is a great river in the ocean that travels northward up the east coast of Mexico, eastward between Florida and Cuba, and northward along the U.S. and Canadian east coast. Water carried north cools and grows denser, with evaporation, leaving behind salt. East of Greenland, this water sinks to form Atlantic deep water, part of the global system of underwater currents. As these currents traverse the deep oceans, they carry organic matter including animal waste and parts of dead plants and animals. The currents also carry carbon dioxide that was absorbed by the water when it was cold. Eventually the water with its nutrients and carbon dioxide upwells along the west coasts of North and South America and Africa, and along parts of the Equator, creating rich areas for sea life.

Surface Ocean Currents

Global winds, such as tropical trade winds, drive oceanic surface currents. The currents carry heat from the tropics to the middle and polar latitudes with important effects on climate.

KEY FACTS

The warmth of the Gulf Stream can strengthen tropical and extratropical cyclones that cross it.

+ fact: Off the U.S. Atlantic coastline, the Gulf Stream moves as fast as 5.6 mph (9 kph).

+ fact: The California current, which moves south along the U.S. West Coast, helps keep coastal waters cool.

Earth's ocean currents form oceanwide gyres—clockwise in the Northern Hemisphere, counterclockwise in the Southern. Through the 20th century, scientists had thought that these currents did most of the work of transporting heat toward the poles. Now, however, there is strong evidence that the atmosphere in the Northern Hemisphere carries 78 percent of the heat moved toward the north, but that ocean currents do most of the work in the Southern Hemisphere, carrying 92 percent of the heat moving toward Antarctica. Researchers are also finding strong evidence that the Gulf Stream doesn't do as much to keep Europe warm as previously thought. Southwest winds also help warm Europe during the winter.

El Niño

El Niño, which happens every few years, occurs when unusually warm tropical surface water flows into parts of the Pacific Ocean. The shifting of warm water has global effects.

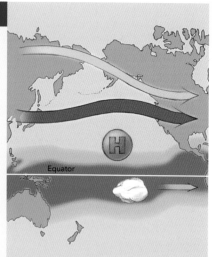

KEY FACTS

El Niño is part of an irregular global climate pattern called the southern oscillation.

+ fact: El Niño usually brings warmer-than-average fall and winter temperatures to the northern U.S. and Canada.

+ fact: El Niño produces high-altitude winds over the Caribbean Sea that can rip hurricanes apart.

Equator

Air flowing out of the tops of Pacific thunderstorms feeds global winds. El Niño pushes these thunderstorms farther east and disrupts jet streams downstream across the Americas. These disruptions, in turn, shift normal patterns of rain, dryness, and storminess as far away as Africa. In North America, the effect is increased rain in normally drier areas and noticeably arid weather in areas that usually get rain. An El Niño occurred in 1957–58, the International Geophysical Year, and scientists began to understand the connection between events that were long assumed to be unconnected. Today, measurements taken in the Pacific Ocean help to predict coming El Niño events, which can have serious economic repercussions.

La Niña

La Niña, the counterpart to El Niño, is a set of global atmospheric events set in motion when the eastern, tropical Pacific cools and the ocean's warmest weather moves to the west.

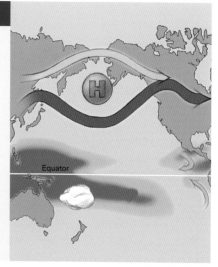

KEY FACTS

Paths of the subtropical and polar jet streams are more variable during La Niña years.

+ fact: La Niña increases the odds of a hurricane hitting the U.S. Atlantic or Gulf coast.

+ fact: La Niña years tend to have warmer-than-normal winters in the southeastern U.S. and cooler-than-normal winters in the northwestern U.S. and Canada.

Equator

Also part of the southern oscillation, La Niña is a Pacific Ocean phenomenon with global implications. El Niño begins with the shift of warm waters in the tropical Pacific to the east, and La Niña begins with enhanced upwelling of deep ocean waters along the South American coast, thus cooling this part of the ocean. Stronger trade winds push the warmest water to the west along with the thunderstorms above it. As with El Niño, winds flowing out of the tops of these storms affect jet streams, but in different patterns. Where El Niño would bring downpours, La Niña generally brings drought. For example, parts of Australia and Indonesia affected by El Niño droughts are wet during a La Niña. In 2011 some scientists linked that year's tornado outbreaks to La Niña, but other scientists disputed this. The question is far from settled.

Arctic Oscillation

The Arctic Oscillation (AO) is an irregular swing between opposite air pressure and wind patterns centered on the Arctic. It strongly affects winter weather in eastern North America.

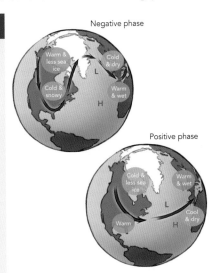

Negative phase

Positive phase

KEY FACTS

Better observations and computers enabled scientists to pin down details of the AO in the 1990s.

+ fact: The most negative AO on record was in February 2010, when three fierce snowstorms hit the United States and Canada.

+ fact: The National Weather Service mentions the AO in winter weather discussions.

The Arctic Oscillation's positive phase features lower air pressures over the Arctic and strong upper air winds around latitude 55° N, which blocks cold outbreaks from hitting the northeastern United States and Canada. The AO's negative phase includes higher air pressure over the Arctic and weaker upper air winds around 55° N, which allows more cold outbreaks to hit the Northeast. The AO can switch between phases in days, but sometimes one phase dominates for long periods. From the early 1960s until the mid-1990s, the AO was positive more often than negative. Since then, the AO has switched phases more often, with extreme negative phases dominating the winters of 2009–10 and 2010–11, which were cold and snowy in the Northeast.

Atlantic Multidecadal Oscillation

The Atlantic Multidecadal Oscillation (AMO) refers to swings in the surface temperature of the Atlantic Ocean between the Equator and Greenland that influence weather widely.

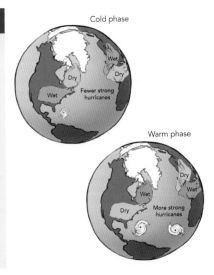

Cold phase

Warm phase

KEY FACTS

During a cold AMO phase, North America and the Caribbean experience more hurricanes.

+ fact: A cold phase from 1971 to 1994 averaged only 1.125 major hurricanes a year; a warm phase from 1995 to 2012 averaged 4 major hurricanes a year.

+ fact: The Dust Bowl drought of the 1930s occurred during a warm phase.

The Atlantic Ocean seems to swing between warm and cool phases lasting 20 to 40 years. Its average temperature during a warm phase is approximately 1°F (0.55°C) above that of a cool phase. Spread out over the ocean, this is a lot of heat, and it can energize hurricanes and affect patterns of high and low atmospheric pressure. Even pressure patterns far from the ocean appear linked to this cycle: African droughts in cold phases, North American droughts in warm phases. Paleoclimatic proxies, such as tree rings and ice cores, show that the AMO has been occurring for at least 1,000 years. It is not an effect of current climate change. Subtle changes in the speed of the Gulf Stream drive the AMO. When it slows, the Atlantic Ocean cools slightly; when it speeds up the Atlantic warms up.

Stationary Fronts

Like all fronts, stationary fronts separate large air masses with different densities—usually caused by temperature differences. Neither air mass is advancing along a stationary front.

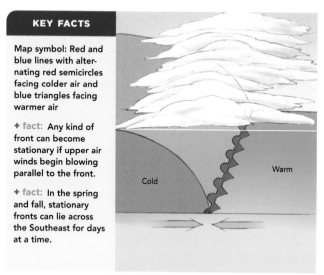

KEY FACTS

Map symbol: Red and blue lines with alternating red semicircles facing colder air and blue triangles facing warmer air

+ fact: Any kind of front can become stationary if upper air winds begin blowing parallel to the front.

+ fact: In the spring and fall, stationary fronts can lie across the Southeast for days at a time.

A stationary front forms when either a cold or a warm front stops moving. Warm, humid air can ride over the front to supply humidity for clouds and precipitation on the cold side of the front. Upper air disturbances can travel along the front, creating clouds and precipitation for days at a time. If an upper air pattern that encourages air to rise moves overhead, a low-pressure area will form on the front. Its counterclockwise winds around the low—in the Northern Hemisphere—begin pushing the warm air toward the north or northwest and the cold air toward the south or southeast to begin organizing an extra-tropical cyclone.

Cold Fronts

A cold front is the leading edge at the surface of a mass of cold air that is replacing warmer air. Showers and thunderstorms and wind shifts accompany most cold fronts.

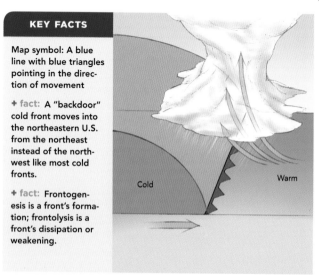

KEY FACTS

Map symbol: A blue line with blue triangles pointing in the direction of movement

+ fact: A "backdoor" cold front moves into the northeastern U.S. from the northeast instead of the northwest like most cold fronts.

+ fact: Frontogenesis is a front's formation; frontolysis is a front's dissipation or weakening.

As a cold front advances, the colder and denser air behind it wedges under the less dense warmer air, lifting it. If the warm air is moist and the atmosphere is unstable—the usual case in North America—this lifting forms showers and thunderstorms. These thunderstorms can be very strong, even severe, especially in the spring when the atmosphere is often unstable. During the winter, when a cold front reinforces dry, cold air already in place, little snow or rain might fall. A reliable sign that a cold front has passed is a wind shift from southwesterly to northwesterly. The coldest air is often a few miles behind the front.

Warm Fronts

A warm front is the boundary where warm air is replacing colder air. The clouds associated with a warm front can be more than 700 miles (1,100 km) ahead of the front.

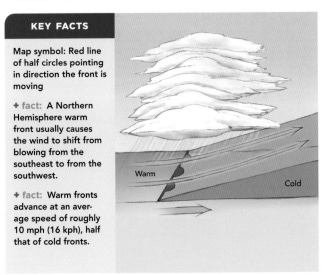

KEY FACTS

Map symbol: Red line of half circles pointing in direction the front is moving

+ fact: A Northern Hemisphere warm front usually causes the wind to shift from blowing from the southeast to from the southwest.

+ fact: Warm fronts advance at an average speed of roughly 10 mph (16 kph), half that of cold fronts.

An advancing warm front doesn't arrive with the drama of a strong cold front, but it affects a much larger area. Because warm air is lighter than cold air, a warm front's air rises over the cold air. The warm air can be 6,000 feet (1,800 m) above the ground and 150–200 miles (249–320 km) ahead of the front. As a warm front approaches, you will first see high cirrus clouds, which become cirrostratus or cirrocumulus. These thicken and descend to become altocumulus and altostratus clouds. Snow or rain could then begin, as when you see nimbostratus clouds. After the surface front passes, the sky will begin clearing and temperatures will warm up.

Occluded Fronts

Unlike stationary, cold, and warm fronts that divide two air masses with contrasting densities, occluded fronts are more complex. They separate three air masses: cold, cool, and warm.

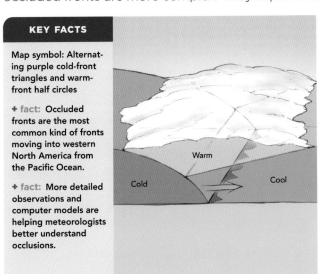

KEY FACTS

Map symbol: Alternating purple cold-front triangles and warm-front half circles

+ fact: Occluded fronts are the most common kind of fronts moving into western North America from the Pacific Ocean.

+ fact: More detailed observations and computer models are helping meteorologists better understand occlusions.

In a "cold" occlusion, the surface boundary separates very cold and cold air with the warm air appearing to have been shoved up by the very cold air to intersect the very cold air aloft. In a "warm" occlusion, cool air is riding over very cold air with the warm air above the cold air. Many textbooks say "warm air catching up with cold air" forms occlusions. Some meteorologists today, using more complete observations and computer models, say a better description involves the warm front and cold front wrapping around the cyclone's low-pressure center after the warm front separates from the low-pressure center.

Dry Line

A dry line, like a front, separates air masses of different densities. These differences are in humidity, not temperature, as with most fronts. Dry lines occur on the Southwestern Plains.

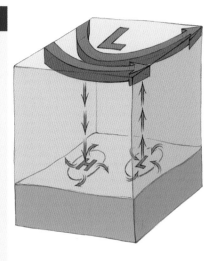

KEY FACTS

Weather map symbol: A gold line with adjacent half circles along the line

+ fact: Dry lines are most often located in western Texas but have been observed as far north as the Dakotas and as far east as the Texas–Louisiana border.

+ fact: Dry lines usually move east in the afternoon and back to the west at night.

The air masses that are in conflict along a dry line are very dry, warm, or hot air moving east from the Southwest and humid hot air moving west from the Gulf of Mexico. Because humid air is less dense than dry air of the same temperature, the dry air pushes under the humid air much as cold air shoves under warm air. This can trigger showers and thunderstorms, sometimes severe thunderstorms with tornadoes, much as advancing cold fronts cause storms. Dry air is denser than humid air because added water molecules are lighter than the nitrogen and oxygen molecules they replace as humidity increases. Thunderstorms tend to form where the dry line bulges and pushes air up. These storms usually are more isolated and severe than those that form elsewhere because they aren't competing with other storms.

Upper Air Troughs

Upper air troughs are elongated areas of low atmospheric pressures relative to adjacent air pressures at particular altitudes. They influence the locations, strengths, and paths of storms.

KEY FACTS

Air rises on the eastern side of a Northern Hemisphere trough aloft, which helps surface storms to intensify.

+ fact: Air sinks on the west side of a Northern Hemisphere trough, creating a dry, high-pressure area below.

+ fact: A trough aloft forms when the air below is colder than air on either side of the trough.

Weather forecasters pay particular attention to troughs aloft because they have a major influence on the weather below by helping storms to form or intensify. When you hear broadcast meteorologists talk about possible bad effects on local weather from "upper air energy" or an "upper air disturbance," they are probably talking about a trough or a "cutoff low." The low formed when the southern end of a trough was pinched off to become an upper-air low pressure area that is disconnected from the upper air wind flow. Cutoffs can hang around for days in the same place or even move to the west, causing cloudy skies and precipitation before they dissipate. A trough and an upper air ridge—where winds aloft turn to the north and back to the south—make one of the three to seven meandering waves that circle Earth.

The Aleutian Low

The Aleutian low is a semipermanent area of low atmospheric pressure that strengthens each fall and fades during spring. From fall through spring, it steers storms into the Pacific Northwest.

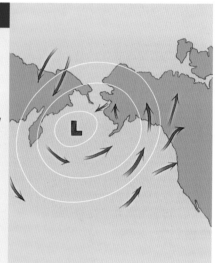

KEY FACTS

In winter, the Aleutian low regularly sends storms into the Pacific Northwest with only brief breaks.

+ fact: During an El Niño in the tropics, the Aleutian low tends to be deeper and spins off stronger storms.

+ fact: A weak Aleutian low can occur any year, but it is especially likely during a strong La Niña.

The Aleutian low forms each winter as Alaska begins turning frigid, leaving the Pacific Ocean around the Aleutian Islands the warmest surface in the region. Because the water is relatively warm, air over it begins rising, forming a low pressure center that continues until spring when the land warms up. Storms that form or strengthen here often have winds faster than 50 mph (80 kph), which create huge waves that surfers in Hawaii love—some as tall as 65 feet (20 m) high when they head south. Waves heading north are the ones that make the Discovery Channel's *Deadliest Catch* show exciting. The Aleutian low's Atlantic Ocean counterpart is the Icelandic low. In summer when the Aleutian low is weak, the North Pacific High moves north so it is west of California, strengthens, and keeps the West Coast mostly dry.

The Bermuda High

While the Bermuda high is centered far out over the Atlantic Ocean, it has several effects on North American weather, including steering Atlantic hurricanes toward or away from the U.S.

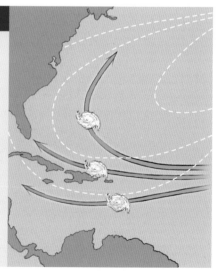

KEY FACTS

The high doesn't always protect Bermuda from hurricanes. One hurricane, on average, hits Bermuda every seven years or so.

+ fact: In 2004 the high was farther west than usual and steered a record four hurricanes to Florida.

+ fact: In winter and spring, the high is centered over the Azores as the Azores high.

Bermuda is in the global belt of high atmospheric pressure and mostly calm winds around latitude 30° N, where air rising in tropical thunderstorms descends to maintain high atmospheric pressure. Winds flowing clockwise out of the high contribute to the easterly tropical trade winds and often feed warm humid air into the eastern U.S. in the summer, especially the Southeast. The clockwise winds around the high steer hurricanes. Slight changes in the strength and position of the high help to determine whether a hurricane heads northward between Bermuda and the U.S., hits the Northeast, the Southeast, or heads into the Gulf of Mexico. Predicting how the Bermuda High and winds around it will change is a major aspect of forecasting the likely paths of hurricanes in the Atlantic Ocean.

Ground/Radiation Fog

Radiation fog forms when heat radiates from the Earth overnight, cooling the air enough for its water vapor to begin condensing into tiny water drops. The fog forms next to the ground.

KEY FACTS

Ground fog is usually thickest before sunrise when evaporation starts.

+ fact: Fog cuts visibility to 0.6 mi (1 km); mist cuts visibility to less than 6 mi (9 km).

+ fact: The NWS reports "fog" when it is less than 20 ft (6 m) deep, and "shallow fog" when it is less than 6 ft (2 m) deep.

Radiation fog, also called ground fog, is most likely to form when the sky has been mostly clear all night, which allows the most heat to radiate away from Earth. Winds should also be calm or nearly calm because stronger winds will mix the coldest air next to the ground with slightly warmer air a few feet higher. Chances of fog are much better when rain has soaked the ground the day before. Water from soaked ground evaporates into the air. The added moisture allows water vapor to began condensing at a higher temperature than it would in drier air. The fog usually begins evaporating shortly after sunrise. Afternoon rain that leaves water on the ground to evaporate, a sky that clears overnight, and 5 to 10 mph (8 to 16 kph) winds combine to make morning fog likely.

Steam Fog

The wisps of "steam" you see rising from ponds or lakes in the fall when the year's first cold air arrives are "steam" fog, which needs a combination of warm water and cold air to form.

KEY FACTS

Steam fog forms when warm water evaporates into cold air, making the air humid enough to form fog.

+ fact: All year, steam fog forms above thermal ponds in Yellowstone National Park.

+ fact: Steam fog begins a few inches above the water because the rising air needs to cool enough to form fog.

Arctic air that begins moving south over North America in the fall is too dry to form fog without some help. This help comes when it flows over ponds, lakes, or rivers, and some of the relatively warm water evaporates into the cold air, giving it enough humidity for condensation to begin at the air's current temperature. Steam fog that forms over an ocean is called "sea smoke." When frigid air moves over much warmer water, the rising fog can create steam devils up to 1,600 feet (500 m) high. In a case studied, the water was 39°F (22°C) warmer than the air. Most fog forms in light winds, but steam devils illustrate the turbulence associated with steam fog. It forms under extreme temperature differences between warm water and frigid air. The Great Lakes are a prime location for steam devils.

Advection Fog

Advection fog forms when winds push humid air over ground or water cold enough to chill the air to a temperature that causes its humidity to condense into tiny fog drops.

KEY FACTS

Coastal advection fog supplies 30 to 40 percent of moisture to California's redwoods.

+ fact: Warm Gulf Stream air advected over the nearby Labrador Current makes Newfoundland's Grand Banks one of Earth's foggiest places.

+ fact: Summer breezes across the cool Great Lakes form persistent advection fog.

Meteorologists use the word "advection" for the horizontal transport of meteorological property such as temperature or moisture. Some of North America's most troublesome fog is the widespread, dense, and long-lasting advection fog across the Midwest created when warm, humid air moves north from the Gulf of Mexico. Southern California's famous "May Gray" and "June Gloom" marine layer is advection fog that is formed when tropical Pacific air flows over the cold California Current near the coast. This type of fog normally rolls in early in the morning and dissipates during the day. Radiation fog does not form over oceans as it does over land, which quickly radiates heat away on clear nights.

Ice Fog

Ice fog is made of tiny ice crystals that float in the air, just as water drops do in ordinary fog. It forms only at temperatures below -30°F (-35°C)

KEY FACTS

Ice fog particles are small enough for ten to fit side by side on the edge of a piece of paper.

+ fact: When the air is -40°F (-40°C), water from a car's tailpipe drops from 250°F (121°C) to the air temperature in less than 10 seconds.

+ fact: Ice fog is sometimes called pogonip, from a Shoshone word for "cloud."

Because ice fog needs such low temperatures, it occurs only in the northern provinces of Canada, in inland and northern Alaska, and in some high elevations in the Pacific Northwest. Ice fog can also be called frozen fog, but it should be distinguished from freezing fog, which is made of water drops that instantly freeze when they touch anything. Ice fog forms only in extremely dry air, but in built-up places such as Fairbanks, Alaska, exhaust from vehicles adds water vapor to the air, and that can instantly become tiny ice fog particles. In Fairbanks, the resulting thick ice fog—with other pollutants added—can drop visibility to near zero and push air quality to unhealthy levels. A few miles from the city's traffic, the air can be clear.

Natural Haze

Haze is a collection of very fine, widely dispersed, solid or liquid particles suspended in the air. They turn the sky milky white and subdue colors. It can be natural or from human activities.

KEY FACTS

The chemical properties of haze particles can change the effects of the haze on the view.

+ fact: Volatile organic compounds from trees help to form a bluish haze: That's what makes the Blue Ridge Mountains blue.

+ fact: Salt particles in the air affect beach scenes, creating a low, white haze in the daytime and red sunsets.

Haze has a direct effect on how far we can see, because its particles are just the right size in relation to wavelengths of light to scatter or absorb some colors, which reduces visibility. We often think of any such obstruction to visibility as pollution, but it's not always the case. Haze is sometimes air pollution—particles or gases in the air added by human activity with the potential for harming life or property. But haze can be natural, coming from a volcanic eruption, or smoke from wildfires started by lightning, or dust. Water vapor can condense on dry haze particles and make them larger, which further reduces visibility by scattering or absorbing more light. This condition most likely occurs in the morning or evening, when relative humidity is higher.

Pollution Haze or Smog

Photochemical smog, a brownish haze, is a mixture of hundreds of hazardous chemicals. It is most often found over and downwind of cities. It is unpleasant and unsightly, and can be deadly.

KEY FACTS

Henry Antoine des Voeux, a London physician, coined "smog" for smoke and fog in 1905.

+ fact: Reactions involving sunlight, nitrogen oxides, and volatile organic compounds produce photochemical smog.

+ fact: Ozone near the Earth's surface is a pollutant, but ozone in the stratosphere blocks dangerous ultraviolet light.

An inversion—air aloft that's warmer than ground-level air—sets the stage for smog: It blocks warm air from rising, so that it cannot be replaced by clean air descending from above. At times, an inversion can trap polluted air over a city for days as more smog brews. In addition to the visible smog, invisible pollutants are at work, including carbon monoxide and extremely small particles so tiny that they travel deep into the lungs, causing damage that can be fatal. Rain and snow wash pollutants out of the air, but civilization often produces so much pollution that this natural cleaning process cannot keep up. Volatile organic compounds (with noticeable odors) from industrial and natural sources, such as trees, are one (but far from the only) component of smog.

Rainbows

When sunlight, a mix of colors, enters and reflects out of a water drop, it bends, with each color bending at a different angle. The light reflects off the drop's back and is visible as separate colors.

KEY FACTS

In 1637, French philosopher René Descartes discovered how refractions in individual drops form rainbows.

+ fact: Rainbows may be partial if it is raining only in part of the sky directly opposite from the sun.

+ fact: Large raindrops create bright rainbows with sharp colors; small drops make washed-out colors.

Rainbows are visible phenomena, but are not actual objects, and your eyes must be in the right relationship between the sun and raindrops to see them. If the sun isn't directly in back of your head, you aren't looking at a rainbow. (You could be seeing iridescence or a circumzenithal arc.) Light hits the raindrops, which act like tiny prisms, dividing it into its component colors as it shines back out. Double rainbows are made by the same raindrops that make the primary rainbow. You often hear that rainbows have seven colors, but a rainbow has the whole range of colors from red at the top to violet at the bottom of a primary bow. Usually we see fewer than seven colors.

Fogbows and Moonbows

A fogbow is created the same way as a rainbow but by water drops in fog, not rain. Rain or a waterfall can disperse the light of a full moon to create a moonbow. Both are rare.

KEY FACTS

Some waterfalls are notable for their moonbows including those in Yosemite National Park and Cumberland Falls near Corbin, Kentucky.

+ fact: Photos taken with the camera on a tripod and long exposure times show moonbows best.

+ fact: You might see a fogbow with a "glory"—a circular bow—at its center.

Fogbows are much fainter than rainbows because light is traveling through much smaller water drops, and their colors aren't as sharp as those of rainbows. They usually appear white, sometimes with faint red and blue. Moonbows will occur only when the moon is full and brightest, but not directly overhead: For a moonbow to occur, the moon cannot be more than 42°, or a little less than halfway up from the horizon to the zenith. As with a sunlit rainbow, the moon must be behind your head as you look into the rain or waterfall, where drops of water are bending the incoming light and dividing it like a prism into its component colors. Full moons set around sunrise, and the hours before sunrise are the best time to look for a moonbow.

Sun Dogs

"Sun dog" is an informal name for a parhelion, a splotch of light seen on one or both sides of the sun. They are the most common visual display caused by ice crystals floating in the air.

Parhelia—the plural of parhelion—are colorful, glowing spots formed by light bending as it enters and leaves plate-shaped, hexagonal ice crystals floating facedown in cirrus clouds. In most parts of the world, sun dogs appear at a 22° angle higher than the sun and on one or both sides of the sun. They can appear as often as a couple of times a week, most visibly when the sun is close to the horizon. Once you start looking for sun dogs when cirrus clouds are in the sky, you might be amazed how often you see them (and you can amaze others by pointing them out, and explaining how ice causes them).

Atmospheric Halos

Arcs or circles shining around the sun or moon, atmospheric halos are the visible effects of light passing through ice crystals in the atmosphere.

Ice crystals hanging in cirrus clouds high in the upper troposphere, 3 to 6 miles (5 to 10 km) above the ground, can create visible arcs, circles, and spots by reflecting and refracting light. As in rainbows, the white light is sometimes split into its component colors. Atmospheric halos have been observed and interpreted for millennia, sometimes as signals of weather to come and sometimes as spiritual messages. Today, the optics of the phenomenon are fully understood, but the marvel of halos remains. On January 11, 1999, for example, those at the South Pole saw 22 different kinds of halos.

Circumzenithal Arc

A circumzenithal arc is a ring directly overhead or almost directly overhead. It is easier to miss than most other halos because you're not likely to look straight up.

KEY FACTS

If you see a sun dog with the sun 15° to 25° above the horizon, look straight up. You might see a circumzenithal arc.

+ fact: When looking at halos, you need to shield your eyes from the sun.

+ fact: Regular skywatchers might see a circumzenithal arc 25 times a year.

Unlike most halos that appear white, maybe with a tinge of color, a circumzenithal arc is colorful. It has been described as an "upside-down rainbow." The colors normally range from blue on the inside to red on the outside—the side toward the sun. Bending of light rays as they enter and leave ice crystals causes all halos. Such refraction separates sunlight into its colors. Hexagonal columns shaped like a pencil cause circumzenithal arcs when light enters the top face and leaves through one of the sides. The arc forms only when the sun is lower than 32.2° above the horizon. The arc can be wider than shown here. Some call them fire rainbows because they sometimes have a fiery look, but they are not rainbows.

Moon Dogs

A moon dog is the lunar version of a sun dog: a glowing spot or pair of spots visible alongside the bright full moon, created by moonlight passing through ice crystals in the atmosphere.

KEY FACTS

The scientific name for a moon dog is paraselene (plural paraselenae), meaning "beside the moon."

+ fact: You see little color in moon dogs because they aren't bright enough to activate your eye's cones, which perceive color.

+ fact: A few nights before and after the full moon, it is bright enough to make halos.

Because the moon is a source of light in the sky, it produces halos, including moon dogs, just as the sun does. If ice crystals are present in the atmosphere, moonlight coming through may reflect and refract that light, making it visible in these attractive patterns. Moon dogs and halos are much dimmer than those generated by the sun, because all of the moon's light is reflected. Moon halos are more common than moon dogs. Folklore says that a ring around the moon means that rain is on the way. Maybe. Cirrus clouds that create halos are sometimes, but not always, a sign that a warm front is arriving with rain.

Coronas

You see a corona when the sun or moon shines through thin clouds. It appears as a bright center (the sun or moon) surrounded by one or more reddish or brownish rings.

KEY FACTS

A safe way to view a corona around the sun is to look at its reflection in water.

+ fact: Coronas are often seen in altocumulus or altostratus clouds, which are often part of a storm that could bring rain or snow.

+ fact: Coronas have colors when the cloud drops or ice crystals in the cloud are the same size.

Rainbows and halos are visible when water or ice causes the refraction, or bending, of light rays. Coronas (and other sky phenomena such as iridescence) are visible when water drops in the atmosphere cause the diffraction, or spreading and rejoining, of light rays. Coronas are created by smaller water particles than halos, and they are more commonly seen. Sometimes the physics of the light and water droplets is such that colors are visible. When that happens, red is always the outermost color and sometimes the only color. This use of the word should not be confused with its astronomical meaning, referring to the sun's outer atmosphere. A corona's size depends on the diameters of the cloud drops. Small drops produce large coronas. A corona is clearest when the drops are mostly the same size.

Crepuscular Rays

Close to sunset or sunrise, beams of light will often appear as if they are radiating from the sun, coming through breaks in the clouds and fanning out as they come down to Earth.

KEY FACTS

"Crepuscular" is derived from the Latin word *crepusculum*, which means "twilight."

+ fact: People sometimes say "the sun is drawing water," harking back to the ancient Greek belief that sunbeams draw water into the sky.

+ fact: Crepuscular rays are often red or yellow because air molecules selectively scatter blue light.

You see the "beams" of light—called crepuscular rays—because haze, dust, or tiny water drops floating in the air scatter the sun's light in all directions, including essentially parallel lines toward you. From your vantage point, the beams appear to converge in the sky because of the principles of perspective—the same reason the edges of a straight road appear to converge at the horizon. While you're looking at crepuscular rays, turn around, put the sun at your back, and if the sky is clear, you may see "anticrepuscular" rays converging on the "antisolar" point opposite the sun. If the air were perfectly clean, you would not see crepuscular rays. Light from the sun is traveling a straight path from the sun to Earth, not toward you. Tiny particles such as dust scatter light in all directions, including toward your eyes.

Sun and Light Pillars

Near sunrise or sunset, bright columns of light appear to shoot above and below the sun: sun pillars. On a frigid night, you may see columns shooting into the sky from streetlights.

Halos become visible when the sun's light rays bend as they pass through ice crystals, but both sun pillars and light pillars emerge when light reflects off ice crystals instead. This is why they are the color of the source of the light—red beaming above the red setting sun near the horizon—instead of being mostly white. Sun pillars usually extend at an angle of only 5° to 10° above the sun. Light pillars form on clear nights that are cold enough to form ice-crystal fog, composed of tiny crystals. This form of fog is also called diamond dust.

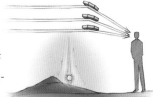

Auroras

Auroras, shimmering curtains of green and brown-red lights seen high in the skies of Earth's far north and far south, are visible evidence of our planet's direct connections to the sun.

Solar winds send energetic charged particles from the sun toward the Earth, attracted especially into the atmosphere over the Arctic and Antarctic by the electromagnetic fields of our planet. Here they smash into the atoms of nitrogen and oxygen, the primary constituents of our atmosphere. These collisions send off photons, which create visible light: the eerie, shimmering, sky-filled curtains of the aurora borealis (northern lights) over the Arctic and aurora australis (southern lights) over Antarctica. The auroras are the only visible aspect of space weather: the many effects on Earth of high-energy particles from the sun, which can disrupt satellites.

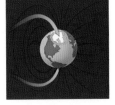

Iridescence

Iridescent clouds have washed-out, mostly pastel colors visible on some or all of the cloud. Most often, iridescence appears as a border of red and green along the edge of a cloud formation.

KEY FACTS

Every cloud does not have a silver lining: Some have colored iridescent linings; most have no lining at all.

+ fact: Iridescence can be seen in cirrus, altocumulus, cirrocumulus, and lenticular clouds.

+ fact: Iridescence is more visible if the sun is shaded, either by a denser cloud or deliberately by the viewer stepping behind a tree or building.

Sometimes a cloud will seem to show colors similar to a rainbow contoured to the shape of the cloud. Iridescence—the appearance of colors within or along the borders of a cloud—is caused by tiny water drops or small ice crystals, each individually scattering and diffracting light. The optics that create an iridescent cloud are similar to those that create colors on the surface of an oily puddle. Iridescence develops primarily in thin clouds positioned relatively close to the sun. The colors can be very subtle or rather bright, but they will shift and change quickly. The colors you see in an iridescent cloud can be considered fragments of a corona. Unlike coronas that form in clouds or parts of clouds with drops of relatively uniform size, iridescence shows that the cloud is made of drops with different sizes.

Glories

From your window seat on an airplane just above the clouds, you're casually looking out when you see the shadow of your airplane with a ring of light around it. You're looking at a glory.

KEY FACTS

Before flying became common, the best way to see a glory was from a mountain above cloud tops.

+ fact: You sometimes see a glory around the shadow of your head on the cloud.

+ fact: The shadow glory is called a "Brocken spectre," for the highest peak in Germany's Harz Mountains, known for its glories.

A glory, a colorful halo encircling a shadow in the clouds below the viewer's eye level, is not as simple as light scattered back from the cloud to create colored rings. A phenomenon that appears at the antisolar point—directly opposite the sun in relation to the viewer—a glory is visible only to a person positioned between the sun and the top layer of clouds. It must be explained by a more complex physical process than simple reflection or diffraction. Physicists are still unsure of the process, and determining the atmospheric optics of glories is a challenge to this day. Their findings could apply to climate science because of the possibility that clouds reflect more sunlight than previously thought.

Green Flash

As you watch the sun set over an open-water horizon, you may see a green flash: a momentary change in the color of the sun to green before it disappears.

KEY FACTS

The best place to see a green flash is from a view at a high elevation above an ocean horizon.

+ fact: Looking for a green flash with binoculars or an SLR camera viewfinder can permanently damage your eyes.

+ fact: In Antarctica, where the sun rises and sets slowly—once a year for each—a green flash can last a half hour.

In locations where a large lake, bay, or the ocean stretches out to the west, far as the eye can see, many people gather in early evening to try to catch a glimpse of the elusive green flash. It may look like a green spot for a second or two or like a green ray shooting up from the sun. There is a physical explanation for this. The atmosphere refracts or bends light's different wavelengths in a standard order of colors. As the sun slips below the horizon, the red disappears first, then a second or so later the yellow, green, and finally blue and violet disappear. Because air scatters blue and violet the most, these colors don't reach your eye. But green might reach you, enhanced as a mirage caused by a layer of warm air over a relatively cool ocean.

Twilight

Twilight is the transition between night and day: from the time light first appears in the morning until sunrise, and from the time the sun reaches the horizon until light fades from the sky.

KEY FACTS

The time from first twilight to dark is 72 minutes.

+ fact: The brightest stars and planets are visible during civil twilight if the sky is clear.

+ fact: The sky turns brilliant colors at twilight because the visible light has traveled through more atmosphere, scattering blue and violet rays, leaving red and yellow.

Twilight—the time between last or first light and the brightness of day—has been defined in many ways. Civil twilight is the period when the sun's center is between the horizon and 6° below the horizon, and the ambient light is usually still bright enough for outdoor activities. Nautical twilight is the period when the sun's center is between 6° and 12° below the horizon, and shapes are visible but not distinct. Astronomical twilight is the period when the sun's center is between 12° and 18° below the horizon. After that, twilight ends, and it is dark. The term "twilight" can also be applied to the same periods of time as the sun is rising.

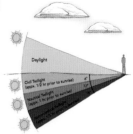

Thermal Inversions

A thermal inversion (usually called just an inversion) is a layer of warm air above colder air. Temperatures normally decrease with altitude, which makes a thermal inversion unusual.

KEY FACTS

Cold air flowing into valleys pushes up warm air, creating inversions.

+ fact: An inversion caused the 1948 pollution episode in Donora, Pennsylvania, that killed 20 people and sickened hundreds.

+ fact: August inversions at the South Pole can make surface temperatures 70°F (39°C) colder than those 1,000 ft (305 m) above.

Surface inversions form in winter when heat radiates away from Earth faster than solar warmth can replace it. The strongest inversions form in polar regions when the sun doesn't rise for weeks or months, and these are sources of cold waves. In warmer places, inversions can develop overnight. Inversions also form where air from aloft is sinking to create surface high pressure. Inversions sometimes form caps that keep warm air from rising, preventing thunderstorm development. When rising air finally breaks such an inversion, the hot air that has been bottled up can quickly rise to form severe thunderstorms.

Blue Sky

Anyone who isn't color-blind and who looks at the sky knows it's blue. But why? The sky's blue has baffled some of the world's best minds, from the ancient Greeks to 19th-century physicists.

KEY FACTS

Air scatters violet more efficiently than blue, but sunlight contains less violet than blue.

+ fact: Scattered light moves in all directions. Reflected light rebounds according to the angle of its arrival in the air.

+ fact: Outer space is black because it contains nothing to scatter light.

From Aristotle to Leonardo da Vinci and Isaac Newton, the greatest minds of science throughout world history have tried to explain why the sky is blue. Many ideas included reflection of one kind or another. Finally the British physicist John William Strutt, usually referred to as Lord Rayleigh, solved the puzzle in the 1870s. Since Newton, it had been known that sunlight is composed of all colors, which can be split apart by a prism. Lord Rayleigh suggested that some colors in light scatter more readily than others. Blue scatters the most; hence, the sky looks blue, except around sunrise and sunset.

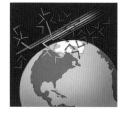

Mirages

In meteorological parlance, a mirage is not a hallucination—it is an optical distortion caused by atmospheric conditions. In fact, you can photograph a mirage, as many people have.

Mirages are caused when layers of air at different temperatures bend light in various ways, distorting an apparent object into an image that does not correspond with physical reality. The greater the temperature differences, the more striking the mirage. Air near the ground that is warmer than air higher can create an "inferior mirage," with objects appearing lower than they really are. For instance, the sky might appear as a pool of water in the desert. A thermal inversion—air near the ground that is colder than air higher—can create a "superior mirage," with objects appearing higher than they are. Striking superior mirages occur especially in the polar regions, including some that show objects that are below the horizon to the viewer's eye.

Fata Morgana

A fata morgana is a complex mirage with elements both compressed and stretched, both inverted and right side up, combining into visions that change quickly.

Steep thermal inversions with warm air over cold water, cold air, or polar ice help create the atmospheric ducts needed to evoke a fata morgana mirage. An atmospheric duct is a layer in the lower atmosphere that guides light along Earth's curvature. The components of a fata morgana can shift between being inferior and superior mirages. Some fata morganas create multiple images, alternately expanded and compressed vertically. At times they appear to be buildings of a city or hills of an island where none exists. Over the years, fata morganas have led polar explorers to map islands or other lands that don't exist. Numerous reports of ghost ships, sometimes in the air, are likely the results of fata morganas.

Thunderstorm Safety

An estimated 100,000 thunderstorms hit the United States yearly. Ten percent of thunderstorms are severe—with winds of 58 mph (93 kph) or faster, large hailstones, or a tornado.

KEY FACTS

When thunderstorms are likely, make sure to stay near shelter in case a storm develops.

+ fact: A severe thunderstorm watch means severe thunderstorms are possible; you should be ready to take shelter.

+ fact: A severe thunderstorm warning means a severe thunderstorm has been spotted; you should take shelter.

Lightning is the big danger in all thunderstorms. Severe thunderstorms add the danger of winds that can topple trees and send debris flying. A sturdy building is the best shelter against both. Your best defense is not being caught far from safe shelter—hiking up a mountain, two hours or more away from safety, for instance—when a thunderstorm hits. Although forecasters can give advance warning about hurricanes and winter storms, they can rarely predict when and where a thunderstorm will hit more than about a half hour ahead. They can, however, pin down general areas where thunderstorms are likely to occur hours ahead of time. Use these general alerts to plan time outdoors to ensure that you won't be caught in the open.

Lightning Safety

Lightning is one of the top three weather killers most years in the United States. If you are anywhere outdoors when you hear thunder or see a lighting flash, lightning could hit you.

KEY FACTS

If you see lightning or hear thunder, no matter how far away, immediately take shelter.

+ fact: Enclosed buildings with electrical wiring and plumbing or enclosed metal vehicles are the best shelters.

+ fact: If lightning catches you outdoors, don't squat or lie on the ground and don't hide under a tree. Run for shelter as fast as you can.

The lightning to be concerned about is a powerful but very brief electrical current seeking a path to the ground with the least resistance. To avoid death or injury, you need to avoid being that easiest path to the ground that it is seeking. When you are inside an enclosed building with electrical service and plumbing, lightning will find the building's wires or pipes to be a better path to the ground. But stay away from water and from anything plugged in, including a telephone with a wire, when you are inside. (A cell phone is safe.) An enclosed vehicle is safe because electricity goes through the metal and to the ground, but a strong lightning strike could blow out the tires.

Heat Exhaustion and Heatstroke

High temperatures and humidity upset the body's systems for regulating temperature. Perspiration is the body's natural cooler, but humidity hinders evaporation, making you hotter.

KEY FACTS

Heatstroke, also called sunstroke, is a medical emergency worth a 911 call.

+ fact: The National Weather Service Heat Index combines temperature and humidity for an "apparent temperature."

+ fact: Many cities have cooling centers where the elderly and others endangered by heat can relax in air-conditioned places.

The dangers presented by extreme heat and humidity are different for healthy men and women than for the elderly and people in ill health who don't have air-conditioning. If you know someone in this latter category, checking on them through a heat spell could be a lifesaver. For those in good health, the dangers arise when a person is so focused on a workout or sport that signs of trouble, such as cramps or thirst, are ignored. Heat exhaustion with heavy sweating, weakness, and pale, clammy skin is serious. Victims should get out of the sun—into air-conditioning if possible—and sip water. Heatstroke with a high body temperature could be next, leading even to unconsciousness. This would be a medical emergency.

Tornado Safety

Tornadoes are the strongest storms on Earth, but they are small in reach and relatively rare, with the strongest ones extremely rare. It's still a good idea to know what to do if one threatens you.

KEY FACTS

Flying debris is a tornado's greatest danger; avoiding it should be your main safety goal.

+ fact: Don't waste time opening windows as a tornado approaches; flying debris will do it for you.

+ fact: Don't even think of sheltering under a highway overpass; wind squeezing through makes it a debris-laden wind tunnel.

Because a tornado's path cannot be predicted until it is on the ground, you and your family need to have a tornado plan. You could even conduct a tornado drill, as schools do. Because your goal is to avoid flying debris, you should take shelter in a low room with no windows. A basement under a strong workbench or table is best. In a high-rise building, the interior stairway is the best shelter. Forget the elevator—tornadoes bring power failures. Many schools and shopping centers on the Great Plains, where tornadoes are more prevalent, have marked tornado shelters. Walk-in coolers have saved many people when tornados hit restaurants or convenience stores. If a tornado threatens, you should seek shelter.

Hurricane Safety

Two good rules for responding when a hurricane threatens your home or where you are staying are: Run from the water and hide from the wind. To do so, you need to know the flood danger.

KEY FACTS

Water is the biggest hurricane killer: both the water that a storm pushes ashore and the flash floods it can create inland.

+ **fact:** Wind can destroy a well-built house if the windows are not protected against flying debris.

+ **fact:** You need to check flood maps to see whether a hurricane could flood your home.

Over the centuries and around the world, hurricanes and tropical cyclones have killed more people with water than with wind. Usually, the highest death tolls are from flash floods from downpours dumped by dying storms far inland. If inundation maps show that your home is well clear of any possible surge, you can think about hiding from the wind at home. But, as Hurricane Andrew in 1992 proved in South Florida—the region of the U.S. where hurricanes are most likely to hit—expensive homes were not necessarily built to withstand storms. If you expect a need to evacuate, plan where you'll go and what you'll take before hurricane season begins. You should have storm centers or plywood window covers ready before hurricane season begins.

Cold Weather Safety

Cold weather brings two unique health threats: frostbite (freezing of tissue such as your fingers) and hypothermia (a lowering of your body's core temperature).

KEY FACTS

The U.S. Antarctic survival schools stress staying hydrated to help avoid hypothermia and frostbite.

+ **fact:** The U.S. National Weather Service and Environment Canada used wind-tunnel tests with volunteers to develop windchill charts.

+ **fact:** Hypothermia can occur at temperatures above 40°F (4.4°C) if a victim is wet.

Frostbite is not life threatening unless it is untreated and leads to gangrene. Hypothermia is always life threatening if not stopped in time. Windchill figures can alert you to the danger of both. Windchill does not change the temperature. If the temperature is above freezing with the windchill below freezing, your fingers and toes cannot freeze. Nevertheless, a frigid windchill means wind is carrying warmth away from your body faster than in calm air. This can speed hypothermia, when your body's core temperature falls below 95°F (35°C). First aid for hypothermia includes removing wet clothing and carefully warming a victim without burning him or her. Mental confusion can be a symptom of hypothermia.

Winter Storm Safety

If a winter storm traps you at home or in your car, be prepared to stay warm to avoid hypothermia. Carbon monoxide poisoning is a major winter storm danger.

KEY FACTS

Roughly 70 percent of deaths related to ice and snow occur in vehicles.

+ fact: A winter storm watch means you should prepare for snow or ice that closes roads and interrupts electric power.

+ fact: About 50 percent of those who die from hypothermia are over age 60.

A winter storm watch should signal you to stock up on food, water, prescription medications, and other things you will need if you find yourself trapped at home with roads closed by ice or snow and possibly having no electrical power for a few days. These threats are worse in the South, where snow and ice are uncommon and people are less well prepared. Anything that burns fuel—a generator, a charcoal grill, or backpacking stove—produces carbon monoxide, so be sure to use these in ventilated places. If you must go anywhere in a car, you should be dressed to survive for a few hours in the conditions outside, so that if you slide off the road or find yourself otherwise immobilized, you can survive the cold.

Flood Safety

The main flood danger, especially in flash floods, is drowning when trying to walk or drive into floodwaters. The danger continues after waters recede; bacterial contamination remains.

KEY FACTS

You shouldn't wade in floodwater if the water is moving faster than you can walk, or if you cannot see the bottom.

+ fact: A mere 18 in (46 cm) of water is enough to lift and carry a car or SUV downstream with a good chance of it rolling over.

+ fact: More than half of those killed in floods are inside vehicles.

An essential element of flood safety is keeping healthy after a flood. When you clean up, you may not want to think about what was in the water that soaked everything, but you must think about it enough to protect yourself from diseases and hazardous chemicals. If possible, find out what could have been in the floodwater. Your hepatitis A and tetanus shots should be current. The Federal Emergency Management Agency recommends wearing boots for flood cleanup and having bleach and water to decontaminate them before getting into your car or going into your home or garage. Also, remember that floods can seriously weaken buildings, so be aware of structural dangers as well. Flooded buildings can harbor venomous snakes.

Surface Weather Observations

Weather data such as temperature, atmospheric pressure, wind speed, and wind direction are raw materials for weather forecasts and the ground truth used to check forecast accuracy.

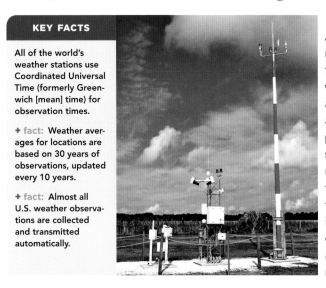

As the computers used to produce weather forecasts become more powerful, they can handle more and more data. Data are being collected and fed into the National Weather Service (NWS) network not only from traditional weather stations but also from automated stations at numerous small airports and locations operated by state, local, and private highway operators. Electronic devices that sense data such as atmospheric pressure make automated stations possible. In addition to traditional observations, the NWS receives data including water levels in streams, dryness of woodland areas, and soil temperatures and moisture levels.

Aviation Weather

Aviation is extremely weather dependent, hence the effort that the Canadian Weather Office and the U.S. National Weather Service put into collecting data for aviation forecasts.

Automated stations at airports provide all sorts of information essential to pilots. The height of the cloud ceiling and current visibility are essential elements in flying and air traffic management. Pilots normally point into the wind for takeoff and landing, so wind speed and direction are necessary knowledge. If the wind is not blowing directly down a runway, pilots prefer to use the runway that is nearest to pointing into the wind. Pilots flying into small airports can use a special radio frequency to hear the airport's automated weather report. The U.S. National Weather Service produces most U.S. aviation forecasts. The Aviation Weather Center in Kansas City supplies nationwide predictions, and local NWS offices make forecasts for airports. Some airlines employ forecasters.

Measuring Snow

Scientists are working on high-tech instruments to measure the amount of snow that falls, but for now, plywood left on the ground and rulers are still the best way to track snowfall.

KEY FACTS

To be considered measurable, at least 0.1 in (0.25 cm) of snow has to fall; less is reported as a "trace."

+ fact: With powder snow, approximately 20 in (51 cm) amounts to an inch of water.

+ fact: In a heavy, wet snow, only 5 in (13 cm) of snow may amount to an inch of water.

National Weather Service volunteer observers use a "snowboard"—a 16-by-16-inch (41 cm) piece of plywood on the ground in a place without drifts or bare spots as snow falls. Each hour, the volunteer uses a ruler marked in tenths of an inch to measure the snow and then brushes the board clean. Another board that is not brushed off measures the snow on the ground after natural settling and melting. In the western mountains, the U.S. Natural Resources Conservation Service uses automated instruments that measure the weight of snow and then convert it into water content. That information helps forecasters to predict water flows in western rivers when the snow melts. Melting snow is also an important source of soil moisture for farmers.

Upper Air Weather Observations

Data about winds, temperatures, humidity, and other atmospheric properties measured as high as 100,000 feet (30,000 m) above Earth are vital for making today's forecasts.

KEY FACTS

Radiosondes rise until the balloon bursts, above 100,000 ft (30,000 m); tiny parachutes lower them to the ground.

+ fact: If you find a radiosonde, directions on the side tell how to return it.

+ fact: From 1898 to 1933, the Weather Bureau used kites to collect data up to 10,000 ft (3,000 m).

Since the 1940s, basic upper air data have come from weather balloons carrying radiosondes, small boxes that collect and radio back the data twice a day. Approximately 800 global weather stations launch balloons once or twice a day. Extra balloons are launched when more data are needed on a storm. The main supplement to balloon data is automated reports that many airliners send every few minutes measuring temperatures, wind speeds, and wind directions. When all U.S. airline flights were grounded September 11–13, 2001, the accuracy of forecasts dropped significantly. Forecasts for three hours ahead were only as accurate as 12-day predictions using constant airline data. The forecasts most affected were those especially intended for aviation, but others were also less accurate.

High-Altitude Turbulence

Although turbulence isn't likely to bring down one of today's airliners, it regularly causes injuries to passengers en route. Researchers are seeking ways to warn pilots of turbulence.

KEY FACTS

Most lower-altitude severe turbulence is in or near clouds; pilots know which clouds to avoid.

+ fact: FAA statistics show that passengers not wearing seat belts account for 98 percent of turbulence injuries.

+ fact: High-altitude turbulence encounters are mostly in clear air with no warning.

Most high-altitude turbulence is in or near jet-stream winds with few clouds. Because radar tracks by sensing cloud water drops or ice crystals, turbulence outside of clouds is invisible to an airliner's radar. The best forecasters can do now is predict where turbulence is likely to occur. An airplane's turbulence encounter is a good detector for other airplanes, and pilots and controllers exchange real-time turbulence information via radio. Researchers are developing a LIDAR (a radar that uses light instead of microwaves) for airplanes, which shows promise of spotting turbulence. Others are testing ways to use data from ground-based Doppler radars, which collect more data than airliners' radars, to spot high-altitude turbulence.

Weather Research Airplanes

Unlike many other scientists, meteorologists can get inside the phenomena they study. Since weather extends to the edge of space, scientists need aircraft to examine the weather in full.

KEY FACTS

NOAA's two WP-3D hurricane hunters began flying into hurricanes in 1977 and are still at it.

+ fact: In 1987, NASA's ER-2, based in Chile, collected Antarctic ozone hole data.

+ fact: In 2012, NASA's Global Hawk, a large unmanned airplane, made the first of many planned flights observing hurricanes from above.

Weather balloons collect needed data, but to study the weather scientists must be able to say: "What's that? Let's take a look." Research flights into hurricanes grew out of the reconnaissance flights by the U.S. military during World War II. Meteorologists on some flights learned new things about hurricanes and saw the value of studying storms from airplanes. This has led to today's research fleet, which includes NASA's ER-2, a civilian version of the U-2 spy plane; two Gulfstream business jets equipped as flying research labs; and NOAA's two WP-3Ds, which are best known for hurricane flights but have helped scientists investigate weather phenomena globally, including El Niño, winter storms, ocean winds, Great Plains thunderstorms, and low-level jet streams.

Tornado Chasing

Since the 1970s, scientists have realized that the only way to improve tornado forecasts is to collect extensive data on what happens inside the supercells that spawn the strongest twisters.

KEY FACTS

VORTEX-1 produced the first full documentation of the life cycle of a tornado.

+ **fact:** The worst tornado in U.S. history happened in 1925, killing 695 people in Missouri, Illinois, and Indiana.

+ **fact:** A person is more likely to fall off a cliff or contract leprosy than be killed by a tornado.

Tornado researchers began using portable Doppler radar devices in 1995 to collect unprecedented close-up views of tornadic supercells. VORTEX (Verification of the Origins of Rotation in Tornadoes Experiment) is the largest tornado research project ever, designed to study how, when, and why tornadoes form. In 1994–95, 18 vehicles collected data as part of VORTEX-1; in 2009–10, roughly 100 men and women in 40 vehicles collected data from 11 supercells as part of VORTEX-2. The goal of these storm chases was to find changes inside supercells that could be detected, measured, and used in the future as indicators that a strong tornado is likely to form. VORTEX-2 scientists will be studying data and presenting analyses for another decade.

Weather Radar

Radar transmitters emit microwaves that objects such as raindrops scatter, some back to the radar. Computers convert the signals into information such as precipitation location.

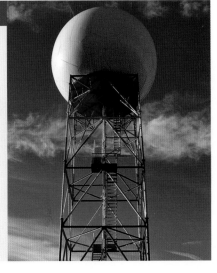

KEY FACTS

Radar is an acronym for RAdio Detection And Ranging.

+ **fact:** Today's Doppler radars detect precipitation intensity and locations plus wind speeds and directions.

+ **fact:** There are nearly 200 weather radars in the United States and Canada.

The U.S. and the United Kingdom developed radar during World War II to detect and track enemy ships and aircraft. When researchers realized that they were encountering interference caused by precipitation—rain or snow—they began applying radar to track weather phenomena as well. The U.S. National Weather Service (NWS) completed its first nationwide radar network in 1967, and radar quickly became important for both forecasters and researchers. Today, radar meteorology is a separate branch of the science. In 2012, NWS finished updating NEXRAD to dual polarization, which supplies more data about the nature of precipitation.

Weather Forecasting

Computerized predictions are the core of all of today's professional weather forecasts.
As computers have gained power and speed since the 1950s, predictions have improved.

KEY FACTS

A 40 percent chance of rain means that any place in the area has a 40 percent chance of at least 0.01 in (0.25 mm).

+ fact: "Isolated showers" should affect 20 percent or less of an area; "scattered showers," 30–50 percent; "numerous showers," 60–70 percent.

+ fact: Wording such as "rain today" means that more than 80 percent of an area will be affected.

Forecasts begin with weather data from around the world flowing into the National Weather Service (NWS) National Centers for Environmental Prediction in College Park, Maryland. Supercomputers use the data to run several "models." These are computer programs that use equations of fluid dynamics and thermodynamics to predict weather around the world, with more detailed forecasts for the U.S. These are sent as maps and data that local NWS offices and other meteorologists use as starting points for their predictions. The results of different models enable forecasters to get what amount to second, or even third, opinions on what's likely to happen. These products are available on the Internet for anyone to use.

Weather Satellites

Weather satellites are such a part of our lives today that we're no longer amazed by their stunning images, such as of hurricanes. They collect other data as well as taking pictures.

KEY FACTS

The U.S. launched the world's first weather satellite, Tiros 1, on April 1, 1960.

+ fact: Since 1960 satellites have had infrared sensors that measure cloud-top temperatures, which indicates their height.

+ fact: Some polar orbiters' infrared sensors "see" water vapor in the air with their "vapor channel."

Almost all the satellite images you see come from geostationary satellites, which orbit 22,238 miles (35,788 km) above the Equator. At this altitude their orbital speed matches Earth's rotation, keeping them above the same spot. They scan an entire disk of Earth. The U.S. has two of these, which with similar satellites of other nations give global coverage. Polar orbiters circle Earth from north to south and back 540 miles (869 km) high. Their closer view allows them to collect more detailed atmospheric data. For example, one polar orbiter's cloud-top temperature sensor detects changes 540 miles below—equivalent to detecting whether a lightbulb 22.6 miles (36 km) away is 100 watts or 101.6 watts.

Weather Radio

Some weather warnings, such as those for tornadoes and flash floods, require immediate action. The consequences of missing a warning could be dire.

The NWS began Weather Radio in 1967. From then until the 1990s, local meteorologists recorded on tape.

+ fact: In the late 1990s a text-to-computer system using "Paul" was introduced, but many didn't like his voice.

+ fact: The text-to-voice system introduced in 2000 has "Donna," "Tom," and a Spanish voice, "Javier."

All National Weather Service (NWS) offices broadcast warnings on NOAA Weather Radio, which requires special receivers to hear. The necessary radios, which pick up the seven VHF frequencies used for weather radio, are available at most stores that sell electronics. Many of these radios can be used also as ordinary radios. Most have a feature that will automatically turn the radio on and sound an alarm when a warning is broadcast. Newer weather radios have Specific Area Message Encoding (SAME), which allows those interested to program the radio to turn on and sound the alarm only for warnings for an area that is specified, not other parts of the region the local NWS office covers.

Ocean Observations

Weather doesn't begin at the ocean's edge. Forecasters need data from along the coast and far out at sea to make forecasts. Automated buoys and coastal stations meet this need.

The U.S. maintains Pacific Ocean buoys collecting data to track El Niño.

+ fact: Hurricane Sandy in 2012 hit a New York Harbor data buoy with a record 32.5-ft (10 m) wave.

+ fact: A 141 mph (198 kph) wind at the Fowey Rocks C-MAN station was the highest recorded when Hurricane Andrew hit Dade County, Florida, in 1982.

The short towers and moored buoys marked as belonging to NOAA are supplying weather data NWS forecasters said they required, as in the 1980s. The U.S. Coast Guard was automating lighthouses, whose keepers were also weather observers. The NWS set up its National Data Buoy Center to maintain the Coastal-Marine Automated Network (C-MAN) stations along the U.S. coast as well as buoys moored in the water to collect data. When a hurricane or extratropical cyclone is battering the shore, forecasters and the public turn to reports from these stations to see how bad the storm is. Even more important, the stations also supply vital information on average winds and waves as well as extreme events in the same way weather stations on land supply the climate data needed for planning.

Short-term Forecasts

A general rule in meteorology is that the further into the future a weather forecast projects, the higher are the odds that the forecast will not be accurate.

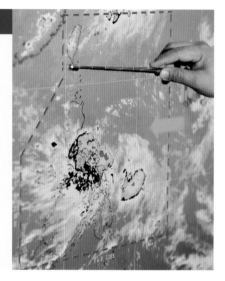

The U.S. National Weather Service (NWS) and the Canadian Weather Office, along with many private forecasters, regularly produce day-to-day forecasts ten days ahead. In general, it is wisest not to make any important decisions based on a forecast for more than two or three days into the future. The atmosphere is an extremely complex system in which minor differences in the present can have major consequences in the near future. Furthermore, the atmosphere is chaotic, with random changes that no model or computer can predict accurately. When a forecast contains a percentage—"a 40 percent chance of rain," for instance—the meteorologist has calculated both physical likelihood and confidence in the prediction.

Long-term Forecasts

The NWS Climate Prediction Center produces outlooks for short and long periods: from the next 6 to 14 days to the 3-month period ending 12 months ahead.

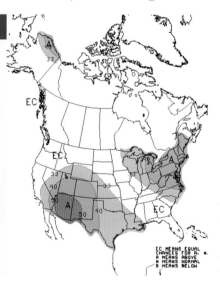

Weather outlooks are intended for rather sophisticated users who are comfortable with probabilities. All outlooks are for averages over monthlong or three-month periods. They will never be specific enough to select a fall weekend for a college homecoming, but they will have general information one could use to estimate fuel oil needs through the period. Still, it's a gamble: If the long-term forecast calls for a cold winter, you can contract to buy oil for the season at a low but fixed price. But if the winter turns out to be warmer, the cost could be below the total price you paid. Nevertheless, by using long-term forecasts, you would be like a professional poker player who loses occasionally but wins more often.

Weather Watches

The terms "watch" and "warning" for potentially dangerous weather may sound somewhat the same, yet they have distinctly different meanings.

KEY FACTS

Severe storm, tornado, and flash-flood watches should make you extra alert for a potential event.

+ fact: Hurricane and winter storm watches give you a small amount of time to get last-minute supplies.

+ fact: You and your family should spend time deciding how to respond to different kinds of weather watches.

The U.S. National Weather Service (NWS) and Canada's Weather Office issue a weather watch when forecasters see a chance that a dangerous event will occur but do not yet consider it a sure thing. A given locale will be put under storm watch, for instance, when the storm is still far enough away not to pose an immediate threat. What you should do when you hear a storm watch depends on the kind of event. A tornado could happen in an hour or so after NWS issues a watch. A hurricane could arrive 48 hours after the watch begins. Depending on the kind of event, it can mean that you should stay alert and be ready to act, or it can mean that you should prepare for an event that could greatly impact your daily life, such as a hurricane or winter storm.

Weather Warnings

A weather warning indicates imminent danger. When it is issued, you should already have used the weather watch period to prepare to keep you and your family safe from the hazard.

KEY FACTS

Each weather warning includes specific advice for the wisest steps to take for safety.

+ fact: A flash-flood warning means leave quickly if you're in a location subject to flash floods.

+ fact: A red flag warning means conditions are right for wildfires to start and spread rapidly, so great care must be taken and no open fires should be lit.

A weather warning means dangerous conditions are occurring or will occur soon. NWS and Canada's Weather Office consider a warning to indicate a threat to life or property. Ideally, you heard the watch and already planned what to do in response. Tornado warnings often advise people to take shelter in a place safe from flying debris. When a winter storm warning is in place, you should stay off the road, remaining safe in your home or elsewhere with the supplies you'll need for a day or two. Hurricane warnings are issued approximately 36 hours before the brunt of the storm is expected, because responses to an oncoming hurricane are more complex and take more time. You should evacuate if you're in a possible flood zone or a home that won't withstand high winds.

The Milky Way and other celestial attractions illuminate
the night sky above Death Valley, California.

5 | Night Sky
Nature's Nocturnal Entertainment

We tend to overlook the night sky, because nighttime itself is something we work around; often nothing but a light switch stands between our day and night activities. But the night sky is worth knowing: It is our window on the past, present, and future. Knowledge of night sky phenomena—moon, stars, planets, comets, and galaxies—draws us in toward the origins of the universe, helps us understand the flow of the seasons, and makes us ponder our planet's destiny. Above all (no pun intended), observing the night sky is the ultimate free pastime, available to all without a reservation.

■ The Basics

The naked eye remains the most important piece of "gear" for sky watching, combined with a vantage point as free as possible from ambient light. Getting your night vision in shape is the first step. Allow your eyes to adapt to the dark (which takes a minimum of 15 to 20 minutes) and then keep them that way. A glance at a car headlight can undo the process. Because you'll need a flashlight to read this chapter or sky charts, tape a piece of red cellophane over a regular flashlight lens or buy a red-lensed light, preferably LED. Dim red light interferes less with night vision.

Choose the darkest location you can find. At home, turn off your indoor and outdoor lights. A hill, field, or park is a better choice, as far from city light as possible. In the country, find an observation point that keeps city lights to your back. A black cloth or jacket over your head can help block out further light.

Look for a clear night; the sky is especially clear of humidity and haze following a cold front or late-afternoon storm. Developing high-pressure systems also create clear skies. Be sure to prepare for the chill and dampness that can creep in as the night wears on. A beach-type lounge chair offers a relaxing viewing angle, especially when settling in for a meteor show.

■ Getting Oriented

Star maps show the night sky as if all the celestial objects revolved around the Earth. It helps to picture the Earth as encased in a sky globe with the stars attached to it. This globe has an equator and poles, just as the Earth does. Stars are assigned celestial coordinates similar to latitude and longitude, and these appear on many sky charts. Declination, like latitude, describes how far the star appears above or below the celestial equator. Right ascension is similar to longitude, divided like

Earth's day into 24 hours, with each hour equal to 15° of circumference. From our point of view, the stars pass above from east to west. It is important to know your approximate latitude and the reference latitude of any star guide you're using. The charts in this book, for example, are based on a view from latitudes close to 40° N.

To use the large seasonal sky charts on pages 394 through 401, choose the appropriate one for the season and try to observe at the time indicated for each month. Turn the map so the name of the direction you are facing appears right side up. Then use this orientation of the chart to help locate constellations and other objects.

The seasonal charts and the smaller maps that accompany constellation descriptions in this book show other objects such as star clusters, nebulae, and galaxies. The meanings of these symbols appear on the seasonal charts. Most objects shown are numbered. Many numbers have an M in front; those were listed and catalogued by 18th-century French astronomer Charles Messier and range from 1 to 110. An overlapping and much larger list is the New General Catalogue; objects from this list carry the letters NGC.

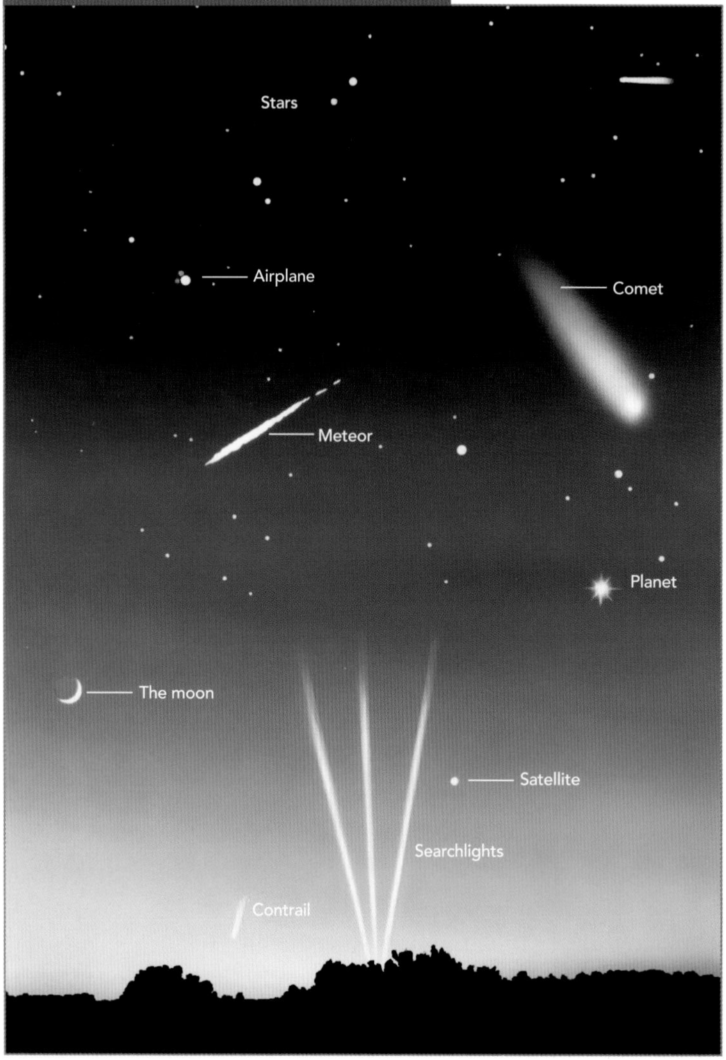

■ What's Out There

So what am I looking at? Are those little lights the stars, planets, planes, meteors, or what? Where are those gods, goddesses, and animals with their connect-the-dot outlines? There are no outlines, just stars. But with time and practice, you can discern the shapes of many constellations and learn to tell apart stars and satellites, comets and spacecraft, planets and airplanes.

Stars move from east to west through the night. They twinkle because their pinpoints of light are perturbed by fluctuations in the atmosphere. Planets don't usually twinkle because they are much closer to us. Unlike starlight, planets' disk shapes reflect light from the sun, and this relatively broader width of light is less affected by the atmosphere. Five planets are visible to the naked eye: Mercury, Venus, Mars, Jupiter, and

ELEMENTS OF THE NIGHT SKY

Stars

Airplane

Comet

Meteor

Planet

The moon

Satellite

Searchlights

Contrail

Time-lapse star trails etch the darkness at Arches National Park in Utah.

Saturn. Mercury is a bit elusive, but can be viewed well at certain intervals during the year. Venus is the second brightest object in the night sky after the moon. It shines a brilliant white. Mars glows orangish red, while Jupiter is a steady white, and Saturn is pale yellow. Planets take the same path that the sun and moon take across the sky.

The blurry band of light running across the sky, high in winter and summer, is the Milky Way. Yes, we also reside in the Milky Way, but we are able to view other parts of it. Small blurry blobs in the sky are star clusters, nebulae, or perhaps even a galaxy, such as the Andromeda. It appears as a little blur in the constellation Andromeda and is the most distant object visible to the naked eye.

Fast-moving objects include meteors ("shooting stars"), which are bits of solar system debris that flame out in Earth's atmosphere. Slower and steadily moving objects with steady, blinking lights are airplanes. Satellites are visible just after sunset or

||||||||||||||||||||||||||||||||||||||||||||||||||||||||||||||||

*But when I follow at my
pleasure the serried multitude
of the stars in their circular course,
my feet no longer touch the earth.*
—PTOLEMY

||||||||||||||||||||||||||||||||||||||||||||||||||||||||||||||||

before sunrise, or during the night depending on latitude; they appear as small, starlike lights that move smoothly before disappearing into Earth's shadow.

■ Equipment and Resources

As with any hobby, there is a range of gear that can take sky watching to the nth degree, if you want to. But beginners need to acquire no more than a basic pair of binoculars to enhance their viewing.

The night sky's wonders are accessible with or without equipment.

your best choice. This size is large enough to let in enough light and small enough to wear around your neck and hold steady.

A telescope is a much larger investment and should be carefully chosen. Avoid the low-end ones sold in toy stores and "big box" stores. The poor quality and results of these mass-produced instruments will only disappoint. Instead, go to a specialized store where experts can help.

Much more of what you would like to know and learn about the night sky is just a few taps away on your computer. There are many wonderful instructional and even interactive websites that offer all kinds of practical information about every aspect of sky watching. An example is the website for *Sky & Telescope* (www.skyandtelescope.com), a print publication that has been guiding amateur astronomers for many decades. Among other features, it allows you to create a custom-made naked-eye map of the sky for any location. And of course there are apps by the dozens for your mobile phone that will show you the stars and planets overhead and help you locate them. Many of these are available free or for only a small fee. But, before you turn to your gadgets, you might want to learn about the night sky in the low-tech way that humans have done for millennia. Your well-trained eyes on their own can take you a long distance into its wonders.

Binoculars can reveal a fantastic array of objects and details such as the larger craters on the moon, the distant planets Uranus and Neptune, and even beautiful star fields along the Milky Way. A pair of 7x50 binoculars (7 being the power of magnification and 50 the diameter of the outer lens) is probably

HOW TO MEASURE DISTANCE

Distance across the night sky is typically measured in degrees. You can roughly measure degrees and distance using only your hands. Outstretch your arms and hold out your hands against the sky. At arm's length, a thumb covers approximately 2° of sky, a fist covers about 10°, and a fully outstretched hand about 20°. These approximations can be used to estimate the width and height of constellations. Each constellation entry in this book gives a width measurement using this system. Of course, hands come in all sizes, so these measurements are just approximations for average adult hands.

NIGHT SKY

The Atmosphere

Any sky watching we do is affected by the atmosphere—miles-high layers of gases, liquids, and solids that envelop the Earth. The first few layers contain most of the atmosphere's mass.

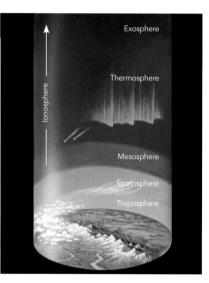

KEY FACTS

The troposphere (to 6–10 mi) is where clouds form and weather occurs.

+ fact: The stratosphere (10–30 mi/16–48 km) is the radiation-absorbing ozone.

+ fact: The mesosphere and thermosphere (30–210 mi/48–338 km) are where temperatures drop, then rise steeply.

+ fact: The exosphere (above 210 mi) contains hydrogen and helium.

Light from the sun, stars, and other celestial objects travels great distances to reach Earth. The portion of light that reaches Earth's atmosphere must navigate the properties and activities that each layer of atmosphere contains. These cause further reductions in the light that will eventually reach the surface. The most challenging transit is through the troposphere, the layer between the surface and six to ten miles above it. This layer holds the largest concentration of atmospheric mass, including more than 97 percent of water vapor. Clouds, dust, chemical pollution, and light pollution combine there to hinder our observations of the night sky.

Nightfall

Sunset initiates nightfall, but the often spectacular colors that accompany the event reveal only part of the interactions among the Earth, the sun, the moon, and the atmosphere.

KEY FACTS

The higher the latitude, the longer twilight lasts because of the angle at which sunlight reaches Earth.

+ fact: In the far north during and near the summer solstice, twilight lasts all night long.

+ fact: The opposite phenomenon occurs in the far north's winter, when little or no sunlight arrives during the day.

As the sun sets in the west, the atmosphere disperses the sun's light like a prism. The long wavelengths of red, orange, and yellow in the spectrum reach us even after the sun has set. Meanwhile, the atmosphere scatters short-wavelength blue and violet at the opposite end of the spectrum, and almost none of these colors reach us. Look to the east and you may see a deep blue band across the horizon—the shadow of Earth, cast on the sky. The moon and stars replace the sun. The amount of moonlight we see at night depends on the reflected light of the sun; changing amounts cause the moon's phases. Atmospheric conditions also affect the moon's brightness.

Noctilucent Clouds

In northern locations, twilight sometimes illuminates high-flying noctilucent clouds that originate far above the region where most clouds occur.

KEY FACTS

Noctilucent clouds were first observed in 1885, two years after the Krakatau Volcano erupted, and initially linked to the eruption.

+ fact: When Krakatau's ashes disappeared, noctilucent clouds remained.

+ fact: Noctilucent clouds (NLCs) are also called polar mesospheric clouds (PMCs).

+ fact: Atmospheric methane increases formation of NLCs.

Clouds typically occur in the troposphere, which ranges from 6 to 10 miles (10 to 16 km) above the Earth. Noctilucent (night-shining) clouds form at much higher altitudes in the mesosphere, which extends to 50 miles (80 km). These thin, wispy clouds appear pale white or an eerie electric blue, and they often occur above the typical orange-red layer of sunset. The arid, frigid mesosphere offers poor conditions for cloud building, but noctilucent clouds assemble from tiny ice crystals and possibly meteor smoke. They are visible in summer when the sun has dipped about 10° below the horizon and usually at altitudes above 40° N. Astronauts on the International Space Station have viewed the clouds from their unique vantage point and have taken scientifically valuable photographs of the phenomenon.

Magnetosphere

A magnetic shield sustained far beyond Earth's atmosphere by the planet's inherent magnetism, the magnetosphere deflects solar and cosmic radiation.

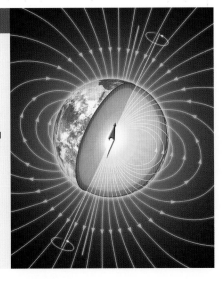

KEY FACTS

On the side of Earth facing the sun, solar wind compresses the magnetosphere to about 40,000 miles.

+ fact: Geologic and other studies show that the magnetic field changes over time.

+ fact: The location of the north magnetic pole has wandered over arctic Canada.

+ fact: In the future, the north magnetic pole will move into arctic Russia.

Earth's molten metal core generates a magnetic field on a cosmic scale. Magnetic current flows outward from the two poles and loops back toward the center, forming a magnetosphere that extends tens of thousands of miles into space. It deflects the steady stream of atomic particles known as the solar wind, which can knock out communications and electronics during intense geomagnetic storms. Buffeting by the solar wind causes the magnetosphere to change shape during the day. It becomes compressed on the side facing the sun, and it elongates into a long tail extending past the orbit of the moon on the opposite side. The sun and the other planets also have magnetospheres, but Earth's is the strongest among rocky planets of the inner solar system.

Van Allen Belts

Consisting of two concentric zones of intense radioactivity that gird the Earth, Van Allen belts are named for James Van Allen, the physicist who discovered them in 1958.

The Van Allen belts harness atomic particles that have penetrated the magnetosphere, including protons and electrons from outside the solar system and helium ions from the sun. The trapped, highly charged particles mingle with the atmosphere and travel a spiral path as they bounce back and forth between Earth's magnetic poles. The belts can be visualized as concentric doughnuts, thickest at the Equator and weaker at the poles. The inner belt begins approximately 600 miles (966 km) above Earth's surface and extends to about 3,000 miles (4,800 km). The outer belt extends from about 10,000 to 25,000 miles (14,000 to 24,000 km). In 2013, NASA announced the detection of a third belt by twin satellites named the Van Allen Probes.

Auroras

A by-product of solar storms, auroras are visible mainly in polar areas. The light show is called aurora borealis in the Northern Hemisphere and aurora australis in the Southern Hemisphere.

During violent solar events, such as mass coronal ejections and solar flares, the discharge of highly charged particles from the sun increases astronomically, shooting out billions of tons of matter at speeds as fast as 2 million miles (3 million km) an hour. This barrage breaches the magnetosphere and Van Allen belts, reaching perhaps as close as 50 miles (80 km) to the Earth's surface. The abundant solar material energizes molecules of oxygen and nitrogen in the atmosphere, causing dazzling curtains of green and red and pink light that intensify in proportion to the intensity of the solar storms in their 11-year cycle.

Stars

From their earliest formation about 13 billion years ago, stars have been a universal sky feature. A few thousand can be seen with the naked eye on a clear, moonless night.

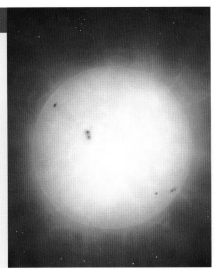

KEY FACTS

The sun is a third-generation star with elements from two previous star explosions.

+ fact: The sunlight currently shining on Earth was generated in the star's core 50 million years ago.

+ fact: Sunlight takes 8.3 seconds to reach Earth.

+ fact: A neutron star fragment the size of a sugar cube would weigh about 100 million tons on Earth.

Elementally simple—composed almost entirely of hydrogen and helium gases—stars have not yet revealed the precise details of their formation. In basic outline, gravity pulls inward dense patches of gases and dust, causing pressure and heat to build at the center of the ball of gas and triggering nuclear fusion. The energy created, which we see as starlight, pushes outward and is balanced by the pull of gravity, establishing equilibrium. Stars are born in nebulae, large gas clouds found throughout the Milky Way. They also come into being from the supernova explosions of older stars that create raw material for new stars. The nuclear reactions within stars form the heavy elements carbon, oxygen, and nitrogen—the essential building blocks of life as we currently understand it.

Types of Stars

Astronomers classify stars by the characteristics of size, temperature, and color. They use hydrogen spectral analysis of stars' hydrogen emission lines to make their determinations.

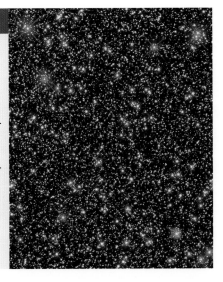

KEY FACTS

Stars are classified by a letter system that runs O, B, A, F, G, K, M (largest/hottest to smallest/coolest).

+ fact: A mnemonic to remember the order is "Oh boy, a flying giraffe kicked me!"

+ fact: A Hertzsprung-Russell diagram shows the relationship between color, temperature, and luminosity of stars.

+ fact: The sun is a modest G star.

A star's size is determined by its mass and not by a linear measurement, such as diameter. Mass regulates the other properties of a star, including temperature, color, and longevity. Smaller, cooler stars burn fuel at a slower pace and live longer than larger, fuel-guzzling hot stars. Red dwarfs are the most common smaller stars and have potential life spans of tens of billions of years. Massive blue giants are the hottest and may burn out in a million years or so. The supergiant is the largest star; Betelgeuse in Orion is a red supergiant, about 1,000 times bigger than the sun. Despite its large size, a red supergiant is a relatively cool star. It burns approximately ten times cooler than hotter and smaller blue O stars.

Star Brightness

We talk about star brightness in terms of magnitude, a number that often represents how bright the star is from our vantage point on Earth.

In the second century B.C., the Greek astronomer Hipparchus created a catalog that ranked the brightest stars as of the first magnitude, less bright stars as the second magnitude, and so on. Adopted by Ptolemy and refined in later centuries, the system became a logarithmic scale. As observation methods improved and fainter stars came into view, the scale was extended at the higher (dimmer) end. The need also arose to accommodate differences among first-magnitude stars. With nowhere to go but down, negative numbers were added. Magnitude is now calculated in many ways, including absolute magnitude, which is brightness if all stars were placed at the same distance from Earth. Under average night sky conditions, the naked eye can see stars to magnitude +6.

Star Death

The state of equilibrium that keeps a star together is a temporary one. As a star's core consumes its fuel, it sets in motion the scenario of its death.

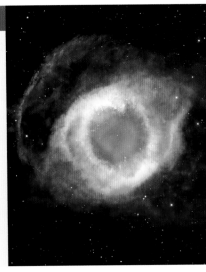

As a star burns up its fuel, its nuclear core begins to shut down. Gravity can no longer be held at bay, and the star contracts. This raises its temperature and triggers renewed fusion, allowing the star to expand. Smaller stars will become pulsing red giants. Eventually, they shed outer layers and become dead stars known as white dwarfs. Larger stars morph into supergiants, with contracting cores that may reach 1,000,000,000°F. More intense gravity in the core leads to an implosion followed by an explosion—a supernova—that blows apart the outer layers. Next follows either a neutron star or a total collapse into a black hole.

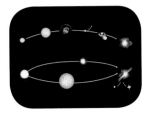

Sun

Our sun came into being the way any star does—in a cauldron of condensed gas and fused atoms. This life-sustaining star sits at the center of our solar system.

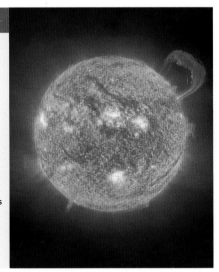

KEY FACTS

For a star, the sun is of middling size and heat—10,000°F (5500°C) at surface.

+ fact: Sun's core temperature: about 27,000,000°F (14,999,820°C)

+ fact: Composition of sun's outer layer: 74 percent hydrogen, 25 percent helium, 1 percent heavy elements

+ fact: The sun provides 1,400 watts of energy to each square yard of Earth.

Despite its seemingly extreme statistics, the sun is the perfect star for our planet. One of about 200 billion stars in the Milky Way, it resides 93 million miles from Earth. Its dimensions, dynamics, and distance provide just what we need in terms of energy and light to support life without overwhelming the planet with violent radiation. The sun is classified as a yellow dwarf star and lies in the middle of the classification scheme for color and temperature. Earth's life-supporting relationship with the sun is jeopardized by depletion of the upper ozone layer in the atmosphere, which allows more ultraviolet light to reach the planet. As with any star, the sun has an expiration date, estimated at about five billion years from now.

Sun Anatomy

The sun is a roiling, complicated body of violent extremes, with its own weather systems and a disturbing hint of unpredictability.

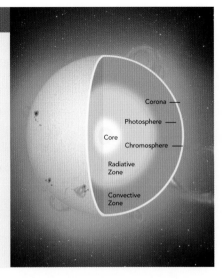

Corona

Photosphere

Core

Chromosphere

Radiative Zone

Convective Zone

KEY FACTS

The sun's equatorial region rotates in 25 days, but its polar regions take 34 days to complete rotation.

+ fact: In the core, about half the sun's mass is packed into 7 percent of its volume.

+ fact: Nuclear fusion in the core generates 400 trillion watts of energy a second.

+ fact: Holes in the corona allow the flow of a steady stream of solar wind.

The sun has a dense, high-energy core, where fusion takes place and hydrogen is converted to helium; emitted photons begin a journey to the surface that can last millions of years. Three-fourths of the way, they reach the convection zone that continues to draw gas and heat outward. They then encounter the 300-mile-thick (483 km) photosphere, which gives the sun the appearance of a solid with gas bubbling at the surface. The chromosphere is a pinkish layer that harbors solar flares. Beyond that lies the corona, a billowing megahot halo with temperatures reaching 2,000,000°F (1,111,093°C) that extends millions of miles. Holes in the corona, caused by the sun's magnetic field, allow a steady stream of particles known as solar wind to break free.

Sunspots

Sunspots peak and ebb in an 11-year cycle that coincides with other solar activity. Chinese astronomers made note of sunspots in 28 B.C.

Disrupted currents in the sun's magnetic field can slow the flow of solar material to the sun's surface, causing it to cool and darken in color. This produces sunspots, dark, visible splotches on the sun's photosphere. The center of a sunspot, depressed slightly below the surrounding gases, can measure some 3600°F (1980°C) cooler than the sun's typical surface temperature of 10,000°F (5500°C). Within a cycle, sunspot activity usually starts in the regions of latitudes 30° N and 30° S and then moves near the sun's equator. The 11-year cycle is part of a 22-year period in which the magnetic fields in the sun's upper and lower hemispheres switch polarities. NASA's Solar Dynamics Observatory is designed to study the variations in solar activity that affect Earth and near-Earth space.

Solar Flares

One thing leads to another: Solar flares, immense eruptions of high-energy radiation from the sun's photosphere, are associated with sunspots in ways not completely understood.

When magnetic energy that is pent up as a result of sunspot activity breaks free, it releases intense bursts of radiation across the entire electromagnetic spectrum, from radio waves to gamma waves. Solar flares extend into the sun's corona and take the form of arcs of highly charged superheated material known as coronal loops. The solar storms produced by flares sabotage electronics, communications, and electrical grids as they unleash their disruptive radiation. The most troublesome kinds of solar flares are known as X-class flares, which produce prolonged radiation storms in the upper atmosphere.

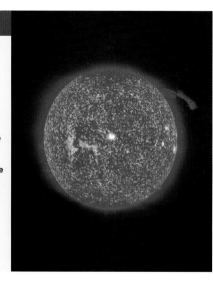

Sun's Path

For centuries, we have known that the Earth orbits the sun, but to conveniently describe all kinds of solar and celestial phenomena, we talk of the sun's apparent path across the sky.

KEY FACTS

Earth rotates from west to east, making the sun's path appear to be from east to west.

+ fact: Perihelion is the point at which a planet or other sky object in orbit is closest to the sun.

+ fact: Aphelion is the point at which a planet or other sky object is farthest from the sun.

We on Earth cannot track in a useful way our planet's daily rotation or yearly orbit without reference to the sun. The sun's apparent daily journey across the sky—and the changes to its route over the course of a year—have long formed the means by which we tell time, mark the seasons, and plan annual celebrations. The path of the sun across the sky is known as the ecliptic, which is also the term given to the Earth's route around the sun. Most sky charts indicate the sun's ecliptic on the night sky as a useful reference for locating constellations and planets and describing their whereabouts. The constellations that form the zodiac follow the path of the sun's ecliptic over the course of a year, the basis of the 12 astrological signs.

Sun & Seasons

Earth and sun are oriented to each other and interact in a way that produces predictable periods of difference in sunlight and temperature throughout the year.

KEY FACTS

During the equinoxes, the division of day and night is about the same around the globe.

+ fact: Summer at the North Pole means nearly 24 hours of daylight.

+ fact: The Earth's axis moves in position a little more than a half degree a century.

+ fact: The change of direction in the Earth's axis is known as precession.

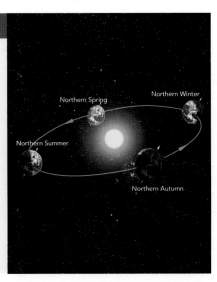

Northern Spring

Northern Winter

Northern Summer

Northern Autumn

As Earth orbits the sun, its rotational axis is tilted 23.5 degrees off the perpendicular. Because of this, the sun's direct light will hit different parts of the Earth at different times of the year, causing the seasons. Astronomical convention marks the transitions in two ways: as equinoxes in the spring and autumn, when the sun at noon shines directly on the Equator; and as solstices in winter and summer, when the sun shines on Earth as it is tilted away from or toward the sun. The Northern and Southern Hemispheres experience seasons at opposite times of the year, and equatorial areas experience little seasonal variation at all. Seasonal differences in sunlight and temperature affect many of the life processes on Earth.

Viewing the Sun

Many of us quickly reach for our sunglasses when we hit the bright sun—a good instinct. Viewing the sun directly with the naked eye can cause permanent damage.

KEY FACTS

Telescopes can be used for safe viewing.

+ fact: A filter on the objective lens allows safety; never use one only on the eyepiece.

+ fact: Use a scope as a projector. Attach cardboard to the objective lens, leaving a 2–4 in (5–10 cm) opening.

+ fact: Stand white paper behind the eyepiece, and view the sun on the paper.

Here is a mantra to repeat over and over for sky watchers young and old: Never, ever look directly at or point your binoculars or telescope toward the sun—not even for a split second. That said, there are safe ways to observe the sun. Welders' goggles rated 14 or higher are one option, as are approved solar filters and "eclipse glasses." These items can be found online. It is also easy to make a pinhole projector by piercing a card with a small hole and placing a whole card below it. Position the top card toward the sun, without looking. A binocular lens can be used the same way; leave the caps on the unused lens. Using exposed film, medical x-rays, or smoked glass as filters is dangerous. Eye damage may be painless, as it occurs because it is caused by invisible infrared rays.

Solar System

We tend to talk of *the* solar system—ours—but there are many of them. Our sun's gravity draws planets and other objects into orbit around it.

KEY FACTS

More than 99 percent of the solar system's mass is contained within the sun.

+ fact: Primitive meteorites provide a measure of the age of the solar system: about 4.5 billion years old.

+ fact: The solar system orbits the center of the Milky Way every 220 million years.

+ fact: The word "planet" means "wanderer."

The sun holds sway over a diverse array of objects ranging from planets and dwarf planets to moons and asteroids—and even to dust. Its influence spreads from its own surface to the distant Oort cloud that encircles the solar system. The planets under the sun's control—including Earth, of course—spread across a distance ranging from Mercury, on average about 36 million miles (58 million km) from the sun, to Neptune, with an orbit averaging about 2.8 billion miles (4.5 billion km) from the sun. Pluto and other dwarf planets range beyond that. Of course, our solar system is only one of many that have formed or at this moment are forming around distant stars, as ongoing discoveries attest.

Moon

The moon is a stargazer's nearly constant companion, visible to the naked eye even in competition with extreme light pollution. Binoculars bring out details of its features.

KEY FACTS

Made of lighter elements than Earth, the moon has $1/4$ the diameter of the Earth, contains about $1/80$ the mass, and has about $1/6$ the gravity.

+ fact: Diameter is 2,159 miles (3,465 km).

+ fact: Distance from Earth is 238,855 miles (384,400 km).

+ fact: Orbital period is 27.3 days; cycle of moon phases is 29.5 days.

Moons are natural satellites that orbit celestial bodies such as planets and asteroids, and exercise their own gravitational pull. Most planets in our solar system have moons, usually multiples. Earth's solitary moon likely formed from debris ejected when an object about the size of Mars struck the young Earth, about 4.5 billion years ago. Evidence from moon missions shows a similar composition of their respective rocks. Characteristics of their orbits also support this theory. Over time, the moon acquired its craters and seas from bombardments by space debris and the eruption of molten rock from the interior. The "man in the moon" is an impression that the unaided eye sees of contrasting lunar surface features. Some people see a woman, a hare, or a toad.

Moon in Motion

The motion of the moon shares complex interactions with that of the Earth. The planet and its satellite regulate and influence each other in various ways.

KEY FACTS

Friction from Earth-moon interaction causes the Earth to slow slightly.

+ fact: The slowdown is offset by a slight pulling away of the moon—about 1.5 in (3.8 cm) a year.

+ fact: The moon's orbital period is 27.3 days.

+ fact: The Earth's rotation accounts for much of the moon's visible motion.

The moon spins on its axis in the same manner as the Earth. But the moon's rotation takes the same amount of time as its orbit around the Earth—27.3 Earth days. This is because the Earth's gravity produces a bulge, or "land tide," in the moon that slows it down. As a result of this synchronization of rotation and orbit, the moon always presents the same side to the Earth. The moon still exerts its own influence: For example, it creates a gravitational tug responsible for the tides on Earth. The moon's gravity requires spacecraft engineers to take it into consideration when calculating trajectories for lunar landings. World cultures have attached other influences to the moon, such as influencing the course of love or initiating the legendary transformation from human to werewolf.

Moon Phases

The moon "borrows" light from the sun to create its shine. The phases, or appearance, of the moon at different times of its orbit, reflect changes in available sunlight.

KEY FACTS

The moon takes 29.5 days to complete the cycle of its phases.

+ fact: Moonrise occurs 12 degrees farther to the east each night.

+ fact: A full moon rises with the setting sun and sets at sunrise the next day.

+ fact: View crescent moons best in the evening in spring and just before sunrise in autumn.

Observers on Earth cannot see the moon during its phase known as new moon. It occurs when the moon lies between the Earth and the sun and is fully backlit. About two weeks later, the Earth lies between the sun and moon, and we see the latter's maximum appearance—a full moon. In the weeks leading to the full moon, the moon comes into view in gradual increments: a waxing crescent, crescent, first quarter, and then waxing gibbous. (Gibbous is from the Latin for "hump.") In the weeks after the full moon, its light starts to diminish, and it becomes a waning gibbous, last quarter, crescent, and then waning crescent. Worldwide, many agricultural tasks as well as religious observations are linked to the phases of the moon.

Moon's Near Side

Collisions with meteors and comets, combined with volcanic activity from the moon's interior, produced the landscape of ridges and craters we see on the near side of the moon.

KEY FACTS

Each part of the moon's surface is visible for about two weeks in a phase cycle.

+ fact: Galileo observed the moon's surface through his telescope in the 1600s.

+ fact: First and last quarter phases show greater detail on the moon's surface.

+ fact: The last major crater to form was the Tycho crater about 109 million years ago.

Earth's moon began in violence that continued for the first half billion or so years. Debris bombarded the surface, creating large, wide, and shallow areas that later filled with dark lava pouring from the moon's core. These areas are called "maria" (MAH-ree-uh), the Latin word for "seas." The most famous is the Mare Tranquillitatis, the Sea of Tranquillity, site of the first moonwalk by Neil Armstrong in 1969. Smaller debris gouged smaller, deeper holes known as craters. These appear much lighter on the moon's surface than the dark maria. Over time, the bombardment decreased dramatically. NASA's Astronomy Picture of the Day (APOD), available at *apod.nasa.gov/apod/astropix.html*, often features amazing images of the moon, including some interactive ones.

Moon's Far Side

Although it is not a visual presence in the night sky, understanding the moon's far side rounds out our appreciation of Earth's satellite.

KEY FACTS

Many far-side features have names that reflect Russian space exploration, such as the Moscow Sea.

+ fact: The far side experienced fierce asteroid bombardment, causing craters.

+ fact: Bombardment on the far side caused volcanic activity on the near side.

+ fact: The far side has a thicker crust with fewer seas than the near side.

The moon's far side is often referred to as its "dark side," but it is by no means dark on the far side; we just cannot see the light because, due to its synchronous orbit with Earth, the moon presents only its near side to our view. The far side experiences phases just as the near side does; there is a time each month when it is in full sunlight and a time when it is in full shadow. The far side of the moon was first mapped from space-based observations by the Russians in 1959, with pictures the spacecraft Luna 3 took. After that, three decades of probes and space missions created detailed maps. The quest continues to provide even more detailed and scientifically advanced topographical images of the moon's far side with the Lunar Reconnaissance Orbiter, launched by NASA in 2009.

Moon Exploration

On July 20, 1969, men on the moon—the astronauts of Apollo 11—met the man in the moon, so to speak. Remote reconnaissance and robotic landings preceded the historic event.

KEY FACTS

Scientists believe that most of the 6 U.S. flags planted on the moon still stand.

+ fact: In the 1960s, orbiters photographed 99 percent of the moon's surface.

+ fact: The last Apollo mission, in December 1972, collected more than 250 pounds of moon rocks.

+ fact: India, China, and Japan are developing lunar exploration projects.

The manned Apollo missions with their moon landings and space buggy rides were preceded by remote data gathering. The Russians sent the first probes to check out the moon in 1959: Luna 1 flew by it, Luna 2 crash-landed, Luna 3 took photographs of the far side; and in 1966, Luna 9 made a soft landing. NASA followed with several series of lunar probes, including orbiters and surveyors that paved the way for six manned Apollo missions and their 12 astronauts.

NASA maintains a Lunar Reconnaissance Orbiter that acquires sophisticated data, and the space agency is committed to returning humans to the moon.

Solar Eclipse

The sun and moon collaborate to produce three kinds of solar eclipses, obscurations of the sun by the moon. These are called annular, partial, and total.

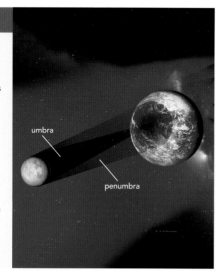

An annular eclipse occurs when the moon is too far from the sun to cover it completely. Instead, a ring or annulus, of the sun's bright edge is visible. A partial eclipse happens when only part of the sun is obscured if viewed from within the moon's penumbra, the outer region of its shadow. Annular and partial eclipses only dim the sun. During totality of a total solar eclipse, the moon's disk completely covers the sun for those on Earth within the moon's umbra, or central shadow. Daylight fades to dark twilight, and stars are visible during totality. Sky watchers can see the sun's corona feathering out around the edges of the moon's shadow.

Lunar Eclipse

Like the sun, the moon experiences three kinds of eclipses: penumbral, partial, and total. Unlike solar eclipses, they can be viewed with the naked eye.

If the moon passes through the outer region of the Earth's shadow, the penumbra, it causes a penumbral lunar eclipse. This will dim the moon a bit. If some of the moon passes through the Earth's dark central shadow, the umbra, a partial eclipse will occur. When the moon moves completely into the Earth's umbra, a total eclipse takes place. Even then, the moon does not become completely dark. Sunlight bends around Earth's surface and shines dimly on the moon. The refracted light casts a reddish or orange glow. If Earth's atmosphere is unusually dusty, the moon might be barely visible for a while.

Satellites

The U.S.S.R. launched the first artificial satellite, Sputnik 1, into orbit on October 4, 1957. Since then, thousands more have been launched for scientific and commercial purposes.

KEY FACTS

Sputnik 2 carried a dog named Laika. Her flight proved that an animal could survive launch and weightlessness, although she perished shortly after.

+ fact: In 1959, U.S. satellite Explorer 6 photographed Earth for the first time.

+ fact: In 1962, Telstar, the first communications satellite, was launched.

A satellite is any object that orbits a planet; Earth's moon is a natural satellite. Artificial satellites were created and sent into space for purposes of data gathering and signal transmission. The launch of Sputnik 1 in 1957 ushered in the space age. The United States followed on January 31, 1958, with the launch of American Explorer 1—and the space race between the U.S.S.R. and the U.S. was under way. In the meantime, satellite launchings have become commonplace. There are roughly 6,000 satellites in orbit now, not all of them functional, with specialized missions such as communications, weather, navigation, and astronomy.

Highly-Elliptical Orbit

Geosynchronous Orbit

Low Earth Orbit

Semi-synchronous Orbit

Cassini-Huygens Spacecraft

A joint venture of NASA, the European Space Agency, and the Italian Space Agency, the Cassini-Huygens spacecraft launched in October 1997 to begin the long journey to Saturn.

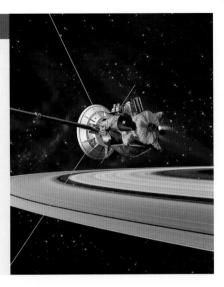

KEY FACTS

The Huygens probe was the first craft to land on a body in the outer solar system.

+ fact: Seventeen countries collaborated in the Cassini-Huygens mission.

+ fact: It takes more than an hour for radio signals from Cassini to reach Earth.

+ fact: The Cassini mission has been extended to 2017.

Saturn, with its rings suggestive of the sun's encircling gases and its many moons representing diverse aspects of planet formation, was an ideal target for the Cassini-Huygens mission. The hyphenated name reflects the craft's two components: the plutonium-powered Cassini orbiter and the wok-shaped Huygens probe, named for Italian-French and Dutch astronomers, respectively. Cassini-Huygens took nearly seven years to reach Saturn. Huygens parachuted to a soft landing on Saturn's moon Titan in January 2005. The duo has made pathbreaking discoveries, including strange weather patterns, new moons, methane lakes on Titan, and ice geysers on the moon Enceladus. Cassini is on a second mission extension until 2017, when the Saturn summer solstice can be observed.

Chandra X-ray Observatory

The Chandra X-ray Observatory, named for Indian American astrophysicist Subrahmanyan Chandrasekhar, provides a view of the universe not possible from Earth's surface.

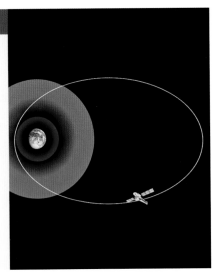

KEY FACTS

Chandra can see the x-rays from particles at the edge of black holes.

+ fact: Its orbit takes it 200 times higher than the Hubble Space Telescope.

+ fact: Its resolving power compares to reading a stop sign from 12 miles away.

+ fact: Chandra needs only the wattage of a hair dryer to run the spacecraft.

Launched in July 1999 by the space shuttle Columbia, NASA's Chandra X-ray Observatory is a telescope that focuses on the high-energy objects in space that radiate energy in x-ray wavelengths, such as supernovae, binary stars, and black holes. The world's most powerful x-ray telescope, it can detect x-ray sources only one-twentieth as bright as earlier telescopes could. To do so, it orbits above Earth's atmosphere, up to an altitude of 86,500 miles (139,000 km). Chandra, a NASA project that is hosted by the Harvard-Smithsonian Center for Astrophysics, has thrilled astronomers with its data, which it shares with scientists around the world.

Hubble Space Telescope

Launched in 1990, the Hubble Space Telescope is the astronomical equivalent of the greatest thing since sliced bread. It looks at the universe in mind-boggling detail.

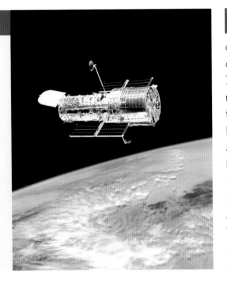

KEY FACTS

The Hubble travels at 17,000 mph (28,000 kph) and orbits 353 miles (569 km) above the Earth.

+ fact: Every year, the Hubble makes more than 20,000 observations.

+ fact: Space shuttle missions have upgraded and repaired equipment, including cameras on the Hubble.

Its ability to view far-flung reaches of the universe without atmospheric interference is the hallmark of the Hubble Space Telescope. Roughly the size of a large bus, the Hubble carries a primary mirror 7.9 feet (2.4 m) in diameter and an array of instruments, including cameras, spectrographs, and fine-guidance sensors used to aim the telescope. In the more than two decades since its launch, a relentless stream of images from the Hubble has helped determine the age of the universe, informed us about quasars, and enlightened us about dark energy. Its successor, the infrared James Webb Space Telescope, is in development. The Webb, planned to launch in 2018, will have a mirror nearly three times the size of the Hubble's and a sun shield the size of a tennis court.

Voyagers 1 & 2

Launched more than 35 years ago, NASA's Voyagers 1 and 2 are now headed out of the solar system, prepared for encounters with intelligent life.

KEY FACTS

The probes' mission originally was to study Jupiter and Saturn and their moons.

+ fact: Early success led to an expansion and continuation of the Voyager mission, with flybys of all outer planets.

+ fact: Voyager 1 is the farthest artificial object in existence—119 times farther from the sun than the Earth is.

The deep-space probes Voyager 1 and 2 were launched in 1977 to study the outer solar system and, eventually, interstellar space. These nuclear-powered craft are headed out of the solar system on two separate trajectories, with Voyager 1 well ahead. Should either encounter intelligent life, each carries a gold-plated 12-inch (30 cm) phonograph record—very "old school," but the go-to durable format at the time. Drawings on the cover demonstrate how to play the record, and they show a map of our solar system relative to 14 beaconlike pulsars. Music included ranges from Bach to Javanese gamelan to Navajo chant to Chuck Berry's "Johnny B. Goode." The record also contains spoken greetings in 55 world languages and natural sounds, such as thunder, birds, and whales.

Space Junk

What goes up into space may come down, but it often stays up and accumulates with other outmoded hardware to become orbiting space junk.

KEY FACTS

The now nonfunctional Vanguard 1 satellite, launched in 1958, is expected to orbit another 150 years.

+ fact: Space junk endangers weather, communications, and GPS satellites.

+ fact: When space junk falls out of orbit, it can cause potential damage on Earth.

+ fact: In 1978, a Soviet satellite spewed radioactive debris into the Arctic.

With the exception of manned spacecraft, most satellites and other equipment sent into space do not make the return journey. Even when no longer functional, these objects remain in orbit to become part of a growing array of space junk. Some 600,000 items ranging from rocket boosters to explosion debris to actual space station garbage float out there, creating hazards to functioning spacecraft and satellites from time to time. At most peril is the International Space Station, or ISS, and its inhabitants. In 2001, the ISS had to be nudged into a slightly higher orbit by the space shuttle *Endeavour* to avoid a collision with the upper stage of a spent Soviet rocket.

Space Station

Sky watchers are often treated to the sight of the International Space Station, or ISS, in the night sky. NASA's website and its SkyWatch 2.0 applet provide information on its whereabouts.

KEY FACTS

Skylab fell from orbit prematurely in 1979, breaking up over the Indian Ocean and Australia.

+ fact: The ISS is the largest space station ever constructed.

+ fact: Human habitation on the ISS has gone on uninterrupted since it began in 2000.

+ fact: Plants are grown experimentally in the ISS lab and service module.

Prolonged sojourns in orbit require a space station. Both the Soviet Union (and later Russia) and the United States committed to this endeavor, beginning with Salyut 1 in 1971. The Soviet Salyut series was succeeded by the Mir space station, which operated from 1986 through 2006. The U.S. launched Skylab in 1973, and it was manned through 1974. Today, the International Space Station is the sole station in orbit, and it serves the exploration and research needs of the U.S., Russia, Japan, Canada, and 11 European countries. The ISS received its first crew in November 2000, and is expected to remain in operation until at least 2020. A yearlong mission is planned for 2015 that would further examine human physiological responses to prolonged space travel.

UFOs

Few topics are as controversial as unidentified flying objects—UFOs. Despite many unexplained sightings, UFOs often turn out to be ordinary phenomena or objects.

KEY FACTS

The term "flying saucer" was coined by a pilot in 1947.

+ fact: Over the decades, NASA astronauts have reported a number of unexplained phenomena.

+ fact: In 2012, Mars rover Curiosity appeared to capture a UFO image that turned out to be dust kicked up by the impact of its sky crane with the surface.

Beginning sky watchers see a lot of objects they cannot identify, but that doesn't make those objects UFOs. The watchers soon learn to sort out planes, larger satellites, and sometimes even the International Space Station from relatively stationary objects in the sky. Otherwise, UFOs in general can usually be accounted for as meteors, clouds, birds, weather balloons, atmospheric phenomena such as sun dogs, or military aircraft. The often secret nature of military aircraft leaves many sightings unexplained. A large number of unresolved UFO sightings may well be unfamiliar atmospheric phenomena with unusual physical properties. The SETI Institute searches for signs of extraterrestrial life using radio and optical telescopes. Many prominent scientific institutions, as well as NASA, have collaborated on the project.

Comets

Dirty balls of frozen gases and dust, comets are leftovers from the formation of the solar system. Nudged out of their original orbits, they streak through the inner solar system.

KEY FACTS

Comets typically are large enough to survive hundreds of trips through the inner solar system.

+ fact: British astronomer Edmond Halley never saw the comet named for him.

+ fact: The most recurrent comet is Encke, cycling past Earth every 3.3 years.

+ fact: A stream of dust shed by a comet's nucleus forms a second tail.

When an icy ball of gases and dust enters a new orbit and approaches the sun, the warming object forms a glowing head—the coma—and a flow of charged ions is shaped by solar wind into a glowing tail. Long-period comets come from the Oort cloud, a belt of icy debris that encircles the solar system. Short-period comets, with orbits easy to predict, originate in the Kuiper belt beyond Neptune or in a far-flung area known as the scattered disk. Comets can be viewed with the naked eye and, of course, through optical equipment. To tell a comet from a star, check a star chart to see if it's there, look for a tail, and look for movement over time. The NASA website offers an Asteroid and Comet Watch that provides up-to-date information.

Meteors

Called "shooting stars," meteors are bits of material, often left behind by a comet, that ignite and burn up as they pass into Earth's atmosphere.

KEY FACTS

More than 60 meteorites found are pieces from the moon, and more than 30 are fragments of Mars.

+ fact: Comets under the pull of Earth's gravity reach speeds of 20,000–160,000 mph (32,000–257,000 kph).

+ fact: Atmospheric friction heats meteors to 2000°F (1093°C).

+ fact: Most meteors vaporize before they get within 50 mi (80 km) of Earth.

Meteors originate as meteoroids, bits of interstellar dust and debris. The particles may be very small or quite large, approaching asteroid size, although the difference between the two is somewhat arbitrary. Larger meteoroids may be broken-off pieces of asteroids, planets, or our moon. Many smaller meteoroids represent burned-off ashes from comets, which are left in massive trails throughout the solar system. As Earth travels in its orbit, it encounters perhaps a half billion meteoroids each year. Objects entering the atmosphere are known as meteors and those that reach the Earth are called meteorites. On a given night, there may be several visible shooting stars each hour. Up to 10,000 tons of meteoric material, much of it dust size, fall on Earth daily.

Meteor Showers

About 30 times a year, Earth passes through a dense dust trail, typically left by a comet, giving rise to a meteor shower or a rare, spectacular meteor storm.

KEY FACTS

The International Meteor Organization depends on information from amateurs to expand its database.

+ fact: Meteor viewing is affected by atmospheric conditions and light pollution.

+ fact: Halley's comet left a dust trail that produces two meteor showers each year.

Meteor showers, unlike the sporadic meteors, occur in steady streams at predictable times from specific locations in the sky known as radiants. Radiants typically are identified with constellations. The Leonid meteor shower, for example, originates in Leo. Meteor showers may produce several dozen meteors an hour; meteor storms can produce that many a second. Meteor showers are well publicized, with tips for best viewing times. No equipment is needed to view a meteor shower, but binoculars help focus on fainter shooting stars. A lounge chair provides comfort for all that looking up. Scan the whole sky; meteors may emerge anywhere. The Great Meteor Storm of November 1833 astonished and awed Western Hemisphere observers with an estimated 240,000 meteors emanating from Leo.

Asteroids

Rocky bodies left from the formation of the solar system about 4.5 million years ago, asteroids mainly occur in a belt between Mars and Jupiter.

KEY FACTS

A NASA-sponsored program tracks larger asteroids near Earth to gain important information.

+ fact: Asteroids are rich in metals and one day could be mined.

+ fact: Asteroids reveal details of solar system formation and planet migration.

+ fact: Asteroid names include Jabberwock and Dioretsa, a backward-orbiting asteroid.

Hundreds of asteroids are theoretically within range of even a 3-inch telescope. But asteroids are notoriously difficult to spot; their light is faint and they are indistinguishable from stars. Comparing the field of view through a telescope with a sky chart may confirm a starlike object not on the map as an asteroid. Or, you can map what you see through the telescope and compare it with the same view several hours later to determine which objects have moved. Those objects likely are asteroids. The millions of fragments in the asteroid belt collide and sometimes break free from the belt, streaking toward Earth as meteors. In February 2013, an asteroid half the size of a football field came harmlessly within 17,200 miles (27,700 km) of Earth, inside the orbits of some satellites.

Kuiper Belt

A major comet breeding ground, the Kuiper belt is a region of ice and rocky debris that begins just past Neptune and extends outward through the solar system.

KEY FACTS

The Kuiper belt has an outer extension of icy and rocky debris—the scattered disk.

+ fact: The Kuiper belt is about 20 times wider than the asteroid belt between Mars and Jupiter.

+ fact: Kuiper belt objects (KBOs) are difficult to measure because they are so distant.

+ fact: KBOs are also called trans-Neptunian objects, or TNOs.

The Kuiper (kind of rhymes with "wiper") is named for 20th-century Dutch astronomer Gerard Kuiper. It is a fertile field of debris, loaded with objects large and small that divulge information about the history of our solar system. The belt also contains a number of almost planet-size objects. Pluto is the biggest object in the Kuiper belt, at about 1,400 miles (2,300 km) wide. The demoted planet shares the belt with other dwarf planets Makemake and Haumea (Eris resides in the scattered disk). A number of Kuiper belt dwarf planet candidates are under investigation. Some of these objects seem to be in unusual orbits.

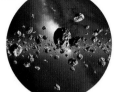

Oort Cloud

The Oort cloud is an enormous spherical cloud that houses a vast population of icy objects and encircles the solar system.

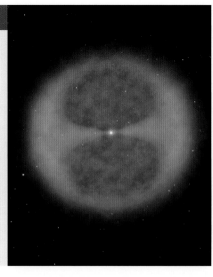

KEY FACTS

The trans-Neptunian object Sedna may be the first observed planetlike object to originate in the Oort cloud.

+ fact: The Oort cloud is so distant that telescopes have never seen it.

+ fact: There are up to 2 trillion icy rocks in the Oort cloud.

+ fact: Dislodged Oort cloud comets may get flung from the solar system.

In the mid-20th century, astronomers Ernst Öpik and Jan Oort theorized the existence beyond Pluto of an even more distant cloud of icy debris left over from the solar system's formation. Now called the Oort cloud, or sometimes the Öpik–Oort cloud, it is one of three solar system nurseries for comets, along with the Kuiper belt and its nearby scattered disk. Comets dislodged from the Oort cloud are long-period comets that can take thousands and even millions of years to orbit the sun. Short-period comets, with orbital periods less than 200 years, originate from the Kuiper belt. The Oort cloud's nearest edge lies some 465 billion miles (750 billion km) away. Its outer edge may extend 20 times farther than that, where the sun has less gravitational influence than nearby stars.

Planets

In 2006, the International Astronomical Union changed the rules for what defines a planet, demoting Pluto and ending more than seven decades of belief in a nine-planet solar system.

KEY FACTS

Current planets are Mercury, Venus, Earth, Mars, Jupiter, Saturn, Uranus, and Neptune; dwarf planets are Pluto, Ceres, Eris, Makemake, and Haumea.

+ fact: In the 1850s, the official list of planets numbered 41.

+ fact: If Earth's moon had an orbit independent of the Earth, it could be a planet.

Discoveries of new objects larger than Pluto, some with their own moons, made it necessary to rethink the definition of a planet. The International Astronomical Union established criteria an object needed to meet to be considered a planet. It must orbit a star, not be a satellite of another object, remain spherical by the pull of its own gravity, and be substantial enough to clear objects from its orbit. Pluto didn't meet the last benchmark. Currently, 13 sky objects qualify as planets or dwarf planets, 8 from the traditional list and 5 dwarf planets, and there is a waiting list of other objects under consideration. The four planets closest to the sun in the inner solar system have solid, rocky natures and are known as terrestrial planets; the four outer ones are known as gas giants.

Planet Viewing

Observing the planets can be done "old school," without gear or charts, or can involve more effort and a large array of equipment and resources.

KEY FACTS

Some planets can be observed with the naked eye; others require optics.

+ fact: Venus, brightest object in the sky after the sun and moon, is highly visible to the naked eye.

+ fact: Jupiter is also visible; its moons are starlike points through binoculars.

+ fact: Saturn can be viewed unaided; viewing its rings requires a telescope.

Both hardware and software greatly aid planet viewing. Good binoculars and a telescope with a lens aperture of 3 inches (7.5 cm) or more are basic tools to be combined with charts, almanacs, and other resources that help determine the best viewing times. The Internet is a tremendous source of detailed information; Space.com is a great place to start, with its monthly summaries and daily updates of planetary locations. Mobile apps provide another level of assistance, and many are available free or for only a small fee. Add a clear night and a location with unobstructed viewing and minimal light pollution, and you're all set. With the demand for commercial space travel, it is only a matter of time before non-astronauts can add Earth to their planet-viewing wish lists.

Mercury

Smallest and innermost of the planets, Mercury is close enough to be seen with the naked eye but is one of the more difficult planets to spot.

KEY FACTS

Mercury, slightly larger than Earth's moon, has a highly eccentric orbit at irregular distances from the sun.

+ fact: Mercury has a diameter of 3,031 mi (4,878 km).

+ fact: The planet's distance from the sun is 29–43 million mi (47–69 million km).

+ fact: Its orbital period is 88 Earth days, and its rotational period is 58.6 days.

The surface of Mercury is barren, rocky, and pocked with craters. A molten iron core makes up about 65 percent of the planet, which essentially has no atmosphere. Mercury circles the sun every 88 days, making it invisible much of the time due to competition with the sun's glare. The planet is best viewed in the west in evenings in March and April and in the east in mornings in September and October. Extremely clear viewing conditions are usually needed to see it, and it can be a challenge even with binoculars or a telescope. Consult resources such as almanacs and astronomy sites on how best to hunt for it. Mercury is named after the fleet Roman messenger god.

Venus

Neither similarities in size, mass, and orbital year with Earth nor having the goddess of love for a namesake can mask the treacherous nature of Venus.

KEY FACTS

+ fact: Earth's closest neighbor, Venus has phases somewhat similar to those of Earth's moon, but has no moon of its own.

+ fact: Venus has a diameter of 7,521 mi (12,103 km).

+ fact: The planet has a distance from the sun of 67.2 million mi (108.2 million km).

+ fact: Its orbital period is 225 Earth days, and its rotational period is 243 days.

Although farther from the sun than Mercury, Venus has a greenhouse-type atmosphere that traps the sun's heat and sends temperatures soaring to 860°F (460°C). The atmosphere is thick with carbon dioxide and is blanketed with a 40-mile (64 km) layer of sulfuric acid, which makes Venus highly reflective and bright, but hides its surface. Known as both the "evening star" and the "morning star," Venus is visible for some months each year, low to the horizon in the western sky at nightfall or rising in the east before sunrise. Telescope users generally get a better view at twilight when there is not so much contrast with the dark sky because the planet's brightness can be overpowering.

Transit of Venus

This rare phenomenon, the passing of Venus across the face of the sun, last took place in June 2012. The next transit does not occur until December 2117.

KEY FACTS

The ancient Baby-lonians may have recorded the transit as early as 3000 B.C.

+ fact: Captain James Cook set up an obser-vation post in Tahiti to take transit measure-ments in 1769.

+ fact: In 1882–83, John Philip Sousa wrote the "Transit of Venus March."

+ fact: The June 2012 transit was celebrated with observing parties, webcasts, and music.

The alignment that places Venus directly between the Earth and sun affords a chance to see Venus as a dot that slowly moves across the face of the sun. The rare transits occur in pairs eight years apart, with more than a hundred years between pairs. In recent centuries, each transit was observed and recorded with increas-ingly sophisticated technology. The anticipated transit of 1769 inspired England's Royal Society to send a team, with Captain Cook at the helm of the H.M.S. *Endeavour*, to observe the transit from southern latitudes. Transits provide an opportu-nity to learn about the size of the solar system. NASA's webcast of the June 2012 transit of Venus as viewed from Mauna Kea, Hawaii, is archived on its website.

Mars

The stuff of alien fantasies, Mars in reality is just as fantastic—and knowable, thanks to concentrated efforts to study the red planet.

KEY FACTS

Mars has two moons called Phobos and Deimos—"fear" and "panic"—and the solar system's largest volcano.

+ fact: Mars has a diameter of 4,222 mi (6,794 km).

+ fact: Mars is 141.6 million mi (227.9 mil-lion km) from the sun.

+ fact: Its orbital period is 687 Earth days, and its rotational period is 24.6 Earth hours.

Mars and Earth share a number of similarities. A day on Mars is just a tad longer than an Earth day. Mars has seasons, an atmosphere, clouds, and polar caps, but it is much smaller than Earth. As numerous remote and on-site investi-gations attest, the surface of Mars is barren and a dusty red, the result of a high iron oxide, sili-con, and sulfur content. Its equator is marked by immense volcanoes and dramatic canyons. Mars is visible to the naked eye, but its apparent size and brightness vary significantly during its 687-day orbit. A modest telescope reveals details such as dark lava, boulder fields, and frozen polar caps. Binoculars will show the planet's reddish color.

Mars Rovers

Mars has had more than its fair share of scientific inspection in the form of flyby probes, orbiters, landers, and rovers, including the latest—the Curiosity rover.

KEY FACTS

The Mars Pathfinder spacecraft delivered rover Sojourner to Mars in 1997.

+ fact: Rover Opportunity, originally expected to travel 2,000 ft (0.6 km), has racked up more than 22 mi (35 km).

+ fact: Rover Opportunity's twin, Spirit, went off-line in March 2010.

+ fact: Rover Curiosity is about the size of a Volkswagen Beetle.

Robotic vehicles festooned with scientific equipment provide invaluable information about the surface of Mars. The rover Opportunity was deployed in 2004, and was expected by some to survive there only 90 days; it completed its ninth year on Mars in January 2013, providing an ongoing bonus of data. The Curiosity rover, a car-size vehicle carrying a payload of ten of the most sophisticated science instruments ever designed, landed there in August 2012. It is looking at the planet's geology, past planetary processes, surface radiation, and biological potential. In early 2013, it sent back the first nighttime photos of Mars. Yet much of the excitement of Curiosity's mission involves the evidence it has detected of water-bearing minerals in the planet's rocks.

Asteroid Belt

A gap of 340 million miles (547 million km) between Mars and Jupiter, the asteroid belt marks the division between the solar system's terrestrial planets and its gas giants.

KEY FACTS

The dwarf planet Ceres comprises one-third of the mass in the asteroid belt.

+ fact: Jupiter's orbit contains a group of asteroids known as the Trojan asteroids.

+ fact: Earth has its own small Trojan asteroid belt located at about 60° east or west of the sun.

+ fact: Each of the Beatles has an asteroid named for him.

Jupiter's massive gravitational field prevents the rocky debris in the asteroid belt—leftovers from the formation of the solar system—from coalescing into planets. The asteroids in the belt are like a warehouse of raw materials that under different circumstances could have made the transition to full-fledged planets. The debris won't achieve that destiny, but larger objects in the belt, such as Ceres, have the chance to be considered dwarf planets. Astronomers have cataloged more than 120,000 bodies in the asteroid belt, and more than 13,000 are named, sometimes very personally for girlfriends and favorite writers. According to the guidelines set out by the International Astronomical Union, near-Earth objects are supposed to have names taken from the mythology of any world culture.

Jupiter

A planet of superlatives, Jupiter is the largest by far and has the largest number of satellites with 65 moons, making it a kind of solar system unto itself.

KEY FACTS

Jupiter has a ring, but it is small and thin compared to other planets' rings.

+ fact: Jupiter has a diameter of 88,846 mi (142,984 km).

+ fact: The planet's distance from the sun is 483.7 million mi (778.4 million km).

+ fact: Its orbital period is 11.9 Earth years, and its rotational period is 9.9 Earth hours.

Jupiter has a massive influence in the solar system with its strong magnetic field and gravitational pull, intense emissions of radio waves, and bursts of radiation. It has a metallic core but otherwise is a swirling mass of gases: hydrogen, helium, methane, and ammonia. Its complete rotation lasts just short of ten hours; at this clip, the atmosphere appears to us as faint pastel bands. Persistent storms, including the Great Red Spot, a perpetual high-pressure zone, add to the spectacle. At the other extreme, Jupiter takes nearly 12 years to orbit the sun, spending nearly a year in each constellation of the zodiac and becoming an easy target for sky watchers.

Jupiter's Moons

In 1610, Galileo spotted the four largest moons of Jupiter with his telescope. Io, Callisto, Ganymede, and Europa became known as the Galilean moons.

KEY FACTS

Jupiter has a chart-topping 65 moons; presently, only 51 of them have names.

+ fact: Callisto is the most heavily cratered object in the solar system and lacks a molten core.

+ fact: Moon Io is the most volcanically active object in the solar system.

+ fact: Many of Io's more than 400 volcanoes erupt almost continuously.

Jupiter's intense gravitational pull, second only to the sun's, has lured many objects into its orbit. It governs these moons like the masterful planet that it is, and its four largest moons do the planet proud. Jupiter's Galilean moons are large enough that if they orbited the sun, they would qualify as planets. Among them, Ganymede is the largest moon in the solar system, larger than Mercury, Ceres, Pluto, and Eris combined. It generates its own magnetic field. Europa may harbor a saltwater ocean under its icy surface, and it may have twice as much water as Earth does. The Galilean moons can be viewed easily from Earth with binoculars. In orbit, all the moons keep the same face toward Jupiter, just as Earth's moon does.

Saturn

Gas giant Saturn, the second largest planet, is best known for its rings, but it has also acquired a multitude of moons. Despite its massive dimensions, Saturn is less dense than water.

KEY FACTS

Saturn with its rings is the most recognizable planet.

+ fact: Saturn has a diameter of 74,898 mi (120,536 km).

+ fact: Its distance from the sun is 885.9 million mi (1.4 billion km).

+ fact: Saturn's orbital period is 29.4 Earth years, and its rotation period is 10.7 Earth hours.

Known to the ancient Mesopotamians, Saturn was dubbed "the old sheep" for taking more than 29 years to orbit the sun. The plodding orbit is countered by speedy rotation; a day on Saturn lasts just under 11 hours. And although the planet's volume could contain 763 Earths, its low density means that it would float in water if a big enough basin could be found. It also means that the planet appears considerably squashed in profile; the diameter through the poles is significantly shorter than its equatorial diameter. Saturn is visible to the naked eye, but binoculars or a telescope are needed to view details. The Cassini orbiter has captured some amazing images.

Saturn's Rings

Saturn's iconic rings are composed of rubble, dust, and ice. The rings may also contain the pulverized remains of moons, comets, and asteroids that strayed too close.

KEY FACTS

Galileo viewed Saturn through his telescope in 1610, but he thought the side bulges he saw were separate bodies.

+ fact: Saturn's rings vary in width and thickness; they extend 170,000 mi (274,000 km) from side to side.

+ fact: Saturn's rings were named alphabetically in the order of their discovery.

+ fact: Saturn's rings may disappear.

A ring around Saturn was first proposed by Dutch astronomer Christiaan Huygens in 1655. We now know that Saturn has multiple rings fashioned from billions of icy particles—some as small as grains of sand and others as big as buildings. The rings are complex and not solid, containing ringlets and gaps between them. The appearance of the rings, visible with a small telescope, changes over time. When the tilt of Saturn is toward Earth, the rings are visible from a broad top-down view. Roughly every 14 years, the rings are tilted edgewise toward Earth, making them all but disappear from view—an event that will next occur in 2023. The Cassini-Huygens spacecraft is named in part for the Dutch astronomer.

Saturn's Moons

Saturn's impressive assemblage of moons includes many that exhibit unusual features. Only 53 of the 62 known moons currently have names.

KEY FACTS

Saturn's largest moon is the massive Titan, larger than the planet Mercury.

+ fact: Titan's atmosphere contains more than 90 percent nitrogen.

+ fact: Moons Titan and Enceladus have the potential for life.

+ fact: Some of Saturn's moons share orbits, traveling the same paths.

Among Saturn's fascinating moons, Titan has surface liquids in the form of two large methane lakes. The poles of Enceladus shoot geysers of water vapor and ice particles that suggest liquid water beneath its frigid surface. Moon Mimas sports a large impact crater of unknown origin. Several moons appear to wrangle the billions of ice particles that comprise Saturn's rings. These "shepherd moons"—Pan, Atlas, Pandora, and Prometheus—straddle two of the rings and keep them intact. Much of what we know about Saturn, its moons, and its rings comes from the data being collected by the Cassini orbiter since its arrival at Saturn in 2004. Cassini transmits many raw images that are made available to the public before they are calibrated and analyzed by NASA scientists.

Uranus

Uranus is a tilted planet, spinning on its side at an angle of 98° to its orbital plane, possibly the result of a massive collision.

KEY FACTS

Uranus has a retrograde rotation, revolving in the opposite direction of most other planets.

+ fact: Uranus has a diameter of 31,764 mi (51,119 km).

+ fact: The planet's distance from the sun is 1.8 billion mi (2.9 billion km).

+ fact: Its orbital period is 84.02 Earth years, and its rotation period is 17.24 Earth hours.

Distant Uranus, barely visible to the naked eye, was incorrectly identified as a star by ancient astronomers before its discovery in 1781 by astronomer William Herschel. Spotted more realistically through binoculars, Uranus appears as a small blue-green disk, the color due to methane gas in its upper atmosphere. Uranus is considered an ice giant, with temperatures reaching down to −357°F (−216°C). Uranus's tilted rotation means that the planet's poles point toward the sun: first one pole and then, 42 years later, the other. In addition to 5 large moons and more than 20 smaller ones, 13 narrow rings surround Uranus. The ring system was imaged by the Hubble Space Telescope.

Neptune

Named for the Roman god of the sea, Neptune displays a vivid blue color, a product of the planet's surrounding clouds of icy methane.

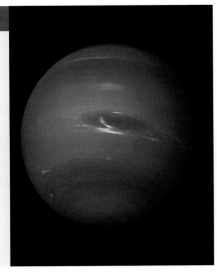

KEY FACTS

Neptune's moon, Triton, has a retrograde orbit, orbiting opposite the direction of Neptune's rotation.

+ fact: Neptune has a diameter of 30,776 mi (49,529 km).

+ fact: The planet's distance from the sun is 2.8 billion mi (4.5 billion km).

+ fact: Its orbital period is 164.79 Earth years, and its rotational period is 16.11 Earth hours.

The outermost major planet in the solar system, Neptune is an ice giant similar in size to Uranus, although denser. A deep atmosphere containing hydrogen, helium, and a small amount of methane surrounds it. The planet seems to have an internal heat source and a puzzling tendency toward very turbulent weather, despite its great distance from the sun. Winds on Neptune can reach more than 1,300 miles (2,100 km) an hour. Neptune is surrounded by a system of rings that are faint and lumpy, and appear to be younger than the planet itself. The rings could be the remnants of a former moon pulled apart by Neptune's gravity.

Dwarf Planets & Plutoids

Dwarf planets are almost planets, which tend to be small and roundish and travel with debris in their orbits. Plutoids are dwarf planets in the trans-Neptunian Kuiper belt.

KEY FACTS

Haumea is named for a mythical Hawaiian sorceress; its moons are named for her daughters.

+ fact: Haumea is sometimes called the "cosmic football" for its shape and motion.

+ fact: Makemake is named for the chief god of Easter Island in the South Pacific.

+ fact: Eris was originally called Xena after the TV warrior princess.

When Pluto was demoted in 2006, it was put into a category of dwarf planet. Besides Pluto, several other objects in the Kuiper belt qualify. Generally, dwarf planets are larger than Mercury, roundish, and are not satellites of other objects. Haumea, a rocky, egg-shaped object, tumbles end over end as it spins every four hours. Makemake seems to have an atmosphere containing nitrogen and methane. It shines red and has no moon. The orbit of Eris and its moon, Dysnomia (Greek for "lawlessness") takes it 10 billion miles (16 billion km) beyond the Kuiper belt. Dwarf planet Ceres is not a plutoid; it appears in the interior asteroid belt beyond Mars. Astronomers are investigating hundreds of objects to determine whether the objects qualify for dwarf planet status.

Pluto

Pluto enjoyed 76 years of planetary status until, one day in 2006, it didn't. Demoted to dwarf planet, the celestial body named for the god of the underworld retains legions of fans.

KEY FACTS

Some astronomers consider Pluto and its moon Charon to be a double planet.

+ fact: Pluto has a diameter of 1,430 mi (2,301 km).

+ fact: The planet's distance from the sun is 3.7 billion mi (5.9 billion km).

+ fact: Its orbital period is 247.92 Earth years, and its rotational period is 6.38 Earth days.

In 1930, an assistant at the Lowell Observatory found a long-anticipated planet beyond Neptune. Excitement about the newbie generated a naming contest, which an 11-year-old British girl won. She proposed the name of the Greek god of the underworld that coincidentally was the name of Mickey Mouse's popular pup. By 2006, the International Astronomical Union, or IAU, had its doubts about Pluto and demoted it to the status of dwarf planet. Pluto did not meet one of the prerequisites the IAU had drawn up: It didn't clear its orbit of debris. Pluto now lends its name to dwarf planets that inhabit the Kuiper belt—the plutoids. Many of Pluto's devoted fans mourn its demotion.

Ceres

Ceres, a dwarf planet, rose out of the asteroid belt between Mars and Jupiter. Since its discovery in 1801, Ceres has worn several solar system hats: planet, asteroid, and now dwarf planet.

KEY FACTS

Ceres has a rocky inner core and a mantle of water ice, and is one of two dwarf planets without a moon.

+ fact: Ceres's diameter is 590 mi (950 km).

+ fact: The dwarf planet's distance from the sun is 258 million mi (415 million km).

+ fact: Its rotational period is 9.075 Earth hours.

In 2006, Ceres was promoted to the new group of dwarf planets. About one-fourth the diameter of Earth's moon, Ceres is the largest object in the asteroid belt, an area holding thousands of objects that are remnants left from the formation of the solar system. Dwarf Ceres is the fifth planet from the sun and is classified among the terrestrial planets because of its rocky nature. It is suspected of having water ice under its crust because it is less dense than Earth and its surface shows evidence of water-bearing minerals. Named for the Roman goddess of agriculture, Ceres takes 4.6 Earth years to complete one orbit around the sun. Ceres was discovered by Italian astronomer Giuseppe Piazza.

The Universe

The word "universe" evokes all the superlatives one can imagine, only to take on more with each breakthrough in cosmic understanding enabled by sophisticated technology.

KEY FACTS

Many physicists and astronomers believe that our universe is just one of an unknown number of universes.

+ fact: The age of our universe is about 13.7 billion years old.

+ fact: The first stars were formed about 100 million years after the big bang.

+ fact: The first galaxies formed less than 600 million years after the big bang.

The unfolding story of the universe involves greater stretches of the imagination than most science fiction tales could devise. Consider the plot: An infinitely small point of infinite density and infinite gravity erupts to create both space and time, unleashing radiation that eventually forms matter that eventually forms stars, galaxies—and us. The initial radiation of the universe's chaotic beginnings was left by hot plasma (ionized gas) that emitted a microwave signal we can detect by space probes such as the Wilkinson Microwave Anisotropy Probe. The probe accomplished a full-sky survey of background microwave radiation between 2001 and 2010, which scientists scrutinize for information about this phenomenon.

Big Bang

More of an expansion than an explosion, the big bang set in motion the formation of the universe about 13.7 billion years ago.

KEY FACTS

Events immediately after the big bang occurred at an incomprehensibly fast speed.

+ fact: In 10^{-35} seconds after the big bang, its energy turned into matter.

+ fact: In 10^{-5} seconds after the big bang, the universe's natural forces took shape.

+ fact: In 3 seconds after the big bang, the nuclei of simple elements formed.

The term "big bang" was coined in the 1950s by British astronomer Fred Hoyle. Ironically, he used it derisively, but it stuck. Current thinking holds that the universe began as a singularity, a point of infinite gravity, where space, time, and all subsequent matter and energy were contracted into an object without size. At the point in which contents of the singularity escaped, a dramatic expansion began that has played out over billions of years. Radiation from the initial event persists as detectable cosmic background radiation. The origin of the singularity remains scientifically elusive, but predicted scenarios of the universe's demise include a big crunch, a big chill, or a big rip. Astronomers have a few tens of billions of years to look for answers.

Galaxy

Gravity holds together the vast assemblage of stars, interstellar dust and gas, and dark matter (little understood matter that emits no light) that form a galaxy.

KEY FACTS

There may be one hundred billion galaxies in the universe.

+ fact: Galaxies occur in clusters throughout the universe.

+ fact: Galaxy clusters form larger associations, giving the universe a "clumpy" nature.

+ fact: The most distant galaxy yet identified is more than 13 billion light-years away.

Shape determines galaxy classification. Spiral galaxies, by far the most numerous, rotate around a bright nucleus. Arms spiral out from that point. Elliptical galaxies may be nearly spherical or stretched to an oblong; stars in them are mostly older red giant stars. Irregular galaxies, which lack a coherent shape, are amorphous collections of stars that include star-forming nebulae. The most recently identified type is the starburst galaxy, so-called because the stars in it seem to burst out. Galaxies are so massive that the strength of one can rip apart another, and they can collide even in the relative emptiness of space. The Hubble Space Telescope's website (*www.hubblesite.org*) features an extensive gallery of amazing, downloadable photos that demonstrate galaxy diversity—and beauty.

Milky Way

Home sweet home, the Milky Way galaxy contains our solar system, just a tiny enclave in the vast spiral galaxy of several hundred billion stars.

KEY FACTS

The Milky Way is composed of some 200 to 500 billion stars.

+ fact: Our solar system is located about 25,000 light-years from the galaxy core.

+ fact: The Milky Way's core contains a supermassive black hole.

+ fact: In ancient times, the Milky Way was a more prominent feature in the sky and many myths developed around it.

Thought to be about 13 billion years old, the Milky Way is a bit younger than its oldest stars, which are estimated to be about 13.5 million years old. The galaxy's disk shape bulges at the center with orange and yellow stars. A corona of old stars and globular clusters extends above and below the disk. The younger, brighter stars and nebulae crowd the arms of the galaxy. The entire spiral structure spins, completing a rotation approximately every 200 million years. Overall, the Milky Way has a diameter of about 100,000 light-years; its thickness ranges from 1,000 to 13,000 light-years. Dark nights and clear skies give the best chance of viewing the Milky Way. In the Northern Hemisphere, the summer months usually offer the best views of the galaxy.

Andromeda Galaxy

A faint oval smudge in the gap between the constellations Cassiopeia and Andromeda, the Andromeda galaxy is the most distant object visible to the naked eye.

KEY FACTS

Astronomer Edwin Hubble's study of the Andromeda galaxy was the first confirmation of the existence of galaxies other than the Milky Way.

+ fact: The light we see from the Andromeda galaxy left the galaxy 2.5 to 3 million years ago.

+ fact: The galaxy has a magnitude of 4.4, making it one of the brighter Messier objects.

Possessing a spiral shape like the Milky Way, the Andromeda galaxy is the larger of the two, containing an estimated one trillion stars. Nevertheless, recent observations indicate that the Milky Way may contain greater mass, if dark matter is included. The Andromeda galaxy is accompanied by several smaller galaxies, including M32 and M110, visible through binoculars. Under good conditions, especially during autumn, the Andromeda galaxy is visible to the naked eye. Astronomers as far back as at least A.D. 964 observed and recorded the feature. Many observers comment that its beauty compares to its namesake, the mythical princess Andromeda. A distance "only" about 2.5 million light-years away turns the Andromeda galaxy into the girl next door.

Galaxy Collisions

Despite the general emptiness of space, galaxies have a tendency to move toward each other and collide, a mechanism that allows galaxies to increase in size.

KEY FACTS

A merger from a galaxy collision can take millions of years.

+ fact: When a large galaxy merges with a smaller one, the larger one usually keeps its shape.

+ fact: Galaxy collisions cause friction that can trigger shock waves that lead to star formation.

+ fact: Galaxy collisions provide vital information about galaxy evolution.

In the scheme of the universe, galaxy collisions occur with different frequencies. More often, larger galaxies collide with dwarf galaxies and incorporate the mass of the dwarf galaxies into their own. Collisions of larger galaxies happen less frequently. Collisions can produce a galaxy of a different shape; two spiral galaxies, for example, can form an elliptical galaxy. The Hubble Space Telescope has revealed details of the approaching collision—4 billion years from now—between the Milky Way and the Andromeda galaxy. The news seems good for our solar system: displacement, perhaps, but not destruction. This future merger of the two galaxies sometimes is referred to as Milkomeda or even the Andromeda Way.

Nova

A spectacular sudden surge in the brightness of a star, a nova is the equivalent of a supermassive atomic explosion, but one from which the star often can recover.

KEY FACTS

Novae are not often spotted, but when they are, you can find them with binoculars along the Milky Way.

+ fact: At times, a nova can be viewed with the naked eye.

+ fact: The density of a white dwarf allows it to draw in hydrogen from the larger star.

+ fact: A classical nova erupts only once.

Typically, a nova occurs when stars in a binary system reach different points in their stellar evolution, with one star collapsed into a dense white dwarf and the other in a red giant phase. The white dwarf siphons off hydrogen from the red giant, and the temperature and pressure rise until the gas erupts in a thermonuclear explosion. This causes the white dwarf to brighten for a period of hours or days by as many as ten magnitudes before dimming over a period of months as the remnants of the blast dissipate. Novae are not necessarily fatal to the stars involved and often occur in cycles in some binary pairs. The Keck Interferometer, a two-telescope system at the W. M. Keck Observatory in Mauna Kea, Hawaii, has captured invaluable data about a nova in the constellation Ophiuchus.

Supernova

Unlike the nova, a supernova is a one-time event. It signals the death of a star, an irrevocable transformation of its material.

KEY FACTS

A supernova can briefly release as much energy as all the stars in the Milky Way.

+ fact: A supernova creates raw material for the formation of new stars.

+ fact: The Crab Nebula (M1) in the constellation Taurus is the remnant of an event in 1054.

+ fact: The brightest supernova in recent history occurred in 1987.

Like the nova, a supernova can also occur in a white dwarf if its core takes in so much hydrogen that it implodes. A supernova also announces the end stage for a red supergiant when its core implodes from intense heat and pressure, unleashing a violent shock wave that blasts away the surrounding cloud of gas. The supernova was first recorded by the Chinese in A.D. 185. Supernovae again were recorded in 1006, 1054, and 1572, and in the modern era. The bright flash of a supernova can occasionally be viewed by the naked eye, but usually requires an 8-inch or larger telescope. More than 25 years after its explosion in the Large Magellanic Cloud, Supernova 1987A continues to glow as it transforms into a supernova remnant, which the Hubble Space Telescope has imaged.

Black Hole

A black hole provides a one-way ticket to visual oblivion for the contents of a supernova explosion. The resulting gravitational field collapses in on itself, allowing no light to escape.

KEY FACTS

The point of no return at the edge of a black hole is called the event horizon.

+ fact: The energy from stellar gases swirl into a black hole like a whirlpool.

+ fact: The black hole in a galaxy in the constellation Virgo seems to have the density of 3 billion suns.

+ fact: The Chandra X-ray Observatory provides critical data on black holes.

Born in one of the universe's most dramatic and awe-inspiring events, a black hole is one of the two options that follow a supernova explosion. The last explosive gasps of a red supergiant create a gravitational field so intense that the force of gravity causes matter to collapse inward. This forms the black hole, in which distance between subatomic particles reduces to zero, and density and gravity expand to their maximum expression. Black holes affect the stars and gas around them, which is one way they can be detected. Small black holes formed at the creation of the universe, and supermassive black holes formed when the galaxies they inhabit were formed. Supermassive black holes spin furiously as they swallow stars and merge with other black holes.

Quasar

The bright light of a quasar is fed by a black hole in the form of a brilliant stream of energy that a black hole releases as it consumes the contents of nearby stars.

KEY FACTS

Coined in the 1960s, the word "quasar," comes from "quasi-stellar," as in quasi-stellar radio source.

+ fact: More than 120,000 quasars have been identified.

+ fact: Quasars are among the universe's very ancient objects.

+ fact: The Sloan Digital Sky Survey catalogs positions and absolute brightness of celestial objects such as quasars.

Bright, deep sky objects with a starlike appearance, quasars puzzled astronomers from the time they were first detected from their radio emissions. When spotted, quasars seemed too bright to be so far away. A boost from association with a black hole was proposed for the anomaly. When quasars were found to be abundant, they came to be considered a part of normal galactic evolution. Though distant, a quasar is within the reach of a skilled amateur astronomer. The quasar 3C 273 can be spotted some 3 billion light-years away through an 8-inch-aperture telescope just 5 degrees northwest of the figure's head in Virgo. Sometimes, though, even the Hubble Space Telescope can struggle to observe a quasar embedded in a galaxy obscured by a large amount of cosmic dust.

Binary & Multiple Stars

"Twinkle, twinkle, little stars"—plural—might be more appropriate, because about 80 percent of stars exist with a companion or companions.

KEY FACTS

The star pair Mizar and Alcor in Ursa Major are about 3 light-years apart and are not gravitationally linked.

+ fact: Mizar is actually a double binary, and Alcor is a binary, making a 6-star system.

+ fact: Binary stars, each star with its own orbiting planet, have been discovered.

+ fact: Binary stars, each with multiple planets in orbit, have also been discovered.

Our singleton sun aside, the massive gas clouds that nurture stars usually produce them in pairs or greater multiples that remain gravitationally attached to each other throughout their lives. Binary stars, triples, and larger groups are the norm, whether the companion is a twin of similar size and luminosity or a group of small siblings that stick close to a more dominant star. Double stars can be optical doubles, which only appear to be close, or true binary stars, with two stars orbiting a common center of gravity. Powerful binoculars or a telescope are often needed to pick out the individual members of a binary pair. Alpha Centauri is a binary star that seems to be bound to Proxima Centauri, which is far from Alpha Centauri but closer to us, as a so-called wide binary.

Nebula

Giant clouds of stellar gas, nebulae provide the raw materials for stars, planets, and galaxies. They are present at both star birth and star death.

KEY FACTS

Despite the name, a planetary nebula does not have an association with planets.

+ fact: Dark nebulae are so dense that they hide the light of stars within them, not only behind them.

+ fact: The Helix Nebula is the closet nebula to Earth.

+ fact: The North American Nebula is a large emission nebula in the constellation Cygnus.

Various types of nebulae represent those giant gas clouds in their different and dynamic roles and relationships with stars. Emission nebulae are star-forming clouds energized by the young stars within them, such as the middle star of Orion's belt. Planetary nebulae are formed of the gas departing from a dying red giant, as seen in the Ring Nebula in the constellation Lyra. Dark nebulae are collections of interstellar dust that obscure the light of the stars behind them. Reflection nebulae shine from the light of nearby stars. Many nebulae are visible with a 4- or 6-inch telescope, maximized with higher magnification and filters. The Orion Nebula, which is 1,500 light-years from Earth, is considered a textbook example of a planetary nebula, showing a number of different star-forming processes.

Eagle Nebula

The Eagle Nebula, located in the tail portion of the constellation Serpens about 7,000 light-years away, is an active region of star formation.

KEY FACTS

The dark towers are 56 trillion mi (90 trillion km) high.

+ fact: Images from the Chandra X-ray Observatory show that the pillars are low in x-ray content.

+ fact: The scant content perhaps signals the pillars' end.

+ fact: Further study suggests that a supernova destroyed the pillars 6,000 years ago, but evidence hasn't yet reached Earth.

The stellar nursery in the Milky Way known as the Eagle Nebula is actually a combination nebula and star cluster. Philippe Loys de Cheseaux discovered it between 1745 and 1746, as did Charles Messier independently in 1764. The nebula has dark pillars of dense material that rise at its center and can be seen with a 12-inch telescope. The Hubble Space Telescope captured these pillars in a now iconic image known as the Pillars of Creation. Newborn stars sculpt the pillars by burning away some of the gas within the nebula. The nebula in general can be viewed with low-powered telescopes and even binoculars. But to view the Eagle Nebula in all its dramatic glory, you will want to check out the online images available on HubbleSite (*www.hubblesite.org*).

Exoplanets

Exoplanets, short for extrasolar planets, are celestial bodies beyond our solar system that orbit stars and share other characteristics of known planets.

KEY FACTS

More than 2,300 exoplanets have been identified, with many more expected.

+ fact: Exoplanets include "hot Jupiters," gas giants that orbit closer to their stars than Mercury does to the sun.

+ fact: Hot Jupiters are also known as "roaster planets."

+ fact: Super-Earths refer to rocky exoplanets about twice the mass of Earth.

Radio astronomers discovered the first extrasolar planet in 1991. Since then, ground-based and space-based telescopes have confirmed the presence of hundreds of objects that orbit stars of other systems or that freelance as nomads—planets "kicked out" of their star systems. Various scientific organizations use different criteria to confirm an exoplanet's status, coming up with different totals, but all concede that the numbers are increasing. The ultimate find among exoplanets would be an Earthlike body in the habitable zone of its solar system that not only had a potential for life, but also signs of life itself. To keep up with the latest exoplanet news, visit the PlanetQuest website (*planetquest.jpl.nasa.gov*). It documents exoplanet research at the Jet Propulsion Laboratory.

Kepler Space Telescope

Launched in 2009, just south of Cape Canaveral, the Kepler Space Telescope is on a mission to monitor more than 150,000 stars, looking for Earthlike exoplanets.

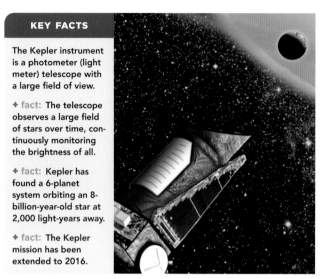

KEY FACTS

The Kepler instrument is a photometer (light meter) telescope with a large field of view.

+ fact: The telescope observes a large field of stars over time, continuously monitoring the brightness of all.

+ fact: Kepler has found a 6-planet system orbiting an 8-billion-year-old star at 2,000 light-years away.

+ fact: The Kepler mission has been extended to 2016.

The Kepler Space Telescope scrutinizes our portion of the Milky Way galaxy for extrasolar planets and the stars they orbit. The big prize would be an Earthlike planet technically capable of supporting life (known as a "just right Goldilocks" planet), but all kinds of exoplanets are given attention: gas giants, hot super-Earths with short-period orbits, and ice giants. Kepler finds potential candidates through the concept of the transit—a planet crossing in front of its star from the observer's point of view. When this happens, the star's brightness dims by a factor related to the size of the object. A periodic occurrence at four equal intervals confirms an object as a planet candidate. Follow-up observations are made to eliminate other possibilities before the planet is verified.

Dark Energy

Einstein rightly realized that space was not just empty space. Though only inferred, dark energy may be the "dark force" involved in rapid expansion of the universe.

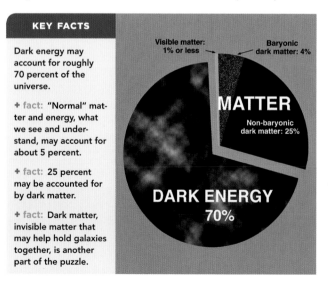

KEY FACTS

Dark energy may account for roughly 70 percent of the universe.

+ fact: "Normal" matter and energy, what we see and understand, may account for about 5 percent.

+ fact: 25 percent may be accounted for by dark matter.

+ fact: Dark matter, invisible matter that may help hold galaxies together, is another part of the puzzle.

Visible matter: 1% or less

Baryonic dark matter: 4%

MATTER

Non-baryonic dark matter: 25%

DARK ENERGY 70%

Too much was going on in the universe for there not to be something in space besides space. When a study of supernovae showed an unexpected acceleration of the expansion of the universe, some force had to be held accountable. That force has been designated as dark energy, a concept that creates more questions than it answers at present. It can be seen as the reverse of gravity: Dark energy pushes objects away from each other rather than pulling them together and coalescing them. How it operates keeps astrophysicists in a quandary, because it calls into question the laws of gravity and the theory of relativity. Observations by the Hubble Space Telescope in 1998 set this discussion in motion, but a lot more data obviously are needed to arrive at some answers.

Constellation

Over the millennia, stargazers have teased out shapes of humans, animals, and objects from patterns of stars in the night sky. A recognized star pattern is called a constellation.

The International Astronomical Union (IAU) recognizes 88 official constellations. These include the original 48 constellations listed by Ptolemy in the second century A.D., which have names and backstories rooted in Greco-Roman mythology. The Greek constellation Argo Navis has since been divided into three separate constellations. The list also includes a series of "modern" constellations that have been added since about 1600 to fill in gaps, especially in the southern sky. According to the IAU, a constellation refers not just to the star pattern itself, but also to an agreed-upon, bounded segment of the sky in its vicinity. The convention helps us locate objects in the night sky and allows astronomers to convey information about deep-space objects that are not part of a constellation.

Asterism

When is a constellation not a constellation? When it's an asterism, a small group of stars that makes a recognizable, attention-grabbing shape in the sky.

A look at the official list of 88 constellations will show that some prominent celestial names, like the Big and Little Dippers, are missing. These well-known star formations are asterisms, small groups of stars that have a distinct shape and may form part of a constellation but are not constellations themselves. The iconic dippers, for example, are contained within Ursa Major and Ursa Minor, respectively. Asterisms are useful points of orientation in the sky and may include stars from multiple constellations. The Winter Triangle, for example, connects the alpha stars Betelgeuse of Orion with Procyon in Canis Minor and Sirius in Canis Major. Similarly, the Summer Triangle asterism connects the bright stars Vega in constellation Lyra, Deneb in Cygnus, and Altair in Aquila.

Cluster

Two kinds of star clusters appear in the deep sky. These dramatic formations are sometimes visible to the naked eye, but a telescope reveals their scale and complexity.

KEY FACTS

In an open cluster, the individual stars often can be resolved through a telescope.

+ fact: Individual stars often escape open clusters and travel on their own.

+ fact: Globular clusters are prominent in the central bulge of the Milky Way.

+ fact: Many of the globular clusters in our galaxy have highly eccentric orbits.

Star clusters are classified as either open or globular clusters. Open clusters are loose concentrations of a few to several thousand associated stars that are bound weakly by gravity. They also share a common star birth "cloud," or source of raw materials. A good example is the Pleiades in the constellation Taurus, a young open cluster that will eventually disperse over time. Another type is the globular cluster, a massive star association that may contain up to a million stars and may be up to 13 billion years old. About 150 of them have been identified, mostly in the outer reaches of the Milky Way. Star clusters in proximity to each other appear to merge, creating larger clusters. Astronomers hope that the James Webb Space Telescope will reveal details of such interactions.

The Pleiades

Any Subaru owner should recognize this asterism, which appears in stylized fashion as the logo for the Japanese auto manufacturer named for the star cluster.

KEY FACTS

The Pleiades cluster (M45) contains hundreds of stars; the 6 or 7 brightest stars make up the famous Seven Sisters asterism, most of them visible to the naked eye.

+ fact: months best viewed: January–February

+ fact: seasonal chart location: Winter, southwest quadrant

+ fact: noted skymark: Taurus

Also known as the Seven Sisters, the Pleiades cluster is one of the most easily identified objects in the night sky. A line traced from alpha stars Betelgeuse in Orion through Aldebaran in Taurus brings the asterism into view. The Pleiades' seven bright stars feature prominently in many world mythologies; they get their names from seven sisters in Greek mythology, daughters of the Titan Atlas and the Oceanid Pleione. Zeus set them in the sky to thwart Orion's advances. The brightest star is Alcyone, with a magnitude of 2.86. A Kiowa Indian tale tells of seven sisters chased by giant bears onto a tall rock and then whisked into the sky. The rock is Devil's Tower and its long cracks are the claw marks of the frustrated bears.

Big Dipper

The first asterism many people learn to locate, the Big Dipper is an abiding feature of the northern sky in all seasons, because of its proximity to the celestial North Pole.

KEY FACTS

This 7-star asterism, which forms a distinct ladle or dipper, is incorporated into the constellation Ursa Major. Its orientation changes according to time of the year.

+ fact: **months best viewed:** March–April

+ fact: **seasonal chart location:** Spring, center of chart

+ fact: **noted skymark:** Ursa Major

To locate the Big Dipper, face the north horizon. As it circles the celestial pole during the year, the asterism faces downward in spring and upward in autumn. The Big Dipper also points to its counterpart in Ursa Minor, the Little Dipper. Follow an imaginary line from the front edge of the Big Dipper's ladle to find Polaris on the end of the Little Dipper's handle. Visible year-round, the Big Dipper served as a constant guide for escaped slaves traveling along the Underground Railroad and is mentioned in many stories and songs, including "Follow the Drinking Gourd." It is known as the Plow in Britain and the Great Cart in Germany. In India it is the Seven Sages. Cultures in much of the world have given locally significant names to this familiar asterism.

Polaris

Not the brightest star in the sky—as many assume—Polaris is actually the 50th brightest star, but its year-round presence makes it an invaluable "skymark."

Polaris

KEY FACTS

Polaris is a yellow supergiant star that varies in brightness throughout the year. It is part of a triple star system, although its other two stars are not visible to the naked eye.

+ fact: **months best viewed:** Year-round

+ fact: **seasonal chart location:** All star charts

+ fact: **noted skymark:** Little Dipper

Located directly above Earth's northern axis, Polaris appears to be the star around which the rest of the sky revolves. The stars in the front edge of the Big Dipper asterism line up with it, making the North Star, or polestar, easy to find. Due to the wobbling of the Earth on its axis as an effect of gravitational pull from the sun, moon, and planets—a motion known as precession—the Earth's pole points to different stars over time. The pole will be closest to Polaris about the year 2100, and after that will move through a series of new polestars and reach Vega in Lyra about 12,000 years from now. In the southern sky, Beta Hydri in the southern constellation Hydrus (different from the northern sky constellation Hydra) is the closest bright star to the celestial south pole.

Northern Sky

As most of the world's population lives in the Northern Hemisphere, the northern sky has gotten the lion's share of attention, from the ancient astronomers onward.

KEY FACTS

Northern sky constellations are named mostly for figures in Greco-Roman mythology.

+ fact: Some constellations are visible in the Northern Hemisphere all year.

+ fact: Ursa Major is an example of a north circumpolar constellation.

+ fact: Others near the celestial equator are visible in northern and southern skies.

When Greek astronomer Ptolemy codified the constellations in the second century A.D., he followed traditions handed down from ancient Mesopotamian observers who had watched the skies at about latitude 35° N. They would have seen the northern sky up to the celestial north pole and, in theory, a good bit of the southern sky, although objects near the horizon are difficult to see. Constellations and features farther south would have been blocked by Earth. This partially accounts for the fact that ancient astronomers limited their constellation list to 48, although they saw some of the southern constellations some of the time. The "Tonight's Sky" feature at HubbleSite (www.hubblesite.org) offers a preview of constellations, deep sky objects, planets, and events visible each month in the skies of the Northern Hemisphere.

Southern Sky

Not fully appreciated until voyages of exploration sailed southern routes beginning in the 1500s, the southern sky harbors extraordinary sights.

KEY FACTS

Crux, or the Southern Cross, is the smallest in area of all the constellations.

+ fact: The Southern Cross appears on the flags of Australia and New Zealand.

+ fact: Many beautiful globular sky clusters are visible in the southern sky.

+ fact: The brightest part of the Milky Way passes over southern skies in winter.

The skies of the Southern Hemisphere contain a glorious array of stars, clusters, galaxies, and nebulae. The newer names for southern constellations often honor the tools of science and art. The iconic Southern Cross, or Crux, includes several notable stars as well as the "Jewel Box" open cluster and the Coalsack Nebula, a dense cloud of gas and dust silhouetted against the Milky Way. The southern sky also contains the huge constellation Centaurus—location of Alpha Centauri, neighbor to the sun and only 4.3 light-years away—as well as the Large and Small Magellanic Clouds, irregular dwarf galaxies that orbit the Milky Way. If you have an opportunity to visit the Southern Hemisphere, check out the wonders of its night sky. Many online resources can guide your tour.

Zodiac

A group of star patterns recognized from ancient times, the 12 zodiac constellations represent animals, people, creatures—and one inanimate object, Libra—that appear in an annual cycle.

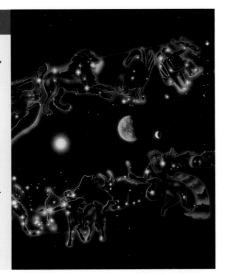

KEY FACTS

The zodiac constellations are Aries, Taurus, Gemini, Cancer, Leo, Virgo, Libra, Scorpius, Sagittarius, Capricornus, Aquarius, and Pisces.

+ fact: Several constellations of the zodiac are always in view.

+ fact: These constellations are located along the ecliptic, the sun's apparent yearly path across the sky.

The word "zodiac" comes from the Greek for "circle of animals," and most of the zodiacal constellations do have animal associations. In the sky, these constellations form an annual procession along the ecliptic, the apparent path taken by the sun across the sky as the Earth orbits the sun. As each of the 12 zodiacal constellations reaches its westernmost phase, it is in alignment with the sun, creating the idea of "sun sign" that forms the basis of the pseudoscience of astrology. Constellation Ophiuchus also reaches into the ecliptic path, but it usually is excluded from the zodiac. In Arab astronomy, even Libra (The Scales) had an animal connection, as reflected in the names of its alpha and beta stars, which were considered the claws of the constellation Scorpius.

Astrology

In the mystical and mind-bending 1970s, "What's your sign?" was a frequent conversation starter, reflecting the influence of astrology—at least superficially—in those changing times.

KEY FACTS

For millennia, astrology was considered a scholarly pursuit, and it remains so in some cultures.

+ fact: The Babylonians developed the first organized system of astronomy.

+ fact: Astrology helped promote the advancement of astronomy.

+ fact: Arab astrologers made many notable contributions to astronomy.

Few people are unaware of their "sign," the constellation of the zodiac where the sun resided when they were born. The correlation between signs and personal characteristics and the alignment of the sun, moon, stars, and planets at a particular point with future events forms the basis of astrology. Astrology is a practice as old as stargazing itself. The ancient Egyptians, for example, looked to the rising of the star Sirius to predict the flooding of the Nile, events that did coincide. Astrologers deem such occurrences to be cause-and-effect situations, using reliable observations to support their claims. In some cultures, predictions based on astrological interpretations are still taken into consideration when planning life events such as marriages and making other important decisions.

Sky & Constellation Charts

The following section of this chapter presents the night sky in the Northern Hemisphere through a series of four seasonal sky charts or maps and following those, through focused charts that highlight discussions of 56 different constellations.

The seasonal sky charts—one each for winter, spring, summer, and autumn—show the visible constellations, the brightest stars, and the locations of many galaxies, nebulae, and other deep sky objects. These maps reflect the fact that the celestial sphere operates on a constantly changing continuum. The maps are prepared from the perspective of an observer at latitude 40° N, at the dates and times indicated on each chart.

Getting Oriented

These charts show the sky at a particular time but are useful any night. Like the sun, the stars and constellations appear to move from east to west, in some cases falling from view below the horizon for several months.

Observers at latitudes above 40° N will find the northern constellations higher in the sky, for longer periods of time, while losing sight of some below the southern horizon. The reverse is true for observers at positions toward the Equator.

The seasonal sky charts include a silhouetted landscape around the border indicating the horizon. Constellations appearing within about

STELLAR MAGNITUDES
- -0.5 and brighter
- -0.4 to 0.0
- 0.1 to 0.5
- 0.6 to 1.0
- 1.1 to 1.5
- 1.6 to 2.0
- 2.1 to 2.5
- 2.6 to 3.0
- 3.1 to 3.5
- 3.6 to 4.0
- 4.1 to 4.5
- 4.6 to 5.0
- Variable star

DEEP SKY OBJECTS
- Star; size indicates apparent magnitude
- Variable star
- Galaxy
- Open star cluster
- Globular star cluster
- Planetary nebula
- Bright nebula

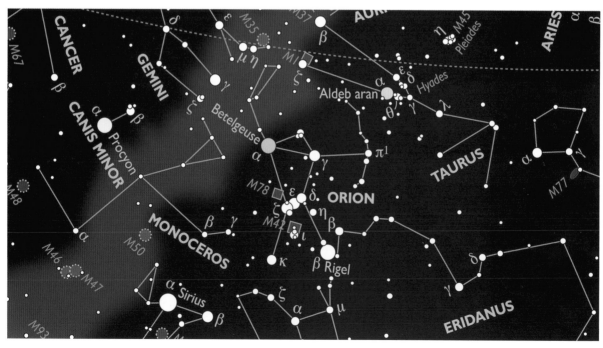

NOTES ABOUT SKY CHARTS

+ **Stars:** The brightest stars have common names, like Antares, which often relate to ancient mythology. But most bright stars that form constellations are typically known by Johann Bayer's 1603 method of calling the brightest star in a constellation "alpha," the second brightest "beta," and so on. For example, Deneb is the common name for the brightest star in the constellation Cygnus (the Swan), but the star is also known as Alpha Cygni, and is often labeled with both names.

+ **Deep sky objects:** Most of the galaxies and other "fuzzy" objects seen by backyard astronomers take their names from either the Messier catalog (M) or the New General Catalog (NGC), where they are assigned a number. Charles Messier

published his catalog of about 100 objects in 1771. The New General Catalog, with 7,840 objects, came in 1888. There are alternative names in these catalogs and others. For example, in addition to the common name Andromeda galaxy, the object is also known as M31 and NGC 224, as well as by other names.

+ **Mythological icons:** These small drawings are designed to help relate the shapes of the constellations to the figures they represent in mythology. They are not necessarily oriented in the direction they are typically seen in the night sky, and their arbitrary line patterns do not always match what is shown on the detailed star charts on the following pages.

Star maps often show line patterns to help illustrate the shape typically recognized as a constellation. These line patterns vary from source to source. Choosing which stars to show as part of a constellation drawing is arbitrary. The International Astronomical Union (IAU) defines constellations as areas of the night sky, not as strings of specific stars. The IAU's well-defined technical boundaries are not shown on these charts.

10 degrees of that edge will be difficult to spot. On the charts, stars with magnitudes down to 5 are connected by lines to form the constellations. Many stars of magnitude 3.5 or brighter (meaning a lower numerical value) are labeled for reference, and some have been tinted to indicate color. A star's size reflects its magnitude, indicated in a table on the margins of the chart, as are the symbols for various deep sky objects. The ecliptic—the sun's apparent path during the year and the line of travel for the zodiacal constellations—is the dotted line across each chart. The faint white field represents the Milky Way.

Constellation Charts

The pages that follow the seasonal sky charts describe 56 of the 88 recognized constellations. Many of the original 48 in Ptolemy's *Almagest,* an astronomical handbook, are included, as are newer constellations added to fill in the sky in the Northern Hemisphere. Those not included are mainly the "deep south" constellations that cannot be viewed from the mid-northern latitudes. You can learn about southern sky constellations online or in astronomy guides that often include such information.

Key Facts

Each constellation heading shows the "Size on the Sky," using a thumb length, closed fist, and outstretched hand or hands held at arm's length to indicate the constellation's approximate size in the sky. The Key Facts for the constellations include the number of brighter stars found in the constellation. "Best viewed" indicates the best months to hunt for the constellation. "Location" indicates the appropriate seasonal sky chart for reference. The constellation's alpha star also is noted.

Each constellation entry features a star map showing the stars that make up a constellation and a portion of the surrounding sky. Light lines connect the brighter member stars of each constellation; prominent asterisms are named. The brightest stars are labeled with letters of the Greek alphabet, beginning with alpha. Some of the brighter and well-known stars are labeled with their proper names and, where appropriate, tinted to indicate the approximate color in the sky. Background stars, neighboring constellations, and nearby deep sky objects also appear on the maps. Messier (M) or New General Catalog (NGC) numbers indicate deep sky objects. See "Notes About Sky Charts," above, for M and NGC histories.

Winter

Orion makes a good orientation point in the winter sky. Bright star Betelgeuse appears on Orion's right shoulder and Rigel forms the left foot. At the sword, you'll find the Orion Nebula (M42). Taurus lies northwest of Betelgeuse and so do two famed star clusters, the Pleiades and the Hyades.

Find Auriga, and draw a line north from Betelgeuse to bright star Capella. Northeast of Orion, Gemini is marked by the stars Castor and Pollux. To the southeast lies alpha star Sirius of Canis Major. Betelgeuse, Sirius, and Procyon in Canis Minor form the Winter Triangle asterism.

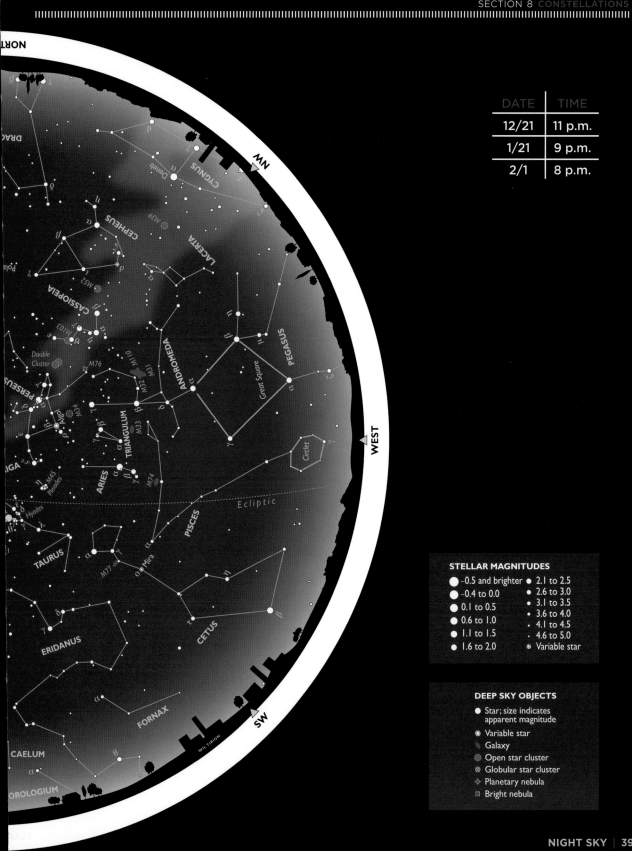

DATE	TIME
12/21	11 p.m.
1/21	9 p.m.
2/1	8 p.m.

STELLAR MAGNITUDES

- −0.5 and brighter
- −0.4 to 0.0
- 0.1 to 0.5
- 0.6 to 1.0
- 1.1 to 1.5
- 1.6 to 2.0
- 2.1 to 2.5
- 2.6 to 3.0
- 3.1 to 3.5
- 3.6 to 4.0
- 4.1 to 4.5
- 4.6 to 5.0
- Variable star

DEEP SKY OBJECTS

- Star; size indicates apparent magnitude
- Variable star
- Galaxy
- Open star cluster
- Globular star cluster
- Planetary nebula
- Bright nebula

Spring

Spring is a good time to hunt for galaxies, as the sky offers a good view of elusive deep sky objects. Earth's position relative to the Milky Way has shifted so that the core of our galaxy lies near the horizon in the west. In spring, Ursa Major swings highest above Polaris, bringing the Big Dipper along with it. Virgo is prominent, together with its dense collection of galaxies, including the Virgo supercluster, the Sombrero galaxy (M104), and many other Messier items.

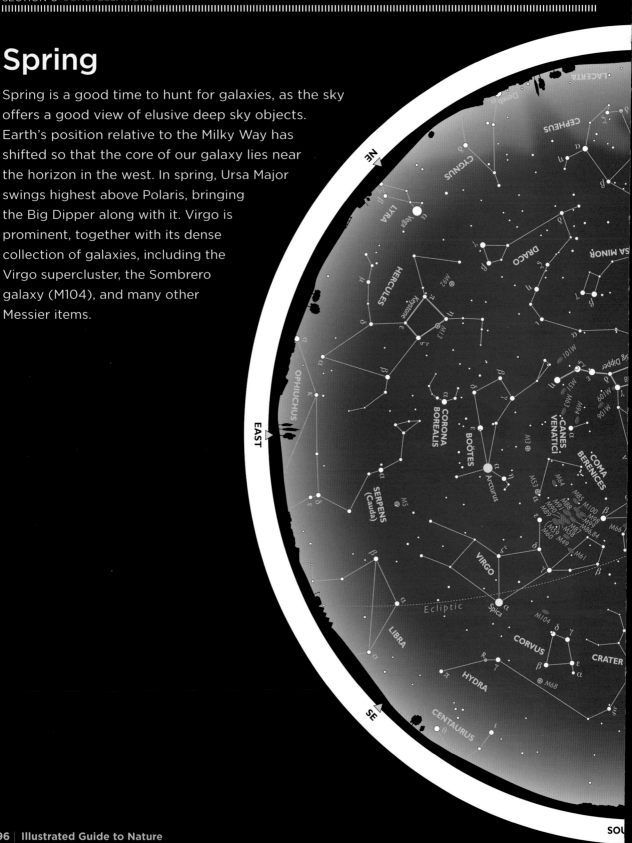

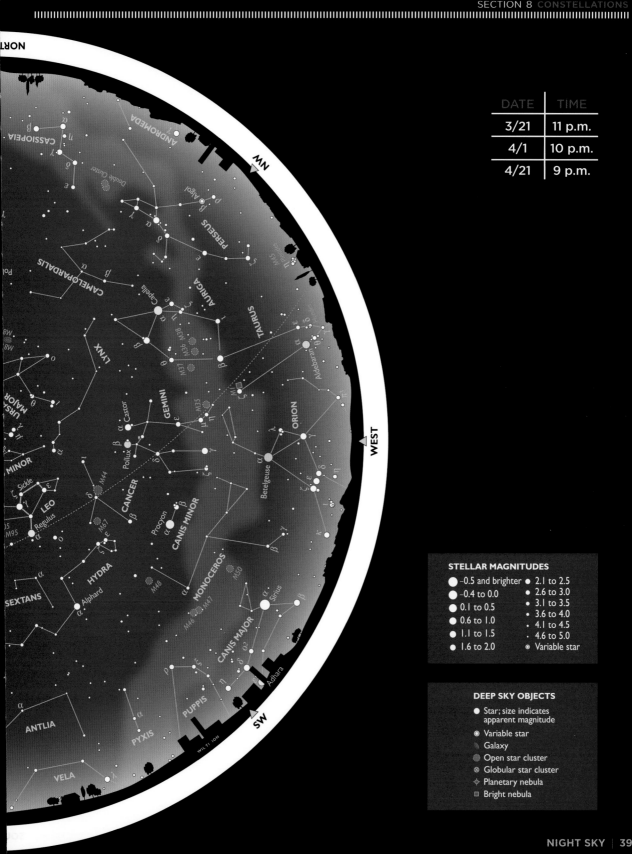

DATE	TIME
3/21	11 p.m.
4/1	10 p.m.
4/21	9 p.m.

STELLAR MAGNITUDES

- -0.5 and brighter
- -0.4 to 0.0
- 0.1 to 0.5
- 0.6 to 1.0
- 1.1 to 1.5
- 1.6 to 2.0
- 2.1 to 2.5
- 2.6 to 3.0
- 3.1 to 3.5
- 3.6 to 4.0
- 4.1 to 4.5
- 4.6 to 5.0
- ⊛ Variable star

DEEP SKY OBJECTS

- ● Star; size indicates apparent magnitude
- ⊛ Variable star
- ◜ Galaxy
- ⊛ Open star cluster
- ⊗ Globular star cluster
- ✧ Planetary nebula
- ◻ Bright nebula

Summer

Summer places the Summer Triangle overhead, highlighting three constellations. Vega in Lyra, Deneb in Cygnus, and Altair in Aquila form the seasonal asterism. Vega will be almost directly overhead, unmistakably bright at magnitude 0. Sagittarius is an area rich in star clusters and nebulae. Locate the Teapot asterism within it to spot notable objects surrounding the lid: the Lagoon (M8) and Trifid (M20) nebulae, and the great Sagittarius star cluster (M22).

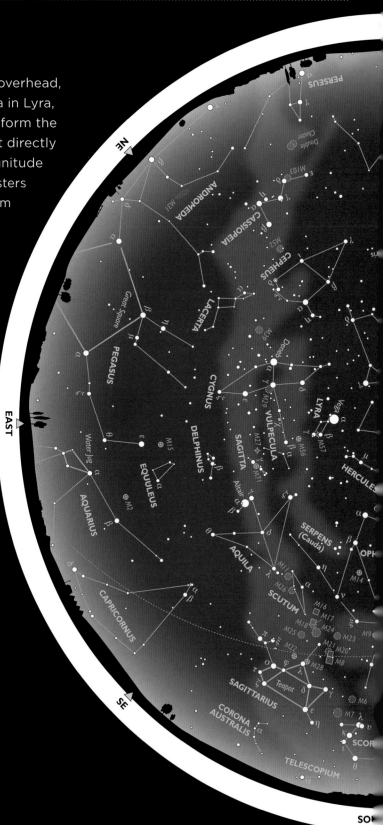

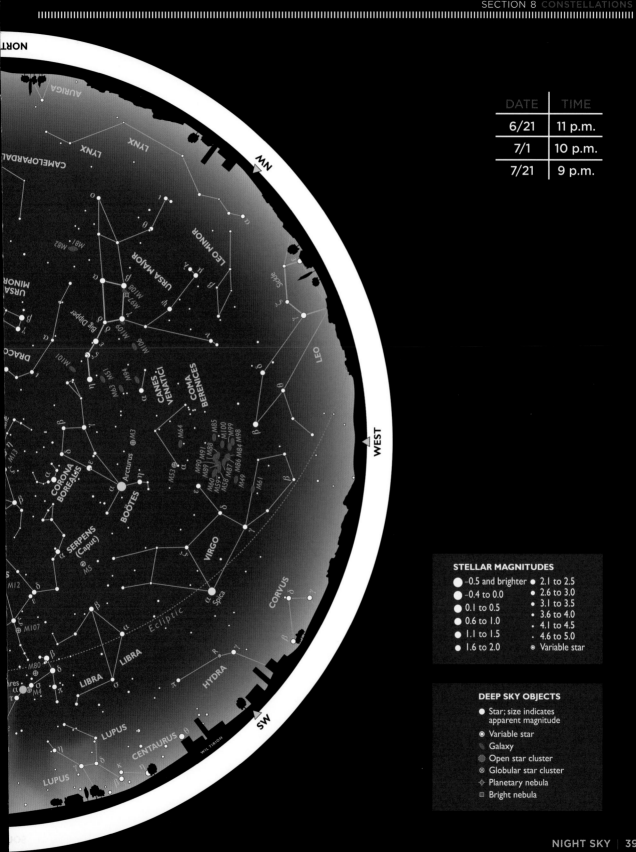

DATE	TIME
6/21	11 p.m.
7/1	10 p.m.
7/21	9 p.m.

STELLAR MAGNITUDES

- −0.5 and brighter
- −0.4 to 0.0
- 0.1 to 0.5
- 0.6 to 1.0
- 1.1 to 1.5
- 1.6 to 2.0
- 2.1 to 2.5
- 2.6 to 3.0
- 3.1 to 3.5
- 3.6 to 4.0
- 4.1 to 4.5
- 4.6 to 5.0
- Variable star

DEEP SKY OBJECTS

- Star; size indicates apparent magnitude
- Variable star
- Galaxy
- Open star cluster
- Globular star cluster
- Planetary nebula
- Bright nebula

WIL TIRION

Autumn

The Great Square of Pegasus, an asterism at the center of constellation Pegasus, is central to locating several other autumn constellations, including Andromeda. Alpha star Alpheratz is shared by Andromeda and the asterism. Autumn brings four major meteor showers: the Orionids in October, the Taurids and Leonids in November, and the Geminids in December. The season also offers rewarding galaxy viewing: The Andromeda galaxy (M31) in that constellation is visible to the naked eye.

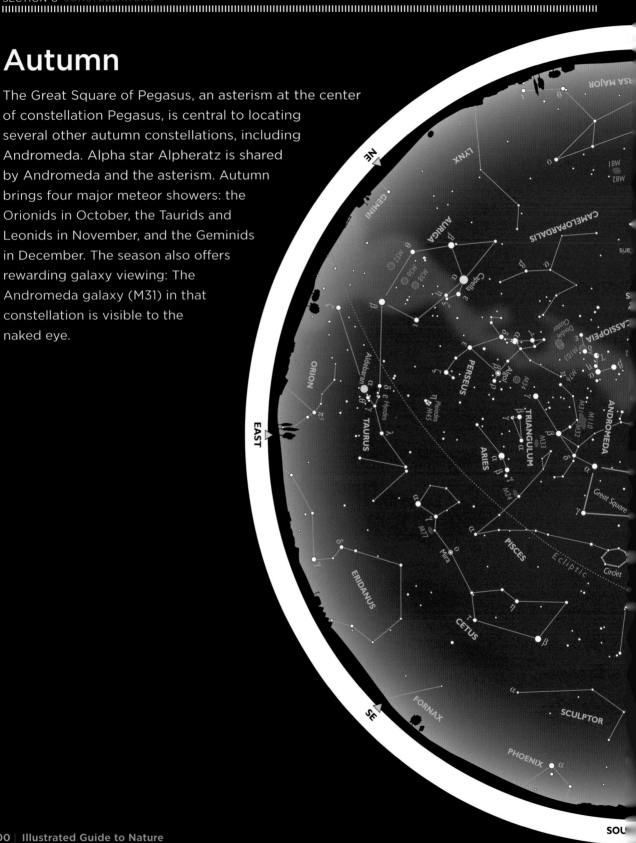

DATE	TIME
9/21	11 p.m.
10/21	10 p.m.
11/1	8 p.m.

STELLAR MAGNITUDES

● −0.5 and brighter	● 2.1 to 2.5
● −0.4 to 0.0	● 2.6 to 3.0
● 0.1 to 0.5	● 3.1 to 3.5
● 0.6 to 1.0	· 3.6 to 4.0
● 1.1 to 1.5	· 4.1 to 4.5
● 1.6 to 2.0	· 4.6 to 5.0
	⊛ Variable star

DEEP SKY OBJECTS

● Star; size indicates apparent magnitude
⊙ Variable star
✺ Galaxy
✳ Open star cluster
⊗ Globular star cluster
✧ Planetary nebula
▫ Bright nebula

Andromeda

The Chained Maiden   Size on the sky:

This constellation notably contains the Andromeda galaxy (M31), a spiral-shaped deep sky galaxy similar to our own Milky Way. Andromeda galaxy is visible to the naked eye.

KEY FACTS

The figure of Princess Andromeda, most often seen upside down, contains 7 main stars. Locate by tracing a line northeast from the northeast corner of the Great Square of Pegasus.

+ months best viewed: October–November

+ seasonal chart location: Autumn, center of chart

+ alpha star: Alpheratz

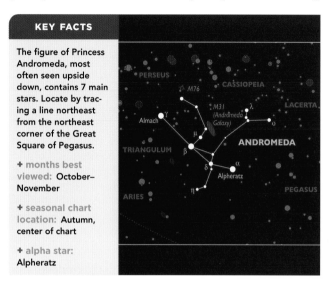

Cassiopeia and Cepheus, the mythical rulers of Ethiopia, angered the Greek gods when Cassiopeia boasted that her daughter Andromeda's beauty surpassed that of the daughters of Nereus, god of the sea and Poseidon's father-in-law. In retaliation, Poseidon sent the sea monster Cetus to destroy the kingdom. Cassiopeia and Cepheus then offered Andromeda as a sacrifice, chained to a rock. At the last moment, the hero Perseus, homeward bound with the head of Medusa, rescued Andromeda, his trophy head turning Cetus to stone in the bargain. Other characters in this famous tale, including Pegasus, the hero's mount, appear nearby in the sky.

Antlia

The Air Pump   Size on the sky:

Northern Hemisphere stargazers can pick out this small constellation in the spring near the southern horizon. It contains two binary stars, one of which can be split with good binoculars.

KEY FACTS

The angular southern constellation Antlia contains 3 main stars; the constellation lies about 5 fist-widths south of the bright alpha star Regulus in Leo.

+ months best viewed: March–April

+ seasonal chart location: Spring, southwest quadrant

+ alpha star: Alpha Antlia

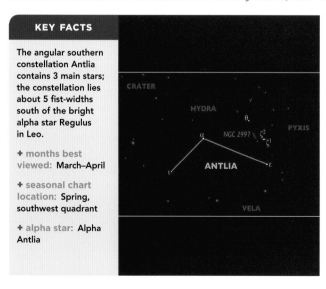

As a southern constellation, Antlia's backstory belongs to the realm of science, not mythology. In the 1750s, French astronomer Nicholas-Louis de Lacaille constructed the constellation from his vantage point at the Cape of Good Hope in South Africa. He named it for Irish chemist and physicist Robert Boyle's discovery of the pneumatic pump, originally calling it Antlia Pneumatica. All together, Lacaille charted some 10,000 southern stars. Antlia's alpha star is its brightest one, but it has no proper name. Antlia harbors the galaxy NGC 2997, faintly visible through a small telescope, just inside its corner.

Aquarius

The Water Bearer   Size on the sky:

A constellation of the zodiac, Aquarius occupies a lot of celestial real estate along the ecliptic—the apparent path of the sun across the night sky—between Pisces and Capricornus.

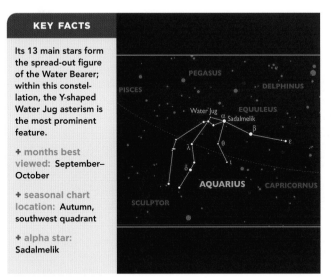

KEY FACTS

Its 13 main stars form the spread-out figure of the Water Bearer; within this constellation, the Y-shaped Water Jug asterism is the most prominent feature.

+ **months best viewed:** September–October

+ **seasonal chart location:** Autumn, southwest quadrant

+ **alpha star:** Sadalmelik

One of the fainter constellations of the zodiac, Aquarius is located south of the Great Square of Pegasus, in the "watery" section of the sky with associated constellations such as Cetus the Whale and Pisces. Aquarius has several mythological associations. The ancient Egyptians linked the constellation to the yearly flooding of the Nile River, the most vital event of the agricultural year, and the hieroglyph for water became the zodiacal symbol for Aquarius. To the Greeks, Aquarius was Zeus's cupbearer, Ganymede. In some versions, he pours water or wine from a jug that maintains the celestial river Eridanus.

Aquila

The Eagle   Size on the sky:

Named by ancient Mesopotamian stargazers, Aquila the Eagle appears close enough to Earth's Equator to be seen from any terrestrial viewing position.

KEY FACTS

This straightforward constellation near the Equator contains 10 stars, with alpha star Altair traditionally representing the eagle's head.

+ **months best viewed:** August–September

+ **seasonal chart location:** Summer, southeast quadrant

+ **alpha star:** Altair

Aquila's alpha star, Altair, is one of the brightest stars in the sky and a good reference point for identifying the constellation. It also forms part of the Summer Triangle asterism along with Vega in constellation Lyra and Deneb in Cygnus. Aquila occurs in the Milky Way in an area of abundant starfields. In Greek myth, the eagle was Zeus's companion and carried the god's thunderbolts for him. He also is thought to have carried the young shepherd Ganymede to the sky to serve as Zeus's celestial cupbearer. Ganymede is immortalized nearby as the constellation Aquarius.

Aries

The Ram   Size on the sky:

Ancient astronomers traditionally placed Aries at the beginning of the celestial zodiac because, millennia ago, the sun was "in" Aries at the time of the vernal equinox in spring.

KEY FACTS

The brightest part of constellation Aries contains 4 stars; the "tail" of the ram is Gamma Arietis, or Mesartim, a double star with a wide separation.

+ months best viewed: November–December

+ seasonal chart location: Winter, southwest quadrant

+ alpha star: Hamal

On evenings in late fall and early winter, Aries appears high in the west between the Great Square of Pegasus and the Pleiades asterism in Taurus. The constellation's double star, Mesartim, was one of the first doubles spotted with a telescope—by astronomer Robert Hooke in 1664. In the view of the ancient Greeks, Aries was the source of the Golden Fleece stolen by Jason and the Argonauts. Jason wrested the fleece from the custody of a dragon that had received it from a king. The king had sacrificed the ram in gratitude for saving his two children from an abusive stepmother.

Auriga

The Charioteer   Size on the sky:

The tidy, elegant form of Auriga appears in the heart of the Milky Way and is on Ptolemy's list of the 48 original ancient constellations.

KEY FACTS

The constellation Auriga, found on a line between Orion and Polaris, has 7 main stars. It shares with Taurus a star (lowest on the map) that otherwise would be Auriga's gamma star.

+ months best viewed: December–January

+ seasonal chart location: Winter, center of chart

+ alpha star: Capella

Auriga's shining glory is its alpha star, the brilliant Capella, the sixth brightest star in the sky. Southwest of Capella lies Epsilon Aurigae, an eclipsing binary star veiled every 27 years by a companion star. The constellation lies along the galactic equator, the great circle that passes through the densest part of the Milky Way and contains several interesting star clusters visible through binoculars, notably M36 and M37. The chariot and rider of Auriga may represent Hephaestus of Greek myth, the crippled blacksmith god who built the chariot to accommodate his handicap.

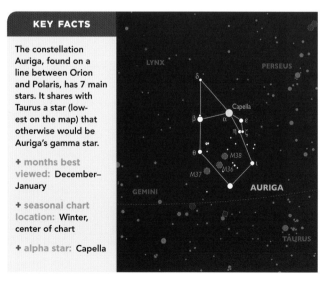

Boötes

The Herdsman   Size on the sky:

The ancient constellation of Boötes is a celestial highlight of early summer. Its alpha star, Arcturus, ranks as fourth brightest star in the sky.

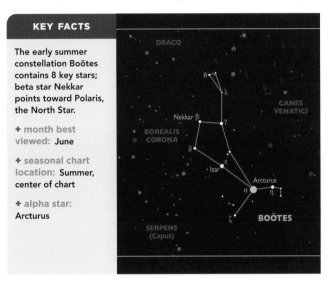

KEY FACTS

The early summer constellation Boötes contains 8 key stars; beta star Nekkar points toward Polaris, the North Star.

+ month best viewed: June

+ seasonal chart location: Summer, center of chart

+ alpha star: Arcturus

Boötes's alpha star, Arcturus, lies on an arc continuing from the handle of the Big Dipper, giving rise to the mnemonic "arc to Arcturus." In a variation of the myth surrounding the constellation, herdsman Boötes keeps his flock moving about the sky in pursuit of Ursa Major and Ursa Minor. In another, he herds a bear; a loose translation of Arcturus is "bear keeper." In the first week in January, the northern part of Boötes hosts the Quadrantid meteor shower, one of the strongest of the year, producing several dozen meteors an hour during its peak. It occurs at the point where Boötes, Hercules, and Draco meet.

Camelopardalis

The Giraffe   Size on the sky:

Camelopardalis is located in a region of the sky where few stars are visible. No star in this faint constellation has a common name.

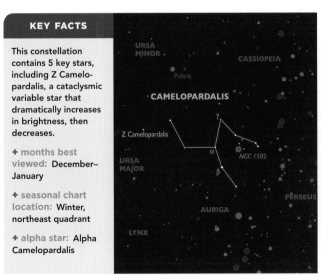

KEY FACTS

This constellation contains 5 key stars, including Z Camelopardalis, a cataclysmic variable star that dramatically increases in brightness, then decreases.

+ months best viewed: December–January

+ seasonal chart location: Winter, northeast quadrant

+ alpha star: Alpha Camelopardalis

Camelopardalis is a modern constellation, created in 1613 to fill the celestial gap between the bears, Ursa Major and Ursa Minor, and Perseus. It neighbors Polaris, the North Star, making it a strictly northern constellation, although a faint one. It contains the star cluster NGC 1502, a good target for a telescope, and spiral galaxy NGC 2403, which lies about 12 million light-years from Earth. Although the ancient Greeks called the giraffe a "camel-leopard" (and gave the animal its species name: *camelopardalis*), the constellation possibly was named for the biblical camel that carried Rebecca to her marriage with Isaac.

Cancer

The Crab   Size on the sky:

As a constellation of the zodiac, tiny Cancer the Crab is far less noticeable than most. It has no stars with a magnitude brighter than 4.

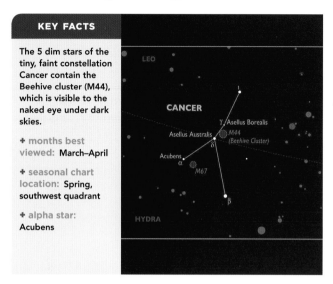

KEY FACTS

The 5 dim stars of the tiny, faint constellation Cancer contain the Beehive cluster (M44), which is visible to the naked eye under dark skies.

+ months best viewed: **March–April**

+ seasonal chart location: **Spring, southwest quadrant**

+ alpha star: **Acubens**

Unassuming Cancer contains some interesting star groups. Viewed with binoculars, the Beehive cluster reveals more than 20 stars, while open cluster M67, with its 500 stars, can be seen through a small telescope. The Tropic of Cancer gets its name from the constellation, although the sun no longer is in it on the summer solstice as it was for early mapmakers. Cancer's story involves Zeus's often jealous wife, Hera. She sent the crab to thwart Hercules, Zeus's son with a mortal, as he battled the multiheaded monster Hydra. Proving no match, the crab perished under the hero's foot, but earned a place in the stars.

Canes Venatici

The Hunting Dogs   Size on the sky:

Location is everything in the night sky: The two stars of Canes Venatici appear just about where two leashed hounds would be expected in relation to Boötes, the Herdsman.

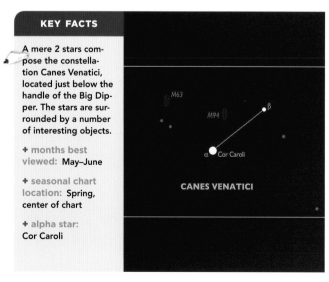

KEY FACTS

A mere 2 stars compose the constellation Canes Venatici, located just below the handle of the Big Dipper. The stars are surrounded by a number of interesting objects.

+ months best viewed: **May–June**

+ seasonal chart location: **Spring, center of chart**

+ alpha star: **Cor Caroli**

An alternative way of viewing Canes Venatici is to look for the two stars running between the legs of Ursa Major, the bear that the dogs are chasing. The small northern constellation was created by Polish astronomer Johannes Hevelius in the 17th century and has been accepted as one of the official 88 constellations. The name of its alpha star, Cor Caroli, stands for the Heart of Charles and was reputedly bestowed on the star by famous English astronomer Edmond Halley in honor of his sponsor, King Charles II. The 500-star globular cluster M3 can be found midway between Cor Caroli and Arcturus in Boötes.

Canis Major

The Larger Dog   Size on the sky:

Canis Major appears just to the southeast of the constellation Orion, not far from the Milky Way. The constellation's alpha star, Sirius, outshines all other stars in the night sky.

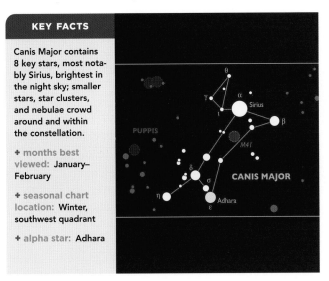

KEY FACTS

Canis Major contains 8 key stars, most notably Sirius, brightest in the night sky; smaller stars, star clusters, and nebulae crowd around and within the constellation.

+ **months best viewed:** January–February

+ **seasonal chart location:** Winter, southwest quadrant

+ **alpha star:** Adhara

Sirius, the Dog Star, shows the way to Canis Major, the constellation representing the larger of Orion's hunting dogs. A line traced through Orion's belt and continuing southeast points directly to Sirius within its constellation. The Dog Star figures in the expression "dog days of summer." In late summer, Sirius rises around the same time as the sun, and the two "conspire" to generate extra warmth in the Northern Hemisphere. Several star clusters and nebulae are located in Canis Major. The brightest, open cluster M41, is easily spotted through binoculars and even more impressive through the lens of a telescope.

Canis Minor

The Smaller Dog   Size on the sky:

Orion's smaller hunting dog, Canis Minor, lies directly northeast of its larger companion, Canis Major. The dim constellation lacks any bright deep sky objects.

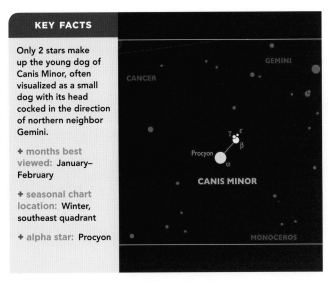

KEY FACTS

Only 2 stars make up the young dog of Canis Minor, often visualized as a small dog with its head cocked in the direction of northern neighbor Gemini.

+ **months best viewed:** January–February

+ **seasonal chart location:** Winter, southeast quadrant

+ **alpha star:** Procyon

Multiple scenarios explain the dim and subdued Canis Minor. One puts the dog beneath the table of Castor and Pollux, the Gemini twins, waiting for scraps. Another makes him Helen of Troy's favorite pup that allowed her to elope with the Trojan prince Paris. Together with Betelgeuse in Orion and Canis Major's Sirius, Canis Minor's alpha star, Procyon, forms the Winter Triangle, a seasonal asterism that helps orient winter stargazers. Procyon lies only 11.2 light-years from Earth and is the eighth brightest star in the night sky. The constellation also hosts the Canis Minorids, a faint meteor shower of early December.

Capricornus

The Sea Goat Size on the sky:

The broad and distinctive triangle of Capricornus stands out in the southern sky, especially in late summer and early fall. Algedi, Arabic for "the goat," is the zodiacal constellation's alpha star.

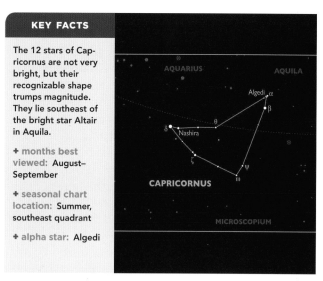

KEY FACTS

The 12 stars of Capricornus are not very bright, but their recognizable shape trumps magnitude. They lie southeast of the bright star Altair in Aquila.

+ **months best viewed:** August–September

+ **seasonal chart location:** Summer, southeast quadrant

+ **alpha star:** Algedi

Long recognized as a goat, Capricornus at some point acquired a fish tail. In one story, the god Pan, a satyr, leaped into the River Nile to escape a monster, and the water transformed him. In an older tale, the animal represents one of Zeus's warriors, discoverer of conch shells with a resounding call that frightened the opposing Titans into retreat. Zeus placed the warrior in the sky with a fish tail and horns to represent his discovery. Alpha star Algedi appears to the naked eye as an optical binary composed of two stars that are closely aligned but separated by more than 500 light-years.

Cassiopeia

The Queen   Size on the sky:

Visible year-round because of its proximity to the north celestial pole, Cassiopeia's iconic W shape (or M, depending on the season) makes identification easy.

KEY FACTS

The 5 stars of this iconic constellation form the figure of Queen Cassiopeia on her throne within a portion of the Milky Way facing Polaris, the North Star; visible much of year.

+ **months best viewed:** October–November

+ **seasonal chart location:** Autumn, northeast quadrant

+ **alpha star:** Shedar

Chained to her throne, Ethiopian Queen Cassiopeia sits between the constellations of her husband, Cepheus, and her daughter, Andromeda. Boasting of her daughter's good looks, the queen drew the ire of the sea god Poseidon, who felt that his own daughters, the Nereids, had been disrespected. Cassiopeia and Cepheus offered to sacrifice Andromeda to save their kingdom. At the last moment, Perseus saved Andromeda, but Cassiopeia was punished nonetheless, forced to hang upside down half the year. Visible through a telescope, the dense open cluster M52, seen off the leg of the W, contains about 100 stars.

Cepheus
The King   Size on the sky:

A circumpolar constellation, Cepheus is visible all year in the Northern Hemisphere, but the brightness of its main stars competes with surrounding stars, often making it hard to spot.

KEY FACTS

The body of King Cepheus, facing the open end of Cassiopeia, contains 5 key bright stars. They take the shape of a small house pointing generally toward Polaris, the North Star.

+ **months best viewed:** September–October

+ **seasonal chart location:** Autumn, center of chart

+ **alpha star:** Alderamin

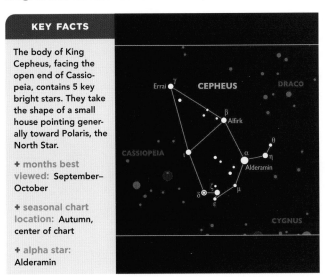

King Cepheus, the hapless consort of Queen Cassiopeia, was forced to put daughter Andromeda at peril because of his wife's unwise boast of their daughter's beauty, which angered the sea god Poseidon. As in his mythological life, the constellation Cepheus plays second fiddle to that of his wife, facing the open end of the W shape of the constellation Cassiopeia. Errai, Cepheus's gamma star, is both a binary star and host to an orbiting planet. Estimated to be almost 1.6 times the size of Jupiter, Errai's planet attests that planets can form in relatively close binary systems, stars that orbit the same center of gravity.

Cetus
The Sea Monster   Size on the sky:

Very large and faint, the constellation Cetus appears in a sky neighborhood known as the Heavenly Waters in the company of constellations Eridanus and Pisces.

KEY FACTS

The 13 stars of the constellation Cetus are divided between the monster's head and body. The parts are connected by two stars; one of them, Mira, is a variable star.

+ **month best viewed:** November

+ **seasonal chart location:** Autumn, southeast quadrant

+ **alpha star:** Menkar

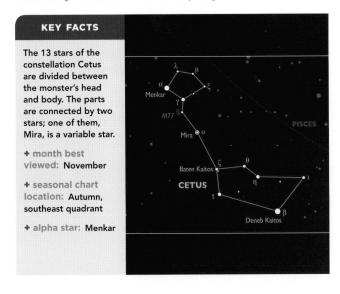

The god Poseidon sent the frightful sea monster Cetus—who gave his name to the mammalian order Cetacea, the whales—to terrorize Ethiopia after Queen Cassiopeia insulted Poseidon's daughters, the Nereids, by boasting that her daughter was more beautiful. Cetus was vanquished by Perseus and turned to stone by the head of Medusa. The constellation is also associated with the whale that swallowed the prophet Jonah in the Old Testament. In autumn, the head of Cetus appears between Taurus and Pisces in the southern sky, with its body bordering Aquarius.

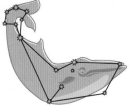

Columba
The Dove   Size on the sky:

Columba is a smallish constellation, one of the modern ones added in the 16th century to fill out the sky chart. It honors the dove Noah sent out from the ark to search for dry land.

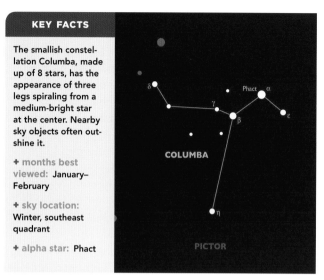

KEY FACTS

The smallish constellation Columba, made up of 8 stars, has the appearance of three legs spiraling from a medium-bright star at the center. Nearby sky objects often outshine it.

+ months best viewed: January–February

+ sky location: Winter, southeast quadrant

+ alpha star: **Phact**

At the end of the 40 days and nights of epic rain that flooded all the Earth, Noah sent out a dove to see if the waters of the great flood had begun to recede. When the dove returned carrying an olive sprig in its bill, Noah knew immediately that it had. The dove's association with the ark and the flood made it a natural companion of the watery constellations of the Heavenly Waters group. The constellation is located very close to brighter objects in the night sky, making the Dove somewhat difficult to make out, especially for novice sky watchers.

Coma Berenices
Berenice's Hair   Size on the sky:

Coma Berenices formed part of other constellations—a wisp of Virgo's hair or tuft of Leo's tail—until 17th-century astronomer Tycho Brahe helped designate it a constellation in its own right.

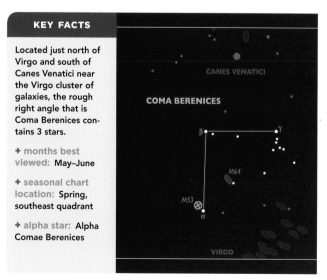

KEY FACTS

Located just north of Virgo and south of Canes Venatici near the Virgo cluster of galaxies, the rough right angle that is Coma Berenices contains 3 stars.

+ months best viewed: May–June

+ seasonal chart location: Spring, southeast quadrant

+ alpha star: **Alpha Comae Berenices**

A rejuvenated tale of an Egyptian queen provides a namesake and story for Coma Berenices. The consort of Ptolemy III, Berenice sought Aphrodite's aid in the safe return of her husband from war, promising the goddess luxurious hair as an incentive. After the wish was granted, an astronomer of the royal court convinced the ruling couple that a grateful Aphrodite had placed the queen's gift in the stars. The constellation contains the deep sky spiral galaxy M64, known as the Black Eye. It lies between the constellation's two outermost stars, along the line that would represent the base of a triangle.

Corona Australis

The Southern Crown   Size on the sky:

Though an original Ptolemaic constellation, Corona Australis is primarily southern, rising only a few degrees above the horizon during midsummer in the mid-northern latitudes.

KEY FACTS

The crown of Corona Australis, which contains 5 key stars, can be found at the feet of Sagittarius and just west of the bright tail star in Scorpius.

+ months best viewed: July–August

+ seasonal chart location: Summer, southeast quadrant

+ alpha star: Alpha Coronae Australis

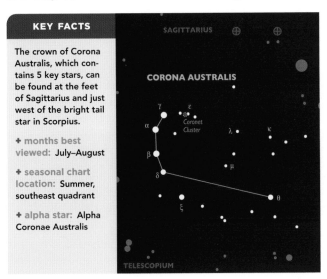

For a small, faint constellation, Corona Australis enjoys a wealth of supporting mythology, much in the form of stories referring to a crown of laurel or fig leaves. One version regards it as the crown of Chiron, the Centaur. Another puts Apollo in the picture, having him fashion the crown from the leaves of his love Daphne, who was changed into a laurel tree to escape the insistent Apollo's advances. The constellation hosts an active star-forming region composed of the Coronet cluster and the nearby Corona Australis Nebula. Though visible only through advanced deep sky technology, this region boasts about 30 "newborn" stars.

Corona Borealis

The Northern Crown   Size on the sky:

Despite having one of the more definitive constellation shapes, small and faint Corona Borealis is squeezed in between its bigger, brighter neighbors Boötes and Hercules.

KEY FACTS

An obvious, semicircular crown shape made of 7 stars in a small and faint constellation is located just south of a line traced between the bright stars Vega in Lyra and Arcturus in Boötes.

+ months best viewed: June–July

+ seasonal chart location: Summer, center of chart

+ alpha star: Alphecca

Another Greek crown-related myth provides a plausible tale for the Northern Crown. In it, good-time god Dionysus threw his crown into the sky to impress Ariadne, princess of Crete and his future wife. She had rebuffed him earlier when he courted in the form of a young mortal. The crown toss changed her mind and they married. In American Indian lore, this celestial shape represents a camp circle. Some of the stars in Corona Borealis vary significantly in brightness. One of them is being investigated as possibly being an extrasolar planet; a body larger than Jupiter already has been found.

Corvus

The Crow   Size on the sky:

Corvus, along with Crater, perch in a crook of Hydra, the serpentine megaconstellation. The Crow is located just west of Spica in nearby Virgo.

KEY FACTS

The constellation Corvus, located near Hydra's tail, contains 5 stars, 4 of which make up its body, while alpha star Alchiba, just outside the crow's body, represents its downward-facing head.

+ **months best viewed:** April–May

+ **seasonal chart location:** Spring, southeast quadrant

+ **alpha star:** Alchiba

In a myth belying the smarts of members of the Corvid family, a crow is sent for a cup of water by the god Apollo. The bird spied a tempting unripe fig and waited for it to ripen. Anticipating the god's ire, the crow made up a tale about being attacked by a snake, which he carried back along with the cup of water. Apollo saw through the deceit and banished the cup, the crow, and the serpent to the sky. A pair of colliding galaxies known as the Ring-tailed galaxy (NGC 4038 and NGC 4039) can be seen with an 8-inch aperture telescope just outside Corvus, with the "tail" near the border of Corvus and Crater.

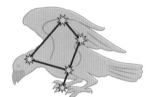

Crater

The Cup   Size on the sky:

The small constellation Crater appears near its mythological associates, Hydra and Corvus. To some, Crater first represented a spike on sea monster Hydra's back.

KEY FACTS

The 8 stars of the constellation Crater make a goblet shape: 4 stars form the base of the goblet and 4 more form the cup, which opens toward the constellation Spica.

+ **months best viewed:** April–May

+ **seasonal chart location:** Spring, southeast quadrant

+ **alpha star:** Alkes

The unmistakable goblet shape of Crater can be found in the spring sky right above Hydra and just west of Corvus, an associated constellation. Crater represents the vessel brought to Apollo by Corvus, the crow. Apollo tossed the two into the sky, along with Hydra, when he realized Corvus had lied about the reason he was late in bringing a requested cup of water. Some sky observers also relate the cup to the constellation Aquarius, the Water Bearer, which occupies the same quadrant of the sky in summer. More recent astronomers associate the cup with the Holy Grail or Noah's wine goblet.

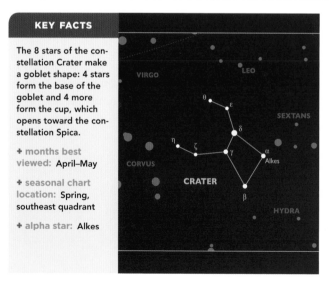

Cygnus

The Swan *Size on the sky:*

Cygnus appears in a dense and visually rewarding portion of the sky. It is sometimes known as the Northern Cross for its shape and bright stars.

KEY FACTS

The 13 stars of the constellation Cygnus form recognizable wings and a neck and head that in late summer and fall point southward like a migratory bird.

+ **months best viewed:** August–September

+ **seasonal chart location:** Summer, northeast quadrant

+ **alpha star:** Deneb

The ancients associated several bird-centric myths with this prominent constellation. It was seen as Zeus, transforming himself into a swan to seduce Leda, and also as Orpheus, who was murdered and turned into a swan for rejecting a group of maidens. He was placed in the sky next to his beloved lyre, the constellation Lyra. Cygnus also represented one of the Stymphalian birds, the quarry Hercules pursued in one of the 12 labors. Alpha star Deneb, along with Altair and Vega, create the Summer Triangle, and a large diffuse gas emission nebula shaped like and named for North America appears on the Swan's southern border.

Delphinus

The Dolphin *Size on the sky:*

The distinct shape of Delphinus, one of the sea creature constellations of the Heavenly Waters sky region, seems to swim toward Pegasus.

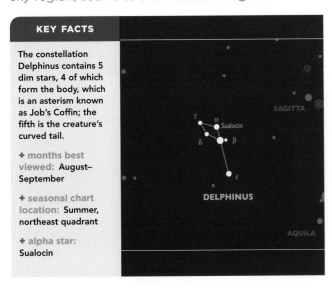

KEY FACTS

The constellation Delphinus contains 5 dim stars, 4 of which form the body, which is an asterism known as Job's Coffin; the fifth is the creature's curved tail.

+ **months best viewed:** August–September

+ **seasonal chart location:** Summer, northeast quadrant

+ **alpha star:** Sualocin

The constellation Delphinus holds its place in the sky because he was a favorite of the Greek sea god Poseidon. The little dolphin succeeded in convincing the Nereid Amphitrite, the sea nymph daughter of Nereus, to marry Poseidon after the god himself had attempted and failed to win her attention. Delphinus can be located just west of a straight line traced between Altair in the constellation Aquila and Deneb in the constellation Cygnus. The Dolphin constellation's gamma star is an optical double star, best viewed through a telescope, whose dimmer star has a greenish tinge.

Draco

The Dragon Size on the sky:

Visible year-round in the Northern Hemisphere, Draco is one of the constellations closest to the north celestial pole. Draco contains the Cat's Eye Nebula (NGC 6543), a dying sun-like star.

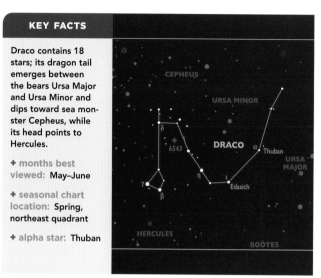

KEY FACTS

Draco contains 18 stars; its dragon tail emerges between the bears Ursa Major and Ursa Minor and dips toward sea monster Cepheus, while its head points to Hercules.

+ **months best viewed:** May–June

+ **seasonal chart location:** Spring, northeast quadrant

+ **alpha star:** Thuban

The constellation Draco represented different scaly beasts to different civilizations. To the ancient Greeks, it was the dragon Ladon, who was slain by Hercules. To the Hindus, it was a celestial alligator, and to the Persians, a giant serpent. Draco's alpha star, Thuban, once was the Earth's polestar, before the phenomenon of precession shifted the celestial position of Earth's axis and made Polaris the polestar. The Quadrantid meteor shower erupts from the region where Draco, Boötes, and Hercules meet in the beginning of January. One of the heaviest meteor showers of the year, the Quadrantids last only a few hours.

Equuleus

The Little Horse Size on the sky:

The second smallest constellation in the area, Equuleus resides in a crowded southern portion of the sky where it is overshadowed by the larger horse constellation, Pegasus.

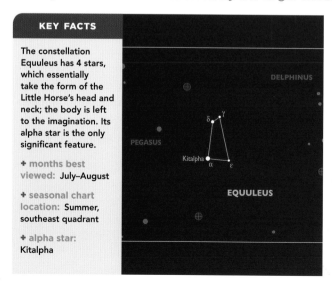

KEY FACTS

The constellation Equuleus has 4 stars, which essentially take the form of the Little Horse's head and neck; the body is left to the imagination. Its alpha star is the only significant feature.

+ **months best viewed:** July–August

+ **seasonal chart location:** Summer, southeast quadrant

+ **alpha star:** Kitalpha

Equuleus is the opening act for Pegasus, rising and setting before its neighbor, a fact that gave it the designation of Equus Prior. The constellation is thought to represent Celeris, the brother-horse of Pegasus. Pegasus is best known as the noble winged steed of Perseus, slayer of Medusa and savior of Andromeda. Celeris belonged to Castor, who along with twin Pollux represents the twins of Gemini. Castor was famed as a skilled equestrian and received the gift of Celeris from the messenger god Hermes. Look for Equuleus between neighbors Pegasus to the northeast and Aquila to the southwest.

Eridanus

The River   Size on the sky:

On a clear night in winter with an open horizon, you have a good chance of seeing Eridanus, although its southern tip disappears below the horizon at about 20° N latitude.

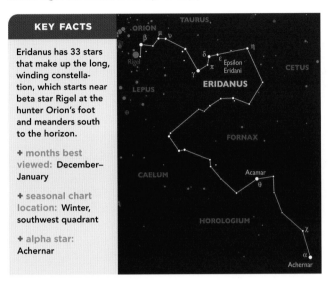

KEY FACTS

Eridanus has 33 stars that make up the long, winding constellation, which starts near beta star Rigel at the hunter Orion's foot and meanders south to the horizon.

+ **months best viewed:** December–January

+ **seasonal chart location:** Winter, southwest quadrant

+ **alpha star:** Achernar

The river of the constellation Eridanus was variously identified as the Euphrates or the Nile. The ancient Greeks saw it as the river into which Phaëthon, son of the sun god Helios, was cast after he failed to control the sun god's chariot and Earth stood in danger of burning up. It is a long river, to be sure, the sixth largest constellation in the sky containing its sixth brightest star, alpha star Achernar. Eridanus bends tightly at its beginning near Rigel and then gradually widens out. It contains a bright star similar to our sun, Epsilon Eridani, which is known to host a confirmed planet.

Gemini

The Twins   Size on the sky:

The twins Castor and Pollux stride hand in hand in the southeast winter sky, northeast of the bright star Betelgeuse on constellation Orion's upraised arm.

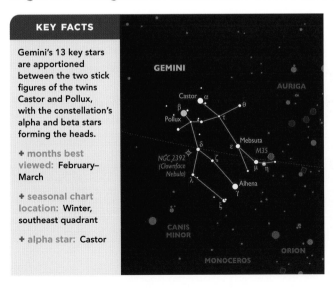

KEY FACTS

Gemini's 13 key stars are apportioned between the two stick figures of the twins Castor and Pollux, with the constellation's alpha and beta stars forming the heads.

+ **months best viewed:** February–March

+ **seasonal chart location:** Winter, southeast quadrant

+ **alpha star:** Castor

The Gemini twins represent Castor and Pollux, sons of the unions of Leda with a swan-disguised, seductive Zeus and her husband Tyndareus, and brothers of Helen of Troy and Clytemnestra—the exact paternity details remaining murky. Castor and Pollux also served as shipmates with Jason on the *Argo*. The impressive Geminid meteor showers originate from the constellation in the middle of December. Binoculars or a telescope yield views of M35, an open star cluster of hundreds of stars that resides near the three "foot stars" of Castor. The constellation also contains the blue-green Clownface Nebula, visible only through a telescope.

Grus

The Crane   Size on the sky:

Grus is a breakout constellation, which was separated from the formation Piscis Austrinus, the Southern Fish. It is visible in the Northern Hemisphere only briefly in the fall.

KEY FACTS

Grus has 11 stars that make up the rough form of a crane in the constellation, including the distinct central shape of an X, which is composed of brighter stars.

+ **month best viewed:** September

+ **seasonal chart location:** Autumn, southwest quadrant

+ **alpha star:** Alnair

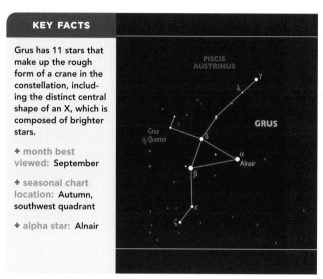

The largely southern constellation Grus is a modern invention, named in 1603 by German celestial cartographer Johann Bayer in honor of the bird the ancient Egyptians made the symbol of the astronomer. Alternate names for the star group include Anastomas (the Stork), den Reygher (the Heron), and Phoenicopterus (the Flamingo). The constellation harbors the Grus Quartet, a group of four spiral galaxies (NGC 7552, 7582, 7590, and 7599) that are located very close together and exert strong influences on each other. Millions of years from now, the four galaxies will likely unite.

Hercules

The Super Hero   Size on the sky:

The constellation Hercules contains a he-man's worth of stars, although none of them are exceptionally bright. Look for him overhead in the summer months.

KEY FACTS

It takes 20 key stars to make up the hefty form of Hercules, including the recognizable backwards K at the center; his "torso" contains the 4-star asterism known as the Keystone.

+ **months best viewed:** July–August

+ **seasonal chart location:** Summer, center of chart

+ **alpha star:** Rasalgethi

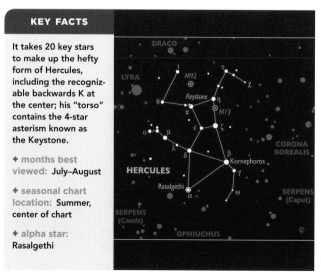

The half-mortal son of Zeus, Hercules was plagued throughout much of his life by the jealousy of Hera, Zeus's wife and queen of the gods, who resented the tangible reminder of her husband's dalliance. She induced a fit of insanity in Hercules that caused him to murder his wife and children. In grief and repentance, he undertook 12 nearly impossible labors, and at their successful completion was granted immortality. The constellation contains M13, considered the best globular cluster visible in the northern sky. It occurs on one side of the Keystone asterism and can be seen as a blur by the naked eye.

Hydra

The Sea Serpent   *Size on the sky:*

Hydra winds across the sky from Cancer to Libra and spans the largest area of any constellation. It used to include today's constellations Sextans, Corvus, and Crater.

KEY FACTS

The constellation Hydra's 17 stars begin with a kite-shaped head, continue past the "heart" at alpha star Alphard, and end with the sea serpent's tail close to the southern horizon.

+ months best viewed: **March–April**

+ seasonal chart location: **Spring, southwest quadrant**

+ alpha star: **Alphard**

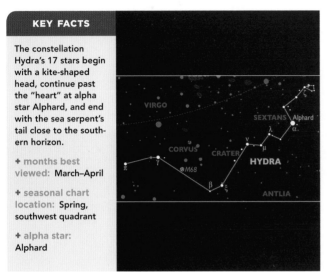

In classical mythology, Hydra lost a decisive battle with Hercules, who fought the multiheaded sea serpent as one of his 12 labors. Hydra put up a good fight, regrowing two heads for each one that Hercules lopped off, and it looked pretty hopeless for Hercules until the stumps of each decapitated head were cauterized to stop the regrowth. Constellation Hydra's enormous span belies the relative dimness of its stars; only two of them surpass a magnitude of 3. Alphard, the brightest star in the formation, can be located by looking just east of a line traced from Regulus in the constellation Leo to the star Sirius in Canis Major.

Lacerta

The Lizard   *Size on the sky:*

A so-called modern constellation, Lacerta was created by 17th-century Polish astronomer Johannes Hevelius. At first depicted as a small mammal, it later morphed into a lizard.

KEY FACTS

The constellation Lacerta, which zigzags between Cassiopeia, Cygnus, and Andromeda, is composed of 8 stars. It forms a W shape like its brighter, queenly neighbor Cassiopeia.

+ month best viewed: **October**

+ seasonal chart location: **Autumn, center of the chart**

+ alpha star: **Alpha Lacerte**

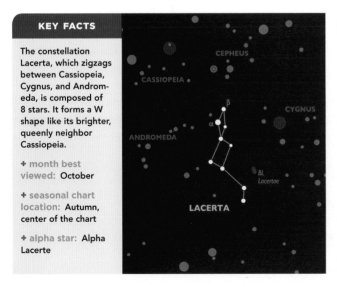

Lacerta is a constellation without a relevant mythology or backstory. At the time it was created to fill in a gap on the celestial map, competing proposals were offered to name it for Louis XIV or Frederick the Great. Although Lacerta's stars are faint, they remain visible to observers for much of the year and appear almost directly overhead in early fall for those in mid-northern latitudes. Lacerta contains an intriguing deep sky object, BL Lacertae, which is the center of an elliptical galaxy visible only through a substantial telescope. Of great interest, its core may harbor a black hole.

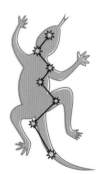

Leo

The Lion   Size on the sky:

One of the most easily visualized constellation shapes, Leo the Lion figured as one of the 13 original zodiacal constellations established by the Babylonians.

KEY FACTS

The kingly constellation Leo has 12 stars, with 8 of them forming the asterism known as the Sickle, which represents the lion's uplifted head.

+ months best viewed: **March–April**

+ seasonal chart location: **Spring, center of chart**

+ alpha star: **Regulus**

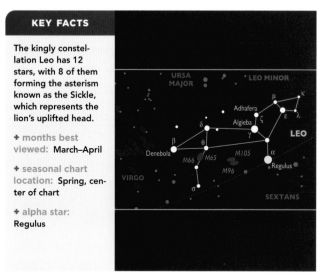

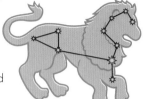

The ancient Egyptians revered the constellation Leo. They associated it with the sun, and may have modeled the Sphinx on its form. Leo also links to Hercules myths as a representation of the Nemean lion that the hero choked to death in the first of his 12 labors. The constellation harbors spiral galaxies M65 and M66, which can be seen with binoculars. Leo conveniently appears at the end of an imaginary line formed from Polaris and through the bowl edge of the Big Dipper. Leo is also the location of the Leonid meteor showers, which are identified with the comet Tempel–Tuttle and peak in mid-November.

Leo Minor

The Small Lion   Size on the sky:

Leo Minor requires a dark night for visibility, but it lies almost directly overhead in the mid-northern latitudes in early spring.

KEY FACTS

Appearing atop Leo, 3 dim key stars make up an indistinct shape of this small constellation; Leo Minor requires a dark night to even be seen.

+ months best viewed: **March–April**

+ seasonal chart location: **Spring, center of chart**

+ alpha star: **none** (beta star: Beta Leonis Minoris)

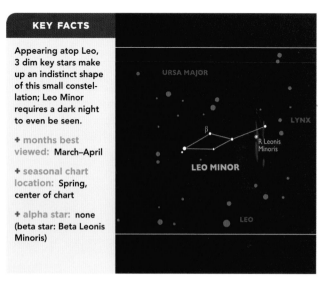

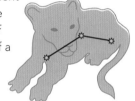

Like other 17th-century additions to the celestial charts, Leo Minor lacks a lot of supporting documentation or mythology. Polish astronomer Johannes Hevelius isolated the formation, which lacks a labeled alpha star, probably as the result of an oversight. Beta star Leonis Minoris is not the constellation's brightest; that distinction belongs to R Leonis Minoris, an oscillating star with more than a five-point range of magnitude during the course of a year. The ancient Egyptians recognized the small Leo Minor group of stars as the hoof prints of a herd of gazelles in flight from neighboring big cat Leo.

Lepus

The Hare   Size on the sky:

Just as Orion keeps close his celestial hunting dogs, he also has his prey nearby in the form of Lepus, the small constellation representing the Hare.

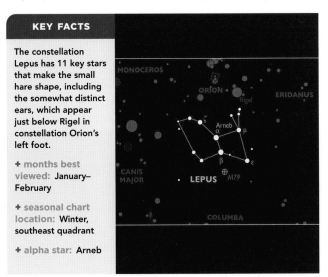

KEY FACTS

The constellation Lepus has 11 key stars that make the small hare shape, including the somewhat distinct ears, which appear just below Rigel in constellation Orion's left foot.

+ months best viewed: January–February

+ seasonal chart location: Winter, southeast quadrant

+ alpha star: Arneb

Lepus the Hare sits crouched and ready to flee just west of Sirius, the bright star in the constellation of a pursuer, Canis Major. Midwinter marks Lepus's highest position in the sky, but it remains close to the horizon for viewers in mid-northern latitudes and dips down below it for part of the year. Lepus contains the globular cluster M79, unusual for the fact that it sits within a southern constellation, while most other globular clusters are found at the center of the Milky Way a significant distance away. It is possible that globular cluster M79 was formed in the neighboring Canis Major dwarf galaxy.

Libra

The Scales   Size on the sky:

Libra falls within the constellations of the zodiac, the only one named for an inanimate object. It appears between zodiac neighbors Scorpius and Virgo.

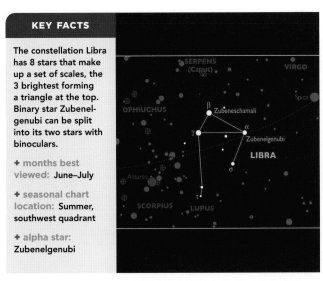

KEY FACTS

The constellation Libra has 8 stars that make up a set of scales, the 3 brightest forming a triangle at the top. Binary star Zubenelgenubi can be split into its two stars with binoculars.

+ months best viewed: June–July

+ seasonal chart location: Summer, southwest quadrant

+ alpha star: Zubenelgenubi

A constellation of the northern summer sky, Libra lacks star magnitude when compared to its neighbors. Arab astronomers, and the Greeks before them, considered the alpha and beta stars of Libra to be the claws of the scorpion of neighboring Scorpius. Their names reflect this designation: Zubenelgenubi means "southern claw" and Zubeneschamali means "northern claw." Later, constellation Libra's scales were associated with those held by Astraea, the Greek goddess of justice. Libra's brightest stars can be found along a line traced from alpha star Antares in Scorpius to alpha star Spica in Virgo.

Lynx

The Lynx   Size on the sky:

The constellation Lynx came into being in the late 17th century to fill an empty celestial area between Ursa Major and Auriga. Polish astronomer Johannes Hevelius introduced it.

KEY FACTS

The 7 key stars of this inconspicuous constellation form a kinked line somewhat suggestive of a rearing lynx, located to the north of stars Castor and Pollux in Gemini.

+ months best viewed: **February–March**

+ seasonal chart location: **Winter, northeast quadrant**

+ alpha star: **Al Fahd**

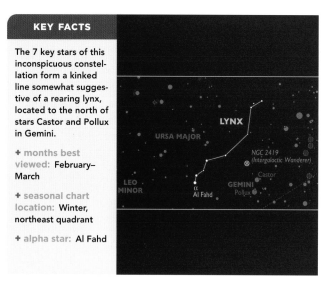

Inconspicuous constellation Lynx makes a difficult sighting for casual stargazers. It is said that Lynx earned its name because it requires the observer to possess the sharp eyesight of the northern forest cat to spot it. As a kind of compensation, the constellation contains a number of double stars to reward the observer who uses even a modest telescope. Lynx also harbors a deep sky object in the form of the Intergalactic Wanderer (NGC 2419). The name "Wanderer" refers to the belief that the object is so far from the core of the Milky Way that it might be expected to break away from its current loose orbit around the galaxy.

Lyra

The Lyre   Size on the sky:

Alpha star Vega steals the show in constellation Lyra, appearing directly overhead in the mid-latitude summer sky. It is ranked the fifth brightest star in the sky.

KEY FACTS

The 6 stars of the easily identified constellation Lyra compose the ancient musical instrument; its alpha star Vega ranks among the brightest in the sky.

+ months best viewed: **July–August**

+ seasonal chart location: **Summer, center of chart**

+ alpha star: **Vega**

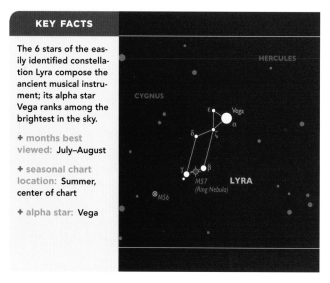

Lyra, the Lyre, belonged to Orpheus, Apollo's tragic son. He lost his beloved wife, Eurydice, not once but twice, when he blew the chance to retrieve her from Hades. Remaining faithful to her memory, Orpheus was killed by a group of scorned women. An impressed Zeus then sent Orpheus's lyre to the sky. With Deneb in Cygnus and Altair in Aquila, alpha star Vega in Lyra forms an asterism known as the Summer Triangle. In 12,000 or so years, alpha star Vega is slated to become the polestar. The Ring Nebula (M57), located at the edge of Lyra, is formed of gas, and it is visible through a telescope.

Monoceros

The Unicorn   Size on the sky:

The name "Monoceros" derives from the Greek for "one horned." A constellation created in the 17th century, the Unicorn was first officially recorded by German astronomer Jakob Bartsch.

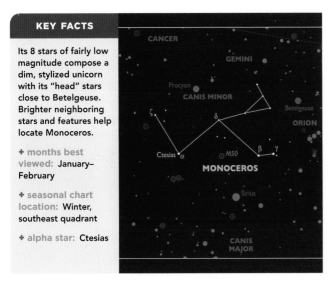

KEY FACTS

Its 8 stars of fairly low magnitude compose a dim, stylized unicorn with its "head" stars close to Betelgeuse. Brighter neighboring stars and features help locate Monoceros.

+ **months best viewed:** January–February

+ **seasonal chart location:** Winter, southeast quadrant

+ **alpha star:** Ctesias

Monoceros appears within the asterism known as the Winter Triangle, which is composed of Betelgeuse in Orion, Procyon in Canis Minor, and Sirius in Canis Major. The constellation is dim in comparison to those very bright stars, but it fills an important gap on the sky chart, and its position on the Milky Way provides many good reference points. One highlight—open star cluster M50—appears between Procyon and Sirius and can be spotted through binoculars. Monoceros may refer to a biblical creature that decided to play in the rain rather than join Noah and the other animals on the ark to escape the great flood.

Ophiuchus

The Serpent Bearer   Size on the sky:

Southern constellation Ophiuchus interrupts neighboring Serpens, splitting it in two. Entwined mythologies perpetually link them.

KEY FACTS

The constellation Ophiuchus has 14 stars that make up the blocky figure of a man. The Serpent Bearer's alpha star, Rasalhague, represents the head.

+ **months best viewed:** June–July

+ **seasonal chart location:** Summer, center of chart

+ **alpha star:** Rasalhague

The constellation Ophiuchus likely honors Asclepius, the Greek god of medicine. He appears between the head and tail of Serpens the snake, who taught the god about the healing properties of plants. Asclepius used this knowledge to raise a fatally wounded Orion. A concerned Hades, god of the dead, fearing irrelevance, had Zeus kill Asclepius. Now healer and serpent watch from the sky, easily identified just to the west of the Milky Way. The constellation's location makes it rich in deep sky objects, including a series of globular clusters— M9, M10, M12, M14, M19, M62, and M107—all visible with binoculars.

Orion

The Hunter   Size on the sky:

Perhaps the most easily recognized constellation in the sky, Orion takes center stage on the celestial equator with more than its fair share of superbright stars.

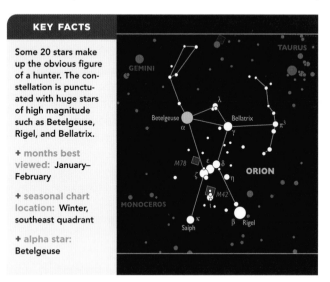

KEY FACTS

Some 20 stars make up the obvious figure of a hunter. The constellation is punctuated with huge stars of high magnitude such as Betelgeuse, Rigel, and Bellatrix.

+ **months best viewed:** January–February

+ **seasonal chart location:** Winter, southeast quadrant

+ **alpha star:** Betelgeuse

Without a doubt, Orion is a stellar rock star. Among many superlatives, three very bright stars form the hunter's iconic belt; red-tinted, variable Betelgeuse boasts a diameter 300 to 400 times wider than the sun's; and blue-white supergiant Rigel shines 57,000 times brighter than the sun. The great hunter of Greek mythology, Orion appears in the vicinity of his canine companions Canis Major and Minor and other associates, such as Lepus the Hare. Located below Orion's belt, the Great Orion Nebula (M42) is visible as a cloud patch to the naked eye. Its gas cloud churns out new stars at a furious pace.

Pegasus

The Winged Horse   Size on the sky:

Pegasus shares the sky with other constellations attached to the same mythology: Andromeda, Cassiopeia, Cepheus, Cetus, and the heroic Perseus.

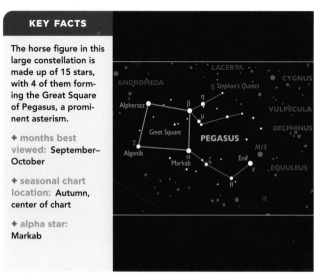

KEY FACTS

The horse figure in this large constellation is made up of 15 stars, with 4 of them forming the Great Square of Pegasus, a prominent asterism.

+ **months best viewed:** September–October

+ **seasonal chart location:** Autumn, center of chart

+ **alpha star:** Markab

Constellation Pegasus is rich in deep sky objects, including the M15 cluster and Stephan's Quintet, a group of five galaxies, four of which are interacting with one another. Pegasus also is a fertile source of exoplanets, planets that circle stars other than the sun. Pegasus shares a star, Alpheratz, and a story with constellation Andromeda (see page 402). Alpheratz forms part of the Great Square asterism with three stars from Pegasus. Pegasus also has associations with the hero Bellerophon, who fought the Chimera with the help of the steed. He had tamed the flying horse with aid from the goddess Athena.

Perseus

The Hero   Size on the sky:

Constellation Perseus, named for a superhero of Greek mythology, has a lot going for it: many bright and easily spotted stars, a superlative meteor shower, and a fascinating double star cluster.

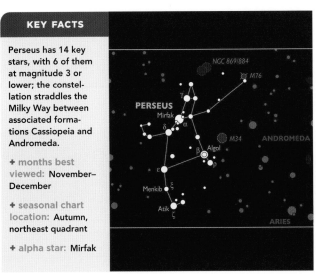

KEY FACTS

Perseus has 14 key stars, with 6 of them at magnitude 3 or lower; the constellation straddles the Milky Way between associated formations Cassiopeia and Andromeda.

+ months best viewed: November–December

+ seasonal chart location: Autumn, northeast quadrant

+ alpha star: Mirfak

Perseus belongs to the same mythological adventure as neighboring constellations Andromeda, Cassiopeia, Cepheus, Cetus, and Pegasus. The hero rescued captive Andromeda, daughter of Cassiopeia and Cepheus, from the predatory sea monster Cetus. The constellation is prominent in the late autumn sky and contains bright and notable signposts. Among them is Algol, the beta star, an eclipsing variable that dims by a full magnitude every three days. Perseus contains double cluster NGC 869 and NGC 884, two separate groups of stars that can be viewed through binoculars. It also is point of origin for the mid-August Perseid meteor showers.

Pisces

The Fish   Size on the sky:

The fish of Pisces are on the small side, but the cord joining them by their tails spans a wide area and crosses the ecliptic, the apparent path taken by the sun across the sky.

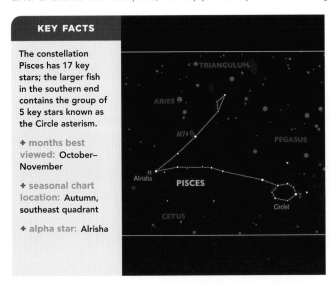

KEY FACTS

The constellation Pisces has 17 key stars; the larger fish in the southern end contains the group of 5 key stars known as the Circle asterism.

+ months best viewed: October–November

+ seasonal chart location: Autumn, southeast quadrant

+ alpha star: Alrisha

Pisces is an ancient constellation of the zodiac that forms a large V in the sky. Its southern asterism, known as the Circle, appears just south of the Great Square of Pegasus. The outside cord attached to the smaller fish is galaxy M74, a faint but elegant spiral galaxy whose arms can be resolved with an 8-inch-aperture telescope. For the ancient Greeks, the two fish of Pisces represented the goddess Aphrodite and her son Eros, who both changed themselves into fish to escape the sea monster Typhon, often equated with Cetus. The cord, tied at alpha star Alrisha, holds mother and son together as they swim.

Piscis Austrinus
The Southern Fish   Size on the sky:

The key to locating this low southern constellation—low to the horizon even at its highest for northern mid-latitude observers—is its formidable alpha star Fomalhaut.

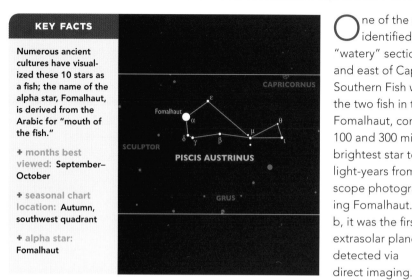

KEY FACTS

Numerous ancient cultures have visualized these 10 stars as a fish; the name of the alpha star, Fomalhaut, is derived from the Arabic for "mouth of the fish."

+ **months best viewed:** September–October

+ **seasonal chart location:** Autumn, southwest quadrant

+ **alpha star:** Fomalhaut

One of the original 48 constellations Ptolemy identified, Piscis Austrinus is located in the "watery" section of the sky just south of Aquarius and east of Capricornus. In Greek mythology, the Southern Fish was considered to be the sire of the two fish in the constellation Pisces. Alpha star Fomalhaut, considered a young star at between 100 and 300 million years old, ranks as the 18th brightest star to the naked eye and is a mere 25 light-years from Earth. In 2008, the Hubble telescope photographed a planet orbiting Fomalhaut. Named Fomalhaut b, it was the first confirmed extrasolar planet detected via direct imaging.

Puppis
The Stern   Size on the sky:

Puppis is one of the three breakout constellations created in 1763 from Argo Navis, one of the original constellations on Ptolemy's list of 48.

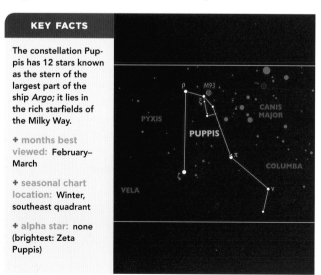

KEY FACTS

The constellation Puppis has 12 stars known as the stern of the largest part of the ship *Argo*; it lies in the rich starfields of the Milky Way.

+ **months best viewed:** February–March

+ **seasonal chart location:** Winter, southeast quadrant

+ **alpha star:** none (brightest: Zeta Puppis)

Puppis, disappointingly to some, doesn't represent a cute little dog, but the stern of the mythological ship *Argo*. The *Argo* carried Jason and the Argonauts on the quest to steal the Golden Fleece, once sported by Aries the Ram, from the clutches of a dragon. When Argo Navis was dismantled during the constellation update of the modern era, its stars were not relabeled within the new constellations and so Puppis was left without a labeled alpha star. Its brightest star is Zeta Puppis, also known as Naos, from the Greek for "ship." A blue supergiant, Zeta Puppis is a spectacular naked-eye star of the second magnitude.

Sagitta

The Arrow   Size on the sky:

Although the shapes associated with many constellations often stretch the imagination, the arrow-like form of the small but distinct Sagitta cleanly hits the mark.

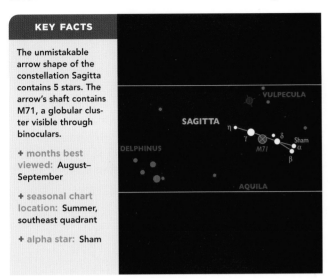

KEY FACTS

The unmistakable arrow shape of the constellation Sagitta contains 5 stars. The arrow's shaft contains M71, a globular cluster visible through binoculars.

+ months best viewed: August–September

+ seasonal chart location: Summer, southeast quadrant

+ alpha star: Sham

The third smallest constellation, Sagitta lies totally within the Summer Triangle asterism formed of Altair in Aquila, Deneb in Cygnus, and Vega in Lyra. Most ancient cultures made associations to the arrow and there are a number of arrow stories to choose from to explain Sagitta's celestial presence. In Greco-Roman myth, it could be the arrow Cupid (Eros) used to spread the love, or the arrow Apollo used to slay the Cyclops, or the one Hercules deployed to dispatch the Stymphalian birds in one of his 12 labors. A relationship to Sagitta, despite the name "arrow," is not widely accepted.

Sagittarius

The Archer   Size on the sky:

Spread across a wide band of the Milky Way, Sagittarius is a treasure trove of celestial superlatives, including two prominent asterisms and eight high-magnitude stars.

KEY FACTS

The complex figure of the constellation Sagittarius contains 22 stars, as well as the clearly defined Teapot asterism and the Milk Dipper asterism.

+ months best viewed: July–August

+ seasonal chart location: Summer, southeast quadrant

+ alpha star: Rukbat

Appearing south of Vega, the Teapot asterism contains the eight central stars of Sagittarius. The Milk Dipper includes the ladle and lambda star handle that form the Teapot's handle. Sagittarius's alpha star is not its brightest; that would be epsilon Kaus Australis. Sigma star Nunki, the second brightest, appeared on the Tablet of the 30 Stars of the ancient Babylonians. The centaur of Sagittarius—half man, half horse—pursues his supposed prey, the neighboring constellation Scorpius, through the sky. Sagittarius is often identified with Chiron of Greek mythology, who paradoxically was a wise and peaceful creature.

Scorpius

The Scorpion   Size on the sky:

The narrow outline of Scorpius harbors 2 of the 25 brightest stars as well as outstanding star clusters, including the spectacular Butterfly cluster (M6).

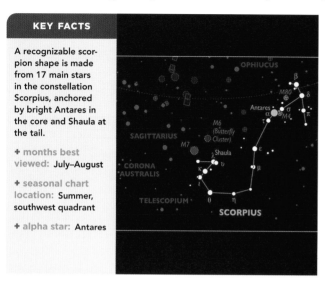

KEY FACTS

A recognizable scorpion shape is made from 17 main stars in the constellation Scorpius, anchored by bright Antares in the core and Shaula at the tail.

+ **months best viewed:** July–August

+ **seasonal chart location:** Summer, southwest quadrant

+ **alpha star:** Antares

A straightforward constellation, Scorpius is located by first-magnitude star Antares, known to the Romans as Cor Scorpionus, or "heart of the scorpion." A red supergiant, about 300 times as large as the sun, Antares's name means "rival of Mars," alluding to its size and color. The Butterfly cluster (M6), one of many open and globular star clusters in the vicinity, reveals its butterfly shape through binoculars. In Greek mythology, Scorpius is tied to the constellation Orion as the animal that inflicted the fatal wound on the hero's leg. The two constellations appear on opposite sides of the sky to keep them separated.

Scutum

The Shield   Size on the sky:

This faint modern constellation was created by Polish astronomer Johannes Hevelius in the late 17th century in the feature-rich Milky Way.

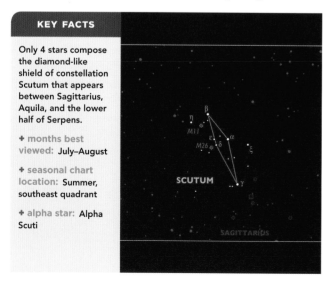

KEY FACTS

Only 4 stars compose the diamond-like shield of constellation Scutum that appears between Sagittarius, Aquila, and the lower half of Serpens.

+ **months best viewed:** July–August

+ **seasonal chart location:** Summer, southeast quadrant

+ **alpha star:** Alpha Scuti

Scutum lacks any bright or otherwise notable stars, but its location on the Milky Way puts it in close proximity to several open star clusters, including M11, the so-called Wild Duck cluster. Binoculars alone will reveal this formation that resembles a dense flock of waterfowl, and an 8-inch telescope will show thousands of glittering stars. The Wild Duck appears just southeast of Scutum's northernmost star. The constellation's original name was Scutum Sobiescianum, or "shield of Sobieski," honoring Polish King (and Hevelius's patron) John III Sobieski, who defeated the Ottoman Empire in the 1683 Battle of Vienna.

Serpens

The Serpent   Size on the sky:

The only split constellation in the night sky, Serpens is divided between head and tail, winding over the shoulder of Ophiuchus, the Serpent Bearer, in the middle.

KEY FACTS

Two groups of stars, the triangular head (Caput) and elongated tail (Cauda), make up the two separate parts of the serpent constellation Serpens.

+ **months best viewed:** June–July

+ **seasonal chart location:** Summer, southern half of chart

+ **alpha star:** Unukalhai

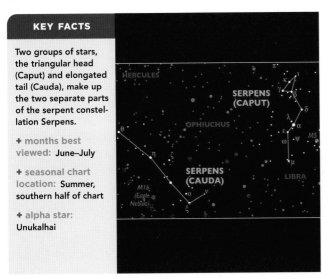

The constellation Serpens formerly encompassed the whole scenario of snake head, tail, and Ophiuchus. They were physically and mythologically united, representing the Greek god of medicine, Asclepius, and the serpent who taught him about the healing power of plants. Serpens Cauda contains the combination nebula and star cluster called the Eagle Nebula (M16), which occurs in an active region of star formation about 7,000 light-years away. Both the nebula and cluster can be spotted with an 8-inch telescope. M16 was famously photographed by the Hubble telescope, which captured a prominent area of star destruction and formation.

Taurus

The Bull   Size on the sky:

Taurus, a zodiacal constellation, is easily located because nearby Orion's belt points directly northwest to Aldebaran, the Bull's alpha star.

KEY FACTS

The renowned constellation Taurus—including the most prominent feature, the horns—contains 13 stars. The southern horn contains Aldebaran, the Bull's "eye."

+ **months best viewed:** January–February

+ **seasonal chart location:** Winter, southwest quadrant

+ **alpha star:** Aldebaran

The Egyptians associated Taurus with Osiris, god of life and fertility. Greek god Zeus disguised himself as a bull to capture Europa and carry her to the continent that bears her name. Taurus contains two prominent star clusters, the Pleiades and the Hyades. The Pleiades (M45), an open cluster, is the more famous and displays 6 or 7 of its hundreds of stars to the naked eye. Vying for celebrity is M1, the Crab Nebula. It was the first deep sky object cataloged by Charles Messier, in 1758. Cultures from the American Southwest to China witnessed and recorded the death of the star in 1054 that created the nebula.

Triangulum

The Triangle   Size on the sky: 🖐

Despite its nonmythological, utilitarian name—a feature of many of the modern constellations—Triangulum was on Ptolemy's list of the original 48.

KEY FACTS

The 3 stars of this geometric constellation form a distinct triangle located between Andromeda, Perseus, and Aries. All 3 are brighter than magnitude 3.

+ months best viewed: November–December

+ seasonal chart location: Autumn, southeast quadrant

+ alpha star: Mothallah

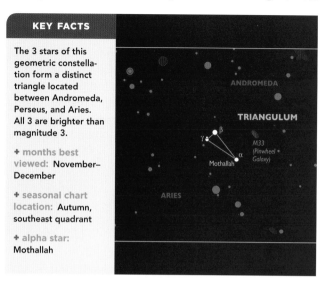

In late autumn, Triangulum is easy to find just east of Andromeda. The Pinwheel galaxy (M33) lies west of the constellation, a spiral galaxy like our own Milky Way. Under good viewing conditions, it appears like a slight glow, although a substantial telescope is needed to view the pinwheel effect. The ancient Hebrews appear to have named this constellation for the small percussion instrument, and the name stuck. The Greeks also took notice that these stars resembled their letter *delta*, and it was known for a time as Deltoton. The theme continues in the name for the alpha star, Mothallah, which means "triangle" in Arabic.

Ursa Major

The Great Bear   Size on the sky: 🖐🖐

To various cultures, the formidable assemblage of stars in Ursa Major represented a bear, a chariot, a horse and wagon, a team of oxen, and even a hippopotamus.

KEY FACTS

The complex figure of the constellation Ursa Major contains 20 stars, including the 7-star asterism of the Big Dipper, representing the bear's rear torso and tail.

+ months best viewed: March–April

+ seasonal chart location: Spring, center of chart

+ alpha star: Dubhe

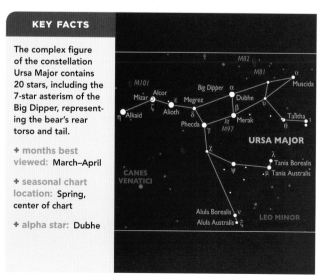

During the year, prominent Ursa Major seems to run in a circle with its back to Polaris, the North Star. The constellation's many named stars include Alcor and Mizar, which appear to be a binary system but are not. Galaxies M81 and M82 lie close together and may be in collision with each other. In Greek mythology, the Great Bear represents Callisto, yet another victim of the goddess Hera's jealousy on finding that her husband, Zeus, had strayed again. When Callisto's son mistakenly tried to kill her while hunting, Zeus placed them both in the sky to protect them.

Ursa Minor

The Little Bear   Size on the sky:

The Little Bear plays second fiddle to none, containing the North Star, Polaris, as its alpha star. The constellation and its polestar are visible year-round.

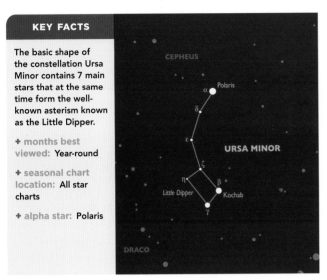

KEY FACTS

The basic shape of the constellation Ursa Minor contains 7 main stars that at the same time form the well-known asterism known as the Little Dipper.

+ months best viewed: **Year-round**

+ seasonal chart location: **All star charts**

+ alpha star: **Polaris**

Navigators, wanderers, and sky gazers throughout history have found their way by the enduring presence of Ursa Minor and Polaris, indicating the celestial pole. This won't always be the case. Over millennia, the celestial pole shifts. It will be closest to Polaris around 2100, and then will start to move away, passing other stars and eventually settling at Vega for a while in about 12,000 years. Ursa Minor appears in the sky as the mythological child of the Great Bear, who represents Zeus's paramour Callisto. Late December brings the Ursid meteor shower, which display about ten meteors an hour at its peak.

Virgo

The Virgin   Size on the sky:

The only zodiacal constellation representing a woman, Virgo is the second largest constellation and a fertile area for viewing star clusters and galaxies.

KEY FACTS

The figure of a seated woman in the constellation Virgo contains 13 main stars, with bright alpha star Spica in the vicinity of her left hand.

+ months best viewed: **May–June**

+ seasonal chart location: **Spring, southeast quadrant**

+ alpha star: **Spica**

Mythologically, Virgo appears to represent a goddess. For many she was Dike, the goddess of justice, or possibly Demeter, the goddess of agriculture. Virgo's bright alpha star, Spica, is a useful point of orientation as part of an alliterative mnemonic, "arc to Arcturus, then speed on to Spica." This directs one's gaze south from the end of the Big Dipper's handle to find Arcturus in Boötes, then south again to Spica just below it. Northwest of Virgo is an area dense in star clusters and galaxies—thousands of them. To the southwest is the Sombrero galaxy (M104), with a dark band across its middle that serves as the brim.

Natural Regions of North America

The continental United States and Canada can be divided into nine physiographic regions.

■ Appalachian Highlands

The oldest mountain chain in North America, the heavily eroded Appalachians extend for about 2,000 miles (3,200 km) from Alabama to Newfoundland, with the highest peak being Mount Mitchell, in North Carolina, at 6,684 feet (2,040 m). This mountain range started forming 480 million years ago, when continental collisions caused volcanic activity and mountain building. Ecologically, this region hosts eastern temperate forests with a great variety of coniferous and deciduous trees and wildflowers, some of them high-elevation specialists.

■ Coastal Plain

This gradually rising flatland spans some 4,000 miles (6,400 km) in total and covers several distantly related regions along the Gulf of Mexico, the southern Atlantic coast, and the northernmost coasts of the Arctic Ocean. In the past, oceans covered these plains, depositing sediment layers over millions of years, until falling sea levels exposed them. The types of vegetation in these widely separated regions range from subtropical trees and flowers in the southernmost coastal plain to marsh plants along the mid-Atlantic to the largely treeless tundra on the northern coast of Alaska.

■ Interior Plains

Ranging from the lowlands of the Saint Lawrence River Valley in the east to the mile-high Great Plains in the west, the vast Interior Plains of North America provide fertile soils, especially for productive prairie farms. A shallow sea covered much of this region as recently as 75 million years ago, and sediments from rivers draining the Appalachians and western mountains were deposited in layers throughout the sea. The Great Plains were once covered by vast and diverse expanses of natural grasses, sagebrush, and a varied suite of wildflowers. Much of this ecosystem has vanished, the land brought into use by modern agriculture and extensive grazing.

■ Interior Highlands

The Ozark Plateau and Ouachita Mountains form the Interior Highlands, which are centered on Arkansas and southern Missouri, with mountains reaching more than 2,600 feet (800 m) high. These ancient eroded highlands were connected to the Appalachians until tectonic activity separated them some 200 million years ago. Ecologically, this relatively small area straddles the southern Interior Plains and the eastern Coastal Plain, with trees and wildflowers representing both regions.

■ Rocky Mountains

The highest mountain system in North America, the Rockies dominate the landscape for some 3,000 miles (4,800 km), from New Mexico to Alaska, with more than 50 peaks surpassing 14,000 feet (4,300 m). Tectonic activity uplifted the Rockies about 50 to 100 million years ago, making them much younger and less eroded than the Appalachians. The ecological hallmarks of the Rockies are its coniferous forests of pines, firs, and spruces, adapted to high elevations, with wildflower species similarly adapted to elevations and temperatures.

■ Intermontane Basins and Plateaus

This region is called *intermontane* because it is situated between the Pacific and the Rocky Mountain systems. Pacific mountains block most moisture-bearing clouds coming from the Pacific Ocean, giving desert climates to places like the Colorado Plateau, 5,000 to 7,000 feet (1,500 to 2,100 m) high, and the Great Basin. In Canada and Alaska, the immense Yukon River Valley and the Yukon-Tanana Uplands are part of this region. The deserts feature cacti and a host of other specialist plants of the arid West. Cottonwoods, ashes, and willows line rivers that run intermittently through the dry plains.

■ Pacific Mountain System

From Alaska to California, mountains and volcanoes tower over the West Coast in an almost unbroken chain. These mountain ranges are geologically young and seismically active, with uplift starting some five million years ago. The highest mountain in North America, Alaska's Mount McKinley (20,320 feet/6,200 m), is still growing at about a millimeter a year—about the thickness of a fingernail. Distantly separated from the Rockies, this mountain range supports its own distinct varieties of trees and wildflowers adapted to higher elevations.

■ Canadian Shield

The geologic core of North America is the Canadian Shield, which contains the continent's oldest rocks.

Landforms are relatively flat, having been eroded and scoured by glaciers over millions of years. The exposed bedrock ranges in age from 570 million to more than 3 billion years old. This is a vast and extensively diverse region of climatic extremes and varied vegetation, from dense boreal forests in the south to frigid tundra in the north, populated by stunted trees, small shrubs, lichens, and ground-clinging herbs.

■ Arctic Lands

Highlands known as the Innuitian Mountains cover most islands. The icy climate is too harsh for most animals and vegetation, and much of the ground is permanently frozen. Nevertheless, low-growing shrubs, small tundra plants, and lichens manage to survive.

Further Resources

▥ Wildflowers

BOOKS

Brandenburg, David M. *National Wildlife Federation Field Guide to Wildflowers of North America.* Sterling, 2010.

National Audubon Society. *Field Guide to North American Wildflowers: Eastern Region.* Alfred A. Knopf, 2001.

National Audubon Society. *Field Guide to North American Wildflowers: Western Region.* Alfred A. Knopf, 1979.

Peterson, Roger Tory, and Margaret McKenny. *A Field Guide to Wildflowers: Northeastern and North-Central North America,* revised ed. Peterson Field Guide Series. Houghton Mifflin Harcourt, 1998.

WEBSITES

Flora of North America
   www.efloras.org

Lady Bird Johnson Wildflower Center
   www.wildflower.org

U.S. Department of Agriculture, National Resources Conservation Service, Plants Database
   www.plants.usda.gov

U.S. Forest Service, Celebrating Wildflowers
   www.fs.fed.us/wildflowers

APPS

Audubon Wildflower Identification App
   www.audubonguides.com/field-guides/wildflowers-north-america.html

National Science Foundation, Project BudBurst
   www.budburst.org/gomobile.php

▥ Trees

BOOKS

Dirr, Michael A. *Manual of Woody Landscape Plants: Their Identification, Ornamental Characteristics, Culture, Propagation and Uses,* 4th ed. Stipes Publishing Co., 1990.

Elias, Thomas S. *The Complete Trees of North America: A Field Guide and Natural History.* Gramercy Publishing Co., 1987.

Harris, James G., and Melinda Woolf Harris. *Plant Identification Terminology: An Illustrated Glossary.* Spring Lake Publishing, 2001.

Rushforth, Keith, and Charles Hollis. *National Geographic Society Field Guide to Trees of North America.* National Geographic Society, 2006.

Rutkow, Eric. *American Canopy: Trees, Forests, and the Making of a Nation.* Scribner, 2012.

Sibley, David Allen. *The Sibley Guide to Trees.* Alfred A. Knopf, 2009.

WEBSITES

Flora of North America
   www.efloras.org

Lady Bird Johnson Wildflower Center
   www.wildflower.org

Missouri Botanical Garden Plant Finder
   www.missouribotanicalgarden.org/gardens-gardening/your-garden/plant-finder.aspx

University of Connecticut Plant Database of Trees, Shrubs, and Vines
   www.hort.uconn.edu/plants/index.html

USDA Forest Service Fire Effects Information System
   www.fs.fed.us/database/feis

Virginia Tech Dendrology Factsheets
   www.dendro.cnre.vt.edu/dendrology/factsheets.cfm

▥ Rocks & Minerals

BOOKS

Dixon, Dougal, and Raymond Bernor, eds. *The Practical Geologist.* Simon and Schuster, 1992.

Geology Underfoot (series of books). Mountain Press Publishing Company.

Klein, Cornelis, and Anthony Philpotts. *Earth Materials: Introduction to Mineralogy and Petrology,* Cambridge University Press, 2013.

Marshak, Stephen. *Essentials of Geology.* W. W. Norton and Company, 2009.

Price, Monica T. *The Sourcebook of Decorative Stone.* Firefly Books Ltd., 2007.

Roadside Geology (series). Mountain Press.

Wenk, Hans-Rudolf, and Andrei Bulakh. *Minerals: Their Constitution and Origin.* Cambridge University Press, 2004.

WEBSITES

Association of American State Geologists
www.stategeologists.org

Dinosaur Society's Dinosaur News
www.dinosaursociety.com/news/category/
dinosaur-news

Geological Society of America
www.geosociety.org

Mindat.org
www.mindat.org

Paleontology Portal
www.paleoportal.org

U.S. Geological Survey
www.usgs.gov

U.S. Geological Survey Volcano Hazards Program
www.volcanoes.usgs.gov

University of California Museum of Paleontology's
Online Exhibits
www.ucmp.berkeley.edu/exhibits/index.php

Volcano World
http://volcano.oregonstate.edu

■ Weather

BOOKS

Ahrens, C. Donald. *Meteorology Today: An Introduction to Weather, Climate, and the Environment.* West Publishing Company, 2008.

Henson, Robert. *The Rough Guide to Climate Change.* (Rough Guide Reference Series). Rough Guides, 2011.

Williams, Jack. *The AMS Weather Book: The Ultimate Guide to America's Weather.* University of Chicago Press and The American Meteorological Society, 2009.

WEBSITES

American Meteorological Society
www.ametsoc.org/aboutams/index.html

American Meteorological Society's education pages
www.ametsoc.org/amsedu

Atmospheric Optics
www.atoptics.co.uk

Canadian Weatheroffice
www.weatheroffice.gc.ca/canada_e.html

Cloud Appreciation Society
www.cloudappreciationsociety.org

National Weather Association
www.nwas.org/about.php

U.S. National Weather Service
www.weather.gov

U.S. National Weather Service JetStream Online
Weather School
www.srh.weather.gov/jetstream

University Cooperation for Atmospheric Research
Spark science education
www.spark.ucar.edu

Weatherwise magazine
www.weatherwise.org

■ Night Sky

BOOKS

Aguilar, David A. *13 Planets: The Latest View of the Solar System.* National Geographic Society, 2011.

Daniels, Patricia. *The New Solar System.* National Geographic Society, 2009.

Ridpath, Ian, and Wil Tirion. *Stars and Planets,* 4th ed. Princeton Field Guides. Princeton University Press, 2007.

Trefil, James. *Space Atlas: Mapping the Universe and Beyond.* National Geographic Society, 2012.

WEBSITES

Astronomy magazine
www.astronomy.com

EarthSky
www.earthsky.org

Harvard-Smithsonian Center for Astrophysics
www.cfa.harvard.edu

Hubble Space Telescope
www.hubblesite.org

International Astronomical Union
www.iau.org

National Aeronautics and Space Administration
www.nasa.gov

Sky & Telescope magazine
www.skyandtelescope.com

Space.com
www.space.com/news

About the Contributors

▥ About the Authors

Bland Crowder is associate director and editor with the Flora of Virginia Project, whose *Flora of Virginia* was published in December 2012. He is also a freelance writer and editor. He lives in Richmond, Virginia.

Sarah Garlick is a writer and educator specializing in earth and environmental science. She holds degrees in geology from Brown University and the University of Wyoming, and she is the author of the award-winning book *Flakes, Jugs, and Splitters: A Rock Climber's Guide to Geology* (2009). She lives with her family in the mountains of New Hampshire.

Catherine Herbert Howell, a former National Geographic staff member, has written extensively on natural history. She explored the relationships between people and plants in *Flora Mirabilis: How Plants Have Shaped World Knowledge, Health, Wealth, and Beauty* (2009) and covered the importance of birds in culture in the *National Geographic Bird-watcher's Bible* (2012). Howell serves as a master naturalist volunteer in Arlington, Virginia.

Jack Williams was founding editor of the *USA TODAY* weather page in 1982 and the USATODAY.com weather section in 1995. After retiring from USA TODAY in 2005, Williams was director of public outreach for the American Meteorological Society through 2009. Since then, he has been a freelance writer and is the author or coauthor of seven books, including the *National Geographic Field Guide to the Water's Edge* (2012).

▥ About the Consultants

George Yatskievych is a curator at the Missouri Botanical Garden. His research involves the wild plants of Missouri and surrounding states. He also has interests in ferns and parasitic flowering plants.

Kay Yatskievych is a retired botanist associated with the Missouri Botanical Garden. She is the author of a field guide to Indiana wildflowers and is working on a catalog of that state's flora.

▥ About the Artists

Fernando G. Baptista graduated in Fine Arts from País Vasco in Spain. He worked as a graphic artist for a newspaper in Bilbao, as a freelancer creating reconstructive pieces for museums, and spent six years as a professor of infographics at the University of Navarra. Baptista has won more than 125 international awards. In 2012, he was named one of the five most influential infographic artists of the last 20 years.

Jared Travnicek is a scientific and medical illustrator. He received his M.A. in Biological and Medical Illustration from the Johns Hopkins University School of Medicine in Baltimore, Maryland. Travnicek is a Certified Medical Illustrator and a professional member of the Association of Medical Illustrators.

Jenny Wang was trained in medical and biological illustration at the Johns Hopkins School of Medicine in Baltimore, Maryland. Her areas of focus include scientific illustration, biomedical visualization, and information design. As a nature lover, she is delighted to have contributed to this guide.

Illustrations Credits

New artwork appearing in this book was created by Fernando G. Baptista, Jared Travnicek, and Jenny Wang.

FRONT COVER: Top Row (Left to Right): Frans Lanting/NG Stock; Renaud Visage/The Image Bank/Getty Images; Yva Momatiuk & John Eastcott/Minden Pictures/NG Stock; Paul Nicklen/NG Stock. Middle Row: Ron Gravelle/NG My Shot; Tim Fitzharris/Minden Pictures/NG Stock; Lincoln Harrison/NG My Shot; Amy White & Al Petteway/NG Stock. Bottom Row: Yva Momatiuk & John Eastcott/Minden Pictures/NG Stock; Paul Nicklen/NG Stock; David Muench/Getty Images; Crisma/Getty Images.

BACK COVER: Top Row: Terry Donnelly; Bret Webster/Science Source; Mike Theiss/NG Stock; Ingo Arndt/Minden Pictures. Middle Row: Frans Lanting/NG Stock; Jack Dykinga; Jim Brandenburg/Minden Pictures; NASA/Science Source. Bottom Row: Carr Clifton/Minden Pictures; Mary Liz Austin; Paolo De Faveri/paolodefaveri.com; Pasieka/Science Source.

INTERIOR: 2-3, Tim Fitzharris/Minden Pictures; 4, Kevin Barry; 6, Carr Clifton; 8, Terry Donnelly; 10, UIG via Getty Images; 14 (UP), Mark Turner/Getty Images; 14 (LO), Judywhite/GardenPhotos.com; 15 (UP), Gerald D. Tang; 15 (LO), Glenis Moore/Science Source; 16 (UP), Gerald D. Tang; 16 (LO), Bob Gibbons/Science Source; 17 (UP), No:veau/Shutterstock; 17 (LO), Visuals Unlimited, Inc./Gerry Bishop/Getty Images; 18 (UP), Arco Images GmbH/Alamy; 18 (LO), Willem Kolvoort/Foto Natura/Getty Images; 19 (UP), Roger Whiteway/iStockphoto; 19 (LO), Robert and Jean Pollock/Science Source; 20 (UP), Stephen P. Parker/Science Source; 20 (LO), Bob Gibbons/Science Source; 21 (UP), Gerald D. Tang; 21 (LO), Michael P. Gadomski/Science Source; 22 (UP), Andy Crawford/Getty Images; 22 (LO), LianeM/Shutterstock; 23 (UP), David Nunuk/Science Source; 23 (LO), karloss/Shutterstock; 24 (UP), Kenneth M. Highfill/Science Source; 24 (LO), Lindasj22/Shutterstock; 25 (UP), Hal Horwitz/Corbis; 25 (LO), Ron Wolf/Tom Stack & Associates; 26 (UP), Sumikophoto/Shutterstock; 26 (LO), Jerry Pavia/Getty Images; 27 (UP), Joshua McCullough, PhytoPhoto/Getty Images; 27 (LO), Clint Farlinger; 28 (UP), Kenneth M. Highfill/Science Source; 28 (LO), Inga Spence/Science Source; 29 (UP), Rod Planck/Science Source; 29 (LO), John Greim; 30 (UP), Geoff Kidd/Science Source; 30 (LO), Gregory K. Scott/Science Source; 31 (UP), Peter Herring; 31 (LO), Maria Mosolova/Getty Images; 32 (UP), Barry Breckling; 32 (LO), Kevin Schafer/Getty Images; 33 (UP), Kenneth M. Highfill/Science Source; 33 (LO), Steve Guttman; 34 (UP), Gerry Bishop/Visuals Unlimited, Inc.; 34 (LO), Michael Wheatley/All Canada Photos/Getty Images; 35 (UP), judywhite/GardenPhotos.com; 35 (LO), William A. Bake/Corbis; 36 (UP), Jeff Lepore/Science Source; 36 (LO), Kenneth M. Highfill/Science Source; 37 (UP), William S. Moye; 37 (LO), Ross Hoddinott/Minden Pictures; 38 (UP), Barry Breckling; 38 (LO), Bill Beatty; 39 (UP), Robert & Jean Pollock/Visuals Unlimited, Inc.; 39 (LO), Laura Berman; 40 (UP), Valerie Giles/Science Source; 40 (LO), Michael P Gadomski/Getty Images; 41 (UP), Roanna Littlefield; 41 (LO), Kenneth M. Highfill/Science Source; 42 (UP), Nature's Images/Science Source; 42 (LO), Dave Welling; 43 (UP), Douglas Craig/iStockphoto; 43 (LO), Artefficient/Shutterstock; 44 (UP), Michael P. Gadomski/Science Source; 44 (LO), Gerald D. Tang; 45 (UP), Michael P. Gadomski/Science Source; 45 (LO), Stephen Dalton/Minden Pictures; 46 (UP), Adrian Bicker/Science Source; 46 (LO), jack thomas/Alamy; 47 (UP), Fotosearch; 47 (LO), David Hall/Alamy; 48 (UP), Mike Theiss/National Geographic Stock; 48 (LO), Mike Comb/Science Source; 49 (UP), David Davis/Science Source; 49 (LO), Will & Deni McIntyre/Science Source; 50 (UP), Luther Linkhart/Visuals Unlimited, Inc.; 50 (LO), Gary Cook/Visuals Unlimited, Inc.; 51 (UP), Brian Barnes/Alamy; 51 (LO), Ron Wolf/Tom Stack & Associates; 52 (UP), Scott Cramer/Getty Images; 52 (LO), Sue Carnahan; 53 (UP), Jerry Pavia/Getty Images; 53 (LO), Roger Hyam/Getty Images; 54 (UP), Visuals Unlimited, Inc./Gerry Bishop/Getty Images; 54 (LO), Scientifica/Visuals Unlimited, Inc.; 55 (UP), John W. Bova/Science Source; 55 (LO), Pi-Lens/Shutterstock; 56 (UP), James Steinberg/Science Source; 56 (LO), 56 (LO) Photo by Jessie Harris; 57 (UP), Douglas Graham/Wild Light Photography, Inc.; 57 (LO), Ventura/Shutterstock; 58 (UP), Nature's Images, Inc./Science Source; 58 (LO), Thomas & Pat Leeson/Science Source; 59 (UP), James Randklev/Getty Images; 59 (LO), Gilbert S. Grant/Science Source; 60 (UP), Scott Camazine/Science Source; 60 (LO), Len Rue Jr./Science Source; 61 (UP), Visuals Unlimited, Inc./John Gerlach/Getty Images; 61 (LO), Richard Bloom/Getty Images; 62 (UP), Gail Jankus/Science Source; 62 (LO), Gerry Bishop/Visuals Unlimited, Inc.; 63 (UP), Brian Gadsby/Science Source; 63 (LO), Martin Ruegner/Getty Images; 64 (UP), Nigel Cattlin/Science Source; 64 (LO), Juan Silva/Getty Images; 65 (UP), Meg Sommers; 65 (LO), Peter Haigh/Getty Images; 66 (UP), Gerry Bishop/Visuals Unlimited, Inc.; 66 (LO), David Schwaegler; 67 (UP), Bill Pusztai; 67 (LO), Bob Gibbons/Science Source; 68 (UP), Len Rue Jr./Science Source; 68 (LO), Tim Graham/Getty Images; 69 (Both), Gerald D. Tang; 70 (UP), Lucy Jones/Visuals Unlimited, Inc.; 70 (LO), Gerald D. Tang; 71 (UP), Nature's Images/Science Source; 71 (LO), Visuals Unlimited, Inc./Nigel Cattlin/Getty Images; 72 (UP), Bob Gibbons/Science Source; 72 (LO), Gabriel Bertilson; 73 (UP), Ingrid Russell; 73 (LO), George Grall/National Geographic/Getty Images; 74 (UP), Gerry Bishop/Visuals Unlimited, Inc.; 74 (LO), Gerald D. Tang; 75 (UP), James Steinberg/Science Source; 75 (LO), Gerald D. Tang; 76 (UP), john t. fowler/Alamy; 76 (LO), Gerald D. Tang; 77 (Both), Gerald D. Tang; 78 (UP), George H. Huey; 78 (LO), Gerald D. Tang; 79 (UP), All Canada Photos/Alamy; 79 (LO), Ron Wolf/Tom Stack & Associates; 80 (UP), Geoff Kidd/Science Source; 80 (LO), Rich Wagner/WildNaturePhotos; 81 (UP), Gregory K. Scott/Science Source; 81 (LO), Rod Planck/Science Source; 82 (UP), James Steinberg/Science Source; 82 (LO), S.J. Krasemann/Getty Images; 83 (UP), judywhite/GardenPhotos.com; 83 (LO), Steffen Hauser/botanikfoto/Alamy; 84 (UP), Joshua McCullough/Getty Images; 84 (LO), Harley Seaway/Getty Images; 85 (UP), Bill Pusztai; 85 (LO), Nature's Images, Inc./Science Source; 86 (UP), Robert and Jean Pollock/Science Source; 86 (LO), Andreas Riedmiller; 87 (UP), Mark Steinmetz; 87 (LO), Dayton Wild/Visuals Unlimited, Inc.; 88 (UP), Rick & Nora Bowers/BowersPhoto.com; 88 (LO), kpzfoto/Alamy; 89 (UP), Shaughn F. Clements/Alamy; 89 (LO), Martin Shields/Science Source; 90 (UP), Nature's Images, Inc./Science Source; 90 (LO), Scott Camazine/Science Source; 91 (UP), Dr. John D. Cunningham/Visuals Unlimited, Inc.; 91 (LO), Gerald D. Tang; 92 (UP), Howard Rice/Getty Images; 92 (LO), imagebroker/Alamy; 93 (UP), Jeffrey Lepore/Science Source; 93 (LO), Sandra Ivany/Getty Images; 94, Floris van Breugel/NPL/Minden Pictures; 97, Tim Fitzharris/Minden Pictures; 100 (UP), Ron & Diane Salmon/Flying Fish Photography LLC; 100 (LO), Photos Lamontagne/Getty Images; 101 (UP), Barrett & MacKay/All Canada Photos/Getty Images; 101 (LO), Richard Thom/Visuals Unlimited/Getty Images; 102 (UP), Ted Kinsman/Science Source; 102 (LO), Terry Donnelly; 103 (UP), Tim Fitzharris/Minden Pictures; 103 (LO), Tony Wood/Science Source; 104 (UP), Perry Mastrovito/First Light/Corbis; 104 (LO), Ron & Diane Salmon/Flying Fish Photography LLC; 105 (UP), Ron Hutchinson Photography; 105 (LO), David Hosking/Alamy; 106 (UP), Bob Gibbons/Science Source; 106 (LO), David Matherly/Visuals Unlimited/Getty Images; 107 (UP), John Hagstrom; 107 (LO), Stephen J. Krasemann/Science Source; 108 (UP), Philippe Clement/NPL/Minden Pictures; 108 (LO), Michael P. Gadomski/Science Source; 109 (UP), Susan Glascock; 109 (LO), Fred Bruemmer/Getty Images; 110 (UP), Michael P. Gadomski/Science Source; 110 (LO), David Hosking/FLPA/Minden Pictures; 111 (UP), Ted Kinsman/Science Source; 111 (LO), Visuals Unlimited, Inc./Rob Kurtzman/Getty Images; 112 (UP), David Hosking/Minden Pictures; 112 (LO), Bob Gibbons/Science Source; 113 (UP), KENNETH W FINK/Getty Images; 113 (LO), Bob Gibbons/Minden Pictures; 114 (UP), Inga Spence/Science Source; 114 (LO), Thomas & Pat Leeson/Science Source; 115 (Both), Gerald D. Tang; 116 (UP), Susan Glascock; 116 (LO), Dennis Flaherty/Getty Images; 117 (UP), Michael P. Gadomski/Science Source; 117 (LO), Keith Rushforth/Minden Pictures; 118 (UP), David Middleton/NHPA/Photoshot; 118 (LO), David Jensen; 119 (UP), Cora Niele/Getty Images; 119 (LO), Jim Zipp/Science Source; 120 (UP), David Winkelman/David Liebman; 120 (LO), Ron & Diane Salmon/Flying Fish Photography LLC; 121 (UP), Susan Glascock; 121 (LO), David Jensen; 122 (UP), Ethan Welty/Aurora Photos; 122 (LO), Ron & Diane Salmon/Flying Fish Photography LLC; 123 (UP), Tim Fitzharris/Minden Pictures; 123 (LO), Ron & Diane Salmon/Flying Fish Photography LLC; 124 (UP), David Woodfall/Photoshot Holdings Ltd/Alamy; 124 (LO), Michael P. Gadomski/Science Source; 125 (Both), Ron & Diane Salmon/Flying Fish Photography LLC; 126 (UP), Carr Clifton/Minden Pictures; 126 (LO), Colin Marshall/FLPA/Minden Pictures; 127 (UP), Ron & Diane Salmon/Flying Fish Photography LLC; 127 (LO), Jim Brandenburg/Minden Pictures; 128 (UP), Inga Spence/Science Source; 128 (LO), David Liebman; 129 (UP), Kenneth Murray/Science Source; 129 (LO), Adam Jones/Science Source; 130 (UP), James Steakley; 130 (LO), Inga Spence/Science Source; 131 (UP), Geoff Bryant/Science Source; 131 (LO), Frank Zullo/Science Source; 132 (UP), Adam Jones/Science Source; 132 (LO), Ron Boardman/Life Science Image/FLPA/Science Source; 133 (UP), Gerald D. Tang; 133 (LO), Dave Watts/Alamy; 134 (UP), Tim Fitzharris/Minden Pictures/Getty Images; 134 (LO), Geoff Kidd/Science Source; 135 (UP), William Weber/Visuals Unlimited, Inc.; 135 (LO), Joel Sartore/National Geographic/Getty Images; 136 (UP), Ron & Diane Salmon/Flying Fish Photography LLC; 136 (LO), Panoramic Images/Getty Images; 137 (UP), DEA/C.SAPPA/De Agostini/Getty Images; 137 (LO), Dane Johnson/Visuals Unlimited, Inc.; 138 (UP), Michael Orton/Getty Images; 138 (LO), Doug Sokell/Visuals Unlimited, Inc.; 139 (UP), John Shaw/Science Source; 139 (LO), Gerald D. Tang; 140 (UP), Mark Oatney/Getty Images; 140 (LO), Eliot Cohen; 141 (UP), Kent Foster/Science Source; 141 (LO), Phillip Merritt; 142 (UP), Marcos Issa/Argosfoto; 142 (LO), John Glover/Alamy; 143 (UP), De Agostini/S. Montanari/Getty Images; 143 (LO), Peter Chadwick LRPS/Getty Images; 144 (UP), Melinda Fawver/Shutterstock; 144 (LO), E. R. Degginger/Science Source; 145 (UP), Lee F. Snyder/Science Source; 145 (LO), Stuart Wilson/Science Source; 146 (UP), Altrendo Nature/Getty Images; 146 (LO), Ron & Diane Salmon/Flying Fish Photography

LLC; 147 (UP), Adam Jones/Science Source; 147 (LO), Michael P. Gadomski/Science Source; 148 (UP), Stan Osolinski/Oxford Scientific/Getty Images; 148 (LO), William Webber/Visuals Unlimited, Inc.; 149 (UP), shapencolour/Alamy; 149 (LO), Ron & Diane Salmon/Flying Fish Photography LLC; 150 (UP), Anne Gilbert/Alamy; 150 (LO), Ron & Diane Salmon/Flying Fish Photography LLC; 151 (UP), Susan A Roth/Alamy; 151 (LO), Mack Henley/Visuals Unlimited, Inc.; 152 (Both), Ron & Diane Salmon/Flying Fish Photography LLC; 153 (Both), Ron & Diane Salmon/Flying Fish Photography LLC; 154 (UP), Ron & Diane Salmon/Flying Fish Photography LLC; 154 (LO), DEA/S. MONTANARI/Getty Images; 155 (UP), Gerald D. Tang; 155 (LO), Ron & Diane Salmon/Flying Fish Photography LLC; 156 (UP), Br. Alfred Brousseau, Saint Mary's College; 156 (LO), Ron & Diane Salmon/Flying Fish Photography LLC; 157 (UP), A. N. T./Science Source; 157 (LO), Charles Bush/Alamy; 158 (UP), Dorling Kindersley/Getty Images; 158 (LO), Jodie Coston/Getty Images; 159 (UP), James Young/Getty Images; 159 (LO), Ron & Diane Salmon/Flying Fish Photography LLC; 160 (UP), Eliot Cohen; 160 (LO), Plantography/Alamy; 161 (UP), David Winkelman/David Liebman; 161 (LO), Karl Magnacca; 162 (UP), M Timothy O'Keefe/Getty Images; 162 (LO), Laura Berman; 163 (UP), Alex Hare/Getty Images; 163 (LO), Richard Shiell; 164 (UP), Q-Images/Alamy; 164 (LO), Eliot Cohen; 165 (UP), Ed Jensen; 165 (LO), Michael P. Gadomski/Science Source; 166 (UP), Gerald & Buff Corsi/Visuals Unlimited, Inc.; 166 (LO), Doug Sokell/Visuals Unlimited, Inc.; 167 (UP), Joseph Sohm/Visions of America/Corbis; 167 (LO), James Steinberg/Science Source; 168 (UP), Dwight Kuhn ; 168 (LO), Eliot Cohen; 169 (UP), Michael P Gadomski/Getty Images; 169 (LO), David Roth/Botanica/Getty Images; 170 (UP), Gary Vestal/Getty Images; 170 (LO), DEA/RANDOM/Getty Images; 171 (Both), Ron & Diane Salmon/Flying Fish Photography LLC; 172 (UP), Ron & Diane Salmon/Flying Fish Photography LLC; 172 (LO), Geoff Kidd/Science Source; 173 (UP), Gilbert Twiest/Visuals Unlimited, Inc.; 173 (LO), Martin Page/Getty Images; 174 (UP), Kathleen Nelson/Alamy; 174 (LO), Jim Brandenburg/Minden Pictures/Getty Images; 175 (UP), Ron & Diane Salmon/Flying Fish Photography LLC; 175 (LO) John Hagstrom; 176 (UP), John Hagstrom; 176 (LO), Nature's Images/Science Source; 177 (UP), Carr Clifton; 177 (LO), Susan Webb/Flying Fish Photography LLC; 178 (UP), Nature's Images, Inc./Science Source; 178 (LO), Kenneth Murray/Science Source; 179 (UP), Kyle Wicomb; 179 (LO), Eliot Cohen; 180, Art Wolfe/www.artwolfe.com; 183, Pasieka/Science Source; 186 (UPLE), Juraj Kovac/Shutterstock; 186 (UPRT), Boris Sosnovyy/Shutterstock; 186 (LOLE), Scientifica/Visuals Unlimited, Inc.; 186 (LORT), DEA/A.RIZZI/Getty Images; 187 (UPLE), Joel Arem/Science Source; 187 (UPRT), Mark A. Schneider/Science Source; 187 (LOLE), Scott Camazine/Science Source; 187 (LORT), 2013 National Museum of Natural History, Smithsonian Institution; 188 (UPLE), Scientifica/Visuals Unlimited, Inc.; 188 (UPRT), The Trustees of the British Museum/Art Resource, NY; 188 (LOLE), Mark Schneider/Visuals Unlimited, Inc.; 188 (LORT), David McLain/Aurora/Getty Images; 189 (UPLE), Mark A. Schneider/Science Source; 189 (UPRT), Biophoto Associates/Science Source; 189 (LOLE), Andrew J. Martinez/Science Source; 189 (LORT), Phil Degginger; 190 (UPLE), Marli Miller/Visuals Unlimited, Inc.; 190 (UPRT), Mark Steinmetz; 190 (LOLE), Biophoto Associates/Science Source; 190 (LORT), 2013 National Museum of Natural History, Smithsonian Institution; 191 (UPLE), Ted Kinsman/Science Source; 191 (UPRT), Biophoto Associates/Science Source; 191 (LOLE), Marli Miller/Visuals Unlimited, Inc.; 191 (LORT), Joel Arem/Science Source; 192 (UPLE), Charles D. Winters/Science Source; 192 (UPRT), The Trustees of the British Museum/Art Resource, NY; 192 (LOLE), 2013 National Museum of Natural History, Smithsonian Institution; 192 (LORT), Joel Arem/Science Source; 193 (UPLE), John Cancalosi/Getty Images; 193 (UPRT), Richard Leeney/Getty Images; 193 (LOLE), RF Company/Alamy; 193 (LORT), Biophoto Associates/Science Source; 194 (UPLE), Gary Cook/Visuals Unlimited, Inc.; 194 (UPRT), James H. Robinson/Science Source; 194 (LOLE), DEA/Photo 1/Visuals Unlimited, Inc.; 194 (LORT), Harry Taylor/Getty Images; 195 (UPLE), Joel Arem/Science Source; 195 (UPRT), Boykung/Shutterstock; 195 (LOLE), Roberto de Gugliemo/Science Source; 195 (LORT), Image copyright © The Metropolitan Museum of Art. Image source: Art Resource, NY; 196 (UPLE), Biophoto Associates/Science Source; 196 (UPRT), Marli Miller/Visuals Unlimited, Inc.; 196 (LOLE), Dirk Wiersma/Science Source; 196 (LORT), Tyler Boyes/Shutterstock; 197 (UPLE), Scientifica/Visuals Unlimited, Inc.; 197 (UPRT), Siim Sepp/Alamy; 197 (LOLE), Manamana/Shutterstock; 197 (LORT), Biophoto Associates/Science Source; 198 (UPLE), 2013 National Museum of Natural History, Smithsonian Institution; 198 (UPRT), Dirk Wiersma/Science Source; 198 (LOLE), Joel Arem/Science Source; 198 (LORT), Scientifica/Visuals Unlimited, Inc.; 199 (UPLE), Ted Kinsman/Science Source; 199 (UPRT), Image copyright © The Metropolitan Museum of Art. Image source: Art Resource, NY; 199 (LOLE), Scientifica/Visuals Unlimited, Inc.; 199 (LORT), WitthayaP/Shutterstock; 200 (UPLE), 2013 National Museum of Natural History, Smithsonian Institution; 200 (UPRT), bpk, Berlin/Aegyptisches Museum, Staatliche Museen, Berlin, Germany/Margarete Buesing/Art Resource, NY; 200 (LOLE), Scientifica/Visuals Unlimited, Inc.; 200 (LORT), Mark A. Schneider/Science Source; 201 (UPLE), Nadezda Boltaca/Shutterstock; 201 (UPRT), 2013 National Museum of Natural History, Smithsonian Institution; 201 (LOLE), 2013 National Museum of Natural History, Smithsonian Institution; 201 (LORT), Scala/Art Resource, NY; 202 (UPLE), 2013 National Museum of Natural History, Smithsonian Institution; 202 (UPRT), Tyler Boyes/iStockphoto; 202 (LOLE), Scientifica/Visuals Unlimited, Inc.; 202 (LORT), E. R. Degginger/Science Source; 203 (UPLE), carlosdelacalle/Shutterstock; 203 (UPRT), Mark Schneider/Visuals Unlimited, Inc.; 203 (LOLE), 2013 National Museum of Natural History, Smithsonian Institution; 203 (LORT), GIPhotoStock/Science Source; 204 (UPLE), 2013 National Museum of Natural History, Smithsonian Institution; 204 (UPRT), Mark Schneider/Visuals Unlimited, Inc.; 204 (LOLE), 2013 National Museum of Natural History, Smithsonian Institution; 204 (LORT), The Trustees of the British Museum/Art Resource, NY; 205 (UPLE), 2013 National Museum of Natural History, Smithsonian Institution; 205 (UPRT), Album/Art Resource, NY; 205 (LOLE), 2013 National Museum of Natural History, Smithsonian Institution; 205 (LORT), Roman Beniaminson/Art Resource, NY; 206 (UPLE), 2013 National Museum of Natural History, Smithsonian Institution; 206 (UPRT), AnatolyM/Shutterstock; 206 (LOLE), Vitaly Raduntsev/Shutterstock; 206 (LORT), Scientifica/Visuals Unlimited, Inc.; 207 (UPLE), 2013 National Museum of Natural History, Smithsonian Institution; 207 (UPRT), Martin Novak/Shutterstock; 207 (LOLE), George Whitely/Science Source; 207 (LORT), 2013 National Museum of Natural History, Smithsonian Institution; 208 (UPLE), Scientifica/Visuals Unlimited, Inc.; 208 (UPRT), Scientifica/Visuals Unlimited, Inc.; 208 (LOLE), Doug Martin/Science Source; 208 (LORT), Dr. Marli Miller/Visuals Unlimited, Inc.; 209 (UPLE), Borislav Marinic; 209 (UPRT), Dr. Marli Miller/Visuals Unlimited, Inc.; 209 (LOLE), Scientifica/Visuals Unlimited, Inc.; 209 (LORT), Marli Miller/Visuals Unlimited, Inc.; 210 (UPLE), Marli Miller/Visuals Unlimited, Inc.; 210 (UPRT), Mark Schneider/Visuals Unlimited, Inc.; 210 (LOLE), Dan Suzio/Science Source; 210 (LORT), Ron Schott; 211 (UPLE), Spring Images/Alamy; 211 (UPRT), Dirk Wiersma/Science Source; 211 (LOLE), Marli Miller/Visuals Unlimited, Inc.; 211 (LORT), Scientifica/Visuals Unlimited, Inc.; 212 (UPLE), Duncan Shaw/Science Source; 212 (UPRT), Doug Martin/Science Source; 212 (LOLE), Peter von Bucher/Shutterstock; 212 (LORT), Ted Kinsman/Science Source; 213 (UPLE), Ron Schott; 213 (UPRT), Michael Szoenyi/Science Source; 213 (LOLE), Alan Majchrowicz; 213 (LORT), E. R. Degginger/Science Source; 214 (UPLE), William H. Mullins/Science Source; 214 (UPRT), Wally Eberhart/Visuals Unlimited, Inc.; 214 (LOLE), Jerry McCormick-Ray/Science Source; 214 (LORT), kavring/Shutterstock; 215 (UPLE), Gerald & Buff Corsi/Visuals Unlimited, Inc.; 215 (UPRT), Adrienne Hart-Davis/Science Source; 215 (LOLE), David Hosking/Science Source; 215 (LORT), Joyce Photographics/Science Source; 216 (UPLE), Images & Volcans/Science Source; 216 (UPRT), DEA/R. APPIANI/Getty Images; 216 (LOLE), Dr. Marli Miller/Visuals Unlimited, Inc.; 216 (LORT), Joyce Photographics/Science Source; 217 (UPLE), John Buitenkant/Science Source; 217 (UPRT), Mark A. Schneider/Science Source; 217 (LOLE), Inga Spence/Visuals Unlimited, Inc.; 217 (LORT), Tyler Boyes/Shutterstock; 218 (UPLE), Francois Gohier/Science Source; 218 (UPRT), michal812/Shutterstock; 218 (LOLE), Richard J. Green/Science Source; 218 (LORT), E. R. Degginger/Science Source; 219 (UPLE), Steve McCutcheon/Visuals Unlimited, Inc.; 219 (UPRT), Ted Kinsman/Science Source; 219 (LOLE), Dr. Marli Miller/Visuals Unlimited, Inc.; 219 (LORT), Biophoto Associates/Science Source; 220 (UPLE), Christian Grzimek/Science Source; 220 (UPRT), Scientifica/Visuals Unlimited, Inc.; 220 (LOLE), Hubertus Kanus/Science Source; 220 (LORT), Scientifica/Visuals Unlimited, Inc.; 221 (UPLE), Len Rue Jr./Science Source; 221 (UPRT), Albert Copley/Visuals Unlimited, Inc.; 221 (LOLE), Gianni Tortoli/Science Source; 221 (LORT), Trevor Clifford Photography/Science Source; 222 (UPLE), Ken M. Johns/Science Source; 222 (UPRT), Joyce Photographics/Science Source; 222 (LOLE), Ken M. Johns/Science Source; 222 (LORT), Joyce Photographics/Science Source; 223 (UPLE), Michael P. Gadomski/Science Source; 223 (UPRT), Scientifica/Visuals Unlimited, Inc.; 223 (LOLE), Ellen Thane/Science Source; 223 (LORT), Biophoto Associates/Science Source; 224 (UPLE), Dr. Marli Miller/Visuals Unlimited, Inc.; 224 (UPRT), Tyler Boyes/Shutterstock; 224 (LOLE), John Arnaldi/Visuals Unlimited, Inc.; 224 (LORT), Marli Miller/Visuals Unlimited, Inc.; 225 (UPLE), Arthur W. Snoke; 225 (UPRT), 2013 National Museum of Natural History, Smithsonian Institution; 225 (LOLE), Marli Miller/Visuals Unlimited, Inc.; 225 (LORT), Marli Miller/Visuals Unlimited, Inc.; 226 (UPLE), Marli Miller/Visuals Unlimited, Inc.; 226 (UPRT), Joel Arem/Science Source; 226 (LOLE), Scientifica/Visuals Unlimited, Inc.; 226 (LORT), Scientifica/Visuals Unlimited, Inc.; 227 (UPLE), Albert Copley/Visuals Unlimited, Inc.; 227 (UPRT), Scientifica/Visuals Unlimited, Inc.; 227 (LOLE), Marli Miller/Visuals Unlimited, Inc.; 227 (LORT), Andrew Alden; 228 (UPLE), William D. Bachman/Science Source; 228 (UPRT), Aaron Haupt/Science Source; 228 (LOLE), John Shaw/Science Source; 228 (LORT), © RMN-Grand Palais/Art Resource, NY; 229 (UPLE), Walt Anderson/Visuals Unlimited, Inc.; 229 (UPRT), Image copyright © The Metropolitan Museum of Art. Image source: Art Resource, NY; 229 (LOLE), Jack Ballard/Visuals Unlimited, Inc.; 229 (LORT), Albert Copley/Visuals Unlimited, Inc.; 230 (UP), Mark A. Schneider/Science Source; 230 (LO), Barbara Strnadova/Science Source; 231 (UP), Colin Keates/Getty Images; 231 (LO), Mark A. Schneider/Science Source; 232 (UP), Mark A. Schneider/Science Source; 232 (LOLE), Scott Camazine/Science Source; 232 (LORT), Trilobite Illustration © Emily S. Damstra; 233 (Both), Mark A. Schneider/Science Source; 234 (UP), Mark A. Schneider/Science Source; 234 (LO), holbox/Shutterstock; 235 (UP), Mark A. Schneider/Science Source; 235 (LO), Scott Camazine/Science Source; 236 (UP), Dirk Wiersma/Science Source; 236 (LO), Geoff Kidd/Science Source; 237 (UP), James L. Amos/Science Source; 237 (LO), Volker Steger/Science Source; 238 (UP), Francois Gohier/Science Source; 238

(LO), Dr. John D. Cunningham/Visuals Unlimited, Inc.; 239 (UP), John Cancalosi/Okapia/Science Source; 239 (LOLE), Victor Habbick Visions/Science Source; 239 (LORT), Francois Gohier/Science Source; 239 (LORT), Laurie O'Keefe/Science Source; 240 (Both), Marli Miller/Visuals Unlimited, Inc.; 241 (Both), Marli Miller/Visuals Unlimited, Inc.; 242 (UP), Michael Szoenyi/Science Source; 242 (LO), Marli Miller/Visuals Unlimited, Inc.; 243 (UP), Doug Sokell/Visuals Unlimited, Inc.; 243 (LO), Ted Kinsman/Science Source; 244 (UP), Dennis Flaherty/Science Source; 244 (LO), Bruce M. Herman/Science Source; 245 (UP), Dr. Ken Wagner/Visuals Unlimited, Inc.; 245 (LO), Marli Miller/Visuals Unlimited, Inc.; 246 (UP), mikenorton/Shutterstock; 246 (LO), EastVillage Images/Shutterstock; 247 (UP), Pierre Leclerc/Shutterstock; 247 (LO), Jim Edds/Science Source; 248 (UP), Ken M. Johns/Science Source; 248 (LO), Marli Miller/Visuals Unlimited, Inc.; 249 (UP), Marli Miller/Visuals Unlimited, Inc.; 249 (LO), Walt Anderson/Visuals Unlimited, Inc.; 250 (UP), Bryan Lowry/SeaPics.com; 250 (LO), Stephen & Donna O'Meara/Science Source; 251 (UP), Explorer/Science Source; 251 (LO), Stephen & Donna O'Meara/Science Source; 252 (Both), Georg Gerster/Science Source; 253 (UP), Francois Gohier/Science Source; 253 (LO), Brenda Tharp/Science Source; 254 (UP), Inga Spence/Visuals Unlimited, Inc.; 254 (LO), Marli Miller/Visuals Unlimited, Inc.; 255 (UP), Robert and Jean Pollock/Science Source; 255 (LO), Marli Miller/Visuals Unlimited, Inc.; 256 (UP), Craig K. Lorenz/Science Source; 256 (LO), William D. Bachman/Science Source; 257 (UP), Douglas Knight/Shutterstock; 257 (LO), Jim W. Grace/Science Source; 258 (UP), ANT Photo Library/Science Source; 258 (LO), Marli Miller/Visuals Unlimited, Inc.; 259 (UP), Marli Miller/Visuals Unlimited, Inc.; 259 (LO), Tim Pleasant/Shutterstock; 260 (UP), Michael Male/Science Source; 260 (LO), Mark Newman/Science Source; 261 (UP), Planet Observer/Science Source; 261 (LO), Michael P. Gadomski/Science Source; 262 (UP), Marli Miller/Visuals Unlimited, Inc.; 262 (LO), William D. Bachman/Science Source; 263 (Both), G. R. 'Dick' Roberts/NSIL/Visuals Unlimited, Inc.; 264 (UP), Ned Therrien/Visuals Unlimited, Inc.; 264 (LO), Andrew J. Martinez/Science Source; 265 (UP), Thomas & Pat Leeson/Science Source; 265 (LO), Bill Kamin/Visuals Unlimited, Inc.; 266, Mike Grandmaison, 268, Paul Marcellini/PaulMarcellini.com; 269, Mike Grandmaison, 272 (UP), Rene Ramos/Shutterstock; 272 (LO), Michael & Patricia Fogden/Minden Pictures/National Geographic Stock; 273 (UP), Ilya Akinshin/Shutterstock; 273 (LO), Monica Schroeder/Science Source; 274 (UP), Nemeziya/Shutterstock; 274 (LO), Henry Lansford/Science Source; 275 (UP), Joyce Photographics/Science Source; 275 (LO), Detlev van Ravenswaay/Science Source; 276 (UP), Joyce Photographics/Science Source; 276 (LO), Robert and Jean Pollock/Science Source; 277 (UP), Adam Jones/Science Source; 277 (LO), G. R. Roberts/Science Source; 278 (UP), Science Source; 278 (LO), Gregory K. Scott/Science Source; 279 (UP), Mark Schneider/Visuals Unlimited, Inc.; 279 (LO), WimL/Shutterstock; 280 (UP), Brenda Tharp/Science Source; 280 (LO), Gregg Schieve/schievephoto.com; 281 (UP), Ralf Broskvar/Shutterstock; 281 (LO), Mike Hollingshead/Science Source; 282 (UP), Jim Reed/Science Source; 282 (LO), Jim Corwin; 283 (UP), David R. Frazier/Science Source; 283 (LO), Christophe Cadiran/Science Source; 284 (UP), Jerry Schad/Science Source; 284 (LO), Pekka Parviainen/Science Source; 285 (UP), Jim Reed/Science Source; 285 (LO), Jim W. Grace/Science Source; 286 (UP), Robert & Jean Pollock/Visuals Unlimited, Inc.; 286 (LO), Giselle Goloy; 287 (UP), Mike Hollingshead/Science Source; 287 (LO), Design Pics/Steve Nagy/Getty Images; 288 (UP), Santanor/Shutterstock; 288 (LO), David Hosking/FLPA/Minden Pictures; 289 (UP), Paul Mansfield Photography/Getty Images; 289 (LO), Kamparin/Shutterstock; 290 (UP), Stan Honda/AFP/Getty Images; 290 (LO), Daniel Friedrichs/AFP/Getty Images; 291 (UP), Tim Laman/National Geographic Stock; 291 (LO), Olga Miltsova/Shutterstock; 292 (UP), Kent Wood/Science Source; 292 (LO), Vortex 2/Science Source; 293 (UP), Howard Bluestein/Science Source; 293 (LO), Mike Hollingshead/Science Source; 294 (UP), Jim Reed/Science Source; 294 (LO), Jim Reed; 295 (UP), NOAA; 295 (LO), Anna Omelchenko/Shutterstock; 296 (UP), Chase Studio/Science Source; 296 (LO), Victor Habbick Visions/Science Source; 297 (UP), muratart/Shutterstock; 297 (LO), Lisa Dearing; 298 (UP), Victor Habbick Visions/Science Source; 298 (LO), Gary Meszaros/Science Source; 299 (UP), Science Source; 299 (LO), Mike Hollingshead/Science Source; 300 (UP), Eric Nguyen/Science Source; 300 (LO), Mike Hollingshead/Science Source; 301 (UP), Howard Bluestein/Science Source; 301 (LO), Dr. Bernhard Weßling/Science Source; 302 (UP), St. Meyers/Science Source; 302 (LO), Will & Deni McIntyre/Science Source; 303 (UP), Jim Edds/Science Source; 303 (LO), Planet Observer/Science Source; 304 (UP), Kaj R. Svensson/Science Source; 304 (LO), Syd Greenberg/Science Source; 305 (UP), James Steinberg/Science Source; 305 (LO), AP Photo/Kiichiro Sato; 306 (UP), Kaj R. Svensson/Science Source; 306 (LO), Jim Reed/Science Source; 307 (UP), Bruce Roberts/Science Source; 307 (LO), Hoa-Qui/Science Source; 308 (LO), Jules Bucher/Science Source; 309 (LO), Jim Edds/Science Source; 320 (UP), David Hosking/FLPA/Minden Pictures; 320 (LO), Dean Krakel II/Science Source; 321 (UP), Alan Copson/Getty Images; 321 (LO), Lowell Georgia/Science Source; 322 (UP), Carr Clifton/Minden Pictures; 322 (LO), Justin Lambert/The Image Bank/Getty Images; 323 (UP), Tony Freeman/Science Source; 323 (LO), Laurent Laveder/Science Source; 324 (UP), Mike Hollingshead/Science Source; 324 (LO), Steve Allen/Science Source; 325 (UP), Mark A. Schneider/Science Source; 325 (LO), Sebastian Saarloos; 326 (UP), Mike Hollingshead/Science Source; 326 (LO), Tim Holt/Science Source; 327 (UP), Mike Hollingshead/Science Source; 327 (LO), Jamen Percy/Shutterstock; 328 (UP), Vera Bradshaw/Science Source; 328 (LO), Lizzie Shepherd/Robert Harding World Imagery/Getty Images; 329 (UP), Ron Wolf/Tom Stack & Associates; 329 (LO), Mike Hollingshead/Science Source; 330 (UP), Wesley Bocxe/Science Source; 330 (LO), dotshock/Shutterstock; 331 (UP), Richard W. Brooks/Science Source; 331 (LO), Patrick Endres/Visuals Unlimited/Corbis; 332 (UP), plampy/Shutterstock; 332 (LO), Robert Jakatics/Shutterstock; 333 (UP), David R. Frazier/Science Source; 333 (LO), Jim Reed/Science Source; 334 (UP), Jim Edds/Science Source; 334 (LO), Sergiy Zavgorodny/Shutterstock; 335 (UP), Bruce M. Herman/Science Source; 335 (LO), Dariush M/Shutterstock; 336 (UP), Science Source; 336 (LO), Mark Horn/Getty Images; 337 (UP), Ted Kinsman/Science Source; 337 (LO), 4FR/Getty Images; 338 (Both), NOAA; 339 (UP), Mike Berger/Science Source; 339 (LO), Science Source; 340 (UP), Hank Morgan/Science Source; 340 (LO), Science Source; 341 (UP), NOAA; 341 (LO), Gregory Ochocki/Science Source; 342 (UP), Noel Celis/AFP/Getty Images; 342 (LO), NOAA; 343 (UP), Mike Theiss/National Geographic/Getty Images; 343 (LO), Ken Gillespie/Getty Images; 344, Marc Adamus; 346, Dorling Kindersley/Getty Images; 347, Grant Ordelheide/National Geographic My Shot; 348, Babak Tafreshi/Science Source; 350 (UPRT), Image produced by F. Hasler, M. Jentoft-Nilsen, H. Pierce, K. Palaniappan, and M. Manyin. NASA Goddard Lab for Atmospheres - Data from National Oceanic and Atmospheric Administration (NOAA). NASA/JPL/GSFC; 350 (LORT), Félix Pharand-Deschênes, Globaïa/Science Source; 350 (UPLE), Davis Meltzer; 350 (LOLE), Paul Leong/Shutterstock; 351 (UP), Pekka Parviainen/Science Source; 351 (LO), Mark Garlick/Science Source; 352 (LORT), Image courtesy NASA; 352 (UP), Detlev van Ravenswaay/Science Source; 352 (LOLE), Zoltan Kenwell/National Geographic My Shot; 353 (UP), Mark Garlick/Science Source; 353 (LO), NASA, ESA, and the Hubble SM4 ERO Team; 354 (UP), John Chumack/Science Source; 354 (LOLE), Don Dixon; 354 (LORT), Don Dixon/National Geographic Stock; 355 (UPLE), SOHO (ESA & NASA); 355 (UPRT), Kenneth Garrett/National Geographic Stock; 355 (LO), National Geographic Stock; 356 (UP), Science Source; 356 (LOLE), ESA/Science Source; 356 (LORT), NASA/SDO/Steele Hill; 357 (UP), Larry Landolfi; 357 (LOLE), Monica Schroeder/Science Source; 357 (LORT), Dieter H/Shutterstock; 358 (LORT), Colorization by Eric Cohen/Science Source; 358 (UP), Babak Tafreshi/Science Source; 358 (LOLE), Detlev van Ravenswaay/Science Source; 359 (UP), Mark Newman/Science Source; 359 (LO), NASA, 360 (UPLE), John Sanford; 360 (UPRT), Detlev van Ravenswaay/Science Source; 360 (LO), Encyclopaedia Britannica/UIG Via Getty Images; 361 (UPRT), Manfred Kage/Science Source; 361 (UP), NASA/JPL/USGS; 361 (LOLE), NASA; 362 (UPRT), Babak Tafreshi/Science Source; 362 (LORT), Philippe Morel/Science Source; 362 (UPLE), Detlev Van Ravensway; 362 (LOLE), Mark Garlick; 363 (UPLE), NASA; 363 (LO), NASA/JPL; 363 (UPRT), NG Maps; 364 (UPRT), NGST; 364 (UPLE), NASA/CXC/SAO (Photo: CXC/M. Weiss); 364 (LO), NASA; 365 (UP), NASA; 365 (LOLE), ESA; 365 (LORT), Victor Habbick Visions/Science Source; 366 (UP), NASA; 366 (LO), Mike Agliolo/Science Source; 367 (UP), Dan Schechter; 367 (LO), James Keenan/National Geographic My Shot; 368 (UP), John Chernack; 368 (LO), Mark Garlick/Science Photo Library/Getty Images; 369 (UPLE), David A. Aguilar; 369 (UPRT), Detlev van Ravenswaay/Science Source; 369 (LO), Mark Garlick/Science Source; 370 (UP), David A. Aguilar; 370 (LO), Frank Zuillo; 371 (UP), Johns Hopkins Univ. Applied Physics Laboratory/Carnegie Institution of Washington; 371 (LO), NASA; 372 (UPLE), Thomas Tuchan/iStockphoto; 372 (UPRT), Colorization by: Mary Martin/Science Source; 372 (LO), NASA; 373 (UP), NASA/JPL-Caltech/Malin Space Science Systems; 373 (LO), Mark Garlick/Science Source; 374 (UP), NASA; 374 (LO), Atlas Photo Bank/Science Source; 375 (UP), NASA/JPL/Space Science Institute; 375 (LO), JPL; 376 (UP), SPL/Science Source; 376 (LO), NASA, ESA, and M. Showalter (SETI Institute); 377 (UP), NASA/JPL; 377 (LO), NASA/JPL-Caltech; 378 (UP), David A. Aguilar; 378 (LO), Chris Butler/Science Source; 379 (UPLE), NASA, ESA, R. Windhorst (Arizona State University) and H. Yan (Spitzer Science Center, Caltech); 379 (UPRT), Colorization by: Jessica Wilson/Science Source; 379 (LO), Mehau Kulyk/Science Source; 380 (UP), NASA, ESA, K. Kuntz (JHU), F. Bresolin (University of Hawaii), J. Trauger (Jet Propulsion Lab), J. Mould (NOAO), Y.-H. Chu (University of Illinois, Urbana), and STScI; 380 (LO), The Milky Way galaxy as conceptualized by Ken Eward and National Geographic Maps; 381 (UP), Adam Evans/sky-candy.ca; 381 (LO), NASA, ESA, and the Hubble Heritage Team (STScI/AURA)-ESA/Hubble Collaboration; 382 (UP), NASA, ESA, and the Hubble Heritage Team (STScI/AURA); 382 (LO), ESA/J. Hester & A. Loll/Arizona State University; 383 (UP), X-ray: NASA/CXC/CfA/R. Kraft et al.; Submillimeter: MPIfR/ESO/APEX/A.Weiss et al., Optical: ESO/WFI; 383 (LO), ESO/M. Kornmesser; 384 (UP), John Chumack/Science Source; 384 (LO), NASA; 385 (UP), T. A. Rector & B. A. Wolpa, NOAO, AURA; 385 (LO), NASA/JPL-Caltech/R. Hurt (SSC); 386 (UP), NASA Ames Research Center/Kepler Mission; 386 (LO), Jon Lomberg/Science Source; 387 (UP), Royal Astronomical Society; 387 (LO), Celestial Image Co./Science Source; 388 (LORT), José Antonio Peñas/Science Source; 388 (UP), ESA/NASA; 388 (LOLE), Reinhold Wittich/Shutterstock; 389 (UP), Jerry Schad/Science Source; 389 (LO), National Geographic Stock; 390 (UP), Daniel McVey/National Geographic My Shot; 390 (LO), Babak Tafreshi/Science Source; 391 (UP), Peter Lloyd/National Geographic Stock; 391 (LO), The Art Archive/British Library/Via Art Resource, NY.

Index

Boldface indicates illustrations.

ILLUSTRATED GUIDE
TO
Nature

CELEBRATING
‹125›
YEARS

Published by the National Geographic Society

John M. Fahey, *Chairman of the Board and Chief Executive Officer*

Declan Moore, *Executive Vice President; President, Publishing and Travel*

Melina Gerosa Bellows, *Executive Vice President; Chief Creative Officer, Books, Kids, and Family*

Prepared by the Book Division

Hector Sierra, *Senior Vice President and General Manager*

Janet Goldstein, *Senior Vice President and Editorial Director*

Jonathan Halling, *Design Director, Books and Children's Publishing*

Marianne Koszorus, *Design Director, Books*

Susan Tyler Hitchcock, *Senior Editor*

R. Gary Colbert, *Production Director*

Jennifer A. Thornton, *Director of Managing Editorial*

Susan S. Blair, *Director of Photography*

Meredith C. Wilcox, *Director, Administration and Rights Clearance*

Staff for This Book

Barbara Payne, *Editor*

Paul Hess, *Text Editor*

Sanaa Akkach, *Art Director*

Catherine Herbert Howell, *Developmental Editor*

Miriam Stein, *Illustrations Editor*

Linda Makarov, *Designer*

Uliana Bazar, *Art Researcher*

Carl Mehler, *Director of Maps*

Marshall Kiker, *Associate Managing Editor*

Judith Klein, *Production Editor*

Galen Young, *Illustrations Specialist*

Katie Olsen, *Production Design Assistant*

Production Services

Phillip L. Schlosser, *Senior Vice President*

Chris Brown, *Vice President, NG Book Manufacturing*

George Bounelis, *Vice President, Production Services*

Nicole Elliott, *Manager*

Rachel Faulise, *Manager*

Robert L. Barr, *Manager*

The National Geographic Society is one of the world's largest nonprofit scientific and educational organizations. Founded in 1888 to "increase and diffuse geographic knowledge," the Society's mission is to inspire people to care about the planet. It reaches more than 400 million people worldwide each month through its official journal, *National Geographic,* and other magazines; National Geographic Channel; television documentaries; music; radio; films; books; DVDs; maps; exhibitions; live events; school publishing programs; interactive media; and merchandise. National Geographic has funded more than 10,000 scientific research, conservation and exploration projects and supports an education program promoting geographic literacy. For more information, visit www.nationalgeographic.com.

For more information, please call 1-800-NGS LINE (647-5463) or write to the following address:
National Geographic Society
1145 17th Street N.W.
Washington, D.C. 20036-4688 U.S.A.

For information about special discounts for bulk purchases, please contact National Geographic Books Special Sales: ngspecsales@ngs.org

For rights or permissions inquiries, please contact National Geographic Books Subsidiary Rights: ngbookrights@ngs.org

ISBN: 978-1-4262-1174-4 (trade)
ISBN: 978-1-4262-1309-0 (regular)
ISBN: 978-1-4262-1255-0 (deluxe)

Printed in China

13/RRDS/1

Dig Into More Nature-Related Books From National Geographic!

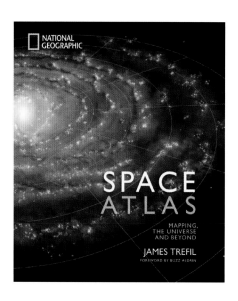

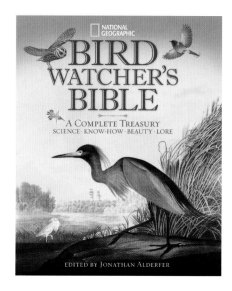

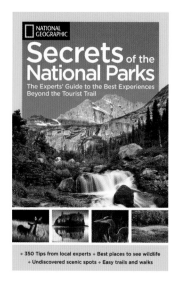